Pathway Design for Industrial Fermentation

Pathway Design for Industrial Fermentation

Walter Koch

WILEY-VCH

Author

Dr. Walter Koch
BASF SE
Carl-Bosch-Str. 38
Ludwigshafen
67056
Germany

Cover Image: © Parilov/Shutterstock

Library of Congress Card No.: applied for

British Library Cataloguing-in-Publication Data
A catalogue record for this book is available from the British Library.

Bibliographic information published by the Deutsche Nationalbibliothek
The Deutsche Nationalbibliothek lists this publication in the Deutsche Nationalbibliografie; detailed bibliographic data are available on the Internet at <http://dnb.d-nb.de>.

Print ISBN: 978-3-527-35275-3
ePDF ISBN: 978-3-527-84393-0
ePub ISBN: 978-3-527-84394-7
oBook ISBN: 978-3-527-84395-4

Typesetting Straive, Chennai, India
Printing and Binding CPI Group (UK) Ltd, Croydon, CR0 4YY

C9783527352753_290124

Contents

Preface

Fermentation technology has often outcompeted organic chemistry in the case of advanced molecular functionality. This refers especially to enzymes and antibodies, though also includes selected organic acids (e.g. lactic acid, citric acid, and itaconic acid), amino acids (e.g. lysine, glutamic acid, and alanine), and various fine chemicals. It offers a great advantage and cost reduction since one fermentation process bundles several steps, which would alternatively be needed with a cascade of single synthesis steps in organic chemistry. The process shortcut is enabled by the multifunctional abilities of the biocatalyst, which handles a complete pathway in one cell. On the other hand, the catalysts applied in organic chemistry, be it a homogeneously or heterogeneously catalyzed process, are usually restricted to single synthesis steps.

The political and societal attempts to replace petrochemistry by solutions with an improved sustainability profile and lower or zero product carbon footprint have drawn management attention toward fermentation products with lower molecular functionality and fuels. These target molecules comprise 3-hydroxypropionic acid or acrylic acid, butanol, isobutanol, isobutene, 1,4-butanediol, and building blocks for the PA6 and PA66 value chains such as adipic acid, hexamethylenediamine, caprolactam, or 6-aminocaproic acid, respectively, and the MDI value chain like anthranilic acid.

The success of fermentative fuels relies on a high energy efficiency in the conversion of feedstock into hydrocarbons with high energy density. Ethanol is the fermentation fuel used as a fuel additive in combustion engines. The rise to a production volume of 100 Mio. t in the last 30 years has been enabled by the national fuel mandates and quota systems and has profited technologically from the existing and highly efficient yeast technology. R&D focus switched to second-generation approaches using lignocellulosic biomass as the feedstock. Significant R&D resources were also allocated to establish butanol as a fermentative fuel.

This book tries to capture more than 20 chapters, respectively, in the technology competition of organic chemistry versus fermentation for molecules with lower molecular functionality and fuels beyond ethanol. Mature organic chemistry is facing increasing competition since fermentation technology is further enhanced by advancements in molecular biology, enzyme screening and optimization, metabolic engineering, and downstream processing. Furthermore, chemolithoautotrophic metabolism with carbon monoxide, carbon dioxide, and hydrogen as feedstocks has successfully been introduced with an acetogenic host to overcome the limitations of the potential volume of feedstocks of the traditional heterotrophic metabolism.

May 2023 *Walter Koch*

Introduction

Fermentation as Production Technology

Fermentation is to be conceived as a production technology competing with natural extraction, organic chemistry, and biocatalysis. All three technologies to access chemical products have their own advantages and disadvantages.

Raw material extraction, which may be regarded as the first production technology, uses natural crops as reactors with controlled cultivation to make chemical products. The production technology is facing limited arable land that could alternatively be used for food production, and an expansion is critical. The cultivation period might be long and define a critical hurdle in case a period of several years is required. Raw material extraction often bears a high extraction cost since the product is mixed with various other crop ingredients. The low content of the target molecule in the biomass might be increased to improve production economics. Subject of raw material extraction might be a natural variety (wild type), a non-natural variety derived by breeding, or a genetically modified variety obtained by plant biotechnology (Chawla 2003; Kempken and Kempken 2004). Most of the agronomic varieties applied for industrial food production were obtained by breeding and possess a higher product content and thereby provide a higher yield per ha than the original natural varieties. Given the inherent disadvantages, natural extraction has in many instances been supplemented or been replaced by organic chemistry, like for natural dyes (Wittke 1984), flavors, and fragrances.

Organic chemistry offers a broad range of potential reaction conditions with respect to temperature, pressure, pH, and organic solvents. It often facilitates high yield (turnover × selectivity), high productivity (space–time-yield), and reasonable investment efforts. Accordingly, the technology has seen a tremendous development and is broadly applied in the petrochemical and chemical industry (Arpe 2007). Organic chemistry has allowed a significant reduction of production costs and enabled the synthesis of many molecules and polymers or plastics, respectively, not found in nature. It has often replaced, supplemented, or at least partially pushed back natural extraction into niche markets, like e.g. in the case of terpenes (Breitmaier 2005) and vanillin (Banerjee and Chattopadhhyay 2018; Kaur and Chakraborty 2013).

Biocatalysis uses enzymes as catalysts and is favored in case an enantioselective synthesis is needed to derive, e.g. chiral alcohols or amines (Hilterhaus et al. 2016). The technology is in some instances applied in high-volume applications, such as the isomerization of glucose into fructose for the soft drink industry (Schmid 2016), but it is often a niche technology for complicated molecular functionality not accessible via organic chemistry within a reasonable number of steps (Faber 2011). Biocatalysis is favored in case the needed raw material conversion is merely an isomerization or a hydrolysis since no cofactors are required. Thus, most of the industrial enzymes applied in the food, feed, or detergent industry are hydrolases, which exclusively rely on water as a cofactor (Sahm et al. 2013). On the other hand, if the needed raw material conversion is a redox reaction in which electrons are transferred, the biocatalyst depends on cofactors like NADH or FAD, which are very costly. For the synthesis of low-volume fine chemicals with sufficient value perspective, suitable cofactor regeneration systems are available (Wu et al. 2021; Grunwald 2018), but synthesis routes for low-value commodity products are often not commercially feasible.

Fermentation has historically covered the microbial conversion of carbohydrates in the absence of oxygen. The term was originally introduced by Louis Pasteur: "La fermentation c'est la vie sans l'air" (Schegel 1992). Fermentation initially happened accidentally if non-sterilized carbohydrate containing food or biomass was stored in the absence of oxygen and was then culturally developed as an on purpose conservation technology. The cultural history of e.g. alcoholic beverages mirrors the long-term development of fermentation technology. The conceptual restriction of fermentation as an exclusively anaerobic process was later given up, and the meaning of the term now comprises the anaerobic and the aerobic conversion of chemical feedstocks by prokaryotic or eukaryotic hosts (Schegel 1992). The application field of fermentation may be differentiated into "White Biotech," the production of fuels and chemical intermediates, "Red Biotech," the production of recombinant proteins and human antibodies for pharma application (Lee and Kildegaard 2020); and "Industrial Enzymes," the production of enzymes for application in industrial processes. The focus of this book in on "White Biotech."

Rising Interest in Fermentation as Production Technology

Production costs and price, respectively, have been up until now, for chemical high-volume products the most important and, in some cases, the single decisive purchase criterion. The products are regarded as "commodities," opposed to "specialties", with low-volume demand, higher molecular functionality, and potentially customization to meet the defined requirements of single customers. The marketing of high-volume chemicals appeals to molecular functionality and specification. The low-cost level of the still dominant petrochemistry is mainly enabled by the circumstance that oil refining to make gasoline for automotive combustion engines generates byproducts like naphtha, which is handed over to chemical companies to feed a cracker (Baerns et al. 2013; Jess and Wasserscheid 2013). The cracker reassembles the C—C bonds of naphtha to derive the main four petrochemical aliphatic intermediates, i.e. ethylene, propylene, isobutene, butadiene, and the three aromatic intermediates, i.e. benzene, toluene, and xylene.

Fermentation is the technology of choice for producing pharma proteins, antibodies ("Red Biotech") and enzymes ("Industrial Enzymes"). Though there is an increasing interest in fermentation technology for high-volume chemicals with much less advanced molecular functionality ("White Biotech"), which receives promotion in national industry development plans. The US bioeconomy strategy pursues the direction of "moving beyond fuels toward biobased and bio-enabled production of chemicals and other products"

(A Bioeconomy Strategy 2022). The growing importance of fermentation as production technology and the increasing share of "white biotechnology" to replace petrochemistry are due to several reasons (Rosales-Calderon and Arantes 2019; Tsuge et al. 2016; Jang et al. 2012).

Biobased raw materials and product carbon footprint considerations are gaining importance, although production costs are still the most important sourcing criterion for chemical products. Chemicals might additionally be marketed with reference to their biobased origin if the raw material is derived from natural sources and the integrated carbon has originally been derived from carbon dioxide captured out of the atmosphere via photosynthesis. The interest in biobased products represents a rising trend for about 20–30 years. Biobased origin was traditionally an advantage for chemicals applied in the food or cosmetic industry with high proximity to the final consumer, which achieved a price premium in the market (Ravenscroft 2019; Ravenscroft 2013). The price premium is required since the production starting with natural extraction and some chemical conversion steps is often more costly than a petrochemical synthesis. It can be observed that in parallel to food ingredients and cosmetics also markets for biobased chemical intermediates are developing, and petrochemistry is facing competition from alternative feedstocks. The reason of this trend reflects the growing customer demand for natural products and the concern that the high carbon dioxide emissions associated with petrochemistry need to be overcome for climate protection. The trend is enforced by national and international regulation and shifting customer attention as well.

The claim of a biobased origin might be achieved via the direct use of a natural feedstock or mediated via a mass balance approach in which the biobased carbon origin of natural feedstocks is allocated to a petrochemically generated product. Fermentation technology profits from the trend toward biobased feedstocks since glucose and sucrose are biobased and derived from, e.g. corn, wheat, sugar beet, sugarcane, or cassava. Fermentation might be the technology providing the lowest production cost, such as in the case of ethanol, lactic acid, citric acid, glutamic acid, and lysine, and has become the leading technology irrespective of its conversion of biobased feedstocks. On the other hand, in case fermentation incurs higher cost than petrochemistry, a price upside in the market is required. Biotech companies and startups were often inclined to overestimate the biobased price premium. Fermentation technology was often able to provide a sound production process, though the marketing success was limited given a cost level above petrochemistry. It turned out that the market success of fermentative products often requires cost equivalence or functional advantages:

"Customers in the electronic and automotive industries will not pay a premium for biobased products unless there is a technical advantage," M. Scheibitz, specialist at BASF automotive (Ravenscroft 2013).

"A cost-advantaged and more sustainable route to adipic acid is expected to offer significant value for the company‘s performance materials business," J. Iademarco, DSM's Vice President Biobased Chemicals (Coons 2011).

"Products won't sell themselves just because they are based on renewables resources [...] Companies need to understand how to bring new biopolymers to market," J. Ravenstijn, bioplastic consultant (Ravenscroft 2013).

"The success of biobased chemicals will rest on competitive economics or premium and uses," M. Hackett, senior manager at IHS Chemicals (Ravenscroft 2013).

"We are not in a situation where [...] any company manufacturing diapers at this moment is ready to pay a 15% premium because it is a biobased raw material [...] This is because no consumer is ready to pay a premium for the diaper" (Tullo 2013).

The second marketing advantage for fermentation technology refers to product carbon footprint considerations (Carruthers and Lee 2022). The original proposal for enhanced sustainability is transforming more and more into the request of politics that the petrochemical and the chemical industry, and the corresponding big consumer brands, need to implement a carbon-neutral business perspective until 2050 (Boulamanti and Moya 2017). Triggered by climate change and the enhanced regulatory burden of carbon dioxide emissions, the conviction was born that no »fresh« carbon in the form of oil and gas must be allowed as feedstock for the petrochemical and chemical industry, and the concept of biobased feedstocks and chemical recycling is gaining momentum (vom Berg et al. 2022). The "endgame" of petrochemistry might have been kicked off. This would imply that the carbon footprint of chemicals needs to be reduced more and more until finally a carbon neutral industry – with net zero carbon dioxide emissions – is established. Ultimately, the carbon content of chemicals, which usually cannot be substituted, needs to be gained via recycling or will be derived from biobased raw materials. Fermentation does often, though not always, allow a lower product carbon footprint than petrochemistry in case the biogenic uptake is considered. It will be interesting to see if fermentation succeeds in providing an alternative supply for the main petrochemical intermediates given the limited availability of carbohydrate as a carbon and energy source (SYSTEMIQ 2022).

Technical Comparison of Fermentation Versus Organic Chemistry

The fermentation process has several restrictions and disadvantages, respectively, though also decisive advantages compared to organic chemistry. The benchmark of both production technologies is discussed with respect to defined criteria:

Coverage of reactions: The outstanding advantage of fermentation technology compared to organic chemistry is the ability of the fermentation biocatalyst to carry out several chemical reactions in one production step without the isolation and purification of intermediates. The substrate is modified by several enzymes within the host cell and channeled through a whole pathway. Fermentation is usually the preferred technology if several steps toward fine chemicals with high molecular functionality can be bundled and the alternative order of steps in organic chemistry is avoided (Vandamme and Revuelta 2016). The coverage of >100 reactions that otherwise would need to be pursued in separate steps of organic chemistry – with yield loss and purification efforts in each step – explains that enzymes or antibodies are more or less exclusively produced via fermentation. The complete field of "Industrial Enzymes," i.e. the production of enzymes for application in industrial processes, and "Red Biotech," i.e. the production of recombinant proteins and human antibodies for pharma application, is dominated by fermentation technology (Lee and Kildegaard 2020).

Feedstock: Fermentation processes usually apply carbohydrates, i.e. glucose as a monosaccharide or more complex carbohydrates, as their sole carbon and energy source. There is a limited number of feedstocks available that can be used as carbon and energy sources by the major industrially used hosts such as *Escherichia coli*, *Saccharomyces cerevisiae*, *Corynebacterium glutamicum*, and *Aspergillus niger*. These comprise, e.g. the monosaccharide glucose (C6) and glycerol (C3). Pending on the specific host also disaccharides or complex carbohydrates may be used. Sucrose, a common table sugar, can usually be metabolized in case an efficient glycosidase is integrated. Otherwise, a chemical hydrolysis is needed. The use of cellulose and hemicellulose requires specific treatment procedures to ensure accessibility of the C5 and C6 sugars. The metabolism of the shorter C5 sugars D-xylose and L-arabinose derived via the digestion of lignocellulosic biomass usually requires the addition of specific pathway genes. Fermentation technology with acetogenic bacteria as hosts using carbon monoxide and hydrogen (Mock et al. 2015; Molitor et al. 2016; Müller 2019) as energy sources is available and industrially applied (Zhu et al. 2020). The methanization of carbon dioxide with exclusively hydrogen as an energy source is evaluated at pilot scale (Bernacchi et al. 2014a, 2014b). Furthermore, methanol as a potential feedstock for fermentation processes is under development, though commercial maturity has not yet been derived. On the other hand, organic chemistry is much more flexible with respect to feedstock: nearly all organic compounds may be regarded as educt. A comparison of fermentation and organic chemistry with respect to defined criteria reveals the main differences (Table 1):

Table 1 Comparison of production technologies for chemical molecules.

	Fermentation	Organic chemistry
Feedstock	Glucose, xylose, glycerol, $CO/CO_2/H_2$	In principle all chemical molecules
Enthalpy	Exothermic overall	Exothermic or endothermic
Temperature	25–60 °C	Up to 800 °C
Pressure	1 barg up to 10 barg	Vacuum up to 300 bar
Reaction medium	Liquid phase	Gas phase or liquid phase
Yield	max. slightly above 90% (including biomass)	Dependent on reaction (may achieve values 99.0-99.9%)
Product titer	10–200 g/l	»200 g/l
Productivity	0.001-0.01 kg/lxh	1–10 kg/lxh heterogen. catalysis 0.01–1 kg/lxh homogen. catalysis
Catalyst	Biocatalyst/Host required	With or without catalyst
Reactor	CSTR/Airlift reactor	Broad portfolio of reactors
Investment	<1000 €/t anaerobic >1000/5 000 €/t aerobic	<1000 €/t commodity >1000/5 000 €/t fine chemical

Temperature and pressure: The fermentation temperature is usually between 25 and 35 °C for the common biocatalysts since the enzyme activity of the host is very sensitive to temperature. A tight window does also apply with respect to pressure, and a fermentation with *E. coli* or *S. cerevisiae* is usually applied at ambient pressure. In exceptional cases, boundary conditions beyond this window are selected. E.g. the fermentative hydrogenation of carbon dioxide with hydrogen is pursued at a temperature of 60–65 °C and a pressure of 4–9 bar (Rusmanis et al. n.d.). Reaction temperatures outside the defined temperature window of the specific host would negatively interfere with the process productivity, which usually cannot be restored. Higher temperatures trigger the partial denaturization of enzymes, which lose their individual three-dimensional structure important for functionality. Since the overall fermentation process is exothermic, sufficient cooling equipment needs to be installed to align with the exothermic heat release of the main reaction. On the other hand, organic chemistry offers a much higher flexibility in the selection of the most suitable temperature and may be pursued at temperatures up to 800 °C in case catalyst and product stability are ensured. However, it is often preferred to avoid very high temperatures to extend the catalyst lifetime, to reduce the investment, and to limit maintenance efforts. Additionally, the chemical reaction may be pursued under vacuum or high pressure to improve kinetics, yield, or selectivity.

Yield: The yield may be defined as the molar ratio between the obtained product amount and the used educt amount and is indicated as percentage. The realistic yield achieved in industrial processes is in almost all cases <100% since the turnover is often not complete (despite recycling loops of non-converted educt) and a compromised selectivity allows by-products whose formation cannot completely be suppressed:

$$\text{Yield}(\%) = \text{Turnover} \times \text{Selectivity}/100$$

With respect to fermentation, the yield is often expressed as "carbon yield", given by the ratio of the mass of the obtained product and the mass of the applied glucose (Arpe 2007). However, this value is especially misleading in case fermentation products with a higher reduction status and higher energy content are considered. The maximum carbon yield in the fermentative production of ethanol amounts to 0.51 t ethanol per ton of glucose (2 × 46 /180 g/mol), which might seem low. However, the industrial process achieves a carbon yield of about 0.46 t ethanol per ton of glucose, which corresponds to about 90% of the theoretical maximum molar yield and is thereby quite high since the formation of biomass is included. Since one glucose derives maximum two mol of ethanol, the theoretical maximum molar yield is 2.0. Thereof 90% or 1.8 mol ethanol per mol glucose are achieved 90%:

$$0.46\,\text{t ethanol per ton glucose} = 90\% \times 2 \times 46/180\,\text{g/mol}$$

Accordingly, the "yield" referred to in this book is defined as the ratio between the achieved molar yield and the theoretical maximum molar yield. Hereby the formation of biomass is included which, pending on the biomass titer and the potential recycling of biomass as inoculum, also consumes a share of the feedstock.

The turnover of the applied glucose might be close to 100% since the reaction is often stopped after the glucose feed in the medium has completely been consumed. A weak selectivity may be caused by competing pathways that channel intermediates into unwanted by-products and the formation of biomass. The carbon yield usually includes the formation of biomass, which may be added as inoculum at the beginning of the reaction or may be formed during the fermentation. The most developed industrial fermentation processes for ethanol and lactic acid achieve a yield of slightly >90%, including the formation of biomass (Jansen et al. 2017; van Maris et al. 2004).

Product titer and productivity: A high yield is important for the reduction of raw material costs. Titer and productivity are important to limit the reactor volume and the scope of the downstream process to reduce the investment. Hereby, it is to be considered that the glucose molecule – the conventional feedstock – needs to be absorbed out of the broth, is converted within the cell by passing a metabolic pathway, and the respective product secreted into the broth. Hereby, specific uptake rates for each host do apply. With respect to the likely most important bacterial host, *E. coli*, the maximum glucose uptake rate for the transfer through the prokaryotic cell wall is about 20 mmol per g DCW (dry cell weight) per h (Yim et al. 2011) and defines a threshold for productivity at a given cell concentration. For in silico simulations, a reasonable average glucose uptake rate of 10 mmol per g DCW per h may be set (Table 2):

Table 2 Maximum glucose uptake (q_S) rates for selected hosts.

Host	Glucose uptake rate
E. coli	20 mmol/g DCW × h (maximum threshold, Ahn et al. 2020; Liebal et al. 2018; Chae et al. 2017)
E. coli	10 mmol/g DCW × h (reasonable average value Ahn et al. 2020; Liebal et al. 2018; Chae et al. 2017)
Corynebacterium glutamicum	4.6 mmol/g DCW × h (Becker et al. 2008)
C. glutamicum	3.6 mmol/g DCW × h (Buschke et al. 2013)
Vibrio natriegens	21 mmol/g DCW × h (aerobically growing cells; Thoma and Blombach 2021)
V. natriegens	43 mmol/g DCW × h (anaerobically growing cells; Thoma and Blombach 2021)
V. natriegens	6 mmol/g DCW × h (anaerobically resting cells; Thoma and Blombach 2021)

The value for *C. glutamicum* refers to a lysine producing strain *C. glutamicum* lysC^{fbr} with feedback resistant aspartate kinase (Becker et al. 2008). For comparison, the xylose uptake rate in *C. glutamicum* is with about 4.0 mmol/g DCW × h in the same range (Buschke et al. 2013).

High figures have been reported for *Vibrio natriegens*, which have triggered several product-specific R&D projects with the target to exceed productivities of the established mainstream hosts. However, the commercial relevance of *V. natriegens* is still limited.

The resulting productivity for an industrial fermentation process may display a level of up to 10 g/l × h, which may be achieved in efficient periods with respect to ethanol or lactic acid (Jansen et al. 2017; van Maris et al. 2004). For laboratory scale experiments, higher values are reported in exceptional cases (Table 3):

Table 3 Very high productivity values achieved in lab-scale.

Productivity	Product	Source	DCW
14.0 g/l × h	Butanol	Nguyen et al. (2019)	29 g DCW/l
21.3 g/l × h	Succinic acid	Ahn et al. (2020)	18 g DCW/l (own estimate)
34.0 g/l × h	Alanine	Hoffart et al. (2017)	(resting cells)

The maximization of productivity is in competition with the goal to achieve a high yield since productivity is enhanced by a higher concentration of the fermentation host (i.e. biomass titer). It is obvious that a higher biomass titer, which consumes glucose feedstock, raises the productivity in case the supply of glucose, oxygen, or other media components is not limited. Though then also the ratio of biomass to product is increasing, which compromises the yield because the biomass can often not be recycled (e.g. used as inoculum of the next batch). A process analysis referring to the maximum product titers of ethanol (Jansen et al. 2017) and lactic acid (van Maris et al. 2004) at batch end and an assumed batch duration of 24 h reveals that a productivity between 5 and 10 g/l × h combined with an average glucose uptake rate of 10 mmol glucose per g DCW and h still allows a yield of 90% of the theoretical maximum yield, facilitated by a rather low biomass byproduct factor (Table 4):

Table 4 Process analysis for the fermentative production of ethanol and lactic acid. In case the carbon content of biomass is estimated with 50% the glucose gap given as difference between the theoretical maximum carbon yield and the realistic carbon yield is sufficient to account for the generation of the needed dry cell weight.

Fermentation	Ethanol	Lactic acid
Titer at batch end	200 g/l	150 g/l
Batch duration	24 h	24 h
Productivity	8.33 g/l × h	6.25 g/l × h
Carbon yield theoretical	0.51 t EtOH/t Glc	1.00 t LA/t Glc
Carbon yield realistic	90%	90%
Carbon yield realistic	0.46 t EtOH/t Glc	0.90 t LA/t Glc
Glucose consumption	18.1 g/l × h	6.94 g/l × h
Glucose consumption	101 mmol/l × h	38.54 mmol/l × h
Glucose uptake assumed	10 mmol/gDCW × h	10 mmol/gDCW × h
Dry cell weight needed	10.1 g/l	3.85 g/l
Dry cell weight	0.05 g/g EtOH	0.03 g/g EtOH
Biomass productivity	0.823 g/g DCW × h	1.6 g/g DCW × h
Process conditions	Anaerobic	Anaerobic
Host	*Saccharomyces cerevisiae*	Lactobacillus

For a bulk chemical to be produced by fermentation, a yield of 80%–90% of the theoretical maximum, a product titer of 100 g/l, and a productivity of 2.5–3.0 g/l × h are regarded as minimum threshold for a commercial feasibility (Thoma and Blombach 2021; Van Dien 2013). A product titer greater than 50 g/l may be defined as the minimum threshold for an industrial process, though a clear-cut borderline is difficult to define (Sun and Alper 2015). These performance data are significantly below the value of a heterogeneous or a homogeneous catalysis in organic chemistry (Klemm 2011). The exemplary productivity for organic chemistry in the acetoxylation of ethylene amounts to 0.3–0.6 kg/l × h (Arpe 2007). It is to be kept in mind that a comparison is difficult since organic chemistry needs to proceed step-by-step, whereas fermentation technology may allow to facilitate a whole pathway, e.g. >10 chemical conversions, in one process steps. However, in case a high-value product is generated that is not accessible via organic chemistry, a fermentation productivity ≪1 g/l × h may be acceptable.

Feeding Strategy

The feeding strategy defines the addition of feedstock during the fermentation covering the carbon source, the energy source, trace elements, and other nutrients. The addition of carbohydrates may be regarded as the most essential part of the feeding strategy; trace elements and other nutrients define media composition. Carbohydrates usually function as combined carbon and energy source in most industrial processes. The most straightforward and easiest approach is the batch process, which foresees that carbohydrate is once added at the start of the fermentation and is not replenished. The host splits the glycosidic bond in case a disaccharide (e.g. sucrose, lactose) is applied and converts glucose into product. The batch is stopped immediately after the glucose has been spent. Since the broth may be disregarded after the product has been isolated, an exhaustive uptake and conversion of carbohydrate is decisive for a high yield. Temperature and pH are controlled and kept at the target value, and oxygen is permanently added in case an aerobic process is pursued.

The batch process does not allow to operate the reactor at a high productivity since a carbohydrate limitation might apply already at the midpoint of the batch. Accordingly, most industrial fermentation processes pursue the fed-batch approach (Sahm et al. 2013), which foresees that the feedstock is permanently be replenished after the initial carbohydrate supply at batch initiation. In case the fed-batch process can be run as a continuous process, the production cost of the product would profit from a high reactor productivity and a high use of capacity of the downstream equipment because the periods with interrupted production for cleaning and batch preparation are minimized. However, most hosts are not suited for a continuous process since a permanent in situ-extraction of product is usually not possible and the avoidance of contamination is challenging. In case a gas fermentation is pursued the carbon source and the energy source are disconnected. Carbon dioxide or carbon monoxide may be applied as carbon source and hydrogen or carbon monoxide as energy source. Industrial experience in the supply of gaseous feedstock is available (Heijstra et al. 2017), though still limited compared to feeding glucose. It may be assumed that a gas mixture is permanently fed into the reactor and that the non-converted components are captured in the head space of the reactor and are recycled.

Metabolic Engineering Strategy

A main part of fermentation R&D refers to the engineering of the metabolic pathway. Until 1980, the strain development programs started with natural overproducers for specific products, like *S. cerevisiae* for ethanol, Lactobacillus for lactic acid, *C. glutamicum* for glutamate and lysine, and *A. niger* for citric acid, and improved performance via media optimization, random mutagenesis, and selection. Accordingly, the industrial fermentative production of ethanol, lactic acid, glutamate, lysine, and citric acid follows natural pathways. Since 1980, the toolbox of molecular biology has allowed more and more rational optimization of the existing pathways or the design and the implementation of artificial pathways (Nielsen and Keasling 2011).

Advanced technological possibilities allowed for innovative recombinant pathways and the development of new industrial processes. The R&D activities of Dupont/Genencor for 1,3-propanediol, the working group of J. C. Liao for butanol and isobutanol, Genomatica for 1,4-butanediol, and Amyris with the working group of J. Keasling for terpenes like artemisinic acid and farnesene define landmark projects in industrial fermentation technology.

The metabolic engineering strategy identifies the most suitable pathway toward the defined target product. Different computer models and tools are available. In this context, the "SymPheny" tool of Genomatica is described as example that significantly contributed to the development of a commercial fermentation technology toward 1,4-butanediol. It consists of five steps: "Reactor operator management" summarizes all potential chemical steps based on known organic chemistry and enzyme classes. The suitable secondary metabolites or cofactors are defined. The toolbox comprises all potential enzymatic steps, even if a suitable enzyme might not yet have been identified. "The Network calculation" starts with the target molecule, figures out potential reaction cascades, and calculates the presumable network reactions. The "Automated integration" adds secondary metabolites, describes charged species at neutral pH, derives balance reactions, and considers thermodynamic feasibility. Unknown biochemical reactions are flagged. In the following module, "Pathway tracing", the starting molecules and the maximum pathway length are set. Additionally, the overall pathway thermodynamics are calculated. Finally, the module "Pathway ranking" pursues a calculation of the maximum theoretical yield and prioritizes pathways according to defined selection criteria such as thermodynamic feasibility, length or number of steps, coverage by known enzymes, or gaps requiring enzyme evolution (Yim et al. 2011). After the in silico-based decision on the recombinant pathway, the real experimental work can be initiated. Native genes from other species are screened, optimized, and integrated into a recombinant host to either transfer a natural pathway into a new organism or to create an artificial pathway not found in nature. The new pathway might be obtained by an orchestrated combination of existing, natural genes, or a new reaction might be implemented based on natural promiscuity or the artificial modeling of enzymes.

The translation of the metabolic engineering strategy into an efficient biocatalyst requires a high number of iteration circles following the general procedure of in silico design, molecular construction and transformation, fermentation at small-scale, analysis of product and process, and re-design. After assembly of the recombinant genes into suitable cassettes and plasmids, respectively, the host is genetically transformed. Different constructs are tested in microtiter plates with limited reaction volume, and the bottlenecks of the pathway are characterized. Competing pathways and undesired by-products are identified. Bottlenecks are overcome via the overexpression of individual genes, and genes encoding enzymes that tend to channel metabolic flux into competing pathways are knocked out or silensed. Additionally, more efficient enzymes might be generated via random or rational replacement and substitution of amino acids to improve enzymatic activity or to adjust substrate specificity in the desired direction. The cofactor specificity of enzymes toward NADH and NADPH might be changed to facilitate a better recycling of reduction equivalents. Suitable membrane proteins for secretion of the target molecule might be added to facilitate a better secretion and to improve the product titer in the broth.

Redox Status of Feedstock and Product to Categorize Metabolic Pathway

In case a new metabolic pathway shall be designed an analysis of the redox status of the product compared to the feedstock is recommended. Glucose, as the most common feedstock, bears 24 electrons without considering the electrons of the reduced oxygen (Dugar and Stephanopoulos 2011; Nielsen et al. 2003). Whereas the typical pathway scheme focuses on the rearrangement of the C-backbone also the 24 electrons are reshuffled and deserve attention in metabolic engineering. The electron density of the feedstock can be calculated, and since the 24 electrons are shared by 6 C atoms, an electron density of 4 is derived. Several conversion types with different electron allocations can be distinguished (Table 5):

Table 5 Electron balance in glucose-based fermentation based on the electron density of feedstock and product. The electron density is also referred to as degree of reduction symbolized with $\kappa = 4.0$ for glucose and expresses a measure of the number of available electrons per unit of carbon (Clomburg and Gonzalez 2013).

Target conversion	Electron / product	Electron density	Comments
Glucose → 2 x Ethanol + 2 CO_2	2 x 12	4 e/C → 6 e/C	Balance of electrons achieved via disproportionation (CO_2 as by-product)
Glucose → Isobutanol + 2 CO_2 + H_2O	24	4 e/C → 6 e/C	
Glucose → Butanol + 2 CO_2 + H_2O	24	4 e/C → 6 e/C	
Glucose → Isobutene + 2 CO_2 + 2 H_2O	24	4 e/C → 6 e/C	
Glucose + 4 e + 4 H^+ → Isoprene + CO_2	28	4 e/C → 5.6 e/C	Gap of electrons implies partial oxidation of glucose
Glucose + 4 e + 4 H^+ → Lysine	28	4 e/C → 4.7 e/C	
Glucose + 8 e + 8 H^+ → 2 x 1,3-Propanediol	16	4 e/C → 5.33 e/C	
Glucose + 2 e + 2 H^+ → Adipic acid	26	4 e/C → 4.33 e/C	
Glucose → 2 Lactic acid	2 x 12	4 e/C → 4 e/C	Balance of electrons with constant electron density
Glucose → 2 x 3-Hydroxypropionic acid	2 x 12	4 e/C → 4 e/C	
Glucose → 1,4-butanediol + 2 CO_2 + 2 e + 2 H^+	22	4 e/C → 5.5 e/C	Surplus electrons propose aerobic process
Glucose → Itaconic acid + CO_2 + 6 e + 6 H^+	18	4 e/C → 3.6 e/C	
Glucose → Glutamic acid + CO_2 + 6 e + 6 H^+	18	4 e/C → 3.6 e/C	
Glucose → Succinic acid + 2 CO_2 + 10 e + 10 H^+	14	4 e/C → 3.5 e/C	
Glucose → Citric acid + 6 e + 6 H^+	18	4 e/C → 3.0 e/C	

Balance of electrons via disproportionation and carbon dioxide as by-product: In the generation of the fermentative fuels ethanol, butanol, and isobutanol the electron density is increasing from 4 e/C in the educt to 6 e/C in the product. These biochemical conversions facilitate a "concentration" of electrons in the respective product. The pursued chemical reaction, irrespective of the chosen pathway, implies the generation of carbon dioxide as byproduct because the disproportionation of glucose and the electron allocation of the product do not allow a less oxidized byproduct. No surplus of reduction equivalents is derived, and the generated carbon dioxide cannot be avoided.

Gap of electrons, which implies oxidation of glucose: In the metabolic pathways toward isoprene, 1,3-propanediol, and adipic acid, the products are in a higher reduction status than glucose, and an electron gap is given. This applies especially to 1,3-propanediol. Two mol of 1,3-propanediol might be formed out of one mol of glucose unless a gap of 8 electrons applies. This is the reason why microbial electrosynthesis has been considered as a suitable tool to improve the carbon yield (Zhou et al. 2015; Kracke and Krömer 2014). Since an electron supply via electrosynthesis is usually not available and technically challenging, the electron gap in this type of conversion triggers the partial aerobic oxidation of the feedstock to provide additional electrons.

Balance of electrons with constant electron density by molecular rearrangement: In some pathways, the glucose molecule might simply be rearranged such that the electron density remains the same. This is the case in the formation of lactic acid and 3-hydroxypropionic acid. The fermentative production toward lactic acid pursues a straightforward pathway: Glycolysis converts glucose into pyruvate and allows the generation of two mol ATP per mol glucose via substrate level phosphorylation. Since oxygen or other inorganics are not available as final electron acceptors in the anaerobic process, the surplus of reduction equivalents needs to be recycled to NAD^+ for further metabolism of glucose. The reduction of the keto function at C2 to obtain lactic acid ensures recycling of NAD^+. This natural pathway is more or less ideal since it achieves a balance of reduction equivalents and allows the generation of ATP to provide energy for maintenance and growth. The 24 electrons carried by glucose (without counting reduced oxygen) are equally transferred to two mol of lactic acid. According to the fermentative pathway to lactic acid, in the generation of 3-hydroxypropionic acid, the 24 electrons of glucose (without counting reduced oxygen) are equally allocated to two mol product. The balance of reduction equivalents in the metabolic pathways explains the circumstance that cell-free technologies for the products ethanol (historical experiments of Buchner), isobutanol (Guterl et al. 2012), and butanol (Krutsakorn et al. 2013) have been evaluated. However, the commercial feasibility of cell-free metabolic engineering also relies on a net balance of ATP, which is not given in the pathways of the aforementioned products. With respect to isobutanol, a modified non-phosphorylative pathway using cell-free metabolic engineering has been proposed (Guterl et al. 2012).

Surplus of electrons proposes an aerobic process: Finally, the biochemical conversion might generate an electron surplus in case the product has a higher oxidation status than glucose and the electron density of the product decreases. This applies to the fermentative conversion of glucose into itaconic acid, succinic acid, and citric acid (Table 3). In these cases, an electron acceptor is needed, and usually an aerobic process is chosen to provide oxygen as an electron acceptor in the formation of water. Alternatively, an anodic electro-fermentation is discussed, which offers the anode as an extracellular electron sink. The surplus of NADH or NADPH may be oxidized and thereby stabilize the redox status of the host (Vassilev et al. 2018).

Succinic acid defines a special case since a rather high electron surplus, i.e. 10 electrons in the case where one mol glucose is converted into one mol succinic acid, would be derived and imply a high oxygen consumption. In order to exploit the surplus of reduction equivalents, carbon dioxide is provided as an additional feedstock. The pathway is then designed in a way such that one mol glucose (24 electrons) and two mol carbon dioxide generate two mol succinic acid (2 × 14 electrons):

$$\text{Glucose} + 2\,CO_2 + 2\,\text{NADH} + 2\,H^+ \rightarrow 2\,\text{Succinic acid} + 2\,\text{NAD}^+$$

The electron gap has been moved to the left side of the reaction equation due to the structural integration of carbon dioxide, and a share of the glucose needs to be oxidized to balance the electron gap. The product carbon footprint has been improved since carbon dioxide is established as educt.

Reduction Potentials in Biochemistry

Biochemical systems are able to facilitate redox reactions in a very efficient way. The cofactor systems $NAD(P)^+/NAD(P)H$ and $FAD/FADH_2$ have evolved as efficient electron carriers that either receive electrons generated in the oxidation of a pathway intermediate (electron acceptor) or provide electrons in the reduction of a pathway intermediate (electron donor). Hereby, it needs to be considered that the established cofactors have defined half-reaction potentials under standard boundary conditions, i.e. pH 7.0 in biochemical systems as opposed to pH 0 in electro-chemistry. The half-reaction potentials for several reductions are summarized (Table 6):

Table 6 Reduction potentials in biochemical systems. All values are defined with pH 7 as reference. The half-reaction potential of hydrogen at pH 0 with 0 mV is defined as reference point in electrochemistry.

Reduction half-reaction potential	E_0' (pH 7.0)	Source
Organic acid + 2e + $2H^+$ ⇆ Ald. + H_2O	−580 mV	Haas et al. 2018; Mock et al. 2015
Formate + 2e + $2H^+$ ⇆ FA + H_2O	−580 mV	(according acid to aldehyde)
CO_2 + 2e + $2H^+$ ⇆ CO + H_2O	−520 mV	Bertsch and Mueller 2015; Katysv 2020
Acetyl-CoA + CO_2 + 2e + $2H^+$ ⇆ Pyruvate	−500 mV	Haas et al. 2018
Fd + 2e ⇆ Fd^{2-}	−500 mV	Katsyv 2020; Schuchmann and Müller 2016; Thauer et al. 1977
CO_2 + 2e + $2H^+$ ⇆ Formate	−432 mV	Katysv 2020; Wang et al. 2013; Thauer et al. 1977
$2H^+$ + 2e ⇆ H_2	−414 mV	Bertsch and Mueller 2015; Thauer et al. 1977
$NAD(P)^+$ + 2e + $2H^+$ ⇆ NAD(P)H	−320 mV	Schuchmann and Müller 2016; Bertsch 2015; Thauer et al. 1977
Methylene-THF + 2e + $2H^+$ ⇆ Methyl-THF	−200 mV	Buckel 2018; Thauer et al. 1977
Acetaldeh. + 2e + $2H^+$ ⇆ EtOH	−200 mV	Voet et al. 2016; Thauer et al. 1977
Pyruvate + 2e + $2H^+$ ⇆ Lactate	−190 mV	Schuchmann and Müller 2016; Thauer et al. 1977
Oxalacetate + 2e + $2H^+$ ⇆ Malat	−170 mV	Voet et al. 2016; Thauer et al. 1977
Fumarate + 2e + $2H^+$ ⇆ Succinate	−30 mV	Voet et al. 2016; Thauer et al. 1977

These values uniformly refer to standard conditions for better comparability. However, the actual half-reaction potential under physiological conditions may be different and thereby allow a redox reaction with sufficient thermodynamic efficiency, which does not seem possible under standard conditions. E.g. the redox potential of NAD^+ to NADH is −320 mV under standard conditions and assumed to be more in the range of −280 mV under physiological conditions, and the redox potential of ferredoxin to reduced ferredoxin is −500 mV under standard conditions and assumed to be more in the range of −400 to −500 mV under physiological conditions (Schuchmann and Müller 2016).

A biochemical reduction is thermodynamically favored in case the half-reaction reduction potential of the used electron donating cofactor for the electron transfer is sufficiently negative to overcome the resistance of the electron acceptor to undergo the reduction. This means that the half-reaction potential of the reduction equivalent needs to be more negative than the half-reduction equivalent of the intermediate foreseen for the reduction.

Some conclusions can be drawn that help to understand the thermodynamic efficiency of natural pathways. E.g. NADH is able to reduce acetaldehyde to ethanol and pyruvate toward lactic acid in the native fermentation pathways to ethanol and lactic acid. The reaction can also proceed if the half-reaction potentials are in the same range, e.g. reduced ferredoxin is able to reduce carbon dioxide to formate in acetogenic bacteria. Additionally, implications for the design and the efficiency of recombinant metabolic pathways do apply. It is obvious that the reduction of organic acids is rather challenging since a suitable cofactor with a reduction potential in the range of −580 mV is not available. Accordingly, the conversion of carboxylic acids toward their respective aldehydes requires, in most cases a specific activation step of the acid. It may proceed with the respective CoA-thioester as an intermediate in the formation of e.g. succinate semialdehyde, 4-hydroxybutyraldehde or 4-hydroxybutanal, adipate semialdehyde, and 6-aminocaproate semialdehyde (Figure 1):

Succinyl-CoA + NADH + H^+ → Succinate semialdehyde + CoA + NAD^+	CoA-dep. succinate semialdehyde DH
4-Hydroxybutyrate + Acetyl-CoA → 4-Hydroxybutyryl-CoA + acetate	CoA-acyl transferase
4-Hydroxybyturyl-CoA + NAD(P)H + H^+ → 4-Hydroxybutanal + CoA	CoA-dep. aldehyde dehydrogenase
Adipyl-CoA + NADH + H^+ → adipate semialdehyde + CoA + NAD^+	Aldehyde dehydrogenase
6-aminocaproate + ATP + CoA → 6-Aminocaproyl-CoA + ADP	6-Aminocaproyl-CoA synthase
6-aminocaproyl-CoA + NADH → 6-Aminocaproate-SA + NAD^+	6-Aminocaproyl-CoA reductase

Figure 1 Pathway steps with acid to aldehyde reduction.

Alternatively, the carboxylic acid function may be activated via the conversion into a carboxylic acid phosphate ester facilitated by the transfer of a phosphate group from ATP:

Acid + ATP → Acid-phosphate + ADP

This approach to overcome the thermodynamically difficult reduction step is exemplified in the biochemical pathway toward lysine and, in particular, the conversion of aspartate into aspartate-phosphate (or aspartyl-ß-phosphate):

Aspartate + ATP → Aspartate-4-P + ADP	Aspartate kinase
Aspartate-4-P + NADH + H^+ → Aspartate semialdehyde + Pi + NAD^+	Aspartate semialdehyde DH

The analogue activation strategy can be found in the metabolic pathway towards the amino acids proline and arginine with glutamate as starting point:

Glutamate + ATP → Glutamate-5-P + ADP	γ-Glutamyl-kinase
Glutamate-5-P + NADH + H^+ → Glutamate-5-semialdehyde + Pi + NAD^+	Dehydrogenase

Estimation of Theoretical Maximum-Molar-Yield (TMY)

The theoretical molar yield is nowadays modeled with digital tools like the "SymPheny" tool of the biotech company Genomatica (Yim et al. 2011). These tools try to reflect the complete biological system, i.e. the fermentation host under real process conditions, to derive the overall equation of the reaction and the molar ratio of glucose to product. It is usually supposed that competing pathways cannot completely be avoided and that, e.g. glucose is partially metabolized by conventional glycolysis and the pentose phosphate pathway. Additionally, the genetic engineering measures for yield improvement are reflected, which might silence or knock out individual genes encoding enzymes for native pathways. Given the high complexity of the complete biological system, the in silico approach is more comprehensive and provides more accurate results than the traditional approach of summing up a cascade of single reactions. The provided TMY reduces system complexity, though it might be counterintuitive and difficult to validate. Accordingly, in this book, the conventional approach has been chosen to regain full transparency. It is acknowledged that a comprehensive in silico analysis, considering the complexity of the whole biological system, provides a more accurate result.

In case the net balance of the modeled pathway reveals a gap in energy equivalents or reduction equivalents, respectively, and no external equivalents besides glucose are supplemented to the medium, an additional amount of glucose needs to be metabolized and taken into consideration in the estimation of the theoretical maximum molar yield to close the gap. Therefore, conversion ratios need to be defined.

In the first instance, an aerobic pathway is considered. For the calculation of the ATP derived by reduction equivalents, the P/O ratio needs to be defined, which expresses the relation of the mol ATP per mol oxygen consumed in the total aerobic oxidation of glucose:

$$\text{Glucose} + 6\,O_2 \rightarrow 6\,CO_2 + 6\,H_2O$$

$$\begin{aligned}\Delta G^{\circ\prime} &= -2850\ \text{GJ/mol glucose (Voet 2016)}\\ &= -15.8\ \text{GJ/t glucose}\end{aligned}$$

In the case where 36 mol of ATP would be generated combined with the consumption of six mol of oxygen a P/O ratio of 3.0 would be obtained. For the estimation of the theoretical molar yield, a more realistic approach is chosen, and for the complete aerobic oxidation of one mol glucose, it is assumed that in total only 32 mol of ATP are generated, applying a P/O ratio of 1.5 mol ATP/mol $FADH_2$ and 2.5 mol ATP/mol NADH (Table 7):

Table 7 Deduction of total energy equivalents in the aerobic metabolism of glucose. ATP and GTP can be regarded as equivalent in the first approximation.

Metabolic pathway	Substrate level phosphorylation	Respiratory chain phosphorylation		
	Energy equivalents	Reduction equivalents	P/O ratio	Energy equivalents
Glycolysis	2 ATP	2 NADH	2.5	5 ATP
Pyruvate dehydrogenase	n.a.	2 NADH	2.5	5 ATP
Citric acid cycle	2 GTP	6 NADH	2.5	15 NADH
		2 $FADH_2$	1.5	3 ATP
Total	4	12		28

Accordingly, the oxidation of one mol of glucose compensates for 32 ATP in an aerobic process. In case a gap in reduction equivalents needs to be compensated, it is assumed that 1 mol glucose provides 12 mol NAD(P)H given the 24 electrons carried by one mol glucose. In order to compensate for a combined gap of energy and reduction equivalents, the coproduction of 12 reduction equivalents and 4 energy equivalents (2 mol ATP and 2 mol GTP) can be considered.

Secondly, a different rule applies for the anaerobic pathway. The generation of ATP is limited to substrate level phosphorylation in case acetogenic bacteria are excluded. A gap in energy or reduction equivalents needs to be compensated via the partial oxidation of glucose, and respective by-products need to be defined. In case ethanol or lactic acid are assumed as by-product, 2 mol of ATP are generated without coproduction of reduction equivalents because the NADH generated during glycolysis is consumed in the reduction of acetaldehyde or pyruvate. The generation of acetate may be assumed in case a gap in energy and reduction equivalents does apply, and one mol glucose is assumed to provide two mol acetate (plus two mol carbon dioxide), two mol of ATP and 4 reduction equivalents. The 24 mol electrons of glucose are transferred to acetate (2×8 without counting reduced oxygen), and the remaining 8 electrons are given as 4 reduction equivalents.

References

A Bioeconomy Strategy (2022). The U.S. Bioeconomy: charting a course for a resilient and competitive future, Schmidt. *Futures* 7.

Ahn, J.H., Seo, H., Park, W. et al. (2020). Enhanced succinic acid production by Mannheimia employing optimal malate dehydrogenase, nature. *Communication* 11: 1970, https://doi.org. doi: 10.1038/s41467-020-15839-z.

Arpe, H.-J. (2007). *Industrielle Organische Chemie. Bedeutende Vor- und Zwischenprodukte*, 179. Wiley 320, 449ff.

Baerns, M., Behr, A., Brehm, A. et al. (2013). *Technische Chemie*, 509. Weinheim: Wliey-VCH.

Banerjee, G. and Chattopadhhyay, P. (2018). Vanillin biotechnology: the perspective and future. *Journal of the Science of Food and Agriculture* https://doi.org/10.1002/jsfa.9303.

Becker, J., Klopprogge, C., and Wittmann, C. (2008). Metabolic responses to pyruvate kinase deletion in lysine producing *Corynebacterium glutamicum*. *Microbial Cell Factories* 7: 8. https://doi.org/10.1186/1475-2859-7-8.

Bernacchi, S., Weissgram, M., Wukovits, W., and Herwig, C. (2014a). Process efficiency simulation for key process parameters in biological methanogenesis. *AIMS Bioengineering* 1 (1): 53–71. https://doi.org/10.3934/bioeng.2014.1.53.

Bernacchi, S., Rittmann, S., Seifert, A.H. et al. (2014b). Experimental methods for screening parameters influencing the growth to product yield ($Y_{[x/CH4]}$) of a biological methane production (BMP) process performed with *Methanothermobacter marburgensis*. *AIMS Bioengineering* 1 (2): 72–87. https://doi.org/10.3934/bioeng. 2014.2.72.

Bertsch, J. and Mueller, V. (2015). Bioenergetic constraints for conversion of syngas to biofuels in acetogenic bacteria. *Biotechnology for Biofuels* 8: 210. https://doi.org/10.1186/s13068-015-0393-x.

Boulamanti, A. and Moya, J.A. (2017). *Energy Efficiency and GHG Emissions: Prospective Scenarios for the Chemical and Petrochemical Industry*, JRC science for policy report, European Union, 17.

Breitmaier, E., *Terpene*. Aromen, Düfte, Pharmaka, Pheromone, Wiley-VCH 2005

Buschke, N., Becker, J., Schäfer, R. et al. (2013). Systems metabolic engineering of xylose-utilizing *Corynebacterium glutamicum* for production of 1,5-diaminopentane. *Biotechnology Journal* 8: 557–570. doi:10.10027biot.201200367.

Carruthers, D.N. and Lee, T.S. (2022). Translating advances in microbial bioproduction to sustainable biotechnology. *Frontiers in Bioengineering and Biotechnology* https://doi.org/10.3389/fbioe.2022.968437.

Chae, T.U., Ko, Y.-S., Hwang, K.-S., and Lee, S.Y. (2017). Metabolic engineering of *Escherichia coli* for the production of four-, five-, and six-carbon lactams. *Metabolic Engineering* 43: 82–91. https://doi.org/10.1016/y.mben.2017.04. 001.

Chawla, H.S. (2003). *Introduction to Plant Biotechnology*. Science Publishers.

Clomburg, J.M. and Gonzalez, R. (2013). Anaerobic fermentation of glycerol: a platform for renewable fuels and chemicals. *Trends in Biochemistry* 31 (1): 20–28. https://doi.org/10.1016/j.tibtech.2012.10.006.

Coons, R. (2011). DSM adds adipic acid to biobased chemical efforts. *Chemical Week* October 17: 14.

Dugar, D. and Stephanopoulos, S. (2011). Relative potential of biosynthetic pathways for biofuels and bio-based products. *Nature Biotechnology* 29 (12): 1074–1078.

Faber, K. (2011). *Biotransformations in organic chemistry. A textbook*. Springer.

Grunwald, P. (2018). *Biocatalysis. Biochemical fundamentals and applications*. World Scientific Publishing.

Guterl, J.-K., Garbe, D., Carsten, J. et al. (2012). Cell-free metabolic engineering: production of chemicals by minimized reaction cascades. *ChemSusChem* 5: 2165–2172.

Haas, T., Krause, R., Weber, R. et al. (2018). Technical photosynthesis involving CO2 electrolysis and fermentation. *Nature Catalysis* 1: 32–39. http://doi.org/10.1038/s41929-017-0005-1.

Heijstra, B.D., Leang, C., and Juminga, A. (2017). Gas fermentation: cellular engineering possibilities and scale up. *Microbial Cell Factories* 16: 60. https://doi.org/10.1186/s12934-017-0676-y.

Hilterhaus, L., Liese, A., Kettling, U., and Antranikian, G. (2016). *Applied Biocatalysis. From fundamentals science to industrial application*. Wiley-VCH.

Hoffart, E., Grenz, S., Lange, J. et al. (2017). High substrate uptake rates empower *Vibrio natriegens* as production host for industrial biotechnology. *Applied and Environmental Microbiology* 83 (22): e01614–e01617.

Jang, Y.-S., Kim, B., Shin, J.H. et al. (2012). Bio-based production of C2-C6 platform chemicals. *Botechnology and Bioengineering* 109 (10): 2437–2459. https://doi.org/10.1002/bit.24599.

Jansen, M.L.A., Bracher, J.M., Papapetridis, I. et al. (2017). *Saccharomyces cerevisiae* strains for second-generation ethanol production: from academic exploration to industrial implementation. *FEMS Yeast Research* 17 (5): https://doi.org/10.1093/femssyr/fox044.

Jess, A. and Wasserscheid, P. (2013). *Chemical technology. An integrated technology*, 604f. Wiley-VCH, Weinheim.

Katsyv, A. and Mueller, V. (2020). Overcoming energetic barriers in acetogenic C1 conversion. *Frontiers in Bioengineering and Biotechnology* 8: article 621166. https://doi.org/10.3389/fbioe.2020.621166.

Kaur, B. and Chakraborty, D. (2013). Biotechnological and molecular approaches for vanillin production: a review. *Applied Biochemistry and Biotechnology* 169: 1353–1373. https://doi.org/10.1007/s12010-012-0066-1.

Klemm, E, *Heterogeneous catalysis engineering*, University of Stuttgart, DFG Priority Program 1362, Stuttgart, 2011, https://www.chm.tu-dresden.de/ac1/spp-mof/workshop_2011_stuttgart/06_Klemm/HetCat Eng Workshop_USTUTT.pdf,

Kempken, F. and Kempken, R. (2004). *Gentechnik bei Pflanzen*. Springer.

Kracke, F. and Krömer, J.O. (2014). Identifying target processes for microbial electrosynthesis by elementary mode analysis. *BMC Bioinformatics* 15: 410. https://doi.org/10.1186/s12859-014-0410-2.

Krutsakorn, B., Honda, K., Ye, X. et al. (2013). In vitro production of n-butanol from glucose. *Metabolic Engineering* 20: 84–91.

Lee, G.M. and Kildegaard, H.F. (ed.) (2020). *Cell culture engineering. Recombinant protein production*. Wiley-VCH.

Liebal, U.W., Blank, L.M., and Ebert, B.E. (2018). CO_2 to succinic acid – estimating the potential of biocatalytic routes, metabolic engineering. *Communications* 7: https://doi.org/10.1016/j.mec.2018.e00075.

Mock, J., Zheng, Y., Mueller, A.P. et al. (2015). Energy conservation associated with ethanol formation from H_2 and CO_2 in *clostridium autoethanogenum* involving electron bifurcation. *Journal of Bacteriology* 197 (18): 2965–2980.

Molitor, B., Kirchner, K., Henrich, A.W. et al. (2016). Expanding the molecular toolkit for the homoacetogen clostridium ljungdahlii. *Scientific Reports* 6: 31518. https://doi.org/10.1038/srep31518.

Müller, V. (2019). New horizons in acetogenic conversion on one-carbon substrates and biological hydrogen storage. *Trends in Biotechnology* 37 (12): 1344–1354. http://doi.org/10.1016/j.tibtech.2019.05.008.

Nguyen, N.-P.-T., Raynaud, C., Meynial-Salles, I., and Soucaille, P. (2019). Reviving the Weizmann process for commercial *n*-butanol production. *Nature Communications* 9: 3682. https://doi.org/10.1038/s41467-018-05661-z.

Nielsen, J. and Keasling, J.D. (2011). Synergies between synthetic biology and metabolic engineering. *Nature Biotechnology* 29 (8): 693–695.

Nielsen, J., Villadsen, J., and Liden, G. (2003). *Bioreaction Engineering Principles*, 104. Springer.

Ravenscroft, M. (2019). Flavors and fragrance sense expansion. Consumer preference is shifting to natural and healthy ingredients. *Chemical Week* 23.

Ravenscroft, M. (2013). Industrial biotechnology. Chemical building blocks from renewable resources, IHS. *Chemical Week* 5 (12): 19–22.

Rosales-Calderon, O. and Arantes, V. (2019). A review on commercial-scale high-value products that can be produced alongside cellulosic ethanol. *Biotechnology for Biofuels* 12: 240. http://doi.org/10.1186/s13068-019-1529-1.

Rusmanis, D., O'Shea, R., Wall, D.M., and Murphy, J.D. Biological hydrogen methanation systems—an overview of design and efficiency. *Bioengineered* 10 (1): 604–634. 10.1080/21655979.2019.1684607.

Schuchmann, K. and Müller, V. (2016). Energetics and application of heterotrophy in acetogenic bacteria. *Applied and Environmental Microbiology* 82 (14): 4056–4069. https://doi.org/10.1128/AEM.00882-16.

Sahm, H., Antranikian, G., Stahmann, K.-P., and Takors, R. (2013). *Industrielle Mikrobiologie*. Heidelberg: Springer Verlag p. 32–33, p. 205–224.

Schegel, H.G. (1992). *Allgemeine Mikrobiologie*, 286. Stuttgart: Georg Thieme Verlag.

Schmid, R.D. (2016). *Taschenatlas der Biotechnologie und Gentechnik*, 180. Wiley-VCH.

Sun, J. and Alper, H.S. (2015). Metabolic engineering of strains: from industrial-scale to lab-scale chemical production. *Journal of Industrial Microbiology and Biotechnology* 42: 423–436. https://doi.org/10.1007/s10295-014-1539-8.

SYSTEMIQ (2022, source: https://plasticseurope.org/de/2022/04/05/reshaping-plastics-neuer-report-mit-pfaden-zu-einer-treibhausgasneutralen-kunststoffindustrie/?msclkid=65d48658b8b811ec8e2965359d4b8942). ReShaping Plastics: Pathways to a circular, climate neutral plastics system in Europe. In: . accessed 10.4.2022.

Thauer, R.K., Jungermann, K., and Decker, K. (1977). Energy conservation in chemotrophic anaerobic bacteria. *Bacterial Reviews* 41: 100–180. https://doi.org/10.1128/MMBR.41.1.100-180.1977.

Thoma, F. and Blombach, B. (2021). Metabolic engineering of *vibrio natriegens*. *Essays in Biochemistry* 65: 381–392. https://doi.org/10.1042/ECB20200135.

Tsuge, Y., Kawaguchi, H., Sasaki, K., and Kondo, A. (2016). Engineering cell factories for producing building block chemicals for bio-polymer synthesis. *Microbial Cell Factories* 15: 19. https://doi.org/10.1186/s12934-016-0411-0.

Tullo, A.H. (2013). Sustainable acrylic acid. Companies vie for a nonpetrochemical route into this lucrative industry. *CEN* 18 (11): 18–19.

Vandamme, E.J. and Revuelta, J.L. (ed.) (2016). *Industrial biotechnology of vitamins, biopigments and antioxidants*. Wiley-VCH.

Van Dien, S. (2013). From the first drop to the first truckload: commercialization of microbial processes for renewable chemicals. *Current Opinion in Biotechnology* 24: 1061–1068. https://doi.org/10.1016/j.copbio.2013.03.002.

vom Berg, C., Carus, M., Stratmann, M., and Dammer, L. (2022, source: http://www.renewable-carbon-initiative.com). Renewable carbon as a guiding principle for sustainable carbon cycles. In: *Renewable Carbon Initiative*, February. .

van Maris, A.J.A., Konings, W.N., van Dijken, J.P., and Pronk, J.T. (2004). Microbial export of lactic and 3-hydroxypropionic acid: implications for industrial fermentation processes. *Metabolic Engineering* 6: 245–255. https://doi.org/10.1016/j.ymben.2004.05.001.

Vassilev, I., Gießelmann, G., Schwechheimer, S.K. et al. (2018). Anodic electro-fermentation: anaerobic production of L-lysine by recombinant *Corynebacterium glutamicum*. *Biotechnology and Bioengineering* 115: 1499–1508. https://doi.org/10.1002/bit.26562.

Voet, D., Voet, J.G., and Pratt, C.W. (2016). *Principles of Biochemistry*. Wiley p. 575–576, p. 453, 466.

Wang, S., Huang, H., Kahnt, J. et al. (2013). NADP-specific electron-bifurcating [FeFe]-dehydrogenase in a functional complex with formate dehydrogenase in *clostridium autoethanogenum* grown on CO. *Journal of Bacteriology* 195 (19): 4373–4386.

Wittke, G. (1984). *Farbstoffchemie*, 41–44. Moritz Diesterweg: Frankfurt.

Wu, S., Snajdrova, R., Moore, J.C. et al. (2021). Biocatalysis: enzymatic synthesis for industrial applications. *Angewandte Chemie* 60: 88–119. https://doi.org/10.1002/anie.202006648.

Yim, H., Haselbeck, R., Niu, W. et al. (2011). Metabolic engineering of *Escherichia coli* for direct production of 1,4-butanediol. *Nature Chemical Biology* 10.1038/NCHEMBIO.580.

Zhou, M., Freguia, S., Dennis, P.G. et al. (2015). *Development of bioelectrochemical activity stimulates mixed-culture reduction of glycerol in a bioelectrochemical system, microbial biotechnology* 8 (3): 483–489. https://doi.org/10.1111/1751-7915.12240.

Zhu, H.-F., Liu, Z.-Y., Zhou, X. et al. (2020). Energy conservation and carbon flux distribution during fermentation of CO or H_2/CO_2 by clostridium ljungdahlii. *Frontiers in Microbiology* 11: article 416. https://doi.org/10.3389/fmicb.2020.00416.

1 Methane

1.1 Application

Methane is a simple gaseous hydrocarbon with the formula CH_4 and a molecular weight of 16 g/mol. It is the main ingredient of natural gas which is defined as "dry gas" with a methane content of >80% and "wet gas" with a methane content of <80% (Thinkstep 2017). Natural gas also consists of ethane, propane, and butane referred to as the "natural gas liquids" (NGLs), which condense gat lower temperatures and can rather easily be removed from natural gas. The NGLs s may be channeled into diverse value chains for producing other chemicals like a cracking process in the ethane cracker to derive, e.g. ethylene and propylene, a dehydrogenation of propane to derive propylene, or a chemical conversion to C4 molecules with higher molecular functionality derived from butane.

Natural gas is mainly used as an energy source to generate heat and power or as a feedstock in the chemical industry like in the synthesis of acetylene, syngas, carbon monoxide, and methanol (Arpe 2007). Its global consumption is estimated about 3.800 bn m^3 in 2020 (Focus 2021). The carbon footprint (scope 3) of methane varies between 6 and 30 g CO_2e/MJ (lower heating value [LHV]), depending on the specific extraction efforts and the distance between source and sink (OCI 2021; Gan et al. 2020; Thinkstep 2017). Natural gas is usually sold according to its energy content, and the price level is about 5–6 €/mmBTU in Europe and 2–3 €/mmBTU in the United States (OCI 2021; Focus 2021; Centi 2020; Capital 2019).

1.2 Conventional Production of Methane

Natural gas and methane, accordingly, are mainly produced via the exploitation of natural reservoirs. The process does not include a chemical conversion since methane is the main component of the gas mixture extracted out of the natural reservoir through some purification steps that comply with the specification of the respective grid. The local gas supplier is usually slightly adjusting the composition of natural gas to comply with the natural Wobbe index (Thema et al. 2019) and to ensure a rather constant heating value despite fluctuations and varying composition in the natural reservoir. For both the application of natural gas as fuel and chemical feedstock, it has been managed to use the gas mixture without costly processing efforts. The use of natural gas for the production of ammonia or syngas

Pathway Design for Industrial Fermentation, First Edition. Walter Koch.

for hydroformylations does usually not require isolating methane, though the natural gas mixture is channeled into the process.

Given the low extraction costs, alternative technologies for methane were traditionally not focused on. The rising interest in academia and the chemical and power industry for synthetic natural gas (SNG) or fermentative methane with carbon dioxide as feed is driven by sustainability considerations. Both technologies would presumably result in higher production costs, though might provide a huge leverage for the chemical use of carbon dioxide as a chemical building block. This shall be accomplished via the generation of green power based on photovoltaic cells and wind energy to provide green hydrogen via electrolysis, which would then be applied for the hydrogenation of carbon dioxide. The thermic use of methane or SNG would still generate carbon dioxide, though the net balance would be zero since each burnt mol of methane has beforehand consumed 1 mol of carbon dioxide.

Furthermore, methane shall be developed as energy storage ("power-to-gas," Götz et al. 2016; Ewald et al. 2015). The shift to renewable production of energy derived from solar power and wind energy – which are not permanently available pending on daytime and season – combined with the prospected phasing out of fossil energy sources will require the development and installation of energy storage technology. Since conventional batteries are not economic for high power capacity, the chemical energy carriers hydrogen, ammonia, methanol, and methane are discussed (Rivard et al. 2019; Burkart et al. 2019; Spurgeon and Kumar 2018; Goeppert et al. 2014). All would have different advantages and disadvantages as energy storage if the criteria production cost, energy density, cost of transportation, and cost of storage and handling security are compared. The great advantage of methane would be that it can easily be stored and transported via the existing national gas grids. The German gas grid has a maximum capacity of 24.6 bn m^3 corresponding to an energy storage potential of about 246 TWh (Handelsblatt 2022; Güssgen et al. 2021; Ewald et al. 2015). This is a major advantage for methane as opposed to hydrogen for which – with some exceptions – a new transportation infrastructure would need to be installed. It is also considered to transport hydrogen in the existing gas grids and to selectively extract hydrogen at defined exit points. The current national grid regulations define a maximum hydrogen content of 0.1 up to 10% (v/v), and it needs to be ensured that the current equipment and material are suitable for hydrogen transportation with respect to potential attack of pipe material and diffusion loss (Rusmanis et al. n.d.). Accordingly, the chemical hydrogenation of carbon dioxide and the fermentative generation of methane with "green" hydrogen as energy source (Thema et al. 2019) have attracted attention.

1.3 Carbon Dioxide as Feedstock

The hydrogenation of carbon dioxide with green hydrogen may consume carbon dioxide from different feedstocks. In particular, the carbon dioxide might be captured at a power plant generating electrical power based on fossil fuels (e.g. the burning of natural gas or coal), at chemical plants that use natural gas as energy source (e.g. ammonia and hydrogen production), maybe captured out of the atmosphere via direct air capture (DAC; Ozkan et al. 2022; House et al. 2011), or be obtained in the upgrading of biogas. The fermentative production of biogas defines a technology field on its own. It applies various agricultural

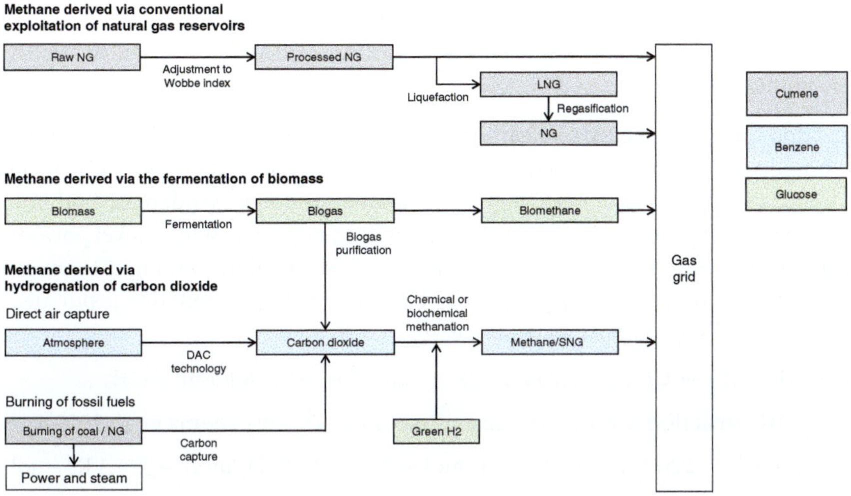

Figure 1.1 Potential role of biological methanation in the generation of methane. Source: Walter Koch.

and organic feedstocks that, when digested by anaerobic bacteria, convert into basically 50–60% (v/v) methane and up to 35% (v/v) carbon dioxide. A broad mixture of hosts or microbial consortia, respectively, is used which are rather complex and characterized by diverse metabolic interactions. After the hydrolysis of the biomass, acidogenesis takes place. The generated organic fatty acids are then further digested by other fermenting bacteria into either acetic acid or carbon dioxide and hydrogen. Finally, methane formation proceeds via acetotrophic methanogenesis (acetate as substrate) and hydrogenotrophic methanogenesis (carbon dioxide and hydrogen as feedstock). The nonindustrial biogas technology is operated in a high number of local small-scale plants, mainly according to the local availability of agricultural waste streams. The obtained biogas is either directly burnt to generate heat or purified to biomethane and channeled into the gas grid (Adnan et al. 2019). Feeding the biomethane share of biogas into the gas grid requires the removal of carbon dioxide and would cause a rather high emission of carbon dioxide. In case hydrogen would be available, the carbon dioxide might be converted into additional methane and the emission of carbon dioxide would be avoided. Accordingly, the hydrogenation of carbon dioxide might also be applied for upgrading the quality of biogas to raise the methane output and avoid the emission of the by-product carbon dioxide. The technological options for the generation of methane, the sources of carbon dioxide, and the potential role of biological methanation may be summarized as follows (Figure 1.1).

Since a commercially reasonable hydrogen supply of the small and often nonindustrial biogas plants is difficult, specific boundary conditions for technology design are to be obeyed. The local generation of hydrogen would avoid high efforts for hydrogen processing and transportation. The direct integration of water electrolysis and hydrogenation of carbon dioxide with in situ conversion of hydrogen at the cathode might avoid the implementation of Environmental, Health and Safety (EHS) efforts to address the explosion risk in hydrogen handling, which might hamper commercial feasibility of local settings (Mayer et al. 2017).

1.4 Conversion of Carbon Dioxide into Methane

1.4.1 The Chemical Sabatier

The target reaction for the conversion of carbon dioxide into methane would be the direct reduction with molecular hydrogen. This reaction is sometimes referred to as the "chemical Sabatier" in honor of the historical discovery by P. Sabatier and J.-B. Senderens. Both are known for their pioneering work on catalytic hydrogenation originally with Nickel powder and were awarded with the Novel Price in 1912. The chemical and biotechnological conversion obtain water in different states of aggregation. The chemical hydrogenation is pursued at temperatures >100 °C and obtains water as steam:

$$CO_2\,(g) + 4\,H_2\,(g) \rightarrow CH_4\,(g) + 2\,H_2O\,(g) \qquad \Delta H_R{}^\circ = -164.8\ \text{KJ/mol}\ CH_4$$

The enthalpy of the reaction is estimated according to Hess's law (Thema et al. 2019):

$$\Delta H_R{}^\circ = \Sigma\Delta H_f{}^\circ{}_P - \Sigma\Delta H_f{}^\circ{}_E = (-74.8\ \text{KJ/mol} + 2 \times -241.8\ \text{KJ/mol}) - 393.3\ \text{KJ/mol}$$
$$= -164.8\ \text{KJ/mol given}$$

$$\Delta H_f{}^\circ CO_2\,(g) = -393.3\ \text{KJ/mol},$$

$$\Delta H_f{}^\circ CH_4\,(g) = -748\ \text{KJ/mol and}$$

$$\Delta H_f{}^\circ H_2O\,(g) = -2418\ \text{KJ/mol}$$

The maximum stoichiometric energy efficiency of the reaction amounts to 83.3% in case the low heating value of hydrogen (120 GJ/t) and methane (50 GJ/t) is considered (50 GJ/t/0.5 × 120 GJ/t; Bernacchi et al. 2014a,b) and the process utilities are omitted. The "chemical Sabatier" is pursued as a heterogeneously catalyzed gas phase reaction. The catalyst is sensitive toward contaminations in the feed stream, such as hydrogen sulfide and siloxanes, which need to be removed. This defines a disadvantage versus the biological methanation, which is more robust with respect to contaminations in the feed stream (Leonzio 2016; Strevett et al. 1995). The selectivity is rather efficient. Partially the typical nickel hydrogenation catalyst has been replaced by a more expensive ruthenium catalyst which is less prone to decomposition (Burkart et al. 2019). The conversion of the reaction is incomplete and a recycling of the non-converted educts is required. The "Le Chatelier" principle favors a low temperature of the exothermic reaction and a high pressure which, however, would raise the EHS requirements and the investment for an industrial plant. On the other hand, kinetics favors high temperature. Accordingly, a compromise of the reaction parameters is needed and several combinations have been tried. The reaction is often pursued at a pressure of about 100 bar to support conversion to the right side of the equilibrium at temperatures between 200 and 500 °C (Rönsch et al. 2016; Götz et al. 2014). Especially, efficient heat removal needs to be considered in the choice of the most suited reactor. Usually, fixed-bed reactors are used, though also fluidized-bed, monolith, foam, micro-channel, membrane-based, and slurry reactors were evaluated. Most methanation plants apply a fixed-bed reactor (Rönsch et al. 2016; Ghaib et al. 2016). The mechanism of the reaction is not yet fully elucidated and might proceed via carbon monoxide or a direct hydrogenation. Catalyst development is focusing on decreasing the reaction temperature to increase conversion of the exothermic reaction and to extend the catalyst's lifetime. The initial chemisorption of carbon dioxide is depending on the catalyst surface. Several metal combinations for catalyst design are tested comprising rhodium, ruthenium, palladium, nickel, alumina-supported nickel, silica-supported nickel, zirconia, and ceria (Frontera et al. 2017). Hereby low catalyst cost enabled by high activity and low metal cost are crucial for process economy and especially the noble metal catalysts Ru, Rh, and Pd shall be avoided (Zhou et al. 2016). A prominent pilot plant is the Audi “e-gas” project which operated a 6 MW power-to-methane plant in 2013. The hydrogen feed is pressurized to a level of 10 bar before it is channeled into the reactor. Heat removal is facilitated with a salt bath reactor (Ghaib et al. 2016).

A broad industrial application of the chemical Sabatier is not yet possible given the much higher cost for producing methane – compared to the exploitation of natural gas reservoirs – and the limited availability of green hydrogen with low or zero carbon footprint. At minimum 2.75 t carbon dioxide is consumed for producing of one-ton methane (44 g/mol/16 g/mol), though about 4.5 tons carbon dioxide might be released given a hydrogen unit consumption rate of 0.5 t hydrogen per ton methane and a carbon footprint of about 9 t carbon dioxide per ton hydrogen (including Scope 3) if generated via conventional steam methane reforming technology (Boulamanti and Moya 2017).

1.4.2 The Biochemical Sabatier

The fermentative conversion of carbon dioxide into methane or the “biochemical Sabatier” is operated at temperatures below 100 °C and obtains liquid water as by-product. Since the evaporation energy is not absorbed in the gasification, the released heat is higher than in the chemical process and needs to be removed by cooling:

$$CO_2\ (g) + 4\ H_2\ (g) \rightarrow CH_4\ (g) + 2\ H_2O\ (l)\ \Delta H_R{}^\circ = -253.1\ \text{KJ/mol}\ CH_4$$

(Thema et al. 2019)

The enthalpy of the reaction is estimated according to Hess's law:

$$\Delta H_R{}^\circ = \Sigma\Delta H_f{}^\circ{}_P - \Sigma\Delta H_f{}^\circ{}_E = (-74.8\ \text{KJ/mol} + 2\text{x} - 285.8\ \text{KJ/mol}) - 393.3\ \text{KJ/mol}$$
$$= -253.1\ \text{KJ/mol given}$$

$$\Delta H_f{}^\circ CO_2\ (g) = -393.3\ \text{KJ/mol}$$

$$\Delta H_f{}^\circ CH_4\ (g) = -748\ \text{KJ/mol and}$$

$$\Delta H_f{}^\circ H_2O\ (l) = -2858\ \text{KJ/mol}$$

The Gibbs free energy of the reaction has been estimated and reveals that the reaction is in principle sufficiently exergonic to pursue a fermentative technology:

$$CO_2\ (g) + 4\ H_2\ (g) \rightarrow CH_4\ (g) + 2\ H_2O\ (l)\ \Delta G^{\circ\prime} = -131\ \text{KJ/mol}\ CH_4$$

(Goyal et al. 2016; Thauer et al. 2008)

A more realistic value of $\Delta G^{\circ\prime} = -126.6$ KJ/mol was obtained with Aspen at ambient pressure and 338 K or 65 °C (Bernacchi et al. 2014a,b). This value should in principle be sufficient in case the biochemical system efficiently manages to channel the energy potential toward the formation of ATP. For the biochemical conversion, it is critical that the Gibbs free

Table 1.1 Comparison of chemical and biological CO_2 methanation.

	Chemical methanation	**Biological methanation**
Reaction	CO_2 (g) + 4 H_2 (g) → CH_4 (g) + 2 H_2O (g)	CO_2 (g) + 4 H_2 (g) → CH_4 (g) + 2 H_2O (I)
Phase	Gas phase reaction/200 °C – 500 °C	Liquid phase reaction/<70 °C
Enthaply	$\Delta H_R{}^\circ = -164.8$ kJ/mol	$\Delta H_R{}^\circ = -253.1$ kJ/mol
Pressure	About 100 bar	<10 bar
Catalyst	Ni-based/nobel metal	Biocatalyst (methanogenic bacteria)
Reactor	Fixed bed/fluidized bed	CSTR
Challenge	Efficient removal of heat/catalyst sensitivity	Low solubility of H_2 Low space time yield

energy is sufficiently negative to allow in parallel the generation of ATP for cellular maintenance in resting cells. In case the same feedstock shall be applied to raise the inoculum also sufficient energy for microbial growth is needed. The Gibbs free energy of the fermentation pathway needs to provide a phosphoryl group transfer potential for the formation of ATP of at least −31.8 KJ/mol plus a surplus to account for realistic energy efficiency in the biological system (Thauer et al. 1977).

The reaction is implemented at moderate temperatures <100 °C pending on the heat resistance of the biocatalyst and is pursued at low pressure. A slow kinetics and a potentially weak mass transfer in the reactor are regarded as disadvantages compared to the chemical process (Ghaib et al. 2016).

The chemical methanation and the biological methanation are two competing technological approaches to convert green power into methane as energy storage (Table 1.1).

1.5 Biochemical Pathway Design

The biochemical pathway for the reduction of carbon dioxide toward methane covers several steps, and the involved co-factors may deviate between various species for which the ability of biological methanation has been reported. With respect to metabolic engineering, it can be concluded that still the native biochemical pathway for methane generation is pursued. The biological methanation is carried out by methanogenic bacteria, which are a diverse group of Archaea microorganisms characterized by their ability to produce methane (Balch et al. 1979). They are found in anoxic habitats, are highly sensitive toward oxygen, and, accordingly, require specific anaerobic cultivation procedures. Pending on the feedstock hydrogenotrophic (hydrogen), aceticlastic (acetic acid) and methylotrophic (methanol) methanogenesis are distinguished. The process exemplified by the archaebacterial model organism *Methanococcus maripaludis*, a strictly anaerobic, hydrogenotrophic methanogen, is reconstructed as follows (Goyal et al. 2016; Table 1.2):

Altogether eight electrons need to be transferred to the fully oxidized C-atom of methane. Carbon dioxide in the first step reacted with methanofuran (MFR) to provide formyl-THF as the first reduction product after having received two electrons from reduced ferredoxin (Figure 1.2).

Table 1.2 Biochemical pathway for the generation of methane.

Reaction	$\Delta G^{\circ\prime}$
$CO_2 + MFR + Fd_{red} \rightarrow$ Formyl-MFR $+ Fd_{ox} + H_2O$	$\Delta G^{\circ\prime} = 0.0$ KJ/mol (Thauer et al. 2008)
$Fd_{ox} + H_2 + PMF \rightarrow Fd_{red}$	
Formyl-MFR + THMPT → Formyl-THMPT + MFR	$\Delta G^{\circ\prime} = -5.0$ KJ/mol (Thauer et al. 2008)
Formyl-THMPT → Methenyl-THMPT + H_2O	$\Delta G^{\circ\prime} = -5.0$ KJ/mol (Thauer et al. 2008)
Methenyl-THMPT + $F420_{red}H_2$ → Methylene-THMPT + $F420_{ox}$	$\Delta G^{\circ\prime} = 6.0$ KJ/mol (Thauer et al. 2008)
Methylene-THMPT + $F420_{red}H_2$ → Methyl-THMPT + $F420_{ox}$	$\Delta G^{\circ\prime} = -6.0$ KJ/mol (Thauer et al. 2008)
$2\ F420_{ox} + 2\ H_2 \rightarrow 2\ F420_{red}H_2$	$\Delta G^{\circ\prime} = -11.0$ KJ/mol (Thauer et al. 2008)
Methyl-THMPT + CoM-SH → Methyl-S-CoM + THMPT	$\Delta G^{\circ\prime} = -30.0$ KJ/mol (Thauer et al. 2008)
Methyl-S-CoM + HS-CoB → CH_4 + CoM-S-S-CoB	$\Delta G^{\circ\prime} = -30.0$ KJ/mol (Thauer et al. 2008)
CoM-S-S-CoB + H_2 → CoM-SH + CoB-SH	$\Delta G^{\circ\prime} = -39.0$ KJ/mol (Thauer et al. 2008)
$CO_2 + 4\ H_2 \rightarrow CH_4 + 2\ H_2O$	$\Delta G^{\circ\prime} = -131.0$ KJ/mol (Thauer et al. 2008)

MFR, methanofuran; Fd, ferredoxin; THMPT, tetrahydrosarcinapterin; HS-CoB, coenzyme B; HS-CoM, coenzyme M. The standard free energy was calculated from equilibrium constants or from the standard free energies for formation at 25 °C with H_2, CO_2, and CH_4 in the gaseous state at 10^5 Pa, H_2O in the liquid state, pH at 7.0 and all other compounds at 1 M activity.
Source: Adapted from Thauer et al. (2008).

This step is facilitated by the Fwd/Fmd enzyme, a tungsten-molybdenum containing formylmethanofuran dehydrogenase. The preceding reduction of ferredoxin may be facilitated via EchA or the Vhu/Hdr bifurcation complex. In the statement of the reaction, the pathway via the EchA enzyme is assumed which additionally uses a proton motive force (PMF) in order to provide the required reduction potential for the reduction of ferredoxin (−500 mV) which cannot be achieved with hydrogen (−414 mV) alone. Alternatively, the reduction of ferredoxin might be facilitated by the Vhu/Hdr bifurcation complex. It consumes 2 mol of hydrogen and additionally accomplishes the reduction of the disulfide bridge in CoM-S-S-CoB to regenerate the cofactors with each carrying a reduced sulfhydryl function. In the second step, the formyl group is transferred to THMPT (tetrahydromethanopterin) arranged by the formyltransferase to obtain formyl-THMPT. Formyl-THMPT is the subject of a dehydration facilitated by the methylene-THMPT cyclohydrolase (Mch) to obtain methenyl-THMPT. Afterward four electrons are transferred by two mol of $F420_{red}$ to obtain methylene-THMPT in the first step and Methyl-THMPT in the second step whereby the electron-providing cofactor is oxidized. The regeneration and reduction, respectively, of F420 are facilitated by the oxidation of hydrogen. The derived methyl function is handed over by a membrane-bound enzyme complex, namely, the methyltransferase (Mtr), to coenzyme M to obtain Methyl-S-CoM with activated thioester

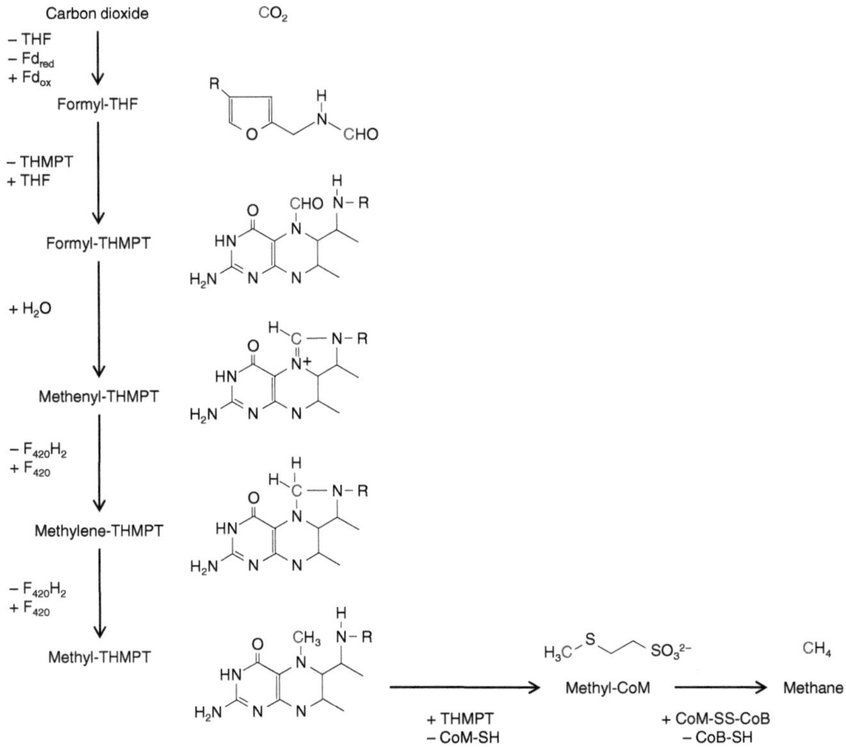

Figure 1.2 Biochemical pathway for the conversion of carbon dioxide into methane (Sowers 2009). Source: Walter Koch.

in a highly exergonic reaction. This methyl transfer is coupled to the translocation of two sodium ions to generate a sodium gradient which drives an ATP synthase. Finally, methyl-S-CoM is subject to a reductive demethylation to obtain methane. The oxidized disulfide bridge between the co-factors CoM and CoB is recycled via a reduction with molecular hydrogen (Goyal et al. 2016).

The limited metabolic engineering tools for metabolic engineering of methanogenic bacteria had blocked the implementation of pathway improvement projects in the past. The recent progress with respect to the model organism *Methanothermobacter* – the microbe of the year 2021 – may change this and might allow the generation of other products than methane (Molitor et al. 2023).

1.6 Integration of Hydrogen Production and the Biochemical Methanation

1.6.1 Conversion of Carbon Dioxide into Methane with Integrated Production of Hydrogen

The fermentative reduction of carbon dioxide to methane as outlined above relies on the external supply of molecular hydrogen. Since the conversion of carbon dioxide into

Table 1.3 Generation of methane coupled with the electrolysis of water.

$4\ H_2O\ (l) \rightarrow 4\ H_2\ (g) + 2\ O_2\ (g)$	$\Delta H_R{}^\circ = 285.8$ KJ/mol H_2O
$CO_2\ (g) + 4\ H_2\ (g) \rightarrow CH_4\ (g) + 2\ H_2O\ (l)$	$\Delta H_R{}^\circ = -253.1$ KJ/mol CH_4
$CO_2\ (g) + 2\ H_2O\ (l) \rightarrow CH_4\ (g) + 2\ O_2(g)$	$\Delta H_R{}^\circ = 890.1$ KJ/mol CH_4 (Thema et al. 2019)

methane is from a sustainability point of view exclusively justified in case hydrogen is generated without significant emissions of carbon dioxide, water electrolysis driven by green power is needed unless rather infinite capacities for Carbon capture and storage (CCS) projects are granted. Thus, the external supply of hydrogen supposes an infrastructure for transportation and presumable storage which is circumvented in case the generation of hydrogen and the hydrogenation of carbon dioxide are integrated (Mayer et al. 2017). The integration of electrical green power to drive the water electrolysis with the conversion of carbon dioxide into methane avoids the generation, storage, and eventually the transportation of hydrogen. Since an infrastructure for the storage and transportation of hydrogen is usually not yet available – with the exemption of some chemical clusters with steam methane reformer (SMR) plants and adjacent hydrogenation plants – the technology integration might be an advantage for future application. The green power generated by wind turbines and photovoltaics could directly be used to obtain methane which would then be used or stored in local gas grids. The overall reaction is derived as the addition of the two single reactions shown below. The high endothermic character of the reaction is due to the high energy input for splitting water (Table 1.3):

The enthalpy of the reaction is confirmed by application of Hess's law:

$$\Delta H^\circ{}_R = \Sigma\Delta H_f{}^\circ{}_P - \Sigma\Delta H_f{}^\circ{}_E = -74.8\ \text{KJ/mol} - (-393.3\ \text{KJ/mol} + 2x - 285.8) = 890.1\ \text{KJ/mol}$$

$$\Sigma\Delta H_f{}^\circ\ CO_2\ (g) = -393.3\ \text{KJ/mol},$$

$$\Sigma\Delta H_f{}^\circ\ CH_4\ (g) = -748\ \text{KJ/mol and}$$

$$\Sigma\Delta H_f{}^\circ\ H_2O\ (l) = -2858\ \text{KJ/mol}$$

In case the electrolysis of water is integrated with the fermentative hydrogenation of carbon dioxide, the field of microbial electrosynthesis (MES) is entered. MES is a developing technology to provide electrical power for fermentation systems providing electrons for the buildup of reducing equivalents, which can be used for the reduction of material feedstocks (Gong et al. 2020; Jafary et al. 2015). The common setup foresees to install an anode and a cathode in a two-chambered system with separated anode and cathode space. Once a sufficient potential is applied, water splitting proceeds at the anode which forms molecular oxygen, protons, and electrons. The potential at the cathode may be controlled with a potentiostat to ensure a sufficient potential for the generation of hydrogen through or to block a short-term high potential that might hurt the host or the biofilm, respectively. This measure is reasonable in R&D but might be impractical for a commercial application. The oxygen is not needed and is released into the reactor headspace. The protons diffuse into the cathode space via a proton-exchange membrane and receive electrons at the cathode for the generation of fermentative reduction equivalents. In order to save purchasing of membranes, which likely need to be replaced in defined intervals, a membrane-less reactor has been proposed (Giddings et al. 2015). The vertical arrangement of both electrodes with the anode above the cathode in membrane-less reactor ensures that the formed and ascending oxygen does not interfere with the oxygen-sensitive host in the cathode space.

1.6.2 Mechanisms at the Cathode for the Uptake of Reduction Equivalents

The uptake of reduction equivalents by the biocatalyst may proceed via different mechanisms and three modes of interaction are distinguished (Enzmann et al. 2018; Mayer et al. 2019). In case a direct electron transfer (DET) is assumed, the organism would be in contact with the cathode via cytochromes, membrane-bound proteins, conductive filaments (so-called "nanowires"), or outer membrane extensions. The microbial host usually forms a biofilm on the cathode. A DET mechanism is assumed for the species *Geobacter sulfurreducens* and *Shewanella oneidensis*. The metabolic abilities of the microorganisms inhabiting the cathode surface are not yet fully understood. Transcriptome analysis of host community members isolated in the supernatant of the cathode or near the cathode surface revealed a distinct expression pattern of soluble hydrogenases, ferredoxins, formate dehydrogenases, and cytochromes which are likely involved in the metabolic management of the electron flow (Marshall et al. 2017; Perona-Vico et al. 2020).

A mediated electron transfer (MET) is assumed in case a redox-active component receives the electrons from the cathode and channels them to the fermentation host. The mediator might be added to the biochemical system or be induced and self-secreted by the host. *S. oneidensis* is known to secrete riboflavin as redox mediator. A MET is assumed for *Escherichia coli* mediated by neutral red; a cationic dye that is used as a histological stain due to its optical sensitivity to pH in the physiological range between pH 6–8. If neutral red is added to the medium, it is protonated with protons out of the cathode medium and receives electrons from the cathode. It diffuses into the inner membrane and transfers the electrons to menaquinone, which diffuses within the membrane to the active site of a terminal reductase or to the arcB enzyme which ultimately delivers the electrons to a terminal electron acceptor like fumarate (Harrington et al. 2015).

Additionally, an indirect electron transfer (IET) might proceed in which the electrons at the cathode are received by the protons generated at the anode and molecular hydrogen is abiotically synthesized without the involvement of the microbial host. The hydrogen would then be absorbed by the microorganism submersed in the cathode medium for the biochemical hydrogenation steps (Enzmann et al. 2018). In case an abiotic hydrogen generation is pursued, it is crucial to establish an efficient uptake of the molecular hydrogen in *statu nascendi* to avoid diffusion of the costly reduction agent into the reactor headspace. With respect to *Sporomusa ovata,* it could be shown that the biofilm at the cathode is highly effective in scavenging the evolving hydrogen (Giddings et al. 2015).

MES is pursued for different interests pending on the target molecule. It might be used to provide additional electrons in case the fermentation feedstock does not provide sufficient electrons for the target pathway, e.g. in the fermentative production of 1,3-propanediol based on glucose (Kracke and Krömer 2014). So far several molecules including acetic acid, butyric acid, isobutyric acid, isopropanol, butanol, and isobutanol could be produced via MES from carbon dioxide and electricity (Cheng et al. 2009; Ganigué et al. 2015; Nevin et al. 2011; Vassilev et al. 2018). MES might alternatively be tested in case a highly or

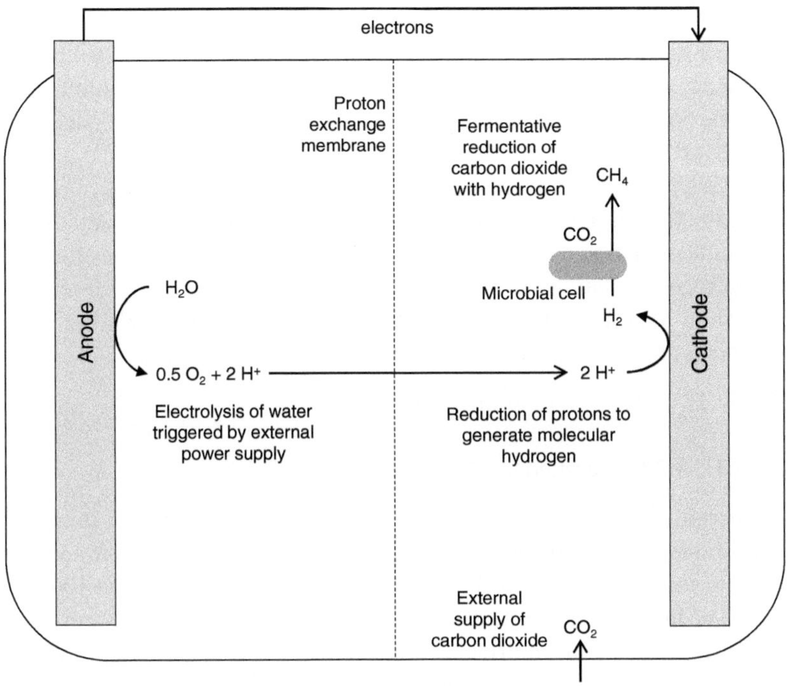

Figure 1.3 Integrated bioelectrochemical system consisting of a water electrolysis combined with the fermentative reduction of carbon dioxide toward methane. Source: Adapted from Kracke et al. (2019).

fully oxidized feedstock shall be reduced to obtain a biofuel. This approach is pursued for producing methane based on carbon dioxide, water, and electrical power.

1.6.3 R&D with Integrated Hydrogen Production and Biochemical Methanation and IET

Proof of concept based on experimental data for the combination of water electrolysis integrated with the fermentative reduction of carbon dioxide via IET has been shown (Kracke et al. 2019). The bio-electrochemical reactor consists of two chambers with a volume of 150 ml separated by a Nafion proton-exchange membrane with a surface area of 7.5 cm^2. It avoids the diffusion of oxygen into the cathode chamber where the oxygen-sensitive hosts are cultivated (Figure 1.3).

In the first step, the formation of molecular hydrogen at the cathode was shown by applying either cobalt-phosphide (CoP), molybdenum-disulfide (MoS_2), or nickel-molybdenum (NiMo) alloy as cathode materials. The commonly used and cheaper carbon-based cathode was ruled out since microbial electrosynthesis is pursued under specific conditions – neutral pH in the presence of media components for the biocatalyst – and instead, cathode materials including transition metals were evaluated, which are known to significantly increase the hydrogen evolution rate in MES. These cathodes were tested with respect to their biocompatibility for the generation of hydrogen. The cathode space was sparged with carbon dioxide at neutral pH in 0.03 M $NaHCO_3$. Applying a constant current of 1 mA/cm^2

and a potential of −1 V versus the standard hydrogen electrode (SHE), a maximum hydrogen production rate of 48 μmol/h × cm^2 observed for the NiMo cathode is achieved with a Coulomb efficiency of 98%. Coulomb efficiency was calculated by dividing the electrons recovered in the target molecule by the electrons supplied as current for a defined incubation time. In the next steps, the cathode efficiency was tested after microbial media were added to the electrolyte suitable for methanogenic and homoacetogenic bacteria. It could be observed that the Coulomb efficiency decreased in the homoacetogenic medium likely due to the unspecific reduction of organic media compounds. After separate analysis and confirmation of hydrogen formation, the cathode space was inoculated with either the homoacetogenic bacterium *S. ovata* or the methanogenic archaeon *M. maripaludis*. Both would use the generated hydrogen for producing either methane or acetic acid.

$$2\,CO_2 + 4\,H_2 \rightarrow CH_3COOH + 2\,H_2 \text{ pursued by } S.ovata$$

$$CO_2 + 4\,H_2 \rightarrow CH_4 + 2\,H_2 \text{ pursued by } M.maripaludis$$

The MES would proceed via an IET and the evolved hydrogen is used by the hosts as electron source either to produce methane or acetic acid. It could be shown that an elevated hydrogen concentration could exclusively be found in close proximity to 50–100 μm of the cathode surface. The hydrogen concentration was 0.2–0.6 μmol/l in the medium containing the biocatalyst and around 220 μmol/l without the biocatalyst. This efficient hydrogen uptake is seen as an advantage since it avoids the development of a DET, which may restrict the repertoire of suitable hosts, and facilitates efficient production rates. With an electron supply rate of 3.73×10^{-2} mmol/h, a methane generation of 9.16×10^{-3} mmol/h (Coulomb efficiency 98.2%) and an acetic acid generation of 4.7×10^{-3} mmol/h (Coulomb efficiency 100.8%), respectively, could be observed (Kracke et al. 2019). The supply of electrons and the generation of methane and acetic acid (indicated as electron equivalents) were plotted against time (batch duration 50 hours). A proportional relation was shown as expected.

1.6.4 Boundary Conditions for Potential Commercial Application

The commercial attractiveness of the integrated hydrogen production depends mainly on the overall energy efficiency of the microbial electro-fermentation. The benchmark for the integrated approach is the external supply of green hydrogen generated by water electrolysis (polymer exchange membrane; PEM), which has an energy usage of 50 MWh/t H_2 (Tountas et al. 2021; Götz et al. 2016). In case the hydrogen yield in the fermentative conversion of carbon dioxide with hydrogen to methane (without microbial electrosynthesis) is assumed with 98% and the corresponding hydrogen unit consumption rate is 0.51 t H_2/t CH_4, the overall power consumption of the process would be 25.5 MWh/t H_2 without including the power demand in the fermentation process itself, e.g. gas compression and downstream purification (50.0 MWh/t H_2 × 0.51 t H_2/t CH_4).

For comparison, the power consumption for the integrated hydrogen production and biochemical methanation can be estimated based on the available information keeping in mind that the mini setup of the experiments is not designed to accomplish a potential commercial efficiency. The estimation outlined below refers to the work of Prof. Spormann at Stanford University (Kracke et al. 2019). In one experiment, a hydrogen evolution rate of

39 μmol/cm^2 × h was achieved with a platin cathode. The constant potential was −1.0 V versus SHE. The cathode surface area is 1 cm^2 and the current density is given with constant 1 mA/cm^2 for all experiments. The intensity of current of the system can be derived with 1 mA. Thus, given that 1 W = 1 V × 1 A and 1 A = 1 C/s, it can be derived that 1 hour of incubation implies an electrical work of 3.6 W. In this 1 h 39 μmol [H]/cm^2, cathode area is generated translating into 39 μmol hydrogen absolute generation within the system:

3.6 Wh corresponds to 39 μmol [H], derive

3.6 Wh corresponds to 39 μg H_2, derive

0.092 Wh corresponds to 1 μg H_2, derive

92 MWh corresponds to 1 kg H_2

The cathode has absorbed 3.6 C in 1 h. The derived 39 μmol [H] requires 3.76 C (96.485 C/mol) and the indicated Coulomb efficiency of 99% is plausible given a realistic measurement accuracy of the system. Accordingly, the derived power consumption in the experimental system is 1840-fold higher than the consumption in the fully optimized stand-alone PEM water electrolysis. The huge gap indicates the ambitious development target of such a system to achieve commercial maturity. Especially, the intensity of current needs to be increased up to a range of 100–1000 mA/cm^2 (Kracke et al. 2021).

The high-power unit consumption rate is caused by several reasons. It is in the first instance not realistic to achieve a commercially interesting usage rate in the experimental setup. The whole system is designed to optimize the biochemical conversion of carbon dioxide. Anode and cathode space for the electrolysis of water is optimized with respect to the requirements of the biocatalyst (i.e. pH 7.0, media components) and not for electrochemical efficiency. The suitability of the integrated system for the methanogenic host *M. maripaludis* allows to use its specific ability for IET and thereby avoids the tedious development of a technology with DET. Specific cathode material is evaluated and the high experience in cathode material development for a conventional alkaline electrolysis (AE) or the PEM technology can likely not be exploited.

1.7 Process Development for the "Biochemical Sabatier" without Integrated Water Electrolysis

The feeding of a gas stream consisting of carbon dioxide and hydrogen and its circulation within the reactor defines a critical hurdle for facilitating an efficient productivity. Carbon dioxide has a reasonable solubility in water via the formation of bicarbonate, though hydrogen might display a limiting accessibility for the host and might negatively impact space-time yield. Solubility of hydrogen is increased with lowered temperature and raised pressure. Whereas the first option is limited given the temperature sensitivity of the biocatalyst the pressure can be raised to some extent although then also the methane pressure in the reactor headspace would be higher (Leonzio 2016). In a process simulation using Aspen, a theoretically optimized pressure of 11 bar could be identified at which about 80% of the total carbon dioxide feed is converted (Bernacchi et al. 2014a,b). The critical mass transfer and the corresponding hydrogen consumption of the biocatalyst are enhanced with a high mass transfer coefficient K_L, an increased surface area between the gas bubbles and the outer host membrane, and the hydrogen concentration gradient in the liquid phase and the gas phase.

Several reactor types have been considered to address this challenge (Thema et al. 2019). In the continuously stirred tank reactor (CSTR), the microorganism is suspended in the liquid phase, and the feed gases are introduced at the bottom of the reactor. The stirrer would permanently be mixing the liquid and the gaseous phase and especially provide a high surface area via a reduction of the size of the gas bubbles. The heat of the exothermic reaction is removed with cooling equipment. The methane product stream including non-converted educts is captured at the reactor head and channeled into the downstream process section. Alternatively, a bubble column reactor (BCR) might be employed in which the pressurized gas stream of carbon dioxide and hydrogen ensures efficient mixing of both phases in the reactor. A stirrer would not be needed. The saved power demand is counterbalanced by the efforts to pressurize the feed gases. An interesting approach to address the difficult accessibility of hydrogen for the biocatalyst would be the trickle-bed reactor (TBR), which is packaged with small solid particles to facilitate a small diameter and, accordingly, a huge overall surface of the hydrogen and carbon dioxide gas bubbles which are fed into the reactor at the bottom. The biocatalyst might be immobilized on the surface of the solid particles or be suspended in the liquid phase containing the medium. The liquid phase is captured at the bottom of the reactor and pumped for recirculation to the head of the reactor. In this step, also the heat of the reaction would be removed (Thema et al. 2019). The published mass transfer coefficients k_L are for all three reactor types in the range of 0.3 up to 4×10^{-4} m/s. With respect to the range of the effective surface area a_{eff}, the CSTR and the BCR are subscribed with 100 up to 1500 m^{-1} higher values than the TBR with 60 up to 640 m^{-1} (Thema et al. 2019). The innovative concepts of a TBR or a BCR might provide a better solution for the biological methanation. However, the applied known reactor of the pilot plant of Electrochaea is a conventional CSTR with a height of 9 m and four impellers (Rusmanis et al. n.d.). It remains to be seen if this reactor concept turns out to be the superior solution for the fermentative generation of methane or if the alternative reactor concepts are finally chosen after additional development work has been done.

1.8 Commercial Application of Fermentative Methane Production

The biochemical conversion of carbon dioxide to methane with external supply of hydrogen is pursued by the companies Electrochaea (www.electrochaea.com) and Krajete (www.krajete.com). Their technology is not yet commercialized in high-volume plants, though a broader application is to be expected. The production of SNG, either via the chemical or the biochemical methanation, is still suffering from the high costs of green hydrogen. Whereas the cost of conventional natural gas amounts to 30 €/MWh in Europe, the cost of SNG derived via the methanation of carbon dioxide is estimated to 400 €/MWh (Vahrenholt 2023).

The performance of the Electrochaea technology is described in this section since detailed information of their pilot plant in Denmark is available (Electrochaea 2014;

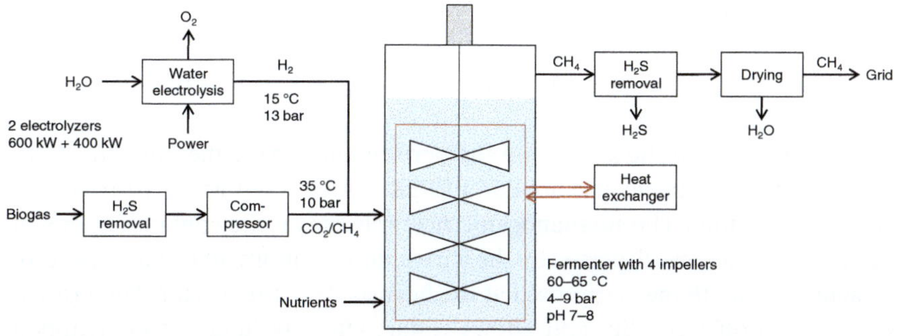

Figure 1.4 Flow diagram for the conversion of carbon dioxide with hydrogen into methane according to Electrochaea. Source: Walter Koch.

Rusmanis et al. n.d.). The technology has been demonstrated in a pilot scale in Denmark at the BIOFOS AvedØre wastewater treatment center and Electrochaea is offering the technology to commercial partners. The first commercial plant shall be constructed in Denmark and a funding of 36 € m has been secured in 2022 (Krapp 2022). Electrochaea applies a nongenetically modified archaea bacterium and combines water electrolysis with the fermentative production of methane. The feedstock is raw biogas, though also carbon dioxide might be used. The methane purity in the reactor headspace is reported higher than 97% and exceeds the critical threshold of 96% methane in compliance with local and national requirements of grid operators (Figure 1.4).

Accordingly, the methane can be injected into the gas grid as SNG. The technical ability to achieve a grid injection level gas quality is crucial for the commercial application of the technology and a major boundary condition for process design. Electrochaea succeeded in meeting the 96% threshold by using a high mixing rate, elevated pressure, and a pure strain archaea culture (Rusmanis et al. n.d.).

The process displays a high on/off flexibility which might be useful in case renewable power is applied for producing hydrogen (http://electrochaea.com 2021). Based on patents (Electrochaea patent 2020), it may be assumed that the reaction is pursued at a temperature around 60–65 °C (Bernacchi et al. 2014a,2014b) and a pH of about 7.0. The dead biomass might be kept in the reactor together with the still-active cells because the nutritional components of the dead biomass are partially released and absorbed by the living cells. Although it is common sense that oxygen inhibits important enzymes in biological methanogenesis and hampers productivity, it is reported that microaerobic conditions and minor residues of H_2S can be tolerated by some modified species. This would especially limit efforts for feedstock processing and purification. A culture of *Methanothermobacter thermoautotrophicus* with a biomass concentration of 10–12 g/l is reported in an Electrochaea patent to display a productivity of 40 l CH_4/l × d which is corresponding to a productivity of 1.2 g/l × h (Electrochaea 2020). In case this level could be achieved in the pilot plant, it would be in the range of typical fermentation productivities up to 4 g/l × h. However, based on conference presentations, a conversion of even 800 l CH_4/l × d is reported which is corresponding to an outstanding productivity of 24 g/l × h (Rusmanis et al. n.d.). This value is corresponding to the methane stream obtained in the reactor headspace, but it needs to be taken into consideration that raw biogas – which a methane content of 50%–60% (v/v) – has been applied as feedstock, and the synthetic productivity in the reactor is likely in the range of about 12 g/l × h. The reactor operates at a temperature of 60–65 °C and a pressure of 4–9 bar (Rusmanis et al. n.d.). The pressure displays a gradient in the reactor. Since the reactor has a height of 8 m the broth pressure at the bottom of the reactor is about 1 bar higher than below the headspace of the reactor.

Less details are published on the status and the performance of the methanation technology executed by Krajete. Krajete also applies a CSTR, which is fed with hydrogen and carbon dioxide at the bottom. The methanogenic host tolerates some contaminations in the feed gas. The gas bubbles are dispersed by the stirrer, and a uniform distribution within the reactor is facilitated. Methane is captured at the headspace of the reactor. The process does display a high on/off flexibility and allows restart after 500 hours of interruption (Krajete homepage 2021).

References

Adnan, A.I., Ong, M.Y., Nomanbhay, S. et al. (2019). Technologies for upgrading biogas to biomethane: a review. *Bioengineering* 6: 92. https://doi.org/10.3390/bioengineering6040092.

Arpe, H.-J. (2007). *Industrielle Organische Chemie. Bedeutende Vor- und Zwischenprodukte.* Wiley.

Balch, W.E., Fox, G.E., Magrum, L.J. et al. (1979). Methanogens: reevaluation of a unique biological group. *Microbiological Reviews* 43: 260–296.

Bernacchi, S., Weissgram, M., Wukovits, W., and Herwig, C. (2014a). Process efficiency simulation for key process parameters in biological methanogenesis. *AIMS Bioengineering* 1 (1): 53–71. https://doi.org/10.3934/bioeng.2014.1.53.

Bernacchi, S., Rittmann, S., Seifert, A.H. et al. (2014b). Experimental methods for screening parameters influencing the growth to product yield ($Y_{[x/CH4]}$) of a biological methane production (BMP) process performed with *Methanothermobacter marburgensis. AIMS Bioengineering* 1 (2): 72–87. https://doi.org/10.3934/bioeng.2014.2.72.

Boulamanti, A. and Moya, J.A. (2017). Energy efficiency and GHG emissions: prospective scenarios for the Chemical and Petrochemical Industry. JRC science for policy report, European Union, p. 17.

Burkart, M.D., Hazari, N., Tway, C.L., and Zeitler, E.L. (2019). Opportunities and challenges for catalysis in carbon dioxide utilization. *ACS Catalysis* 9: 7937–7956. https://doi.org/10.1021/acscatal.9b02113.

Capital (2019). LNG. Edition 4e, pp. 18–20. www.capital.de.

Centi, G., Perathoner, S., Salladini, A., and Iaquaniello, G. (2020). Economics of CO_2 utilization: a critical analysis. *Frontiers in Energy Research* 8: https://doi.org/10.3389/fenrg.2020.567986.

Cheng, S., Xing, D., Call, D.F., and Logan, B.E. (2009). Direct biological conversion of electrical current into methane by electromethanogenesis. *Environmental Science and Technology* 43: 3953–3958.

Electrochaea (2014). Power-to-gas via biological catalysis (P2G-Biocat). Final report. https://energiforskning.dk/sites/energiforskning.dk/files/slutrapporter/12164_final_report_p2g_biocat.pdf.

Electrochaea homepage (2021). www.electrochaea.com (accessed 14 November 2021).

Electrochaea patent (2020). Method to use industrial CO_2 containing gas for the production of a methane enriched gas composition. 7.5.2020, WO 2020/089181 A1.

Enzmann, F., Mayer, F., Rother, M., and Holtmann, D. (2018). Methanogens: biochemical background and biotechnological applications. *AMB Express* 8: 1. https://doi.org/10.1186/s13568-017-0531-x.

Ewald, S., Koschany, F., Schlereth, D. et al. (2015). Power-to-gas. *Chemie in unserer Zeit* 49: 270–278. https://doi.org/10.1002/ciuz.201500715.

Focus (2021). Faktenreport: Erdgas, primary data from Statistisches Bundesamt, Statista and DIW, p. 59.

Frontera, P., Macaria, A., Ferraro, M., and Antonucci, P. (2017). Supported catalysts for CO_2 methanation: review. *Catalysts* 7: 59. https://doi.org/10.3390/catal7020059.

Gan, Y., El-Houjeiri, H.M., Badahdah, A. et al. (2020). Carbon footprint of global natural gas supplies in China. *Nature Communications* 11: 824. https://doi.org/10.1038/s41467-020-14606-4.

Ganigué, R., Puig, S., Batle-Vilanova, P. et al. (2015). Microbial electrosynthesis of butyrate form carbon dioxide. *Chemical Communications* 51: 3235–3238.

Ghaib, K., Nitz, K., and Ben-Fares, F.-Z. (2016). Chemical methanation of CO_2: a review. *Chemical Bioengineering Review* 3 (6): 266–275.

Giddings, C.G.S., Nevin, K.P., Woodward, T. et al. (2015). Simplifying microbial electrosynthesis reactor design. *Frontiers in Microbiology* 6: 468. https://doi.org/10.3389/fmicb.2015.00468.

Goeppert, A., Czaun, M., Jones, J.-M. et al. (2014). Recycling of carbon dioxide to methanol and derived products – closing the loop. *The Royal Society of Chemical*. https://doi.org/10.1039/c4cs00122b.

Gong, Z., Yu, H., Zhang, J. et al. (2020). Microbial electro-fermentation for synthesis of chemicals and biofuels driven by bi-directional extracellular electron transfer. *Synthetic and Systems Biotechnology* 5: 304–313. https://doi.org/10.1016/j.synbio.2020.08.004.

Götz, M., McDaniel Koch, A., and Graf, F. (2014). State of the art and perspectives of CO_2 methanation process concepts for power-to-gas application. Conference paper. https://www.researchgate.net/publication/273139805_State_of_the_Art_and_Perspectives_of_CO2_Methanation_Process_Concepts_for_Power-to-Gas_Applications.

Götz, M., Lefebvre, J., Mörs, F. et al. (2016). Renewable power-to-gas: a technological and economic review. *Renewable Energy* 85: 1371–1390. https://doi.org/10.1016/j.renene.2015.07.066.

Goyal, N., Zhou, Z., and Karimi, I.A. (2016). Metabolic processes of *Methanococcus maripaludis* and potential applications. *Microbial Cell Factories* 15: 107. https://doi.org/10.1186/s12934-016-0500-0.

Güssgen, F., Tutt, C., and Wettach, S. (2021). Moskaus monopoly. *Wirtschaftswoche* 24 (9): 46–49.

Handelsblatt (2022). Im Notfall Staatshilfe. p. 33.

Harrington, T.D., Tran, V.N., Mohamed, A. et al. (2015). The mechanism of neutral red-mediated microbial electrosynthesis in *Escherichia coli*: menaquinone reduction. *Bioresource Technology* 689–695. https://doi.org/10.1016/j.biortech.2015.06.037.

House, K.Z., Baclig, A.C., Ranjan, M. et al. (2011). Economic and energetic analysis of capturing CO_2 from ambient air. *PNAS* 108 (51): 20428–20433. https://doi.org/10.1073/pnas.1012253108.

Jafary, T., Daud, W.R.W., Ghasemi, M. et al. (2015). Biocathode in microbial electrolysis cell: present status and future prospects. *Renewable and Sustainable Energy Reviews* 47: 23–33. https://doi.org/10.1016/j.rser.2015.03.003.

Kracke, F. and Krömer, J.O. (2014). Identifying target processes for microbial electrosynthesis by elementary mode analysis. *BMC Bioinformatics* 15: 410. https://doi.org/10.1186/s12859-014-0410-2.

Kracke, F., Wong, A.B., Maegaard, K. et al. (2019). Robust and biocompatible catalysts for efficient hydrogen-driven microbial electrosynthesis. *Communications Chemistry*. https://doi.org/10.1038/s42004-019-0145-0.

Kracke, F., Deutzmann, J.S., Jayathilake, B.S. et al. (2021). Efficient hydrogen delivery for microbial electrosynthesis via 3D-printed cathodes. *Frontiers in Microbiology* 12: https://doi.org/10.3389/fmicb.2021.696473.

Krajete homepage (2021). www.krajete.com (accessed 14 November 2021).

Krapp, C. (2022). Electrochaea 36 Millionen Euro für power-to-gas. *Handelsblatt* 7 (1): 19.

Leonzio, G. (2016). Process analysis of biological Sabatier reaction for bio-methane production. *Chemical Engineering Journal* 290: 490–498. https://doi.org/10.1016/j.cej.2016.01.068.

Marshall, C.W., Ross, D.E., Handley, K.M. et al. (2017). Metabolic engineering and modeling microbial electrosynthesis. *Scientific Reports* 7: 8391. https://doi.org/10.1038/s41598-017-08877-z.

Mayer, F., Enzmann, F., Lopez, A.M., and Holtmann, F. (2017). Methanogene in der Elektrobiotechnologie. Die mikrobielle Elektrosynthese von Methan. *BIOspektrum* 4 (17): 471–473. https://doi.org/10.1007/s12268-017-0825-1.

Mayer, F., Enzmann, F., Lopez, A.M., and Holtmann, D. (2019). Performance of different methanogenic species for the microbial electrosynthesis of methane from carbon dioxide. *Bioresource Technology*, https://doi.org/10.1016j.biortech.2019.121706.

Molitor, B., Casini, I., Fink, C. et al. (2023). Power-to-gas: neuer Rückenwind für ein altes Modellsystem. *Biospektrum* 1 (23): 94–96. https://doi.org/10.1007/s12268-023-1880-4.

Nevin, K.P., Hensley, S.A., Franks, A.E. et al. (2011). Electrosynthesis of organic compounds from carbon dioxide is catalyzed by a diversity of acetogenic microorganisms. *Applied and Environmental Microbiology* 77 (9): 2882–2886. https://doi.org/10.1128/AEM.02642-10.

OCI (2021). Q3 2021 results presentation. p. 17. https://www.oci.nl/investor-centre/results-and-presentations (accessed 23 December 2021).

Ozkan, M., Nayak, S.P., Ruiz, A.D., and Jiang, W. (2022). Current status and pillars of direct air capture technologies. *iScience* 25 (103): 990.

Perona-Vico, E., Feliu-Paradeda, L., Puig, S., and Bañeras, L. (2020). Bacteria coated cathodes as an in-situ hydrogen evolving platform for microbial electrosynthesis. *Scientific Reports* 10: 19852. https://doi.org/10.1038/s41598-020-76694-y.

Rivard, E., Trudeau, M., and Zaghib, K. (2019). Hydrogen storage for mobility. *Materials* 12: https://doi.org/10.3390/ma.12.121.973.

Rönsch, S., Schneider, J., Matthischke, S. et al. (2016). Review on methanation. *Fuel* 166: 276–296. https://doi.org/10.1016/j.fuel.2015.10.111.

Rusmanis, D., O'Shea, R., Wall, D.M., and Murphy, J.D. Biological hydrogen methanation systems – an overview of design and efficiency. *Bioengineered* 10 (1): 604–634. https://doi.org/10.1080/21655979.2019.1684607.

Sowers, K.R. (2009). Methanogenesis. In: *Encyclopedia of Microbiology*, 3e (ed. M. Schaechter), 265–286. eBook ISBN 9780123739445.

Spurgeon, J.M. and Kumar, B. (2018). A comparative technoeconomic analysis of pathways for commercial electrochemical CO_2 reduction to liquid products. *Energy & Environmental Science* 11: 1536–1551.

Strevett, K.A., Vieth, R.F., and Grasso, D. (1995). Chemo-autotrophic biogas purification for methane enrichment-mechanism and kinetics. *Chemical Engineering Journal and the Biochemical Engineering Journal* 58 (1): 71–79.

Thauer, R.K., Jungermann, K., and Decker, K. (1977). Energy conservation in chemotrophic anaerobic bacteria. *Bacterial Reviews* 41: 100–180. https://doi.org/10.1128/MMBR.41.1.100-180.1977.

Thauer, R.K., Kaster, A.-K., Seedorf, H. et al. (2008). Methanogenic archaea: ecologically relevant differences in energy conservation. *Nature Reviews Microbiology* 6: 579–591.

Thema, M., Weidlich, T., Hörl, M. et al. (2019). Biological CO_2 methanation: an approach to standardization. *Energies* 12: 1670.

Thinkstep. (2017). GHG intensity of natural gas transport. https://www.globallnghub.com/wp-content/uploads/attach_380.pdf.

Tountas, A.A., Ozin, G.A., and Sain, M.M. (2021). Solar methanol energy storage. *Nature Catalysis* 4: 934–942. https://doi.org/10.1038/s41929-021-00696-w.

Vahrenholt, F. (2023). *Die grosse Energiekrise und wie wir sie bewältigen*, 92. München: Langen Müller Verlag.

Vassilev, I., Geißelmann, G., Schwechheimer, S.K. et al. (2018). Anodic electro-fermentation: anaerobic production of L-lysine by recombinant *Corynebacterium glutamicum*. *Biotechnology and Bioengineering* 115: 1499–1508. https://doi.org/10.1002/bit.26562.

Zhou, R., Rui, N., Fan, Z., and Liu, C. (2016). Effect of the structure of Ni/TiO_2 catalyst on CO_2 methanation. *International Journal of Hydrogen Energy* 41: 22017–22025. https://doi.org/10.1016/j.ihydene.2016.08.093.

2

Ethanol Ex Glucose

2.1 Application

Ethanol is a linear alcohol with the formula C_2H_6O and a molecular weight of 46.1 g/mol. It is mainly applied as a fuel additive for conventional combustion engines. The quota of ethanol blending in national and regional fuel mixtures is enforced by political legislation and varies from 3% up to 28%. National governments define mandates to be obeyed by the mineral oil companies when serving the national markets. Politics is thereby supporting the application of biobased fuels with improved carbon footprint compared to conventional gasoline based on petrochemical resources. The application of ethanol as fuel for automotive combustion engines did not require the buildup of a new infrastructure for storage and transportation. This is a main advantage opposed to hydrogen and facilitated the significant volume increase since 2000 (Herrera 2006). Additionally, ethanol is used as chemical building block for the synthesis of ethyl-esters and ethyl-ethers. Producers also take the opportunity of the strong demand of hand sanitizers due to the Covid-19 epidemic and started production and marketing of consumer products like hand sanitizers (FAZ 2020).

The growth rate for bioethanol is driven by the increasing chemical conversion of ethanol into ethylene. The dehydration is pursued as a heterogeneously catalyzed gas phase reaction:

$$CH_3 - CH_2 - OH \rightarrow CH_2 = CH_2 + H_2O \quad \Delta H^\circ = 44.9\ \mathrm{KJ/mol}$$

It is reported that a ZSM-5 zeolite catalyst modified through dealumination achieved a yield of 98.5% and a sensitivity of 100% which is a good basis for a commercial process (Wu and Wu 2017; Autthanit et al. 2021). The obtained ethylene is marketed as "green" product and replaces petrochemical ethylene in the established high-volume value chains. The share of bio-ethylene based on bioethanol amounted to 0.4% in 2010, namely, 600 kt bio-ethylene of a total production of 142,000 kt ethylene pa (Hüsing 2012), though has seen a significant growth until 2020. Braskem was the first industrial mover and commissioned a 200 kt/pa ethylene in Triunfo (Brazil) in 2010 (de Guzman 2010), and Dow is running a 350 kt/pa ethanol to ethylene plant in Brazil. Both plants use low-cost sugarcane as feed. Braskem is expanding its ethanol to ethylene capacity up to 260 kt/pa with expected finalization end of 2022 (Braskem 2022). Braskem is partnering with Lummus, the leading ethylene technology provider with a market share of 40%, in the commercialization of Braskem's registered ethanol dehydration technology (Croteau 2021). Solvay Indupa is

Pathway Design for Industrial Fermentation, First Edition. Walter Koch.

running an ethanol-to-ethylene plant with a capacity of 220 kt/pa in Argentina feeding an adjacent PVC plant (Rosales-Calderon and Arantes 2019; Bozell and Petersen 2010). Ethylene may be further converted into ethylene oxide (EO) and Clariant, Ineos, and Croda offer bio-based EO derivatives in the market (Bettenhausen 2023).

2.2 Production of Ethanol

Ethanol is mainly produced via fermentation. As carbon and energy source predominantly corn in the US, wheat and corn in Europe and sugarcane in Brazil is used which reflects regional agricultural preferences. The feed is hydrolyzed to the monosaccharide level and then fermented with customized yeast strains typically of *Saccharomyces cerevisiae* family. Most of the ethanol fermentation plants do not source glucose or sucrose in the market, as usual for the standard non-ethanol fermentation plant, though are backward-integrated and source the grain or respective carbohydrate externally and pursue the digestion of the polysaccharide into the monosaccharide in their own plant. The obtained by-products are independently sold in the market. The backward integration into the feedstock provides a cost advantage.

These activities covering the hydrolysis of cellulose and the fermentation part are subsumed under the header first-generation ethanol technology. In parallel, alternative raw material sources are accessed. Lignocellulosic biomass like corn stover or wheat straw is applied as fermentation feed. The endeavor to employ lignocellulosic biomass is driven by several reasons: Besides economic and political competition on the use of agricultural acreage either for "food or fuel," the more sustainable use of lignocellulosic biomass for the production of chemicals opposed to the traditional burning of biomass for heat generation also a reduction of production cost is envisaged. Last, but not least, ethanol is produced based on industrial off-gas consisting of primarily CO, CO_2, and H_2. This developing technology has for the first time been implemented by Lanzatech in China (Shougang). The technology applies acetogenic bacteria like *Clostridium autoethanogenum* and allows efficient conversion of industrial waste streams (Heijstra et al. 2017). The chemical production of ethanol is minor and can be neglected.

2.3 Pathway Design

2.3.1 Glycolysis as Natural Fermentation Pathway

The glycolysis or the Embden–Meyerhof–Parnas is to be regarded as the main biochemical pathway for carbon oxidation in heterotroph organisms besides the Entner–Doudoroff pathway, the pentose-phosphate pathway, and the phosphoketolase pathway. It facilitates the conversion of glucose into pyruvate and thereby generates two ATP and two reduction equivalents NADH (Table 2.1).

Glycolysis may either be pursued in anaerobe or aerobe metabolism. In anaerobe metabolism, the generated ATP is available to allow the formation of the main building blocks for cellular growth, e.g. amino acids and proteins, sugars and carbohydrates, and lipids. Oxygen as an electron acceptor is not available for the generation of water and

Table 2.1 G6P, Glucose-6-phosphate; F6P, fructose-6-phosphate; FBP, fructose-1,6-bisphosphate; GAP, glyceraldehyde-3-phosphate; DHAP, dihydroxyacetone-phosphate; 1,3-BPG, 1,3-bisphosphoglycerate; 3-PG, 3-phosphoglycerate; 2-PG, 2-phosphoglycerate; PEP, phosphoenolpyruvate; GAPDH, glyceraldehyde-3-phosphate-dehydrogenase; PGK, phosphoglycerate-kinase; PGM, phosphoglycerate-mutase.

Pathway step	Enzyme	$\Delta G^{\circ\prime}$ (KJ/mol)
Glucose + ATP → G6P + ADP	Hexokinase	−20.9
G6P → F6P	G6P isomerase	+2.2
F6P + ATP → FBP + ADP	Phosphofructokinase (PFK)	−17.2
FBP → DHAP + GAP	Aldolase	+22.8
DHAP → GAP	Triose-P-isomerase (TIM)	+7.9
GAP + NAD^+ → 1,3-BPG + NADH	GAPDH	−16.7
1,3-BPG + ADP → 3-PG + ATP	PGK	(incl.)
3-PG → 2-PG	PGM	+4.7
2-PG → PEP+ H_2O	Enolase	−3.2
PEP + ADP → Pyruvate + ATP	Pyruvate-kinase	−23.0
Glucose + 2 NAD^+ + 2 ADP + 2 Pi → 2 Pyruvate + 2 NADH + 2 ATP + 2 H_2O + 2 H^+		

Source: Voet et al. (2016)/John Wiley & Sons.

the two reduction equivalents generated via substrate chain phosphorylation cannot be recycled in the respiratory chain. Pyruvate needs to be reduced to ensure an efficient recycling of NADH. This is exemplified in the simplest fermentation pathways toward ethanol or lactic acid whose products display, accordingly, a more reduced state than pyruvate since two electrons per NADH have been transferred.

The theoretical molar yield amounts to 2 mol of ethanol per mol glucose which translates into a carbon yield of 2.0 t ethanol per t glucose. The calculated Gibbs free energy for the conversion of glucose into ethanol and carbon dioxide allows the estimation of the energetic efficiency of the reaction:

$$\text{Glucose} \rightarrow 2\,\text{Ethanol} + 2\,\text{CO}_2 \quad \Delta G^{\circ\prime} = -235\ \text{KJ/mol glucose}$$

The negative free Gibbs energy indicates that the reaction is thermodynamically highly favored. Since the formation of 1 mol of ATP based on ADP and P_i requires a Gibbs free energy of −30.5 KJ/mol, the overall energetic efficiency of the reaction can be estimated with 26% in the standard state (Voet et al. 2016; Folch et al. 2021). Under physiological conditions, it is assumed that about 50–60 KJ/mol of ATP are required for the synthesis of 1 mol of ATP (Flamholz et al. 2012) and the energy efficiency of the pathway is likely more in the range of 47%.

The generation of 1 mol of carbon dioxide per mol ethanol, which is increasing the carbon footprint of the technology and is therefore in some contexts described as a disadvantage, cannot be avoided since the 24 electrons carried by glucose are equally allocated on the 2 mol of ethanol each carrying 12 electrons (without counting reduced oxygen). A higher yield providing 3 mol of ethanol derived from 1 mol of glucose might be possible in case only the C-balance is taken into consideration, though then 12 electrons from an additional

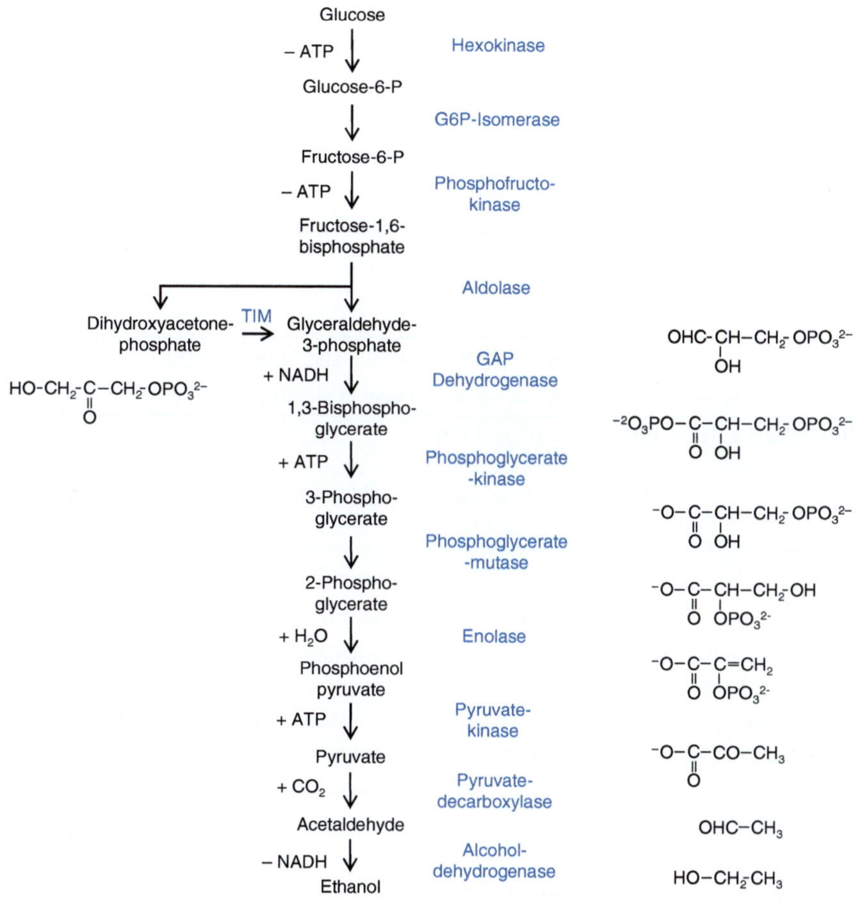

Figure 2.1 Glycolysis with the conversion of glucose into ethanol in *S. cerevisiae*. Source: Walter Koch.

electron donor are required and 6 mol of hydrogen need to be supplemented:

$$\text{Glucose} + 6\,H_2 \rightarrow 3\,\text{Ethanol} + 3\,H_2O \quad \Delta G^{\circ\prime} = -337\,\text{KJ/mol}$$

The straightforwardness of the existing biochemical pathway and the spontaneous formation of ethanol by *S. cerevisiae* in the absence of oxygen has triggered the development of fermentation as originally food conservation technology (Katsyv and Mueller 2020). Opposed to nonnatural pathways and designed enzymes not described in nature, ethanol fermentation technologies have been “discovered” and merely optimized for industrial production. The standard metabolic pathway pursued by *S. cerevisiae* proceeds via the oxidative glycolysis from glucose to pyruvate (Figure 2.1).

Pyruvate is the subject of a non-oxidative decarboxylation and the keto function is converted into an aldehyde – opposed to the oxidative decarboxylation pursued by the pyruvate dehydrogenase complex which transforms the keto function into a carboxylic acid function – and the obtained acetaldehyde is hydrogenated with the alcohol dehydrogenase to ethanol and the following reaction scheme is derived (Table 2.2).

Table 2.2 Biochemical conversion of glucose into ethanol.

Reaction	Enzyme
Glucose + 2 P_i + 2 ADP + 2 NAD^+ → 2 Pyruvate + 2 ATP + 2 NADH + 2 H^+ + 2 H_2O	
2 Pyruvate → 2 Acetaldehyde + 2 CO_2	Pyruvate decarboxylase
2 Acetaldehyde + 2 NADH + 2 H^+ → 2 Ethanol + 2 NAD^+	Alcohol dehydrogenase
Glucose + 2 ADP + 2 P_i → 2 Ethanol + 2 CO_2 + 2 ATP + 2 H_2O	

Table 2.3 Reduction of acetaldehye to ethanol.

Reaction	Potential
NADH → NAD^+ + H^+ + 2 e	$E°' = 0.320$ V (Bertsch and Mueller 2015)
Acetaldehyde + 2 H^+ + 2 e → Ethanol	$E°' = -0.200$ V (Voet et al. 2016)
Acetaldehyde + NADH + H^+ → Ethanol + 2 NAD^+	$\Delta E°' = 0.120$ V
	$\Delta G°' = -n \times F \times \Delta E°' = -2 \times 96\,485$ J/V × mol × 0.12 V = −23.1 KJ/mol

The conventional route toward ethanol thereby provides 2 mol of ATP for cell growth and cell maintenance and achieves a balance of reduction equivalents since the 2 mol of NADH generated in the synthesis of pyruvate are consumed in the reduction of acetaldehyde to ethanol. The pathway is very efficient which is also supported by the negative free Gibbs energy of the last step. The reductive potential of the energy-rich electron transfer carrier NADH is sufficiently strong to render the combined reaction into a thermodynamically downhill path (Table 2.3).

2.3.2 *S. cerevisiae* as Fermentation Host

The industrial anaerobic fermentation is mainly pursued with the eukaryotic host *S. cerevisiae*, a facultative anaerobic species. *S. cerevisiae* is clearly the preferred host since it displays a high tolerance toward ethanol and naturally secrets ethanol under anaerobic growth conditions to block the growth of competing microorganisms. Although ethanol is used as the standard antibiotic in the laboratory, an industrial titer of 150–200 g/l is achieved. The overall reaction equation reveals that 2 of mol ethanol are obtained from 1 mol of glucose; the theoretical molar yield is 2 mol of ethanol per mol glucose. This translates into a theoretical carbon yield of 0.51 t ethanol/t glucose (2 × 46.1 g/mol/180.2 g/mol). The reaction has a very high theoretical energy efficiency of about 97% defined as the ratio of the output heating value of ethanol in relation to the invested high heating value of glucose:

$$\text{Energy efficiency (theory)} = 0.51\ \text{t ethanol/t glucose} \times 29.7\ \text{GJ/t}/15.6\ \text{GJ/t} = 97\%$$

The achieved carbon yield is in the range of about 90–95% of the theoretical carbon yield and amounts to 0.46–0.48 t ethanol/t glucose. Thus, the energy efficiency of the glucose to ethanol conversion shrinks to 88% in case a realistic carbon yield of 0.46 t ethanol/t glucose is applied. The gap between theoretical and achieved carbon yield is explained by the required formation of biomass and by-products like glycerol. It is assumed that about

4% of the carbohydrate substrate is lost due to the generation of glycerol (Papapetridis et al. 2017). The achieved ethanol titer is in the range of 200 g/l and the productivity is in the range of at least 2–3 g/l × h (Jansen et al. 2017). It is also to be taken into consideration that the fermentation process itself is energy consuming. It may be assumed that the stirrer has an energy input of 2 kW/m^3 (Meyer et al. 2017). This derives an energy demand of 0.5 MWh/t or 1.8 GJ/t ethanol in case a batch time of 50 h and a titer of 200 g/l is assumed. Additionally, the isolation of ethanol out of the broth requires that at least the evaporation enthalpy of ethanol of 40.5 KJ/mol or 0.88 GJ/t of ethanol is invested if one distillation step is sufficient for purification. The energy efficiency of the whole process may then be estimated as follows:

$$\text{Energy efficiency (practice)} = 0.46\,\text{t ethanol/t glucose} \times 29.7\,\text{GJ/t}/15.6\,\text{GJ/t} + 2.7\,\text{GJ} = 75\%$$

Energy efficiency of 75% of the whole process is still a high value and explains the high-volume production of fermentative ethanol as fuel for combustion engines. In case the generated biomass – which is often sold as feed – is given an energy credit, the efficiency would even be higher.

The ethanol producers apply natural strains for producing non-GMO beverage alcohol in Europe or further optimized genetically modified organism (GMO) strains for producing ethanol as a fuel additive. Accordingly, the ethanol pricing of ICIS differentiates a beverage grade (96%) and a fuel grade (99.3–99.9%), which are reflecting the limited GMO acceptance with respect to food of the European consumer and politics (ICIS 2020a, b). Most of the ethanol plants do not pursue their own host development, though source the inoculum from a dedicated yeast supply industry.

2.3.3 Generation of Carbon Dioxide as By-Product

The high energy efficiency of the process is enabled by the partial reduction (i.e. generation of the methyl-group in ethanol) and partial oxidation (i.e. generation of carbon dioxide) of glucose. The fermentative production of ethanol is formally a disproportionation. The generation of carbon dioxide implies that per ton ethanol at least 0.96 t carbon dioxide are obtained which traditionally have been released into the atmosphere. For the United States with 216 existing biorefineries, a total emission of 45 Mio. t carbon dioxide is estimated (Sanchez et al. 2018). In order to address the critical role of carbon dioxide in global warming and the tightened political and regulatory framework, measures are evaluated to avoid the release of the by-product into the atmosphere. These measures include in first instance the capturing of the carbon dioxide. This step is gaining interest since the capturing cost of carbon dioxide off-gases out of an ethanol plant is in the range of 10 €/t and thereby very low compared to other processes in which carbon dioxide is generated (Naims 2016).

The captured carbon dioxide may then be converted into valuable downstream products (CCU, carbon capture and utilization) or handed over to long-term storage projects filling suitable geologic reservoirs (CCS, carbon capture and sequestration).

2.3.4 *Zymomonas mobilis* as Fermentation Host

The prokaryote *Z. mobilis* has been proposed as an alternative production host for ethanol. The rationale refers to the classical Entner–Doudoroff (ED) pathway or the

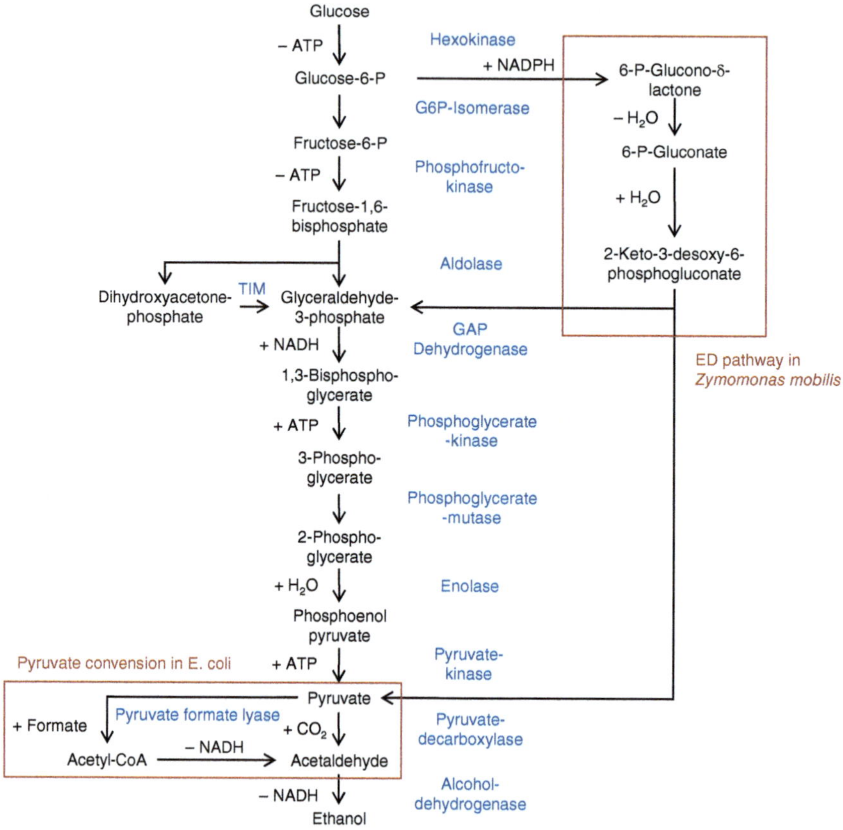

Figure 2.2 Glycolysis toward ethanol in *S. cerevisiae*. The alternative ED pathway via KDPG partially pursued by *Z. mobilis* is indicated in red box on the right site. Source: Walter Koch.

2-keto-3-desoxy-6-phosphogluconate (KDPG) pathway for the conversion of glucose into pyruvate (Spaans et al. 2015; insert 1). In deviation to conventional glycolysis, it converts glucose-6-phosphate into glyceraldehyde-3-phosphate and pyruvate which might then also be converted into ethanol (Figure 2.2).

The difference is given in the low energy efficiency of the ED pathway which provides exclusively 1 mol of ATP per mol glucose instead of 2 mol of ATP achieved by glycolysis:

$$\text{Glucose} + \text{ADP} + \text{Pi} + \text{NAPD}^+ + \text{NAD}+ \rightarrow 2\,\text{Pyruvate} + \text{ATP} + \text{NADPH} + \text{NADH} + \text{H}_2\text{O}$$

Since abundant ATP availability is associated with cell growth and biomass formation competing with the desired formation of ethanol, the ED pathway with inferior energy efficiency might allow a higher carbon yield than the current ethanol technology based on yeast (Kang and Lee 2015). Already in 1985, it has been published that *Z. mobilis* achieves a sucrose to ethanol conversion efficiency of 95–98% and with a titer of 95.5 g/l given 200 g/l sucrose feed after 24–30 h an impressing productivity (Doelle and Greenfield 1985). Additionally, *Z. mobilis* has been engineered to convert also mannose and xylose and thereby allow to switch feed to lignocellulosic biomass. However, this theoretical advantage has so far not been materialized.

Table 2.4 Biochemical conversion of glucose into ethanol and formic acid.

Glucose + 2 Pi + 2 ADP + 2 NAD^+ → 2 Pyruvate + 2 ATP + 2 NADH + 2 H^+ + 2 H_2O	
2 Pyruvate + 2 CoA → 2 Acetyl-CoA + 2 formic acid	Pyruvate formate lyase
2 Acetyl-CoA + 2 NADH → 2 Acetaldehyde + 2 CoA + 2 NAD^+	Aldehyde dehydrogenase
2 Acetaldehyde + 2 NADH → 2 Ethanol + 2 NAD^+	Alcohol dehydrogenase
Glucose + 2 ADP + 2 NADH + 2 P_i → 2 Ethanol + 2 formic acid +2 ATP + 2 NAD^+ + 2 H_2O	

2.3.5 *Escherichia coli* as Fermentation Host

Since more than 20 years, it has been evaluated to use *E. coli* as an alternative host for producing ethanol. It has a higher growth rate than yeast, it displays simple nutritional requirements and a highly developed molecular biology toolbox for strain improvement is available. Its native anaerobe metabolism is characterized as a mixed-acid fermentation and converts sugars into several products comprising formic acid, carbon dioxide, ethanol, acetic acid, lactic acid, and hydrogen (Clark 1989). The pyruvate dehydrogenase activity is hardly detectable (Snoep et al. 1993). It is to be kept in mind that the main pyruvate conversion proceeds via the pyruvate formate lyase (formate acetyltransferase) to obtain acetyl-CoA and formic acid. *E. coli* expresses the pyruvate formate lyase under anaerobic conditions in the absence of respiratory electron acceptors. Thus, glucose is via pyruvate mainly converted into acetyl-CoA and formic acid (insert 2). The generated acetyl-CoA can be converted into acetyl-phosphate and is further processed into acetate and ATP. Formic acid is secreted out of the cell. In case the external pH would generate physiological stress, formic acid is taken back into the cell and is converted into carbon dioxide and molecular hydrogen via the formate hydrogen lyase (Pinske and Sargent 2016; Wood n.d.). In case a metabolic pathway toward ethanol is to be designed, the various by-products would need to be suppressed and the acetyl-CoA as C2 intermediate would need to be converted into ethanol. Acetyl-CoA would therefore be reduced from the acid to the aldehyde level and the obtained acetaldehyde would be further reduced to ethanol according to the final step in the pathway of *S. cerevisiae*. This step is in bacterial systems catalyzed by the bifunctional acetaldehyde/alcohol dehydrogenase encoded by the *adhE* gene. The direct conversion of the intermediate aldehyde without release into the cytosol might reflect the high activity of aldehydes which might trigger harmful non-intended reactions (Table 2.4).

This pathway would not provide a balance of reduction equivalents and not be applicable for an industrial fermentation process. If compared to the conventional pathway in yeast, it is obvious that the avoidance of carbon dioxide and the lower oxidation status of formic acid requires two additional reduction equivalents, which cause the metabolic imbalance of this approach. *E. coli* cells with silenced pyruvate formate lyase and chromosomal integration of the pyruvate decarboxylase have been developed to pursue the same pathway as yeast under anaerobic conditions. However, these strains have not replaced *S. cerevisiae* as an industrial fermentation host.

The fermentative production of ethanol with *S. cerevisiae* does not generate significant amounts of by-products besides fusel oils. In some studies minor amounts of ethyl acetate have been observed. This applies especially to the *E. coli* host KO11 which has been

Table 2.5 Biochemical conversion of ethanol into acetaldehyde and hydrogen.

Glucose + 2 Pi + 2 ADP + 2 NAD^+ → 2 Pyruvate + 2 ATP + 2 NADH + 2 H^+ + 2 H_2O	
2 Pyruvate + 2 CoA → 2 Acetyl-CoA + 2 formic acid	Pyruvate formate lyase
2 Formic acid → 2 CO_2 + H_2	Pyruvate hydrogen lyase
2 Acetyl-CoA + 2 NADH + 2 H^+ → 2 Acetaldehyde + 2 CoA + 2 NAD^+	Acetyl-CoA reductase
Glucose + 2 ADP + 2 P_i → 2 Acetaldehyde + 2 ATP + 2 CO_2 + 2 H_2 + 2 H_2O	

designed to potentially replace the dominant *S. cerevisiae* host conventionally applied for ethanol. The native fermentation pathway of *E. coli* has been replaced by the homo-ethanol pathway of *Z. mobilis* in the KO11 strain which is able to ferment a diverse array of sugars present in lignocellulosic biomass. In this host, a competing ethyl acetate titer of 2 g/l has been observed and triggered some metabolic engineering approaches to achieve a higher selectivity. Since the known acetyltransferases might have other important functions in metabolism, the activation of an esterase has been chosen as the preferred metabolic engineering strategy compared to the knock-out of the respective acetyltransferase genes. Finally, the expression of the short-chain aliphatic ester *estZ* gene of *P. putida* in *E. coli* reduced the ethyl acetate concentration in the broth to a level below 20 mg/l (Hasona et al. 2002).

2.3.6 Digression: *E. coli* as Host for Acetaldehyde and Hydrogen

It has been proposed to use the established metabolic pathways in *E. coli* to generate acetaldehyde and hydrogen based on glucose as raw material source (Zhu et al. 2011). The idea is to proceed via the reaction scheme indicated above and to convert glucose into acetyl-CoA and formic acid. Formic acid would then be further converted via the formate hydrogen lyase into carbon dioxide and hydrogen. The interesting reaction to be pursued with recombinant strains would be the conversion of acetyl-CoA via a reductive reaction into acetaldehyde. The thioester of acetyl-CoA would be cleaved, the CoA cofactor be released, and the carboxyl-function be reduced to the aldehyde level (Table 2.5).

This reaction would be pursued by the acetyl-CoA reductase (ACR) and the respective *eutE* gene has been identified in *S. enterica*. The respective conversion has a sufficiently negative Gibbs free energy to allow the generation of 2 mol of ATP in parallel:

$$\text{Glucose} \rightarrow 2\ \text{Acetaldehyde} + 2\ \text{H}_2 + 2\ \text{CO}_2 \quad \Delta G'^{\circ} = -151\ \text{KJ/mol}$$

The overall net reaction has a partial overlap with the ethanol fermentation pathway with the major difference that the product has a higher oxidation level (aldehyde versus alcohol), which allows to generate a reduction equivalent or molecular hydrogen (Zhu et al. 2011). The reductive acetyl-CoA conversion would also facilitate a redox balance. For comparison, the generation of acetaldehyde has also been installed in *Z. mobilis* by using the pyruvate decarboxylase resulting in acetaldehyde, carbon dioxide, and one excess NADH (Bongers et al. 2005; Wecker and Zall 1987). Redox balance was achieved via the addition of oxygen which requires more technical efforts in the fermentation process and would likely not allow to generate hydrogen as a valuable by-product due to process security reasons. With respect to the experiments applying the ACR, a production rate of 0.68 g/h × g dry cell weight was achieved with an acetaldehyde generation at 86% of the maximum theoretical yield.

The isolation of acetaldehyde relies on an efficient gas separation. It has a low boiling point of 20 °C and would allow a direct in situ-product recovery out of the fermentation broth and bypass potential issues caused by a high aldehyde titer in the medium. It would directly evaporate out of the broth since it can be assumed that the fermentation would be pursued at temperatures of at least 30 °C. The three off-gases acetaldehyde, carbon dioxide, and hydrogen in the off-gas fraction would then need to be separated. It remains to be seen if this approach gains commercial interest in the future. Acetaldehyde is usually generated by the partial oxidation of ethylene.

2.3.7 Ethanol Production in Cell-Free Systems

Cell-free systems are "ensembles of catalytic proteins prepared from purified enzymes or crude lysates of cells for the production of target products" (Dudley et al. 2015). These systems are also referred to as synthetic *in vivo* metabolic networks and are prospected to expand the technology options of conventional fermentation in the future (Erb 2019; Ullah et al. 2016). They are especially interested in case the target product is toxic to the host and the productivity remains very low. It might then be possible by artificial strain evolution to raise tolerance of the host or to install an in situ-product recovery such that the actual concentration of the toxic target product in the broth remains very low. However, in case these process adjustments are not successful, a cell-free system might be the outstanding approach.

The generation of ethanol in a yeast-based cell-free system by Buchner at the end of the nineteenth century defines a landmark in the development of biochemistry and supported the view that chemistry and biochemistry are basically one academic discipline (Buchner 1897). In the historical experiments of Buchner, a complete conversion of glucose was not achieved because the ATP surplus of the overall pathway did not allow to replenish the cofactor ADP. This limitation was overcome later on by Welch and Scopes who added an ATPase or arsenate to the cell-free extract and could thereby increase glucose conversion (Welch and Scopes 1985). These attempts might gain industrial relevance in case the activity of the cell-free enzyme extract can be restored for a long time, and the catalyst cost of the system is minor compared to the value of the produced ethanol. This will require a technology to efficiently separate the catalyst – the enzyme mix – and the product after the reaction via enzyme immobilization or encapsulation (Ullah et al. 2014). Otherwise, the catalyst replenishment would render the overall cost position of ethanol rather unattractive.

2.4 Process Development

2.4.1 Technology Scope

There are two technologies applied in the United States for the fermentative production of ethanol. The original "wet milling" technology for corn processing is operated by the sugar producers, e.g. Cargill and Tate & Lyle. It pursues the strategy to separate maize into four components starch, germ, fiber, and protein (gluten), which are then further processed to serve different market segments. The starch fraction may be used for the production of glucose and then ethanol. Pending on actual demand and price levels, the starch fraction might also be hydrolyzed and partially be isomerized to fructose to obtain a syrup mixture of glucose and fructose sold in the soft drink industry. For the producers of sugar and syrup, ethanol is more an opportunistic product with second priority. The so-called dry milling process has primarily been developed for the production of ethanol and it has reached a market share of about 80% based on the high-volume growth of fuel ethanol from 2000 to 2020. It is pursued to primarily derive ethanol as the target product. It is pursued by the major players like ADM, POET, and Great Plains, which conceive themselves as ethanol producers. In order to fully exploit the value of the feedstock, maize, or corn starch, the generation of the by-products DDGS (distilled dried grains) and corn oil has been optimized.

2.4.2 Dry Milling Process

The corn, wheat, or tapioca starch feed is grinded in the first step and the obtained powder is put in water to obtain a slurry. The slurry of grinded starch is then channeled into the liquefaction step, which enhances the physical access of the starch carbohydrate chain for the following saccharification process. The powder is solved in water at a temperature of about 80 °C and a thermo-stable α-amylase is added to initiate the digestion of internal α-1,4-glycosidic bonds and thereby reduce the chain length and the corresponding viscosity. A short heating above 100 °C for a duration of about 5–8 min in the jet cooker is needed although it weakens the activity of the α-amylase. The process stream is then entering the secondary liquefaction at a temperature of about 95 °C and a duration of 90 minutes. In the following saccharification step, a glucoamylase is added and the starch is further digested at 60 °C for 8–10 h (Schubert 2006; Figure 2.3).

The applied α-amylase enzyme and the glucoamylase are usually delivered by dedicated enzyme providers like Novozymes and subjected to continued R&D activities to improve thermo-stability or to allow a reduction of the duration of the reaction or the reaction temperature. The α-amylase might alternatively be expressed in the EtOH fermentation host and obtained via secretion, though the liquefaction is pursued at 80 °C and the yeast strain is not viable and efficiently working at this temperature. The thermostable α-amylase has an optimum enzymatic activity at pH 5 and a temperature of 85 °C. The product stream with the glucose mono- and disaccharides is then channeled into the fermentation reactor and the yeast biocatalyst is added. The by-product carbon dioxide is removed via the off-gas derived in the reactor head. After the fermentation the biomass and the remaining starch constituents are removed and sold as DDGS into the feed industry, ethanol is isolated out of the broth via a two-step distillation. The evaporation temperature of 78 °C of the "light-boiler" ethanol allows an efficient extraction with limited energy efforts. Since water and ethanol form an azeotropic mixture at a ratio of 96% (w/w) ethanol and 4% (w/w) water, a complete separation via distillation is not possible and mol sieves are applied to achieve a purity of 99.9% (w/w).

The process has achieved significant yield improvements in the last 15 years if the overall corn input per t ethanol is considered. The calculation of the corn-to-ethanol input ratio has been estimated with reference to a publication of 2009 and 2022:

1 bushel corn yields 2.5 gal ethanol (Sheehan 2009), derives
25.4 kg corn yield 2.5 gal ethanol (1 bushel equals 25.4 g)
1 kg corn yields 98 gal ethanol

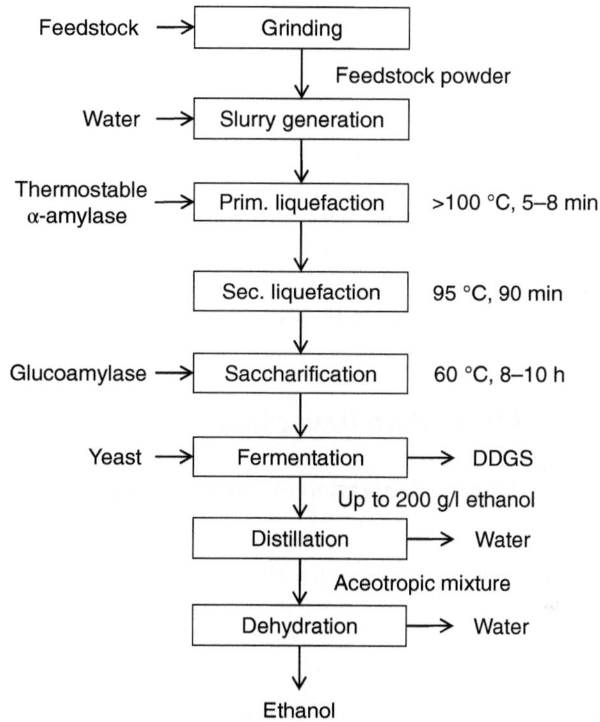

Figure 2.3 Process flow of the dry milling process for fermentative ethanol. Final ethanol titers at batch end up to 21% (w/w) are reported. Source: Adapted from Jansen et al. (2017).

1 kg corn yields 373 l ethanol (3.79 l ethanol/gal)
1 kg corn yields 295 kg ethanol (0.79 kg ethanol/l)
3.39 kg corn yield 1 kg ethanol (input ratio before 2009), compared to
1 bushel corn yields 2.8 gal ethanol (Cropwatch 2022), derives
3.03 kg corn yield 1 kg ethanol

The reduction of the corn unit consumption is likely due to improvements in the fermentation part, though likely mainly achieved in the processing of the biomass and the downstream processing after the fermentation.

2.4.3 Sterile or Non-sterile Fermentation

Fermentation processes usually require sterilization of media and feedstock to avoid contamination by competing microorganisms. Sterilization efforts and handling protocols gain importance proportional to the value of the fermentation products as an "insurance" to protect the product which can be seen in the rather high efforts for the fermentative production of value enzymes. A contamination incident often requires the withdrawal of the complete batch given regulatory and quality guidelines. The consequences depend on the type of contamination and a fungus with the risk of releasing toxic compounds is likely the most stringent case.

Usually, the media are sterilized via steam heating (e.g. 15 min, 121 °C) and the fermenter is sterilized before each batch cycle with steam with a pressure of 1–2 bar. In case the fermentation conditions are discriminating for competing hosts, sterilization efforts may be reduced or completely omitted. This applies especially for brewing beer and the fermentative production of ethanol (*S. cerevisiae*) and citric acid (*Aspergillus niger*). Additionally, the fermentative production of acids with acid-tolerant hosts like lactic acid with Lactobacillus at low pH allows the reduction of sterilization efforts. The anaerobe fermentation process for ethanol and lactic acid with very limited availability of oxygen excludes all obligatory aerobe microorganisms. A growth advantage of the production host is ensured with a high cell load of the inoculum. It is also reported that the ethanol industry supplements the feed with antibiotics or hops-derived extracts to suppress competing microorganisms. The missing sterilization requirements for ethanol production translated into a significantly lower specific investment since seals and valves do not need to be designed to stand steam treatment (Meyer 2017, Stanbury 2017).

2.5 Alternative Raw Material Source

2.5.1 Second-Generation Ethanol Technology

First-generation ethanol comprises technologies that derive ethanol from food crops with high level of starch and sugar. The raw material source for current ethanol production varies by region. In the United States, predominantly corn (maize) is applied and sugar beet in the EU. In Brazil, the early mover in the ethanol-to-fuel approach, often a mixture of seasonal "green juice" derived in sugarcane processing and molasse is used. The "green juice" contains carbohydrates and dilutes inhibitory components contained in the molasse. These fermentative ethanol technologies are subsumed under the header of first-generation technology (G1).

Despite the high availability of corn starch in the United States and globally with a production volume above 1000 Mio. t, alternative carbon sources for the fermentative production of ethanol are explored. This endeavor was driven by mainly three arguments comprising the limitation of arable land, the optimization of the carbon footprint, and the reduction of production costs (Knocke 2009). In the first instance, the political and social "food versus fuel" argumentation is raising the concern that the production of gasoline should not limit food production given the limited availability of agricultural acreage. About 9% of the global grain production is allocated to biofuels, which is mainly ethanol (Table 2.6; Economist 2022).

Table 2.6 Global grain production and consumption.

Grain production 2019		Grain allocation 2019	
Maize	1153 Mio. t	Food	1337 Mio. t
Rice	734 Mio. t	Animal feed	987 Mio. t
Wheat	725 Mio. t	Biofuel	265 Mio. t
Others	294 Mio. t	Processing & losses	244 Mio. t
		Seeds	73 Mio. t
Total	2906 Mio. t	Total	2906 Mio. t

Table 2.7 Technology for the digestion of lignocellulosic biomass.

Technology	Auxiliary chemicals	Mode of action
Chemical treatment		
Dilute-acid treatment	H_2SO_4, H_3PO_4, strong acids	Hydrolysis of glycosidic bonds
Alkali treatment	NaOH, Na_2CO_3, lime Other alkaline compounds	Extraction of lignin
Ionic liquids	Large organic cation/inorganic anion	Fractionation of polymers
Deep-eutectic solvents	Mixture of Lewis and Bronsted acids	Fractionation of polymers
Organosolv	Organic solvents (ethanol, butanol)	Extraction of lignin
Physicochemical treatment		
Steam treatment	Hot/supersaturated steam Catalyst may be added	Hydrolysis of hemicellulose Fiber separation
Physical treatment		
Milling/grinding	(Machine equipment)	Surface increase Improved access
Biological treatment	Biologicals (e.g. brown/white fungi)	Degradation of polymers

Since the global food demand is increasing, the occupation of arable land for producing fuels implies second-order land substitution and indirectly triggers deforestation. Opposed to the starch fraction of conventional food crops in first-generation ethanol (G1), for second-generation ethanol (G2) lignocellulosic biomass shall be employed which is not suitable for food production, though often burnt to exploit the intrinsic energy. These approaches based on lignocellulosic biomass use exclusively the inedible part of food or non-food crops and are referred to as second-generation technology (G2) (Ning et al. 2021; Robak and Balcerek 2018; Peters et al. 2014). The use of livestock waste materials (i.e. manure) has been considered, though has not yet have commercial relevance (Jung et al. 2021). Secondly, besides the avoided cannibalization of food resources second-generation ethanol has an improved carbon footprint and thereby a better sustainability profile than first-generation ethanol (Jeswani et al. n.d.). Finally, the second-generation ethanol technology was prospected to achieve a reduction of production costs for fermentable sugars.

Lignocellulosic biomass contains cellulose, hemicellulose, lignin, and ash (Kumar and Sharma 2017). The cellulose fraction consists of glucose molecules tight together with β-1.4-glycosidic bonds. Hemicellulose contains C5 sugars like D-xylose and L-arabinose which in principle both can be converted into pyruvate via metabolic engineering of the fermentation host. However, the high physical strength of lignocellulosic biomass reflecting the natural stability functionality requires significant chemical, physicochemical, physical, or biological pretreatment efforts for decomposition. A broad technology, repertoire has been developed to access the fermentable C6 and especially C5 sugars in lignocellulosic biomass (Galbe and Wallberg 2019; Kucharska et al. 2018; Kumar and Sharma 2017; Table 2.7).

The subsequent steps in the G2 fermentation technology depend on the efficiency and duration of the pretreatment step and the obtained molecular portfolio of decomposition products. This includes the application of hydrolytic enzymes to fully convert the oligo- and polysaccharides into monomers (i.e. glucose ex cellulose, D-xylose, and L-arabinose ex hemicellulose), which can easily be taken up by the fermentation host. The applied enzyme mix required to efficiently convert the oligo- and polysaccharides into monosaccharides may be supplied by an independent enzyme producer or be generated on-site. The "on-site cellulase production" has several advantages including less activity loss (just in time), limited logistics and handling efforts, and especially a higher efficiency. It has been found that the enzymes are more efficient if they were produced and generated by a host, which uses the same biomass as carbon and the energy source the enzymes are supposed to digest (Siqueira et al. 2020). It is assumed that the respective cellulase isoenzymes are in this circumstance ideally customized to their target substrate.

To fully exploit the fermentable C5 and C6 sugars (on the monosaccharide level), the fermentation host needs to be equipped with the respective enzyme portfolio. The conventional *S. cerevisiae* strains are not able to hydrolyze cellulose and hemicellulose and can also not utilize the C5 sugars D-xylose and L-arabinose contained in hemicellulose (Knocke and Vogt 2009). Different strategies have been evaluated. The most straightforward idea would be to apply anaerobic bacteria, which are able to break down cellulose and hemicellulose such as *Clostridium*, *Thermoanaerobacterium*, and *Ruminococcus*. These species secrete multi-enzyme complexes, called cellulosome, into their natural habitat to convert cellulose into glucose. The well-characterized model organism *Clostridium cellulolyticum* is able to grow on crystalline cellulose (Gaida et al. 2016). However, the natural specialists for the breakdown of cellulose are no suitable biocatalysts for fermentation technology and the genes encoding the cellulosome could not yet efficiently be transferred to other bacterial species. Another option would be to develop microbes such as the bacteria *E. coli* and *Klebsiella oxytoca* and the yeast *Pichia stipites* as hosts which naturally utilize pentoses for the fermentative production of ethanol. On the other hand, ethanologenic microorganisms like *S. cerevisiae* or *Z. mobilis* might be transformed with genes for a pentose utilization pathway. Several pathways exist for the channeling of D-xylose and L-arabinose into central metabolism (Figure 2.4).

D-xylose can directly be converted into D-xylulose via the xylose isomerase. D-xylulose is then phosphorylated with a kinase and the obtained D-xylulose-5-phosphate can then further be converted into pyruvate via the pentose phosphate pathway. The route via the xylose isomerase is referred to as the "bacterial pathway." The alternative route reduces D-xylose to xylitol in the first step, converting the terminal aldehyde function into primary alcohol, followed by the xylitol dehydrogenase pursuing the reverse step to obtain D-xylulose. This route is referred to as the "yeast pathway" (Guo et al. 2018). Additionally, D-xylose and L-arabinose may also be metabolized in a non-phosphorylative manner (not shown here), which might be interesting for a biocatalytic approach to avoid consumption of ATP (Watanabe et al. 2019).

It should be pointed out that the bacterial pathway and the yeast pathway – both converting D-xylose into D-xylulose – are neutral with respect to reduction equivalents, but the two-step route often consumes 1 mol of NADPH and provides 1 mol of NADH. This imbalance needs to be avoided in the metabolic engineering strategy. This might be facilitated with a transhydrogenase or by identifying an aldose reductase accepting

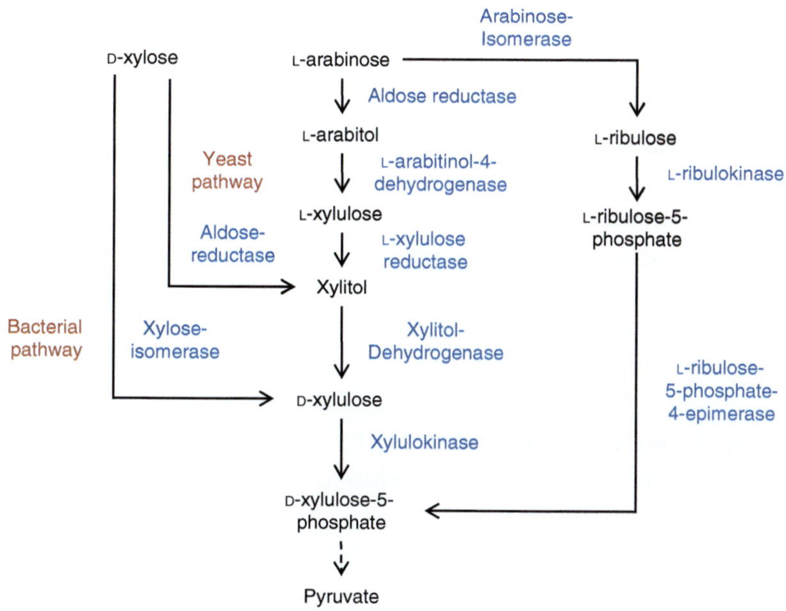

Figure 2.4 Artificial metabolic pathways for the conversion of D-xylose and L-arabinose. Source: Adapted from Nevoigt, 2008.

NADH as a cofactor. The overall reaction equation reveals that 3 mol of D-xylose can be converted into 5 mol of ethanol and 5 mol of carbon dioxide. The partial reduction of the hydroxy-methylene groups of the pentose to the level of ethanol is thermodynamically counterbalanced by the partial oxidation to carbon dioxide (Table 2.8).

D-xylose has a molecular weight of 150.1 g/mol. The molar yield of 1.66 mol of ethanol per mol D-xylose translates into a theoretical carbon yield of 0.51 t ethanol/t D-xylose ((5 × 46.1 g/mol)/(3 × 150.1 g/mol)) which is identical to the theoretical carbon yield with glucose as feed. In case it is assumed that about 90% of the theoretical carbon yield is achieved, the realistic carbon yield amounts to 0.46 t ethanol/t D-xylose.

The carbon yield based on lignocellulosic biomass depends on the specific composition which varies by source (Kumar and Sharma 2017). For the estimation of the unit consumption rate of biomass for producing ethanol, an exemplary composition of lignocellulosic biomass representing an average of different wood types is assumed. Based on 100 kt biomass, the conversion factor from poly- to monosaccharides (mass gain due to integrated water molecules in the hydrolysis of glycosidic bonds) and an extraction yield of 70% of the accessible fermentable sugar is derived (Table 2.9).

It is assumed that about 30% of the lignocellulosic biomass is hemicellulose and that about two-thirds of the hemicellulose contain D-xylose (including L-arabinose). Accordingly, it may be assumed that about 21 kt ethanol can be obtained from 100 kt lignocellulosic biomass (0.2 t EtOH/t lignocellulosic biomass) representing a composition of wood types. This yield is further reduced in case a minor share of the fermentable sugars is used for the production of an enzyme mix used in the digestion of the biomass especially the hydrolysis of polysaccharides. If one ton of lignocellulosic biomass is charged with 100 €/t including local collection and transportation (Brown 2015), the raw material cost per ton ethanol add

Table 2.8 Conversion of D-xylose into ethanol.

3 D-xylose + 3 NAD(P)H + 3 H^+ → 3 Xylitol + 3 $NAD(P)^+$
3 Xylitol + 3 NAD^+ → 3 D-Xylulose + 3 NADH + 3 H^+
3 D-Xylulose + 3 ATP → 3 D-xylulose-5-P + 3 ADP
Xylulose-5-P → Ribose-5-P
Xylulose-5-P + Ribose-5-P → Seduheptulose-7-P + GAP
Seduheptulose-7-P + GAP → Fructose-6-P + Erythrose-4-P
Xylulose-5-P + Erythrose-4-P → Fructose-6-P + GAP
2 Fructose-6-P + 2 ATP → 2 Fructose-1,6-BP + 2 ADP
2 Fructose-1,6-BP → 2 DHAP +2 GAP
2 DHAP → 2 GAP
5 GAP + 5 NAD^+ → 5 × 1,3-Bisphosphoglycerate + 5 NADH + 5 H^+
5 × 1,3-Bisphosphoglycerate + 5 ADP → 5 × 3-Phosphoglycerate + ATP
5 × 3-Phosphoglycerate → 5 × 2-Phosphoglycerate
5 × 2-Phosphoglycerate → 5 PEP + 5 H_2O
5 PEP + 5 ADP → 5 Pyruvate + 5 ATP
5 Pyruvate → 5 Acetaldehyde + 5 CO_2
5 Acetaldehyde + 5 NADH + 5 H^+ → 5 EtOH + 5 NAD^+
3 D-Xylose + 5 ADP + 5 Pi → 5 EtOH + 5 CO_2 + 5 ATP + 5 H_2O

Table 2.9 Composition of lignocellulosic biomass and conversion into ethanol.

Composition	Polysaccharides	Monosaccharide	Extraction	Ethanol yield
40% cellulose	40 kt cellulose	44.4 kt glucose	31.1 kt	14.3 kt
30% hemicellulose	20 kt D-xylose	22.6 kt D-xylose	15.8 kt	7.3 kt
20% lignin	(as polysaccharide)			
10% Others (e.g. ash)				

already up to 500 €/t ethanol with a market price of about 600–800 €/t in Europe (ICIS 2020a, ICIS 2020b) and 600–800 $/t in the United States (ICIS 2020a,b). The lignin by-product may be sold at fuel value or be subjected to more advanced valorization technologies and thereby provide an additional by-product credit.

When the second-generation technology approach was started around 2000, it was clear that the new technology will have a cost disadvantage in the initial phase though it was prospected that ethanol second-generation will achieve the same cost level of ethanol first-generation once the technological hurdles of the pretreatment step are solved. The cost comparison relies to a great extent on the raw material prices for corn and biomass and the assumed development of conversion cost toward ethanol as well. The following estimation refers to the cost and price level as of 2008 (Table 2.10).

The production cost of ethanol first- and second-generation were stated about 2 USD/gal for first-generation and 5 USD/gal for second-generation in 2007 (Sandmair

Table 2.10 Feedstock, fermentable sugar, and ethanol G1/G2 production cost (3.79 l/gal, 0.785 kg/l). The ethanol market price is highly fluctuating. For the years 2010 and 2020, a reasonable average has been chosen. A market price for ethanol 2G is not available though the improved carbon footprint might allow a price premium versus ethanol G1.

	G1 Ethanol		G2 Ethanol		Source
Feedstock					
Raw material type	Corn		Biomass		
Raw material price	91.7 USD/t		50.0 USD/t		
Energy content	16.3 USD/t		17.4 USD/t	(LHV)	
Fermentable sugars					
Cost no coproducts	152 USD/t	2007	166 USD/t	2007	Lynd et al. (2008)
Cost with coproducts	148 USD/t	2007	194 USD/t	2007	Lynd et al. (2008)
			83 USD/t	mature	Lynd et al. (2008)
			79 USD/t	mature	Lynd et al. (2008)
Ethanol cost/price					
Production costs	672 USD/t	2007	1.681 USD/t	2007	
	2.0 USD/gal	2007	5.0 USD/gal	2007	Sandmair and Zeitalter (2010)
	672 USD/t	2010	1.008 USD/t	2010	
	2.0 USD/gal	2010	3.0 USD/gal	2010	Sandmair and Zeitalter (2010)
Investment only			3.361 USD/t	2015	Brown, (2015)
			10 USD/gal	2015	Brown, (2015)
Market price	672 USD/t	2010			
	2.0 USD/gal	2010			Trading Economics (n.d.)
	588 USD/t	2020			
	1.75USD/gal	2020			Trading Economics (n.d.)

and Zeitalter 2010). Given the technology advancements, the costs in 2010 were stated about 2 USD/gal for first-generation and 3 USD/gal for second-generation ethanol and are predicted to reach the same order of magnitude in the future (Sandmair and Zeitalter 2010). However, the cost reduction could not be achieved and the huge cost disadvantage remained. It is reported that exclusively the investment for cellulosic biofuel is in the range of 2.64 USD/l (10 USD/gal), which translates into 3.360 USD/t (Brown 2015). This figure would translate into depreciation costs of 1 USD/gal in case a linear depreciation over a 10 years period is assumed. These rough estimations indicate the economic hurdles of biorefineries (Behr and Seidensticker 2018), though also the potential in case the lignocellulosic biomass – regarded as a low-value agricultural by-product – can be sourced at low prices. The minimum share of second-generation biofuel based on lignocellulosic biomass was defined in the US Renewable Fuel Standard (RFS2), though the originally mandated goals were significantly revised in the years 2010–2014 (Brown, 2015). The cost level needs to be further improved to allow a broader market penetration.

2.5.2 Third-Generation Ethanol Technology

The third-generation biorefinery or ethanol technology aims to use microbial cell factories, carbon dioxide captured out of the air or from industrial processes, and renewable energy for producing ethanol or other biofuels (Liu et al. 2020; Tong et al. 2022). The renewable energy may comprise sunlight, artificial light, and electricity generated from sustainable sources, e.g. photovoltaic cells and wind power and compounds extracted from waste streams. In particular, autotrophic microbial organisms capable to use energy in the form of light for photosynthesis are fed with carbon dioxide as a carbon source and other media components to derive ethanol or products with higher functionality and value. As an example of the diverse innovation field of startups and R&D programs the company Cyano Biofuels (later on renamed as Cyano Biotech GmbH) is mentioned which was founded in 2007 and focused on the optimization of cyanobacteria to generate ethanol (Humboldt Innovation n.d.). The large-scale implementation of the technology was planned together with Algenol (www.algenol.com) and an area of 100 000 m^2 in Texas was prepared for the cultivation of cyanobacteria applying a photobioreactor developed in cooperation with Dow. A second project together with Biofields was envisaged comprising a cultivation area of 40 000 ha in the Mexican desert for the conversion of 8 Mio. t carbon dioxide pa (Müller 2010). However, these plans have not materialized so far or a delayed. Cyano Biotech is offering a technology service portfolio around the cultivation and processing of cyanobacterial strains (Cyano Biotech n.d.). The technology of Algenol based on microalgae is still in the pilot scale. Based on current estimation given boundary conditions of the technology the price of algae-based biofuels is estimated with 200 USD/gal compared to 2.6 USD/gal (equals 0.69 USD/l) for petroleum diesel (Liu et al. 2020).

2.6 Industrial Production and Capacity

2.6.1 Ethanol Production First Generation

The fermentative production of ethanol has experienced a boom in the last 20 years and the global consumption has increased from 20 bn liters in 2002 up to 129 bn liters in 2020. Consumption is predicted to grow even further reaching a level of 143 bn liters in 2028. The United States had enforced in 2005 a new RFS (Renewable Fuel Standard) partially in reaction to the attack on September 11, 2001, to provide energy security, reduce carbon dioxide emissions, and ensure income in rural areas. The RFS program requires that biofuels, in particular ethanol, are blended into the gasoline fuel supply. Accordingly, the United States is the world leader in biofuel production and generated 47% of global output over the last decade (Lark et al. 2022) and is the biggest producing region based on their cost-efficient agriculture of GMO corn, which is still the predominant feedstock in the United States, followed by Brazil based on their sugarcane industry (Witsch 2020; Table 2.11).

The consumption in 2020 corresponds to a global production of more than 100 Mio. t given a density of 0.785 kg/l. The product is applied as fuel additive (about 85%), for beverages (about 7.5% as additive for herbal extracts) and as a building block for chemical synthesis (about 7.5% for e.g. ethyl-ester, ethyl-ether). It should be indicated that the ethanol generation of the brewing industry and wine growing is not covered.

Table 2.11 Consumption of ethanol 2020 and 2028.

Region	Consumption 2020, bn liters	Consumption 2028, bn liters
USA	59.0	58.8
Brazil	30.0	36.6
China	10.0	15.5
EU	8.0	7.4
Others	22.0	24.9
Total	129.0	143.2

Table 2.12 Planned capacity for plants using lignocellulosic biomass.

Company	Startup	Capacity	Location	Feedstock	Status
BP	n.a.	36 m gal	H. County, Florida	Biomass	Stopped 2012
POET-DSM	Sept. 2014	20 m gal	Emmetsburg, Iowa	Corn stover	Phased out
Abengoa	Oct. 2014	25 m gal	Hugoton, Kansas	Corn stover	Phased out
Dupont	Oct. 2015	30 m gal	Nevada, Iowa	Corn stover	Phased out
GranBio	n.a.	22 m gal	Brazil	Bagasse	Running
Raízen	2014/2105	11 m gal	Brazil	Bagasse	Running

2.6.2 Ethanol Production Second Generation

The RFS protocol in the United States has been extended in 2007 to support the development of ethanol second generation with further improvement of the carbon footprint. Several attempts were pursued in the United States to commercially establish the fermentative production based on lignocellulosic biomass. These R&D programs pursue different approaches for decomposition and are testing the innovative technology in mid-size capacity with the option to increase afterward once technical issues have been resolved (Sheridan 2013; Jansen et al. 2017; Greenwood 2018; Ko and Lee 2018; da Silva et al. 2020; Table 2.12).

The biofuel volume goals of the initiative have not been achieved in the United States. The overall target for biofuel in 2022 was originally set with 36 billion gallons (=136 bn liters, first and second generation) compared to a demand of 200 billion gallons of motor fuel in the United States. The achieved volume in 2019 amounts to 20 billion gallons (=75 bn liters, first and second generation) mainly comprising corn ethanol and soybean biodiesel. The target for cellulosic biofuel of 16 billion gallons (=61 bn liters, second generation) in 2022 will not be met since only the tiny volume of 10 million gallons (=38 million liters) has been achieved in 2019 (DeCicco 2021). In a comprehensive review published in *Nature*, the reasons are listed and an industry expert that all second-generation plants (at least in the United States) are significantly reducing their volume: "Alas, the flow of cellulosic fuel is barely a trickle. Processing equipment proved hard to operate, petrol prices fell and governments eased mandates designed to force the pricier cellulosic biofuels into the market.

'Ultimately all of those facilities struggled. Most are either producing at very low levels today or not producing at all,', says John Field, who studies the climate mitigation potential of bioenergy systems at Oak Ridge National Laboratory in Tennessee" (Fairley n.d.).

Accordingly, in the Bioeconomy Strategy for the US bioeconomy, it is stated that the second-generation technology accessing cellulose and hemicellulose as fermentation feedstock has not yet achieved its goals: "Unfortunately, releasing the sugars tied up in cellulose, a major structural component of all plants, is not as easy as liberating it from corn kernels, sugar cane, or sugar beets" [...] "more research is needed to address the technical challenges of converting most biomass into desired bioproducts" (A Bioeconomy Strategy 2022). The high expectation and the technical issues of the developing technology are also reflected in the funding experience of Bill Gates, who acquired a 50 Mio. USD share in an US-based biofuel plant which finally had to be closed: "A few years ago, I learned about a U.S. company that had a proprietary process for converting biomass, such as trees, into fuels. I went to visit the plant and was impressed by what I saw, and after doing due diligence, I invested $50 million in the company. But its technology just didn't work well enough - various technical challenges meant the plant couldn't produce at nearly the volume it needed to be economical - and the plant I visited eventually shut down. It was a $50 million dead end, but I'm not sorry I did it. We need to be exploring lots of ideas, even knowing that many of them will fail" (Gates 2021).

The different stories behind the selected projects mirror the great potential and promise to switch to a renewable raw material source, though also reveal that technical hurdles were underestimated. Abengoa pursued ethanol second-generation activities with a plant in Kansas though declared bankruptcy in 2016 (Rosales-Calderon and Arantes 2019). The major players which still pursue ethanol second-generation activities are Clariant, Granbio, Pray, Raízen, and Versalis. Granbio pursues a project with bagasse in Brazil with a planned capacity of 82.5 Mio. l (Jansen et al. 2017). Raízen pursues a project in Brazil with a planned capacity of 40.3 Mio. l based on sugarcane bagasse and straw (Jansen et al. 2017). In 2022, Raízen announced that additional capacity is under preparation with a startup in 2023/24 to derive an overall capacity of 280 Mio. l (Bettenhausen 2023; AgriBusiness, Raízen n.d.). Praj opened a cellulosic ethanol plant in India in 2023 with a target capacity of 30 Mio. l based on rice straw (Bettenhausen 2023). The Indian company applies catalysts for the breakdown of biomass into cellulose, hemicellulose, and lignin. On the other hand, Clariant and Raízen employ steam explosion (Bettenhausen 2023). In Europe, Versalis pursues a second-generation ethanol technology with a plant in Crescentino. The activities of POET & DSM, Dupont, and Clariant shall be described in more detail. There are other R&D initiatives pursued and the failure of the described examples does not conclude the case of second-generation biofuel.

2.6.3 The Project "LIBERTY" of POET and DSM

The R&D activities of POET-DSM, Abengoa, and Dupont were trying lignocellulosic biomass such as the abundant corn stover in the US derived as by-product in the processing of corn. The sugar fraction of cellulose and hemicellulose is extracted and used for the fermentation, the remaining lignin is mostly burnt to use the heat or to generate power. The established ethanol producer POET and the Dutch chemical and food ingredient company DSM joined forces in 2012 after several years of independent R&D activities.

The project "LIBERTY" as a joint venture between POET and DSM planned to develop an ethanol 2nd generation plant in Iowa and foresaw to license the technology after commercial maturity. DSM has a position in the market for yeast hosts and contributed to the yeast host and the enzyme cocktail to break down the bigger macromolecules into C5 and C6 sugars. In 2011, DSM acquired C5 Yeast Company B.V. to strengthen its R&D pipeline for second-generation biofuels (DSM press release 2011). POET contributed their engineering and production know how. The overall costs of the facility were indicated at 250 Mio. $ (Bomgardner 2014). It was also reported that the plant is designed to provide maximum of 72 gal ethanol per ton of biomass (Merritt, 2017). In case 70 gal per ton of biomass are assumed as average output 0.21 t ethanol/t biomass (70 gal = 272 l = 0.21 ethanol/t biomass) would be obtained. This unit consumption rate is according to the figure estimated above for a wood-based lignocellulosic biomass. In 2017, POET announced that the plant has achieved a breakthrough and solved several pre-treatment issues hampering to reach full scale in the last years. Pre-treatment would now be running at 80% of its capacity. Furthermore, an on-site enzyme production had been implemented to directly feed the enzyme mix into the pre-treatment section (Dale 2017; POET press release 2017). However, in 2019 POET declared that activities were put on hold with the reason that the Renewable Fuel Standard (RFS) has not been adjusted to allow a commercially viable business for 2nd generation ethanol (Godwin, 2019). DSM sold its second-generation enzyme technology Prosea to Versalis in 2022 (Sapp 2022).

2.6.4 The Ethanol Second Generation Activities of Dupont

Dupont has a business stake in enzymes and was interested in expanding activities in enzymes for the breakdown of lignocellulosic biomass. In order to be a frontrunner in this promising field it engaged in the ethanol business with a plant in Iowa. However, Dupont invested about 200 $m and phased out activities in 2018 with the rationale that Dupont is focusing on specialties according to their new strategy and the long-term ethanol commodity business would not match that strategy. The CEO of DowDupont at that time, E. Breen, classified second-generation ethanol as a "moonshot project" (Tullo 2018; Lane 2017). DuPont declared to further continue the enzyme business for the cellulosic biofuel industry and to keep the share in Butamax. The plant was sold to Verbio which planned to use the plant to produce methane from biomass (Verbio press release 2018). Verbio has refurbished the plant and plans to convert 100 kt corn stover into renewable natural gas (RNG) which is offered as a climate-friendly biofuel and can be injected into the existing gas grids (*Verbio launches the first cellulosic RNG plant in the USA* n.d.).

2.6.5 The "Sunliquid" Technology of Clariant

Clariant has developed a proprietary fermentation technology for the production of second-generation ethanol which is marketed under the brand name "sunliquid." The initial R&D stage has been acquired in the context of the acquisition of Südchemie in 2011, which has been upgraded to reach commercial maturity. A demonstration plant has been operated in Germany (Straubing) since 2012 and the first commercial plant to be operated by Clariant is envisioned for startup in 2022 in Romania (Podari). The raw material source of the "sunliquid" technology is wheat straw. The availability of wheat straw in Europe is

Table 2.13 Composition of wheat straw and conversion into ethanol.

Composition	Polysaccharides	Monosaccharide	Extraktion	Ethanol yield
32% cellulose	32 kt cellulose	35.5 kt glucose	28.4 kt	13.1 kt (0,46 t/t EtOH)
29% hemicellulose	20 kt D-xylose	22.6 kt D-xylose	18.1 kt	8.3 kt (0.46 t/t EtOH)
18% lignin			18.0 kt	

estimated about 144 Mio. t (Cordis 2019; Rarbach 2016). The market value is 63 USD/t, though the price will display a broad range pending on region, volume, distance to the plant, and alternative conversion technologies (Tan et al. 2018). The generation of wheat straw on small farms implies rather high effort for collection and logistics and Clariant has signed contracts with more than 300 farmers to ensure supply (Lane 2022). For the estimation of the unit consumption rate of the wheat straw feed per ton ethanol, the specific composition needs to be considered. The calculation refers to a starting volume of 100 kt wheat straw and applies the mathematical mean of the wheat straw cellulose, hemicellulose, and lignin composition range (Kumar and Sharma 2017; Bharathiraja et al. 2017; Table 2.13).

The applied technology is a proprietary version of "steam explosion," which basically applies hot steam to break down the biomass into its major three components cellulose, hemicellulose, and lignin. Reported extraction yields are in the range of 35–71% (da Silva et al. 2020). The extraction yield referred to in this context is assumed with 80% given the less recalcitrant physical matter of wheat straw and taking the technology maturity after 10 years of pilot scale into account. For the conversion of the C6 and C5 sugars (i.e. D-xylose and L-arabinose), a carbon yield of 0.46 t/t EtOH according to the yield in the first-generation ethanol technology has been assumed. The estimation derives that 100 kt wheat straw would finally provide 21.4 kt ethanol. This corresponds to an input factor of 4.7 t wheat straw per ton ethanol. Clariant states that the commercial plant in Romania has a capacity of 50 kt pa ethanol and converts a feedstock of 250 kt wheat straw (IV Internationales Vekehrswesen 2018; Rarbach 2019). This translated into a unit consumption rate of 5 t wheat straw per ton ethanol corresponding to the figure estimated above. It should be considered that Clariant produces the enzyme mix for the hydrolysis on site applying the same feedstock and, thus, a minor share of the wheat straw is allocated for the production of the enzyme mix. Based on the published content of 18% lignin in wheat straw, the process might potentially also provide 18 kt lignin or 0.84 t lignin per ton ethanol.

References

A Bioeconomy Strategy (2022). The U.S. bioeconomy: charting a course for a resilient and competitive future. Schmidt Futures. p. 68.

AgriBusiness, Raízen (n.d.). Novozymes strengthen partnership for enhancing production of second generation ethanol. http://agribusinessglobal.com/sustainability/raizen-novozymes-strengthen-partnership-for-enhancing-production-of-second-generation-ethanol (accessed 22 September 2022).

Autthanit, C., Likitpiriya, N., Praserthdam, P., and Jongsomjit, B. (2021). Development of a new Al_2O_3-HAP-pd catalyst for diethyl ether and ethylene production using the preferential dehydration of ethanol. *ACS Omega* 6: 19911–19923. https://doi.org/10.1021/acsomega.1c02818.

Behr, A. and Seidensticker, T. (2018). *Einführung in die Chemie nachwachsender Rohstoffe*, 339ff. Springer.

Bertsch, J. and Mueller, V. (2015). Bioenergetic constraints for conversion of syngas to biofuels in acetogenic bacteria. *Biotechnology for Biofuels* 8: 210. https://doi.org/10.1186/s13068-015-0393-x.

Bettenhausen, C. (2023). Ethanol rising. *C&EN* 20–26.

Bharathiraja, B., Jayamuthunagai, J., Sudharsanaa, T. et al. (2017). Biobutanol – an impeding biofuel for future: a review on upstream and downstream techniques. *Renewable and Sustainable Energy Reviews* 68: 788–807. https://doi.org/10.1016/j.rser.2016.10.017.

Bomgardner, M. (2014). Cellulosic biofuel opens in the US. *C&EN* 8 (9): 7.

Bongers, R.S., Hoefnagel, M.H., and Kleerebezem, M. (2005). High-level acetaldehyde production in lactobacillus lactis by metabolic engineering. *Applied and Environmental Microbiology* 71: 1109–1113.

Bozell, J.J. and Petersen, G.R. (2010). Technology development for the production of biobased products from biorefinery carbohydrates – the US department of energy's "top 10" revisited. *Green Chemistry* 12: 539–554.

Braskem (2022). www.braskem.com.br/europe/combatingclimatechange (accessed 12 August 2022).

Brown, T.R. (2015). A techno-economic analysis of thermochemical cellulosic biofuel pathways. *Bioresource Technology* 178: 166–176. https://doi.org/10.1016/j.biortech.2014.09.053.

Buchner, E. (1897). Alkoholische Gärung ohne Hefezellen. *Berichte der Deutschen Chemischen Gesellschaft* 30: 117–124.

Clark, D. and (1989). The fermentation pathway of *Escherichia coli*. *FEMS Microbiology Review* 5: 223–234. https://doi.org/10.1016/0168-6445(89)90033-8.

Cordis (2019). Wie man Weizenstroh in grüne Chemikalie umwandeln kann. https://cordis.europa.eu/article/id/124991-how-to-convert-wheat-straw-waste-into-green-chemicals/de (accessed 23 April 2021).

Cropwatch (2022). Corn grain as an ethanol feedstock. https://cropwatch.unl.edu/bioenergy/corn (accessed 16 October 2022).

Croteau, J. (2021). *Lummus, Braskem Partner for Green Ethylene Projects*, 9. Chemical Week.

Cyano Biotech (n.d.). http://www.cyano-biotech.com/content/biomasstocompounds/index.php (accessed 9 December 2021).

Dale, B. (2017). Real progress in cellulosic biofuel. *Biofuels, Bioproducts & Biorefining* 11: 943–944. https://doi.org/10.1002/bbb.1837.

DeCicco, J. (2021). Biofuel isn't as neutral as we want it to be. https://www.greenbiz.com/article/biofuel-isnt-carbon-neutral-we-want-it-be (accessed 2 January 2022).

Doelle, H.W. and Greenfield, P.F. (1985). The production of ethanol from sucrose using *Zymomonas mobilis*. *Applied Microbiobiolgy and Biotechnology* 22: 405–410.

DSM press release (2011). DSM strengthens yeast technology leadership with G2 biofuels. https://www.dsm.com/corporate/news/news-archive/2011/45-11 (accessed 15 April 2021).

Dudley, Q.M., Karim, A.S., and Jewett, M.C. (2015). Cell-free metabolic engineering: biomanufacturing beyond the cell. *Biotechnology Journal* 10: 69–82.

Economist (2022). Against the grain. Graphic Detail World Food, p. 81

Erb, T. (2019). Structural organization of biocatalytic systems: the next dimension of synthetic metabolism. *Emerging Topics in Life Sciences* 3: 579–586.

Fairley, P. The biofuel course correction. *Nature* 611: S15–S17.

FAZ, Rettung nach vom Acker (2020). Für den Grundstoff Alkohol in Desinfektionsmitteln springen nun auch Bioethanol-Hersteller in die Bresche. No. 72, p. 22.

Flamholz, A., Noor, E., Bar-Even, A., and Milo, R. (2012). eQuilibrator – the biochemical thermodynamics calculator. *Nucleic Acids Research* 40: https://doi.org/10.1093/nar/gkr874.

Folch, P.L., Bisschops, M.M.M., and Weusthuis, R.A. (2021). Metabolic energy conservation for fermentative product formation. *Microbial Biotechnology* 14 (3): 829–858. https://doi.org/10.1111/1751-7915.13746.

Gaida, S.M., Liedtke, A., Jentges, A.H.W. et al. (2016). Metabolic engineering of *clostridium cellulolyticum* for the production of *n*-butanol from crystalline cellulose. *Microbial Cell Factories* 15: 6. https://doi.org/10.1186/s12934-015-0406-2.

Galbe, M. and Wallberg, O. (2019). Pretreament for biorefineries: a review of common methods for efficient utilization of lignocellulosic materials. *Biotechnology for Biofuels* 12: 294. https://doi.org/10.1186/s13068-019-1634-1.

Gates, B. (2021). *How to Avoid a Climate Disaster: The Solutions We Have and the Breakthroughs We Need*, 138–139. Allen Lane.

Godwin, J. (2019). POET-DSM idling project Liberty cellulosic plant. https://chemweek.com/CWDocument/107511/POET-DSM-idling-Project-Liberty-cellulosic-plant (accessed 15 April 2021).

Greenwood, A. (2018). US left with one large-scale cellulosic ethanol plant. *ICIS* 29: 11.

Guo, J., Huang, S., Chen, Y. et al. (2018). Heterologous expression of *Spathaspora passalidarum* xylose reductase and xylitol dehydrogenase genes improved xylose fermentation ability of *Aureobasidium pullulans*. *Microbial Cell Factories* 17: 64. https://doi.org/10.1186/s12934-018-0911-1.

de Guzman, D. (2010). *Latin America Outlook Good for Bioplastic Expansion*, 22. ICIS Chemical Business.

Hasona, A., York, S.W., Yomano, L.P. et al. (2002). Decreasing the level of ethyl acetate in ethanolic fermentation broths of *Escherichia coli* KO11 by expression of *pseudomonas putida estZ* estetrase. *Applied and Environmental Microbiology* 68 (6): 2651–2659. https://doi.org/10.1128/AEM.68.6.2651-2659.2002.

Heijstra, B.D., Leang, C., and Juminga, A. (2017). Gas fermentation: cellular engineering possibilities and scale up. *Microbial Cell Factories* 16: 60. https://doi.org/10.1186/s12934-017-0676-y.

Herrera, S. (2006). Bonkers and biofuels. *Nature Biotechnology* 24 (7): 755–760.

Humboldt Innovation (n.d.). https://humboldt-innovation.de/en/startupserivice/startup-portfolio/cyano-biofuels-gmbh (accessed 9 December 2021).

Hüsing, B. (2012). Strategien und Geschäftsmodelle in der industriellen Biotechnologie, Innovationsreport Weisse Biotechnologie, TBA Brief 40.

ICIS (2020a). *Chemical Profile: Europe Ethanol*. ICIS www.ecn.com.

ICIS (2020b). *Chemical Profile: US Ethanol*. ICIS www.ecn.com.

IV Internationales Vekehrswesen (2018). Biokraftstoff aus Stroh: Auszeichnung für sunliquid-Technologie. https://www.internationales-verkehrswesen.de/biokraftstoff-aus-stroh (accessed 2 February 2023).

Jansen, M.L.A., Bracher, J.M., Papapetridis, I. et al. (2017). *Saccharomyces cerevisiae* strains for second-generation ethanol production: from academic exploration to industrial implementation. *FEMS Yeast Research* 17 (5): https://doi.org/10.1093/femssyr/fox044.

Jeswani, H.K., Chilvers, A., and Azapagic, A. (n.d.). Environmental sustainability of biofuels: a review. *Proceedings of the Royal Society* 476: 20200351. https://doi.org/10.1098/rspa.2020.0351.

Jung, S., Shetti, N.P., Reddy, K.R. et al. (2021). Synthesis of different biofuels from livestock waste materials and their potential as sustainable feedstocks – a review. *Energy Conversion and Management* 236: https://doi.org/10.1016/j.enconman.2021.114038.

Kang, A. and Lee, T.S. (2015). Converting sugar to biofuels: ethanol and beyond. *Bioengineering* 2: 184–203. https://doi.org/10.3390/bioengineering2040184.

Katsyv, A. and Mueller, V. (2020). Overcoming energetic barriers in acetogenic C1 conversion. *Frontiers in Bioengineering and Biotechnology* 8: 621166. https://doi.org/10.3389/fbioe.2020.621166.

Knocke, C. and Vogt, J. (2009). Biofuels – challenges & chances: how biofuel development can benefit from advanced process technology. *Engineering in Life Sciences* 9 (2): 96–99. https://doi.org/10.1002/elsc.200700064.

Ko, J.K. and Lee, S.-M. (2018). Advances in cellulosic conversion to fuels: engineering yeasts for cellulosic bioethanol and biodiesel production. *Current Opinion in Biotechnology* 50: 72–80. https://doi.org/10.1016/j.copbio.2017.11.007.

Kucharska, K., Rybarczyk, P., Holowacz, I. et al. (2018). Pretreatment of lignocellulosic materials as substrates for fermentation processes. *Molecules* 23: 2937. https://doi.org/10.3390/molecules23112937.

Kumar, A.K. and Sharma, S. (2017). Recent update on different methods of pretreatment of lignocellulosic feedstocks: a review. *Bioresources and Bioprocessing* 4: 7. https://doi.org/10.1186/s40643-017-0137-9.

Lane, J. (2017). DowDuPont to exit cellulosic biofuels business, www.biofuelsdigest.com

Lane, J. (2022). 5 minutes with Thomas Hoppe, Clariant's Head of Operations Business Line Biofuels & Derivatives. www.biofuelsdigest.com (accessed 8 September 2022).

Lark, T.J., Hendricks, N.P., Smith, A. et al. (2022). Environmental outcomes of the US renewable fuel standard. *Proceedings of the National Academy of Sciences of the United States of America* 119 (9): https://doi.org/10.1073/pnas.2101084119.

Liu, Z., Wang, K., Chen, Y. et al. (2020). Third-generation biorefineries as the means to produce fuels and chemicals from CO_2. *Nature Catalysis* 3: 274–288. https://doi.org/10.1038/s41929-019-0421-5.

Lynd, L.R., Laser, M.S., Bransby, D. et al. (2008). How biotech can transform biofuels. *Nature Biotechnology* 26 (2): 169–172.

Merritt, M. (2017). Real progress in cellulosic biofuel. *Biofuels, Bioproducts and Biorefining* 11: 943–944.

Meyer, H.P., Minas, W., and Schmidhalter, D. (2017). Industrial-scale fermentation. In: *Industrial Biotechnology. Products and Processes* (ed. C. Wittmann and J.C. Liao), 30, 3–53. Wiley.

Müller, E. (2010). Die CO_2-fresser. *Manager Magazine* 12: 138–144.

Naims, H. (2016). Economics of carbon dioxide capture and utilization – a supply and demand perspective. *Environmental Science and Pollution Research* 23: 22226–22241. https://doi.org/10.1007/s11356-016-6810-1.

Nevoigt, E. (2008). Progress in metabolic engineering of *Saccharomyces cerevisiae. Microbiology and Molecular Biology Reviews* 379–412.

Ning, P., Yang, G., Hu, L. et al. (2021). Recent advances in the valorization of plant biomass. *Biotechnology for Biofuels* 14: 102. https://doi.org/10.1186/s13068-021-01949-3.

Papapetridis, I., van Dijk, M., van Maris, A.J.A., and Pronk, J.T. (2017). Metabolic engineering strategy for optimizing acetate reduction, ethanol yield and osmotolerance in *Saccharomyces cerevisiae. Biotechnology for Biofuels* 10: 107. https://doi.org/10.1186/s13068-017-0791-3.

Peters, D., Ulber, R., and Wagemann, K. (2014). Die deutsche Roadmao. Bioraffinerien. *Chemie in unserer Zeit* 48: 46–59. https://doi.org/10.1002/ciuz.201400622.

Pinske, C. and Sargent, F. (2016). Exploring the directionality of Escherichia coli formate hydrogenlyase: a membrane-bound enzyme capable of fixing carbon dioxide to organic acid. *Microbiology Open* 5 (5): 721–737.

POET press release (2017). POET-DSM achieves cellulosic biofuel breakthrough. New pretreatment system clears path for increased cellulosic production. https://poet.com/pr/poet-dsm-achieves-cellulosic-biofuel-breakthrough (accessed 15 April 2021).

Rarbach, M. (2016). Stroh ist nicht gleich Stroh, Interview. https://www.pflanzenforschung.de/de/pflanzenwissen/journal/stroh-ist-nicht-gleich-stroh-interview-mit-dr-markus-ra-10584 (accessed 2 February 2023).

Rarbach, M. (2019). Clariant sunliquid®: Vom Stroh in die Chemie, Company presentation. www.clariant.com (accessed 15 April 2021).

Robak, K. and Balcerek, M. (2018). Review of second-generation bioethanol production from residual biomass. *Food Technology and Biotechnology* 56 (2): https://doi.org/10.17113/ftb.56.02.18.5428.

Rosales-Calderon, O. and Arantes, V. (2019). A review on commercial-scale high-value products that can be produced alongside cellulosic ethanol. *Biotechnology for Biofuels* 12: 240. https://doi.org/10.1186/s13068-019-1529-1.

Sanchez, D.L., Johnson, N., McCoy, S.T. et al. (2018). Near-term deployment of carbon capture and sequestration from biorefineries in the United States. *Proceedings of the National Academy of Sciences of the United States of America* 115 (19): 4875–4880. https://doi.org/10.1073/pnas.1719695115.

Sandmair, K. and Zeitalter, N. (2010). Biosprit aus Pflanzenabfällen statt aus wertvollem Getreide. Novozymes schafft sich einen Milliardenmarkt. *Focus-Money* 17: 48–49.

Sapp, M. (2022). Versalis acquires DSM's Prosea technology. https://www.biofuelsdigest.com/bdigest/2022/12/19/versalis-acquires-dsms-proesa-technology (accessed 20 December 2022).

Schubert, C. (2006). Can biofuels finally take center stage? *Nature Biotechnology* 24 (7): 777–784.

Sheehan, J. (2009). Engineering direct conversion of CO_2 into fuels. *Nature Biotechnology* 27 (12): 1128–1129.

Sheridan, C. (2013). Big oil turns on biofuels. *Nature Biotechnology* 31 (10): 870–873.

da Silva, A.S., Espinheira, R.P., Teixeira, R.S.S. et al. (2020). Constraints and advances in high-solids enzymatic hydrolysis of lignocellulosic biomass: a critical review. *Biotechnology for Biofuels* 13: 58. https://doi.org/10.1186/s13068-020-01697-w.

Siqueira, J.G.W., Rodrigues, C., de Souza Vandenberghe, L.P. et al. (2020). Current advances in on-site cellulase production and application on lignocellulosic biomass conversion to biofuels: a review. *Biomass and Bioenergy* 132: 105419. https://doi.org/10.1016/j.biombioe.2019.105419.

Snoep, J.L., de Graef, M.R., Westphal, A.H. et al. (1993). Differences in sensitivity to NADH of purified pyruvate dehydrogenase complexes of *Enterococcus faecalis, Lactobacillus lactis, Azetobacter vinelandii* and *Escherichia coli*: implication for their activity in vivo. *FEMS Microbiology Letters* 114: 279–283.

Spaans, S.K., Weusthuis, R.A., van der Oost, J., and Kengen, S.W.M. (2015). NAPDH-generating systems in bacteria and archaea. *Frontiers in Microbiology* 6: article 742. https://doi.org/10.3389/fmicb.2015.00742.

Stanbury, P.F., Whitaker, A., and Hall, S.J. (2017). *Principles of Fermentation Technology*, 3e. Elsevier section on sterilization pp. 273–333.

Tan, J.P., Luthfi, A.A.I., Manaf, S.F.A. et al. (2018). Incorporation of CO_2 during the production of succinic acid from sustainable oil palm frond juice. *Journal of CO_2 Utilization* 26: 595–601. https://doi.org/10.1016/j.jcou.2018.06.006.

Tong, K.T.X., Tan, I.S., Foo, H.C.Y. et al. (2022). Advancement of biorefinery-derived platform chemicals from macroalgae: a perspective for bioethanol and lactic acid. *Biomass Conversion and Biorefinery*. https://doi.org/10.1007/s13399-022-02561-7.

Trading Economics (n.d.). Ethanol. https://tradingeconomics.com/commodity/ethanol (accessed 25 november 2022).

Tullo, A. (2018). DowPuPont CEO favors short-term R&D, C&EN. www.https://pubs.acs.org/doi/full/10.1021/cen-09624-buscon4 (accessed 15 April 2021).

Ullah, M.W., Khattak, W.A., Ul-Islam, M. et al. (2014). Bio-ethanol production through simultaneous saccharification and fermentation using an encapsulated reconstituted cell-free enzyme system. *Biochemical Engineering Journal* 110–119.

Ullah, M.W., Khattak, W.A., Ul-Islam, M. et al. (2016). Metabolic engineering of synthetic cell-free systems: strategies and applications. *Biochemical Engineering Journal* 105: 391–405.

Verbio press release (2018). Verbio erwirbt eine Zellulose-Ethanolanlage von DuPont in Nevada/Iowa, US. https://www.verbio.de/investor-relations/aktuelles-publikationen/ad-hoc/verbio-erwirbt-eine-zellulose-ethanolanla-ge-von-dupont-in-nevadaiowa-usa (accessed 15 April 2021).

Verbio press release (2022). Verbio launches the first cellulosic RNG plant in the USA. https://www.verbio.de/en/press/news/press-releases/detail/verbio-betreibt-erste-stroh-biomethan-anlage-der-usa (accessed 11 May 2022).

Voet, D., Voet, J.G., and Pratt, C.W. (2016). *Principles of Biochemistry*. Wiley p. 501, 503, 520.

Watanabe, S., Fukumori, F., Nishiwaki, H. et al. (2019). Novel non-phosphorylative pathway of pentose metabolism from bacteria. *Scientific Reports* 9: 155. https://doi.org/10.1038/s41598-018-36774-6.

Wecker, M.S. and Zall, R.R. (1987). Production of acetaldehyde by *Zymomonas mobilis*. *Applied and Environmental Microbiology* 53: 2815–2820.

Welch, P. and Scopes, R.K. (1985). Studies on cell-free metabolism: ethanol production by a yeast glycolytic system reconstituted from purified enzymes. *Journal of Biotechnology* 2: 257–273.

Witsch, K. (2020). Die Rückkehr des Biofuels. *Handelsblatt* 16 (1): 16.

Wood, W.A. (n.d.). Fermentation of carbohydrates and related compounds. In: *The Bacteria: A Treatise on Structure and Function*, vol. 2 (ed. I.C. Gunsalus and R.Y. Stanier), 59–149. New York: Academic Press.

Wu, C.-Y. and Wu, H.-S. (2017). Ethylene formation from ethanol dehydration using ZSM-5 catalyst. *ACS Omega* 2: 4287–4296. https://doi.org/10.1021/acsomega.7b00680.

Zhu, H., Gonzalez, R., and Bobik, T.A. (2011). Coproduction of acetaldehyde and hydrogen during glucose fermentation by *Escherichia coli*. *Applied and Environmental Microbiology* 77: 6441–6450.

3

Acetate and Ethanol Ex CO/H_2

3.1 The Wood–Ljungdahl Pathway

3.1.1 Methyl-Branch and CO-Branch

Acetogenic bacteria grow in aqueous media containing gaseous mixtures of carbon dioxide, carbon monoxide, and hydrogen and are able to produce acetate (Schuchmann and Müller 2016). They comprise about 60 species which are characterized with respect to substrate, product, and growth conditions (Takors et al. 2018) and follow the Wood–Ljungdahl pathway which is pursued by methanogenic archaea and acetogens (Drake et al. 2008; Drake et al. 2013; Ragsdale and Pierce 2008; Thauer et al. n.d.). The pathway generates acetyl-CoA derived by condensation of carbon monoxide and methyl-tetrahydrofolate (methyl-THF). Hereby, carbon dioxide is metabolized down to the reduced level of the methyl function by the transfer of six electrons. In the first step, carbon dioxide is reduced with reduction equivalents or molecular hydrogen to formate which is then coupled with tetrahydrofolate (THF) enabled by the investment of 1 mol of ATP. The obtained formyl-THF is converted with water into methylene-THF and the C1-moiety is in two consecutive reduction steps converted into methyl-THF ("methyl-branch"). In parallel, carbon dioxide or carbon monoxide ("CO-branch") is prepared for the condensation with the activated methyl function to derive acetyl-CoA. The thioester of acetyl-CoA is then converted into a phosphoric acid ester to obtain acetyl-phosphate which is highly active and bears the energy potential to enable the generation of ATP. The final product of the pathway, namely, acetate, is also the compound for naming organisms applying this pathway ("acetogens") (Figure 3.1).

The reductive steps of the pathway with electron transfer are calculated with either molecular hydrogen (left column) or NADH (middle column) as electron donors. The Gibbs free energy values demonstrate that the formation of formyl-THF requires the investment of 1 mol of ATP and that the formation of acetate allows the generation of 1 mol of ATP (right column). The reduction of carbon dioxide to carbon monoxide in the CO branch is done with reduced ferredoxin to provide the needed reduction power.

3.1.2 ATP Generation via Generation of Ion Gradient

The Wood–Ljungdahl pathway is to be regarded as a metabolic invention of ancestral forms of life to survive in an atmosphere mainly containing carbon dioxide, carbon monoxide, and hydrogen. Energy-rich substrates for heterotrophic glycolysis were not yet available.

Pathway Design for Industrial Fermentation, First Edition. Walter Koch.

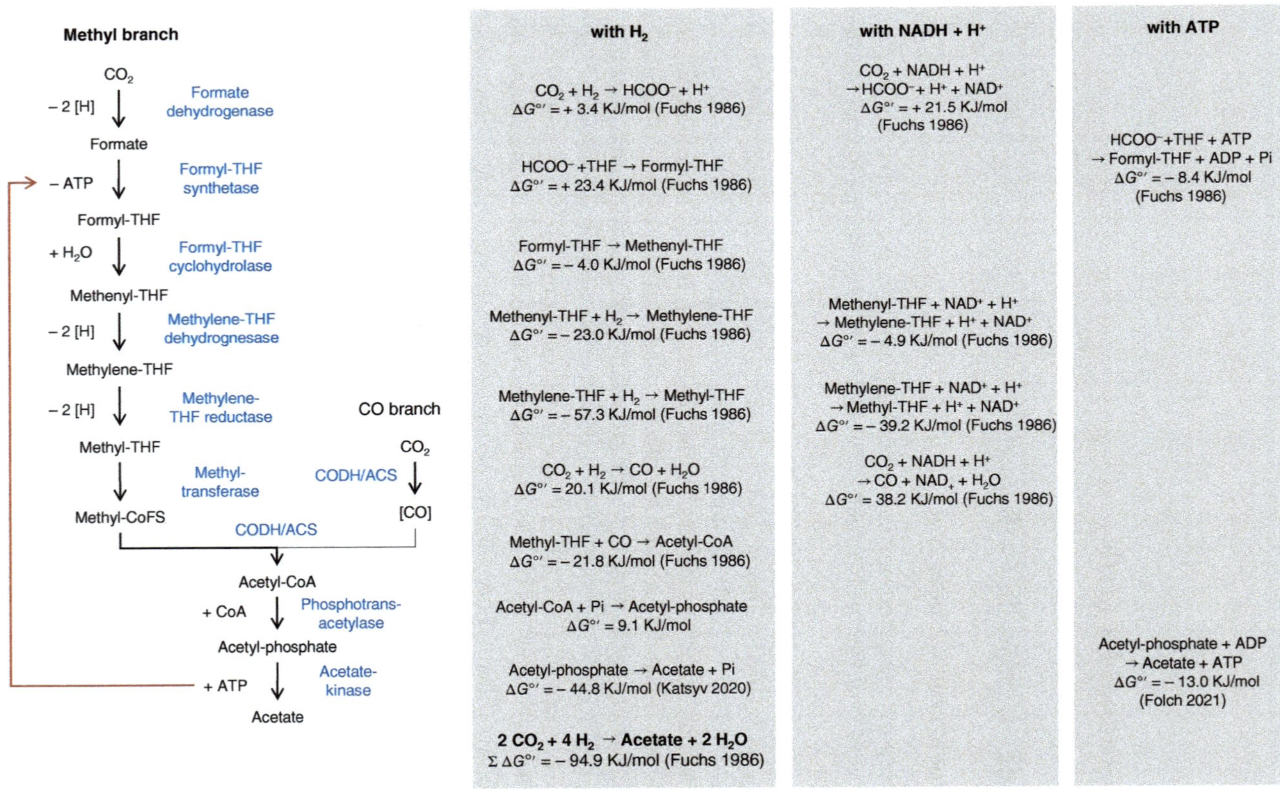

Figure 3.1 The Wood–Ljungdahl pathway for the conversion of carbon dioxide, carbon monoxide, and hydrogen into acetate. CODH/ACS: CO dehydrogenase or acetyl-CoA synthase. The Gibbs free energy for the separate steps is according to literature (Fuchs 1986; Katsyv and Mueller 2020; Folch et al. 2021). The Gibbs free energy for the reduction of acetate with molecular hydrogen is also indicated with −3 KJ/mol (Müller 2019). The value for the phosphotrans-acetylase has been calculated to match the overall Gibbs free energy of the pathway (Fuchs 1986; Katsyv and Mueller 2020). Source: Walter Koch.

However, several acetogens are nevertheless able to metabolize sugars and can utilize a vast number of electron sources (Schuchmann and Müller 2016). Substrate-level phosphorylation in the Wood–Ljungdahl pathway is restricted to the conversion of acetyl-phosphate into acetate. The formation of 1 mol of ATP via the phosphorylation of ADP requires a phosphoryl group transfer potential of at least −31.8 KJ/mol (Thauer et al. 1977). Including unavoidable dissipation in living systems, the minimum threshold for ATP synthesis via substrate-level phosphorylation is estimated with a minimum Gibbs free energy of −40 to 50 KJ/mol (Folch et al. 2021; Voet et al. 2016) which restricts substrate-level phosphorylation to creatine-phosphate, arginine-phosphate, carbamoyl-phosphate, butyryl-phosphate, 1,3-BPG, PEP, and acetyl-phosphate:

$$\text{Acetyl-phosphate} + H_2O \rightarrow \text{Acetate} + \text{Pi} \quad \Delta G^{\circ\prime} = -44.8\ \text{KJ/mol}$$

This match is exploited in the ATP regeneration of the WLP (Katsyv and Mueller 2020; Voet et al. 2016). The gained ATP is in the first instance needed as a cofactor of the formyl THF synthetase which transfers formate on THF to obtain formyl-THF. It is crucial for the understanding of acetogenic bacterial life that the formation of acetate based on carbon dioxide, carbon monoxide, and hydrogen also needs to provide a surplus of energy for cell maintenance and bacterial growth. Thus, it is not sufficient to replenish ATP for the formation of formyl-THF though in parallel an energy surplus needs to be generated to drive the endergonic reactions for cell maintenance and biomass formation with the exergonic hydrolysis of ATP. Acetogens manage to provide additional ATP via membrane-bound respiratory enzymes that use reduced ferredoxin as a reductant. The oxidation of reduced ferredoxin is coupled to the secretion of positively charged ions through the bacterial membrane to generate a gradient which is exploited by an F_1F_O-ATP synthase to convert ADP into ATP. The cumulative energy mechanism for ATP generation does not suppose a strict stoichiometry, which might require that 1 mol of ATP is generated by 1 mol product. The generation of an ion gradient with a membrane-bound enzyme complex allows to generate ATP with a step-wise buildup of the gradient without strict coupling to the stoichiometry of a defined reaction (Fuchs 1986).

3.1.3 Taxonomy of Acetogenic Bacteria

Acetogens can be differentiated by the type of enzyme complex facilitating the oxidation of reduced ferredoxin coupled with the formation of an ion gradient, which may be managed by either the Ech-complex or the Rnf-complex. The model organisms representing the Ech-complex acetogens are *Thermoanaerobacter kivui* (Jain et al. 2020) and *Moorella thermoacetica*. Reduced ferredoxin is oxidized by an 8–9 subunit membrane complex which transfers ions out of the cell to generate a gradient. The electrons are handed over to protons, and molecular hydrogen is obtained as the final respiration product. It should be pointed out that the volatile final product might be removed via diffusion to restore the thermodynamic possibility of this step. In case the hydrogen concentration or the corresponding partial pressure should increase, the actual free Gibbs energy might turn positive and block further conversion.

The second type of acetogens is using the Rnf-complex for the generation of the ion gradient and is represented by *Acetobacterium woodii*, *Clostridium ljungdahlii* (Köpke et al. 2010; Zhang et al. 2020), *Clostridium autoethanogenum*, and *Clostridium ragsdalei*. The Rnf-complex employs reduced ferredoxin to generate the ion gradient and the electrons are transferred to NAD^+ to provide NADH, which feeds reduction steps in the formation of acetyl-CoA. The co-production of reduction equivalents should in principle allow the addition of recombinant pathways toward other products besides acetate and explains the commercial interest in this subgroup of acetogens (Katsyv and Mueller 2020). The metabolic engineering toolbox is not state of the art, though has seen major improvements in the last years (Bourgade et al. 2012).

A further distinction is based on genetic data and the type of ion gradient. The three acetogenic bacterial species *C. autoethanogenum*, *C. ljungdahlii* and *C. ragsdalei* share several unique features. All three species grow on carbon monoxide as the sole carbon and energy source and are producing ethanol and acetic acid. Since the 16S RNA sequences are 99% identical and the gene arrangement in the Wood–Ljungdahl operon is very similar, the species are believed to be close relatives. They grow ideally in a pH window of 5.0–5.5 and are able to generate acetate and ethanol if grown on hydrogen and carbon dioxide. A proton gradient is formed during the execution of the Wood–Ljungdahl pathway feeding an ATPase to obtain ATP (Mock et al. 2015). These characteristics distinguish them from *Ceropegia woodii* which generates a sodium gradient and feeds the ATPase with sodium ions. The species is not able to grow on carbon monoxide as the sole carbon and energy source. The type of ions exploited by the F_1F_O-ATP synthase are deduced from the conserved binding sites of the enzyme. Protons are bound to the carboxyl function of aspartate or glutamate residues, whereas sodium ions are additionally bound by a glutamine and a serine or threonine residue (Katsyv and Mueller 2020). The stoichiometry of the reaction has been estimated based on bioenergetic calculations and the relation of ion-binding sites to ATP synthesizing centers. Experimental data are available for *A. woodii* which displays 10 Na^+-binding sites and three ATP synthesizing centers in the ATP synthase (Matthies 2014), which proposes that 3.3 Na^+/ATP are required. Since no experimental data for the ATP synthase of other acetogens are reported the ATP synthase of *Clostridium paradoxum* is referred to which displays 11H^+-binding sites and three ATP synthesizing centers. Thus, a stoichiometry of 3.66H^+/ATP is assumed (Katsyv and Mueller 2020; Zhu et al. 2020).

3.1.4 Hydrogen as Energy Source

Acetogenic bacteria use carbon monoxide and molecular hydrogen as reducing equivalents, and fermentation technology with acetogenic hosts may be referred to as "syngas" usually comprising a mixture of carbon monoxide and hydrogen. Nevertheless, the two different energy sources need to be distinguished. In the first instance, carbon dioxide and hydrogen may be applied as feed. In this case, hydrogen is the sole energy source and carbon dioxide is the sole carbon source. Given the reduction potential of hydrogen at standard conditions, hydrogen is not capable to accomplish the reduction of carbon dioxide to carbon monoxide without further support. This is indicated by the respective reduction potentials of the two redox pairs:

$$2\,H^+ + 2\,e \rightarrow H_2\; E_0' = -414\,\text{mV (at pH 7.0)}$$

$$CO_2 + 2\,e + 2\,H+ \rightarrow CO + H_2O\; E_0' = -520\,\text{mV (at pH 7.0)}$$

Accordingly, the conversion of carbon dioxide into carbon monoxide facilitated by the CO dehydrogenase (CODH) requires reduced ferredoxin whose reduction level is in the same range of the redox couple carbon dioxide/carbon monoxide (Bertsch and Mueller 2015; Bertsch and Mueller 2015):

$$\text{Fd} + 2\,\text{e} \rightarrow \text{Fd}^{2-}\ E_0' = -500\,\text{mV (at pH 7.0)}$$

The theoretical reduction potential needed for the generation of carbon monoxide is not fully achieved by reduced ferredoxin, though the difference between the reduction potentials is minor and it can be assumed that the reduction of carbon dioxide with reduced ferredoxin is managed in case the CO concentration is kept sufficiently low (Katsyv and Mueller 2020):

$$\text{CO}_2 + 2\,\text{e} + 2\,\text{H}^+ \rightarrow \text{CO} + \text{H}_2\text{O}\ E_0' = -520\,\text{mV (at pH 7.0)}$$

$$\text{Fd}^{2-} \rightarrow \text{Fd} + 2\,\text{e}\ E_0' = 500\,\text{mV (at pH 7.0)}$$

$$\text{CO}_2 + \text{Fd}^{2-} + 2\,\text{H}^+ \rightarrow \text{CO} + \text{Fd} + 2\,\text{H}_2\text{O} \quad \Delta E^{\circ\prime} = -0.020\,\text{V}$$

The combination of the two redox pairs CO_2/CO and Fd/Fd^{2-} derives a slightly positive Gibbs free energy:

$$\Delta G^{\circ\prime} = -n \times \text{F} \times \Delta E^{\circ} = -2 \times 96{,}485\,\text{J/V} \times \text{mol} \times -0.020\,\text{V} = 3.8\,\text{KJ/mol}$$

It should be pointed out that the calculation of the Gibbs free energy depends on the assumed redox potentials which differ slightly by source, e.g. in literature also a Gibbs free energy of 14 KJ/mol for the reaction above can be found (Zhu et al. 2020).

After carbon dioxide has been reduced with ferredoxin and two electrons have been transferred, the thermodynamic hurdle is not yet taken. The remaining question is how acetogenic bacteria manage to generate reduced ferredoxin if hydrogen as the sole energy source does not provide the required reduction power. This is facilitated via electron bifurcation (Buckel and Thauer 2018; Wang et al. 2013). Electron bifurcation may be defined as the process in which an electron pair of a hydride is split into two single electrons which reduce two separate electron acceptors with reduction potentials above and below the oxidized hydride-donor. One electron receives a higher reduction potential – i.e. to reduce ferredoxin – paid by the other electron receiving a weaker reduction potential. This may also proceed in the opposite direction which is expressed by the corresponding concept of electron configuration. The reduction of carbon dioxide to carbon monoxide is pursued by the CODH in acetogenic bacteria which applies reduced ferredoxin. The enzymes involved in electron bifurcation depend on the specific acetogenic bacterium, e.g. in *C. autoethanogenum* reduced ferredoxin is obtained in the oxidation of hydrogen to yield NADPH ("downhill reduction") and reduced ferredoxin ("uphill reduction") and the reduction of methylene-THF to methyl-THF with NADH providing methyl-THF ("downhill reduction") and reduced ferredoxin ("uphill reduction"). Thus, electron bifurcation allows the reduction of carbon dioxide with hydrogen despite the nonmatching reduction potentials.

Acetate and ethanol are the most straightforward native fermentation products of most acetogenic bacteria and their Gibbs free energy is given as follows:

$$2\,\text{CO}_2 + 4\,\text{H}_2 \rightarrow \text{CH}_3 - \text{COOH} + 2\,\text{H}_2\text{O} \quad \Delta G^{\circ\prime} = -95\,\text{KJ/mol}$$

$$\text{CO}_2 + 6\,\text{H}_2 \rightarrow \text{CH}_3 - \text{CH}_2 - \text{OH} + 3\,\text{H}_2\text{O} \quad \Delta G^{\circ\prime} = -105\,\text{KJ/mol}$$

These reactions do not generate carbon dioxide as by-product and are favored from a sustainability point of view (Katsyv and Mueller 2020; Katsyv and Mueller 2020). It is additionally to be taken into consideration that the energy efficiency in the generation of ethanol based on hydrogen is 80% (HHV) rather high. Hereby, the HHV heating value for ethanol (29.7 GJ/t) and hydrogen (143 GJ/t) is put into relation given a hydrogen unit consumption rate of 0.26 t hydrogen/t ethanol (29.7 GJ/t/0.26 × 143 GJ/t). The theoretical value will likely not be achieved on a commercial scale and the efficiency would be compromised by some utility efforts to run the process and to purify the product. However, the theoretical value at least indicates that the pathway can be translated into an efficient commercial process.

3.1.5 Carbon Monoxide as Energy Source

In case carbon monoxide is applied as energy and carbon source, carbon dioxide is obtained as by-product. According to the disproportionation of glucose in conventional fermentation, carbon monoxide is then partially reduced for product formation and partially oxidized to carbon dioxide to provide the required energy. The respective overall reactions are given as follows:

$$4\,CO + 2\,H_2O \rightarrow CH_3\text{–}COOH + 2\,CO_2 \quad \Delta G^{\circ\prime} = -175\ KJ/mol$$

$$6\,CO + 3\,H_2O \rightarrow CH_3\text{–}CH_2\text{–}OH + 4\,CO_2 \quad \Delta G^{\circ\prime} = -216\ KJ/mol$$

From a thermodynamic point of view, the application of carbon monoxide as feed may be favored given its higher reduction potential compared to hydrogen and thereby the improved energy balance of the host (Hermann et al. 2020; Katsyv and Mueller 2020; Liew et al. 2016). On the other hand, from the point of view of industrial biotechnology and the high attention on sustainability, the feed of carbon dioxide and hydrogen is clearly favored since acetogenic fermentation may then be used as efficient carbon capture and use technology (CCU). The political request to reduce carbon dioxide emissions has triggered the search for technologies suitable for a high-volume conversion and recycling of carbon dioxide into chemical building blocks, final products, or fuels.

In order to limit the generation of carbon dioxide as by-product and to restrict the metabolic function of carbon monoxide on carbon replenishment, the idea of a mixed hydrogen and carbon monoxide feed has been explored. The availability of such a gas mixture might profit from the circumstance that several industrial processes generate off-gas containing carbon monoxide and hydrogen. These available off-gas streams might then directly be applied as fermentation feed. Hydrogen supplementation is discussed with respect to the product ethanol which is commercially more interesting than acetic acid. The supplementation of hydrogen in the ratio of 1:1 allows to improve the carbon footprint and only 1 mol of carbon dioxide per mol ethanol is formed. Pending on the hydrogen-to-carbon monoxide ratio, the following equations might be pursued:

$$3\,CO + 3\,H_2 \rightarrow CH_3\text{–}CH_2\text{–}OH + CO_2 \quad \Delta G^{\circ\prime} = -158\ KJ/mol$$

$$2\,CO + 4\,H_2 \rightarrow CH_3\text{–}CH_2\text{–}OH + H_2O \quad \Delta G^{\circ\prime} = -138\ KJ/mol$$

Furthermore, in case the hydrogen to carbon monoxide ratio is raised to 2:1, the generation of carbon dioxide can completely be avoided (Wilkins and Atiyeh 2011). Hereby the feed preference of the host needs to be considered. It is known that in most batch cultures of acetogenic bacteria fed with carbon monoxide and hydrogen carbon monoxide – the energy source with higher reduction potential – is consumed before co-utilization is initiated. This is different in continuous cultures which display a simultaneous utilization of both substrates (Valgepea et al. 2018). The experiments also showed that hydrogen supplementation of a carbon monoxide feedstock enhances the carbon flux toward ethanol compared to the native by-products acetate and 2,3-butanediol. Hereby, it is to be considered that already a low CO concentration may inhibit the Fe-based hydrogenases in acetogens and the productivity of the technology might be challenging (Wang et al. 2013; Shima and Thauer 2013; Adams 1990). However, the CO sensitivity might be overcome by genetic engineering.

In summary, the growth of acetogenic bacteria with either hydrogen or carbon monoxide as energy source is possible in case the Wood–Ljungdahl pathway toward a defined product provides a surplus of ATP for cell maintenance and bacterial growth. Hereby, the different pathway designs in acetogenic bacteria including the enzymes capable of electron bifurcation, the needed reduction equivalents, and the type and efficiency of the ATPase driven by a proton or a sodium gradient need to be considered. Out of the high number of acetogenic strains, four hosts are regarded as potential industrial workhorses, which are *A. woodii*, *C. ljungdahlii*, *M. thermoacetica* and *C. autoethanogenum*(Humphreys et al. 2015; Takors et al. 2018). Although genetic modification of acetogenic bacteria is still challenging and the molecular toolbox is underdeveloped compared to *E. coli* or *S. cerevisiae*, several genetic modifications could already be achieved (Liew et al. 2016; Molitor et al. 2016). A detailed analysis of metabolism has been pursued the acetogenic species *A. woodii* and *C. autoethanogenum* in the working groups of Volker Müller and Peter Dürre. The following pathway reconstruction and estimation of surplus ATP is mainly based on their publications.

3.2 Formation of Acetate in *A. woodii* Based on Carbon Dioxide and Hydrogen

The Wood–Ljungdahl pathway and the possible ATP generation have been estimated for *A. woodii* on a detailed level. In general, *A. woodii* is able to generate acetate if fed with carbon dioxide and hydrogen. The formation of acetate based on carbon dioxide and hydrogen as the sole energy source is reconstructed as follows; an ATP surplus is derived to sustain bacterial growth (Katsyv and Mueller 2020; Müller 2019; Mock et al. 2015; Table 3.1).

Acetyl-CoA may also be involved in the acetylation of lysine as a common reversible post-translational protein modification that regulates diverse cellular functions (Weinert et al. 2013). This potential side-reaction can be neglected from a volume point of view and is not considered in the pathway calculations. The estimation reveals that the formation of acetate based on carbon dioxide and hydrogen is in principle possible and that about 0.3 mol of ATP are generated per mol acetate (Bertsch and Mueller 2015; Mock et al. 2015; Katsyv and Mueller 2020). For *A. woodii* – exceptional for acetogens – the ion/ATP stoichiometry of the ATP synthase is known. In particular, 3.3 Na^+ are needed for the formation of 1 mol of ATP. Thus, in case one sodium ion is translocated, 0.3 mol of ATP can be obtained.

The reduction of carbon dioxide requires 8 electrons per mol acetate which are provided via the oxidation of 4 mol of hydrogen. The overall reaction for the formation of acetate can be stated as follows:

$$2\,CO_2 + 4\,H_2 \rightarrow CH_3\text{–}COOH + 2\,H_2O \quad \Delta G^{\circ\prime} = -95\ \text{KJ/mol}$$

Table 3.1 Metabolic pathway toward acetate based on CO_2 and H_2.

Synthesis of methyl-group	
$CO_2 + H_2 \rightarrow$ Formate	Formate dehydrogenase
Formate + ATP + THF → 10-Formyl-THF + ADP + Pi	Formyl-THF synthetase
10-Formyl-THF → 5,10-Methenyl-THF + H_2O	Methenyl-THF dehydrogenase
5,10-Methenyl-THF + NADH + $H^+ \rightarrow$ 5,10-Methylene-THF + NAD^+	Methylene-THF dehydrogenase
5,10-Methylene-THF + NADH + $H^+ \rightarrow$ 5-Methyl-THF + NAD^+	Methylene-THF reductase
5-Methyl-THF + CoA → 5-Methyl-CoFeS + THF	Methyltransferase
Generation of ATP and reduction equivalents	
$3\ H_2 + 1.5\ Fd + 1.5\ NAD^+ \rightarrow 1.5\ NADH + 4.5H^+ + 1.5\ Fd^{2-}$	Hydrogenase HydABCD
$0.5\ Fd^{2-} + 0.5\ NAD^+ + 0.5\ H^+ + Na^+ \rightarrow 0.5\ NADH + 0.5\ Fd + Na^+$ (outside)	Rnf complex
$0.3\ ADP + 0.3\ Pi + Na^+ \rightarrow 0.3\ ATP + 0.3\ H_2O + Na^+$ (inside)	ATP synthase
Synthesis of acetate	
$CO_2 + Fd^{2-} + 2\ H^+ \rightarrow CO + H_2O + Fd$	Acetyl-CoA Synthase / CODH
CO + 5-Methyl-CoFeS → Acetyl-CoA	Acetyl-CoA Synthase / CODH
Acetyl-CoA + Pi → Acetyl-P + CoA	Phosphotransacetylase
Acetyl-P + ADP → Acetate + ATP	Acetate kinase
$2\ CO_2 + 4\ H_2 + 0.3\ ADP + 0.3\ Pi \rightarrow$ Acetate + 0.3 ATP + 2.3 H_2O	

The overall energetic efficiency of the reaction can be estimated if the assumed ATP gain is put in relation to the Gibbs free energy of the reaction (Katsyv and Mueller 2020). The generation of 1 mol of ATP requires about −30.5 KJ/mol under normal conditions, though under physiological conditions it is assumed that about 50 KJ/mol are needed to generate an additional phosphoric acid anhydride bond to derive ATP ex ADP. This would translate into energy efficiency of about 17% in case it is assumed that the generation of 1 mol of ATP requires about 55 KJ/mol (Flamholz et al. 2012) given a Gibbs free energy of −95 KJ/mol of acetate.

The overall pathway including the generation of reduction equivalents and ATP via substrate-level phosphorylation and electrochemical gradient can be reconstructed. It is to be considered that *A. woodii* employs a hydrogen-dependent formate dehydrogenase for the reduction of carbon dioxide and the methylene-THF dehydrogenase and the methylene-THF reductase both apply NADH as co-factor (Figure 3.2)

3.3 Formation of Acetate in *A. woodii* Based on Carbon Monoxide

Resting cells of *A. woodii* are able to generate acetate if fed with carbon monoxide but so far no growth could be observed. The formation of acetate based on carbon monoxide as

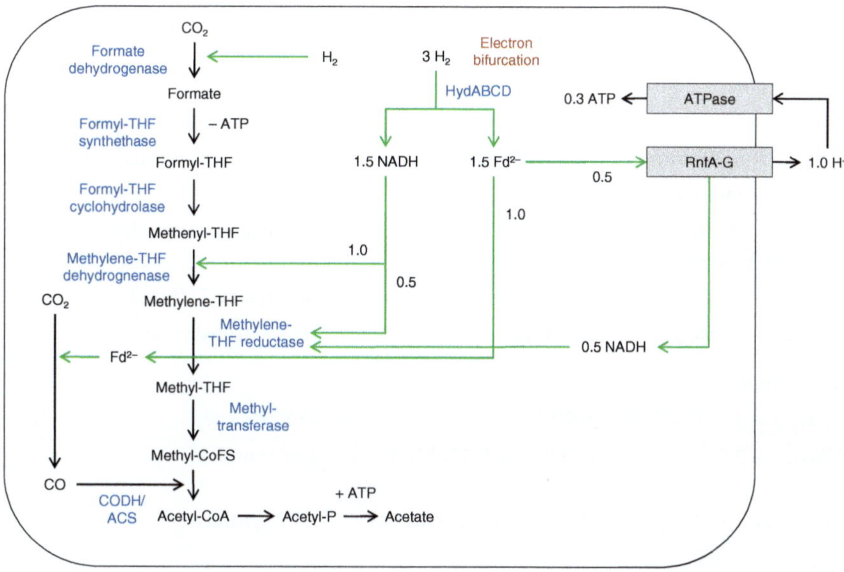

Figure 3.2 Pathway reconstruction for the generation of acetate in *A. woodii* based on carbon dioxide and hydrogen. The acetate formation proceeds via acetyl-phosphate and does allow substrate-level phosphorylation to replenish the ATP invested in the formation of formyl-THF. The chemical conversion is indicated in black arrows. The electron flow is indicated in green arrows. HydABCD: electron-bifurcating and NADH-dependent [FeFe]-hydrogenase, Rnf: membrane-associated and energy-conserving reduced ferredoxin NAD^+ oxidoreductase (Mock et al. 2015; Müller 2019). Source: Walter Koch.

the sole energy source is reconstructed as follows (Katsyv and Mueller 2020; Bertsch and Mueller 2015; Table 3.2).

The estimation reveals that the formation of acetate based on carbon monoxide is in principle possible and that about 1.5 mol of ATP are generated per mol acetate (Bertsch and Mueller 2015; Mock et al. 2015; Katsyv and Mueller 2020). It should be pointed out that the metabolic pathway does not include the water gas shift (WGS) reaction as known in organic chemistry:

$$CO + H_2O \rightarrow CO_2 + H_2 \quad \Delta H^\circ = -41\,\text{KJ/mol}$$

In the metabolic pathway of *A. woodii*, carbon monoxide and water are reacted to obtain reduction equivalents in the form of reduced ferredoxin, though molecular hydrogen is not provided (Atsbha et al. 2021). The reduction power of CO is sufficient to generate reduced ferredoxin based on the reduction of half potentials. This defines a difference to the former pathway employing hydrogen as an energy source which requires electron bifurcation. The energetic efficiency of the reaction can be estimated if the assumed ATP gain is put in relation to the Gibbs free energy of the reaction. The overall reaction for the formation of acetate can be stated as follows:

$$4\,CO + 2\,H_2O \rightarrow CH_3\text{–}COOH + 2\,CO_2 \quad \Delta G^{\circ\prime} = -175\,\text{KJ/mol}$$

The generation of 1 mol of ATP requires about −30.5 KJ/mol under normal conditions, though under physiological conditions it is assumed that about 50–60 KJ/mol are needed to generate an additional phosphoric acid anhydride bond to derive ATP ex ADP (Flamholz

Table 3.2 Metabolic pathway toward acetate based on CO_2 and H_2 in *A. woodii*.

Synthesis of methyl-group	
$CO_2 + H_2 \rightarrow$ Formate	Formate dehydrogenase
Formate + ATP + THF → 10-Formyl-THF + ADP + Pi	Formyl-THF synthetase
10-Formyl-THF → 5,10-Methenyl-THF + H_2O	Methenyl-THF dehydrogenase
5,10-Methenyl-THF + NADH + $H^+ \rightarrow$ 5,10-Methylene-THF + NAD^+	Methylene-THF dehydrogenase
5,10-Methylene-THF + NADH + $H^+ \rightarrow$ 5-Methyl-THF + NAD^+	Methylene-THF reductase
5-Methyl-THF + CoA → 5-Methyl-CoFeS + THF	Methyltransferase
Oxidation of carbon monoxide	
$3\ CO + 3\ Fd + 3\ H_2O \rightarrow 3\ CO_2 + 3\ Fd^{2-} + 6H^+$	CODH/ACS
$2.5\ Fd^{2-} + 2.5\ NAD^+ + 2.5\ H^+ + 5\ Na^+ \rightarrow 2.5\ Fd + 2.5\ NADH + 5\ Na^+$ (outside)	Rnf complex
$1.5\ ADP + 1.5\ Pi + 5\ Na^+ \rightarrow 1.5\ ATP + 1.5\ H_2O + 5\ Na^+$ (inside)	ATP synthase
Synthesis of acetate	
CO + 5-Methyl-CoFeS → Acetyl-CoA	Acetyl-CoA Synthase / CODH
Acetyl-CoA + Pi → Acetyl-P + CoA	Phosphotransacetylase
Acetyl-P + ADP → Acetate + ATP	Acetate kinase
Generation of hydrogen	
$0.5\ NADH + 0.5H^+ + 0.5\ Fd^{2-} \rightarrow 0.5\ NAD^+ + 0.5\ Fd + H_2$	Hydrogenase HydABCD
$4\ CO + 1.5\ ADP + 1.5\ Pi + 0.5\ H_2O \rightarrow$ Acetate $+ 1.5\ ATP + 2\ CO_2$	

et al. 2012; Katsyv and Mueller 2020). This would translate into an energy efficiency of about 47% in case it is assumed that the generation of 1 mol of ATP requires about 55 KJ/mol given a Gibbs free energy of −175 KJ/mol of acetate.

3.4 Formation of Ethanol in *A. woodii* Based on Carbon Dioxide and Hydrogen without AOR

From a commercial point of view, the generation of ethanol is much more attractive than the generation of acetic acid (Bengelsdorf and Dürre 2017). Thus, in academia and commercial biotechnology also, the option to provide ethanol has been taken into consideration. The WLP provides acetate though it might be extended with two optional steps toward ethanol via the CoA-dependent acetaldehyde dehydrogenase or the acetaldehyde:ferredoxin oxidoreductase (Figure 3.3).

It is known from feeding experiments that *A. woodii* is not able to generate ethanol when grown on carbon dioxide and hydrogen. This is explained by the circumstance that *A. woodii* lacks the aldehyde ferredoxin oxidoreductase (AOR) (Poehlein et al. 2012), which converts acetate into acetaldehyde. The reduction of acetate to acetaldehyde – the acid to the

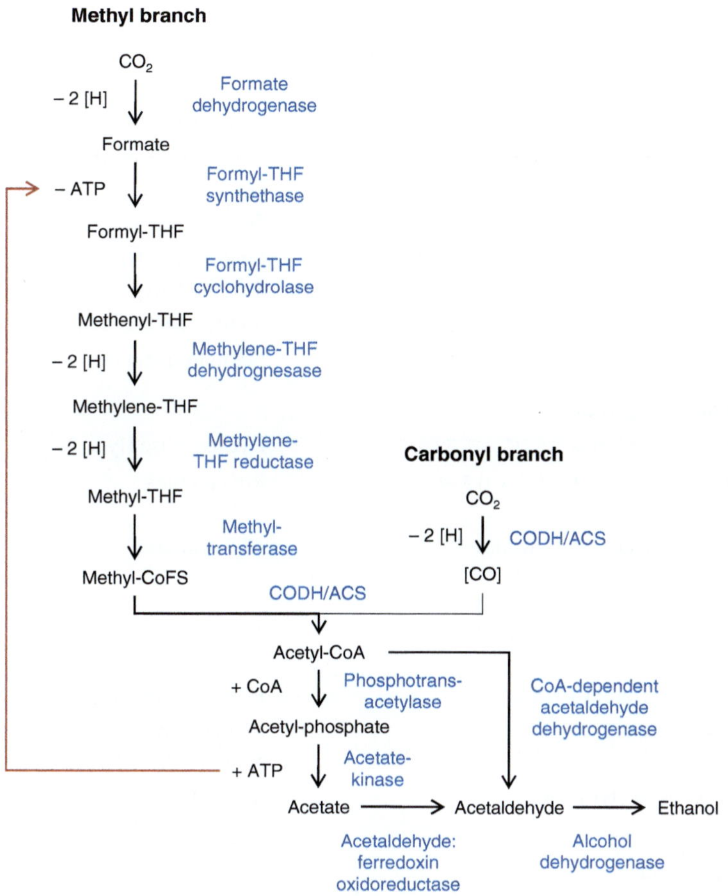

Figure 3.3 The Wood–Ljungdahl pathway for the conversion of carbon dioxide, cabon monoxide, and hydrogen into acetate. Source: Walter Koch.

aldehyde level – is biochemically not straightforward since the electron transfer potential of carbon monoxide and NAD(*P*)H is not sufficient. Accordingly, the acetaldehyde ferredoxin oxidoreductase needs to operate with electron bifurcation. The option to generate ethanol via the alternative pathway employing the CoA-dependent acetaldehyde dehydrogenase has been reconstructed in *A. woodii,* though the ATP gas reveals that this pathway is not feasible since the minimum ATP requirement for growth and cell maintenance can not be provided. The resulting lack of ATP generation causes a net ATP consumption which renders the whole pathway as thermodynamically not feasible (Bertsch and Mueller 2015; Mock et al. 2015). The demand of 1 mol of ATP in the formation of 10-formyl-THF cannot be compensated since exclusively 0.9 mol of ATP are generated (Table 3.3).

Acetaldehyde is assumed to be toxic to the host cell and might be generated in subcellular, carboxysome-like compartments as in *Clostridium kluyveri* (Köpke et al. 2010). The metabolic engineering strategy needs to consider that the recombinant synthesis of acetaldehyde might suppose a respective compartment.

Table 3.3 Hypothetical metabolic pathway toward ethanol based on CO_2 and H_2 in *A. woodii*.

Synthesis of methyl-group	
$CO_2 + H_2 \rightarrow$ Formate	Formate dehydrogenase
Formate + ATP + THF → 10-Formyl-THF + ADP + Pi	Formyl-THF synthetase
10-Formyl-THF → 5,10-Methenyl-THF + H_2O	Methenyl-THF dehydrogenase
5,10-Methenyl-THF + NADH + H^+ → 5,10-Methylene-THF + NAD^+	Methylene-THF dehydrogenase
5,10-Methylene-THF + NADH + H^+ → 5-Methyl-THF + NAD^+	Methylene-THF reductase
5-Methyl-THF + CoA → 5-Methyl-CoFeS + THF	Methyltransferase
Generation of ATP and reduction equivalents	
$3\ H_2 + 1.5\ Fd + 1.5\ NAD^+ \rightarrow 1.5\ NADH + 4.5\ H^+ + 1.5\ Fd^{2-}$	Hydrogenase HydABCD
$0.5\ Fd^{2-} + 0.5\ NAD^+ + Na^+ \rightarrow 0.5\ NADH + H^+ + 0.5\ Fd + Na^+$ (outside)	Rnf complex
$0.3\ ADP + Na^+ \rightarrow 0.3\ ATP + 0.3\ H_2O + Na^+$ (inside)	ATP synthase
Synthesis of acetyl-CoA and ethanol	
Acetyl-CoA + NADH + H^+ → Acetaldehyde + NAD^+	Acetaldehyde dehydrogenase
Acetaldehyde + NADH + H^+ → Ethanol + NAD^+	Ethanol dehydrogenase
$CO_2 + Fd^{2-} + 2\ H^+ \rightarrow CO + H_2O + Fd$	Acetyl-CoA Synthase/CODH
CO + 5-Methyl-CoFeS + CoA → Acetyl-CoA	Acetyl-CoA Synthase/CODH
Generation of reduction equivalents	
$2\ H_2 + 1.0\ Fd + 1.0\ NAD^+ \rightarrow 1.0\ NADH + 3\ H^+ + 1.0\ Fd^{2-}$	Hydrogenase HydABCD
$1.0\ Fd^{2-} + 1.0\ NAD^+ + H^+ + 2\ Na^+ \rightarrow 1.0\ NADH + 1.0\ Fd + 2\ Na^+$ (outside)	Rnf complex
$0.6\ ADP + 0.6\ Pi + 2\ Na^+ \rightarrow 0.6\ ATP + 0.6\ H_2O + 2\ Na^+$ (inside)	ATP synthase
$2\ CO_2 + 6\ H_2 + 0.1\ ATP \rightarrow$ Ethanol + 0.1 ADP + 0.1 Pi + 2.9 H_2O	

3.5 Formation of Ethanol in *A. woodii* Based on Carbon Dioxide and Hydrogen with AOR

As outlined above, *C. woodii* is not able to generate ethanol when grown on carbon dioxide and hydrogen. However, the AOR (acetaldehyde ferredoxin oxidoreductase) might be introduced via genetic engineering and the resulting recombinant *C. woodii* host should be able to achieve an ATP surplus in the production of ethanol if grown on carbon dioxide and hydrogen (Table 3.4).

The estimation reveals that the formation of ethanol based on carbon dioxide and hydrogen should in principle be possible in case an AOR is available (Bertsch and Mueller 2015; Mock et al. 2015; Katsyv and Mueller 2020). The energetic efficiency of the reaction can be

Table 3.4 Potential metabolic pathway toward ethanol based on CO_2 and H_2 in *A. woodii* with AOR.

Reaction	Enzyme
Synthesis of methyl-group	
$CO_2 + H_2 \rightarrow$ Formate	Formate dehydrogenase
Formate + ATP + THF → 10-Formyl-THF + ADP + Pi	Formyl-THF synthetase
10-Formyl-THF → 5,10-Methenyl-THF + H_2O	Methenyl-THF dehydrogenase
5,10-Methenyl-THF + NADH + $H^+ \rightarrow$ 5,10-Methylene-THF + NAD^+	Methylene-THF dehydrogenase
5,10-Methylene-THF + NADH + $H^+ \rightarrow$ 5-Methyl-THF + NAD^+	Methylene-THF reductase
5-Methyl-THF + CoA → 5-Methyl-CoFeS + THF	Methyltransferase
Generation of ATP and reduction equivalents	
$3\ H_2 + 1.5\ Fd + 1.5\ NAD^+ \rightarrow 1.5\ NADH + 4.5\ H^+ + 1.5\ Fd^{2-}$	Hydrogenase HydABCD
$0.5\ Fd^{2-} + 0.5\ NAD^+ + H^+ + Na^+ \rightarrow 0.5\ NADH + 0.5\ Fd + Na^+$ (outside)	Rnf complex
$0.3\ ADP + Na^+ \rightarrow 0.3\ ATP + 0.3\ H_2O + Na^+$ (inside)	ATP synthase
Synthesis of acetyl-CoA and ethanol	
$CO_2 + Fd^{2-} + 2\ H^+ \rightarrow CO + H_2O + Fd$	Acetyl-CoA synthase/CODH
CO + 5-Methyl-CoFeS → Acetyl-CoA	Acetyl-CoA synthase/CODH
Acetyl-CoA + Pi → Acetyl-P + CoA	Phosphotransacetylase
Acetyl-P + ADP → Acetate + ATP	Acetate kinase
Acetate + $Fd^{2-} + 2\ H^+ \rightarrow$ Acetaldehyde + Fd + H_2O	AOR
Acetaldehyde + NADH + $H^+ \rightarrow$ Ethanol + NAD^+	Ethanol dehydrogenase
$2\ H_2 + 1.0\ Fd + 1.0\ NAD^+ \rightarrow 1.0\ NADH + H^+ + 1.0\ Fd^{2-}$	
$2\ CO_2 + 6\ H_2 + 0.3\ ADP \rightarrow$ Ethanol + $0.3\ ATP + 3.3\ H_2O$	

estimated if the assumed ATP gain is put in relation to the Gibbs free energy of the reaction. The overall reaction for the formation of acetate can be stated as follows:

$$2\,CO_2 + 6\,H_2 \rightarrow CH_3\text{–}CH_2\text{–}OH + 3\,H_2O \quad \Delta G^{\circ\prime} = -105\ \text{KJ/mol}$$

The generation of 1 mol of ATP requires about −30.5 KJ/mol under normal conditions, though under physiological conditions it is assumed that about 50–60 KJ/mol are needed to generate an additional phosphoric acid anhydride bond to derive ATP ex ADP (Flamholz et al. 2012; Katsyv and Mueller 2020). This would translate into a potential energy efficiency of about 16% in case it is assumed that the generation of 1 mol of ATP requires about 55 KJ/mol given a Gibbs free energy of −105 KJ/mol ethanol.

3.6 Formation of Ethanol in *C. woodii* Based on Carbon Monoxide

On the other hand, *C. woodii* is able to generate ethanol if fed with carbon monoxide as carbon and energy source. The oxidation of 4 mol of carbon monoxide provides sufficient reduction equivalents in form of NADH and allows the translocation of 9 mol of sodium ions such that a high surplus of ATP is achieved. The resulting generation of ATP has been estimated (Katsyv and Mueller 2020; Bertsch and Mueller 2015; Table 3.5).

The overall reconstruction of the pathway assumes that *A. woodii* is able to generate 1.7 mol of ATP per mol ethanol. The energetic efficiency of the reaction can be estimated if the assumed ATP gain is put in relation to the Gibbs free energy of the reaction. The overall reaction for the formation of acetate can be stated as follows:

$$6\,CO + 3\,H_2O \rightarrow CH_3\text{–}CH_2\text{–}OH + 4\,CO_2 \quad \Delta G^{\circ\prime} = -216\,KJ/mol$$

The generation of 1 mol of ATP requires −30.5 KJ/mol under normal conditions, though under physiological conditions it is assumed that about 50–60 KJ/mol are needed to generate an additional phosphoric acid anhydride bond to derive ATP ex ADP (Flamholz et al.

Table 3.5 Metabolic pathway toward ethanol based on CO/CO_2 and H_2 in *A. woodii*.

Reaction	Enzyme
Synthesis of methyl-group	
$CO_2 + H_2 \rightarrow$ Formate	Formate dehydrogenase
Formate + ATP + THF → 10-Formyl-THF + ADP + Pi	Formyl-THF synthetase
10-Formyl-THF → 5,10-Methenyl-THF + H_2O	Methenyl-THF dehydrogenase
5,10-Methenyl-THF + NADH + H^+ → 5,10-Methylene-THF + NAD^+	Methylene-THF dehydrogenase
5,10-Methylene-THF + NADH + H^+ → 5-Methyl-THF + NAD^+	Methylene-THF reductase
5-Methyl-THF + CoA → 5-Methyl-CoFeS + THF	Methyltransferase
Oxidation of carbon monoxide	
$5\,CO + 5\,Fd + 5\,H_2O \rightarrow 5\,CO_2 + 5\,Fd^{2-} + 10\,H^+$	CODH/ACS
$4.5\,Fd^{2-} + 4.5\,NAD^+ + 4.5\,H^+ + 9\,Na^+ \rightarrow 4.5\,Fd + 4.5\,NADH + 9\,Na^+$ (outside)	Rnf complex
$2.7\,ADP + 2.7\,Pi + 9\,Na^+ \rightarrow 2.7\,ATP + 2.7\,H_2O + 9\,Na^+$ (inside)	ATP synthase
Synthesis of acetyl-CoA and ethanol	
CO + 5-Methyl-CoFeS → Acetyl-CoA	Acetyl-CoA synthase/CODH
Acetyl-CoA + NADH + H^+ → Acetaldehyde + NAD^+ + CoA	Acetaldehyde dehydrogenase
Acetaldehyde + NADH + H^+ → Ethanol + NAD^+	Ethanol dehydrogenase
Generation of hydrogen	
$0.5\,NADH + 0.5\,Fd^{2-} + 2\,H^+ \rightarrow 0.5\,NAD^+ + 0.5\,Fd + H_2$	Hydrogenase HydABCD
$6\,CO + 1.7\,ADP + 1.7\,Pi + 1.3\,H_2O \rightarrow$ Ethanol $+ 1.7\,ATP + 4\,CO_2$	

2012; Liew et al. 2016). This would translate into an energy efficiency of about 43% in case it is assumed that the generation of 1 mol of ATP requires about 55 KJ/mol given a Gibbs free energy of −216 KJ/mol ethanol.

3.7 Formation of Acetate in *C. autoethanogenum* Based on Carbon Dioxide and Hydrogen

C. autoethanogenum does also belong to the group of bacteria which display the ability to reduce carbon dioxide with molecular hydrogen to obtain acetic acid ("acetogens"). The original strain JA1-1 was isolated in 1994 from rabbit feces cultivated on CO as sole energy and carbon source. It is a mesophilic, acetogenic gram-positive bacterium related to *C. ljjungdahlii* (Abrini et al. 1994). The metabolic pathway and the related ATP generation has been estimated on the basis of cell extracts of *C. autoethanogenum* which were fed with carbon dioxide and hydrogen. The metabolic model assumes that the methylene-THF dehydrogenase is electron-bifurcating and NADH specific (Mock et al. 2015; Table 3.6).

The overall reconstruction of the pathway assumes that *C. autoethanogenum* is able to generate 1 mol of ATP per mol acetate. It is hereby assumed that the methylene-THF reductase (i.e. methylene-H4F reductase) is involved in electron bifurcation and NADH specific (Fuchs 1986; Mock et al. 2015; Katsyv and Mueller 2020). The redox potential of methylene-THF/methyl-THF is −200 mV and might allow electron bifurcation in case 1 mol of NADH per mol methylene-THF is invested. Thus, it is assumed that the formation of methyl-THF with 2 mol of NADH does also facilitates the generation of 1 mol reduced ferredoxin (Valgepea et al. 2018):

$$\text{Methylene-THF} + 2\,\text{NADH} + \text{Fd} \rightarrow \text{Methyl-THF} + 2\,\text{NAD}^+ + \text{Fd}^{2-} \quad \Delta G^{\circ\prime} = -5\,\text{KJ/mol}$$

It is assumed that 3.66 Na^+ are required for 1 mol of ATP; the obtained 1 mol of ATP is rounded up (Buckel and Thauer 2018). The energetic efficiency of the reaction can be estimated if the assumed ATP gain is put in relation to the Gibbs free energy of the reaction. The overall reaction for the formation of acetate can be stated as follows:

$$2\,CO_2 + 4\,H_2 \rightarrow CH_3\text{–}COOH + 2\,H_2O \quad \Delta G^{\circ\prime} = -95\,\text{KJ/mol}$$

The generation of 1 mol of ATP requires −30.5 KJ/mol under normal conditions (Katsyv and Mueller 2020). Under physiological conditions, it is assumed that about 50–KJ/mol are needed to generate an additional phosphoric acid anhydride bond to derive ATP ex ADP (Flamholz et al. 2012). This would translate into an energy efficiency of about 58% in case it is assumed that the generation of 1 mol of ATP requires about 55 KJ/mol given a Gibbs free energy of −95 KJ/mol of acetate.

3.8 Formation of Ethanol in *C. autoethanogenum* Based on Carbon Dioxide and Hydrogen

Within the group of acetogenic bacteria the species *C. autoethanogenum*, *C. ljungdahlii* and *C. ragsdalei* are also able to reduce carbon dioxide with molecular hydrogen to ethanol (Zhu et al. 2020). Especially for the commercially applied strain *C. autoethanogenum* the potential

Table 3.6 Metabolic pathway toward acetate based on CO/CO_2 and H_2 in *C. autoethanogenum*. The conversion of carbon dioxide into formate and the reduction of ferredoxin and $NADP^+$ are pursued by the same enzyme.

Synthesis of acetyl-CoA	
$CO_2 + Fd^{2-} + 2\ H^+ \rightarrow CO + H_2O + Fd$	Acetyl-CoA synthase/CODH
CO + 5-Methyl-CoFeS → Acetyl-CoA	Acetyl-CoA synthase/CODH
Synthesis of methyl-group	
$CO_2 + H_2 \rightarrow$ Formate	Formate dehydrogenase HytA-E
Formate + ATP + THF → 10-Formyl-THF + ADP + Pi	Formyl-THF synthetase
10-Formyl-THF → 5,10-Methenyl-THF + H_2O	Methenyl-THF DH
5,10-Methenyl-THF + NADPH + H^+ → 5,10-Methylene-THF + $NADP^+$	Methylene-THF DH
5,10-Methylene-THF + 2 NADH + Fd → 5-Methyl-THF + 2 NAD^+ + Fd^{2-}	Methylene-THF reductase
5-Methyl-THF + CoA → 5-Methyl-CoFeS + THF	Methyltransferase
Generation of ATP and reduction equivalents	
$3\ H_2 + 1.5\ Fd + 1.5\ NADP^+ \rightarrow 1.5\ NADPH + 4.5\ H^+ + 1.5\ Fd^{2-}$	HytA-E
$0.5\ NADPH + 0.25\ NAD^+ + 0.25\ Fd \rightarrow 0.5\ NADP^+ + 0.25\ NADH + 0.25\ Fd^{2-}$	Nfn
$1.75\ Fd^{2-} + 1.75\ NAD^+ + 1.75\ H^+ \rightarrow 1.75\ NADH + 1.75\ Fd + 3.5\ H^+$ (outside)	RnF A-G
$3.5\ H^+ + ADP + Pi \rightarrow ATP + H_2O + 3.5\ H^+$ (inside)	F_0F_1 ATPase
Conversion of acetyl-CoA into acetate	
Acetyl-CoA + Pi → Acetyl-P + CoA	Phosphotransacetylase
Acetyl-P + ADP → Acetate + ATP	Acetate kinase
$2\ CO_2 + 4\ H_2 + ADP + Pi \rightarrow Acetate + 3\ H_2O + ATP$	

Source: Adapted from Mock et al. (2015).

ATP gain has been estimated. In the reconstruction of the pathway and the estimation of the ATP gain two options are compared which derive an ATP gain of 0.5 mol of ATP or 1.2 mol of ATP per mol ethanol, respectively. The main difference of both models is the conversion of acetyl-CoA into ethanol, which may proceed via acetaldehyde (first pathway) or, alternative, via acetyl-phosphate, acetate, and acetaldehyde (second pathway). The first pathway via acetaldehyde proceeds as following (Mock et al. 2015; Table 3.7).

The overall reconstruction of the pathway assumes that *C. autoethanogenum* is able to generate 0.5 mol of ATP per mol ethanol (Mock et al. 2015). It is hereby assumed that the methylene-THF reductase (i.e. methylene-H_4F reductase) is involved in electron bifurcation and NADH specific. It should be pointed out that the respective ATP gain in other studies has been estimated with 0.4 mol of ATP per mol ethanol (Katsyv and Mueller 2020). The energetic efficiency of the reaction can be estimated if the assumed ATP gain is put in

Table 3.7 Metabolic pathway toward ethanol based on CO/CO_2 and H_2 in *C. autoethanogenum*.

Reaction	Enzyme
Synthesis of acetyl-CoA	
$CO_2 + Fd^{2-} + 2\,H^+ \rightarrow CO + H_2O + Fd$	Acetyl-CoA synthase/CODH
CO + 5-Methyl-CoFeS → Acetyl-CoA	Acetyl-CoA synthase/CODH
Synthesis of methyl-group	
$CO_2 + H_2 \rightarrow$ Formate	Formate dehydrogenase HytA-E
Formate + ATP + THF → 10-Formyl-THF + ADP + Pi	Formyl-THF synthetase
10-Formyl-THF → 5,10-Methenyl-THF + H_2O	Methenyl-THF DH
5,10-Methenyl-THF + NADPH + H^+ → 5,10-Methylene-THF + $NADP^+$	Methylene-THF DH
5,10-Methylene-THF + 2 NADH + Fd → 5-Methyl-THF + 2 NAD^+ + Fd^{2-}	Methylene-THF reductase
5-Methyl-THF + CoA → 5-Methyl-CoFeS + THF	Methyltransferase
Generation of ATP and reduction equivalents	
$5H_2 + 2.5\,Fd + 2.5\,NADP^+ \rightarrow 2.5\,NADPH + 7.5\,H^+ + 2.5\,Fd^{2-}$	HytA-E
$0.5\,NADPH + 0.25\,NAD^+ + 0.25\,Fd \rightarrow 0.5\,NADP^+ + 0.25\,NADH + 0.25H^+ + 0.25\,Fd^{2-}$	Nfn
$2.75\,Fd^{2-} + 2.75\,NAD^+ + 2.75H^+ + 5.5\,H^+ \rightarrow 2.75\,Fd + 2.75\,NADH + 5.5H^+$ (outside)	RnF A-G
$5.5H^+ + 5.5\,ADP + 5.5\,Pi \rightarrow 1.5\,ATP + 1.5\,H_2O + 5.5\,H^+$ (inside)	F_0F_1 ATPase
Conversion of acetyl-CoA into ethanol	
Acetyl-CoA + NADPH + H^+ → Acetaldehyde + CoA + $NADP^+$	Acetaldehyde dehydrogenase
Acetaldehyde + NADH + H^+ → Ethanol + NAD^+	Alcohol dehydrogenase
$2\,CO_2 + 6\,H_2 + 0.5\,ADP + 0.5\,Pi \rightarrow$ Ethanol + 0.5 ATP + 3.5 H_2O	

relation to the Gibbs free energy of the reaction. The overall reaction for the formation of acetate can be stated as follows:

$$2\,CO_2 + 6\,H_2 \rightarrow CH_3\text{–}CH_2\text{–}OH + 3\,H_2O \quad \Delta G^{\circ\prime} = -105\ \text{KJ/mol}$$

The generation of 1 mol of ATP requires −30.5 KJ/mol under normal conditions, though under physiological conditions it is assumed that about 50–60 KJ/mol are needed to generate an additional phosphoric acid anhydride bond to derive ATP ex ADP (Flamholz et al. 2012; Katsyv and Mueller 2020). This would translate into an energy efficiency of about 26% in case it is assumed that the generation of 1 mol of ATP requires about 55 KJ/mol given a Gibbs free energy of −105 KJ/mol of ethanol. The overall reaction scheme expresses the metabolic complexity in the arrangement of electron transfer (Figure 3.4).

The ethanol formation proceeds via acetaldehyde and does not allow substrate chain phosphorylation. It is to be seen that *C. autoethanogenum* has a higher metabolic flexibility than *A. woodii* given the three reactions used for electron bifurcation. Whereas electron bifurcation in *A. woodii* seems to be limited to the oxidation of hydrogen,

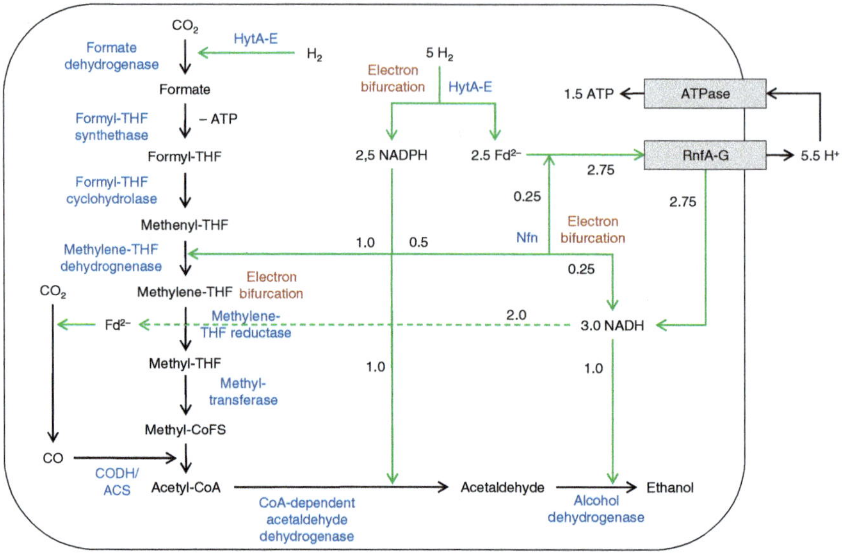

Figure 3.4 Pathway reconstruction for the generation of ethanol in *C. autoethanogenum* based on carbon dioxide and hydrogen. The chemical conversions are indicated in black arrows. The electron flow is indicated in green arrows. HytA-E: electron-bifurcating and NADP-dependent [FeFe]-hydrogenase, Nfn: electron-bifurcating and ferredoxin-dependent transhydrogenase, Rnf: membrane-associated and energy-conserving reduced ferredoxin NAD^+ oxidoreductase. Source: Walter Koch.

C. autoethanogenum is also able to oxidize NADPH and NADH for the generation of reduced ferredoxin:

A. woodii.

$$2\,H_2 + NAD^+ + Fd \rightarrow NADH + 3\,H^+ + Fd^{2-} \text{ HydABCD}$$

C. autoethanogenum.

$$2\,H_2 + NADP^+ + Fd \rightarrow NADPH + 3\,H^+ + Fd^{2-} \text{ HytA-E}$$

$$2\,NADPH + NAD^+ + Fd \rightarrow 2\,NADP^+ + NADH + H^+ + Fd^{2-} \text{ Nfn}$$

$$\text{5,10–Methylene-THF} + 2\,NADH + Fd \rightarrow \text{5-Methyl-THF} + 2\,NAD^+ + Fd^{2-} \text{ Methylene-THF reductase}$$

The second pathway toward proceeds via acetyl-phosphate and acetate and proceeds as follows (Mock et al. 2015; Table 3.8).

It is assumed that the methylene-THF reductase is electron bifurcating and allows the generation of reduced ferredoxin. The energetic efficiency of this pathway would amount to 63% and thereby be significantly high in case; it is also assumed that under physiological conditions about 55 KJ/mol are required for the generation of 1 mol of ATP. This pathway allows the generation of ATP via substrate-level phosphorylation in the conversion of acetyl-P to acetate with the acetate kinase. Hereby especially the first reduction step from the acid to the aldehyde level is critical given the high reduction potential needed to

Table 3.8 Metabolic pathway toward ethanol based on CO/CO_2 and H_2 in *C. autoethanogenum*.

Synthesis of acetyl-CoA	
$CO_2 + Fd^{2-} + 2\,H^+ \rightarrow CO + H_2O + Fd$	Acetyl-CoA synthase/CODH
CO + 5-Methyl-CoFeS → Acetyl-CoA	Acetyl-CoA synthase/CODH
Synthesis of methyl-group	
$CO_2 + H_2 \rightarrow$ Formate	Formate dehydrogenase HytA-E
Formate + ATP + THF → 10-Formyl-THF + ADP + Pi	Formyl-THF synthetase
10-Formyl-THF → 5,10-Methenyl-THF + H_2O	Methenyl-THF dehydrogenase
5,10-Methenyl-THF + NADPH + H^+ → 5,10-Methylene-THF + $NADP^+$	Methylene-THF dehydrogenase
5,10-Methylene-THF + 2 NADH + Fd → 5-Methyl-THF + 2 NAD^+ + Fd^{2-}	Methylene-THF reductase
5-Methyl-THF + CoA → 5-Methyl-CoFeS + THF	Methyltransferase
Generation of ATP and reduction equivalents	
$5H_2 + 2.5\,Fd + 2.5\,NADP^+ \rightarrow 2.5\,NADPH + 7.5\,H^+ + 2.5\,Fd^{2-}$	HytA-E
$1.5\,NADPH + 0.75\,NAD^+ + 0.75\,Fd \rightarrow 1.5\,NADP^+ + 0.75\,NADH + 0.75H^+ + 0.75\,Fd^{2-}$	Nfn
$2.25\,Fd^{2-} + 2.25\,NAD^+ + 2.25H^+ + 4.5\,H^+ \rightarrow 2.25\,Fd + 2.25\,NADH + 4.5\,H^+$ (outside)	RnF A-G
$4.5\,H^+ + 1.2\,ADP + 1.2\,Pi \rightarrow 1.2\,ATP + 1.2\,H_2O + 4.5\,H^+$ (inside)	F_0F_1 ATPase
Conversion of acetyl-CoA into ethanol	
Acetyl-CoA + Pi → Acetyl-P + CoA	Phosphotransacetylase
Acetyl-P + ADP + Pi → Acetate + ATP	Acetate kinase
Acetate + Fd^{2-} + 2 H^+ → Acetaldehyde + Fd + H_2O	AOR
Acetaldehyde + NADH + H^+ → Ethanol + NAD^+	Ethanol dehydrogenase
$2\,CO_2 + 6\,H_2 + 1.2\,ADP + 1.2\,Pi \rightarrow$ Ethanol $+ 1.2\,ATP + 4.2\,H_2O$	

transfer two electrons on the acid. This step is facilitated with the AOR (aldehyde ferredoxin oxidoreductase), which activates acetate and uses reduced ferredoxin as an electron source:

$$\text{Acetate} + 2\,\text{e} + 2\,\text{H}^+ \rightarrow \text{Acetaldehyde} + \text{H}_2\text{O} \quad E_0 = -580\,\text{mV (at pH 7.0)}$$

$$\text{Fd}^{2-} \rightarrow \text{Fd} + 2\,\text{e}\,E_0 = 500\,\text{mV (at pH 7.0)}$$

$$\text{Acetate} + \text{Fd}^{2-} + 2\,\text{H}^+ \rightarrow \text{Acetaldehyde} + \text{Fd} + 2\,\text{H}_2\text{O} \quad \Delta E° = -0.080\,\text{V}$$

The overall Gibbs free energy of the net reaction is still positive given the higher reduction potential needed for the reduction of the carboxylic acid function to the aldehyde function

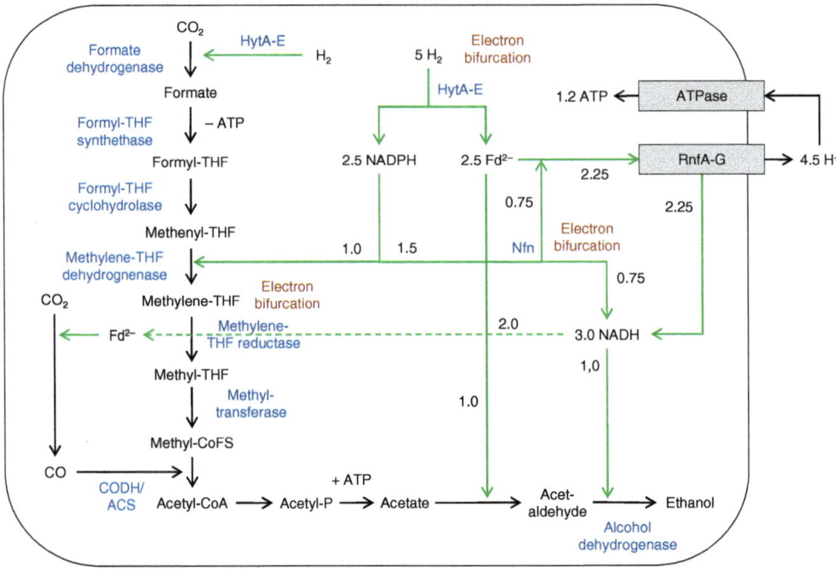

Figure 3.5 Pathway reconstruction for the generation of ethanol in *C. autoethanogenum* based on carbon dioxide and hydrogen. The chemical conversions are indicated in black arrows. The electron flow is indicated in green arrows. HytA-E: electron-bifurcating and NADP-dependent [FeFe]-hydrogenase, Nfn: electron-bifurcating and ferredoxin-dependent transhydrogenase, Rnf: membrane-associated and energy-conserving reduced ferredoxin NAD^+ oxidoreductase. Source: Walter Koch.

though nevertheless seems to allow an efficient conversion:

$$\Delta G^{\circ\prime} = -n \times F \times \Delta E^{\circ} = -2 \times 96{,}485\ J/V \times mol \times -0.08\ V = 15.4\ KJ/mol.$$

Since the reduction potential of ferredoxin to reduced ferredoxin is sometimes estimated with −400 mV, the Gibbs free energy might derive a value of 35 KJ/mol further raising the thermodynamic hurdle (Mock et al. 2015). The organism obviously needs to operate with a "pull" of the next downstream reaction converting acetaldehyde into ethanol which has a negative Gibbs free energy of −23.1 KJ/mol. The importance of a fast aldehyde removal has been shown for the activity of the AOR from *C. thermoaceticum*. The acid reduction was increased when an aldehyde removal system was present (White et al. 1989). Additionally, it needs to be considered that the intracellular acetate concentration is significantly higher than the intracellular acetaldehyde concentration given the exergonic character of the reduction of acetaldehyde with NADH (Mock et al. 2015). Additionally, the picture is changing if the actual pH of 6 is considered which supports the reaction to proceed in the right direction. The reaction is consuming three protons and is driven to the right side with decreasing pH:

$$Fd^{2-} + CH_3COO^- + 3\ H^+ \rightarrow Fd + CH_3CHO + H_2O$$

The overall metabolic and electron flow of the pathway is shown in Figure 3.5:

3.9 Industrial Fermentation and Capacity

With respect to commercial application, the limited solubility of the gaseous feedstock might turn out as a technical hurdle to achieve a high productivity. The solubility of CO, H_2, and CO_2 is reported with 28 mg/l, 1.6 mg/l, and 1.7 mg/l at 293 K and 1 atm which is very low compared to the solubility of glucose with about 900 g/l (Liew et al. 2016). Though the low solubility does not necessarily translate into a weak productivity and the CO uptake rate is more important, which is reported with 18–120 mmol CO/g DCW × h (Gaddy et al., 2003) or 33–100 mol CO/g DCW × h (Hermann et al. 2021), respectively. In case we assume the conversion of carbon monoxide with hydrogen toward ethanol as exemplary reaction with theoretically a yield and a selectivity of 100%, a carbon monoxide unit consumption rate of 1.21 g CO/g ethanol would be derived (2 × 28 g/mol/46 g/mol). The maximum productivity the fermentative conversion might achieve depends on the biomass titer which may be estimated with 5 g/l. The average limit of the carbon monoxide uptake rate covering the complete batch may be assumed with 100 mmol CO/g DCW × h. Under these boundary conditions, considerable productivity might be achieved:

100 mmol CO uptake/g DCW × h and a biomass titer of 5 g/l derive,
500 mmol CO uptake/l × h, and a CO molecular weight of 28 g/mol derive,
14 g CO uptake/l × h, and a stipulated 100% conversion and selectivity derive,
11.5 g ethanol/l × h.

A productivity of 11.5 g/l × h would be in the same range or even be slightly higher than the productivity achieved in the industrial production of ethanol and lactic acid. It is interesting to note that the highest published productivity for a syngas fermentation with a feedstock of 62% H_2, 31% CO, and 5% C_2H_6 at a pressure of about 5 bar achieved a remarkable value of 15.4 g/l × h at a biomass titer of 2 g/l with the *C. ljungdahlii* strain C-01 as host (Gaddy et al. 2003). As an additional reference point, the fermentative production of isopropanol and acetone with *C. autoethanogenum* is reported to achieve a productivity of 3 g isopropanol and acetone/l × h (Liew et al. n.d.) which is according to the reasonable target level of a commercial fermentation of a high-volume chemical (Thoma and Blombach 2021; Van Dien 2013). The actually achieved productivities in the first industrial syngas fermentation plants are not published.

Several R&D approaches with a focus on different feedstocks have been pursued. One of the early players in the field was the company Coskata which had been founded in 2006. The original technology envisaged to apply methane as feedstock for the production of syngas with a H_2/CO ratio of 2:1 or 3:1. The selected feedstock reflected the low natural gas prices in the United States enabled by the shale gas boom. The syngas was then supposed to be converted into ethanol via acetogenic bacteria. However, in 2015 Coskata announced to phase out its semi-commercial pilot plant in Madison (Pennsylvania) and later on went out of business (Heijstra et al. 2017). The technology has been taken over by the company Synata Bio in 2016 (Lane 2016a, b) which acquired a former Abengoa cellulosic biofuel plant in Hugoton to further pursue R&D activities with a focus on natural gas and biomass as feedstock (Lane 2016a, b).

Ineos Bio initiated its activities with the acquisition of Bioengineering Resources Inc. in 2008 which already produced biofuels on a smaller scale in Arkansas based on wood chips. Ineos indicated in its state grant application as of 2008 to broaden the raw material base including carpets and tires. It pursued the idea to gasify biomass to generate syngas composed of carbon monoxide, carbon dioxide, and hydrogen fed to anaerobic bacteria for conversion into ethanol. A pilot plant with a target capacity of 8 m gal was operated in Vero Beach, Florida which was run commercially and setup as a JV with New Planet Energy Holdings (Heijstra et al. 2017). First minor volumes of ethanol were announced in 2013, though technical issues were also reported including a higher moisture content in the Florida wood, the installment of scrubbers to remove hydrogen cyanide, which was hampering the fermentation host, and scale-up issues in the main fermenter due to limited run time. Biomass gasification may generate minor amounts of hydrogen cyanide which negatively impacts host productivity (Lin et al. 2010; Patersen et al. 2005). Finally, the R&D activities were given up in 2016 (Daprile 2017; Greenwood 2018).

Lanzatech was founded in 2005 with the target to produce ethanol and other value products like acetone, isopropanol, acetic acid, and 2,3-butanediol based on carbon dioxide, carbon monoxide, and hydrogen as feedstock (Köpke and Simpson 2020). The company managed to develop a successful technology platform for gas fermentation with *C. autoethanogenum* as host. Lanzatech partnered up with a Chinese steel mill in Chougang which can provide a suitable off-gas stream derived in steel mills. The usual steel production process implies a by-product stream containing hydrogen and carbon monoxide which may be fed to acetogenic bacteria (Bender et al. 2018). The first ethanol plant using the gas fermentation technology of Lanzatech is running since 2018 in Chougang in China. The plant has a pre-commercial capacity of 100,000 gal pa and is running continuously without technical issues. The technology is able to cope with varying H_2/CO ratios and an additional processing to achieve a constant molar ratio in the feedstock is not required. Two further plants in China had startup in 2021 and 2022 and the overall ethanol capacity is reported with >150 kt pa (Lanzatech Investor Day presentation 2023). The H_2/CO ratio is in the range of 1.3–1.5 (Köpke and Simpson 2020). Furthermore, a project was initiated with ArcelorMittal and a plant with a capacity of 21 million gallons pa has been constructed in Ghent (Belgium) receiving feedstock from the nearby steel production of ArcelorMittal (Heijstra et al. 2017; Köpke and Simpson 2020). The startup is foreseen in 2023 with the operation of four reactors with a height of 30 m. Additional plants are under preparation (Handelsblatt 2022).

References

Abrini, J., Naveau, H., and Nuns, E.-J. (1994). *Clostridium autoethanogenum*, sp., nov., an anaerobic bacterium that produces ethanol from carbon monoxide. *Archives of Microbiology* 161: 345–351.

Adams, M.W. (1990). The structure and mechanism of iron-hydrogenases. *Biochimika et Biophysica Acta* 1020 (2): 115–145. 10/1016/0005-2728(90)90044-5.

Atsbha, T.A., Yoon, T., Seongho, P., and Lee, C.-L. (2021). A review on the catalytic conversion of CO_2 using H_2 for synthesis of CO, methanol, and hydrocarbons. *Journal of CO2 utilization* 44: 101413. https://doi.org/10.1016/j.ycou.2020.101413.

Bender, M., Roussiere, T., Schelling, H. et al. (2018). Coupled production of steel and chemicals. *Chemie Ingenieur Technik* 11: 1–25. https://doi.org/10.1002/cite.201800048.

Bengelsdorf, F.R. and Dürre, P. (2017). Gas fermentation for commodity chemicals and fuels. *Microbial Biotechnology* 10 (5): 1167–1170. https://doi.org/10.1111/1751-7915.12763.

Bertsch, J. and Mueller, V. (2015). Bioenergetic constraints for conversion of syngas to biofuels in acetogenic bacteria. *Biotechnology for Biofuels* 8: 210. https://doi.org/10.1186/s13068-015-0393-x.

Bourgade, B., Minton, N.P., and Islam, A. (2012). Genetic and metabolic engineering challenges of C1-gas fermenting acetogenic chassis organisms. *FEMS Microbiology Reviews* fuab008, 45: 1–20. https://doi.org/10.1093/femsre/fuab008.

Buckel, W. and Thauer, R.K. (2018). Flavin-based electron bifurcation, ferredoxin, flavodoxin, and anaerobic respiration with protons (Ech) or NAD^+ (RnF) as electron acceptors: a historical review. *Frontiers in Microbiology* 9: https://doi.org/10.3389/fmicb.2018.00401.

Daprile, L. (2017). *Ineos failed despite $129 million in taxpayer duties.* https://eu.tcpalm.com/story/news/2017/01/17/ineos-closes-vero-beach-biofuel-plant/96412616/

Drake, H.L., Gößner, A.S., and Daniel, S.L. (2008). Old acetogens, new light. *Annual New York Academy of Sciences* 1125: 100–128. https://doi.org/10.1196/annals.1419.016.

Drake, H.L., Küsel, K., and Matthies, K. (2013). Acetogenic prokaryotes. In: *The Prokaryotes* (ed. E. Rosenberg, E.F. DeLong, S. Lory, et al.). doi.org/10.1007/978-3-642-30141-4-61.

Flamholz, A., Noor, E., Bar-Even, A., and Milo, R. (2012). eQuilibrator – the biochemical thermodynamics calculator. *Nucleic acid Research* 40 (database issue): D770–D775.

Folch, P.L., Bisschops, M.M.M., and Weusthuis, R.A. (2021). Metabolic energy conservation for fermentative product formation. *Microbial Biotechnology* 14 (3): 829–858. https://doi.org/10.1111/1751-7915.13746.

Fuchs, G. (1986). CO_2 fixation in acetogenic bacteria: variations on a theme. *FEMS Microbiology Reviews* 39: 181–213.

Gaddy, J.L., Arora, D.K., Ko, C.-W. et al. (inventors) (2003). Methods for increasing the production of ethanol from microbial fermentation. US 2003/0211585 A1.

Greenwood, A. (2018). US left with one large-scale cellulosic ethanol plant. *ICIS* 29: 11.

Handelsblatt (2022). Genug Abgase für alle, p. 24.

Heijstra, B.D., Leang, C., and Juminga, A. (2017). Gas fermentation: cellular engineering possibilities and scale up. *Microbial Cell Factories* 16: 60. https://doi.org/10.1186/s12934-017-0676-y.

Hermann, M., Teleki, A., Weitz, S. et al. (2020). Electron availability in CO_2, CO and H_2 mixtures constrains flux distribution, energy management and product formation in *Clostridium ljungdahlii*. *Microbial Biotechnology* 13 (6): 1831–1846. https://doi.org/10.1111/1751-7915.13625.

Hermann, M., Teleki, A., Weitz, S. et al. (2021). Identifying and engineering bottlenecks of autotrophic isobutanol formation in recombinant *C. ljungdahlii* by systemic analysis. *Frontiers in Bioengineering and Biotechnology* 9: article 647853. https://doi.org/10.3389/fbioe.2021.647853.

Humphreys, C.M., McLean, S., Schatschneider, S. et al. (2015). Whole genome sequence and manual annotation of Clostridium autoethanogenum, an industrially relevant bacterium. *BMC Genomics* 16: 1085. https://doi.org/10.1086/s12864-015-2287-5.

Jain, S., Dietrich, H.M., Müller, V., and Basen, M. (2020). Fomate is required for growth of the thermophilic acetogenic bacterium *Themoanaerobacter kivui* lacking hydrogen-dependent carbon dioxide reductase (HDCR). *Frontiers in Microbiology* 11: article 59. https://doi.org/10.3389/fmicb.2020.00059.

Katsyv, A. and Mueller, V. (2020). Overcoming energetic barriers in acetogenic C1 conversion. *Frontiers in Bioengineering and Biotechnology* 8: article 621166. https://doi.org/10.3389/fbioe.2020.621166.

Köpke, M. and Simpson, S.D. (2020). Pollution to products: recycling of 'above ground' carbon by gas fermentation. *Current Opinion in Biotechnology* 65: 180–189, https://doi.org/10/1016/j.copbio.2020.02.017.

Köpke, M., Held, C., Hujer, S. et al. (2010). *Clostridium ljungdahlii* represents a microbial production platform based on syngas. *Proceedings of National Academy of Sciences of the United States of America* 107 (34): 13087–13092. https://doi.org/10.1073/pnas.1004716107.

Lane, J. (2016a). Coskata's technology re-emerges as Synata Bio. www.biofuelsdigest.com (accessed 24 Aug 2021).

Lane, J. (2016b). Sinatra Bio: Ol' brew eyes is back. www.biofuelsdigest.com (accessed 24 Aug 2021).

Lanzatech Investor Day presentation (2023). Lanzatech. Transforming carbon. Making products, p. 22. https://ir.lanzatech.com/static-files/9eed6c4f-0d20-42f8-9227-5a622c-0103b5 (accessed 25 March 2023).

Liew, F.-M., Martin, M.E., Tappel, R.C. et al. (2016). Gas fermentation – a flexible platform for commercial scale production of low-carbon-fuels and chemicals from waste and renewable feedstocks. *Frontiers in Microbiology* 7: article 694. https://doi.org/10.3389/fmicb.2016.00694.

Liew, F.E., Nogle, R., Abdalla, T. et al. Carbon-negative production of acetone and isopropanol by gas fermentation at industrial pilot scale. *Nature Biotechnology* 40: 335–344. https://doi.org/10.1038/s41587-021-01195-w.

Lin, J.-Y., Zhang, S., Zhang, L. et al. (2010). HCN and NH3 formation during coal/char gasification in the presence of NO. *Environmental Science and Technology* 44: 3719–3723.

Matthies, D., Zhou, W., Klyszejko, A.L. et al. (2014). High-resolution structure and mechanism of an F/V hybrid rotor ring in a Na^+-coupled ATP synthase, Nature Communications, doi: 10.1038/ncomms6286.

Mock, J., Zheng, Y., Mueller, A.P. et al. (2015). Energy conservation associated with ethanol formation from H_2 and CO_2 in *Clostridium autoethanogenum* involving electron bifurcation. *Journal of Bacteriology* 197 (18): 2965–2980.

Molitor, B., Kirchner, K., Henrich, A.W. et al. (2016). Expanding the molecular toolkit for the homoacetogen Clostridium ljungdahlii. *Scientific Reports* 6: 31518. https://doi.org/10.1038/srep31518.

Müller, V. (2019). New horizons in acetogenic conversion on one-carbon substrates and biological hydrogen storage. *Trends in Biotechnology* 37 (12): 1344–1354. https://doi.org/10.1016/j.tibtech.2019.05.008.

Patersen, N., Zhuo, Y., Dugwell, D., and Kandiyoti, R. (2005). Formation of hydrogen cyanide and ammonia during the gasification of sewage sludge and bituminous coal. *Energy Fuels* 19: 1016–1022.

Poehlein, A., Schmidt, S., Kaster, A.-K. et al. (2012). An ancient pathway combining carbon dioxide fixation with the generation and utilization of a sodium ion gradient for ATP synthesis. *PLOS One* 7 (3): e33439.

Ragsdale, S. and Pierce, E. (2008). Acetogenesis and the Wood-Ljungdahl pathway of CO_2 fixation. *Biochimica et Biophysica Acta* 1784 (12): 1873–1898. https://doi.org/10.1016/j.bbapap.2008.08.012.

Schuchmann, K. and Müller, V. (2016). Energetics and application of heterotrophy in acetogenic bacteria. *Applied and Environmental Microbiology* 82 (14): 4056–4069. https://doi.org/10.1128/AEM.00882-16.

Shima, S. and Thauer, R.K. (2013). A third type of hydrogenase catalyzing H2 activation. *The Chemical Record* 7: 37–46. https://doi.org/10.1002/tcr.20111.

Takors, R., Kopf, M., Mampel, J. et al. (2018). Using gas mixtures of CO, CO2 and H2 as microbial substrates: the do's and don'ts of successful technology transfer from laboratory to production scale. *Microbial Biotechnology* 11 (4): 606–625. https://doi.org/10.1111/1751-7915.13270.

Thauer, R.K., Jungermann, K., and Decker, K. (1977). Energy conservation in chemotrophic anaerobic bacteria. *Bacterial Reviews* 41: 100–180. https://doi.org/10.1128/MMBR.41.1.100-180.1977.

Thauer, R.K., Kaster, A.-K., Seedorf, H. et al. Methanogenic archaea: ecologically relevant differences in energy conservation. *Nature Reviews Microbiology* 6: 579–591.

Thoma, F. and Blombach, B. (2021). Metabolic engineering of *Vibrio natriegens*. *Essays in Biochemistry* 65: 381–392. https://doi.org/10.1042/ECB20200135.

Valgepea, K., de Souza Pinto Lemgruber, R., Abdalla, T. et al. (2018). H_2 drives metabolic rearrangements in gas-fermenting *Clostridium autoethanogenum*. *Biotechnology for Biofuels* 11: 55. https://doi.org/10.1186/s13068-018-1052-9.

Van Dien, S. (2013). From the first drop to the first truckload: commercialization of microbial processes for renewable chemicals. *Current Opinion in Biotechnology* 24: 1061–1068. https://doi.org/10.1016/j.copbio.2013.03.002.

Voet, D., Voet, J.G., and Pratt, C.W. (2016). *Principles of Biochemistry*, 501–503. Wiley.

Wang, S., Huang, H., Kahnt, J. et al. (2013). NADP-specific electron-bifurcating [FeFe]-dehydrogenase in a functional complex with formate dehydrogenase in *Clostridium autoethanogenum* grown on CO. *Journal of Bacteriology* 195 (19): 4373–4386.

Weinert, B.T., Lesmantavicius, V., Wagner, S.A. et al. (2013). Acetyl-phosphate is a critical determinant of lysine acetylation in *E. coli*. *Molecular Cell* 51: 265–272.

White, H., Strobl, G., Feight, R., and Simon, H. (1989). Carboxylic acid reductase: a new tungsten enzyme catalyzes the reduction of non-activated carboxylic acids to aldehydes. *European Journal of Biochemistry* 184: 89–96.

Wilkins, M.R. and Atiyeh, H.K. (2011). Microbial production of ethanol from carbon monoxide. *Current Opinion in Biotechnology* 22: 326–330. https://doi.org/10.1016/j.copbio.2011.03.005.

Zhang, L., Zhao, R., Jia, D. et al. (2020). Engineering *Clostridium ljungdahlii* as the gas-fermenting cell factory for the production of biofuels and biochemicals. *Current Opinion in Chemical Biology* 59: 54–61.

Zhu, H.-F., Liu, Z.-Y., Zhou, X. et al. (2020). Energy conservation and carbon flux distribution during fermentation of CO or H_2/CO_2 by *Clostridium ljungdahlii*. *Frontiers in Microbiology* 11: article 416. https://doi.org/10.3389/fmicb.2020.00416.

4

Lactic Acid

4.1 Application

Lactic acid is an α-hydroxy-acid with the formula $C_3H_6O_3$ and a molecular weight of 90.1 g/mol. The asymmetric second C-atom with four different ligands implies the distinction between the two optically active isomers L-lactic acid and D-lactic acid. The natural lactic acid isomer abundant in biological systems is L-lactic acid, which also defines the major market demand. L- and D-lactic acids are commonly produced by fermentation, and strains with a selectivity >99% for either L- or D-lactic acids are available. The efficient fermentative production based on glucose, saccharose, or lactose provided as whey has prepared the path to a high-volume downstream chemistry with several applications in different industries. Lactic acid is used in food applications as an acidulant and flavor enhancer and facilitates the extension of shelf life in food and beverages. The sodium and potassium salts are used as antimicrobials. Other applications are accessed by esterification with methanol, ethanol, and butanol to derive methyl, ethyl- and butyl-lactate. Hereof ethyl-lactate has the highest importance, which is used as an environmentally friendly solvent. Furthermore, low-volume derivatives such as cetyl-lactate and myristyl-lactate are available.

The major and growing application of lactic acid is the polymer polylactic acid (PLA) whose application scope profits from its positioning as a biobased and biodegradable plastic (Rosenboom et al. 2022; Metz et al. 2016). PLA may be derived via the polycondensation of monomeric lactic acid. However, the intermolecular esterification during the polymerization process releases water, which triggers the backward reaction, and polymers with a high molecular weight would be difficult to achieve. The rising viscosity of the growing polymer does not allow for the efficient removal of water anymore. Accordingly, the initial dimerization of lactic acid to the lactide and subsequent ring-opening polymerization with metal-based catalysts are preferred. This technology was first shown by Carothers in 1932 with low molecular weight, though did not reach industrial maturity until Dupont, in 1954, managed to install an efficient lactide purification protocol and patented PLA with a higher molecular weight (Yang et al. 2013). Lactic acid has a chiral center, and the lactide appears accordingly in three stereoisomers. Since specific PLA grades display specific properties pending on the chirality of their monomer building blocks, there is an increasing demand for all stereoisomers (Van Wouwe et al. 2016). The demand for PLA is growing since

Pathway Design for Industrial Fermentation, First Edition. Walter Koch.

it has been established as biobased and biodegradable plastic, besides PHAs (Polyhydroxyalkanoates) and PHBs (Polyhydroxybutyrates) in various application (Metz et al. 2016). The production capacity of PLA was about 190 kt in 2019 (Balla et al. 2021). It supports the trend toward sustainability in packaging and textiles and is gaining interest of several consumer brands. The application of PLA as bottle plastic is still hampered by its oxygen permeability.

Lactic acid has also been considered as a suitable building block for other chemicals and is included in the list of top chemical molecules from biorefinery carbohydrates (Bozell and Petersen 2010). Potential downstream products include acrylic acid, which can be obtained via dehydration to introduce the terminal double bond. However, the limited selectivity of the catalytic gas-phase dehydration is still low (Wojcieszak et al. 2020):

$$CH_3\text{–}CH(OH)\text{–}COOH \rightarrow CH_2 = CH\text{–}COOH + H_2O$$

R&D efforts to improve the selectivity in the generation of acrylic acid are underway. Cargill (biobased materials), IFPEN (catalyst development), and Axens (catalyst manufacturing) are collaborating in the commercialization of the lactic-to-acrylic-acid technology owned by Procter & Gamble (Cargill Press Release 2020). Furthermore, methyl lactate as precursor for acrylic acid and acrylates is discussed (Pang et al. 2022) and might prepare a major volume outlet for lactic acid in the future. The higher selectivity in the chemical conversion favors the use of 3-hydroxypropionic acid as precursor for acrylic acid:

$$HO\text{–}CH_2\text{–}CH_2\text{–}COOH \rightarrow CH_2 = CH\text{–}COOH + H_2O$$

The conversion of lactic acid into acetaldehyde and propylene glycol would be chemically possible, but both molecules can be obtained with lower production costs, e.g. acetaldehyde via partial oxidation of ethylene and propylene glycol via the hydrolysis of propylene oxide. Lactic acid is also considered as precursor toward pyruvate via oxidation of lactate esters. Additionally, it is considered to use lactic acid for the synthesis of 1,2-propylene glycol via catalytic hydrogenation. This path would especially be interesting since calcium lactate, the primary product of fermentation technology given the neutralization with $Ca(OH)_2$, could directly be used as educt for 1,2-propylene glycol. The conversion of calcium lactate into the free acid could be omitted (Binczarski et al. 2016).

4.2 Chemical Synthesis of Lactic Acid

L-Lactic acid is mainly produced by the fermentation of glucose in industrially equipped Lactobacillus strains. The strains are selected to produce exclusively L-lactate and do not possess a lactate racemase. Lactose fermentation is one of the oldest fermentation technologies besides ethanol, with high industrial maturity, exploiting the natural ability of Lactobacillus to convert C6 sugars into lactic acid. The cultural application of food conservation has been transferred into industrial technology, with the production of lactic acid already starting in 1880 in Massachusetts.

Alternative technology based on organic chemistry toward L-lactic acid or D-lactic acid employing petrochemical resources have been developed. A chemical route is pursued by the Japanese company Musashino, which started production of lactic acid in 1949

(Musashino Corporate Profile 2020). Musashino is besides Sterling Chemicals one of last manufacturers which chemically produce lactic acid (Balla et al. 2021). Acetaldehyde based on the partial oxidation of ethylene can be reacted with hydrocyanic acid at high pressure in the liquid phase in an alkaline medium to obtain the α-hydroxynitrile intermediate, which is then hydrolyzed with either hydrochloric acid or sulfuric acid to lactic acid and ammonium chloride.

$$CH_2 = CH_2 + 0.5\,O_2 \rightarrow CH_3\text{–}CHO$$

$$CH_3\text{–}CHO + HCN \rightarrow CH_3\text{–}CH(OH)\text{–}CN$$

$$CH_3\text{–}CH(OH)\text{–}CN + HCl + 2\,H_2O \rightarrow CH_3\text{–}CHOH\text{–}COOH + NH_4Cl$$

This route is unable to provide specifically either the one or the other stereoisomer, and instead a mixture of L- and D-lactic acid is obtained, which is then processed to offer a pure L-lactic acid or D-lactic acid grade. The chemical technology based on acetaldehyde is disadvantaged from an economic point of view given the rather high cost of hydrocyanic acid and the inconvenient generation of ammonium chloride as by-product. Accordingly, this technology serves only a limited market share of less than 10% (Boontawan et al. 2011). The minor importance of organic chemistry is reflected in the rather high efficiency of the industrial fermentative production of lactic acid with respect to carbon yield, selectivity, and productivity.

4.3 Pathway Design

4.3.1 Lactic Acid Production with Homofermentative Lactobacillus Strains

Lactobacilli may be categorized according to their sugar metabolism. The strictly homofermentative pathway is pursued by the genera *Lactococcus*, *Enterococcus*, *Streptococcus*, *Pediococcus*, and group I and II lactobacilli (Pörtner et al. 2016). Representatives of these genera are applied for industrial fermentation and provide either L- or D-lactic acid. The pathway proceeds via pyruvate as the final product of conventional glycolysis, whose prochiral keto function is reduced to a secondary alcohol to obtain either L- or D-lactate (insert 4). The industrial strains possess a native stereoselective lactate dehydrogenase. Racemase activity with lactic acid as substrate – which might generate a mixture of L- and D-lactate – is silenced. The stereoselective reduction of the keto function defines a major advantage of biotechnology versus organic chemistry. For comparison, a stereo-selective reduction of pyruvate to lactic acid by means of organic chemistry is possible, though rather complex and costly catalysts would be needed (Wang and Yang 2020; Figure 4.1).

In the generation of lactic acid, the surplus of reduction equivalents derived in the glycolytic oxidation of glucose is rebalanced in the final hydrogenation of pyruvate (Table 4.1):

The calculated free Gibbs energy for the conversion of glucose into lactic allows the estimation of the energetic efficiency of the reaction:

$$\text{Glucose} \rightarrow 2\,\text{Lactate} + 2\,H^+ \quad \Delta G^{\circ\prime} = -217\ \text{KJ/mol glucose}$$

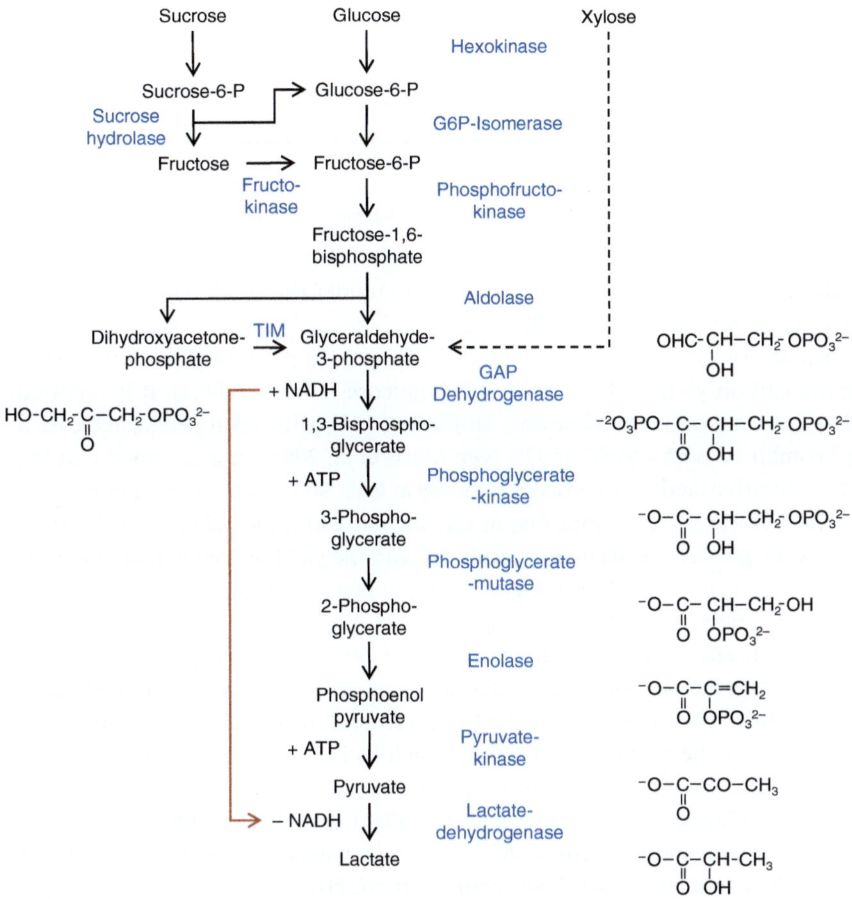

Figure 4.1 Pathway of homofermentative Lactobacillus toward lactate. Source: Walter Koch.

Table 4.1 Biochemical conversion of glucose into lactic acid.

Glucose + 2 ADP + 2 Pi + 2 NAD^+ → 2 Pyruvate + 2 ATP + 2 H_2O + 2 NADH + 4 H^+	
2 Pyruvate + 2 NADH + 4 H^+ → 2 Lactate + 2 NAD^+ + 2 H^+	
Glucose + 2 ADP + 2 Pi → 2 Lactate + 2 H^+ + 2 ATP + 2 H_2O	Net reaction

The indicated $\Delta G°'$ value refers to a pH of 7.0, a product/educt concentration of 1 mM and an ionic strength of 0.1 (Flamholz et al. 2012; Voet et al. 2016; Folch et al. 2021). The negative free Gibbs energy indicates that the reaction is thermodynamically highly favored. Since the formation of one mol of ATP based on ADP and P_i requires a free Gibbs energy of −30.5 KJ/mol, the overall energetic efficiency of the reaction can be estimated at 28% in the standard state. Under physiological conditions, it is assumed that about 50–60 KJ/mol ATP are required for the synthesis of one mol of ATP (Flamholz et al. 2012), and the energy efficiency of the pathway is likely more in the range of 50%. The pathway is supported by the negative free Gibbs energy of the last step. The reductive potential of the energy-rich

Table 4.2 Gibbs free energy for the hydrogenation of pyruvate.

NADH → NAD^+ + H^+ + 2 e	$E°' = 0.320$ V (Voet et al. 2016)
Pyruvate + 2 H^+ + 2 e → Lactate	$E°' = -0.190$ V (Voet et al. 2016)
Pyruvate + NADH + H^+ → Lactate + NAD^+	$\Delta E°' = 0.130$ V

$\Delta G°' = -n \times F \times \Delta E°' = -2 \times 96{,}485$ J/V × mol × 0.130 V = −25.1 KJ/mol (Schuchmann and Müller 2016).

electron transfer carrier NADH is sufficiently strong to render the combined reaction into a thermodynamically downhill path (Table 4.2):

The theoretical maximum molar yield of two mol lactic acid per mol glucose translates into a maximum carbon yield of 1.0 t lactic acid/t glucose (2×90.1/180.2). It is reported that a Lactobacillus lactis strain at laboratory conditions (5 l) cultured at pH 6.2 achieves a titer of 210 g/l combined with a yield of 97% (van Maris et al. 2004). It is assumed that the commercial fermentative lactic acid process achieves at least 90% of the maximum theoretical carbon yield. Additionally, a minor loss in the downstream processing of the broth to isolate and purify the product needs to be considered, and the yield in downstream processing might be assumed to be 98%. Thus, the glucose input factor including product isolation and purification may be estimated in the range of 1.13 t glucose/t lactic acid. The industrial fermentation process is run at a temperature of 45–50 °C and no oxygen needs to be supplied. The balance of reduction equivalents does not require oxygen as final electron acceptor. Lactobacillus species are aerotolerant and can endure some oxygen contamination in the off-gas above the fermentation broth, which facilitates an easier process setup. The achieved productivities are high and reach up to 4.0 g/lxh. For *Lactobacillus delbruckii* a carbon yield of 0.97 t lactic acid/t glucose, a titer of 90 g/l, and a productivity of 3.8 g/lxh have been reported (Abedi and Hashemi 2020). The most advanced technology achieved in the industry will likely manage a significantly higher productivity.

Besides carbon restructuring also the energy balance in the metabolic engineering strategy deserves attention. Given the standard reaction equation of lactic acid fermentation, the parallel synthesis of two mol of ATP per mol lactate for growth, maintenance, and pH management (pending on neutralization of the broth) might represent an ideal constellation. Otherwise, the build-in of an ATP futile cycling element (e.g. overexpression of PEP synthase or adenylate kinase) might help to replenish the ADP pool for additional carbon consumption and hence additional formation of lactate. However, such measures would presuppose a full transparency of the energy balance during the fermentation process.

4.3.2 Lactic Acid Production With Heterofermentative Lactobacillus Strains

The heterofermentative pathway is followed by some Lactobacillus species such as *Leuconostoc*, *Oenococcus*, *Weissella* and group III lactobacilli, which are lacking the aldolase enzyme, which splits fructose-1,6-bisphosphate into glyceraldehyde-3-phosphate and dihydroxyacetone-phosphate (Pörtner et al. 2016). Instead, the heterotrophic glucose metabolism proceeds via the oxidative section of the pentose phosphate pathway and oxidizes glucose-6-phosphate to 6-P-gluconate, which is in the following step decarboxylated to obtain ribulose-5-phosphate. Ribulose-5-phosphate is finally converted into lactate, acetate, and ethanol (Figure 4.2).

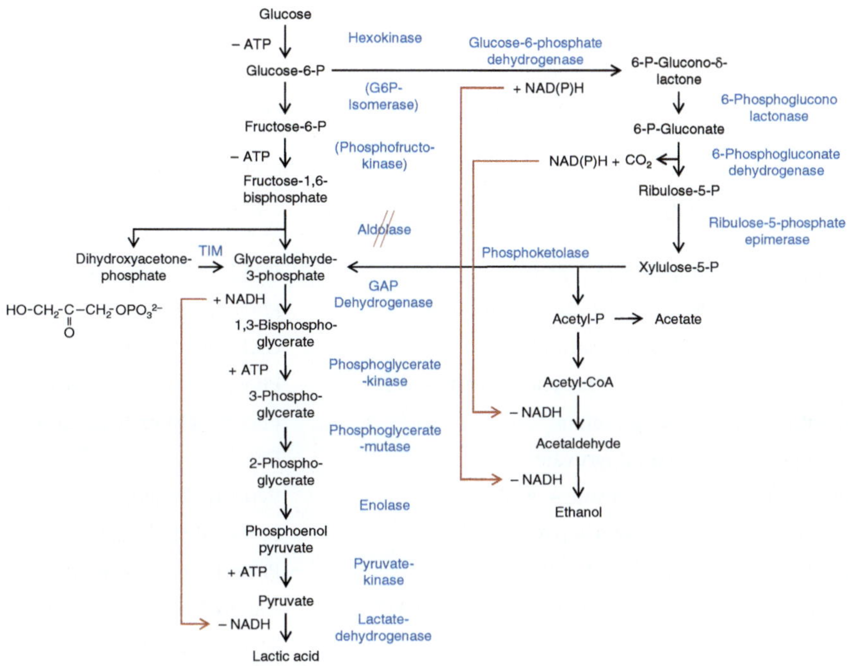

Figure 4.2 Pathway of heterofermentative Lactobacillus. Source: Walter Koch.

The net reaction of the pathway reveals that with lactic acid and ethanol, two products are obtained, and only one mol of lactic acid is derived from one mol of glucose (Table 4.3):

The decisive enzyme of the pathway is the split of xylulose-5-P into glyceraldehyde-3-phosphate and acetyl-phosphate via the phosphoketolase. The generation of lactic acid and ethanol ensures a balance of reduction equivalents since the 24 electrons carried by glucose (without counting reduced oxygen) are allocated to lactic acid and ethanol. Alternatively, the highly active acetyl-phosphate might be used to generate further ATP if converted into acetate (Figure 4.2). The three distinct products lactic acid, ethanol, and acetate, render the heterofermentative pathway as an unattractive route since the needed separation efforts would not be justified given the rather low value of the three products. The maximum molar yield is 0.5 mol lactic acid/mol glucose, and the maximum carbon yield of the heterofermentative pathway via 6-phosphogluconate is 0.5 t lactic acid/t glucose (90.1/180.2). The heterofermentative lactic acid technology has no industrial relevance.

The carbon and energy feed might alternatively be provided via sucrose or xylose. Sucrose is known as common table sugar and consists of glucose and fructose bound together via a glycosidic bond. Its uptake depends on the uptake system of the respective host and might be linked to the conversion of PEP into pyruvate to obtain sucrose-6-phosphate. Alternatively, the phosphorylation and activation might be performed within the cell. Sucrose-6-phosphate is split into glucose-6-phosphate and fructose by the sucrose hydrolase. Glucose-6-phosphate is directly channeled into the glycolytic pathway and fructose is absorbed after its phosphorylation via the fructokinase. Xylose as C5 sugar feed, might

Table 4.3 Biochemical conversion of glucose into lactic acid and ethanol.

Glucose + ATP → Glucose-6-phosphate + ADP	Hexokinase
Glucose-6-phosphate + $NAD(P)^+$ → 6-Phosphoglucono-δ-lactone + NAD(P)H + H^+	G6P DH
6-Phosphoglucono-δ-lactone + H_2O → 6-Phosphogluconate	6-Phosphogluconolactonase
6-Phosphogluconate + $NAD(P)^+$ → Ribulose-5-phosphate + CO_2 + NAD(P)H + H^+	6-Phosphogluconate DH
Ribulose-5-phosphate → Xylulose-5-phosphate	Ribulose-5-P epimerase
Xylulose-5-phosphate → GAP + Acetyl-phosphate + H_2O	Phosphoketolase
GAP + NAD^+ → 1,3-Bisphosphoglycerate + NADH	GAP dehydrogenase
1,3-Bisphosphoglycerate + ADP → 3-Phosphoglycerate + ATP	Phosphoglycerate kinase
3-Phosphoglycerate → 2-Phosphoglycerate	Phosphoglycerate mutase
2-Phosphoglycerate → Phosphoenolpyruvate	Enolase
Phosphoenolpyruvate + ADP → Pyruvate + ATP	Pyruvate kinase
Pyruvate + NADH + H^+ → Lactic acid + NAD^+	Lactate dehydrogenase
Acetyl-phosphate + CoA → Acetyl-CoA + Pi	Phosphotransacetylase
Acetyl-CoA + NADH + H^+ → Acetaldehyde + CoA + 2 NAD^+	Aldehyde dehydrogenase
Acetaldehyde + NADH + H^+ → Ethanol + NAD^+	Alcohol dehydrogenase
Glucose + ADP + Pi → Lactic acid + Ethanol + CO_2 + ATP + H_2O	Net reaction

also be phosphorylated during the cellular uptake or afterwards inside the cell. It would be metabolized via several steps including seduheptulose-7-phosphate and enter glycolysis as glyceraldehyde-3-phosphate.

It has also been considered to convert the lactic acid salt after the fermentation into the respective methyl- or ethylester, which could then easily be isolated from the fermentation broth via distillation. Although these approaches are not known to be applied in industrial fermentation.

4.3.3 Lactic Acid Production with *Lactobacillus paracasei*

Based on the stipulated requirements of an "ideal" production host, further strain improvement efforts are pursued by academia. These requirements were defined as a titer >200 g/l, a productivity of >5 g/lxh and a carbon yield >0.95 g lactic acid/g glucose. Additionally, a fermentation temperature of 45 °C is targeted to reduce cooling efforts and to limit sterilization efforts (Tian et al. 2021). In order to achieve these goals, the conventional repertoire of metabolic engineering technology has been extended to cover mutagenesis screening and adaptive evolution. Tolerance toward a higher fermentation temperature was achieved with adaptive evolution: after activation of the strain at 37 °C the cells were transferred to a medium with 40 °C for 12 hours. This process was repeated 34-times for 17 days before the temperature was raised to 42 °C, and the process was repeated for 23 days. From day 41 onwards the temperature was further increased to 45 °C and the culture was continued for 50 days. The final strain was adapted to the elevated temperature level. The metabolic engineering toolbox, including plasmid-based homologous recombination, Red/RecET-mediated double-stranded DNA recombination, and single-stranded DNA recombination, was broadened to include CRISPR/Cas9 gene editing technique. This technology has already been applied to the established hosts *Lactobacillus lactis* and *Lactobacillus casei* and was in this study customized to *Lactobacillus paracasei*. Three lactate dehydrogenase enzymes were found in the genome sequence of *Lactobacillus paracasei*: L-lactate dehydrogenase (*ldhL1*), L-lactate dehydrogenase 2 (*ldhL2*) and the D-lactate dehydrogenase (*ldhD*). Thereof *ldhL1* is the most relevant gene based on RT-PCR analysis. Since lactic acid production is closely related to cell growth, the *ldhL1* gene was chosen as phenotypic gene to evaluate the CRISPR/Cas9 technology. The *ldhD* gene was knocked out to maximize optical purity. The fermentation medium contained 90 g/l glucose, 10 g/l yeast extract and other additives to account for usual metabolic and growth requirements. The pH was maintained at 6.0 using NH_4OH and $Ca(OH)_2$ for neutralization. Finally, convincing performance data were achieved: a titer of 221 g/l, a productivity of 7.5 g/lxh, and a carbon yield of 0.96 g lactic acid/g glucose. The optical purity of the product exceeded 99.1%. The engineered strain also tolerates a high initial glucose concentration of 250 g/l (Tian et al. 2021).

4.3.4 D-lactate Production with *E. coli*

It has been pursued to develop a heterologous *E. coli* strain for the production of D-lactate (Zhou et al. 2016). One advantage would be that *E. coli* has a *ldh* gene product that primarily generates D-lactate. Under anaerobic conditions, *E. coli* pursues a mixed acid fermentation with D-lactate, acetate, succinate, formate, ethanol, carbon dioxide, and hydrogen as products. The share of D-lactate is in the range of 40%, and metabolism needs to be streamlined to channel pathways into one target product. The pyruvate formate lyase is the main pyruvate processing enzyme under anaerobic conditions with formate and acetyl-CoA as products (Pinske and Sargent 2016). This reaction also generates acetyl-CoA according to the pyruvate dehydrogenase complex though does not release carbon dioxide as by-product. The released C1 unit is kept in a lower oxidation state and absorbs the two electrons that allow the generation of NADH in the pyruvate dehydrogenase complex. The knock-out of the genes *pflA*, *pflB*, *ackA*, and *pta* facilitated a significant decrease in the formation of acetate and formate. This manipulation of the native metabolic network also triggered higher D-lactate production, which is explained by the diminished ATP production based on the conversion of acetyl-P into acetate under oxygen-limited conditions. A deletion of the fumarate reductase *frdA* gene, encoding the enzyme to convert fumarate into succinate, provided a significant reduction of the by-product succinate. On the other hand, deletions of the genes *pps* and *adhE* did not support though weakened lactate accumulation. The inactivation of the flavoprotein pyruvate oxidase (*poxB* gene) did not help to further limit the generation of acetate (Figure 4.3).

The reported data in academia prove the general feasibility of a fermentative D-lactate technology of *E. coli* for industrial application. The reported titers are above 100 g/l, molar yields are above 86%, and the achieved productivities are ranging from 3–6 g/lxh (Grabar et al. 2006; Zhu et al. 2007). The production of D-lactic acid has also been tried with *C. glutamicum* which generates lactate under oxygen deprivation (Okino et al. 2008).

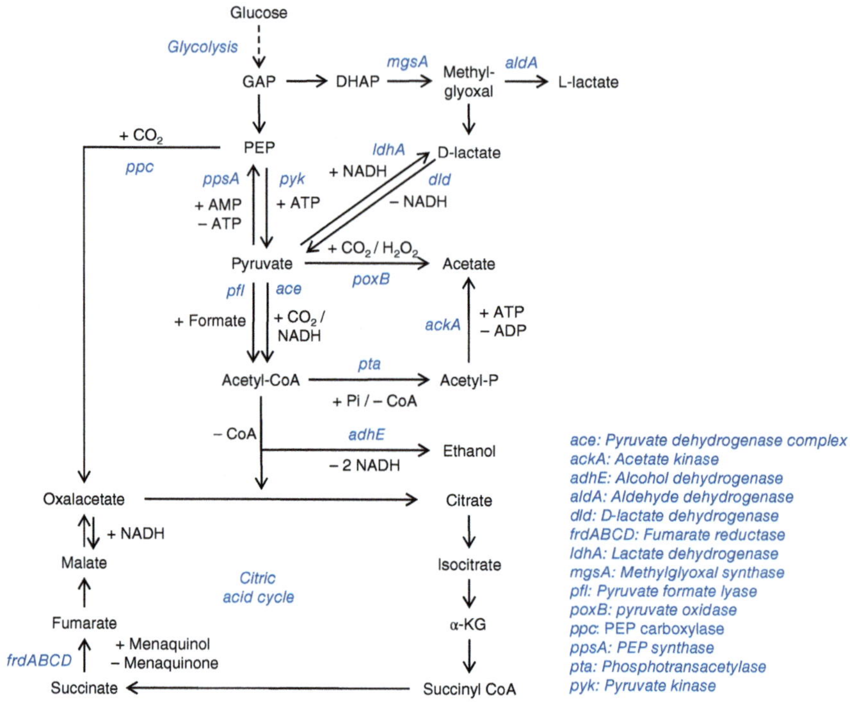

Figure 4.3 Metabolic engineering of *E. coli* for the production of D-lactate. Source: Adapted from Zhou et al. (2016).

4.4 Process Development

Glucose, sucrose, or polysaccharides might be applied as carbon and energy sources. In the latter case, a saccharification step to convert the carbon source into mono- and disaccharides, triggered by a suitable enzyme mix is preparing the fermentation. A stirring device is operated in the main reactor to ensure a homogeneous portioning of the biocatalyst and limits pH gradients within the fermenter. Oxygen is usually not added to the anaerobic process since the glycolysis derives an ATP surplus, which is sufficient for cell maintenance. The neutralization during the fermentation implies that a lactic acid salt is obtained which needs to be deprotonated with an acid, and a salt is inevitably generated as by-product. R&D on the fermentative production of lactic acid focuses on several fields. Only a few studies are published on pathway design given the natural and highly efficient pathway of Lactobacillus species. Instead, activities address Lactobacillus or eukaryotic yeast hosts with even higher acid tolerance to reduce the input factor of auxiliary chemicals and to reduce the co-production of salts (e.g. Na_2SO_4 or $CaSO_4$/gypsum).

4.4.1 pH Management and Calcium Sulfate as By-product

Most commercial production plants apply Lactobacillus strains with high acid tolerance as host. These bacterial strains, like *Lactobacillus delbrueckii*, preferentially for glucose-containing media and *Lactobacillus helvetii* and *Lactobacillus bulgaricus* for whey, have received the regulatory GRAS status ("generally recognized as safe") and leverage their natural ability to convert glucose into lactic acid. Hosts have, in the last decades, been improved by various strain engineering programs to facilitate a product titer in the range of 100 g/l or higher. This implies that the pH is decreasing with increasing product formation and the broth is getting more and more acidic. The pK_a value for lactic acid of 3.9 (Dusselier et al. 2013) corresponds to an equilibrium constant of $K_a = 1.25 \times 10^{-4}$. Given an assumed titer of, e.g. 100 g/l lactic acid (1.11 M), the virtual pH – without neutralization of the broth and hypothetical full productivity – can be estimated:

$$1.25 \times 10^{-4} = \mathrm{c}\left[\mathrm{H_3O^+}\right] \mathrm{c}\,[\text{lactate}^-]\,/\mathrm{c}\left[\text{protonated lactic acid}\right] \text{ and}$$
$$\mathrm{x} = \mathrm{c}\left[\mathrm{H_3O^+}\right] \text{ translates into}$$
$$1.25 \times 10^{-4} = x^2/(1.11 - x) \text{ and derives } x = 0.0131\text{ M and a pH} = 1.88$$

Since commercial plants achieve titers up to 200 g/l (2.22 M) the virtual pH might even be lower:

$$1.25 \times 10^{-4} = \mathrm{c}\left[\mathrm{H_3O^+}\right] \mathrm{c}\,[\text{lactate}^-]\,/\mathrm{c}\left[\text{protonated lactic acid}\right] \text{ and}$$
$$\mathrm{x} = \mathrm{c}\left[\mathrm{H_3O^+}\right] \text{ translates into}$$
$$1.25 \times 10^{-4} = \mathrm{x}^2/\,(2.22 - \mathrm{x}) \text{ and derives x} = 0.0175\text{ M and a pH} = 1.75$$

In general, a low pH level helps to limit sterilization efforts and might even allow a non-sterile process if the operation protocol includes suitable precautionary measures. Nevertheless, at pH figures below 4.5, the lactic acid would more and more enter the bacterial Lactobacillus cells and lower the cytosolic pH within the host. The undissociated lactic acid is membrane-soluble and can pass the cytoplasmic membrane via simple diffusion. It would then dissociate inside the cell and eventually affect the pH management of the host. Thus, pH adjustment is recommended, and it may be assumed that industrial plants neutralize and keep the pH at a level of 5.5 to 6.0. For the neutralization of the broth, NaOH, $CaCO_3$, or NH_3 may be applied. At batch end, sulfuric acid is added to convert the lactic acid salt into the free acid, and a salt as by-product cannot be avoided:

$$2\,\mathrm{CH_3{-}CHOH{-}COOH} + 2\,\mathrm{NaOH} \rightarrow 2\,\mathrm{CH_3{-}CHOH{-}COONa} + 2\,\mathrm{H_2O}$$

$$2\,\mathrm{CH_3{-}CHOH{-}COONa} + \mathrm{H_2SO_4} \rightarrow 2\,\mathrm{CH_3{-}CHOH{-}COOH} + \mathrm{Na_2SO_4}$$

The alternative application of $Ca(OH)_2$ would result in the production of gypsum as by-product:

$$2\,\mathrm{CH_3{-}CHOH{-}COOH} + \mathrm{Ca(OH)_2} \rightarrow \left(\mathrm{CH_3{-}CHOH{-}COO}\right)_2\mathrm{Ca} + 2\,\mathrm{H_2O}$$

$$\left(\mathrm{CH_3{-}CHOH{-}COO}\right)_2\mathrm{Ca} + \mathrm{H_2SO_4} \rightarrow 2\,\mathrm{CH_3{-}CHOH{-}COOH} + \mathrm{CaSO_4}$$

As usually not the anhydrous $CaSO_4$ though the dihydrate gypsum ($CaSO_4$ x 2 H_2O) is obtained about 1 ton gypsum by-product is obtained per ton lactic acid (Dusselier et al. 2013). The pure stoichiometry corresponds to a by-product factor of 0.96 t gypsum/t lactic acid (172.1/2 × 90.1). The complete process based on carbohydrates with enzymatic

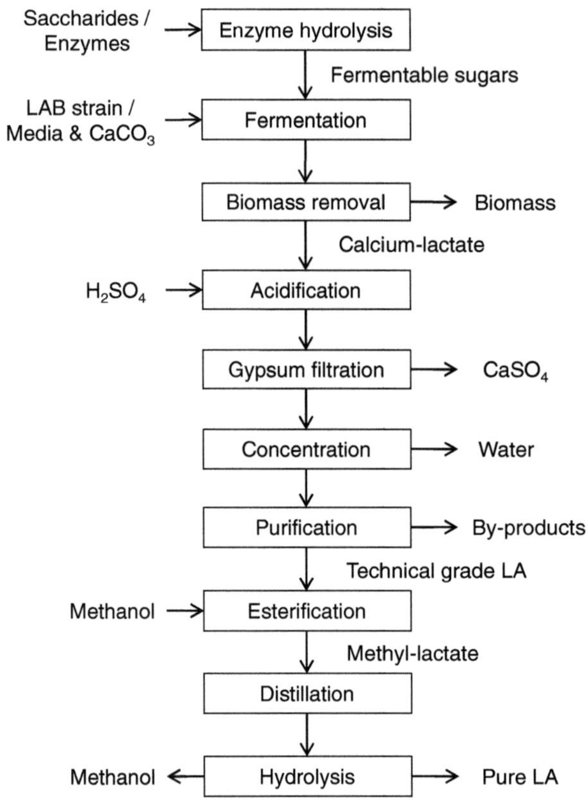

Figure 4.4 Process steps for the fermentative production of lactic acid. Source: Adapted from Dusselier et al. (2013).

breakdown of the carbohydrate monomer and calcium carbonate as an exemplary neutralization agent can be described as follows (Figure 4.4):

The protocol for product isolation and purification might be customized depending on the required product quality and foreseen application. Different technical and non-technical grades of lactic acid are offered in the market (Kubicek 2001). The protonated lactic acid may be esterified with methanol or ethanol to obtain the respective ester as an intermediate which allows an efficient final purification to obtain a food grade. The ester is cleaved, and methanol or ethanol are recycled.

4.4.2 pH Management and Ammonium Sulfate as By-product

ThyssenKrupp and Uhde have developed a proprietary lactic acid and PLA technology which can be constructed as an integrated value chain or as separate units (Thyssen Krupp Industrial Solutions 2016). Thyssen Krupp Industrial Solutions has built a pilot plant for producing lactic acid in Leuna (Germany) to demonstrate the maturity of the technology, which is offered on a license business model ("one-stop technology provider"). The process is regarded as a 2nd generation lactic acid process since it does not generate $CaSO_4$ or gypsum, though the more valuable by-product is ammonium-sulfate $(NH_4)_2SO_4$, which is sold as fertilizer:

$$2\,CH_3\text{–}CHOH\text{–}COOH + 2\,NH_3 \rightarrow 2\,CH_3\text{–}CHOH\text{–}COO^-\,NH_4^+$$

$$2\,CH_3\text{–}CHOH\text{–}COO^-NH_4^+ + H_2SO_4 \rightarrow 2\,CH_3\text{–}CHOH\text{–}COOH + \left(NH_4\right)_2SO_4$$

Based on the equation of the reaction, it may be assumed that about 0.2 t NH_3 and 0.55 t H_2SO_4 (100%) are needed as input factors for the neutralization process per ton lactic acid. The ammonium-sulfate by-product factor will amount to at least 0.74 t per ton lactic acid.

4.4.3 Eukaryotic Host as Biocatalyst with Enhanced Acid Tolerance

Cargill has changed the host in their process applied in Blair (Nebraska) and is using an eukaryotic yeast strain with advanced acid tolerance. More than 1200 yeast strains have been screened before a suitable strain could be identified. In order to suppress the native formation of ethanol the pyruvate decarboxylase needs to be knocked out. The competing pyruvate dehydrogenase complex toward acetyl-CoA needs to be silenced. A complementary lactate dehydrogenase with stereoselectivity for either L- or D-lactic acid needs to be added via transformation. The new process allows to reduce the unit consumption rate for $Ca(OH)_2$ and the by-product factor of gypsum. This reveals that the medium is still buffered, though the process is run at a lower pH, which is not disclosed (Food Ingredients 2010). An increased titer of lactic acid does also reduce the water consumption of the process. For the Cargill technology a reduction in water consumption of 21.1% enabled by a titer increase of 19.7% has been reported (Pothakos et al. 2018).

4.4.4 pH Management with Co-production

An interesting idea has been proposed to absorb the generated protons with a coupled biocatalytical consumption of protons (Zhao et al. 2015). In particular the fermentative production is combined with the biochemical decarboxylation of mono sodium glutamate to γ-aminobutyric acid (GABA) facilitated by the glutamate decarboxylase (GAD). Glutamate decarboxylation is known as the final enzymatic step in the fermentative production of GABA with glutamate as an intermediate (Dhakal et al. 2012; Lee et al. 2013). In this case the glutamate decarboxylation would simply be combined with the generation of lactate and two valuable products would be obtained (Table 4.4):

The reaction avoids a decrease of the pH induced by the formation of lactate since one carboxyl function of glutamate – which is applied in the deprotonated form as sodium salt – is converted into carbon dioxide and a primary amine. Thereby, one proton out of the medium is needed for the formation of the $R\text{-}CH_2\text{-}NH_2$ group. Lactate is obtained as neutral sodium lactate. It is obvious that the desired impact on the pH supposes an efficient removal of carbon dioxide. Otherwise, carbonic acid would be formed which would release protons and thereby decrease the pH.

The experiments were performed with an immobilized GAD and *Rhizopus oryzae* as fungal fermentation hosts for the production of lactic acid. This host has gained interest since it tolerates an acidic pH window of pH 4.0 to 6.0 and produces lactic acid in a single

Table 4.4 pH management with organic salt.

Glucose $\rightarrow$ 2 $CH_3\text{-}CHOH\text{-}COO^-$ + 2 H^+	
2 $NaOOC\text{-}CH(NH_2)\text{-}CH_2\text{-}CH_2\text{-}COO^-$ + 4 H^+ $\rightarrow$ 2 $H_2N\text{-}CH_2\text{-}CH_2\text{-}CH_2\text{-}COO^-$ + 2 H^+ + 2 Na^+ + 2 CO_2	
Glucose + 2 $NaOOC\text{-}CH(NH_2)\text{-}CH_2\text{-}CH_2\text{-}COO^-$ + 2 H^+	
$\rightarrow$ 2 $CH_3\text{-}CHOH\text{-}COO^-$ + 2 Na^+ + 2 $H_2N\text{-}CH_2\text{-}CH_2\text{-}CH_2\text{-}COO^-$ + 2 H^+ + 2 CO_2	Net reaction

isomeric form in low-cost media (Zhao et al. 2015). At a pH range of about 4.7–5.0, a lactic acid concentration of 80.2 g/l with a yield of 67% from glucose has been achieved. In parallel, a GABA titer of 44.8 g/l with a conversion rate of 96.7% from sodium glutamate was observed. In case the glucose consumption could be raised to about 98%–99% a commercially interesting level could be achieved. However, besides technology issues also the volume streams need to match. It is expected that the market for GABA is much smaller than that of lactic acid, and GABA volumes would face a more and more challenging demand limitation in case the combined technology would be scaled up.

4.4.5 Overcome Neutralization by Adjusted Product Concept

The issue of pH management might also be addressed with a modification of the product concept. In case the acid would be esterified with ethanol or butanol inside the cell the pH of the fermentation broth would be kept rather constant. However, these approaches require an extended metabolic pathway, likely the addition of the respective alcohol to the broth and an additional step in downstream processing to reconvert the ester to obtain the "free" lactic acid and a recycling of the alcohol.

4.4.6 *In Situ* Recovery During the Fermentation

In parallel, *in situ*-recovery technologies are evaluated with the main target to limit the efforts for pH management. *In situ*-recovery comprises the immediate removal of the product out of the fermentation broth to avoid interference with the host cell or the medium (Freeman et al. 1993). Interference with the host cell might cause weak productivity, depending on the host's resistance toward the product. Additionally, the product might be volatile and partially get lost into the off-gas phase. An *in situ*-recovery of lactic acid might be achieved in various manners. Ion exchange or electrodialysis technology – functioning with electrostatic surfaces to capture and bind positively or negatively charged cations and anions – are primarily suited given the anionic nature of the compound to be isolated. They do in principle work though might cause product loss. Ion exchange resins need to be regenerated for continuous extraction. It is reported that electrolysis has until the 1970s been applied for the isolation of lactic acid out of the broth. The power demand for a two-stage electrolysis is indicated at 1.5 MWh per t lactic acid, which is high, and electrolysis has been replaced by more cost-efficient technologies (Hábová et al. 2004). Solvent extraction might require toxic solvents and a back-extraction. An efficient partition coefficient needs to be identified to limit the number of steps (Kim et al. 1999). Also, membranes for product removal via size selective permeation might be applied. Activated carbon absorption is pH sensitive and an efficient protocol for desorption needs to be developed.

In order to raise fermentation productivity, it has been tested to efficiently remove calcium lactate via crystallization (Xu and Xu 2014). This approach addresses specifically the derivative calcium lactate as product which is an intermediate of the conventional process – given the required neutralization of the medium with $Ca(OH)_2$ – though is usually reconverted into the free acid. However, it might directly be derived in the process in case the medium is transferred to a crystallization unit to crystallize calcium lactate. To avoid that glucose is trapped in the crystal structure, it is ensured that the glucose is completely exhausted before parts of the broth are transferred to the crystallization unit. The reported lactic acid concentration in the broth is 120 g/l. This titer was also observed to support the formation of granular crystals, which are easier to remove than the amorphous crystals generated with the broth containing 130 g/l lactic acid or higher. The crystals are removed from the crystallization unit, and the remaining supernatant is rechanneled into the fermentation process. Glucose is applied as a powder opposed to the typical 70% solution to limit the addition of water. *Bacillus coagulans* has been employed as fermentation host. It is reported that a lactic acid titer of 168.8 g/l after 36 hours and a corresponding productivity of 4.7 g/lxh has been measured. The calculated carbon yield is indicated with 0.93 t lactic acid/t glucose, corresponding to an input factor of 1.08 t glucose/t lactic acid (Xu and Xu 2014). The potential commercial maturity relies on the product purity of the crystallized product. For cost efficiency reason it will be crucial that the product crystals display a high purity to suffice the food grade specification. Otherwise, a second crystallization step would be needed which requires heat for the evaporation of water or for cooling. It is also known that conventional glucose grades contain a residual share of disaccharides derived from the initial starch hydrolysis. These residues are usually not critically important since standard hosts are able to hydrolyze the disaccharide to the glucose monosaccharide. Though if entrapped in the crystals, they would negatively impact product quality.

4.4.7 Reactor Design and Biocatalyst Immobilization

A high productivity of the fermentation process is usually enabled by a high cell density. A rather high cell density can be achieved with a conventional culture of suspended cells circulated in a continuously stirred tank reactor. In order to keep a high concentration of active cells and to avoid a compromised carbon yield a cell recycling step may be added. The separation of product and biomass might be facilitated with perfusion technology. A share of the medium is permanently removed and channeled into a separation unit in which the lactic acid partially passes a membrane whereas the biomass is retained. These devices do their principal work, though membrane fouling cannot be avoided. Cell immobilization has been evaluated to overcome this obstacle, which might be applied in a fluidized bed reactor with microbial cells entrapped in gel particles. Going one step further, the cells might be embedded in polymeric networks or be immobilized onto synthetic materials to fill a fermentative fixed-bed reactor. The medium might be channeled through the reactor with an axial flow, and the pH impact of the deprotonated lactic acid on the biocatalyst would be decreased. In experiments with *Lactobacillus lactis* in a 100 ml fixed-bed reactor a productivity of 8.1 g/lxh after 52 days operation could be obtained. These attempts might succeed in the future to develop an innovative reactor concept for the fermentative production of lactic acid (Pörtner et al. 2016).

4.5 Evaluation of Alternative Feedstocks

4.5.1 Lignocellulosic Biomass as Carbon and Energy Source

Sustainability considerations drive the application of alternative raw material sources based on lignocellulosic biomass. Biomass from agriculture and forest is partially burnt since suitable processes for the use in chemical processes are rare. A non-thermal use of biomass would avoid the direct emission of carbon dioxide and reduce the demand for petrochemical raw materials. Since the current mainstream hosts for lactic acid do not sufficiently digest biomass and the respective C5 and C6 sugars as feed alternative production hosts are explored. Academic activities provided convincing results with promising values for titer, productivity, and yield (Ma et al. 2022).

These attempts were also pursued with respect to the fermentative production of lactic acid (Pranaw et al. 2021). A fermentation with xylan as feedstock and the host *Thermoanaerobacterium aotearoense* was tried and provided a product with 99.3% optical purity. The formation of acetic acid has been blocked. The strictly anaerobic and thermophilic characteristics of the host suppress the contamination from competing organisms and might reduce sterilization efforts. The host achieved a yield of 0.93 t lactic acid/t glucose and achieved a titer of 3.2 g/l after 48 hours using 10 g/l xylan as the single substrate (Yang et al. 2013). The data indicate a proof of concept. A comprehensive summary of attempts to use innovative feedstocks, including lignocellulosic biomass is given by Yankow, which indicate that often an attractive yield up to 0.99 t lactic acid/t glucose combined with productivities between 6-12 g/lxh has been shown (Yankow 2022). It remains to be seen inasmuch these promising results are finally transferred into the stage of commercial application. Corresponding R&D efforts are also pursued with respect to D-lactic acid (Alexandri et al. 2019).

4.5.2 D-lactate Production with *L. delbrueckii* with Rice Straw Hydrolysate

The fermentative production of D-lactic acid with promising performance data could be shown using rice straw as feedstock and *L. delbrueckii* as host (Ma et al. 2022). After pretreatment of the rice straw with liquid hot water at a temperature of 185 °C under high pressure with 25 minutes incubation time, the reactor was cooled down below 80 °C and the water-insoluble solids were separated from the liquid fraction (prehydrolysate) by vacuum filtration. The water-insoluble solids were then treated with a commercially available cellulase at a dose of 15 FPU/g dry weight at a temperature of 50 °C and an incubation time of 72 hours. After a membrane filtration to detoxify and concentrate the joint fractions of the rice straw hydrolysate, a feedstock containing 50.85 g/l glucose, 4.37 g/l xylose, and 0.82 g/l arabinose was obtained. The fermentation was pursued as continued process with permanent removal of a share of the broth to be channeled into a hollow-fiber microfiltration module. In this unit, the biomass was retained and recycled into the fermentation reactor. The media fraction passing the membrane was stored in the product tank. The experiment was pursued for a duration of 350 hours, during which the biomass increased up to 20 g/l. The titer of D-lactic acid reached a threshold of 46.6 g/l after 20 hours and was kept on this level. The analysis of the process monitoring data revealed a lactic acid productivity of 18.56 g/lxh or 200 mmol/lxh. In case a reasonable glucose uptake rate for Lactobacillus is estimated according to *E. coli* with 10 mmol per g DCW per h (Ahn et al. 2020; Liebal et al. 2018; Chae et al. 2017), a minimum biomass titer of 10 g/l can be deduced. This calculated minimum threshold is according to the reported cell concentration of about 20 g/l. It is interesting to note that the high productivity has been reached after 30 hours incubation time when the cell density was still below 10 g/l. The constant productivity of 18.56 g/lxh after 30 hours incubation time despite the increasing biomass concentration up to 20 g/l after 350 hours seems to indicate a decreasing metabolic performance of the biomass.

The glucose concentration was kept constant at about 5 g/l during the complete incubation time (Ma et al. 2022).

4.5.3 Lactic Acid Based on Methane

A breakthrough technology has been pursued by Calysta who elucidated the aerobic fermentative production of lactic acid based on methane. The project has been co-funded and jointly be pursued with NatureWorks, the biggest producer of PLA and thereby a major consumer of lactic acid. The project shall demonstrate the capability of Calysta's proprietary biological "gas-to-chemicals synthetic biology platform" to create new manufacturing pathways using methane, started in 2013 and achieved proof of concept in 2014 (Chemical Fibers International 2013; Calysta Press Release 2014). About 1 $ million was invested by NatureWorks in a laboratory at its global headquarter in 2016 and an investment of about 50 $ million envisaged for the construction of a pilot plant (Chemical Fibers International 2016). The overall statement of the reaction can be summarized as follows:

$$3\,CH_4 + 3\,O_2 \rightarrow CH_3\text{–}CHOH\text{–}COOH + 3\,H_2O$$

The application of methane as fermentation feed would not primarily be a milestone in the synthesis of lactic acid though industrial fermentation in general. Oxygen would not exclusively function as an electron acceptor in the complete oxidation of glucose in the respiratory chain though also be engaged as an oxygen source in the initial conversion of methane. The pathway might be designed to proceed via methanol, formaldehyde (i.e. the biochemical intermediates with respective oxidation levels), and methylene-THF to finally generate an oxygenated carbohydrate for overall metabolism (Kang and Lee 2016). However, since 2014 no further progress was reported and Calysta seems to focus on its proprietary FeedKind® technology for producing nutritional protein based on methane (Calysta 2021).

4.5.4 Digression: Lactic Acid Based on Carbon Dioxide and Light in *Synechocystis* sp.

The experiments with *Synechocystis sp.* as innovative host pursue the use of sunlight as an abundant and cheap energy source and carbon dioxide as a carbon source for producing fuels or valuable chemicals. The host is supposed to filter carbon dioxide out of the air and to use sunlight for the fixation and reduction of carbon dioxide via the Calvin-Benson cycle. The technology is occasionally referred to as "photo-fermentation" and subsumed under the header of third generation fermentation technology (Hellingwerf and Teixeira de Mattos 2009). It would couple the photosynthetic machinery for carbon fixation applied by plants and photosynthetic bacteria with the production know-how for fermentation. Metabolism would be redirected from the buildup of carbohydrate storage to the respective target product via metabolic engineering. Lactic acid is an exemplary product to demonstrate proof of concept, and also other products like butanol are pursued in academia and the industry (Lan and Liao 2012). The theoretical target reaction, without inclusion of the unavoidable production of biomass and the generation of byproducts, can be stated as follows:

$$3\,CO_2 + 3\,H_2O \rightarrow CH_3\text{–}CHOH\text{–}COOH + 3\,O_2$$

The corresponding reaction enthalpy of the endothermic reaction is estimated according to Hess's law:

$$\Delta H^{\circ}_{R} = \Sigma\Delta H^{\circ}_{P} - \Sigma\Delta H^{\circ}_{E} = -675.21\ \text{KJ/mol} - (3x - 393.5\ \text{KJ/mol} + 3x - 285.8\ \text{KJ/mol}) = 1362.6\ \text{KJ/mol}$$

The species *Synechocystis sp.* strain PCC6803 is recommended for R&D projects because it is characterized by a high stability for batch culturing, natural transformability, and does not show lactate formation. With respect to lactic acid as target product, it could be shown that the expression of the lactate dehydrogenase from *B. subtilis* into *Synechocystis sp.* strain PCC6803 triggers the production of lactic acid (Angermayr et al. 2012). The lactate dehydrogenase usually applies NADH as cofactor for the reduction of the keto-function of pyruvate. Since cyanobacteria derive electrons via the electrolysis of water and store them in the form of NADPH the availability of the needed cofactor needs to be addressed. The lactate dehydrogenase of *B. subtilis* is reported to use NADH and NADPH, though the NADPH share is neglected given the much higher K_M values of 13 μM for NADH and 288 μM for NADPH (Angermayr et al. 2012). Thus, a transhydrogenase is needed to partially convert NADPH into NADH and thereby to supply the lactate dehydrogenase. The co-expression of a non-energy linked soluble transhydrogenase leads indeed to an enhanced production of lactic acid. The best-performing engineered strain SAA017 of *Synechocystis sp.* transformed with the *lactate dehydrogenase* (*ldh* gene) from *B. subtilis* and the soluble transhydrogenase (*sth* gene) from *P. aeruginosa* achieved a maximum activity (V_{Max}) of 0.24 mmol lactic acid per g DW (dry weight) per h which corresponds to 22 mg lactic acid per g DW per h. In case it is hypothetically assumed that the biomass concentration in a potential industrial high-volume fermentation would amount to 20 g DW per l a virtual productivity of 0.4 g/lxh would be derived. This would be much below a reasonable commercial threshold of about 2–4 g/lxh. It is assumed that this limitation can partially be overcome in case the intracellular pyruvate concentration could be raised via modifications of the central metabolic pathways (Angermayr et al. 2012).

The best performing strain SAA017 of *Synechocystis sp.* achieved a final concentration of 3.2 mM or 0.29 g/l after 14 days. This titer would need to be increased for a commercially interesting technology, though the impact of the acidity needs to be considered. Inhibitory effects of lactic acid on the growth rate of wild-type *Synechocystis* were observed at concentrations above 100 mM or 9 g/l (Angermayr et al. 2012). The economic viability will depend to a great extent on the light input of the photo-fermentation reactor. It is assumed that a commercial technology would use electricity to provide the light, and the option of open fermentation ponds with sunlight as energy source is not considered (Shuba and Kifle 2018). According to the reaction equation, the minimum energy input of the endothermic reaction amounts to 4.2 MWh per t lactic acid based on the reaction enthalpy of 1362.58 KJ/mol (1362.6 KJ/mol/3.6 GJ/MWh/90 g/mol). For the estimation of the actual power demand, two factors need to be considered. In the first instance, the efficiency in the conversion of power into light might be estimated at 35% which implies that the minimum energy input increases to 12 MWh per t lactic acid. Additionally, the exposure of the cells will imply a loss of photons such as by diffusion and an assumed efficiency of 50% would further raise the minimum power demand up to 24 MWh per t lactic acid without considering photosynthesis and pathway metabolism. The rough estimation shows the challenging economic attractiveness of such a technology in case power cost of 50 €/MWh is assumed. Alone, the minimum power cost of such a photosynthetic fermentation technology would be in the range of the current market value of lactic acid.

4.6 Production Cost and Market Price

The price for lactic acid is reported at 1300 $/t up to 2300 $/t, reflecting the usual market fluctuation (Rosales-Calderon and Arantes 2019), which translates into a minimum price of about 1200 €/t (1.1 $/€). Given an estimated long-term global glucose price of 350 €/t (Kölbl 2021) and a unit consumption rate of 1.13, the glucose cost would amount to about 400 €/t, which seems reasonable given the lowest market price of about 1200 €/t lactic acid. The target production costs of lactic acid for further development of PLA downstream applications are indicated at 800 USD/t, which shall allow for a reduction of the current industrial PLA market price of 2.200 USD/t by about 50%. At a price level of about 1.100 USD/t PLA a full competition with fossil-based plastics is expected (Okano 2014). However, the high carbon yield of 90% of the theoretical maximum does not allow a significant further reduction in the production cost of lactic acid.

4.7 Industrial Application and Capacity

The global consumption of lactic acid is estimated at 518 kt/pa in 2017, growing to 681 kt/pa in 2023. The biggest producers of lactic acid are Corbion, Cargill/NatureWorks, Galactic/Anhui COFCO Biochemical, and Henan Jindan (Chemical Week 2018).

4.7.1 Corbion

Corbion (formerly Purac) started historically as a sugar producer and entered the L-lactic acid market in the period of 1968–1984 (Corbion 2021). Corbion has a total capacity of 225 kt/pa lactic acid with plants in Brazil, Thailand, and the United States. The product portfolio comprises sodium-, potassium, and calcium-lactate, ethyl-lactate, and polymers as well. Corbion's technology base has been broadened with the acquisition of D-lactic acid technology from Myriant. Corbion has developed a PLA polymerization process jointly with Sulzer (Sulzer 2018). The process is designed as a continuous polymerization using the Sulzer Chemtech's proprietary SMR™ technology and equipment. It uses static mixer technology and allows the flexible production of different PLA grades with different molecular weights due to low residence times. Sulzer has been running a 1000 t/pa PLA pilot plant since 2012. On this basis, Corbion pursued a forward integration into the lactic acid dimer - lactide - in the first step and has built a production facility for lactide in Thailand. The company then decided in 2014 to move one step further down the value chain and entered the market for PLA (c&en 2014) and constructed a 75 kt/pa PLA plant in Rayong (Thailand), which is run as a 50/50 joint venture of Corbion and Total (Laird 2014). Corbion declared in 2020 that an additional PLA plant with a capacity of 100 kt/pa together with Total is planned in Grandpuits (France) with startup in 2024 (Corbion 2020).

4.7.2 Cargill/NatureWorks

Cargill's lactic acid business started internally and has been strengthened with the acquisition of the respective businesses of ADM and Malladi specialties in 2015. A main share of Cargill's lactic acid production in the US was transferred to the JV NatureWorks, the largest producer of PLA globally. NatureWorks was founded in 2002 and was originally a JV of Cargill and Dow and is now a JV of Cargill and PTT Global Chemical after Dow sold its share to PTT Global Chemical. NatureWorks was the first PLA producer on a commercial scale, with operations starting in 2002. The PLA plant of NatureWorks in Blair (Nebraska) has a capacity of 150 kt/pa. Capacities shall be increased by about 10% based on an improved lactide purification process (Kunststoff Information 2021). NatureWorks declared that it is planned to build a new PLA plant with a 75 kt/pa capacity in Malaysia with startup scheduled for 2024. The plant includes the production of lactic acid and lactide and would be the first fully integrated PLA plant (Natureworks 2021). An overall investment of 600 $m is planned (Chemical Week 2018).

4.7.3 Galactic/Anhui COFCO Biochemical

Galactic and Anhui COFCO Biochemical, which have partnered up in the production of lactic acid, own a capacity of 50 kt/pa with plants in Belgium and China. COFCO is since 2018 running a PLA plant in Changchun, China, which is using the proprietary PLA neo technology licensed by ThyssenKrupp. ThyssenKrupp developed the polymerization process of lactic acid based on its broad engineering competencies and markets the technology via licenses and engineering services. The Chinese producer Henan Jindan has a capacity of 100 kt/pa.

References

Abedi, E. and Hashemi, S.S.B. (2020). Lactic acid production – producing microorganisms and substrate-sources – state of art. *Heliyon* 6: e049974.

Ahn, J.H., Seo, H., Park, W. et al. (2020). Enhanced succinic acid production by Mannheimia employing optimal malate dehydrogenase. *Nature Communications* 11: 1970. https://doi.org/10.1038/s41467-020-15839-z.

Alexandri, M., Schneider, R., Mehlmann, K., and Venus, J. (2019). Recent advances in D-lactic acid production from renewable resources: Case studies on agro-industrial waste streams. *Food Technology and Biotechnology* 57 (3): 293–304. https://doi.org/10.17113/ftb.57.03.19.6023.

Angermayr, S.A., Paszota, M., and Hellingwerf, K.J. (2012). Engineering a cyanobacterial cell factory for production of lactic acid. *Applied and Environmental Microbiology* 78 (19): 7098–7106.

Balla, E., Daniilidis, V., Karlioti, G. et al. (2021). Poly(lactic Acid): A versatile biobased polymer for the future with multifunctional properties – from monomer synthesis, polymerization techniques and molecular weight increase to PLA applications. *Polymers* 13: 1822. https://doi.org/10.3390/polym13111822.

Binczarski, M., Berlowska, J., Stanishevsky, A., and Witonska, I. (2016). Biologically synthesized calcium lactate as a substrate for propylene glycol production. *RSC Advances* 6: 92420. https://doi.org/10.1039/c6ra20722g.

Boontawan, P., Kanchanathawee, S., and Boontawan, A. (2011). Extractive fermentation of l-(+)-lactic acid by *Pediococcus pentosaceus* using electrodeionization (EDI) technique. *Biochemical Engineering Journal* 54: 192–199.

Bozell, J.J. and Petersen, G.R. (2010). Technology development for the production of biobased products from borefinery carbohydrates – the US department of Energy's "Top 10" revisited. *Green Chemistry* 12: 539–554.

c&en (2014). Corbion plans to be second polylactic acid producer. Magazine issue 10.11.2014, p. 14.

Calysta (2021). Calysta.com/our-technology/ (accessed 23 December 2021).

Calysta Press Release (2014). Calysta demonstrates lab-scale production of lactic acid from methane – Completes successful first stage of multi year R&D development collaboration with NatureWorks. www.calysta.com (14 June 2014, accessed 29 December 2020).

Cargill Press Release (2020). Cargill, IFPEN and Axens collaborate to advance lactic-to-acrylic-acid technology, 03 Dec 2020. Source: https://www.cargill.com/2020/cargill-ifpen-axens-collaborate-to-advance (accessed 3 February 2023).

Chae, T.U., Ko, Y.-S., Hwang, K.-S., and Lee, S.Y. (2017). Metabolic engineering of *Escherichia coli* for the production of four-, five-, and six-carbon lactams. *Metabolic Engineering* 43: 82–91. https://doi.org/10.1016/y.mben.2017.04.001.

Chemical Fibers International (2013). NatureWorks: R&D collaboration for PLA building block, 63: 3, p. 148; http://www.chemical-fibers.com

Chemical Fibers International (2016). NatureWorks: Development of methane to lactic acid conversion technology, June 2016, 66:2, p. 74; http://www.chemical-fibers.com

Chemical Week (2018). Lactic acid to grow 5.6%/year. https://chemweek.com/document/show/Chemweek/99002/Lactic-acid-use-to-grow-56year?connectPath=&searchSessionId=143fe1e9-08c8-4bae-a517-628c17cdff0d (accessed 28 July 2023).

Corbion press release (2020). Corbion and Total to build first world-scale PLA plant in Europe, 24.9.2020, https://www.globenewswire.com/news-release/2020/09/24/2098442/0/en/Corbion-and-Total-to-build-first-world-scale-PLA-plant-in-Europe.html (accessed 28 July 2023).

Corbion (2021). www.corbion.com/about-corbion/our-history (accessed 9 January 2021).

Dhakal, R., Bajpai, V.K., and Baek, K.-H. (2012). Production of GABA (γ-aminobutyric acid) by microorganisms: a review. *Brazilian Journal of Microbiology* 1230–1241. ISSN 1517-8382.

Dusselier, M., Van Wouwe, P., Dewaele, A. et al. (2013). Lactic acid as a platform chemical in the biobased economy: the role of chemocatalysis. *Energy and Environmental Science* 6: 1415–1442.

Flamholz, A., Noor, E., Bar-Even, A., and Milo, R. (2012). eQuilibrator – the biochemical thermodynamics calculator. *Nucleic Acids Research* 40 (database issue): D770–D775. It should be mentioned that the published free Gibbs energy differs by source pending on the chosen boundary conditions, e.g. a standard textbook indicates a value $\Delta G'^\circ = -196$ KJ/mol (Voet, 2016).

Folch, P.L., Bisschops, M.M.M., and Weusthuis, R.A. (2021). Metabolic energy conservation for fermentative product formation. *Microbiol Biotechnology* 14 (3): 829–858. https://doi.org/10.1111/1751-7915.13746.

Food Ingredients (2010). Cargill awarded for innovation in lactic acid production, 24 March 2010. https://www.foodingredientsfirst.com/news/cargill-awarded-for-innovation-in-lactic-acidproduction.html#:~:text=The%20innovation%20is%20a%20yeast%20engineered%20by%20Cargill,to%20the%20process%20to%20extract%20the%20lactic%20acid (accessed 29 December 2020).

Freeman, A., Woodley, J.M., and Lilly, M.D. (1993). *In situ* product removal as a tool for bioprocessing. *Nature* 11: 1007–1012. The paper is the first publication on *in situ* product recovery in fermentation processes.

Grabar, T.B., Zhou, S., Shanmugam, K.T. et al. (2006). Methylglyoxal bypass identified as source of chiral contamination in L-(+) and D-(-)-lactate fermentations by recombinant *Escherichia coli*. *Biotechnology Letters* 1527: 1527–1535.

Hábová, V., Melzoch, K., Rychtera, M., and Sekarová, B. (2004). Electrodialysis as a useful technique for lactic acid separation from a model solution and a fermentation broth. *Desalination* 162: 361–372.

Hellingwerf, K.J. and Teixeira de Mattos, M.J. (2009). Alternative routes to biofuels: light-driven biofuel formation from CO_2 and water based on the "photanol" approach. *Journal of Biotechnology* 142: 87–90.

Kang, T.J. and Lee, E.Y. (2016). Metabolic versality of microbial methane oxidation for biocatalytic methane conversion. *Journal of Industrial and Engineering Chemistry* 35: 8–13.

Kim, J.K., Iannotti, E.L., and Bajpai, R. (1999). Extractive recovery of products from fermentation broths. *Biotechnology and Bioprocess Engineering* 4: 1–11.

Kölbl, T. (2021). German Corporate Conference, UniCredit/Kepler Cheuvreux, Südzucker company presentation, 18.-20.1.2021, p. 61, www.suedzucker.de/de/investor-relations/publikationen (accessed 2 March 2021).

Kubicek, C.P. (2001). Organic acid. In: *Basic Biotechnology*, 2e (ed. C. Ratledge and B. Kristiansen), 305–324. Cambridge University Press.

Kunststoff Information (2021). NatureWorks: Errichtung einer PLA-Anlage in Thailand. https://www.kunststoffweb.de/branchen-news/natureworks_errichtung_einer_pla-anlage_in_thailand_t247786 (accessed 28 July 2023).

Laird, K. (2014). Corbion Purac to enter the PLA arena, 5.11.2014. https://www.plasticstoday.com/content/corbion-purac-enter-pla-arena/694160721428 (accessed 28 December 2020).

Lan, E.I. and Liao, J.C. (2012). ATP drives direct photosynthetic production of 1-butanol in cyanobacteria. *PNAS* 109 (16): 6018–6023. https://doi.org/10.1073/pnas.1200074109.

Lee, S., Ahn, J., Kim, Y.-G. et al. (2013). Gamma-aminobutyric acid using immobilized glutamate decarboxylase followed by downstream processing with cation exchange chromatography, International Journal of Molecular. *Science* 14: 1728–1739. https://doi.org/10.3390/ijms14011728.

Liebal, U.W., Blank, L.M., and Ebert, B.E. (2018). CO_2 to succinic acid – Estimating the potential of biocatalytic routes. *Metabolic Engineering Communications* 7: https://doi.org/10.1016/j.mec.2018.e00075.

Ma, K., Cui, Y., Zhao, K. et al. (2022). D-Lactic acid production from agricultural residues by membrane integrated continuous fermentation coupled with B vitamin supplementation. *Biotechnology for Biofuels and Bioproducts* 15: 24. https://doi.org/10.1186/s13068-022-02124-y.

van Maris, A.J.A., Konings, W.N., van Dijken, J.P., and Pronk, J.T. (2004). Microbial export of lactic and 3-hydroxypropionic acid: implications for industrial fermentation processes. *Metabolic Engineering* 6: 245–255. https://doi.org/10.1016/j.ymben.2004.05.001.

Metz, A., Hoffmann, A., Hock, K., and Herres-Pawlis, S. (2016). Katalysatoren für die Produktion von Biokraftstoffen. *Chemie in unserer Zeit* 50: 1–11.

Musashino Corporate Profile (2020). www.musashino.com (accessed 28 December 2020).

Natureworks (2021). www.natureworks.com, Natureworks announces key milestone for global manufacturing expansion with 75 ktTa facility for producing Ingeo polymer. https://www.natureworksllc.com/News-and-Events/Press-Releases/2021/2021-06-02-Thailand-Ingeo-PLA-plant-announcement (accessed 12 June 2021).

Okano, K. (2014). Lactic acid. In: *Bioprocessing of Renewables Resources to Commodity Bioproducts* (ed. V. Bisaria and A. Kondo), 353–380. Wiley.

Okino, S., Suda, M., Fujikura, K. et al. (2008). Production of D-lactic acid by *Corynebacterium glutamicum* under oxygen deprivation. *Applied Microbiology and Biotechnology* 78: 449–454. https://doi.org/10.1007/s00253-007-1336-7.

Pang, Y., Lee, C., Vlaisavljevich, B. et al. (2022). Multifunctional amine modifiers for selective dehydration of methyl lactate to acrylates. *JACS* https://doi.org/10.1021/jacsau.2c00513.

Pinske, C. and Sargent, F. (2016). Exploring the directionality of *Escherichia coli* formate hydrogenlyase: a membrane-bound enzyme capable of fixing carbon dioxide to organic acid. *Microbiology Open* 5 (5): 721–737.

Pörtner, R., Faschian, R., and Goelling, D. (2016). Fermentation of lactic acid bacteria: State of the art and new perspectives. In: *Applied Biocatalysis. From fundamental science to industrial applications* (ed. L. Hilterhaus, A. Liese, U. Kettling, and G. Antranikian), 317–342. Wiley-VCH.

Pothakos, V., Deeber, V., Debonne, I. et al. (2018). Fermentation titer optimization and impact on energy and water consumption during downstream processing. *Chemical Engineering and Technology* 41 (12): 2358–2365.

Pranaw, K., Dutta, D., Singh, S., and Khare, S.K. (2021). Lactic acid bacteria for production of platform chemicals: A dark horse in the field of industrial biotechnology. In: *Advances in the Domain on Environmental Biotechnology. Microbiological Developments in Industry, Wastewater Treatment and Agriculture* (ed. N.R. Maddela, L.C.G. Cruzatty, and S. Chakraborty), 3–26. Springer.

Rosales-Calderon, O. and Arantes, V. (2019). A review on commercial-scale high-value products that can be produced alongside cellulosic ethanol. *Biotechnology for Biofuels* 12: 240. https://doi.org/10.1186/s13068-019-1529-1.

Rosenboom, J.-G., Langer, R., and Traverso, G. (2022). Bioplastics for a circular economy. *Plastics* 7: 117–137. https://doi.org/10.1038/s41578-021-00407-8.

Schuchmann, K. and Müller, V. (2016). Energetics and application of heterotrophy in acetogenic bacteria. *Applied and Environmental Microbiology* 82 (14): 4056–4069. https://doi.org/10.1128/AEM.00882-16.

Shuba, E.S. and Kifle, D. (2018). Microalgae to biofuels: 'promising' alternative and renewable energy. *Renewable and Sustainable Energy Reviews* 81: 743–755. https://doi.org/10.1016/j.rser.2017.08.042.

Sulzer (2018). Process Technology. https://www.sulzer.com/-/media/files/products/process-techology/processes-and-applications/brochures/process_plants.ashx?la=en (accessed 6 January 2022).

Thyssen Krupp Industrial Solutions (2016). Planeo Polylactide Technology. https://ucpcdn.thyssenkrupp.com/legacy/UCPthyssenkruppBAIS/assets.files/products_services/chemical_plants_processes/polymer_plants/brochure_pla.pdf (accessed 9 January 2021).

Tian, X., Liu, X., Zhang, Y. et al. (2021). Metabolic engineering coupled with adaptive evolution strategies for the efficient production of high-quality L-lactic acid by *Lactobacillus paracasei*. *Bioresource Technology* 323: 124549.

Van Wouwe, P., Dusselier, M., Vanleeuw, E., and Bert, S. (2016). Lactide synthesis and chirality control for polylactide acid production. *ChemSusChem* 9: 907–921.

Voet, D., Voet, J.G., and Pratt, C.W. (2016). *Principles of Biochemistry*. Wiley p. 501, 520.

Wang, W. and Yang, X. (2020). Computational prediction of chiral iron complexes for asymmetric transfer hydrogenation of pyruvic acid to lactic acid. *Molecules* 25: 1892. https://doi.org/10.3390/molecules25081892.

Wojcieszak, R., Bonnotte, T., Paul, S. et al. (2020). Lactic acid conversion to acrylic acid over fluoride-substituted hydroxyapatites. *Frontiers in Chemistry* 8: article 421, https://doi.org/10.3389/fchem.2020.00421.

Xu, K. and Xu, P. (2014). Efficient calcium lactate production by fermentation coupled with crystallization-based *in situ* product removal. *Bioresource Technology* 163: 33–39.

Yang, J.E., Choi, S.Y., Shin, J.H. et al. (2013). Microbial production of lactate-containing esters. *Microbial Biotechnology* 6: 621–636.

Yang, X., Lai, Z., Lai, C. et al. (2013). Efficient production of L-lactic acid by an engineered *Thermoanaerobacterium aotearoense* with broad substrate specificity. *Biotechnology for Biofuels* 6 (124): http://www.biotechnoloyforbiofuels.com/content/6/1/124.

Yankow, D. (2022). Fermentative lactic acid production from lignocellulosic feedstocks: From source to purified product. *Frontiers in Chemistry* 10: https://doi.org/10.3389/fchem.2022.823005.

Zhao, W., Huang, J., Lv, C. et al. (2015). pH stabilization of lactic acid fermentation via the glutamate decarboxylation reaction: Simultaneous production of lactic acid and γ-aminobutyric acid. *Process Biochemistry* 50: 1523–1527.

Zhou, L., Cui, W., Liu, Z., and Zhou, Z. (2016). Metabolic engineering strategies for D-lactate over production in *Escherichia coli*. *Journal of Chemical Technology and Biotechnology* 91: 576–584.

Zhu, Y., Eiteman, M.A., DeWitt, K., and Altman, E. (2007). Homolactate fermentation by metabolically engineered *Escherichia coli* strains, Applied. *Environmental Microbiology* 73: 456–464.

5
Alanine

5.1 Application

L-Alanine is a non-essential amino acid with the formula $C_3H_7NO_2$ and a molecular weight of 89.1 g/mol. It is used as a chemical building block, feed, and food additive. L-alanine is the only amino acid with a sweet taste (Solms et al. 1965). The product is also applied to pharmaceuticals, e.g. as the main ingredient in Travasol®. An important application of L-alanine is its use as a raw material for producing methylglycine-diacetic acid (MGDA), which is an artificial chelating agent (Liu et al. 2022). MGDA is part of the formulation of several cleaning products and complexes bivalent cations. Given its two functional groups L-alanine is also evaluated as a monomer for tailored thermoplastics (Mallakpour and Dinari 2011; Mallakpour and Behranvand 2014). The worldwide production volume was estimated at 500 t pa in 2002 (Ideka 2003), and it is estimated that the production volume will increase to about 50 kt pa in 2022 (Liu et al. 2022).

It has recently been discovered that acetogenic bacteria (i.e. *Acetobacterium woodii*) are able to use alanine as their sole carbon and energy source. L-Alanine is discussed in this context as an alternative feedstock for fermentation processes that can easily be channeled into the main metabolism via the reversible alanine dehydrogenase (*alaD*). Pyruvate would then be converted into acetyl-CoA via pyruvate: ferredoxin-oxidoreductase (*nifJ*). The obtained acetyl-CoA would, via acetyl-phosphate, finally be converted into acetate. The released carbon dioxide would be recycled via the Wool-Ljungdahl pathway to acetate, such that the following net reaction formula is proposed (Döring and Müller 2018):

$$2\,\mathrm{HOOC{-}CH(NH_2){-}CH_3} + 2\,\mathrm{H_2O} \rightarrow 3\,\mathrm{CH_3COOH} + 2\,\mathrm{NH_3}$$

However, the production costs of L-alanine are likely too high to compete with conventional fermentation feedstocks like glucose or sucrose. It remains to be seen if new application fields are established in the future.

5.2 Chemical Production of L-alanine

Adolph Strecker first created alanine while attempting to create lactic acid, but ended up with alanine instead. The Strecker synthesis with acetaldehyde, cyanic acid, and ammonia

Pathway Design for Industrial Fermentation, First Edition. Walter Koch.

is in principle a feasible alanine technology, though the reported yield of 50% is rather low (Becker et al. 1976):

$$CH_3CHO + HCN + NH_3 \rightarrow H_2N\text{-}CH(CH_3)\text{-}CN + H_2O$$
$$H_2N\text{-}CH(CH_3)\text{-}CN + 2\,H_2O \rightarrow H_2N\text{-}CH(CH_3)\text{-}COOH + NH_3$$

It is to be pointed out that alanine has a central and chiral C-atom, which implies the distinction between an L- and D-amino acid. This is a major disadvantage of the Strecker synthesis, which is not designed to provide stereoselectivity. Glycine, as the easiest amino acid, defines an exception since the central C-atom harboring two hydrogen atoms does not allow chirality. Accordingly, glycine is very well suited for the chemical Strecker synthesis since downstream efforts to separate the racemic mixture of an L- and D-amino acid are not required. Thus, glycine is produced by chemical means with an indicated conversion yield of 78% (D'Este et al. 2018). Formaldehyde, cyanic acid, and ammonia are reacted in the first step to form α-aminoacetonitrile, which is then converted into glycine in the second step:

$$HCHO + HCN + NH_3 \rightarrow H_2N\text{-}CH_2\text{-}CN$$
$$H_2N\text{-}CH_2\text{-}CN + 2\,H_2O \rightarrow H_2N\text{-}CH_2\text{-}COOH + NH_3$$

The Strecker technology may be applied to derive L-alanine through a kinetic resolution, but an additional step would need to be pursued (Grunwald 2018). In the first instance, the amino-function would be acetylated to obtain a mixture of L-*N*-acetylalanine and D-*N*-acetylalanine, which would then, in the second step, be enzymatically reacted with a selective hydrolase, cleaving only the acetyl-moiety of L-*N*-acetylalanine. D-*N*-acetylalanine must not be hydrolyzed, and L-alanine and D-*N*-acetylalanine would then be separated. The efficiency of a kinetic resolution depends on the discrimination selectivity of the hydrolase, the ease of separating the hydrolyzed isomer from the non-hydrolyzed isomer, and the efforts to recycle the non-hydrolyzed isomer. However, the Strecker synthesis does not have commercial relevance for the synthesis of L-alanine. Instead, the enzymatic decarboxylation of L-aspartic acid and the fermentative production based on glucose are the most important technologies (Schäfer 2007). Amino acids may also be derived by extraction from protein hydrolysates. Hereby, natural protein sources, including hair, de-oiled rice bran, or feathers, might be treated with harsh reagents like, e.g. hydrochloric acids or sodium hydroxide to split the peptide bonds of the protein backbone and generate a mixture of amino acids. But for the amino acid cysteine, extraction from natural hydrolysates matters more than for alanine (D'Este et al. 2018).

5.3 Pathway Design

The biochemical pathway towards L-alanine is very easy since suitable enzymes for the direct conversion of the main pathway intermediate pyruvate towards L-alanine are available. Here, the integration of nitrogen is interesting, and two distinct reaction modes need to be distinguished. Pyruvate might be converted into L-alanine via alanine aminotransferase (also called glutamic acid-pyruvic acid transaminase). According to a typical biochemical transamination, the amino function is transferred by an amino acid to be converted into an α-keto acid, i.e. L-glutamate would be employed as a nitrogen donor to generate α-ketoglutarate, and pyruvate would function as a nitrogen acceptor:

$$CH_3\text{–}CO\text{–}COOH + HOOC\text{–}CH_2\text{–}CH_2\text{–}CH(NH_2)\text{–}COOH \\ \rightarrow CH_3\text{–}CH(NH_2)\text{–}COOH + HOOC\text{–}CH_2\text{–}CH_2\text{–}CO\text{–}COOH$$

The alanine aminotransferase has high importance in clinical biochemistry since it is a useful indicator for liver diseases (Liu et al. 2014). The enzyme is also used for the biocatalytic conversion of acetophenone into 1-phenylethylamine (Grunwald 2018). With respect to metabolic engineering strategies, the alanine aminotransferase might be disadvantaged given the efforts to provide L-glutamate as the nitrogen carrier needed in stoichiometric amounts. L-glutamate would not be added as feed, though cytosolic recycling would need to be installed to avoid stoichiometric amounts of α-ketoglutarate as by-product. It is also to be considered that *Escherichia coli* possesses three L-alanine aminotransferases, encoded by *alaA*, *alaC*, and *avtA*, which connect the major biochemical pathways of glycolysis, oxidative pyruvate decarboxylation, and nitrogen metabolism towards aspartate and glutamate (Peña-Solar et al. 2014). Metabolic engineering would likely interfere with the native regulation pattern and cause imbalances in essential pathways.

Members of the hyperthermophilic order *Thermotogales* were found to ferment glucose into L-alanine, and this ability was proposed as an ancestral metabolic characteristic. The conversion of pyruvate towards L-alanine is assumed to be facilitated by transamination with the alanine aminotransferase. Though the academic work was primarily interested in taxonomy and the chosen metabolic pathway was not further elucidated (Ravot et al. 1996).

Alternatively, pyruvate might be aminated with free ammonium to obtain L-alanine via the alanine dehydrogenase. This step does not require a biochemical nitrogen carrier, though it allows the direct addition of ammonia into the medium:

$$CH_3\text{–}CO\text{–}COOH + NH_3 + NADH + H^+ \rightarrow CH_3\text{–}CH(NH_2)\text{–}COOH + H_2O + NAD^+$$

The reaction converts the keto-function of pyruvate at the C2-position in a -$HCNH_2$- moiety, applying NADH as an electron donor and the C-atom as an electron acceptor. Formally a reductive amination is pursued. The overall reaction pathway can be stated as following (Table 5.1):

Table 5.1 Alanine via direct amination.

Glucose + 2 ADP + 2 P_i + 2 NAD^+ → 2 Pyruvate + 2 ATP + 2 NADH + 2 H^+ + 2 H_2O	Glycolysis
2 Pyruvate + 2 NH_3 + 2 NADH + 2 H^+ → 2 L-alanine + 2 NAD^+ + 2 H_2O	Alanine dehydrogenase
Glucose + 2 NH_3 + 2 ADP + 2 P_i → 2 L-alanine + 2 ATP + 4 H_2O	Net reaction

The generation of two mol of ATP concurrent with the generation of two mol of alanine is reasonable from a thermodynamic point of view since the reaction has a sufficiently negative Gibbs free energy (Folch et al. 2021):

$$\text{Glucose} + 2\,NH_3 \rightarrow 2\,\text{L-alanine} + 2\,H_2O \quad \Delta G^{\circ\prime} = -209\ \text{KJ/mol glucose}$$

The generation of ATP from ADP and Pi requires a Gibbs free energy of 30.5 KJ/mol. Since the standard conditions chosen for the calculation of this value are often not present in biochemical systems, it is assumed that under realistic conditions, a Gibbs free energy of about 50–60 KJ/mol is required (Flamholz et al. 2012). The generated surplus of ATP allows cell growth and provides energy for cell maintenance. A balance of reduction equivalents is given, and oxygen supply into the medium is not obvious to allow regeneration of NAD^+. One mol of glucose provides two mol of L-alanine, translating into a theoretical maximum carbon yield of 0.99 t L-alanine/t glucose ($2 \times 89.1/180.1$). This translates into a glucose consumption rate of 1.01 t glucose per ton L-alanine. After the metabolic pathway has been designed and implemented with molecular biology, the potential for racemization needs to be monitored. In many eubacteria, peptidoglycan is a critical component of the cell wall backbone, and D-alanine is a building block of peptidoglycan and needed for the formation of peptidoglycan crosslinks to maintain cell integrity (Marshall et al. 2019). Thus, most eubacteria have an alanine racemase encoded by the *alr* gene, which converts L-alanine into D-alanine. This gene might need to be silenced in order to allow a high molar yield without compromising cell wall integrity. Racemization does not appear to be a significant problem, according to the majority of academic small-scale fermentation studies that have been published (Zhang et al. 2007). The metabolic pathway towards L-alanine based on glucose as carbon and energy feed proceeds via conventional glycolysis (Figure 5.1):

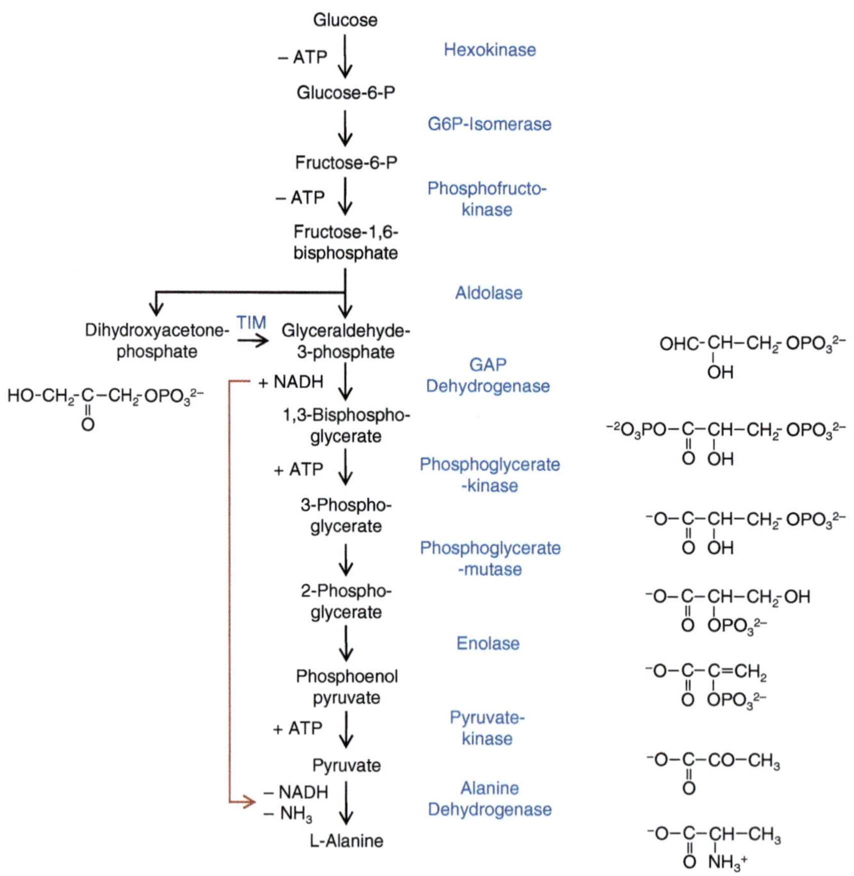

Figure 5.1 Glycolysis with subsequent conversion of pyruvate into L-alanine via the alanine dehydrogenase. Source: Walter Koch.

5.4 Metabolic Engineering

5.4.1 Initial Fermentative Production of L-alanine in *Bacillus subtilis*

The straightforward pathway towards L-alanine via the alanine dehydrogenase has initially been implemented in *Zymomonas mobilis* (Uhlenbusch et al. 1991). The plasmid-based alanine dehydrogenase gene *alaD* of *Bacillus sphaericus* was expressed. Since the gram-negative, ethanologenic bacterium *Z. mobilis* is naturally optimized for ethanol production, and the pyruvate decarboxylase competes with the alanine dehydrogenase for the common substrate pyruvate, the competing allocation of precursor was blocked. For this purpose the cofactor of the pyruvate decarboxylase, thiamin pyrophosphate (TPP), was limited, and a TPP auxotroph mutant of *Z. mobilis* was designed. Thereby, the natural ethanol producer could be remodeled for alanine production. An L-alanine titer of 7.5 g/l after 24 h could be achieved. It was speculated that a further titer increase was hampered by weak secretion efficiency because the intracellular L-alanine concentration was measured at 260 mM (Uhlenbusch et al. 1991).

5.4.2 Fermentative Production of L-alanine in *Escherichia coli*

A more efficient fermentation technology has been developed for the host *E. coli* based on the most straightforward pathway for converting pyruvate via the alanine dehydrogenase (*alaD*) into L-alanine (Zhang et al. 2007). The alanine dehydrogenase of *Geobacillus stearothermophilus* has been chromosomally been integrated into the genome of *E. coli* under the control of the native D-lactate dehydrogenase (*ldhA*) promoter. The competing native D-lactate dehydrogenase (*ldhA*), which might channel pyruvate into the lactate pool, has been knocked out. In order to enhance selectivity and to avoid by-products, respectively, the alanine racemase (*dadX*) and the methyl-glyoxal synthase (*mgsA*) genes were also silenced. The strain provided L-alanine with a chiral purity greater than 99.5%. Strains were grown aerobically for the generation of the inoculum. For fermentative production of L-alanine, cells were grown at 37 °C without antibiotics in neurobasal (NBS) mineral salts medium supplemented with 100 mM ammonium sulfate, 1 mM betaine, and 2% (w/v) glucose. The pH of the broth was controlled with NH_4OH and kept at 7. However, it is not clearly indicated which aeration has been chosen. Since the pathway does not generate a surplus of reduction equivalents and does not suffer an ATP gap, oxygen as the final electron acceptor should not be required. In a review of the fermentative production of alanine, it has been claimed that the respective *E. coli* host XZ132 referred to in the mentioned publication (Zhang et al. 2007) is pursued without oxygen (Liu et al. 2022). After several rounds of process optimization, a maximum titer of 1.3 mol L-alanine/l, or 114 g/l, could be achieved. Given a batch period between 24 and 36 h, a productivity of about 4 g/lxh is reported, which should already be sufficient for commercial maturity. The biomass titer at batch end is reported at 2 g/l corresponding to a specific productivity of 2 g (product)/g (biomass)xh (Zhang et al. 2007). With 120 g/l glucose as feedstock, 1279 mmol L-alanine per l was produced, which indicates that 95% of the theoretical maximum carbon yield was achieved (Zhang et al. 2007).

An alanine exporter was not covered in the metabolic engineering strategy of the *E. coli* host. The export systems were often not the focus of metabolic engineering since most hosts display the natural ability to secrete a specific intermediate in case a defined internal threshold is passed. Though after the identification of the LysE exporter for lysine in *C. glutamicum* (Vrljic et al. 1996), the topic gained attention, and several other exporters have been identified. Also, with respect to *E. coli* some progress could be made. Based on systematic screening of mutants hypersensitive to extracellularly added L-alanyl-alanine, the *ygaW* gene could be identified, which is probably responsible for L-alanine efflux in *E. coli* (Hori et al. 2011). Former experiments derive the conclusion that the exporter functions as a safety valve in the maintenance of the intracellular L-alanine pool and prevents the toxic-level accumulation of L-alanine (Kim et al. 2015).

The fermentative production of L-alanine has been further improved by introducing a thermosensitive promoter for the alanine dehydrogenase. In the first phase, the fermentation is run at 33 °C as an aerobe process for the formation of biomass (up to 50 g biomass/100 g glucose). The ATP generated in glycolysis is available for cell maintenance and growth. The surplus of reduction equivalents is mainly used for the generation of ATP and oxygen as final electron acceptor is available. The expression of alanine dehydrogenase is switched off at this lower temperature. In the second phase, the temperature is raised to 42 °C, and the oxygen supply is limited. The alanine dehydrogenase is activated and the alanine formation is initiated, whereas the formation of biomass is minor. The surplus of reduction equivalents is consumed by the alanine dehydrogenase and not channeled into the respiratory chain phosphorylation. This shrinks the ATP pool, which is restricted to substrate-level phosphorylation and may be completely needed for cell maintenance. The final alanine titer amounted to 120.8 g/l, and a productivity of 3.09 g/lxh (overall process) or 4.18 g/lxh (second phase only) using glucose as the sole carbon source has been achieved. The alanine yield during the second stage with limited oxygen availability achieved 97% of the maximum carbon yield given an output of 96 g L-alanine derived from 100 g glucose (Zhou et al. 2016).

5.4.3 Initial Fermentative Production of L-alanine in *Pediococcus acidilactici*

An outstanding titer and productivity could be achieved with the gram-positive homo-fermentative lactic acid bacterial strain *Pediococcus acidilactici* (Sharma et al. 2021). The cloning of the *alaD* gene was assisted by computational biology, and an auto-inducible P289 promoter was employed for efficient expression. In fed-batch experiments at 37 °C under oxygen depression conditions, an L-alanine titer of 229 g/l after 42 h was measured, corresponding to a productivity of 5.45 g/lxh. The enantiomeric purity was 97%, which indicates that the competing formation of D-alanine could be efficiently suppressed. The downstream processing comprises the removal of biomass and product extraction and purification by crystallization, and 217 g/l L-alanine were recovered (Sharma et al. 2021).

5.4.4 Fermentative Production of L-alanine in *V. natriegens* in a High Cell Density Fermentation

In the last few years, the host *Vibrio natriegens* has been evaluated as a potential industrial host given its high growth rates and outstanding glucose uptake rate. *V. natriegens* is a facultatively anaerobic, non-pathogenic gram-negative γ-proteobacterium that was discovered in 1958 and isolated from salt marsh mud. Growing cells display a glucose uptake rate of 7.81 g (43 mmol) glucose per g dry cell weight (DCW) per h, and resting cells display 1.08 g (6 mmol) glucose per g DCW per h under anaerobic conditions (Thoma and Blombach 2021). In an anaerobic fermentation with resting cells of *V. natriegens*, a productivity of 34 g/lxh could be shown between 0 and 30 min after batch start (Hoffart et al. 2017). The carbon yield amounted to 0.81 t alanine per t glucose, which corresponds to 82% of the maximum theoretical yield of 0.99 t alanine per t glucose. It can be deduced that the average glucose conversion (uptake) rate is 41.9 g glucose per l and h, which translates into a cell density of 38.8 g/l DCW. It needs to be seen whether these promising lab-scale performance results can be applied to a commercial process.

5.4.5 Fermentative Production of D-alanine with *Corynebacterium fascians*

Besides L-alanine, D-alanine might also be produced via fermentation. A strain of *C. fascians* could be isolated that is able to ferment glucose, fructose, or glycerol into D-alanine. The fermentation feed contained 5% glycerol, 4% $(NH_4)_2HPO_4$, and 0.3% corn steep liquor, and after 96 h, a D-alanine titer of 7 g/l was obtained. A productivity of 0.07 g/lxh could be achieved merely by representing the native abilities of the host (Yamada et al. 1973). In case a dedicated strain development program is started, commercially attractive productivity seems realistic.

5.4.6 Biocatalytic (Whole Cell) Production of L-alanine Based on Fumaric Acid

The industrial production of L-alanine relies to some extent on the biocatalytic conversion of fumarate or aspartic acid towards L-alanine. L-aspartate may be sourced on the market, though the use of fumarate provides a cost advantage since aspartate is usually produced via the amination of fumarate (Appleton and Rosentrater 2021). Thus, a biocatalytic two-step approach is recommended:

$$\text{Fumaric acid} + NH_3 \rightarrow \text{L-aspartic acid} \quad \text{Aspartase}$$
$$\text{L-aspartic acid} \rightarrow \text{L-alanine} + CO_2 \quad \text{L-aspartate ß-decarboxylase}$$

The process uses the aspartase activity of immobilized *E. coli* cells and the L-aspartate ß-decarboxylase activity of immobilized *Pseudomonas dacunhae* cells. The decarboxylase is dependent on pyridoxal phosphate (PLP), which needs to be added in small amounts. The minimum concentration is indicated as 0.0001% (w/w) up to 0.0005% (w/w) (EP 0386476 1990). PLP availability in biocatalysis has been identified as an enhancement factor, and process optimization might avoid the requirement of external medium supplementation. However, it is not reported that this lever has already been addressed in the biocatalytic production of L-alanine (Han et al. 2020). Additionally, it was elucidated that the enzyme is allosterically activated by α-keto acids, and therefore pyruvate is added (Takamatsu et al. 1986; Shibatani et al. 1979). In the earlier generations of the technology, L-malic acid (derived from fumaric acid via a fumarase) and D-alanine (via racemization of L-alanine) were identified as competing by-products. Later on, the immobilized cells of *E. coli* and *P. dacunhae* were subjected to a defined pH treatment, and the generation of the by-products could be suppressed (Takamatsu et al. 1985). The immobilization of the cells was supported with κ-carrageenan, which is a suitable polymer for entrapping cells into the gel lattice. Furthermore, hexamethylenediamine (HMD) and glutaraldehyde (GA) are applied as auxiliary chemicals or cross-linking agents, respectively (Çalik et al. 1999). The use of immobilized whole-cell catalysts has the advantage that the enzymes are not subject to a costly isolation and purification process, though they are exclusively overexpressed in the applied host. In addition, compared to an isolated enzyme outside of the native cytosol, the whole-cell biocatalyst may shield the enzymes and restore activity.

Technology development has focused on further process improvements addressing enzyme stability and operational setup. Since both enzymatic steps may display different reaction optimums with respect to pH and temperature, the initial technology was operated as a two-step process. The technology is regarded as the first industrial application of sequential enzyme reactions using two kinds of immobilized enzymes (Takamatsu et al. 1986). Afterwards improvements could be implemented to pursue the two-step reaction in a single reaction tank (EP 0386476 1990). The decarboxylation of L-aspartic acid is further optimized based on screening for alternative decarboxylases. Instead of the standard enzymes from *Alcaligenes faecalis* and *P. dacunhae*, an isomer of *Acinetobacter radioresistens* has been characterized. Structure elucidation with crystal analysis allowed selective replacement of amino acids in the catalytic center, and finally a decarboxylase with a 99.9 conversion yield at pH 6.0 and high optical purity could be constructed (Liu et al. 2021).

5.4.7 Biocatalytic Production of L-alanine

As an alternative to conventional fermentation, a biocatalytic reaction cascade has been proposed (Gmelch et al. 2019). The system employs a modified version of the Entner-Doudoroff (ED) pathway, which avoids ATP as an activating agent (Ahmed et al. 2005) and converts glucose in the first step with the glucose dehydrogenase into gluconate. Contrary to earlier work for the biocatalytic generation of isobutanol (Guterl et al. 2012), the glucose dehydrogenase was chosen from *Bacillus subtilis*, which is 14x more active towards glucose than the previously applied isoenzyme from *Sulfolobus solfataricus*. In the next step, gluconate is dehydrated to 2-Keto-3-deoxy-gluconate (KDG). After several trials and solubility issues with tested archaeal dehydratases, the dihydroxyacid dehydratase (DHAD) from *Caulobacter crescentus* was chosen. The split of KDG into pyruvate and glyceraldehyde is achieved by an aldolase isolated from *Picrophilus torridus*. Furthermore, the alanine dehydrogenase was selected from *Archaeoglobus fulgidus* and the aldehyde dehydrogenase (ALDH) from *Methanocaldococcus jannaschii* (Figure 5.2).

The reaction cascade does not apply ATP since the usual activation of glucose via the formation of phosphoric acid ester bonds with the hydroxyl groups of the sugar is bypassed. Since every ATP regeneration approach would imply at least a minor loss, the addition of fresh ATP (e.g. acetyl-phosphate or polyphosphate) would compromise the cost efficiency of the biocatalytic technology, although the reloading of ADP might be pursued with acetyl-phosphate (Crans 1983). The pathway is also balanced with respect to reduction equivalents. Thus, the application of costly cofactors is not needed, and a potential hurdle for commercial application of the technology has been removed. Since all of the necessary enzymes must be synthesized separately and are not produced by the host as in conventional fermentation, enzyme stability or catalyst cost may be a crucial

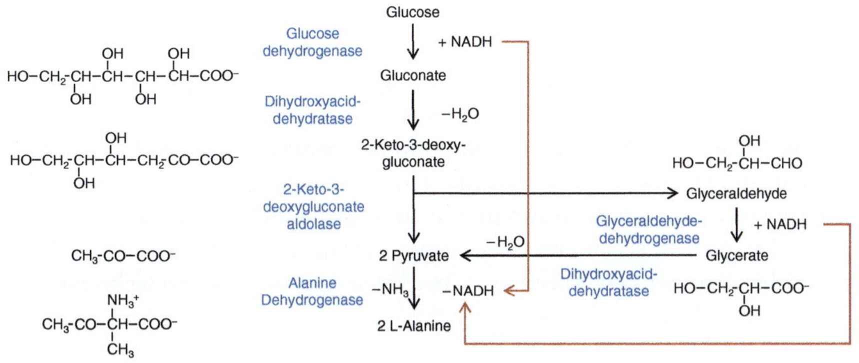

Figure 5.2 Biocatalytic reaction cascade for the synthesis of L-alanine based on glucose. The 2 mol pyruvate are derived via KDG and glycerate. Source: Adapted from Gmelch et al. (2019).

consideration. If the enzymes can be stabilized and efficiently be recycled, the catalyst cost might be minor. However, it is not yet reported that the biocatalytic technology has achieved commercial maturity. The figures of the Michaelis Menten kinetics of the applied enzymes are listed below. Enzyme kinetics have been measured at 50 °C in 100 mM HEPES (4-(2-hydroxyethyl)-1-piperazineethanesulfonic acid) at pH 7.35 and 100 mM ammonium sulfate as a nitrogen source (Gmelch et al. 2019; Table 5.2):

Table 5.2 Kinetic data for the applied enzymes.

Enzyme		**Origin**	K_M **(substrate) (mM)**	K_M **(cofactor) (U/mg)**	**Vmax (mM)**
Glucose dehydrogenase	BsGDH	*B. subtilis*	7.9	0.5 (NAD^+)	218
Dihydroxy acid dehydrogenase	CcDHAD	*C. crescentus*	3.2	n.a.	29
Aldolase	PtKDGA	*P. torridus*	1.1	n.a.	12
Aldehyde dehydrogenase	MjALDH	*M. jannaschii*	0.3	0.3 (NAD^+)	2
Alanine dehydrogenase	AfAlaDH	*A. fulgidus*	0.2	0.06 (NADH)	26

The biocatalysis is reported to provide a yield of >95%. After 12 h of batch incubation, a productivity of 0.17 g/lxh with a catalyst loading of 0.05 g catalyst per g of L-alanine is achieved (Gmelch et al. 2019).

5.5 Industrial Production and Application

The industrial production of L-alanine pursues biocatalysis and fermentation as well. Biocatalysis has played a dominant role, led by the Japanese company Tanabe Seiyaku, which started the biocatalytic conversion of fumarate via L-aspartic acid to L-alanine in 1982 (Takamatsu et al. 1986). Additionally, fermentation technology based on glucose as a carbon and energy source has gained market share and is now the leading technology. It accounts for about 60% of the whole market, compared to the 40% remaining for biocatalysis (Liu et al. 2022). It is reported that the Chinese company Anhui Huaheng Biotechnology has licensed the *E. coli* production host XZ132 developed in the laboratory of Dr. Ingram (Liu et al. 2022). Applying this technology Anhui Huaheng Biotechnology was the first alanine manufacturer in 2014 using fermentation technology (Liu et al. 2022). An additional plant in Hefei was announced to be operated jointly with a partner in the chemical industry (Ramesh 2014). According to reports, two production lines with a combined capacity of 23 kt pa will be running in 2022, using fermenters with a design volume of 250 m^3. The process is claimed to be anaerobic (Liu et al. 2022). Limited public knowledge is available on other major producers, market shares, and applied technology.

References

Ahmed, H., Ettema, T.J.G., Tjaden, B. et al. (2005). The semi-phosphorylative Entner-Doudoroff pathway in hyperthermophilic archaea: a re-evaluation. *Biochemical Journal* 390: 529–540.

Appleton, H. and Rosentrater, K.A. (2021). Sweet dreams (are made of this): A review and perspectives on aspartic acid production. *Fermentation* 7: 49. https://doi.org/10.3390/fermentation7020049.

Becker, H.G.O. et al. (1976). *Organikum*, 446–447. Deutscher Verlag der Wissenschaften.

Çalik, G., Savaçi, H., Çalik, P., and Özdamar, T.H. (1999). Growth and κ-carrageenan immobilization of *Pseudomonas dacunhae* cells for *Pseudomonas dacunhae* cells for L-alanine production. *Enzyme and Microbial Technology* 24: 67–74.

Crans, D.C. and Whitesides, G.M. (1983). A convenient synthesis of disodium acetyl phosphate for use in *in situ* ATP cofactor regeneration. *Journal of Organic Chemistry* 48: 3130–3132.

D'Este, M., Alvarado-Morales, M., and Angelidaki, I. (2018). Amino acids production focusing on fermentation technologies – A review. *Biotechnology Advances* 14–25.

Döring, J. and Müller, V. (2018). Alanine, a novel growth substrate for the acetogenic bacterium *Acetobacterium woodii*. *Applied and Environmental Microbiology* 84 (23): https://doi.org/10.1128/AEM.02023-18.

EP 0386476 (1990). *Process for Producing L-Alanine*. Proprietor: Mitsubishi Petrochemical Co.

Flamholz, A., Noor, E., Bar-Even, A., and Milo, R. (2012). eQuilibrator – the biochemical thermodynamics calculator. *Nucleic Acid Research* 40 (database issue): D770–D775.

Folch, P.L., Bisschops, M.M.M., and Weusthuis, R.A. (2021). Metabolic energy conservation for fermentative product formation. *Microbiol Biotechnology* 14 (3): 829–858. https://doi.org/10.1111/1751-7915.13746.

Gmelch, T.J., Sperl, J.M., and Sieber, V. (2019). Optimization of a reduced enzymatic reaction cascade for the production of L-alanine. *Scientific Reports* 9: 11754. https://doi.org/10.1038/s41598-019-48151-y.

Grunwald, P. (2018). *Biocatalysis: Biochemical Fundamentals and Applications*. World Scientific p. 304ff, p. 770.

Guterl, J.-K., Garbe, D., Carsten, J. et al. (2012). Cell-free metabolic engineering: production of chemicals by minimized reaction cascades. *ChemSusChem* 5: 2165–2172.

Han, H., Xu, B., Zeng, W., and Zhou, J. (2020). Regulating the biosynthesis of pyridoxal 5′-phosphate with riboswitch to enhance L-DOPA production by *Escherichia coli* whole-cell biotransformation. *Journal of Biotechnology* 321: 68–77. https://doi.org/10.1016/j.biotec.2020.05.009.

Hoffart, E., Grenz, S., Lange, J. et al. (2017). High substrate uptake rates empower *Vibrio natriegens* as production host for industrial biotechnology. *Applied and Environmental Microbiology* 83 (22): e01614-17.

Hori, H., Yoneyama, H., Tobe, R. et al. (2011). Inducible L-alanine exporter encoded by the novel gene *ygaW* (*alaE*) in *Escherichia coli*. *Applied and Environmental Microbiology* 77 (12): 4027–4034. https://doi.org/10.1128/AEM.00003-11.

Ideka, M. (2003). Amino acid production processes. In: *Microbial Production of l-Amino Acids. Advances in Biochemical Engineering/Biotechnology*, vol. 79 (ed. R. Fauriem et al.), 1–35. Springer https://doi.org/10.1007/3-540-45989-8_1.

Kim, S., Ihara, K., Katsube, S. et al. (2015). Characterization of the L-alanine exporter AlaE of *Escherichia coli* and its potential role in protecting cells from a toxic-level accumulation of L-alanine and its derivatives. *MicrobiologyOpen* 4 (4): 632–643. https://doi.org/10.1002/mbo3.269.

Liu, Z., Que, S., Xu, J., and Peng, T. (2014). Alanine aminotransferase-old biomarker and new concept: A review. *International Journal of Medical Sciences* 11: 925–935.

Liu, Z., Wang, X., Yu, J., and Han, L. (2021). Characterization and rational modification of aspartate 4-decarboxylase from *Acinetobacter radioresistens* for the production of L-alanine. *Biotechnology and Bioengineering* 118: 2493–2502.

Liu, P., Xu, H., and Zhang, X. (2022). Metabolic engineering of microorganisms for L-alanine production. *Journal of Industrial Microbiology and Biotechnology* 49: kuab057. https://doi.org/10.1093/jimb/kuab057.

Mallakpour, S. and Behranvand, V. (2014). The influence of acid-treated multi-walled carbon nanotubes on the surface morphology and thermal properties of alanine-based poly(amide-imide)/MWCNT nanocomposites system. *Colloid and Polymer Science* 293: 333–339.

Mallakpour, S. and Dinari, M. (2011). Progress in synthetic polymers based on natural amino acid. *Journal of Macromolecular Science* 48: 644–679.

Marshall, D.D., Halouska, S., Zinniel, D.K. et al. (2019). Assessment of metabolic changes in *Mycobacterium smegmantius* wild type and alr mutant strains: evidence for a new pathway of D-alanine biosynthesis. *Journal of Proteome Research* 16 (3): 1270–1279.

Peña-Solar, E., Fernandez, F.J., López-Estepa, M. et al. (2014). Structural analysis and mutant growth properties reveal distinctive enzymatic and cellular roles for the three major L-alanine transaminases of *Escherichia coli*. *PLoS One* 9 (7): https://doi.org/10.1371/journal.pone.0102139.

Ramesh, D. (2014). BASF and Anhui Huafeng mull L-alanine production JV in China. IHS Chemical Week, 24.11.2014, p. 15.

Ravot, G., Ollivier, B., Fardeau, M.-L. et al. (1996). L-Alanine production from glucose fermentation by hyperthermophilic members of the domain bacteria and archaea: a remnant of ancestral metabolism? *Applied and Environmental Microbiology* 62 (7): 2657–2659.

Schäfer, B. (2007). *Naturstoffe in der chemischen Industrie*. Elsevier, p. 139ff.

Sharma, A., Noda, M., Sugiyama, M. et al. (2021). Metabolic engineering of *Pediococcus acidilactici* BD16 for heterologous expression of synthetic *alaD* gene cassette and L-alanine production in the recombinant strain using fed-batch fermentation. *Foods* 10: 1964. https://doi.org/10.3390/foods10081964.

Shibatani, T., Kakimoto, T., and Chibata, I. (1979). Stimulation of L-aspartate ß-decarboxylase by L-glutamate in *Pseudomonas dacunhae* and improved production of L-alanine. *Applied and Environmental Microbiology* 38 (3): 359–364.

Solms, J., Vuataz, L., and Egli, R.H. (1965). The taste of L- and D-amino acids. *Experientia* 692–694.

Takamatsu, S., Tosa, T., and Chibata, I. (1985). Production of L-alanine from ammonium fumarate using two microbial cells immobilized with κ-carrageenan. *Journal of Chemical Engineering of Japan* 18 (1): 66–70.

Takamatsu, S., Tosa, T., and Chibita, I. (1986). Industrial production of L-alanine from ammonium fumarate using immobilized microbial cells of two kinds. *Journal of Chemical Engineering of Japan* 19 (1): 31–36.

Thoma, F. and Blombach, B. (2021). Metabolic engineering of *Vibrio natriegens*. *Essays in Biochemistry* 65: 381–392. https://doi.org/10.1042/ECB20200135.

Uhlenbusch, I., Sahm, H., and Sprenger, G.A. (1991). Expression of an L-alanine dehydrogenase gene in *Zymononas mobilis* and excretion of L-alanine. *Applied and Environmental Microbiology* 57: 1360–1366.

Vrljic, M., Sahm, H., and Eggeling, L. (1996). A new type of transporter with a new type of cellular function: L-lysine export from *Corynebacterium glutamicum*. *Molecular Microbiology* 22: 815–826.

Yamada, S., Maeshima, H., Wada, M., and Chibata, I. (1973). Production of D-alanine by *Corynebacterium fascians*. *Applied Microbiology* 25 (4): 636–640.

Zhang, X., Jantama, K., Moore, J.C. et al. (2007). Production of L-alanine by metabolically engineered *Escherichia coli*. *Applied Genetics and Molecular Biotechnology* 77: 355–366.

Zhou, L., Deng, C., Cui, W.-J. et al. (2016). Efficient L-alanine production by a thermo-regulated switch in *Escherichia coli*. *Applied Biochemistry and Biotechnology* 178: 324–337. https://doi.org/10.1007/s12010-015-1874-x.

6
3-Hydroxypropionic Acid

6.1 Application

3-Hydroxypropionic acid is a C3 building block with the chemical formula $C_3H_6O_3$ and a molecular weight of 90.1 g/mol. It carries a terminal hydroxy function and a carboxylic acid function. Its current use as a chemical block is limited. It is used for the synthesis of various low-volume chemicals, including acrylates. It has a pK_a value of 4.5 due to its carboxylic acid function. The main strategic intention behind the endeavor to develop a fermentative 3-hydroxypropionic technology is the synthesis of acrylic acid (Wagemann 2014). Acrylic acid can be polymerized to obtain acrylic esters (e.g. butyl acrylate, ethyl acrylate, and methyl acrylate) or polyacrylic acid for superabsorbers like diapers. The global consumption of acrylic acid amounts to 6.5 Mio. t in 2022 and is expected to grow at 3–4% pa (mcgroup 2022). 3-Hydroxypropionic acid is considered as chemical intermediate and biobased precursor for acrylic acid since it can be dehydrated to introduce a double bond between C2 and C3, and acrylic acid would be derived:

$$HO\text{–}CH_2\text{–}CH_2\text{–}COOH \rightarrow CH_2 = CH\text{–}COOH + H_2O$$

Acrylic acid is usually produced in a gas-phase reaction with propylene and oxygen as educts. The reaction proceeds via acrolein as an intermediate, which might be isolated or directly converted into acrylic acid. The first step is pursued at temperatures of 330–370 °C and a pressure of 1–2 bar. The second step is pursued at temperatures of 260–300 °C. The complete oxidation achieves a propylene yield of 85–90% with a turnover above 95% (Arpe 2007). The exothermic reaction allows to derive a steam credit to capture the released heat:

$$CH_2 = CH\text{–}CH_3 + O_2 \rightarrow CH_2 = CH\text{–}CHO + H_2O$$
$$CH_2 = CH\text{–}CHO + 0.5\,O_2 \rightarrow CH_2 = CH\text{–}COOH$$

Since the petrochemical production of propylene in conventional steam crackers might decrease in the future and, in general, the carbon dioxide emissions of petrochemistry shall be avoided, the exploration of a biobased and renewable feedstock for acrylic acid is recommended. These considerations have triggered the evaluation of fermentative technology, directly or indirectly, toward acrylic acid. The efficient and cheap petrochemical production of propylene favors the chemical production of acrylic acids, which nevertheless might be counterbalanced by the biobased feedstock and the usually lower carbon footprint of fermentative products, or in this case, the potential fermentative precursor 3-hydropropionic acid. It was assumed in the last decade that the major consumer brands applying acrylic

Pathway Design for Industrial Fermentation, First Edition. Walter Koch.

acid for diapers, such as P&G, would prefer a biobased acrylic acid versus a petrochemically derived product as the basis for a "green marketing campaign," but the required price premium of a biobased product turned out to be minor (Tullo 2013).

Since the direct fermentative production of acrylic acid is regarded as difficult, suitable precursors were identified with 2-hydroxypropionic (i.e. lactic acid) acid and 3-hydroxypropionic acid as the most obvious candidates. Both hydroxy acids with a C3 backbone are competing as precursors for the synthesis of acrylic acid. Lactic acid is advantageous given its established, mature, and cost-efficient high-volume fermentative production. The chemical gas-phase dehydration of lactic acid at temperatures in the range of 300–400 °C is technically not yet very efficient and suffers from decarboxylation with acetaldehyde or propanoic acid as by-products. The yield might be limited to about 80–90%. A selectivity of 92% combined with a lactic acid turnover of 91% based on laboratory research has been published using a phosphate catalyst (Wojcieszak et al. 2020; Zhang et al. 2008), but the incurred catalyst costs need to be considered and might counterbalance the progress in selectivity. An alternative liquid phase technology at temperatures of 120–180 °C has been developed by the University of Erlangen and funded by the German Federal Office for Agriculture and Food and P&G, which is reported to achieve a yield of 80% (Abschlussbericht 2017).

It has also been tried to synthesize acrylic acid based on ethylene and carbon dioxide as feedstock, and commercial production with a startup in 2017 had been envisaged by Novomer (Boswell 2013). However, these plans did not materialize. Thus, 3-hydroxypropionic acid has gained attention, and fermentation technology has been pursued in academia and industry.

6.2 Chemical Synthesis

The most straightforward synthesis of 3-hydroxypropionic acid proceeds via the hydration of acrylic acid, which is a high-volume intermediate for the synthesis of superabsorber:

$$CH_2 = CH\text{–}COOH + H_2O \rightarrow HO\text{–}CH_2\text{–}CH\text{–}COOH$$

As an alternative, ethylene chlorohydrin might be reacted with cyanic acid to obtain the nitrile intermediate, which can be hydrolyzed with water to convert the nitrile function into a carboxylic acid function:

$$HO\text{–}CH_2\text{–}CH_2\text{–}Cl + HCN \rightarrow HO\text{–}CH_2\text{–}CH_2\text{–}C \equiv N + HCl$$

$$HO\text{–}CH_2\text{–}CH_2\text{–}C \equiv N + 2\,H_2O \rightarrow HO\text{–}CH_2\text{–}CH\text{–}COOH + NH_3$$

The second route is likely economically disadvantaged since cyanic acid is used as a raw material and hydrochloride and ammonia are derived as by-products. More complex downstream processing for the isolation of the product and purification of the by-products for recycling is to be expected.

Furthermore, innovative routes are under discussion that might combine different technologies and try to establish a renewable feedstock for producing 3-hydroxypropionic acid or acrylic acid, respectively. These activities include the evaluation of value chains based on furfural to be converted via photooxygenation into hydroxybutenolide, which would then be chemically oxidized to maleic anhydride. After conventional hydrolysis of the anhydride maleic acid, free carboxylic-acid functions are derived, which could then, via an innovative ethenolysis, be converted into acrylic acid (Hermens et al. 2021). Levulinic acid has been explored as an alternative biobased feedstock for the chemical synthesis of 3-hydroxypropionic acid. Academic research has claimed that the oxidation of levulinic acid with hydrogen peroxide under acid conditions produces succinic acid and under basic conditions produces 3-hydroxypropionic acid (CEN 2015).

6.3 Pathway Design

6.3.1 Synthesis Based on Glucose via Malonyl-CoA

Several routes toward 3-hydroxypropionic acid based on glucose have been proposed (Henry et al. 2010). A prominent short route proceeds via acetyl-CoA, which is converted to malonyl-CoA with the aid of ATP. The hydrolysis of the thioester bond with CoA obtains 3-oxopropionic acid (also referred to as malonic semialdehyde), which is finally reduced to obtain the target product (Table 6.1).

The ATP-dependent carboxylation of acetyl-CoA is facilitated by the acetyl-CoA CO_2 ligase or the acetyl-CoA carboxylase. The latter enzyme name is used in this context. The conversion of malonyl-CoA to 3-oxopropionic acid is catalyzed by either – depending on the reference in literature – acetyl-CoA hydrolase or the malonyl-CoA reductase (Henry et al. 2010; Kildegaard et al. 2016). In this context, the enzyme is referred to as malonyl-CoA reductase to emphasize the consumption of reduction equivalents. It should also be pointed out that 3-oxopropionic acid is an aldehyde and is, therefore, most likely toxic to the host due to its high chemical activity, so intra- and extracellular leveling up should be avoided. The reduction of 3-oxopropionic acid is pursued by lactate dehydrogenase or 3-hydroxypropionic acid dehydrogenase. It is interesting to note that 3-hydroxypropionic acid is not an artificial intermediate in biological systems, though it plays a natural role in the metabolism of thermophilic archaea. In these species, the conversion of malonyl-CoA into 3-hydroxypropionic acid might be executed by two distinct enzymes or a bi-functional malonyl-CoA reductase (Kildegaard et al. 2016).

Table 6.1 Synthesis pathway via malonyl-CoA.

Reaction	Enzyme
Glucose + 2 ADP + Pi + 2 NAD^+ → 2 Pyruvate + 2 ATP + 2 H_2O + 2 NADH + 2 H^+	Glycolysis
2 Pyruvate + 2 CoA + 2 NAD^+ → 2 Acetyl-CoA + 2 CO_2 + 2 NADH + 2 H^+	Pyruvate dehydrogenase
2 Acetyl-CoA + 2 ATP + 2 H_2O + 2 CO_2 → 2 Malonyl-CoA + 2 ADP + 2 Pi	Acetyl-CoA carboxylase
2 Malonyl-CoA + 2 NADH + 2 H^+ → 2 3-Oxopropionic acid + 2 CoA + 2 NAD^+	Malonyl-CoA reductase
2 3-Oxopropionic acid + 2 NADH + 2 H^+ → 2 3-Hydroxypropionic acid + 2 NAD^+	Lactate oxidoreductase
Glucose → 2 3-Hydroxypropionic acid	Net reaction

The overall reaction equation reveals that the reaction in theory does not generate any by-products in the form of reduction equivalents, or ATP. The 24 electrons carried by glucose (without counting reduced oxygen) are equally distributed on the two mol product, each carrying 12 electrons (without counting reduced oxygen). Thus, neither the generation of carbon dioxide nor a surplus of reduction equivalents do apply, which would be observed in case the product has a higher oxidation state than glucose. The reaction should proceed as an anaerobic process, though the formation of biomass might be supported under aerobic conditions. Additionally, the two mol of ATP derived in glycolysis are invested in the carboxylation of acetyl-CoA to malonyl-CoA, which indicates that in the first instance a balance is given. The pathway promises to deliver a molar yield of two mol 3-hydroxypropionic acid per mol glucose and a theoretical carbon yield of 1 t product per ton glucose (2 × 90.1 g/mol/180.1 g/mol). However, some ATP will be needed for the formation of biomass and cellular maintenance, which will impact the realistic carbon yield.

6.3.2 Synthesis Based on Glucose via Acetate and Malonyl-CoA

A proof of concept for the metabolic route via malonyl-CoA was originally been provided by Cargill, who expressed the malonyl-CoA reductase of *Chloroflexus aurantiacus* in *Escherichia coli* (Gokarn et al. 2012). A base is added during fermentation to keep the pH in the neutral range. To limit acid consumption and salt generation in the downstream process, the pathway has also been tested in the eukaryotic host *Saccharomyces cerevisiae* with a higher acid tolerance. The academic experiments were pursued at pH 5 and finally provided a titer of 9.8 g/l with a molar yield of 13% based on glucose after 100 h in a carbon-limited fed-batch cultivation (Kildegaard et al. 2016). The results are corresponding to a productivity of about 0.1 g/l × h and were achieved based on several measures. In the first instance, the additional compartment of the eukaryotic host needs to be considered. The pathway relies on an efficient supply of acetyl-CoA, and the pyruvate dehydrogenase is usually localized in the mitochondrion to feed the citric acid cycle. The metabolic engineering strategy may then be designed to pursue a cytosolic route, and the genes encoding the native pyruvate decarboxylase (*pdc*) and the native aldehyde dehydrogenase (*ald*) were overexpressed. These measures ensure that a substantial share of the pyruvate pool is processed in the cytosol and channeled into the recombinant pathway with acetaldehyde and acetate as intermediates (Figure 6.1).

The generation of cytosolic acetyl-CoA is ensured by introducing the gene encoding the acetyl-CoA synthase (*acs*) from *Salmonella enterica*. The establishment of a metabolic pathway based on cytosolic acetyl-CoA within an eukaryotic host is also referred to as the pyruvate dehydrogenase bypass and was originally invented for the mevalonate (MVA) pathway toward isoprene in *S. cerevisiae* (Shiba et al. 2007). Additionally, the recombinant gene encoding the bi-functional malonyl-CoA reductase (*mcr*) from *C. aurantiacus* and the native, but mutated gene encoding the acetyl-CoA carboxylase (*acc1p*) were overexpressed in *S. cerevisiae*. The gene *acc1p* encoding the acetyl-CoA carboxylase bears a double mutation shown to enhance activity and improve the production of malonyl-CoA. A third element of the metabolic engineering strategy addresses the dependency of malonyl-CoA reductase on NADPH. Since exclusively the aldehyde dehydrogenase generates NADPH, a second enzyme replenishing the NADPH pool needs to be added. Therefore, the

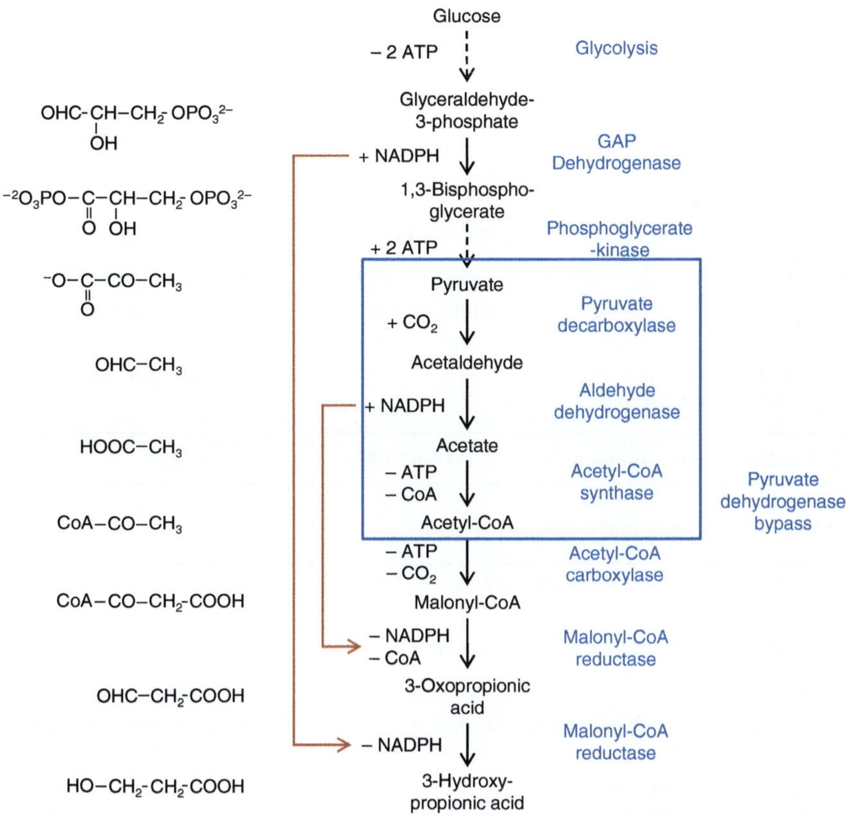

Figure 6.1 Recombinant biochemical pathway toward 3-hydroxypropionic acid. The reduction equivalents are balanced since the 24 electrons of glucose are equally allocated on two mol 3-hydroxypropionic acid. Source: Walter Koch.

glycerol-aldehyde-3-phosphate dehydrogenase (GADPH) of *Clostridium acetobutylicum* was introduced, which consumes $NADP^+$. The activity of the native GAPDH, which consumes NAD^+, was silenced (Kildegaard et al. 2016).

The reconstruction of the overall pathway reveals that the pyruvate dehydrogenase bypass implies an additional consumption of one mol ATP per mol 3-hydroxypropionic acid compared to the direct conversion of pyruvate into acetyl-CoA. This is due to the circumstance that the acetyl-CoA synthase consumes ATP to build the active thioester of the acetate function and CoA (Table 6.2).

6.3.3 Synthesis Based on Glucose via α-Alanine and β-Alanine

An alternative route proceeds via α-alanine and β-alanine. In the first step, pyruvate is subjected to a transamination with glutamate as an amino donor. The obtained α-alanine is isomerized to β-alanine, which can then be converted into 3-hydroxypropionic acid (Table 6.3).

The net reaction equation indicates a rather straight conversion without a surplus or the gap of energy in the form of ATP and reduction equivalents. According to the route via malonyl-CoA, the 24 electrons carried by one mol glucose (without counting reduced oxygen) are transferred to two mol 3-hydroxypropionic acid, each carrying 12

Table 6.2 Synthesis of 3-hydroxypropionic acid via acetate and malonyl-CoA.

Reaction	Enzyme
Glucose + 2 ADP + Pi +2 $NADP^+$ → 2 Pyruvate + 2 ATP + 2 H_2O + 2 NADPH + 2 H^+	Glycolysis
2 Pyruvate → 2 Acetaldehyde + 2 CO_2	Pyruvate decarboxylase
2 Acetaldehyde + 2 $NADP^+$ + 2 H_2O → 2 Acetate + 2 NADPH + 2 H^+	Aldehyde dehydrogenase
2 Acetate + 2 ATP + 2 CoA → 2 Acetyl-CoA + 2 AMP + 2 PPi	Acetyl-CoA synthase
2 Acetyl-CoA + 2 ATP + 2 CO_2 → 2 Malonyl-CoA + 2 ADP + 2 Pi	Acetyl-CoA carboxylase
2 Malonyl-CoA + 2 NADPH + 2 H^+ → 2 3-Oxopropionic acid + 2 CoA + 2 $NADP^+$	Malonyl-CoA reductase
2 3-Oxopropionic acid + 2 NADPH + 2 H^+ → 2 3-Hydroxypropionic acid + 2 $NADP^+$	Malonyl-CoA reductase
Glucose + 2 ATP → 2 3-Hydroxypropionic acid + 2 AMP + 2 PPi	Net reaction

Table 6.3 Synthesis via α-alanine and β-alanine.

Reaction	Enzyme
Glucose + 2 ADP + Pi + 2 NAD^+ → 2 Pyruvate + 2 ATP + 2 H_2O + 2 NADH + 2 H^+	Glycolysis
2 Pyruvate + Glutamate → 2 α-Alanine + α-ketoglutarate	Aspartate transaminase
2 α-Alanine → 2 β-Alanine	Isomerase
2 β-Alanine + 2 α-ketoglutarate → 2 3-Oxopropionic acid + 2 glutamate	Aspartate transaminase
2 3-Oxopropionic acid + 2 NADH + H^+ → 2 3-Hydroxypropionic acid + 2 NAD^+	Lactate oxidoreductase
Glucose → 2 3-Hydroxypropionic acid	Net reaction

electrons (with counting reduced oxygen). It is reported that a metabolic pathway toward 3-hydroxypropionic acid via β-alanine was demonstrated in *S. cerevisiae* displaying a titer of 13.7 g/l in a controlled fed-batch fermentation with mineral medium at pH 5 (Borodina et al. 2015).

6.3.4 Synthesis Based on Glycerol

It is obvious from a chemical point of view that glycerol would be a suitable feedstock for the fermentative production of 3-hydroxypropionic acid. The given C3 building block could be kept, but exclusively the C2-hydroxyl function would need to be removed, and the C1-hydroxyl function would need to be oxidized to the acid level. These steps can be pursued with two enzymes since the glycerol dehydratase facilitates the elimination of one molecule of water and the transient enol-intermediate is directly isomerized to 3-hydroxypropionaldehyde. This step is equal to the first step in the synthesis of 1,3-propanediol based on glycerol as a feedstock (Table 6.4).

Table 6.4 Synthesis of 3-hydroxypropionic acid based on glycerol.

Glycerol → 3-Hydroxypropionaldehyde + H_2O	Glycerol dehydratase
3-Hydroxypropionaldehyde + NAD^+ + 0.5 O_2 → 3-Hydroxypropionic acid + NADH + H^+	Hypothetical monooxygenase
Glycerol + 0.5 O_2 + NAD^+ → 3-Hydroxypropionic acid + H_2O + NADH + H^+	Net reaction

The second step might be facilitated by a hypothetical monooxygenase, which incorporates an oxygen atom to obtain the carboxylic acid function. The pathway would not generate carbon dioxide and avoid an additional scope 1 carbon footprint, which supports the sustainability profile of the product. However, it can also be seen that a surplus of reduction equivalents would be generated since glycerol carries 14 electrons (without counting reduced oxygen) and 3-hydroxypropionic acid carries 12 electrons (without counting reduced oxygen). Thus, for each mol of 3-hydroxypropionic acid, one mol of NADH would be obtained. In order to balance the reduction equivalents, oxygen would need to be added as an electron acceptor and water would be obtained as a by-product (Voet et al. 2016):

$$\text{NAD (P) H} + \text{H}^+ + 0.5\,\text{O}_2 \rightarrow \text{NAD(P)}^+ + \text{H}_2\text{O} \quad \Delta G^{\circ\prime} = -218\ \text{KJ/mol}$$

In case it is assumed that under realistic biochemical conditions about 2.5 mol of ATP per mol of NAD(P)H are obtained, the net reaction equation would be stated as follows:

$$\text{Glycerol} + \text{O}_2 + 2.5\,\text{ADP} + 2.5\,\text{Pi} \rightarrow \text{3-Hydroxypropionic acid} + 2.5\,\text{ATP} + 2\,\text{H}_2\text{O}$$

The surplus of reduction equivalents translates into an energy surplus, which might impact the carbon yield of the fermentation. On the one hand, the energy might be needed to manage the potentially high-maintenance metabolism of the biocatalyst. On the other hand, the host might invest the energy surplus into growth and the cell density during the fermentation might increase. This supports high productivity, though it also implies a high by-product factor of biomass to product, which limits the carbon yield.

Alternatively, the coproduction of 3-hydroxypropionic acid and 1,3-propanediol has been proposed. Both products rely on propionaldehyde as a common precursor, which may either be oxidized to 3-hydroxypropionic acid or reduced to 1,3-propanediol. A coproduction with a molar ratio of 1:1 would ensure a balanced pool of reduction equivalents. Combined production of both products is reported with titers of up to 24.4 g/l 3-hydroxypropionic acid and 49.3 g/l 1,3-propanediol with a yield of 50–80% based on glycerol (Ashok et al. 2011; Huang et al. 2013).

Lactobacillus reuteri is a natural producer of 3-hydroxypropionic acid and might be a potential biocatalyst for industrial applications. However, its slow growth has so far limited its application (Li and Tian 2015). The host has been applied for producing the pathway intermediate 3-propionaldehyde with glycerol as feed. Since aldehyde is toxic, the cells were immobilized and the productivity was increased. A share of the aldehyde is subject to spontaneous dismutation, and 1,3-propanediol and 3-hydroxypropionic acid appear as by-products. The application of the aldehyde scavengers, semi-carbazide and carbohydrazide, reduces the toxic impact of the product on the host and a molar yield of

90% could be achieved. The measured productivity amounted to 11.3 g/l × h (batch) and 21.6 g/l × h (fed-batch), respectively (Krauter 2013).

An interesting host for commercial application is *Klebsiella pneumoniae*, which grows on glycerol as the sole carbon source under anaerobic and microaerobic conditions. Another advantage would be the native capability of *K. pneumoniae* to produce vitamin B_{12}, which is the cofactor of the glycerol dehydratase. This circumstance would define a disadvantage for *E. coli* as a biocatalyst that is not able to synthesize vitamin B_{12} and costly supplementation of the medium would be needed (Li and Tian 2015). The wild type of *K. pneumoniae* generates more 1,3-propanediol than 3-hydroxypropionic acid because the 1,3-propanediol oxidoreductase has a higher activity than the aldehyde dehydrogenase. Accordingly, to obtain a commercially interesting strain, the activity of the respective oxidoreductase would need to be silenced and the activity of the aldehyde dehydrogenase toward 3-hydroxypropionic acid would need to be enhanced. Fermentation protocols for 3-hydroxypropionaldehyde were developed, and a specific toxicity attenuator (in particular, semi-carbazide) was applied to sustain the minimum resistance of the host (Krauter et al. 2012; Krauter 2013).

6.3.5 Digression: Synthesis of Acrylic Acid Based on Glycerol

An obvious route directly to a biobased acrylic acid has been pursued by Arkema. Glycerol is chemically dehydrated and converted into acrolein, which is then, in a second step, oxidized to obtain acrylic acid (Tullo 2013):

$$CH_2OH\text{–}CHOH\text{–}CH_2OH \rightarrow CH_2=CH\text{–}CHO + 2\,H_2O$$

$$CH_2=CH\text{–}CHO + 0.5\,O_2 \rightarrow CH_2=CH\text{–}COOH$$

The technology does, in principle, work. The R&D activities were especially supported in the mid-2000s by the low glycerol price in the market. The rising demand for biodiesel as a fuel additive triggered a significant production of glycerol, which is a by-product in the process of converting plant-based triglycerides and methanol into the fatty acids methylester and glycerol. Since there was no sufficient market for the high glycerol volume, it was available at very low prices. Though the biodiesel boom weakened, additional outlets for glycerol were developed, and, as a consequence, the price recovered. The biobased acrylic acid was available but had a higher production cost than the petrochemically derived acrylic acid, and the original marketing plan of Arkema was given up. Arkema's group vice president for acrylics declared that customers are interested in biobased products but are not willing to accept a price premium: "We are not in a situation where [...] any company manufacturing diapers at this moment is ready to pay a 15% premium because it is a biobased raw material [...]. This is because no consumer is ready to pay a premium for the diaper" (Tullo 2013).

6.4 Industrial Application

It is not known whether a fermentative technology toward 3-hydroxypropionic acid is commercially applied. Based on a patent analysis, it may be assumed that Cargill has pursued the development of a technology based on the recombinant expression of the malonyl-CoA reductase of *C. aurantiacus* in *E. coli*. This technology was later applied and further developed at pilot scale by OPX Technologies (Kildegaard et al. 2016), though the activities have been phased out.

References

Abschlussbericht (2017). Flüssigphasen-Dehydratisierung von fermentativ gewonnener Milchsäure zur Produktion von biobasierter Acrylsäure. https://www.fnr.de/index.php?id=11150&fkz=22009614 (accessed 11 March 2022).

Arpe, H.-J. (2007). *Industrielle Organische Chemie. Bedeutende Vor- und Zwischenprodukte*, 323–325. Wiley.

Ashok, S., Raj, S.M., Rathnasingh, C., and Park, S. (2011). Development of recombinant *Klebsiella pneumoniae* ΔdhaT strain for the co-production of 3-hydroxypropionic acid and 1,3-propanediol from glycerol, applied. *Microbial Biotechnology* 90: 1253–1265. https://doi.org/10.1007/s00253-011-3148-z.

Borodina, I., Kildegaard, K.R., Jensen, N.B. et al. (2015). Establishing a synthetic pathway for high-level production of 3-hydroxypropionic acid in *Saccharomyces cerevisiae* via β-alanine. *Metabolic Engineering* 27: 57–64. https://doi.org/10.1016/j.ymben.2014.10.003.

Boswell, C. (2013). Novomer targets ethylene-to-acrylic acid on US gulf by 2017. *IHS Chemical Week* 12.

CEN (2015). Flipping a switch on biobased chemicals, 26 March 2015, p. 27.

Gokarn, R.R., Selifonova, O.V., Jessen, H.J., et al. (2012). 3-Hydroxypropionic acid and other organic compounds. Patent WO 2002042418 A3.

Henry, C.S., Broadbelt, L.J., and Hatzimanikatis, V. (2010). Discovery and analysis of novel metabolic pathways for the biosynthesis of industrial chemicals: 3-hydroxypropionic acid. *Biotechnology and Bioengineering* 106 (3): 462–473. https://doi.org/10.1002/bit.22673.

Hermens, J.G.H., Jensma, A., and Feringa, B.L. (2021). Highly efficient biobased synthesis of acrylic acid. *Angewandte Chemie* 60: 2–8. https://doi.org/10.1002/ange.202112618.

Huang, Y., Li, Z., Shimizu, K., and Ye, Q. (2013). Co-production of 3-hydroxypropionic acid and 1,3-propanediol by *Klebsiella pneumoniae* expressing *aldH* under microaerobic conditions. *Bioresource Technology* 128: 505–512. https://doi.org/10.1016/j. biortech.2012.10.143.

Krauter, H., Willke, T., and Vorlop, K.D. (2012). Production of high amounts of 3-hydroxypropionaldehyde from glycerol by *lactobacillus reuteri* with strongly increased biocatalyst lifetime and productivity. *New Biotechnology* 29 (2): 211–217. https://doi.org/10.1016/j.nbt.2011.06.015.

Krauter, H. (2013). Optimierung der biotechnologischen Herstellung von 3-Hydroxypropionaldehyde aus Glyzerin. Erhöhung der Standzeit und der Produktivität der Biokatalysators *Lactobacillus reuteri* SD 2112 durch den Fängerstoff Carbohydrazid, Dissertation Universität Braunschweig, 2013.

Kildegaard, K.R., Jensen, N.B., Schneider, K. et al. (2016). Engineering and systems-level analysis of *Saccharomyces cerevisiae* for production of 3-hydroxypropionic acid via malonyl-CoA reductase-dependent pathway. *Microbial Cell Factories* 15: 53. https://doi.org/10.1186/s12934-016-0451-5.

Li, Y. and Tian, P. (2015). Contemplating 3-hydroxypropionic acid biosynthesis in *Klebsiella pneumoniae*. *Indian Journal of Microbiology* 55 (2): 131–139. https://doi.org/10.1007/s12088-015-0513-0.

mcgroup (2022). Acrylic acid market dynamics. https://mcgroup.co.uk/news/20221005/acrylic-acid-market-dynamics.html (accessed 6 Oct 2022).

Shiba, Y., Paradise, E.M., Kirby, J. et al. (2007). Engineering of the pyruvate dehydrogenase bypass in *Saccharomyces cerevisiae* for high-level production of isoprenoids. *Metabolic Engineering* 9: 160–168.

Tullo, A.H. (2013). Hunting for biobased acrylic acid. Companies vie for a nonpetrochemical route into this lucrative industry. *CEN* 18 (11): 18–19.

Voet, D., Voet, J.G., and Pratt, C.W. (2016). *Principles of Biochemistry*. Wiley.

Wagemann, K. (2014). Herstellung von Grundchemikalien auf Basis nachwachsender Rohstoffe als Alternative zur Petrochemie. *Chemie Ingenieur Technik* 86 (12): 2115–2134.

Wojcieszak, R., Bonnotte, T., Paul, S. et al. (2020). Lactic acid conversion to acrylic acid over fluoride-substituted hydroxyapatites. *Frontiers in Chemistry* 8: article 421. https://doi.org/10.3389/fchem.2020.00421.

Zhang, J., Lin, J., Xu, X., and Cen, P. (2008). Evaluation of catalysts and optimization of reaction conditions for the dehydration of methyl lactate to acrylates. *Chinese Journal of Chemical Engineering* 16: 263–269. https://doi.org/10.1016/S1004-9541(08)60073-7.

7

1,3-Propanediol

7.1 Application

1,3-Propanediol is a linear diol with the chemical formula $C_3H_8O_2$ and a molecular weight of 76.1 g/mol. As a chemical intermediate, it offers a broad scope of target products given its two functional groups. It is partially marketed as a single compound, though it is mainly applied as a monomer for polymer applications with polytrimethylene-terephthalic acid (PTT) as the main outlet.

$$n\ \mathrm{HO{-}CH_2{-}CH_2{-}CH_2{-}OH} + n\ \mathrm{HOOC{-}C_6H_4{-}COOH}$$
$$\rightarrow [\mathrm{{-}O{-}CH_2{-}CH_2{-}CH_2{-}O{-}CO{-}C_6H_4{-}CO{-}}]_n + n + 1\ \mathrm{H_2O}$$

PTT is structurally analogous to PET, with the minor difference that the diol-building block bears three instead of two methylene groups. PTT is used for apparel, residential flooring, sporting goods, and other applications and displays good tensile properties, material resilience, outstanding elastic recovery, colored facility, high birefringence, and luminous transmittance (Li and Yang 2016). Polymer properties partially depend on the odd or even number of methylene groups between the repeating ester units, and 1,3-propanediol allows a finetuning of polymer functionality by mixing ethylene glycol and 1,3-propanediol with terephthalic acid as the acid-monomer building block. Development work in polyester chemistry is underway to explore further applications of biobased monomers, including 1,3-propanediol (Lomelí-Rodríguez et al. 2018).

7.2 Alternative Production of 1,3-Propanediol

Until 2007, 1,3-propanediol was exclusively produced using organic chemistry and petrochemical precursors. The first route was originally been developed by Degussa and later taken over by DuPont. It is based on acrolein and pursues an acid-catalyzed hydration at 50–70 °C to obtain 3-hydroxypropionaldehyde in the first step. The aldehyde is then, with Nickel as catalyst, hydrogenated at 135 bar and 75–140 °C to yield 1,3-propanediol (Steverding 2010):

$$\mathrm{CH_2{=}CH{-}CHO} + \mathrm{H_2O} \rightarrow \mathrm{HO{-}CH_2{-}CH_2{-}CHO}$$

$$\mathrm{HO{-}CH_2{-}CH_2{-}CHO} + \mathrm{H_2} \rightarrow \mathrm{HO{-}CH_2{-}CH_2{-}CH_2{-}OH}$$

Pathway Design for Industrial Fermentation, First Edition. Walter Koch.

The second chemical route toward 1,3-propanediol is based on ethylene oxide (EO), which is subjected to hydroformylation at 100 bar and 100 °C to obtain 3-hydroxypropionaldehyde. This technology has been developed by Shell and is operated in one plant in the US with a capacity of 70 kt pa. The intermediate 3-hydroxypropionaldehyde is catalytically hydrogenated according to the reaction step outlined above (Röper 2006; Arpe 2007):

$$\text{EO} + \text{CO}/\text{H}_2 \rightarrow \text{HO–CH}_2\text{–CH}_2\text{–CHO}$$

$$\text{HO–CH}_2\text{–CH}_2\text{–CHO} + \text{H}_2 \rightarrow \text{HO–CH}_2\text{–CH}_2\text{–CH}_2\text{–OH}$$

The development of the technology with EO as the starting material was driven by the attempt to offer a new polymer derived from 1,3-propanediol and terephthalic acid, which was introduced under the brand name "Conterra" (Biebl et al. 1999). The chemical synthesis of 1,3-propanediol might also be based on 3-hydroxypropionic acid as a suitable C3-precursor, followed by catalytic hydrogenation and reduction of the carboxylic acid function, though this route is not applied on an industrial scale.

7.3 Pathway Design Toward 1,3-Propanediol

7.3.1 Glucose as Carbon and Energy Source

The metabolic pathway based on glucose starts with the Embden-Meyerhof-Parnas pathway, or glycolysis, to glyceraldehyde-3-phosphate (GAP) and dihydroxyacetone-phosphate (DHAP), obtained in the cleavage of fructose-1,6-bisphosphate (FBP). The molecular structure of DHAP is already rather similar to that of 1,3-propanediol. Accordingly, the pathway engineering strategy would foresee silencing the downstream pathway toward pyruvate, isomerizing GAP into DHAP, and converting DHAP into product. The medium keto function of DHAP at the C2-position needs to be reduced in the first step to obtain glycerol-3-phosphate, which is then converted into glycerol via hydrolytic cleavage of the phosphoric acid ester. Glycerol is subjected to dehydration, and the enol-intermediate of the enzymatic mechanism isomerizes to 3-hydroxypropionaldehyde (3-HPA), which is reduced to obtain 1,3-propanediol (Figure 7.1).

The reconstruction of the overall net equation of the pathway is divided into conventional glycolysis and the recombinant metabolic pathway toward 1,3-propanediol. The split-up of the monosaccharide carbon backbone derives, in the first instance, a molar yield of two mol C3 intermediates per mol glucose (Table 7.1).

In case two mol of 1,3-propanediol shall be derived from one mol of glucose four reduction equivalents or eight electrons are needed - as can be seen on the left side of the reaction scheme - since glucose carries 24 electrons (without counting reduced oxygen) and the two mol of 1,3-propanediol carry together 32 electrons (2×16, without counting reduced oxygen). The cofactor used in the reductive conversion of 3-hydroxypropionaldehyde may be NADH or NADPH. The replenishment of the NAD(P)H pool might be less complex if the same cofactor is used in the two hydrogenation steps of the recombinant pathway. Since the 3-HPA dehydrogenase encoded by the *yqhD* gene of *Escherichia coli* applies NADPH, a redox balance is not straightforward. The metabolic engineering strategy might ensure that a transhydrogenase is present or may install an isoenzyme of the 3-HPA dehydrogenase, which uses NADH as a cofactor.

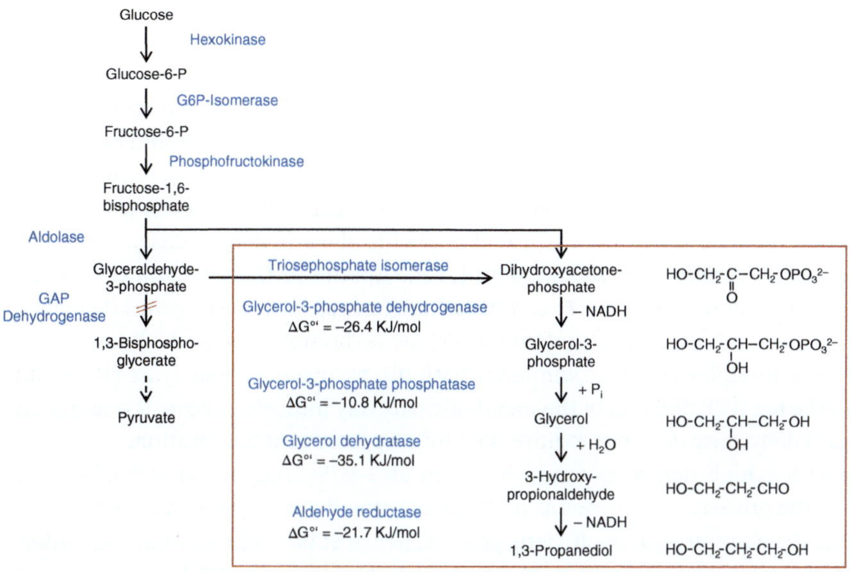

Figure 7.1 Recombinant pathway for the fermentative production of 1,3-propanediol with glucose as feedstock. The Gibbs free energy values for the engineered pathway steps from DHAP to 1,3-propanediol are indicated. Source: Adapted from Frazão et al. (2019).

Table 7.1 Chemical conversion of glucose into 1,3-propanediol.

Conventional glycolysis	
Glucose + ATP → Glucose-6-phosphate + ADP	Hexokinase
Glucose-6-phosphate → Fructose-6-phosphate	G6P Isomerase
Fructose-6-phosphate + ATP → Fructose-1,6-bisphosphate + ADP	Phosphofructokinase (PFK)
Fructose-1,6-bisphosphate → DHAP + GAP	Aldolase
GAP → DHAP	Triose-P-isomerase (TIM)
Engineered pathway	
2 DHAP + 2 NADH + 2 H^+ → 2 Glycerol-3-phosphate + 2 NAD^+	Dehydrogenase
2 Glycerol-3-phosphate + 2 H_2O → 2 Glycerol + 2 P_i	Phosphatase
2 Glycerol → 2 3-HPA + 2 H_2O	Dehydratase
2 3-HPA + 2 NAD(P)H + 2 H^+ → 2 1,3-Propanediol + 2 $NAD(P)^+$	Aldehyde reductase / dehydrogenase
Glucose + 4 NAD(P)H + 4 H^+ + 2 ATP → 2 1,3-Propanediol + 4 $NAD(P)^+$ + 2 ADP + 2 P_i	Net reaction

Note: DHAP: Dihydroxyacetonphosphate. GAP: Glyceraldehyde-3-phosphate.

It has been reported that the glycerol dehydratase usually operates with coenzyme B_{12} as a cofactor and undergoes inactivation during the catalytic step. The enzyme can be activated again in case where specific auxiliary enzymes are present that require ATP and Mg^{2+} (Ma et al. 2012; Tobimatsu et al. 2000). These auxiliary enzymes with their respective ATP demands might be integrated into the metabolic engineering strategy and would need to be considered in the estimation of the maximum carbon yield. Furthermore, the fermentation media might need to be supplemented with coenzyme B_{12} cofactors to ensure a sufficient supply. Metabolic engineering favors the use of a specific glycerol dehydratase, which has been identified in *Clostridium butyricum*. The 1,3-PD operon of *C. butyricum* covers the genes *dhaB1*, *dhaB2*, and *dhaT*, and *E. coli* cells transformed with these genes displayed glycerol dehydratase activity. The related proteins show no homology with the known glycerol dehydratases, though they share similarity with the pyruvate formate lyase (Raynaud et al. 2003). In the net ATP demand of the metabolic pathway indicated above, it is assumed that the glycerol dehydrase does not require ATP for its permanent reactivation.

Aldehydes with a high degree of reactivity, such as 3-HPA, may negatively affect the effectiveness of the process.. Aldehydes as pathway intermediates may not in general cause obstacles in the development of an efficient process, like acetaldehyde as a pathway intermediate in the fermentative production of ethanol, but the topic should be kept in mind in the development of the metabolic engineering strategy. In this respect, the provision of sufficient reduction equivalents and a high efficiency of the aldehyde reductase are key to avoiding a critical cytosolic concentration of 3-HPA. One might conclude that the process should preferentially be pursued as an anaerobic process since oxygen might consume the reduction equivalents needed for an efficient conversion of 3-HPA. According to this consideration, it has been reported that under aerobic conditions, glycerol might be converted into 3-HPA, which is toxic to the cells (Slininger and Bothast 1985). On the other hand, it has also been reported that the production of 1,3-propanediol under micro-aerobic conditions with *Klebsiella pneumoniae* improved the productivity from 0.8 to 1.57 g/l × h (Chen et al. 2003). In case 3-HPA is the target fermentation product that is secreted into the medium, product toxicity is reported as a process issue, and bisulfite as medium supplementation is proposed as a scavenger for *in situ* product removal (Stevens et al. 2013).

Two mol of 1,3-propanediol might be derived from one mol of glucose in a theoretical best-case consideration if exclusively the carbon backbone would be counted. This would imply a carbon yield of 0.84 t 1,3-propanediol/t glucose (2 × 76.1/180.2). However, also reduction and energy equivalents need to be taken into consideration. Hereby an anaerobic process is assumed. Four reduction equivalents need to be provided according to the overall equation of the reaction. These require about 0.33 mol glucose per two mol 1,3-PDO if it is assumed that one mol of glucose provides 12 reduction equivalents. The net reaction equation also reveals that per mol of glucose, two mol of ATP need to be invested. It may be assumed that the ATP demand is partially derived from coproduction via substrate-level phosphorylation. In case 0.33 mol glucose per two mol 1,3-propanediol is invested to provide the reduction equivalents, also 1.32 mol ATP/GTP (0.33 mol glucose × 4 mol ATP/GTP) would be generated as coproduction. The still missing 0.68 mol ATP per mol 1,3-propanediol requires an additional 0.02 mol glucose per mol 1,3-propanediol (0.68 mol ATP/GTP/32 mol ATP/GTP per mol glucose). Altogether 1.35 mol glucose is needed to obtain two mol 1,3-propanediol (Table 7.2)

Table 7.2 Estimation of the molar yield for the syntheisis of 1,3-propanediol.

Glucose demand for...	Energy equivalents	Glucose demand
Carbon backbone	n.a.	1.00 mol Glc/2 mol 1,3-PDO
Reduction equivalents	4.0 mol NADH/2 mol 1,3-PDO	0.33 mol Glc/2 mol 1,3-PDO
ATP demand	1.32 mol ATP coproduction	(no additional demand)
ATP demand	0.68 mol ATP	0.02 mol Glc/2 mol 1,3-PDO
		1.35 mol Glc/2 mol 1,3-PDO

This translates into a maximum molar yield of 1.48 mol 1,3-propionic acid per mol glucose (two mol product/1.35 mol glucose), which is according to values found in the literature (Zhang et al. 2017). The maximum carbon yield amounts to about 0.625 t 1,3-propanediol per t glucose (1.48 × 76.1/180.2).

The glucose based fermentative production of 1,3-propanediol has been implemented by DuPont and Tate & Lyle and the main performance criteria of the developed technology have been published. The reported carbon yield is 0.51 t 1,3-propanediol per t glucose (Nakamura and Whited 2003). This figure reveals that about 82% of the maximum theoretical yield is achieved, though downstream processing and product purification are not yet included. The remaining share of the feed is needed for the production of the biocatalyst (biomass) and potential by-products.

7.3.2 Glycerol as Carbon and Energy Source

The fermentative conversion of glycerol into 1,3-propanediol has already been observed as a natural process since the end of the 19th century. Anaerobe bacterial strains of *Citrobacter* spp., *Clostridium* spp., *Klebsiella* spp. and *Lactobacillus* spp. convert glycerol to 1,3-propanediol under anaerobic conditions. Thus, these hosts were considered industrial producers after the strong volume increase in the production of biodiesel generated a high glycerol by-product volume and the prices dropped significantly (Clomburg and Gonzalez 2013). Thereof *Klebsiella pneumoniae* and *C. butyricum* are facultative anaerobes, which are easier to handle in a high-volume production. These two strains are considered the most suitable target hosts for a process with glycerol as feedstock given their substrate tolerance, yield, and productivity (Gonzales-Pajuelo et al. 2006). Nevertheless, the strains are opportunistic pathogens, and the feasibility and sustainability of commercial high-volume production might be questioned.

The production of first-generation biodiesel uses plant-based triglycerides and petrochemical methanol to produce fatty acid methylester (biodiesel) and glycerol via chemical transesterification. Depending on the processing of the by-product, different glycerol grades are available. It has been reported that the prices for refined glycerol dropped from 1.15 to 0.66 \$/kg glycerol, and those for crude glycerol dropped from 0.44 to 0.11 \$/kg glycerol (Syzmanowska-Powalowska and Bialas 2014). These price trends might provide a good opportunity for glycerol-based technologies that are fully developed and can be immediately implemented, though counterbalancing effects in the market need to be considered. The adoption of glycerol as a feedstock will increase demand and might restore the former price level. Additionally, first-generation biodiesel is facing political pressure referring to deforestation is South/East Asia to provide the required plant oils.

In the metabolic pathway two branches need to be considered. In the first instance, one mol of glycerol is dehydrated by the glycerol dehydratase, and 3-hydroxypropionaldehyde is formed. In the second step, the aldehyde function of 3-hydroxypropionaldehyde is reduced to 1,3-propanediol via an NADH-dependent reaction catalyzed by the aldehyde reductase (Jiang et al. 2016). Both reactions have negative Gibbs free energy and are thermodynamically favored. These two reaction steps are referred to as the "reductive branch" given the consumption of one reduction equivalent (Figure 7.2).

From the point of view of the metabolic engineering scientist, the oxidative branch oxidizes glycerol to provide the reduction equivalents for the reductive branch. Though it is more accurate to rationalize the reductive branch as an "emergency" electron sink for the abundant reduction equivalents generated in the anaerobic metabolism of glycerol. Hereby several options apply since pyruvate might be converted into e.g. carbon dioxide, formate, ethanol, acetate, 2,3-butanediol, butyrate and succinate. Several metabolic engineering attempts have been made to block one or more pathways to reduce the number of by-products. The most straightforward solution would be the generation of lactic acid by Lactobacillus as a single by-product given the direct reduction of pyruvate. Accordingly, the coproduction of 1,3-propanediol and lactic acid has been proposed (Xin et al. 2017) and

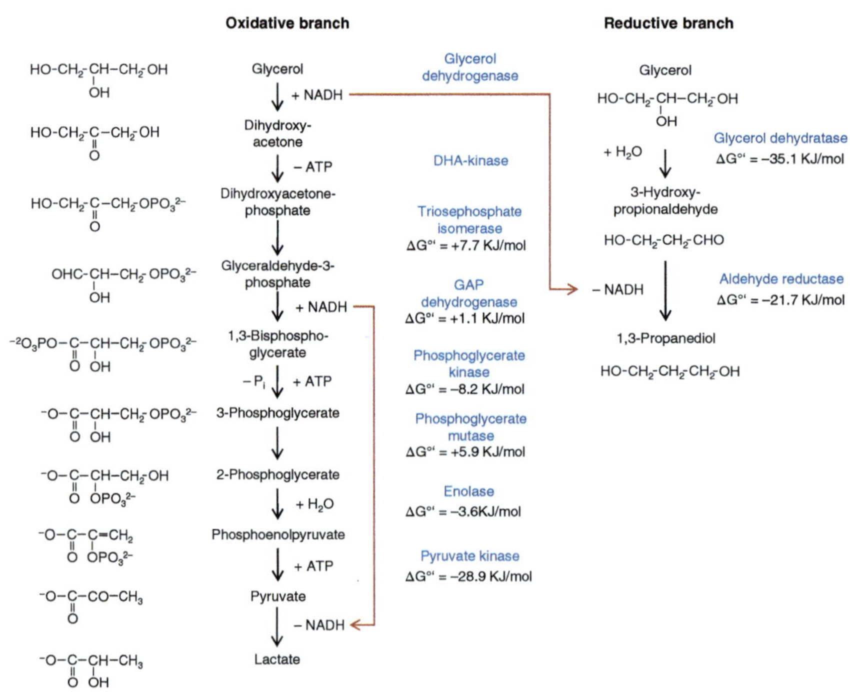

Figure 7.2 Reductive and oxidative branch in the fermentative production of 1,3-propanediol based on glycerol with Gibbs free energy values (Frazão et al. 2019; Alberti 2001 with ionic strength of 0.25 M). Source: Walter Koch.

Table 7.3 Estimation of the molar yield for the synthesis of 1,3-propanediol.

Conversion of glycerol into lactate	
Glycerol + NAD^+ → Dihydroxyacetone + NADH + H^+	
Dihydroxyacetone + ATP → DHAP + ADP	
DHAP → GAP	
GAP + NAD^+ → 1,3-Bisphosphoglycerate + NADH + H^+	
1,3-Bisphosphoglycerate + ADP → 3-Phosphoglycerate + ATP	
3-Phosphoglycerate → 2-Phosphoglycerate	
2-Phosphoglycerate → PEP + H_2O	
PEP + ADP → Pyruvate + ATP	
Pyruvate + NADH + H^+ → Lactate + NAD^+	
Conversion of glycerol into 1,3-propanediol	
Glycerol → 3-Hydroxypropionaldehyde + H_2O	Glycerol dehydratase
3-Hydroxypropionaldehyde + NAD(P)H + H^+ → 1,3-propanediol + NAD^+	Aldehyde dehydrogenase
2 Glycerol + ADP + Pi → 1,3-propanediol + lactate + ATP + 2 H_2O	Net reaction

is also assumed in this context. The overall (chemical) reaction scheme can be stated as following:

$$2\ \text{Glycerol} \rightarrow 1,3-\text{propanediol} + \text{lactic acid} + H_2O$$

The energy balance is revealed if the net reaction is derived from single biochemical steps (Table 7.3).

The pathway would generate a surplus of one mol of ATP per mol of 1,3-propanediol. However, this surplus might be consumed in the regeneration of the aldehyde reductase, which is reported to undergo a mechanism-based inactivation that can be regenerated by an activation factor that needs ATP (Ma et al. 2012). The 28 electrons carried by the two mol of glycerol (2 × 14 without counting reduced oxygen) are allocated to 1,3-propanediol (16 electrons) and lactic acid (12 electrons). The metabolic constraint of balanced reduction equivalents forbids aerobic conditions, which would consume some reduction equivalents for respiratory chain phosphorylation and limit the NADH pool for the reduction of propionaldehyde. It renders all dual pathways rather unattractive since the overall molar yield will inevitably remain below 1.0 mol 1,3-propanediol per mol glycerol. The attractiveness of this approach with two branches depends on the number of reduction equivalents that are generated in the oxidation of glycerol. 0.5 mol 1,3-propanediol maximum would be obtained in case one mol reduction equivalent is generated in the oxidation of glycerol, which is exemplified in the coproduction of lactic acid.

A pathway analysis revealed that the theoretical molar yield may reach up to 0.64 mol 1,3-propanediol per mol glycerol (without the formation of biomass) in the case that acetate with a high oxidation level is formed as a by-product (Biebl et al. 1999). This estimation is reasonable since about two mol of 1,3-propanediol may be obtained from altogether three mol of glycerol in case where acetate and formic acid are generated as by-products, which corresponds to a maximum theoretical molar yield of 0.67 mol of 1,3-propanediol per mol of glycerol:

$$3\ \text{Glycerol} \rightarrow 2 \times 1{,}3-\text{propanediol} + \text{acetic acid} + \text{formic acid} + H_2O$$

The 42 electrons carried by three mol glycerol (3 × 14) are allocated to two mol 1,3-propanediol (2 × 16), one mol acetic acid (8 electrons), and one mol formic acid (2 electrons in all instances without counting reduced oxygen).

7.3.3 Glucose as Carbon Source and Provision of Reduction Equivalents via Electrical Power

The estimation of the maximum carbon yield discussed above has revealed that a significant share of the overall glucose feed would be needed for providing reduction and energy equivalents and does not contribute to the construction of the carbon-backbone of 1,3-propanediol. This circumstance opens up the development potential for microbial electrosynthesis. If the reaction equation is stated with the specific reduction demand – not disguised as biochemical reduction equivalents – the potential impact of a direct electron supply is obvious:

$$\text{Glucose} + 8\,e^- + 8\,H^+ \rightarrow 2\,1{,}3-\text{propanediol} + 2\,H_2O$$

The 24 electrons carried by glucose (without counting reduced oxygen) are not sufficient to provide the 32 electrons (2 × 16 without counting reduced oxygen) carried by two mol of 1,3-propanediol. The unbalanced demand for reduction equivalents might be supplied via microbial electrosynthesis to avoid or at least limit the generation of by-products. The direct feed of electrons might allow for the reduction of NAD^+ and the generation of ATP via the generation of a proton gradient. The overall equation of the reaction for the electro-fermentative production of 1,3-propanediol might be combined with the electrolysis of water (Table 7.4):

The possible application of electrical current designed for microbial electrosynthesis represents a second solution to provide the required electrons for the reduction of glucose to 1,3-propanediol. According to the maximum molar yield estimated above, about 0.165 mol glucose per mol 1,3-propanediol is needed to provide the electrons for the needed reduction equivalents. This would translate into a glucose unit consumption rate of about 0.39 t glucose per t 1,3-propanediol (0.165 × 180.1/76.1) or 137 €/t 1,3-propanediol if a glucose price of 350 €/t is assumed (Kölbl 2021).

Table 7.4 1,3-Propanediol with oxygen as byproduct.

$4\,H_2O \rightarrow 2\,O_2 + 8\,H^+ + 8\,e$	Anode reaction
$4\,NAD(P)^+ + 8\,H^+ + 8\,e \rightarrow 4\,NAD(P)H + 4\,H^+$	Cathode reaction
$4\,NAD(P)H + 4\,H^{++}$ glucose $\rightarrow 2 \times$ 1,3-propanediol $+ 4\,NAD(P)^+ + 2\,H_2O$	
Glucose $+ 2\,H_2O \rightarrow 2$ 1,3-Propanediol $+ 2\,O_2$	Net reaction

These costs need to be compared with the power cost of the alternative approach to providing electrons via microbial electrosynthesis. The estimation of the respective power cost depends on the current in Volt to be applied. In the case where a current of 1 V is assumed, the electron flow in Coulomb is derived in the case where an electrical work of 1 kWh is provided:

$$1\ \text{W} = 1\ \text{V} \times \text{A} = 1\ \text{V} \times 1\ \text{C/s, derives}$$

$$1\ \text{kWh} = 1\ \text{V} \times 3\,600\,000\ \text{C/h}$$

Since one mol of electrons equals 96 485 Coulomb the power input of 1 kWh at a current of 1 V is derived as follows:

$$1\ \text{kWh} = 1\ \text{V} \times 37.3\ \text{electrons/h}$$

Accordingly, since 4 electrons are needed to replenish the gap of reduction equivalents for the synthesis of one mol of 1,3-propanediol (i.e. eight electrons for two mol of 1,3-propanediol) a (theoretical) power input of 0.107 kWh per mol of 1,3-propanediol would be needed, which translates into a (theoretical) power input of about 1.4 MWh power per ton of 1,3-propanediol. In case a power price of 100 €/MWh is assumed, the cost of providing the electrons would amount to about 140 €/t 1,3-propanediol. Thus, the cost of filling the electron gap via glucose oxidation or microbial electrosynthesis would be in the same range. However, in case the required electrical current would amount to 2 V the power cost would double to about 280 €/t 1,3-propanediol. It should be kept in mind that for the microbial electro-fermentation, exclusively the power costs are considered. The technical setup, including microbial interfaces with anode and cathode and the respective mechanisms to facilitate the direct or indirect use of electrons and the energy efficiency of the system, is not reflected.

7.3.4 Glycerol as Carbon Source and Provision of Reduction Equivalents via Electrical Power

The potential direct feed of electrons via a microbial bio-electrochemical system with Citrobacter as host and glycerol as carbon feed has been demonstrated. The molar yield could be increased from 0.27 mol 1,3-propanediol/mol glycerol up to 0.57 mol 1,3-propanediol/mol glycerol when the cathodic current switched from 1 to 10 A/m^2 (Zhou et al. 2015). This suggests that the biofilm receives electrons from the cathode to generate reduction equivalents for the reduction of 3-hydroxypropionaldehyde to 1,3-propanediol. The maximum theoretical molar yield in the biochemical conversion of glycerol to 1,3-propandiol amounts to 0.5 or 0.66 mol 1,3-propanediol per mol glycerol, respectively, in cases where lactate or acetate are generated as by-products as outlined above. Since this threshold has not been exceeded, the 0.57 mol 1,3-propanediol per mol glycerol achieved does not provide conclusive proof that electrons provided by electrical power were taken up by the biochemical system. Nevertheless, the experiments provide a good indication for a proof of concept, and technical maturity has not yet been achieved (Kracke and Krömer 2014). Pending on the efficiency of biochemical power conversion and the electron transfer mechanisms this idea might turn out as a realistic process improvement in the future.

7.4 Metabolic Engineering

7.4.1 Commercial *E. coli* Strain with Glucose as Carbon Source

A plant for the fermentative production of 1,3-propanediol is operated by Dupont and Tate & Lyle. The technology was originally developed by Genencor, which started in 1992. Genencor had been acquired by Danisco in 2005, and Danisco had been acquired by Dupont in 2011. The commercial use of the technology started in 2007. It is reported that the R&D efforts of the projects required about 575 person years, which would imply that up to 40 people on average p.a. contributed to the project (Karim et al. 2017). The development of an *E. coli* strain capable of converting glucose into 1,3-propanediol with commercial maturity is a landmark in metabolic engineering. The idea to use glucose as a raw material has already been discussed in the 1990s, though technical feasibility was regarded as low. A suitable microbial wild-type strain capable of converting glucose directly into 1,3-PDO was not available at that point in time (Biebl et al. 1999).

The metabolic engineering strategy of Genencor and Dupont has been presented at a conference and revealed in several patents (Laffend et al. 1997; Nair et al. 1999; Diaz-Torres et al. 2000; Emptage et al. 2003). The original attempts may have focused on the introduction of the *dha* regulon of *Klebsiella pneumoniae* into *E. coli* combined with a recombinant G3P dehydrogenase and a G3P phosphatase (WO2001/12833A2), and the host has been improved in various iteration rounds. Several gene knockouts and recombinant expressions were required to convert a common strain of *E. coli* K12 strain into an efficient producer of 1,3-propanediol (Figure 7.3).

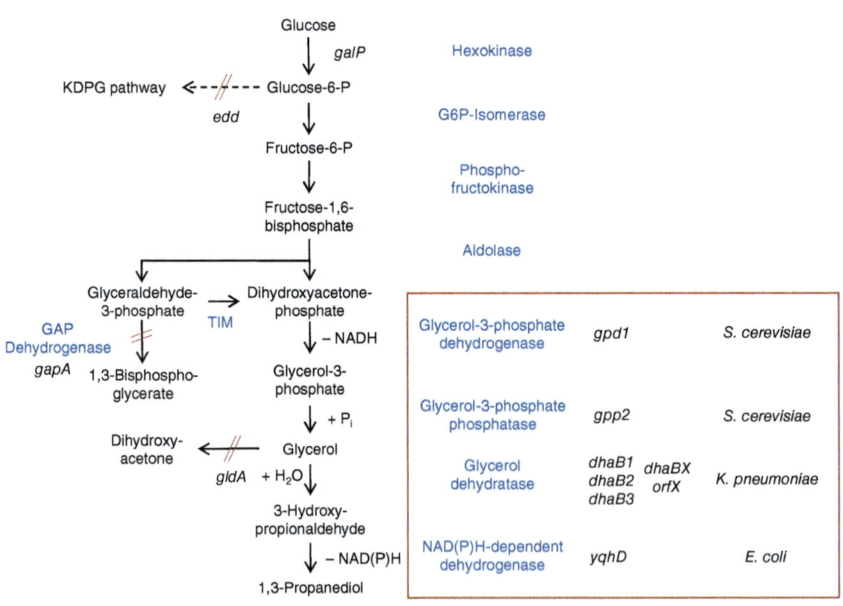

Figure 7.3 Recombinant metabolic pathway of Genencor/Dupont for the fermentative production of 1,3-propanediol in *E. coli* based on glucose. Source: Walter Koch.

In particular, the native phosphoenolpyruvate (PEP)-dependent glucose-uptake system of *E. coli* encoded by the *ptsHI* gene has been knocked out to decouple glucose uptake from ATP consumption. Therefore, the galactose-proton symporter encoded by the *galP* gene has been introduced, which accepts both galactose and glucose as substrate. Competing pathways like the 2-keto-3-deoxy-6-phosphogluconate (KDPG) pathway, the downstream section of glycolysis, or the formation of dihydroxyacetone were silenced or blocked via genetic modification of the *edd* gene, the *gapA* gene and the *gldA* gene, respectively. The formation of glycerol is facilitated by the glycerol-3-phosphate dehydrogenase (*gpd1* gene) and the glycerol-3-phosphate phosphatase (*gpp2* gene). Both genes were obtained from *Saccharomyces cerevisiae* to efficiently generate glycerol. It has been found later on that the construction of a fusion protein of both enzymes provides a significant increase in the production of glycerol given the direct channeling of the intermediate product glycerine-3-phosphate to the next catalytic center (Meynial-Salles et al. 2007), though this approach has not been pursued. In the original work of Genencor and Dupont, the oxidation of glycerol toward 3-hydroxypropionaldehyde was arranged via the glycerol dehydratase of *K. pneumoniae* encoded by the genes *dhaB1*, *dhaB2* and *dhaB3*. The glycerol dehydratase is dependent on the cofactor B12, and two B12 activation factors encoded by *dhaBX* and *orfX* are required. The glycerol dehydratase of *C. butyricum* is not dependent onB12 and the respective metabolic engineering would be easier, though it turned out that the enzyme is not sufficiently active in *E. coli*. The alcohol dehydrogenase encoded by the ydhD gene of *E. coli* facilitates the final conversion of 3-hydroxypropionaldehyde into 1,3-propanediol with NADPH as a cofactor. The alternative 1,3-propanediol dehydrogenase of *K. pneumoniae* using NADH as a cofactor was not effective under aerobic conditions (Meynial-Salles et al. 2007).

7.4.2 Glycerol as Carbon Source with Carbon Dioxide as By-product (*Klebsiella pneumoniae*)

The fermentative synthesis of 1,3-propanediol with glycerol as carbon and energy source facilitates a disproportionation of glycerol toward 1,3-propanediol and by-products like acetate and lactate with higher oxidation status than glycerol. The theoretical molar yield is determined by the number of reduction equivalents generated in the partial oxidation of glycerol. It can be maximized if the glycerol molecule is partially fully oxidized to carbon dioxide. This strategy has been pursued with the expression of the *fdh1* gene encoding the formate dehydrogenase from *Candida boidinii* in *Klebsiella pneumoniae* (Ma et al. 2012), which is supposed to oxidize formate, obtained in the conversion of pyruvate into acetyl-CoA and formate via the pyruvate formate lyase, to obtain one further mol of NADH and carbon dioxide (Figure 7.4):

The net pathway combination of the oxidative and reductive branches with NADH transfer between the two branches can be stated as follows:

$$4\,\text{Glycerol} \rightarrow 3 \times 1,3 - \text{propanediol} + \text{acetic acid} + \text{carbon dioxide} + 2\,H_2O$$

The 56 electrons carried by four mol glycerol (4×14) are transferred to three mol 1,3-propanediol (3×16) and one mol acetate (8 electrons in each instance without counting reduced oxygen). The theoretical molar yield amounts to 0.75 mol 1,3-propanediol per mol glycerol, or a maximum carbon yield of 0.62 g 1,3-prpopamediol per g glycerol

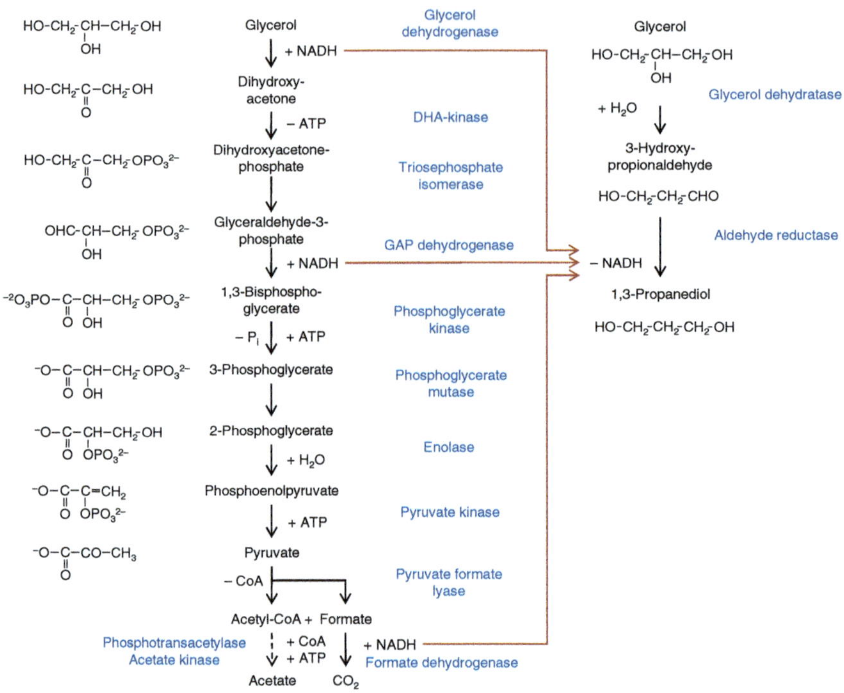

Figure 7.4 Recombinant metabolic pathway toward 1,3-propanediol with glycerol as carbon and energy source. Source: Walter Koch.

Products of the oxidative branch	Electron transfer per mol glycerol	Theoretical molar yield (mol 1,3-PDO/mol glycerol)
Lactate	2	0.5 (1:2)
Acetate + formate	4	0.67 (2:3)
Acetate + CO_2	6	0.75 (3:4)

Figure 7.5 Theoretical molar yield in the conversion of glycerol to 1,3-propanediol in dependency of the by-products generated in the oxidative branch.

(0.75 × 76.1 g/mol/92.1 g/mol). The generation of acetate and carbon dioxide facilitates the transfer of 6 electrons per mol of glycerol and thereby achieves a higher theoretical molar yield than the alternative generation of lactic acid, acetate, and formate as oxidation products of the oxidative branch (Figure 7.5).

The experiments were done with LB medium supplemented with 50 μg/ml kanamycin required to maintain the plasmids carrying the recombinant genes, and the reactor was operated at 30 °C with stirring at 150 r/min. The fermentation was operated under aerobic conditions allowed by the circumstance that *K. pneumoniae* is facultatively anaerobe. The recombinant strain S6-PD without formate dehydrogenase activity achieved a 1,3-propanediol titer of 23 g/l after 30 hours compared to the strain S6-FPD with formate dehydrogenase activity, which displays a 1,3-propanediol titer of 27 g/l (Ma et al. 2012). The glycerol concentration in the medium is reported at 50 g/l, which translates into a carbon yield of 0.54 g 1,3-propanediol per g glycerol. A general proof of concept for the enhanced oxidation of glycerol could be provided through the results are likely not sufficient for a commercial application. It is to be kept in mind that *K. pneumoniae* is a pathogen, and its use in a high-volume commercial production is critical. However, the host has the ability to synthesize coenzyme B_{12}, which is needed for the activity of the glycerol dehydratase (Celińska 2010).

7.4.3 Glycerol as Carbon and Energy Source (*Vibrio natriegens*)

The fermentative production of 1,3-propanediol has also been successfully achieved in the non-mainstream host *V. natriegens*, which is moderately halophile and requires NaCl for optimum growth. This host is characterized by a minimal generation time of 7–10 minutes and has an exceptionally high glucose uptake rate of 3.9 g per g DCW per h under aerobic conditions and 7.81 g per g DCW per h under anaerobic conditions. These abilities qualify *V. natriegens* as a potential host for fermentation processes with very high productivities (Zhang et al. 2021). The host is also able to grow on glycerol as its sole carbon source, and the native metabolism is facilitated by the glycerol kinase (encoded by the *glpk* gene) to convert glycerol into glycerol-3-phosphate and the flavin-dependent glycerol-3-phosphate dehydrogenase (encoded by the *glpD* gene) to convert glycerol-3-phosphate into dihydroxy-acetone-phosphate. Growth on glycerol relies on these two genes, and the knockout of either *glpk* or *glpD* stops microbial growth. Since this route is exclusively pursued under aerobic conditions – according to similar results observed with *E. coli* – aerobic conditions were applied in the fermentation study.

Since the native host *V. natriegens* is not able to convert glycerol into the target product 1,3-propanediol, a suitable recombinant pathway had to be introduced. Accordingly, the glycerol dehydratase of *K. pneumoniae* (encoded by the *dhaBCE* gene), its reactivation factor (encoded by the *orfXY* gene), and the 1,3-propanediol dehydrogenase of *K. pneumoniae* (encoded by the *dhaT* gene) were overexpressed. This strain was exclusively able to produce 1,3-propanediol in small quantities, so the *dhaT* gene of *K. pneumoniae* was replaced by the more efficient *yphD* gene of *E. coli,* which encodes a dehydrogenase to convert 3-hydroxypropionaldehyde into 1,3-propanediol. Glycerol was nearly completely consumed after 12 hours, and a molar yield of 0.22 mol 1,3-propanediol per mol glycerol could be achieved. The applied strain might generate various by-products hampering a high molar yield in the conversion of glycerol into 1,3-propanediol. Thus, the competing pathways toward ethanol (*adhE*), lactate (*ldhA*), formate (*pfl*), and acetate (*pta-ackA*) were deleted. The molar yield could finally be increased up to 0.61 mol 1,3-PDO per mol glycerol by optimization of the promoter for the glycerol dehydratase (J23100), improvement of tolerance toward 3-hydroxypropionic aldehyde via overexpression of the *phaP* gene, increased availability of NADPH via overexpression of the *pntAB* gene (transhydrogenase), and knockout of the *shtA* gene (transhydrogenase). It should be pointed out that the 1,3-propanediol dehydrogenase of *E. coli* operates with NADPH as a cofactor, which favors raising the intracellular NADPH level:

$NADP^+ + NADH \rightarrow NAD^+ + NADPH$	*pntAB* gene (NADPH dependent transhydrogenase)
$NAD^+ + NADPH \rightarrow NADH + NADP^+$	*shtA* gene (soluble pyridine nucleotide transhydrogenase)

Table 7.5 Expressed genes and achieved molar yield.

Molecular engineering	Molar yield
OvEx of *dhaT*, *dhaBCE* & *orfXY*	0.22 mol 1,3-PDO/mol glycerol
OvEx of *dhaT* and *dhaBCE* & J23100	0.37 mol 1,3-PDO/mol glycerol
OvEx of *dhaT* and *dhaBCE* & J23100, *phaP*	0.38 mol 1,3-PDO/mol glycerol
OvEx of *dhaT* and *dhaBCE* & J23100, *phaP* OvEx *pntAB*	0.44 mol 1,3-PDO/mol glycerol
OvEx of *dhaT* and *dhaBCE* & J23100, *phaP* OvEx *pntAB* and KO of *shtA*	0.52 mol 1,3-PDO/mol glycerol
OvEx of *dhaT* and *dhaBCE* & J23100, *phaP* OvEx *pntAB* and KO of *shtA* + optimized media	0.61 mol 1,3-PDO/mol glycerol

Source: Adapted from Zhang et al. (2021).

Table 7.6 Selection of achieved titer and productivity for the fermentative production of 1,3-propanediol based on glycerol. A more comprehensive overview is given elsewhere.

Strain	Titer (g/l)	Yield (mol PDO/mol gly)	Productivity (g/l×h)	Source
C. butyricum	66.2	0.62	1.38	Zhang et al. (2019)
C. butyricum	65.0	0.56	1.21	Saint-Amans et al. (1994)
C. freundii	35.6	0.49	0.23	Celińska et al. (2015)
C. pasteurianum	55.0	0.63	2.30	Kaeding et al. (2015)
M. sludge	81.4	0.63	0.99	Jiang et al. (2017)
K. pneumoniae	76.8	0.66	2.56	Lin et al. (2016)
	80.2	0.45	1.16	Jun et al. (2010)
V. natriegens	56.2	0.61	2.36	Zhang et al. (2021)
	44.6	0.52	1.76	Zhang et al. (2021)

Source: Adapted from Saxena et al. (2009).

The final increase in the molar yield was achieved by media optimization. The optimization steps to maximize the molar yield are summarized; competing pathways toward ethanol, lactate, formate, and acetate are blocked in all cases (Table 7.5).

Summarizing, several attempts in different hosts have been pursued to establish a metabolic pathway toward 1,3-propanediol with glycerol as a carbon source. The reported achieved molar yields are in the range of 0.45–0.66 mol 1,3-propanediol/mol glycerol (Zhang et al. 2021; Table 7.6).

7.5 Process Development

The first commercial fermentation technology developed by Genencor and Dupont uses *E. coli* as the fermentation host and glucose as the carbon source. The technology is pursued by the joint venture of DuPont and Tate & Lyle (DuPont 2021) and achieves a reported titer of 135 g/l; the corresponding productivity is indicated with 3.5 g/l×h (Nakamura and Whited 2003). These performance data reflect the technology maturity level as of 2001/2002, and it should be kept in mind that actual titer and productivity might already be higher based on permanent incremental improvements. The process is pursued as an aerobe fermentation to improve the carbon yield.

The protocol for the isolation of the product out of the broth is not published, though it may be assumed that a conventional liquid-liquid extraction with an organic solvent is pursued. The very high solubility of 1,3-propanediol in water does not favor a crystallization approach, and the evaporation point of 213 °C under atmospheric pressure conditions would require a high energy input for a distillation approach with water as the head-fraction. Thus, the broth might be subjected to filtration to remove the biomass, channeled through an ion exchange step for decolorization, and then isolated with a suitable solvent. The solvent would ideally display a rather exhaustive portioning coefficient and have an evaporation point below 1,3-propanediol. On the other hand, it is reported that the hydrophilic character of 1,3-propanediol negatively impacts an efficient liquid-liquid extraction, and other technologies, including reactive extraction, ion exchange, and molecular sieve adsorption, were evaluated. The processes do often require a pretreatment of the broth (e.g. desalination), and the isolation and purification of the product out of the broth are regarded as hindrances for commercial scale-up (Saxena et al. 2009).

The most advanced strains for the production of 1,3-propanediol might also be used for the fermentative production of acrylic acid. It has been revealed that *Gluconobacter oxydans* possesses a natural pathway to convert 1,3-propanediol into acrylic acid (Zhao et al. 2022). In the first step, 1,3-propanediol is oxidized to 3-hydroxypropionaldehyde (3-HPA) by the alcohol dehydrogenase enzyme (ADH). 3-HPA is then dehydrated into acrolein or partially oxidized to 3-hydroxypropionic acid (3-HP) by the aldehyde dehydrogenase (ALDH). Acrolein is also a substrate for the ALDH enzyme, from which acrylic acid is derived (Zhao et al. 2022). Since acrylic acid is toxic and not common to microorganisms, its commercial suitability is questioned given its inherent weak productivity.

7.6 Industrial Application and Capacity

The first plant for fermentative 1,3-propanediol has been built as a joint venture of Dupont and Tate & Lyle (50/50) in Loudon (Tennessee), with startup in 2007 and a capacity of 50 kt pa. The fermentative production of 1,3-propanediol was not primarily driven by a reduction of production cost but the biobased carbon content of the molecule. In 2018, Dupont and Tate & Lyle announced plans to increase the capacity of the plant by 35 Mio. Pounds, up to 66 kt pa, with startup in 2019 (Wang et al. 2022). It has also been reported that the Shenghong Group (China) has started the construction of a 50 kt pa plant for biobased 1,3-propanediol and a 20 kt pa plant for PTT plant in Suzhou (Jiangsu, China) (Rosales-Calderon and Arantes 2019). Since the chemical technology of Shell based on EO as a starting material is operated in a 70 kt pa plant in the US (Arpe 2007) the overall production capacity of 1,3-propanediol may be assumed to be about 150 kt pa.

The biobased product of the JV of DuPont and Tate & Lyle is marketed under specific brand names (Morales-Cerrada et al. 2021). The Sorona® brand refers to the PTT polymer with unique performance-based advantages in apparel fibers and carpet markets. Susterra® refers to the monomer polyol applied in polyurethanes, fluids, and other applications. Zemea® is a preservative-boosting humectant used in personal care, flavors, pharmaceuticals, laundry, and household cleaning (DuPont 2021).

References

Alberti, R.A. (2001). *Thermodynamics of Biochemical Reactions*, 82. Wiley.

Arpe, H.-J. (2007). *Industrielle Organische Chemie. Bedeutende Vor- und Zwischenprodukte*, 179, 320. Wiley.

Becker, J. and Wittmann, C. (2015). Biotechnologie von morgen: Metabolisch optimierte Zellen für die bio-basierte Produktion von Chemikalien und Treibstoffen. *Materialien und Gesundheitsprodukten, Angewandte Chemie* 127: 3383–3407.

Biebl, H., Menzel, K., Zeng, A.-P., and Deckwer, W.-D. (1999). Microbial production of 1,3-propanediol. *Applied Microbial Biotechnology* 52: 289–297.

Celińska, E. (2010). Debottlenecking the 1,3-propanediol pathways by metabolic engineering. *Biotechnology Advances* 28: 519–530. https://doi.org/10.1016/j.biotechadv.2010.03.003.

Celińska, E., Drożdżyńska, A., Jankowska, M. et al. (2015). Genetic engineering to improve 1,3-propanediol production in an isolated *Ctrobacter freundii* strain. *Process Biochemistry* 50: 48–50.

Chen, X., Xiu, Z., Wang, J. et al. (2003). Stoichiometric analysis and experimental investigation of glycerol bioconversion to 1,3-propanediol by *Klebsiella pneumoniae* under microaerobic conditions. *Enzyme and Microbial Technology* 33 (4): 386–394. https://doi.org/10.1016/S0141-0229(03)00135-2.

Clomburg, J.N. and Gonzalez, R. (2013). Anaerobic fermentation of glycerol: a platform of renewable fuels and chemicals. *Trends in Biotechnology* 31 (1): 20–28.

Diaz-Torres, M., Dunn-Coleman, N.S., Chase, M.W., and Trimbur, D. (2000). Method for the production of 1,3-propanediol. United States Patent 6,136,976.

DuPont (2021). The Sorona story, www.sorona.com/our-story (accessed 10 April 2021).

DuPont (2021). www.duponttateandlyle.com (accessed 11 April 2021).

DuPont patent (n.d.). Process for the biological production of 1,3-propanediol with high titer. WO2001/12833A2, Investors: Emptage, M., Haynie, S., Laffend, L., Pucci, J., and Whited, G.

Emptage, M., Haynie, S.L., Laffend, L.A. et al. (2003). Process for the biological production of 1,3-propanediol with high titer. United States Patent No. 6,514,733.

Frazão, C.J.R., Trichez, D., Serrano-Bataille, H. et al. (2019). Construction of a synthetic pathway for the production of 1,3-propanediol from glucose. *Scientific Reports* 9: 11576. https://doi.org/10.1038/s41598-019-48091-7.

Gonzales-Pajuelo, M., Meynial-Salles, I., Mendes, F. et al. (2006). Microbial conversion of glycerol to 1,3-propanediol: physiological comparison of a natural producer, *Clostridium* butyricum VPI 3266, and an engineered strain, *Clostridium acetobutylicum* DGI (pSPD5). *Applied and Environmental Microbiology* 72 (1): 96–101.

Jiang, W., Wang, S., Wang, Y., and Fang, B. (2016). Key enzymes catalyzing glycerol to 1,3-propanediol. *Biotechnology for Biofuels* 9: 57. https://doi.org/10.1186/s13068-016-0473-6.

Jiang, L., Liu, H., Mu, Y. et al. (2017). High tolerance to glycerol and high production of 1,3-propanediol in batch fermentation by microbial consortium from marine sludge. *Engineering in Life Sciences* 17: 635–644.

Jun, S.A., Moon, C., Kang, C.H. et al. (2010). Microbial fed-batch production of 1,3-propanediol using raw glycerol with suspended and immobilized Klebsiella pneumoniae. *Applied Biochemistry and Biotechnology* 161: 491–501.

Kaeding, T., Daluz, J., and Zeng, A.P. (2015). Integrated study of fermentation and downstream processing in a miniplant significantly improved the microbial 1,3-propanediol production from raw glycerol. *Bioprocess and Biosystems Engineering* 38: 575–586.

Karim, A.S., Dudley, Q.M., and Jewett, M.C. (2017). Cell-free systems for metabolic engineering and biosynthetic pathway prototyping, p. 126. In: *Industrial Biotechnology. Microorganisms* (ed. C. Wittmann and J.C. Liao), 125–148. Wiley.

Kölbl, T. (2021). German Corporate Conference, UniCredit /Kepler Cheuvreux. Südzucker company presentation, p. 61, www.suedzucker.de/de/investor-relations/publikationen (accessed 2 March 2021).

Kracke, F. and Krömer, J.O. (2014). Identifying target processes for microbial electrosynthesis by elementary mode analysis. *BMC Bioinformatics* 15: 410. https://doi.org/10.1186/s12859-014-0410-2.

Laffend, L.A., Nagarajan, V., and Nakamura, C.E. (1997). Bioconversion of a fermentable carbon source to 1,3-propanediol by a single microorganism. US Patent No. 5,686,276.

Li, P. and Yang, G. (2016). Synthesis, morphology, and nonisothermal crystallization behavior of poly (trimethylene terephthalate)/poly (propylene glycol) segmented random copolymers. *Polymer Advanced Technology* 27: 1029–1037.

Lin, J., Zhang, Y., Xu, D. et al. (2016). Deletion of *poxB*, *pta*, and *ackA* improves 1,3-propanediol production by Klebsiella pneumoniae. *Applied Microbiology and Biotechnology* 100: 2775–2784.

Lomelí-Rodríguez, M., Corpas-Martínez, J.R., Willis, S. et al. (2018). Synthesis and characterization of renewable polyester coil coatings from biomass-derived isosorbide, FDCA, 1,5 pentanediol, succinic acid, and 1,3-propanediol. *Polymers* https://doi.org/10.3390/polym10060600.

Ma, Z., Shentu, X., Bian, Y., and Yu, X. (2012). Effects of NADH availability on the *Klebsiella pneumoniae* strain with 1,3-propanediol operon over-expression. *Journal of Basic Microbiology* 53: 348–354. https://doi.org/10.1002/jobm.201100580.

Meynial-Salles, I., Forchhammer, N., Croux, C. et al. (2007). Evolution of a *Saccharomyces cerevisiae* metabolic pathway in *Escherichia coli*. *Metabolic Engineering* 9: 152–159.

Morales-Cerrada, R., Tavernier, R., and Caillol, S. (2021). Fully bio-based thermosetting polyurethanes from bio-based polyols and isocyanates. *Polymers* 13: 1255. https://doi.org/10.3390/polym13081255.

Nair, R.V., Payne, M.S., Trimbur, D.E., and Valle, F. (1999). Method for the production of glycerol by recombinant organisms. PCT International application 1999, WO 99/28480.

Nakamura, C.E. and Whited, G.M. (2003). Metabolic engineering for the microbial production of 1,3-propanediol. *Current Opinion in Biotechnology* 14: 454–459.

Raynaud, C., Carcabal, P., Meynial-Salles, I. et al. (2003). Molecular characterization of the 1,3-propanediol (1,3-PD) operon of *Clostridium butyricum*. *PNAS* 100 (9): 5010–5015. https://doi.org/10.1073/pnas.0734105100.

Röper, M. (2006). Selektivität, Aktivität und Standzeit: Homogene Katalyse in der chemischen Industrie. *Chemie in unserer Zeit* 40: 126–135. https://doi.org/10.1002/ciuz.200600373.

Rosales-Calderon, O. and Arantes, V. (2019). A review on commercial-scale high-value products that can be produced alongside cellulosic ethanol. *Biotechnology for Biofuels* 12: 240. https://doi.org/10.1186/s13068-019-1529-1.

Saint-Amans, S., Perlot, P., Goma, G., and Soucaille, P. (1994). High production of 1,3-propanediol from glycerol by *Clostridium butyricum* VPI 3266 in a simply controlled fed-batch system. *Biotechnology Letters* 16: 832–836.

Saxena, R.K., Anand, P., Saran, S., and Isar, J. (2009). Microbial production of 1,3-propanediol: recent developments and emerging opportunities. *Biotechnology Advances* 27: 895–913. https://doi.org/10.1016/j.biotechadv.2009.07.003.

Slininger, P.J. and Bothast, R.J. (1985). Optimizing aerobic conversion of glycerol to 3-hydroxypropionaldehyde. *Applied and Environmental Microbiology* 64: 1444–1450.

Stevens, M.J.A., Vollenweider, S., Mertes, P., and Lacroix, C. (2013). Bisulfite as scavenger for enhanced biotechnological production of 3-hydroxypropionaldehyde by *Lactobacillus reuteri*. *Biochemical Engineering Journal* 79: 239–245. https://doi.org/10.1016/j.bej.2013.08.002.

Steverding, D. (2010). Fermentative Biotechnologie. Mikrobielle Herstellung von 1,3-Propandiol. *Chemie in unserer Zeit* 44: 384–398. https://doi.org/10.1002/ciuz.201000531.

Syzmanowska-Powalowska, D. and Bialas, W. (2014). Scale-up of anaerobic 1,3-propanediol production by *Clostridium butyricum* DSP1 from crude glycerol. *BMC Microbiology* 14: 45. http://www.biomedcentral.com/1471-2180/14/45.

Tobimatsu, T., Kajiura, H., and Toraya, T. (2000). Specificities of reactivating factors for adenosylcobalamin-dependent diol dehydratase and glycerol dehydratase. *Archives of Microbiology* 174: 81–88. https://doi.org/10.1007/s002030000179.

Wang, X.-L., Zhou, J.-J., Liu, S. et al. (2022). In situ carbon dioxide capture to co-produce 1,3-propanediol, biohydrogen and micro-nano calcium carbonate from crude glycerol by *Clostridium butyricum*. *Biotechnology for Biofuels* 15: 91. https://doi.org/10.1186/s13068-022-02190-2.

Xin, B., Tao, F., Wang, Y. et al. (2017). Coordination of metabolic pathways enhanced carbon conservation in 1,3-propanediol production by coupling with optically pure lactate biosynthesis. *Metabolic Engineering* 41: 102–114.

Zhang, Y., Liu, D., and Chen, Z. (2017). Production of C2-C4-diols from renewable resources: new metabolic pathways and metabolic engineering strategies. *Biotechnology for Biofuels* 10: 299. https://doi.org/10.1186/s13068-017-0992-9.

Zhang, A.H., Zhuang, X.Y., Chen, K.N. et al. (2019). Adaptive evolution of *Clostridium butyricum* and scale-up for high-concentration 1,3-propanediol production. *AIChE* 65: 32–39.

Zhang, Y., Li, Z., Liu, Y. et al. (2021). Systems metabolic engineering of *Vibrio natriegens* for the production of 1,3-propanediol. *Metabolic Engineering* 65: 52–65. https://doi.org/10.1016/j.ymben.2021.03.008.

Zhao, L., Zhu, J., Ro, K.-S. et al. (2022). Discovery of a novel acrylic acid formation pathway in *Gluconobacter oxydans* and its application in biosynthesis of acrylic acid from glycerol. *Process Biochemistry* 182–189. https://doi.org/10.1016/j.procbio.2022.04.027.

Zhou, M., Freguia, S., Dennis, P.G. et al. (2015). Development of bioelectrochemical activity stimulates mixed-culture reduction of glycerol in a bioelectrochemical system. *Microbial Biotechnology* 8 (3): 483–489. https://doi.org/10.1111/1751-7915.12240.

8

Butanol

8.1 Application

Butanol (1-butanol) is a linear C4 alcohol with the formula $C_4H_{10}O$ and a molecular weight of 74.1 g/mol. It is applied as raw material for chemical synthesis such as for butylesters and butylethers. The most common butyl-esters comprise butylacrylate, butylmethylacrylate, and butylacetate. It is additionally used as technical solvent or additive in various applications (Mascal 2012). Its chemical application is well established but limited with respect to volume growth. The global market value for 2015 was estimated with over 5 $ bn in 2012, with a growth rate of about 5% (Mascal 2012). The main interest in butanol from a fermentation point of view is its potential high-volume application as biofuel (Bharathiraja et al. 2017).

Fuels may be regarded as chemical energy storage. The controlled chemical reaction of liquid fuels with air or oxygen ("burning") respectively, in combustion engines or turbines toward carbon dioxide and water releases heat and provides energy for transportation and aviation. Petroleum derived liquid transportation fuels are composed of hydrocarbons with different chain lengths from C5 to C20 with an average energy content of around 40 GJ/t. Jet fuels bear different property requirements given their specific application context like lubricity, freezing point, and energy density, which are met by aromatic compounds along with linear branched and cyclic paraffins with a chain length of C8-C16 (Zhou et al. 2018). Fermentation products might also be applied as fuel and promise a high-volume application perspective with greater business value compared to fermentation products with sophisticated molecular structures and a smaller volume perspective for niche applications (Peralta-Yahya and Keasling 2010). Accordingly, the by far biggest fermentation product with respect to volume is ethanol, with an annual output of about 100 Mio. t (without alcoholic beverage). The significant volume growth since 2000 is mainly due to the increasing utilization of ethanol as fuel (Witsch 2020).

In the last 20 years, tremendous R&D efforts were pursued to also establish butanol as additional fermentation product (Birgen et al. 2019; Ekas et al. 2019; Trindade, da, and dos Santos 2017), besides isobutanol (Becker et al. 2012), hydrogenated farnesene (Jiang et al. 2018; Tippmann et al. 2013), and hydrogenated bisabolene (Jiang et al. 2018; Phelan et al. 2015; Peralta-Yahya et al. 2011). Fuel evaluation is less referring to molecular functionality, i.e. the structure and the spatial arrangements of functional groups, though mainly the price for energy content expressed as low heating value (LHV). Fuels are furthermore

Pathway Design for Industrial Fermentation, First Edition. Walter Koch.

Table 8.1 Energy content and density of fuels.

Fuel	MeOH	EtOH	BuOH	iBuOH	Gasoline	Diesel	Farnesane
Density (g/ml)	0.79	0.79	0.81	0.80	0.74–0.80	0.83	0.77
Density (MJ/l)	16.0	19.6	27–29.2	29.0	32.0	39–47	
LHV (GJ/t)	20.1	26.9	33.1	33.3	43.1	42.8	44.2
HHV (GJ/t)	22.7	29.7	36.1	36.0	47.3	44.8	
Boiling point (°C)	65	78	118	108	27–225	187–343	

ranked based on their logistic properties, like energy density, and their suitability for existing infrastructure and grids. Table 8.1 compares the respective properties of fuels (Fu et al. 2021; Trindade, da, and dos Santos 2017; Tippmann et al. 2013).

In this respect, butanol defines an interesting target for fermentation technology given its lower heating value of 33.1 GJ/t, which is significantly above the respective value of ethanol with a lower heating value of 26.9 GJ/t. It might simply be mixed with conventional fuel for automotive combustion engines like ethanol and partially replace conventional gasoline derived from petrochemical resources or even allow a complete substitution of gasoline. With respect to tank capacity of liquids, especially the energy density in terms of GJ per liter is decisive and the advantage of butanol is further enhanced given its slightly higher density of about 0.81 kg/l compared with 0.79 kg/l for ethanol (Ndaba et al. 2015; Kolesinka et al. 2019). Furthermore, butanol is not hygroscopic, whereas ethanol attracts water, and butanol is much less corrosive and does not require an upgrade in gasoline storage capacity. Fuel application of butanol would allow a high-volume production and thereby justify significant development efforts for the respective R&D in metabolic engineering and downstream processing.

8.2 Conventional Production of Butanol

8.2.1 ABE Fermentation as First Commercial Butanol Technology

The conventional ABE fermentation was the first butanol technology applied on an industrial scale. This natural fermentation process with *C. acetobutylicum* was discovered and developed by Chaim Weizmann at the beginning of the 20th century (Sahm et al. 2013). It was started in 1913 primarily to obtain butanol as a building block for rubber and acetone as solvent, and achieved a technology market share of about 65% for butanol and 33% for acetone in 1945 (Birgen et al. 2019). Despite the formerly significant industrial application, the technology bears several disadvantages if compared to other fermentation processes: The limited tolerance of the applied *Clostridium* strains does not allow raising the butanol titer above 15–20 g/l and the selectivity is weak since with ethanol and acetone two major byproducts are generated. The usual product mixture comprises 6:3:1 (w/w) butanol, acetone, and ethanol. The ABE fermentation pursued with *C. acetobutylicum*, *C. beijerinckii*, and *C. saccharobutylicum* proceeds in two phases: In the first acidogenic phase, glucose is converted into primarily butyrate and acetate, and small amounts of carbon dioxide, hydrogen, and ethanol are formed. In the second solventogenic phase, the secreted acids are recaptured and converted into butanol and acetone. The conversion of acids into alcohols and the ketone stabilizes the external acidity, which would endanger the internal pH management of the clostridia, which do not possess the physiological tools for a robust pH management. With increasing external concentrations of butyrate and acetate, undissociated molecules would pass the membrane and thereby lower the cytoplasmic pH. This is avoided with the initiation of the solventogenic phase.

The weak selectivity limits the carbon yield with respect to butanol and implies a high investment in the downstream processing. A defined isolation protocol is required to capture all three products and to provide pure products channeled into the different markets. All three products need to be isolated and purified, which incurs additional processing costs. Especially the power and steam consumption needed for the isolation and purification of the three products diminishes the energy efficiency of the process, and a significant share of the lower heating value (LHV) of butanol needs to be invested. Accordingly, technology is under development to process the obtained product mixture "as is" as a suitable feedstock for a fuel generation technology. These technologies may comprise a dehydration of alcohols to obtain alkenes, which would then be channeled via an oligomerization reactor to derive carbohydrate for fuel. Alternatively, the alcohols may be dehydrogenated to aldehydes, coupled via aldol condensation, and then be hydrogenated (Nahreen 2015; Ketabchi et al. 2020).

Butanol has, with about 1.200 €/t the highest value compared to acetone and ethanol (Nguyen et al. 2018; Haas et al. 2018). The commercial boundary conditions of the ABE fermentation changed in the 20th century. The value of the by-products has diminished in the second half of the last century given the increasing technology maturity of fermentative ethanol and the rising production volume of phenol via the Hock process, which derives acetone as by-product. Butanol could cost-efficiently be produced via organic chemistry given the development of petrochemistry. Accordingly, most of the ABE fermentation plants were closed due to missing competitiveness, and the technology lost industrial relevance (Tracy 2012). The ABE technology regained attention in the industry and academia in the 1980s with rising oil prices and the accelerated search for biobased fuels. R&D programs focused on a higher butanol selectivity, titer increase, and toxicity management, respectively, either by strain improvement or the development of *in situ* removal technology (Outram et al. 2017) and continuous flow reactors (Vees et al. 2020).

8.2.2 Hydroformylation of Propylene Toward oxo C4 Alcohols

Butanol is currently mainly produced by the hydrogenation of butyraldehyde. The reaction is done with molecular hydrogen and pursued either with Ni-catalysts at a pressure of 2–3 bar and a temperature of 115 °C or with Cu-catalysts at a pressure of 30–50 bar and a temperature of 130–160 °C. In most of the cases the isolated aldehyde is applied, though also a mixture of raw aldehydes obtained in the preceding hydroformylation may be used (Arpe 2007):

$$CH_3\text{–}CH_2\text{–}CH_2\text{–}CHO + H_2 \rightarrow CH_3\text{–}CH_2\text{–}CH_2\text{–}CH_2\text{–}OH$$

Butyraldehyde is obtained in the hydroformylation of propylene. The hydroformylation uses syngas, a defined molar mixture of carbon monoxide and hydrogen, pending on the target product. Since the carbon monoxide of the syngas may form a covalent bond with each of the two C-atoms of the propylene double bond, the selectivity of the reaction is low, and two products are obtained: the linear butyraldehyde as target product and the branched isobutyraldehyde as by-product:

$$CH_3–CH=CH_2 + CO + H_2 \rightarrow CH_3–CH_2–CH_2–CHO$$

$$CH_3–CH=CH_2 + CO + H_2 \rightarrow CH_3–CH(CH_3)–CHO$$

The share of the undesired by-product isobutyraldehyde could be more and more suppressed via catalyst improvement with focus on selectivity (Arpe 2007). Especially the low propylene price enabled by the emerging petrochemistry in the 1950s and the advancement of catalysts for hydroformylation technology triggered the replacement of the formerly dominating ABE fermentation.

8.2.3 Butanol via the Dimerization of Ethanol

Since a very efficient fermentation technology toward ethanol is available and commercially applied, it is also considered to derive butanol via the dimerization of ethanol. The Guerbet coupling process proceeds via three chemical steps. Ethanol is dehydrogenated to the reactive acetaldehyde, which pursues an aldol condensation to obtain the C4 intermediate crotonaldehyde. In the subsequent chemical reaction, indeed, two different hydrogenations occur, whereby the double bond is saturated and the aldehyde function is converted into an alcohol function:

$$CH_3–CH_2–OH \rightarrow CH_3–CHO + H_2$$

$$2\,CH_3–CHO \rightarrow CH_3–CH=CH–CHO + H_2O$$

$$CH_3–CH=CH–CHO + 2\,H_2 \rightarrow CH_3–CH_2–CH_2–CH_2–OH$$

The technology is evaluated in R&D projects and academia. Hereby a gas phase and a liquid phase process are under evaluation. The overall process describes a redox-neutral rearrangement with dehydrogenation and subsequent hydrogenation. It will be crucial to limit the reactor setup since, in total, four chemical steps are needed if the final hydrogenation is separated by the saturation of the double bond and the reduction of the aldehyde. It will also be important to achieve a full recycling of molecular hydrogen. Both the gas phase and the liquid phase technology have not yet reached technical maturity since conversion and selectivity are still weak (Choi et al. 2021; Vlasenko et al. 2019).

8.2.4 Butanol via Catalytic Alkane Hydroxylation

As an alternative to glucose or carbohydrate, butane might be perceived as a cheap and abundant raw material with suitable C4 backbone for producing butanol. Butane is a component of "wet" natural gas and can be separated via liquefaction (thinkstep 2017). However, the conversion of butane is difficult since the reactions often display a weak selectivity. A functionalization toward a primary or secondary alcohol via organic chemistry is in general possible, though it does not yet provide a commercially interesting efficiency. A hydroxylation of butane has been pursued using a P450 monooxygenase (Staudt et al. 2012). The biocatalytic reaction would be pursued with molecular oxygen at temperatures below room temperature. One of the oxygen atoms is added to butane and creates the hydroxy function, and the second oxygen atom is hydrogenated to yield water:

$$CH_3\text{–}CH_2\text{–}CH_2\text{–}CH_3 + NADPH + H^+ + O_2 \rightarrow CH_3\text{–}CH_2\text{–}CH_2\text{–}CH_2\text{–}OH + NADP^+ + H_2O$$

A stoichiometric consumption of the cofactor NADPH would be too expensive for a commercial biocatalytic application, and a recycling system is needed. Thus, the established glucose dehydrogenase regeneration system has been chosen, which converts D-glucose into D-gluconolactone and thereby transfers two electrons on $NADP^+$ to obtain NADPH in the reduced form.

$$\text{D-glucose} + NADP^+ \rightarrow \text{D-gluconolactone} + NADPH + H^+$$

The recycling system is working and applied in other industrial biocatalytic conversions. However, still one mol of glucose would be invested to obtain one mol of butanol and a downstream application for D-gluconolactone with sufficient value perspective is recommended. Pending on the reaction conditions and the chosen methane P450 monooxygenase of *Methylocystis* sp. a 58:42 mixture of 1-butanol and 2-butanol has been obtained. For some monooxygenases and defined reaction conditions, a high regioselectivity toward 2-butanol has been obtained, though a selective synthesis of 1-butanol has not yet been achieved (Staudt et al. 2012). Biocatalysis remains a competing technology with high potential, though regioselectivity toward 1-butanol is a prerequisite. Enzyme optimization via directed evolution might in the future provide a solution.

8.3 Pathway Design Based on Glucose

8.3.1 Formation of Acetyl-CoA via the Pyruvate Dehydrogenase Complex

The easiest route toward fermentative butanol may simply assume condensation of two mol of ethanol with the release of water. Although this reaction would have a Gibbs free energy of -30 KJ/mol and would thereby be highly exergonic, it is not found in the broad toolbox of biocatalytic activity. There is no enzyme known to catalyze this reaction (Bar-Even et al. 2012).

There is one standard biochemical pathway toward butanol, which is followed in the conventional ABE fermentation with the natural producers' *C. acetobutylicum* and *C. beijerinckii* as biocatalyst and the recombinant concepts with *E. coli* as biocatalyst. It proceeds via acetyl-CoA as abundant cellular starting point and may be referred to as CoA-dependent route opposed to eukaryotic fatty acid synthesis with acyl carrier protein bound intermediates. It proceeds via acetoacetyl-CoA as the first C4 intermediate building block. The following steps convert the ß-keto function at the C3-position into a methylene-group according to conventional fatty acid synthesis to derive butyryl-CoA. The activated carboxylic acid function is in the following steps reduced down to the alcohol-level and butanol is obtained (Figure 8.1):

It is interesting to note that the biochemical Claisen condensation of two mol of acetyl-CoA to acetoacetyl-CoA facilitated by a thiolase is thermodynamically not favored. The equilibrium constant is $K_{eq} = 1.1 \times 10^{-5}$, and the estimated Gibbs free energy amounts

Glucose + 2 ADP + 2 Pi + 2 NAD^+ → 2 Pyruvate + 2 ATP + 2 NADH + 2 H^+ + 2 H_2O
2 Pyruvate + 2 NAD^+ + 2 CoA → 2 Acetyl-CoA + 2 NADH + 2 H^+ + 2 CO_2
2 Acetyl-CoA → Acetoacetyl-CoA + CoA
Acetoacetyl-CoA + NAD(P)H + H^+ → 3-Hydroxybutyral-CoA + $NAD(P)^+$
3-Hydroxybutyral-CoA → Crotonyl-CoA + H_2O
Crotonyl-CoA + NADH + H^+ → Butyryl-CoA + NAD^+
Butyral-CoA + NADH + H^+ → Butyraldehyde + CoA + NAD^+
Butyraldehyde + NAD(P)H + H^+ → Butanol + $NAD(P)^+$

Glucose + 2 ADP + 2 Pi → Butanol + 2 CO_2 + 2 ATP + 3 H_2O

Figure 8.1 Metabolic pathway for the fermentative production of butanol.

to about +26.1 KJ/mol (Vögeli et al. 2018; Dellomonaco et al. 2011; Lan and Liao 2012). Thus, the reaction can exclusively proceed in the target direction in case the product, acetoacetyl-CoA, is directly converted into 3-hydroxybutyryl-CoA. Different solutions to overcome this hurdle might be implemented in biochemical systems. It is reported that the thiolase of *Methanothermococcus thermolithotrophicus* is non-covalently linked to the HMG-CoA synthase, which was elucidated in native purification methods. The interlink is assumed to directly channel acetoacetyl-CoA to the active site of the HMG-CoA reductase to react again with acetyl-CoA to obtain HMG-CoA (Vögeli et al. 2018). Both enzymes share a CoA-binding site, which is conserved in most archaea and many bacteria. On the other hand, the synthesis of fatty acids based on acetyl-CoA usually proceeds via malonyl-CoA or malonyl-ACP (acyl carrier protein). ATP is invested in the formation of malonyl-CoA or malonyl-ACP, and the hydrolysis of the active phosphoric acid anhydride bond ensures a supportive thermodynamical framework in the formation of fatty acids (McMurry and Begley 2006):

$$CH_3\text{–CO–S–CoA} + ATP + H_2O + CO_2 \rightarrow HOOC\text{–}CH_2\text{–CO–S–CoA} + ADP + Pi$$

The pathway is continued via the condensation of acetyl-CoA and either malonyl-CoA or malonyl-ACP, respectively. Hereby carbon dioxide is released, which thermodynamically favors the reaction. This is the reason why the metabolic pathway toward butanol according to the synthesis of fatty acids provides one ATP less per mol of butanol than the reversed ß-oxidation pathway, which directly uses acetyl-CoA as donor of two-carbon units (Dellomonaco et al. 2011). The metabolic engineering strategy addressing this thermodynamical hurdle may also try to provide a more suitable ratio of the reduced and the oxidized forms of the cofactor in the reduction of acetoacetyl-CoA to 3-hydroxybutyryl-CoA. Since the NADPH/$NADP^+$ ratio in *A. acetobutylicum* is at least 70 times higher than the NADH/NAD^+ ratio, a strictly NADPH-dependent 3-hydroxybutyryl-CoA dehydrogenase has been searched and was identified in *C. kluyveri*. The replacement of the native NADH-dependent enzyme by the NADPH-dependent 3-hydroxybutyryl-CoA dehydrogenase facilitates a significant flux increase toward butanol (Nguyen et al. 2018).

The overall equation of the reaction indicates that one mol of glucose is needed to derive one mol of butanol. Thereby, two mol of carbon dioxide are generated and two mol of ATP are obtained for cell growth and maintenance. Glucose is via glycolysis split up in two mol of pyruvate, which are disproportionated to obtain butanol and carbon dioxide. The 24 electrons of glucose (without counting the electrons allocated to the reduced oxygen atoms) end up in butanol with a higher ratio of electrons per C-atom compared to glucose (i.e. four electrons/C-atom in glucose and six electrons/C-atom in butanol). The electron balance implies that no surplus reduction equivalents are obtained. Thus, the fermentation can be pursued as an anaerobic process, and oxygen as final electron acceptor is not needed.

The reduction equivalents are balanced in case all steps would apply NADH as cofactor. The reduction equivalents generated in the glycolysis of glucose toward pyruvate and the conversion of pyruvate into acetyl-CoA are consumed in the downstream steps of the pathway. However, since the conversion of acetoacetyl-CoA towards 3-hydroxybutyryl-CoA and of butyraldehyde toward butanol is in some cases – pending on the cofactor specificity of the isoenzyme in the respective host – associated with the usage of NADPH as cofactor, the net equation of the reaction may not be balanced with respect to reduction equivalents, although the electron allocation on the two products is balanced:

$$\begin{aligned}&\text{Glucose} + 2\,\text{ADP} + 2\,\text{Pi} + 2\,\text{NADPH} + 2\,\text{NAD}^+\\&\quad\rightarrow \text{Butanol} + 2\,\text{CO}_2 + 2\,\text{ATP} + 2\,\text{NADP}^+ + 2\,\text{NADH} + 3\,\text{H}_2\text{O}\end{aligned}$$

The unbalance between the chemically slightly different types of reduction equivalents needs to be addressed in the metabolic engineering strategy (Figure 8.2).

The reduction of a carboxylic acid function to the corresponding aldehyde – in this pathway the reduction of butyric acid to butyraldehyde – is from a thermodynamic point of view, difficult in biochemical systems. This is due to the reduction potential of organic acids to aldehydes of $E_0 = -580$ mV, which is not in the range of the standard cofactor for electrons transfer reactions, NAD(P)H/NAD(P)$^+$, with a reduction potential of $E_0 = -320$ mV. A respective reduction might be achieved with reduced ferredoxin ($E_0 = -500$ mV), as e.g. in *C. autoethanogenum* for the reduction of acetate toward acetaldehyde (Mock et al. 2015). As an alternative, the carboxylic acid function may be presented as CoA-thioester in an

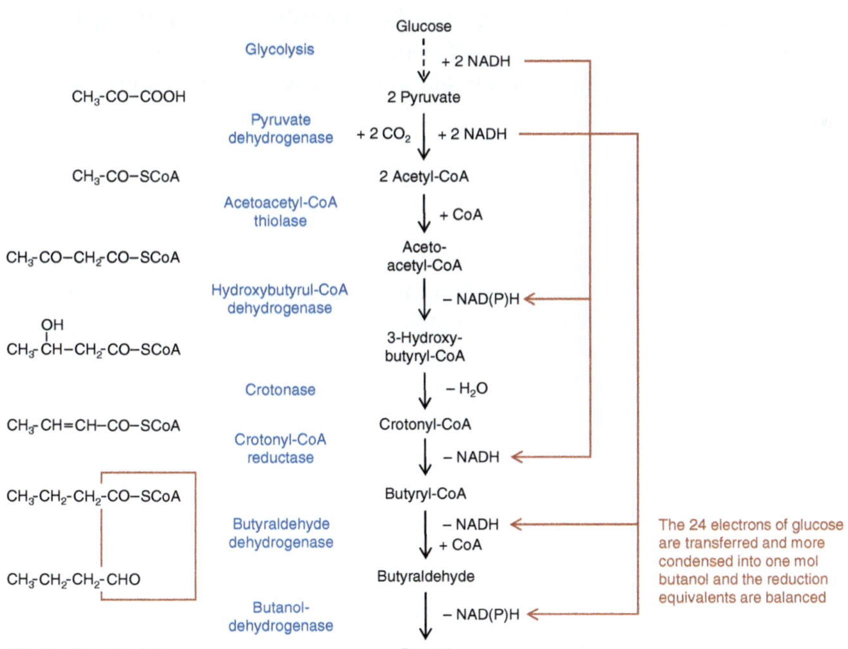

Figure 8.2 Biochemical pathway for the conversion of glucose into butanol with pyruvate dehydrogenase complex for the generation of acetyl-CoA. The reduction of the carboxylic acid function to the aldehyde level proceeds via the activated thioester. Source: Walter Koch.

activated form, as exemplified in the metabolic pathway toward butanol. Thus, the reduction of the carboxylic acid function is achieved as the reduction of butyryl-CoA toward butyraldehyde.

The reaction would generate a minimum carbon dioxide scope 1 emission of 1.19 t CO_2/t butanol (2 × 44 g/mol/74.1 g/mol). The carbon loss via the formation of carbon dioxide does not define a development potential, though mirrors the limits of thermodynamics. The maximum molar yield of one mol of butanol per mol of glucose translates into a maximum carbon yield of 0.41 t butanol per ton glucose (74.1/180.2). The heating value of butanol is estimated given the following formula:

$$\text{Butanol} + 6\,O_2 \rightarrow 4\,CO_2 + 5\,H_2O\ \text{(liquid)}$$

The enthalpy of the reaction is estimated according to Hess's law (webbook.nist.gov 2022):

$$\Delta H°_R = \Sigma\Delta H°_P - \Sigma\Delta H°_E = (4 \times -393.5\ \text{KJ/mol} + 5 \times -285.8\ \text{KJ/mol})$$
$$- -328.0\ \text{KJ/mol} = -2675.0\ \text{KJ/mol given}$$

$$\Delta H_f°\ CO_2\ (g) = -393.3\ \text{KJ/mol},$$
$$\Delta H_f°\ H_2O\ (l) = -285.8\ \text{KJ/mol, and}$$
$$\Delta H_f°\ \text{butanol}\ (l) = -328.0\ \text{KJ/mol}$$

This translates into an estimated higher heating value of 36.1 GJ/t (HHV) for butanol, which corresponds to the corresponding higher heating value of 36.0 GJ/t published in literature (Fu et al. 2021). The fermentative production of butanol based on glucose would provide a significant "energy concentration" given a heating value of 15.6 GJ/t (HHV) for glucose. The reaction has a theoretical energy efficiency of about 95%, defined as the ratio of output high heating value of butanol in relation to the invested high heating value of glucose:

$$\text{Energy efficiency (theory)} = 0.41\ \text{t butanol/t glucose}$$
$$= 0.41 \text{ x } 36.0\ \text{GJ/t}/15.6\ \text{GJ/t} = 95\%$$

It is also to be taken into consideration that the fermentation process itself is energy consuming. It may be assumed that the stirrer – often the fermentation process step with the highest power demand – has an energy input of about 1 kW/m^3 for an anaerobic process (Meyer et al. 2017). In case a technology development with a future productivity of 6 g/lxh is assumed – that is ambitious, though according to the high productivity level for e.g. mature fermentation technology for ethanol and lactic acid – the production of 1 t of butanol would imply a power consumption of 0.17 MWh/t of butanol or 0.6 GJ/t of butanol. The maximum energy efficiency of the process based on a hypothetical carbon yield of 100% may then be estimated as follows:

$$\text{Energy efficiency (HHV) with } 0.41\ \text{t butanol/t glucose}$$
$$= 0.41 \text{ x } 36.0\ \text{GJ/t}/\ (15.6\ \text{GJ/t} + 0.6\ \text{GJ/t}) = 91\%$$

It should be kept in mind that this value still relies on the assumption that an exhaustive glucose turnover is facilitated and the selectivity is 100%, which is not realistic for a commercial process. Furthermore, a share of 5–10% of the glucose feed will likely be needed for the generation of biomass. Since the biomass might be marketed as animal feed or protein source, this yield drag might not be counted. Additionally, the isolation and purification of butanol out of the fermentation broth likely bears a steam consumption, and the realistic energy efficiency would be lower.

8.3.2 Formation of Acetyl-CoA via the Pyruvate Dehydrogenase Bypass

Since acetyl-CoA is the major intermediate in the fermentative formation of butanol via the reversed ß-oxidation pathway, the host specific boundary conditions to supply this intermediate need to be considered. In case a yeast like *S. cerevisiae* is used the mitochondrial acetyl-CoA pool – generated via the pyruvate dehydrogenase – is not available for a cytosolic reversed ß-oxidation. On the other hand, the location of the pathway in the mitochondrion might imply other issues and, thus, a sufficient cytosolic supply needs to be designed and implemented. In the first instance, pyruvate is converted into acetaldehyde via the non-oxidative pyruvate decarboxylation. The formation of ethanol via the ethanol dehydrogenase is typical for *S. cerevisiae*, and the respective *adh* gene needs to be knocked out or silenced. Acetyl-CoA may then be formed via the aldehyde dehydrogenase and the acetyl-CoA synthase. This reaction cascade is referred to as pyruvate dehydrogenase bypass. Alternatively, acetaldehyde may be activated in a non-ATP demanding approach and be oxidized with parallel esterification to the thiol function of CoA (Figure 8.3):

Although a proof of concept for the fermentative anaerobic production of butanol in *S. cerevisiae* via the reversed ß-oxidation pathway has been shown, the results published in academia are far below a commercial threshold. Metabolic engineering with improved

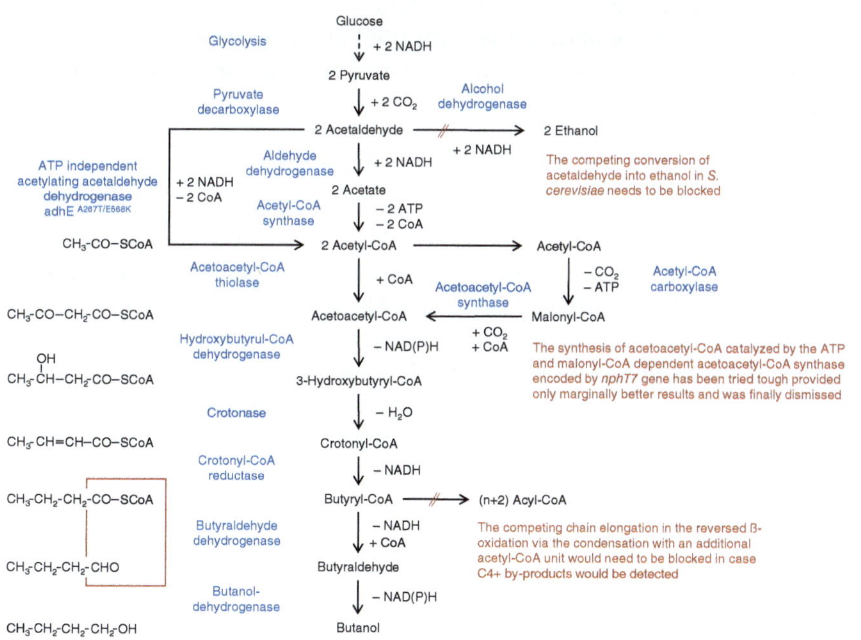

Figure 8.3 Biochemical pathway toward butanol via the pyruvate dehydrogenase bypass (straight route in the middle via acetate). The reduction of the carboxylic acid function to the aldehyde level proceeds via the activated thioester. Source: Adapted from Schadeweg and Boles (2016a).

isoenzymes provided an initial titer of 15 mg butanol/l (Schadeweg and Boles 2016a). The increase of the CoA production by overexpression of the pantothenate kinase gene *coaA* from *E. coli* combined with pantothenate media supplementation provided a titer of 34 mg/l. The deletion of several competing alcohol dehydrogenases genes further raised the titer to 71 mg/l. The additional introduction of a mutant version of the ATP-independent acetylating acetaldehyde dehydrogenase $\text{adhE}^{\text{A267/E568K}}$ resulted in a titer of 95 mg/l. Finally, the stable integration of the *coaA* and the $\text{adhE}^{\text{A267/E568K}}$ genes into the yeast genome, the deletion of a further alcohol dehydrogenase gene, *adh6*, supported by the media supplementation with pantothenate, raised the butanol titer to 130 mg/l after a batch time of 74 h. The glucose yield was 0.012 kg/kg glucose (Schadeweg and Boles 2016a). The switch to an aerobic process allowed a titer of 235 mg/l (Schadeweg and Boles 2016b). It could be shown that especially the conversion of crotonyl-CoA to butyryl-CoA is a rate limiting step, and with overexpression of the *ter* gene encoding the trans-2-enoyl-CoA reductase (i.e. crotonyl-CoA reductase) and some other measures, a titer of 0.86 g/l was achieved. The carbon yield increased to 0.071 kg/kg glucose (Schadeweg and Boles 2016b).

8.4 Pathway Design Based on Carbon Dioxide, Carbon Monoxide and Hydrogen

8.4.1 Syngas Fermentation with Carbon Monoxide and Hydrogen

A different route toward butanol based on CO and H_2 as feedstock has been proposed by Lanzatech, which proceeds via the Wood-Ljungdahl pathway (Lanzatech 2012). The metabolic pathway in the patent application WO 2012/053905 (Lanzatech 2012) proposes that exclusively carbon monoxide is used as a carbon source and the reaction equation may be stated as follows:

$$4\,CO + 8\,H_2 \rightarrow CH_3CH_2CH_2CH_2OH + 3\,H_2O \quad \Delta G^\circ = -334\ \text{KJ/mol}$$

According to this reaction equation, carbon monoxide would exclusively be used as carbon source and hydrogen as energy source (Rückel et al. 2022; Fernández-Naveira and Veiga 2017). In case thermodynamic limitations are exceeded, carbon monoxide might also partially be used as energy source, and carbon dioxide would be generated as byproduct, which would be detrimental for the sustainability profile of the technology. A high selectivity without carbon dioxide as byproduct might be possible in case the metabolic pathway does, in parallel, allow a sufficient generation of ATP needed for growth and cellular maintenance. For acetogenic bacteria in general, ATP might be generated as byproduct of the main fermentation route, e.g. directly with substrate level phosphorylation or indirectly via the generation of a proton or sodium gradient, which can be used to feed an ATP synthase. It would be beneficial for a high selectivity and the avoidance of carbon dioxide as byproduct that ATP is generated along with the main metabolic pathway such that at least the ATP investment in the formation of Formyl-THF can be recycled. This ATP might be replenished in case the processing of butyryl-CoA toward butanol allows the generation of ATP since the thioester formed between butanol and CoA has a sufficient energy level. An ATP might indeed be formed in case butyryl-CoA is converted into butyryl-phosphate, which can be converted into butyrate and ATP facilitated by the butyrate kinase (Figure 8.4):

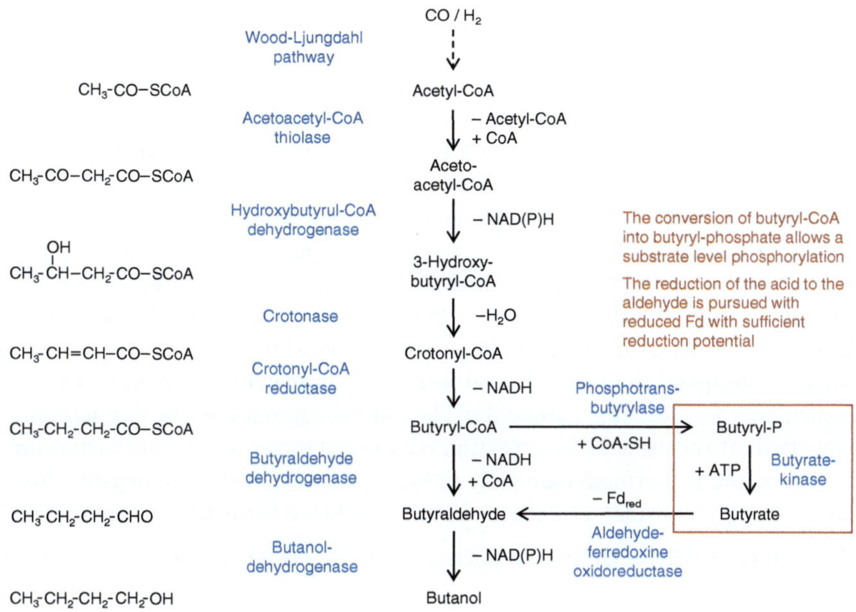

Figure 8.4 Biochemical pathway for the conversion of glucose into butanol (Lanzatech 2012). Source: Adapted from Fernández-Naveira and Veiga (2017).

The hydrolysis of the phosphoric acid ester bond of butyrylphosphate displays a sufficiently high group transfer potential that is more negative than −30.5 KJ/mol (Flamholz et al. 2012), and should thereby allow the generation of one mol of ATP by substrate level phosphorylation:

$$\text{Butyrylphosphate} + H_2O \rightarrow \text{Butyrate} + \text{Pi} \qquad \Delta G^{o\prime} = -35.6\ \text{KJ/mol}$$

However, this route supposes that reduced ferredoxin is generated along with the metabolic pathway since the reduction of butyrate toward butyraldehyde requires reduced ferredoxin given the reduction potential of $E_0{}' = -580$ mV (Katsyv and Mueller 2020). The reduced ferredoxin might be generated via electron bifurcation (Buckel and Thauer 2018; Wang et al. 2013). On the other hand, in case butyryl-CoA is reduced to butyraldehyde, no reduced ferredoxin is needed, but also no ATP can be generated via substrate level phosphorylation.

8.4.2 Syngas Fermentation with Carbon Dioxide, Carbon Monoxide and Hydrogen

A butanol technology with potential "breakthrough" potential is under development by Evonik and Siemens. The overall technology – covering an electrochemical part and a fermentation – pursues to convert carbon dioxide and hydrogen into butanol and hexanol. In the initial step, a share of the carbon dioxide is electrochemically reduced to carbon monoxide since it is not regarded as realistic that a biochemical system with hydrogen as sole energy source is able to develop the required reduction power for the conversion of an organic acid into an aldehyde without compromising selectivity. The reduction of an organic acid toward the respective aldehyde requires reduced ferredoxin, which has a half-reaction potential of $E_0' = -500$ mV, which is on the same level as the half-reaction potential for carbon monoxide toward carbon dioxide. On the other hand, the half-reaction potential of hydrogen with $E_0' = -414$ mV indicates that hydrogen is not an efficient reduction agent to convert oxidized ferredoxin into reduced ferredoxin. In the design of the technology, it was expected that only a minor share of the product spectrum turns out to be butanol and hexanol in case exclusively hydrogen is offered as reduction agent. This implies that a significant share of the product spectrum would be e.g. acetic acid, butyric acid, and hexanoic acid with higher oxidation levels, which are thermodynamically easier to generate. In order to allow a high selectivity toward butanol and hexanol, a share of the carbon dioxide feedstock is electrochemically reduced to carbon monoxide.

The technology is designed to pursue several fermentation reactions in parallel in one reactor. In particular, it foresees that carbon dioxide and hydrogen are in the first fermentation step converted into acetic acid. The reaction is executed by the acetogenic bacterium *C. autoethanogenum* and is thermodynamically rather straightforward if the negative free Gibbs reaction enthalpy is considered. Carbon dioxide is reduced to the level of acetic acid:

$$10\,CO_2 + 20\,H_2 \rightarrow 5\,CH_3COOH + 10\,H_2O, \qquad \Delta G' = -94.6\ KJ/mol$$
$$\Delta H^{\circ\prime} = -310.3\ KJ/mol$$

The second reaction of the first fermentation step is also executed by *C. autoethanogenum* and foresees that acetic acid is partially further converted to ethanol using carbon monoxide and hydrogen as reduction agents (Haas et al. 2018):

$$3\,CH_3COOH + 3\,CO + 3\,H_2 \rightarrow 3\,CH_3CH_2OH + 3\,CO_2 \qquad \Delta G^{\circ\prime} = -29.7\ KJ/mol$$
$$\Delta H^{\circ\prime} = -34.3\ KJ/mol$$

The two products of the first fermentation step are then channeled into a second reactor, in which both intermediates are converted into butyric acid and hexanoic acid (Haas et al. 2018). These condensation and chain prolongation steps, respectively, are pursued by *C. kluyveri*:

$$2\,CH_3COOH + 2\,CH_3CH_2OH \rightarrow 2\,CH_3CH_2CH_2COOH + 2\,H_2O$$
$$CH_3CH_2CH_2COOH + CH_3CH_2OH \rightarrow CH_3CH_2CH_2CH_2CH_2COOH + H_2O$$

Finally, the two acids are in the third fermentation step reduced to butanol and hexanol with carbon monoxide and hydrogen as reaction partners. It should be pointed out that carbon monoxide is not used as carbon source, though oxidized to carbon dioxide to donate the reducing power for the conversion of the acids into alcohols. This conversion is again pursued by *C. autoethanogenum*:

$$CH_3CH_2CH_2COOH + CO + H_2 \rightarrow CH_3CH_2CH_2CH_2OH + CO_2$$
$$CH_3CH_2CH_2CH_2CH_2COOH + CO + H_2 \rightarrow CH_3CH_2CH_2CH_2CH_2CH_2OH + CO_2$$
$$5\,CO_2 + 5\,CO + 25\,H_2 \rightarrow CH_3CH_2CH_2CH_2OH + CH_3CH_2CH_2CH_2CH_2CH_2OH + 13\,H_2O$$

The overall equation of the reaction can be split in two separate reactions providing either butanol or hexanol:

$$2\,CO_2 + 2\,CO + 10\,H_2 \rightarrow CH_3CH_2CH_2CH_2OH + 15\,H_2O$$
$$\Delta G^{\circ\prime} = -283.9\ KJ/mol \quad \Delta H^{\circ\prime} = -747.4\ KJ/mol$$

$$3\,CO_2 + 3\,CO + 15\,H_2 \rightarrow CH_3CH_2CH_2CH_2CH_2CH_2OH + 8\,H_2O$$
$$\Delta G^{\circ\prime} = -453.4\ KJ/mol \quad \Delta H^{\circ\prime} = -1152.4\ KJ/mol$$

The negative reaction enthalpy and especially the free Gibbs energy reveal that the reaction should, in principle, be a suitable target for a fermentation process development (Haas et al. 2018; Haas et al. 2018). The process is sufficiently negative to replenish the ATP pool needed for the formation of formyl-THF and to generate some ATP for cell maintenance and growth.

A decisive evaluation criterion for the efficiency of the proposed fermentation technology is the relation of energy output to energy input. Since both hydrogen and carbon monoxide shall be provided by electrical power, the power consumption is estimated in the first instance. The future power demand for the electrolysis of water to provide H_2 is estimated at 50 MWh/t H_2 according to the alkaline electrolysis (AE) technology (Tountas et al. 2021; Dechema 2019; Rambhujun et al. 2020; Götz et al. 2016). The overall equation of the reaction reveals the hydrogen input rate for butanol and hexanol, and the resulting power demand can be calculated if it is supposed that the hydrogen is provided via the electrolysis of water:

$$0.27\ t\ H_2/t\ \text{butanol}\ \left(10 \times 2\ g/mol/74.1\ g/mol\right)\ \text{which require 13.5 MWh/t butanol}$$
$$0.29\ t\ H_2/t\ \text{hexanol}\ \left(15 \times 2\ g/mol/102.2\ g/mol\right)\ \text{which require 14.5 MWh/t hexanol}$$

The power demand for the electrochemical reduction of carbon dioxide to carbon monoxide is estimated with 6 kWh/Nm^3, or 4.8 MWh/t CO, according to the eCOs^{TM} technology offered by Haldor Topsoe (Topsoe 2022). This power consumption would assume an energy efficiency of 58% in the electrochemical conversion of CO_2 to CO, referring to the reaction enthalpy of the endothermic reaction:

$$CO_2 \rightarrow CO + 0.5\,O_2\ \Delta H^{\circ\prime} = 282{,}8\ KJ/mol\ CO\ \text{or}\ 10.1\ GJ/t\ CO\ \text{or}\ 2.8\ MWh/t\ CO$$

The enthalpy of the reaction is estimated according to Hess's law (webbook.nist.gov 2022):

$$\Delta H_R{}^{\circ} = \Sigma\Delta H_f{}^{\circ}{}_P - \Sigma\Delta H_f{}^{\circ}{}_E$$
$$= -110.5\ KJ/mol - (-393.3\ KJ/mol) = 282.8\ KJ/mol,\ \text{given}$$
$$\Delta H_f{}^{\circ}CO_2\ (g) = -393.3\ KJ/mol,\ \text{and}$$
$$\Delta H_f{}^{\circ}CO\ (g) = -110.5\ KJ/mol.$$

On this basis, the power demand for the generation of butanol and hexanol based on the carbon monoxide demand can be calculated, and the total power consumption (without process utilities) is derived:

$$0.76\ t\ CO/t\ \text{butanol}\ \left(2 \times 28\ g/mol/74.1\ g/mol\right)\ \text{which require 3.65 MWh/t butanol}$$
$$0.82\ t\ CO/t\ \text{hexanol}\ \left(3 \times 28\ g/mol/102.2\ g/mol\right)\ \text{which require 3.94 MWh/t hexanol,}$$

Which derives the total power consumption for the raw material supply of butanol and hexanol:

17.15 MWh/t butanol, and
18.44 MWh/t hexanol.

This estimation is somehow according to the estimation given in the original publication of Evonik, which refers to the overall system and does not differentiate between the electrolysis of water and the electrochemical reduction of carbon dioxide. The energy input is estimated differently, starting with the applied tension and the electron demand for the reduction of carbon dioxide into the two products. At a tension of 2.45 V, 1 Watt electrochemical work is needed to provide a current of 0.41 A: 1 W = 2.45 V × 0.41 A. Since 1 A = 1 C per second, it can be concluded that 1 kWh at a voltage of 2.45 would provide 1.475.000 C, or 15.29 mol electrons, since one mol electron equals 96,485 Coulomb. As one mol of butanol requires 24 electrons if synthesized from carbon dioxide as carbon source, 0.64 mol of butanol (15.29/24) would be obtained in case 1 kWh is invested. Accordingly, 0.42 mol of hexanol (15.29/36) would be obtained in case 1 kWh is invested. This translates into the following energy demand:

0.64 mol butanol with 1 kWh derive 21.09 MWh/t butanol, and
0.42 mol hexanol with 1 kWh derive 23.30 MWh/t hexanol.

The power consumption estimated in the original publication of Evonik and Siemens amounts to 22 MWh/t "product" (mixture of butanol/hexanol; Haas et al. 2018). The figures derived in this study and the original publication cannot directly be compared. The deviation might be linked to the efficiency of the electrolyzer, which might not exactly match the 50 MWh/t H_2 assumed for the water electrolysis.

The overall process setup foresees to apply an electrolyzer, two fermentation reactors, and two separation units, which are continuously operated (Figure 8.5). The electrolyzer unit converts water and carbon dioxide into hydrogen, oxygen, and carbon monoxide. Carbon dioxide, hydrogen, and carbon monoxide are channeled into the first fermenter, whereas the oxygen is released. In both reactors, an agitation of the broth is needed to sustain a balanced local concentration of biomass and feedstock. The medium agitation may be achieved via a conventional stirrer or – in case of the first fermenter – partially with the injection of the pressurized gas stream:

In the first fermenter, CO_2, CO, and H_2 are converted into ethanol and acetic acid by *C. autoethanogenum*. The first fermenter does also convert butyrate and hexanoate into butanol and hexanol. A microfiltration unit is installed at the outlet pipe of both fermenters to retain the host cells. Additionally, a small purge stream is taken, and fresh cells are added

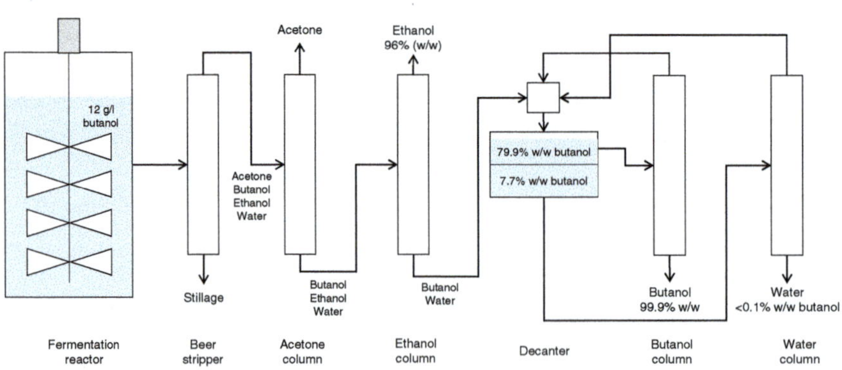

Figure 8.5 Simplified process flow for the butanol & hexanol technology of Evonik. Source: Adapted from Haas et al. (2018).

to some extent to keep the vitality of the fermentation host. The product stream from the first fermenter is channeled into the extraction unit, where butanol and hexanol are extracted with a solvent like oleyl alcohol. The remaining acetate and ethanol fraction (the raffinate of the extractor) is channeled into the second fermenter, in which acetate and ethanol are converted into butyrate and hexanoate by *C. klyveri*. Then the products butanol and hexanol in the solvent are finally isolated in the distillation unit. The butanol and hexanol may be subject of an additional purification step pending on the needed product quality and the solvent is recycled into the extraction unit.

The butanol and hexanol technology is jointly developed by Evonik and Siemens, who bundle their know how in fermentation (Evonik) and electrochemistry (Siemens). The project is referred to as "Rheticus" and is part of the Kopernikus initiative of the German energy transformation. It received a funding of 2.8 € m from the German Federal Ministry of Research and Technology (Evonik Press Release 2018). The technology is not yet commercially applied and has so far been pursued in a 2,000 l fermenter. The bacteria are fed with a gas stream consisting of CO_2, CO, and H_2 derived as industrial waste stream. The production in 2022 is limited to 10–20 t pa. The commercial setup of the process foresees to use a given CO_2 feed, to generate CO and H_2 via electrolysis and then to produce butanol and hexanol. The next stage is already under development, and the first plant will have a capacity of 10 kt pa butanol and hexanol are planned (wiwo 2022; van der Weiden 2020).

8.5 Process Development for Fermentative Butanol

8.5.1 The Conventional ABE Fermentation in the 20th Century

The conventional ABE fermentation with *C. acetobutylicum* is pursued at a temperature of about 37 °C and operated as batch or a fed-batch process, which provides a higher solvent yield. The process may achieve an ABE titer up to 15–20 g/l pending on the host tolerance against butanol. The three products are usually obtained in a ratio of 6:3:1, translating into a titer of about 12 g/l butanol, 6 g/l acetone, and 2 g/l ethanol (Morone and Pandey 2014). Since a higher butanol titer would negatively impact productivity, a share of the broth needs to be removed *in situ* and is channeled into the downstream section (Freeman and Woodley 1993). The butanol capture per h (i.e. the content of the respective volume) contains the same butanol content, which is additionally generated by the fermentation to sustain a constant titer in the broth. The *in situ* recovery approach may be pursued with immobilized cells or free cells, as is usually done in submerged fermentation. Immobilized host cells would allow a permanent medium flow through the fermentative "fixed bed" reactor. All known industrial processes are working with free cells, and the concept of immobilized cells is not yet sufficiently advanced. Accordingly, the following considerations focus exclusively on advancements with non-immobilized cells.

After removal of the biomass, which may be recycled into the fermentation reactor, the commercial plants operated until the 1980s used five consecutive distillation columns and a decanter to isolate the ABE products (Mariano et al. 2011b). The ABE product mixture is subject to a staggered distillation process for separation and purification (Figure 8.6). According to the evaporation temperature of acetone with 56 °C and ethanol with 78 °C below the evaporation temperature of water both are removed in the first two columns via

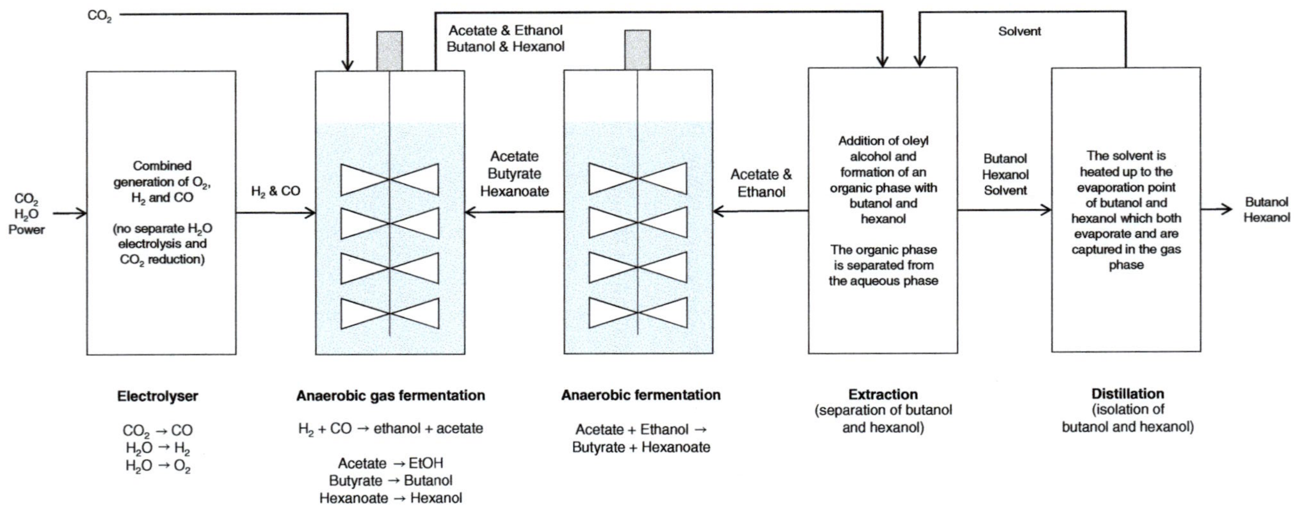

Figure 8.6 Simplified process flow in conventional ABE fermentation. Source: Adapted from Mariano and Filho (2012).

head. Since water and ethanol form an azeotropic mixture at 96% (w/w) ethanol and 4% (w/w) water, a complete separation via distillation is not possible, and mol sieves are applied to achieve a purity of 99.9% (w/w). The further process exploits the circumstance that the maximum butanol solubility in water amounts to 7.7% (w/w). This threshold is exceeded in the decanter, which is fed with the bottom stream of the ethanol column, the head stream of the butanol column, and the head stream of the water column. A second phase is formed in the decanter, resulting in an organic phase and a water phase. The organic phase with a butanol content of 79.9% (w/w) is channeled into the butanol column to separate the remaining water share, and the water phase with a butanol content of 7.7% (w/w) is channeled into the water column (Figure 8.6; Vane 2008).

The phase separation in the decanter is decisive for the commercial viability because it significantly reduces the amount of water, which needs to be evaporated to obtain pure butanol. The total energy demand for an ABE fermentation has been estimated in relation to the butanol concentration in the broth. A reciprocally proportional relation has been found (Mariano and Filho 2012; Mariano et al. 2011a):

Butanol titer	Energy demand
10 g/l	~30 MJ/kg
15 g/l	~20 MJ/kg
20 g/l	~15 MJ/kg

The energy consumption can be decreased, but additional capital and operating costs are incurred. The generated waste-water footprint of the process is rather high, with 97 l stillage / l butanol compared to 11 l stillage /l ethanol (Trindade, da, and dos Santos 2017). Since the overall energy demand for the separation and purification of the three ABE products is too high for a commercially interesting application and the waste-water footprint is significant, alternative downstream processes have been evaluated in academia and the industry. These processes comprise a continuous production mode, *in situ* recovery attempts and butanol isolation, screening of alternative hosts, and high cell density fermentation.

8.5.2 Continuous Production Mode

The switch to a continuous production mode would provide several advantages compared to the formerly pursued batch and fed-batch processes (Vees et al. 2020). A continuous process would allow for higher productivity and reactor output. It would also offer a permanent broth stream with constant quality, which is helpful in case an *in situ* product recovery would be installed to maintain the reactor titer below the critical threshold. However, a continuous operation modus with solventogenic clostridia might be difficult to establish given the two distinct metabolic phases of acidogenesis and solventogenesis. A constant quality of the broth extracts for the *in situ* recovery process might suppose a steady state of host cells either in the first or the second production phase. This potential obstacle has been addressed in several R&D approaches, like the one-stage chemostat with *C. acetobutylicum* ATCC 824, which achieved a stable production of 70 days at pH 6. Mainly butanol was produced with a carbon yield of 0.34 t butanol per ton of glucose combined with a productivity of 0.42 g/l×h (Andrade and Vasconcelos 2003). Also, a process with *C. beijerinkii*, which claims that a mixture of acidogenic-phase cells and solventogenic-phase cells is maintained in a single vessel has been patented. The cells are cultured until a critical butanol titer is achieved. At that point in time, the pressure is released, and a share of the broth is boiling and evaporates into the reactor head space (interval vacuum fermentation). This gas stream is captured and transferred into a second reactor for further processing. The first reactor is flushed with a purging fluid, and new substrate is added to initiate further solvent production (Butler 2010). Alternatively, a two-stage fermentation process might be designed in which the acidogenesis and the solventogenesis are pursued in two different fermenters. Such a process with *C. tyrobutyricum* in the first reactor converting glucose into acids and *C. acetobutylicum* in the second reactor converting acids into acetone, butanol, and ethanol has been patented by ButylFuel (Ramey 1998).

8.5.3 *In Situ* Recovery Attempts and Butanol Isolation

The fermentative production of butanol is strongly hampered by the limited tolerance of the usual hosts, *C. acetobutylicum* and *C. beijerinkij*, toward butanol and does not allow a titer above 20 g/l (Birgen et al. 2019). A higher product concentration in the broth would be detrimental to productivity. The activities addressing host tolerance to decrease toxicity did not yet provide a satisfying solution. The industrial fermentative production with a maximum butanol titer below 20 g/l implies a high energy demand for the dehydration and the isolation of butanol. Accordingly, high R&D efforts were allocated to reduce the energy input for the dewatering or the butanol concentration process, respectively, comprising gas stripping, vacuum fermentation, pervaporation (liquid/gas extraction with membrane), liquid-liquid extraction, perstraction (liquid/liquid extraction with membrane), and adsorption. The discussion of the approaches follows a comprehensive comparison and evaluation in the literature (Outram et al. 2017).

Gas stripping: The ABE products might be extracted directly out of the fermentation broth via gas stripping, which comprises the removal of molecules out of a liquid phase via solution into a gas phase (Outram et al. 2017). The gas phase may be provided as a gas stream passing the liquid phase from bottom to top and may consist of the gases captured in the reactor headspace or be provided from external sources. It is crucial that the intermediate products, acetic acid and butyric acid need to remain in the broth to protect the carbon yield. This is to be expected since the partial deprotonation of the carboxylic acid function raises the hydrophilic character of the intermediates. The efficiency of the system relies on the share of ABE products removed per volume unit broth in relation to the volume of the gas phase passing the fermenter. The gas phase is condensed after it has been passed through the fermentation reactor, and the ABE products are further processed to derive pure acetone, butanol, and ethanol. Since the concentration of ABE products in the gas condensate might be low, a two-stage gas stripping process has been proposed. The first run would be performed on the fermentation broth to reduce the toxic impact on the host. The second run would be performed on the water phase of the condensate to further increase the concentration of ABE products. It is crucial that the butanol concentration in the condensed gas is exceeding the maximum solubility of butanol in water of 7.7% (w/w). In this case, a phase separation will occur to derive an organic phase with a butanol content of 80% (w/w) and a water phase with a butanol content of 7.7% (w/w) (Mariano and Filho 2012; Xue et al. 2014b; Mariano et al. 2011a).

The suitability of gas stripping is not obvious given the boiling point of water (100 °C) and butanol (118 °C) at ambient pressure. In case the evaporation behavior is equal under vacuum conditions, the gas phase and the resulting gas condensate would not be expected to display a higher butanol concentration than the broth. Though this is not the case as butanol and water form a heteroazeotropic mixture, which at butanol concentrations below 7.7% (w/w) boils at a lower temperature than the boiling point of either butanol or water. The vapor has a higher butanol concentration than the broth, and thereby a vacuum fermentation allows a product concentration crucial to reduce the energy input in the isolation and purification of ABE products (Mariano et al. 2011a; Gmehling and Onken 1997).

It has been estimated that the energy demand for a two-stage gas stripping process amounts to 7–15 MJ/kg butanol, which would be a 50% reduction compared to a one-stage gas stripping process (Xue et al. 2014a, 2016). The carbon yields, achieved in testing various *in situ* recovery concepts range from 0.30–0.41 t butanol per ton of glucose. It should be pointed out that a value of 0.41 t of butanol per ton of glucose – which would be 100% of the theoretical maximum – is not realistic, and the figures are not directly comparable due to the different boundary conditions for the calculation. It has also been considered to combine gas stripping with pervaporation. The pervaporation approach pursues the extraction of the ABE products out of the fermentation broth into a gas phase. The liquid phase of the broth would not be in direct contact with the gas phase, though both are separated with a membrane, which selectively allows gases to pass. However, the energy demand of the more complex gas stripping/pervaporation process is estimated at about 20 MJ per kg butanol, and thereby higher than the energy demand of the two-stage stripping process (Xue et al. 2016).

A comprehensive comparison of 21 technical gas stripping variations (e.g. batch, fed-batch, continuous, butanol titer) with consistent estimation of energy demand for gas stripping has been provided (Naidoo et al. 2018). The energy demand is reported for the gas stripping step, ranging from about 27–373 MJ/kg butanol, and for the total process, ranging from 34–380 MJ/kg butanol. The energy demand for the distillation and purification of butanol is in the range of 7–8 MJ/kg butanol, profiting from the enriched ABE concentration in the condensate compared to the fermentation broth (Naidoo et al. 2018). The study concludes that gas stripping for *in situ* product removal does not provide an advantage over the conventional fermentation process with permanent removal of broth and a separation and purification of the three products via distillation (Naidoo et al. 2018).

Vacuum fermentation: The head space in the reactor might be operated under vacuum compared to ambient pressure in conventional fermentation. The vacuum is supposed to pull the ABE products out of the fermentation broth, which then accumulate in the reactor headspace. The gas phase of the reactor headspace is captured and cooled to condensate the ABE products for separation into acetone, butanol, and ethanol and purification. The vacuum might be applied constantly or in defined cycles, which has been found to limit the overall utility requirements. The butanol concentration in the condensed gas phase would be higher than the butanol concentration in the broth, given the heteroazeotropic mixture of butanol and water already referred to in the section on gas stripping. It could be shown that the butanol concentration in the reactor can be kept below 1% (10 g/l) via vacuum fermentation (Mariano et al. 2011a). Five experimental data sets for batch processes with intermittent and continuous vacuum fermentation are reported, which indicate an average carbon yield of 0.3 t of butanol per ton of glucose, corresponding to an average yield decrease of 12% in the representative cases (Outram et al. 2017). The yield loss is likely caused by the inefficient condensation of butanol out of the gaseous phase captured in the reactor head space. Additionally, minor amounts of acetic acid and butyric acid were found in the condensate, which also explain the yield loss. The intermediate products, acetic acid and butyric acid, are further metabolized to butanol in the solventogenic phase of the fermentation and need to remain in the broth. The overall energy requirement for a vacuum distillation was estimated at 32.4 MJ/kg for a continuous and 22 MJ/kg for an intermittent vacuum compared to an overall energy consumption of 26.8 MJ/kg for a comparable batch process without *in situ* product recovery (Outram et al. 2017). It is noted that conclusive data for a final evaluation are not available. The product loss in the condensation step compromises the gain in productivity and substrate utilization.

Pervaporation: The pervaporation pursues the extraction of the ABE products out of the broth into a gas phase. A share of the broth is permanently removed out of the reactor and channeled into a second chamber under vacuum conditions. A specific unit for the removal of biomass is not foreseen. The liquid phase (broth) in the second chamber is not in direct contact with the gas phase, though both are separated by a membrane that selectively allows gases to pass. The ABE products in the gas phase of the second chamber are liquefied by condensation and further treated in the downstream process to finally obtain pure acetone, butanol, and ethanol. As for all *in situ* removal technologies the selectivity is decisive. Whereas the three ABE products should efficiently pass the membrane and enter the gas phase, the nutrients and the two intermediate products, acetic acid and butyric acid, should remain in the broth phase. The second chamber might be operated at a higher temperature level (e.g. 65–78 °C) than the fermentation reactor to accomplish a faster flux of ABE products through the membrane. This would imply a cooling step before the broth with reduced ABE content is rechanneled into the fermentation reactor. As membranes, e.g. silicone tubing, silicone, PDMS/ceramic composite, polypropylene, and Zeolite-PDMS were tested. The carbon yield achieved in the experiments ranges from 0.3 up to values >0.4 ton butanol per ton glucose and is attractive (Outram et al. 2017). The concentration of product in the condensed gas phase, the permeate, achieves values of about 100 g/l butanol, which indicates an efficient extraction and increases the energy efficiency for the isolation and purification of the three products in the downstream process.

Liquid-liquid extraction: The liquid-liquid extraction is designed to pull the products out of the fermentation broth into the extractant by exploiting the differences in relative solubility of a compound in two immiscible components (i.e. the extractant and the aqueous broth). It is recommended to pursue the extraction in a second chamber to safeguard viability of the biomass and to avoid a partial transfer of biomass into the extractant. It is crucial for a commercial process that the extractant displays a high partition coefficient, is immiscible with water, is non-toxic to the host (since small amounts might be transferred into the fermentation reactor), has favorable physical properties, is sterilizable (i.e. temperature stability at 121 °C), allows an easy product recovery, is fully recovered in the process, and is available at a low cost (Outram et al. 2017). It is obvious that the identification of the most suitable extractant, which fulfills several or all of the required characteristics requires significant R&D efforts. Several liquid-liquid extractions have been tested in R&D, though the average carbon yield of 0.25 t butanol per ton of glucose was rather low (0.41 t butanol per ton of glucose theoretical maximum; Outram et al. 2017). The product titer in the organic phase after the extraction is rarely stated, which makes a commercial evaluation difficult.

Especially a high concentration factor compared to the broth titer might be an important decision criterion for a liquid-liquid extraction (Outram et al. 2017). In order to avoid any efforts for product isolation, it has been considered to use biodiesel (fatty acid methylester derived from the esterification of plant-based fatty acids with methanol) as extractant (Yen and Wang 2013; Li et al. 2010; Ishizaki et al. 1999). The biodiesel enriched with butanol might then directly be used as fuel additive. However, it is not entirely clear inasmuch as acetone, acetic acid, and butyric acid might enter the biodiesel phase and disturb the final combustion process.

Perstraction: This technology may be regarded as a specific version and upgrade of liquid-liquid extraction, with the exception that a membrane is installed between the fermentation broth and the extractant. In case the most suitable extractant has a toxic impact on the host the liquid-liquid extraction is disadvantaged. This obstacle might be solved with the perstraction approach since the membrane avoids a direct contact. However, in some experiments, a detrimental effect on productivity has nevertheless been observed, and the potential advantage might require specific membranes, which completely suppress the transfer of the extractant. The membrane needs to discriminate between the solvent products and the acids, acetic acid, and butyric acid, on the other hand. The carbon yield based on ABE fermentation combined with perstraction achieves values up to 0.37 t of butanol per ton of glucose. As membrane silicone tubing, PTFE and polypropylene were combined with oleyl alcohol, 1-dodecanol, and polypropylene glycol as extractant (Outram et al. 2017).

Adsorption: This approach foresees that the product is binding to a solid adsorbent, and thereby the concentration in the fermentation broth is decreased below the critical threshold. The adsorbent might directly be put into the fermentation reactor or be placed in a second chamber to enable the removal of the biomass with a membrane or filter. It is crucial that all three fermentation products are removed from the broth since the accumulation of acetone or ethanol would, at some point in time, also have a detrimental impact on productivity. The intermediate products, acetic acid, and butyric acid must not be adsorbed, though they need to remain in the broth. In the second step desorption would be arranged via, e.g. temperature increase, displacement with steam or another solvent, or vacuum evaporation. The desorption might imply that a batch process is to be pursued unless a second adsorbing column is provided to be used in a cyclic manner with the first column. An overview of different working groups reveals that the approach could increase the productivity up to 323% versus a fermentation protocol without *in situ* recovery and achieve a carbon yield between 28% and 32% (41% theoretical maximum; Outram et al. 2017). The recovered butanol titers after the desorption range from 43 g/l up to 167 g/l, and an additional energy input will be needed to isolate the products and to remove the water. As adsorbent PVP-based resins, polystyrene-co-divinylbenzene or active carbons were tested. The unit consumption rate of the adsorbent per ton of butanol and the energy input needed for the desorption are not disclosed. It is not known that *in situ* product recovery using adsorption is commercially applied.

Gas/gas separation with membrane: An interesting approach has been proposed by the Dalian University of Technology (Xue et al. 2019). It foresees to evaporate a share of the fermentation broth and to channel the obtained gas stream into a second chamber. In this second chamber, the vaporized butanol, acetone, ethanol, and water are selectively permeated through a membrane, and the permeated vapor is collected and condensed in a third chamber. The gas particles in the second chamber, which did not pass the membrane are recycled and used as carrier gas for a gas stripping process in the fermentation reactor. It is claimed that the butanol concentration in the condensate side of the membrane can reach a content of up to 20–30% (w/v), which would imply a concentration increase up to factor 20. This factor might finally allow a commercial application.

8.5.4 Selection of Alternative Hosts

The conventional ABE fermentation is pursued with *C. acetobutylicum*, *C. beijerinckii*, and *C. saccharobutylicum* and applies starch or molasses as feedstock. These natural producers achieve a reasonable technology performance, though do not yet allow the high-volume production of butanol. These specific Clostridium hosts have some disadvantages, like a complex life cycle with acidogenic phase, solventogenic phase, and sporulation; a slow growth; the generation of the by-products acetic acid and butyric acid; and a medium molecular biology toolbox for strain improvement. Furthermore, they are not able to consume cellulose or hemicellulose as feedstock, with the exception that *C. pasteurianum* uses glycerol as feedstock. The attempts to construct a *C. acetobutylicum* strain with the ability to consume cellulose were not successful because the heterologous expression of genes encoding the respective cellulases is difficult (Kovács et al. 2013).

This is the reason why alternative hosts are under development that are not natural butanol producers but offer advantages with respect to growth rate, raw material flexibility, and especially a versatile molecular biology toolbox for strain engineering (Nawab et al. 2020). A remarkable butanol titer of 15 g/l or 30 g/l with *in situ* product removal, respectively, could be shown in a recombinant *E. coli* strain carrying a modified clostridial 1-butanol pathway after alternative NADH-consuming pathways toward lactate, ethanol, and succinate were blocked. Consequentially, the strain lost its ability to grow anaerobically. Additionally, *E. coli's* native acetyl-CoA acetyltransferase (AtoB) was chosen instead of the clostridial acetoacetyl-CoA thiolase (Thl) given the higher specific activity of AtoB with 1078 U/mg compared to 216 U/mg of Thl. The saturation of the double bond of crotonyl-CoA was accomplished with the *trans*-enoyl-CoA reductase of *Treponema denticola* (Ter), which has been shown to possess a high specificity toward crotonyl-CoA using NADH as cofactor (Shen et al. 2011). With another recombinant *E. coli* strain, a butanol titer of 18–20 g/l was achieved (Ohtake et al. 2017; Dong et al. 2017). With a recombinant *S. cerevisiae* strain, a butanol titer of 835 mg/l was achieved (Shi et al. 2016). A fermentation with the cyanobacterium *Synechocystis* PCC 6803, which can grow with carbon dioxide as carbon source and sunlight as energy source, might allow a significant cost reduction compared to the current conventional ABE fermentation. Strains equipped with a synthetic pathway achieved proof of concept with a butanol titer of 4.8 g/l and a productivity of 12.6 mg/l×h. Commercial maturity has not yet been achieved (Liu et al. 2019). Also, the application of the acetogenic hosts *C. ljungdahlii* and *C. autoethanogenum*, which use carbon monoxide, carbon dioxide, and hydrogen as feedstock, is under evaluation (WO 2012/053905 A1). It remains to be seen inasmuch these promising results will be sufficient to establish a commercial production.

8.5.5 High Cell Density Fermentation

The productivity defines a further level for process optimization. Industrial fermentations may usually achieve a productivity of up to 6–8 g/l×h, which is achieved e.g. in the fermentative production of ethanol and lactic acid. The published productivity of 14 g/l×h for a butanol process with *C. acetobutylicum* is remarkable and relies on a high cell density fermentation with a biomass titer in the reactor of 29 g/l (Nguyen et al. 2018). The indicated value is plausible in case the reported productivity and the maximum glucose uptake rate are considered. It should be pointed out that the productivity value does not reflect the accumulated titer at batch end. Since butanol is permanently removed in the *in situ* recovery process the productivity refers to the calculated virtual titer, which is obtained if the product volume of the actual titer and the removed product volume are added for a defined incubation period. Since the carbon yield is reported at 0.35 g/g, it can be concluded that the glucose conversion rate amounts to 40 g/l×h or 220 mmol/l×h. With respect to *E. coli*, the maximum glucose uptake rate is 20 mmol per g DCW per h, and a reasonable average glucose uptake rate is estimated at 10 mmol per g DCW per h (Ahn et al. 2020; Liebal et al. 2018; Chae et al. 2017). This uptake rate may also be attributed to *C. acetobutylicum*. Accordingly, the minimum biomass titer amounts to 22 g/l, which is according to the reported biomass titer of 29 g/l, (Nguyen et al. 2018). The high cell density process allows a reduction of the reactor volume and an optimization of fixed costs, though it supposes that the biomass can be recycled and keeps the productivity for a rather long time. Otherwise, a significant share of the glucose would be consumed for the formation of biomass.

8.5.6 Digression: Derivatization of Butanol

Since butanol is toxic to most hosts, the biochemical derivatization of butanol toward butyl acetate might provide a solution. Butyl acetate is not reported to display a critical toxicity and might thereby allow to achieve a higher titer and a higher productivity, respectively. Butyl acetate would be generated within the host via an acetyl-transfer mediated by the alcohol acetyltransferase because the direct biochemical esterification of butanol and acetate is thermodynamically difficult. Accordingly, the acetyl function needs to be activated in the form of a CoA thioester to facilitate an efficient esterification:

$$\text{Butanol} + \text{acetyl-CoA} \rightarrow \text{Butyl acetate} + \text{CoA}$$

Acetyl-CoA would also be derived from glucose, and the overall reaction equation can be stated as follows:

$$1.5\ \text{Glucose} + 2\ \text{NAD}^+ \rightarrow \text{Butyl acetate} + 3\ \text{CO}_2 + 2\ \text{NADH} + 2\ \text{H}^+ + \text{H}_2\text{O}$$

Acetyl-CoA would preferentially be derived via the pyruvate dehydrogenase, and two reduction equivalents would be obtained since butyl acetate does absorb 32 electrons, whereas 1.5 mol of glucose carry 36 electrons (without counting reduced oxygen) (Ku et al. 2022). The theoretical maximum molar yield would be 0.66 mol butanol per mol glucose and thereby be more than 30% lower compared to the fermentative production of butanol. However, the product would be chemically cleaved into butanol and acetic acid, and the byproduct would provide a credit, which would at least partially compensate the lower yield. The market volume for butyl acetate is minor compared to butanol and estimated at 1 Mio. t with a price of around 1.5 USD/kg (Ku et al. 2022). The idea has already been pursued by academia, and a recombinant *E. coli* strain with the overexpressed acetyltransferase ATF1 from *Saccharomyces cerevisiae* has been generated. The strain provided a butyl acetate titer of 22.8 g/l after an incubation period of 96 h, corresponding to a productivity of 0.24 g/l×h (Ku et al. 2022). The derivatization of target products, which have a detrimental impact on the host and the process efficiency is not entirely new and has already been proposed for other products like geraniol and geranyl acetate (Chacón et al. 2019; Zada et al. 2018).

8.6 Alternative Raw Material Sources

8.6.1 Use of Lignocellulosic Biomass as Second-Generation Feedstock

According to the high efforts to establish lignocellulosic biomass as commercial feedstock for ethanol equivalent, efforts were pursued with respect to butanol. These R&D activities try to access the C5 and C6 sugar fractions of lignocellulosic biomass, e.g. rice straw, corn straw, corncob, wheat straw, wheat bran, rice bran, switchgrass, and cassava flour have been tested (Zhang et al. 2016). Additionally, specific raw materials like empty fruit bunches were evaluated (Salleh et al. 2019). The pursued approaches for the pretreatment of biomass include acid treatment, alkali treatment, enzymatic hydrolysis, and Organosolv treatment. Butanol titers in the range of 10 g/l are reported (Zhang et al. 2016).

Since the application of independently produced and purchased enzymes for the pretreatment of lignocellulosic biomass is expensive and the use of a cellulase may account for 35% of the overall cost (Jiang et al. 2015), the alternative co-production of enzymes during the fermentation is evaluated. Whereas conventional solventogenic clostridia are not able to efficiently secrete cellulases, cellulolytic clostridia can natively degrade lignocellulose and produce ethanol, acetate, butyrate, and to some extent butanol. The pretreatment might partially or completely be pursued by the host in case the suitable hydrolytic enzyme machinery is available. The approach is referred to as consolidated bioprocessing (CBP) and comprises the generation of cellulases (or other hydrolytic enzymes needed for the accessibility of pentoses and hexoses), the hydrolysis of lignocellulosic biomass, and the co-fermentation of pentoses and hexoses in one step. Accordingly, it is pursued to equip solventogenic clostridia with the ability to secrete cellulolytic enzymes and to establish a butanol pathway in cellulolytic clostridia. The fermentative production of butanol has been pursued with *C. cellulolyticum*, which is able to grow on crystalline cellulose because it produces more than 90 glycoside hydrolases. These complex multi-enzyme complexes, the cellulosome, are secreted into the natural habitat or the fermentation medium, respectively, to break down carbohydrate into mono- and oligosaccharides that can be absorbed. A butanol titer of 40 mg/l and 120 mg/l was achieved from cellobiose and crystalline cellulose as feedstock (Gaida et al. 2016). The various attempts have been summarized and did not yet provide the thresholds for a commercial application. The maximum butanol titer is indicated as 11.8 g/l, with most values below 5 g/l combined with a maximum productivity of 0.1 g/l×h. Most of the productivities were significantly below 0.1 g/l×h (Wen et al. 2019).

The promise of reduced raw material costs based on lignocellulosic biomass to allow a competitive fermentative production of butanol has so far not materialized from a high-volume business perspective. The difficulties in accessing fermentable sugars contained in lignocellulosic biomass are comparable as for ethanol fermentation, and technology development addressing butanol has yet to provide the expected results because of the high physical and chemical processing efforts and the required harsh conditions for the generation of C5 and C6 sugar monomers (Birgen et al. 2019; Morone and Pandey 2014).

8.6.2 Digression: Celtic Renewables with Pot Ale as Feedstock

A new butanol plant applying the conventional ABE fermentation technology is being constructed by Celtic Renewables in Scotland (celtic-renewables.com, 2022). The plant shall use pot ale as a cheap feedstock, a by-product in the fermentative production of malt whisky, to allow low butanol production costs. Pot ale contains dead yeast cells, yeast residues, soluble proteins, soluble nutrients, and carbohydrates (Celtic Renewables 2020, patent). It has a low market value and is usually disposed of in the sea. According to a patent of Celtic Renewables, the whisky industry is facing regulatory pressure due to its high copper content. The copper content does not negatively impact productivity if it is diluted below a defined threshold of 20 μM. This invention has been patent protected by Celtic Renewables (Celtic Renewables 2020, patent). In the preparation of the medium for the ABE fermentation the pot ale is diluted with other feedstocks, comprising municipal organic waste, industrial organic waste, agricultural crops, and crop residues, and may require a specific treatment to ensure bioavailability of the sugars. In fermentation experiments with *C. acetobutylicum* at a pH of 4.8, an ABE titer of 15.2 g/l combined with a yield of 0.34 g ABE per g sugar has been achieved (Celtic Renewables 2020, patent).

8.7 Industrial Application

8.7.1 Fermentative Butanol Production with Conventional ABE Fermentation

Butanol is mainly produced via the hydroformylation of propylene and the subsequent hydrogenation of butyraldehyde. The major chemical producers are BASF, Dow, OXO Corporation, Sasol, Formosa Plastics Corporation, Eastman Chemical Company, Oxichimie SAS, KH Neochem, and NPC (Trindade, da, and dos Santos 2017). The ABE fermentation has lost competitiveness against petrochemistry in the second half of the 20th century and has been pushed into a niche technology. It is reported that in 2010, only about 10 plants were still operated in China, with Jilin Ji'an Biochemical, Jilin Cathay Industrial Biotechnology, and Jiangsu Lianhai Biological Technology as major producers. These plants use corn, corn residues, and cassava as feedstock (Dürre 2013). Though the ABE fermentation has gained significant attention since the beginning of the 21st century and various approaches to improve the first-generation technology and to develop a second-generation technology were pursued. However, the expectations for butanol as the next biofuel besides ethanol and the former plans of Arbor Fuel, Butalco, Green Biologics/ButylFuel, Cobalt Technologies, METabolic Explorer, and TetraVitae Bioscience have not yet materialized (Lane 2015).

Arbor Fuel pursued a "consolidated bioprocessing" (CBP), which uses cellulosic biomass as feedstock and covers several steps: removal of lignin, depolymerization of cellulose and hemicellulose to soluble sugars, the fermentation of mixed C5 and C6 sugars, generation of butanol via the solventogenesis pathway, and shutting down the competing pathways to ethanol and acetone (Arbor Fuel 2010).

Butalco was founded in 2007 with the target to develop an innovative yeast strain capable to digest lignocellulosic biomass for conversion into fuels. The product focus was on ethanol, butanol, and isobutanol. Butalco was acquired in 2014 by LeSaffre (Bioökonomie.de 2021; Bharathiraja et al. 2017; Morone and Pandey 2014).

Green Biologics has developed a proprietary thermophilic bacterium, which releases enzymes for the use of cellulose and hemicellulose as feedstock. It is reported that the own strain has a butanol tolerance of 4% (Bharathiraja et al. 2017). The company has been merged with ButylFuel in 2012 and continued business under the name Green Biologics (Lane 2012; Biofuels 2012). ButylFuel has developed a prerequisite for a continuous fermentation, and the company had patented a two-stage fermentation process with operation of the acidogenesis with *C. tyrobutyricum* in one fermenter and the solventogenesis with *C. acetobutylicum* in a second fermenter (Ramey 1998). Green Biologics had bought a corn-based ethanol plant in Minnesota in 2014 to be refurbished for the fermentative production of butanol and acetone. The retrofitted plant started production in 2016, and several products based on renewable resources were launched. However, the board of Green Biologics declared in 2019 that it would cease operation since the required funding could no longer be secured (Voegele 2019).

Cobalt Technologies was founded in 2005 with the target to produce fermentative butanol based on cellulosic feedstocks. The company with headquarter in California used a non-modified *Clostridium* strain in a continuous process (Bharathiraja et al. 2017). For the evaluation of the commercial application, a partnership with the Colorado State University was started. However, in 2015, the assets of Cobalt Technologies were liquidated in a forced sale (Wikipedia 2021).

METabolic Explorer is headquartered in France and originally focused on designed microorganisms for the production of butanol based on plant-derived raw materials (Bharathiraja et al. 2017). The company is still in business though has shifted business interests on methionine, 1,3-propanediol and butyric acid (Metabolic Explorer 2021).

TetraVitae Bioscience had in 2010 taken over a corn-dry mill facility originally designed for ethanol and retrofitted the plant for the production of butanol. The proprietary technology relied on a nongenetically modified strain of *Clostridium beijerinckii* suited for the fermentative production of butanol and acetone. The assets were in 2011 taken over and integrated by Eastman Chemical (Sims 2011).

The difficult scale-up and the postponed market introduction of several companies are likely due to several reasons, and technical and economical hurdles still have to be removed. The broadest survey of issues has been provided by the working group of Professor Dürre, referring to high substrate cost, low selectivity, low titer, low yield, low productivity, culture degeneration, and phage infection as the main challenges of fermentative butanol (Birgen et al. 2019). The future industrial buildup of fermentation technology will mainly rely on improved *in situ*-product recovery to allow higher productivities and to reduce the isolation and purification efforts required for a conventional distillation process. The cost position of fermentative butanol might improve compared to conventional synthesis based on glucose or molasses in case the approach to apply lignocellulosic biomass achieves further advancements. However, the limitations of carbon source second-generation technology (G2) are not yet overcome, and the application of lignocellulosic biomass does not yet provide a cost advantage. Nevertheless, the national legislation of fuel standards may include a minimum butanol threshold to be met, which would further drive fuel market development.

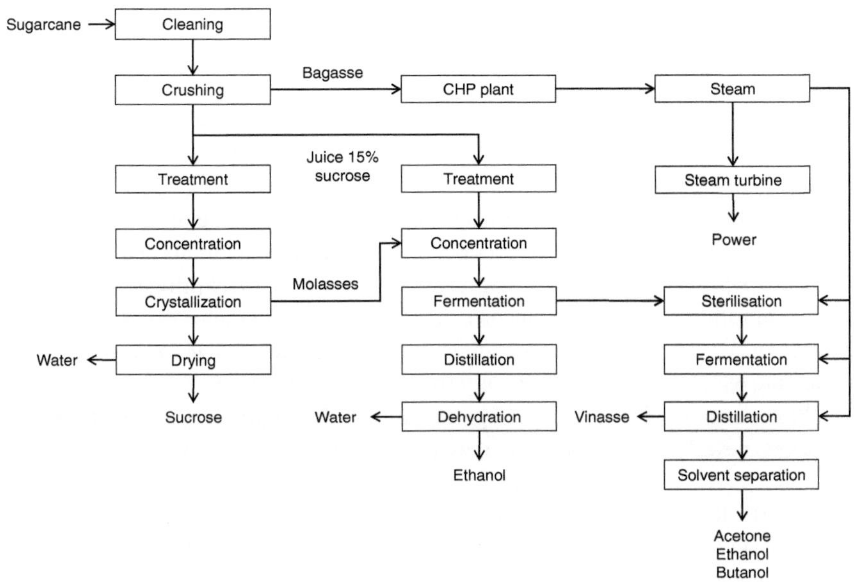

Figure 8.7 Simplified process flow of a sugarcane refinery in Brazil including the fermentative production of ethanol and butanol (Mariano et al. 2013; Sahm et al. 2013). Source: Walter Koch.

8.7.2 Butanol Production Based on Sugarcane in Brazil

The fermentative production of butanol is to some extent pursued in Brazil, with close interlink to the sugar industry. A sugar mill converts the feedstock sugarcane with a sugar (sucrose) content of 15% into pure sucrose. Hereby the feedstock is split into bagasse and the sugar-containing juice after the initial cleaning and crushing step (Mariano et al. 2013). The bagasse is burned together with other sugarcane trash to generate steam for the biorefinery and to convert steam into power in case a steam surplus does apply (Dürre 2013). The juice has a sucrose content of about 15% and is used either for extracting the sucrose or to make ethanol and butanol. It is interesting to note that the sucrose is purified and obtained by crystallization, and the derived mother liquor, the molasses, is used as feedstock for producing ethanol or butanol (Figure 8.7).

A new butanol plant has been announced by GranBio and Rhodia with sugarcane and bagasse as raw material with startup in 2015 (Chemical Week 2013). This plant will likely operate with the same technology already applied in Brazil since 2006. It is located nearby a sugar mill and uses sugarcane as a feedstock.

References

Ahn, J.H., Seo, H., Park, W. et al. (2020). Enhanced succinic acid production by Mannheimia employing optimal malate dehydrogenase. *Nature Communications* 11: 1970. https://doi.org/10.1038/s41467-020-15839-z.

Andrade, J.C. and Vasconcelos, I. (2003). Continuous cultures of *Clostridium acetobutylicum*: culture stability and low-grade glycerol utilization. *Biotechnology Letters* 25: 121–125. https://doi.org/10.1023/a:1021911217270.

Arbor Fuel (2010). Methods for the production of N-butanol. Patent US 2010/0129885.
Arpe, H.-J. (2007). *Industrielle Organische Chemie. Bedeutende Vor- und Zwischenprodukte*, 141–146. Wiley.
Bar-Even, A., Flamholz, A., Noor, E., and Milo, R. (2012). Rethinking glycolysis: on the biochemical logic of metabolic pathways. *Nature Chemical Biology* 8: 509–517. https://doi.org/10.1038/NCHEMBIO.971.
Becker, J., Kind, S., and Wittmann, C. (2012). Systems metabolic engineering of *Corynebacterium glutamicum* for biobased production of chemicals, materials and fuels. In: *Systems Metabolic Engineering* (ed. C. Wittmann and S.Y. Lee), 151–191. Springer.
Bharathiraja, B., Jayamuthunagai, J., Sudharsanaa, T. et al. (2017). Biobutanol – An impeding biofuel for future: A review on upstream and downstream techniques. *Renewable and Sustainable Reviews* 68: 788–807. https://doi.org/10.1016/j.rser.2016.10.017.
Biofuels (2012). Biofuels-news.com, Green Biologics merges with Butylfuel, 20.1.2012. https://biofuels-news.com/news/green-biologics-merges-with-butylfuel/ (accessed 25 August 2022).
Bioökonomie.de (2021). Konzern Lesaffre kauft Biosprit-Pionier Butalco, 29.7.2014. https://biooekonomie.de/nachrichten/neues-aus-der-biooekonomie/konzern-lesaffre-kauft-biosprit-pionier-butalco (accessed 25 December 2021).
Birgen, C., Dürre, P., Preisig, H.A., and Wentzel, A. (2019). Butanol production from lignocellulosic biomass: revising fermentation performance indicators with exploratory data analysis. *Biotechnology for Biofuels* 12: 167. https://doi.org/10.1186/s13068-019-1508-6.
Buckel, W. and Thauer, R.K. (2018). Flavin-based electron bifurcation, ferredoxin, flavodoxin, and anaerobic respiration with protons (Ech) or NAD^+ (RnF) as electron acceptors: a historical review. *Frontiers in Microbiology* 9: https://doi.org/10.3389/fmicb.2018.00401.
Butler, E.T. (Applicant) (2010). Continuous single vessel butanol synthesis by fermentation. Patent WO2010/096065 A1.
Celtic Renewables (Applicant) (2020). Process for the manufacture of butanol or acetone. US2020/0149074 A1.
Chacón, M.G., Marriott, A., Kendrick, E.G. et al. (2019). Esterification of geraniol as a strategy for increasing product titre and specificity in engineered *Escherichia coli*. *Microbial Cell Factories* 18: 105. https://doi.org/10.1186/s12934-019-1130-0.
Chae, T.U., Ko, Y.-S., Hwang, K.-S., and Lee, S.Y. (2017). Metabolic engineering of *Escherichia coli* for the production of four-, five-, and six-carbon lactams. *Metabolic Engineering* 43: 82–91. https://doi.org/10.1016/y.mben.2017.04.001.
Chemical Week (2013). GranBio, Rhodia plan biomass-based n-butanol plant, 2.9.2013, p. 59.
Choi, H., Han, J., and Lee, J. (2021). Renewable butanol production via catalytic routes. *International Journal of Environmental Research and Public Health* 18: 11749. https://doi.org/10.3390/ijerph182211749.
Dechema (2019). Roadmap Chemie 2050, Auf dem Weg zu einer treibhausgasneutralen chemischen Industrie in Deutschland, ISBN: 978-3-89746-223-6, p. 31.
Dellomonaco, C., Clomburg, J.M., Miller, E.N., and Gonzalez, R. (2011). Engineered reversal of the ß-oxidation cycle for the synthesis of fuels and chemicals. *Nature* 476: 355–359. https://doi.org/10.1038/nature10333.
Dong, H., Zhao, C., Zhang, T. et al. (2017). A systematically chromosomally engineered *Escherichia coli* efficiently produces butanol. *Metabolic Engineering* 44: 284–294.
Dürre, P. (2013). Technische Alkohole und Ketone. In: *Industrielle Mikrobiologie* (ed. H. Sahm, G. Antranikian, K.-P. Stahmann, and R. Takors), 71–89. Heidelberg: Springer Verlag.

Ekas, H., Deaner, M., and Alper, H.S. (2019). Recent advancements in fungal-derived fuel and chemical production and commercialization. *Current Opinion in Biotechnology* 57: 1–19. https://doi.org/10.1016/j.copbio.2018.08.014.

Evonik Press Release(2018). Evonik und Siemens erzeugen aus Kohlendioxid und Ökostrom wertvolle Spezialchemikalien, 18.1.2018. https://corporate.evonik.de/de/presse/pressemitteilungen/corporate/evonik-und-siemens-erzeugen-aus-kohlendioxid-und-okostrom-wertvolle-spezialchemikalien-105490.html (accessed 21 June 2022).

Fernández-Naveira, Á., Veiga, M.C., and Kennes, C. (2017). H-B-E (hexanol-butanol-ethanol) fermentation for the production of higher alcohols from syngas/waste gas. *Journal of Chemical Technology & Biotechnology* 92: 712–731. https://doi.org/10.1002/jctb.5194.

Flamholz, A., Noor, E., Bar-Even, A., and Milo, R. (2012). eQuilibrator – the biochemical thermodynamics calculator. *Nucleic Acid Research* 40 (database issue): D770–D775.

Freeman, A., Woodley, J.M., and Lilly, M.D. (1993). *In situ* product removal as a tool for bioprocessing. *Nature* 11: 1007–1012. The paper is the first publication on *in situ* product recovery in fermentation processes.

Fu, C., Li, Z., Jia, C. et al. (2021). Recent advances on bio-based isobutanol separation. *Energy Conversion and Management: X* 100059: https://doi.org/10.1016/j.ecmx.2020.100059.

Gaida, S.M., Liedtke, A., Jentges, A.H.W. et al. (2016). Metabolic engineering of Clostridium cellulolyticum for the production of n-butanol from crystalline cellulose. *Microbial Cell Factories* 15: 6. https://doi.org/10.1186/s12934-015-0406-2.

Gmehling, J. and Onken, U. (1997). *Vapor-liquid equilibrium data collection - Organic hydroxy compounds: Alcohols; DECHEMA Chemistry Data series*, vol. 1. DECHEMA, Frankfurt: part 2a.

Götz, M., Lefebvre, J., Mörs, F. et al. (2016). Renewable Power-to-Gas: A technological and economic review. *Renewable Energy* 85: 1371–1390.

Haas, T., Krause, R., Weber, R. et al. (2018). Technical photosynthesis involving CO_2 electrolysis and fermentation. *Nature Catalysis* 1: 32–39. https://doi.org/10.1038/s41929-017-0005-1.

Ishizaki, A., Michiwaki, S., Crabbe, E. et al. (1999). Extractive acetone-butanol-ethanol fermentation using methylated crude palm oil as extractant in batch culture of *Clostridium saccharoperbutylacetonicum* N1-4 (ATCC 13564). *Journal of Bioscience and Bioengineering* 87: 352–356.

Jiang, Y., Liu, J.L., Jiang, W.H. et al. (2015). Current status and prospects of industrial bio-production of n-butanol in China. *Biotechnology Advances* 33: 1493–1501.

Jiang, W., Gu, P., and Zhang, F. (2018). Steps towards 'drop-in' biofuels: focusing on metabolic pathways. *Current Opinion in Biotechnology* 53: 26–32. https://doi.org/10.1016/j.copbio.2017.10.010.

Katsyv, A. and Mueller, V. (2020). Overcoming energetic barriers in acetogenic C1 conversion. *Frontiers in Bioengineering and Biotechnology* 8: article 621166, https://doi.org/10.3389/fbioe.2020.621166.

Ketabchi, E., Pastor-Pérez, L., Arellano-Garcia, H., and Reina, T.R. (2020). Influence of reaction parameters on the catalytic upgrading of an acetone, butanol, and ethanol (ABE) mixture: Exploring new routes for modern biorefineries. *Frontiers in Chemistry* 7: article 906, https://doi.org/10.3389/fchem.2019.00906.

Kolesinka, B., Fraczyk, J., Binczarski, M. et al. (2019). Butanol synthesis routes for biofuel production: Trends and perspectives. *Materials* 12 (350): https://doi.org/10.3390/ma12030350.

Kovács, K., Willson, B.J., Schwarz, K. et al. (2013). Secretion and assembly of functional mini-cellulosomes from synthetic chromosomal operons in *Clostridium acetobutylicum* ACTT 824. *Biotechnology for Biofuels* 6: 117.

Ku, J.T., Chen, A.Y., and Lan, E.I. (2022). Metabolic engineering of *Escherichia coli* for efficient biosynthesis of butyl acetate. *Microbial Cell Factories* 21: 28. https://doi.org/10.1186/s12934-022-01755.

Lan, E.I. and Liao, J.C. (2012). ATP drives direct photosynthetic production of 1-butanol in cyanobacteria. *PNAS* 109 (16): 6018–6023. https://doi.org/10.1073/pnas.1200074109.

Lane, J. (2012). Merger mania gets underway: Green Biologics, Butylfuel merge, 20.1.2012. www.biofuelsdigest.com/bdigest/2012/01/20/2012-merger-mania-gets-underway-green-biologics-butylfuel-merge/ (accessed 25 December 2021).

Lane, J. (2015). RIP Cobalt Technologies or…How commercializing butanol technology is like riding tour de France, www.biofuelsdigest.com, 24.6.2015 (accessed 25 December 2021).

Lanzatech (2012). M. Koepke, L. Fung-Min (inventors), Production of butanol from carbon monoxide by a recombinant microorganism. Patent WO 2012/053905.

Li, Q., Cai, H., Hao, B. et al. (2010). Enhancing clostridial acetone-butanol-ethanol (ABE) production and improving fuel properties of ABE-enriched biodiesel by extractive fermentation with biodiesel. *Applied Biochemistry and Biotechnology* 162: 2381–2386.

Liebal, U.W., Blank, L.M., and Ebert, B.E. (2018). CO_2 to succinic acid – Estimating the potential of biocatalytic routes. *Metabolic Engineering Communications* 7: https://doi.org/10.1016/j.mec.2018.e00075.

Liu, X., Miao, R., Lindberg, P., and Lindblad, P. (2019). Modular engineering for efficient photosynthetic biosynthesis of 1-butanol from CO_2 in cyanobacteria. *Energy and Environmental Science* 12: 2765–2777.

Mariano, A.P. and Filho, R.M. (2012). Improvements in biobutanol fermentation and their impacts on distillation energy consumption and wastewater generation. *Bioenergy and Resource* 5: 504–514. https://doi.org/10.1007/s12155-011-9172-0.

Mariano, A.P., Qureshi, N., Filho, R.M., and Ezeji, T.C. (2011a). Bioproduction of butanol in bioreactors: new insights from simultaneous *in situ* butanol recovery to eliminate product toxicity. *Biotechnology and Bioengineering* 108: 1757–1765.

Mariano, A.P., Keshtkar, M.J., Atala, D.I.P. et al. (2011b). Energy requirements for butanol recovery using the flash fermentation technology. *Energy & Fuels* 25: 2347–2355. https://doi.org/10.1021/ef200279v.

Mariano, A.P., Dias, M.O.S., Junqueira, T.L. et al. (2013). Butanol production in a first-generation Brazilian sugarcane biorefinery: Technical aspects and economics of greenfield projects. *Bioresource Technology* 135: 316–323. https://doi.org/10.1016/j.biortech.2012.09.109.

Mascal, M. (2012). Chemicals from biobutanol: technologies and markets. *Biofuels, Bioproducts & Biorefining* 6: 483–493. https://doi.org/10.1002/bbb.

McMurry, J. and Begley, T. (2006). *Organische Chemie der biologischen Stoffwechselwege.* Munich, Spektrum: Elsevier p. 112–114.

Metabolic Explorer (2021). Company presentation: The alternative to petrochemicals for manufacturing everyday products, 15.5.2017. https://www.metabolic-explorer.com/annual-and-financial-reports/ (accessed 25 December 2021).

Meyer, H.P., Minas, W., and Schmidhalter, D. (2017). Industrial-scale fermentation, p. 30. In: *Industrial Biotechnology. Products and Processes* (ed. C. Wittmann and J.C. Liao), 3–53. Wiley.

Mock, J., Zheng, Y., Mueller, A.P. et al. (2015). Energy conservation associated with ethanol formation from H_2 and CO_2 in *Clostridium autoethanogenum* involving electron bifurcation. *Journal of Bacteriology* 197 (18): 2965–2980.

Morone, A. and Pandey, R.A. (2014). Lignocellulosic biobutanol production: Gridlocks and potential remedies. *Renewable and Sustainable Energy Reviews* 37: 21–35. https://doi.org/10.1016/j.rser.2014.05.009.

Nahreen, S. (2015). Catalytic upgrading of biomass fermentation products and bio-methane to energy dense hydrocarbon fuels. Dissertation submitted to the Graduate Faculty of Auburn University, Alabama.

Naidoo, M., Tai, S.L., and Harrison, S.T.L. (2018). Energy requirements for the *in-situ* recovery of biobutanol via gas stripping. *Biochemical Engineering Journal* 139: 74–84. https://doi.org/10.1016/j.bej.2018.07.022.

Nawab, S., Wang, N., Ma, X., and Huo, Y.-X. (2020). Genetic engineering of non-native hosts for 1-butanol production and its challenges: a review. *Microbial Cell Factories* 19: 79. https://doi.org/10.1186/s12934-020-01337-w.

Ndaba, B., Chiyanzu, I., and Marx, S. (2015). *n*-Butanol derived from biochemical routes: A review. *Biotechnology Reviews* 8: 1–9. https://doi.org/10.1016/j.btre.2015.08.001.

Nguyen, N.-P.-T., Raynaud, C., Meynial-Salles, I., and Soucaille, P. (2018). Reviving the Weizmann process for commercial *n*-butanol production. *Nature Communications* 9: 3682. https://doi.org/10.1038/s41467-018-05661-z.

Ohtake, T., Pontrelli, S., Laviña, W.A. et al. (2017). Metabolomics-driven approach to solving a CoA imbalance for improved 1-butanol production in *Escherichia coli*. *Metabolic Engineering* 41: 135–143.

Outram, V., Lalander, C.-A., Lee, J.G.M. et al. (2017). Applied *in situ* product recovery in ABE fermentation. *AIChE* 563–579. https://doi.org/10.1002/btpr.2446.

Peralta-Yahya, P.P. and Keasling, J.D. (2010). Advcanced biofuel production in microbes. *Biotechnology Journal* 5: 147–162. https://doi.org/10.1002/biot.200900220.

Peralta-Yahya, P.P., Ouellet, M., Chan, R. et al. (2011). Identification and microbial production of a terpene-based advanced biofuel. *Nature Communication* 2: 482. https://doi.org/10.1038/ncomms1494.

Phelan, R.M., Sekurova, O.N., Keasling, J.D., and Zotchev, S.B. (2015). Engineering terpene biosynthesis in Streptomyces for production of the advanced biofuel precursor bisabolene. *ACS Synthetic Biology* 4: 393–399. https://doi.org/10.1021/sb5002517.

Rambhujun, N., Salman, M.S., Wang, T. et al. (2020). Renewable hydrogen for the chemical industry. *MRS Energy and Sustainability* 7: https://doi.org/10.1557/mre.2020.33.

Ramey, D.E. (inventor) (1998). Continuous two-stage, dual path anaerobic fermentation of butanol and other organic solvents using two different strains of bacteria. US 5753464.

Rückel, A., Oppelt, A., Leuter, P. et al. (2022). Conversion of syngas from entrained flow gasification of biogenic residues with *Clostridium carboxydivorans* and *Clostridium autoethanogenum*. *Fermentation* 8: 465. https://doi.org/10.3390/fermentation8090465.

Sahm, H., Antranikian, G., Stahmann, K.-P., and Takors, R. (ed.) (2013). *Industrielle Mikrobiologie*. Heidelberg: Springer Verlag p. 73, 85.

Salleh, M.S.M., Ibrahim, M.F., Roslam, A.M., and Abd-Aziz, S. (2019). Improved butanol production in 2-L simultaneous saccharification and fermentation with delayed yeast extract feeding and *in situ*-recovery. *Scientific Reports* 9: 7443. https://doi.org/10.1038/s41598-019-43718-1.

Schadeweg, V. and Boles, E. (2016a). n-Butanol production in *Saccharomyces cerevisiae* is limited by the availability of coenzyme A and cytosolic acetyl-CoA. *Biotechnology for Biofuels* 9: 44. https://doi.org/10.1186/s13068-016-0456-7.

Schadeweg, V. and Boles, E. (2016b). Increasing n-butanol production with *Saccharomyces cerevisiae* by optimizing acetyl-CoA synthesis, NADH levels and trans-2-enoyl-CoA reductase expression. *Biotechnology for Biofuels* 9: 257. https://doi.org/10.1186/s13068-016-0673-0.

Shen, C.R., Lan, E.I., Dekishima, Y. et al. (2011). Driving forces enable high-titer anaerobic 1-butanol synthesis in *Escherichia coli*. *Applied and Environmental Microbiology* 77 (9): 2905–2915. https://doi.org/10.1128/AEM.03034-10.

Shi, S., Si, T., Liu, Z. et al. (2016). Metabolic engineering of a synergistic pathway for n-butanol production in *Saccharomyces cerevisiae*. *Scientific Reports* 6: 25675.

Sims, B. (2011). Eastman Chemical captures TetraVitae Bioscience Assets. *Biomass Magazine*, 8 Nov 2011, http://biomassmagazine.com/articles/7440/eastman-chemical-acquires-tetravitae-bioscience-assets/?ref=brm.

Staudt, S., Müller, C., Marienhagen, A.J. et al. (2012). Biocatalytic hydroxylation of *n*-butane with insitu cofactor regeneration at low temperature and under normal pressure. *Beilstein Journal of Organic Chemistry* 8: 186–191. https://doi.org/10.3762/bjoc.8.20.

Thinkstep (2017). GHG intensity of natural gas transport, 24.3.2017. https://www.globallnghub.com/wp-content/uploads/attach_380.pdf

Tippmann, S., Chen, Y., Siewers, V., and Nielsen, J. (2013). From flavor and pharmaceuticals to advanced biofuels: Production of isoprenoids in *Saccharomyces cerevisiae*. *Biotechnology Journal* 8: 1435–1444. https://doi.org/10.1002/biot.201300028.

Haldor Topsoe (2022). https://www.topsoe.com/processes/carbon-monoxide (accessed 18 June 2022).

Tountas, A.A., Ozin, G.A., and Sain, M.M. (2021). Solar methanol energy storage, Nature. *Catalysis* 4: 934–942. https://doi.org/10.1038/s41929-021-00696-w.

Tracy, B.P. (2012). Improving butanol fermentation to enter the advanced biofuel market. *mBIO* 3 (6): e00518-12.

Trindade, da, S.W.R. and dos Santos, R.G. (2017). Review on the characteristics of butanol, its production and use as fuel in internal combustion engines. *Renewable and Sustainable Energy Reviews* 69: 642–651. https://doi.org/10.1016/j.rser.2016.11.213.

Vane, L.M. (2008). Separation technologies for the recovery and dehydration of alcohols from fermentation broths. *Biofuels, Bioproducts & Biorefining* 2: 553–588. https://doi.org/10.1002/bbb.

Vees, C.A., Neuendorf, C.S., and Pflügl, S. (2020). Towards continuous industrial bioprocessing with solventogenic and acetogenic clostridia: challenges, progress and perspectives. *Journal of Industrial Microbiology & Biotechnology* 47: 753–787. https://doi.org/10.1007/s10295-020-02296-2.

Vlasenko, N.V., Kyriinko, P.I., Valihura, K.V., and Kosmambetova, G.R. (2019). Yttria-stabilized zirconia as a high-performance catalyst for ethanol to n-butanol Guerbet coupling. *Omega* 4: 21469–21476. https://doi.org/10.1021/acsomega.9b03170.

Voegele, E. (2019). Green Biologics to cease operation, Ethanol producer magazine, 16.7.2019. http://ethanolproducer.com/articles/16362/green-biologics-to-cease-operations (accessed 25 December 2021).

Vögeli, B., Engilberge, S., Girard, E. et al. (2018). Archaeal acetoacetyl-CoA thiolase/HMG-CoA synthase complex channels the intermediate via a fused CoA-binding site. *PNAS* 115 (13): 3380–3385. https://doi.org/10.1073/pnas.1718649115.

Wang, S., Huang, H., Kahnt, J. et al. (2013). NADP-specific electron-bifurcating [FeFe] -dehydrogenase in a functional complex with formate dehydrogenase in Clostridium autoethanogenum grown on CO. *Journal of Bacteriology* 195 (19): 4373–4386.

Webbook.nist.gov (2022). Webbook.nist.gov/chemistry/name-ser/ National Institute of Standards and Technology, US. Department of Commerce (accessed 25 August 2022).

van der Weiden, S. (2020). Ein Weg zum Klimaparadies, Die Welt, 22.10.2020, Nr. 247, p. WR1.

Wen, Z., Li, Q., Jin, M., and Yang, S. (2019). Consolidated bioprocessing for butanol production of cellulolytic Clostridia: development and optimization. *Microbial Biotechnology* 13 (2): 410–422. https://doi.org/10.1111/1751-7915.13478.

Wikipedia (2021). Cobalt Technologies. https://en.wikipedia.org/wiki/Cobalt_Technologies (accessed 25 December 2021).

Witsch, K. (2020). *Die Rückkehr des Biofuels*. Handelsblatt 16.1.2020, p. 16.

wiwo (2022). *Sprit aus Sonnenlicht*. Wirtschaftswoche 44/2022, p. 64.

WO 2012/053905 A1 (2012). Koepke, M., Fungmin, L. (inventors), Production of butanol from carbon monoxide by a recombinant microorganism.

Xue, C., Du, G.Q., Sun, J.X. et al. (2014a). Characterization of gas stripping and its integration with acetone-butanol-ethanol fermentation for high-efficient butanol production and recovery. *Biochemical Engineering Journal* 83: 55–61.

Xue, C., Zhao, J.B., Chen, L.J. et al. (2014b). Integrated butanol recovery for an advanced biofuel: current state and prospects. *Applied Microbial Biotechnology* 98: 1–12.

Xue, C., Liu, F., Xu, M. et al. (2016). A novel in situ gas stripping-pervaporation process integrated with acetone-butanol-ethanol fermentation for hyper n-butanol production. *Biotechnology and Bioengineering* 113: 120–129.

Xue, C., Zhu, C., and Chen, L. (inventors) (2019). Dalian University of Technology (applicant), Method and device for fermentation integrated with separation and purification of butanol, US 2019/0093143.

Yen, H.W. and Wang, Y.C. (2013). The enhancement of butanol production by in situ butanol removal using biodiesel extraction in the fermentation of ABE (acetone-butanol-ethanol). *Bioresource Technology* 145: 224–228.

Zada, B., Wang, C., Park, J.-B. et al. (2018). Metabolic engineering of *Escherichia coli* for production of mixed isoprenoid alcohols and their derivatives. *Biotechnology for Biofuels* 11: 210. https://doi.org/10.1186/s13068-018-1210-0.

Zhang, J., Wang, S., and Wang, Y. (2016). Biobutanol production from renewable resources: Recent advances. In: *Advances in Bioenergy*, vol. 1. Elsevier, chapter 1, https://doi.org/10.1016/bs.aibe.2016.09.001.

Zhou, Y.J., Kerkhoven, E.J., and Nielsen, J. (2018). Barriers and opportunities in bio-based production of hydrocarbons, Nature. *Energy* 3: 925–935. https://doi.org/10.1038/s41560-018-0197-x.

9

Isobutanol

9.1 Application

Isobutanol (2-methyl-1-propanol) is a branched short-chain alcohol with the chemical formula $C_4H_{10}O$ and a molecular weight of 74.1 g/mol. The traditional markets of isobutanol are solvents, coatings, and low-volume intermediates which rely on the specific molecular structure and functionality such as isobutyl-esters and isobutyl-ethers. Isobutanol may be dehydrated to isobutene, which opens the door for further downstream products obtained via polymerization or adducts of the double bond. The respective markets include isooctane, jet fuel, plastics, rubbers, and lubricants (Fu et al. 2021).

The molecule has gained interest in the scientific community since the perspective of a high-volume, biobased fuel component not derived from petrochemical resources gained shape, mirroring the energy content and the low hygroscopicity. Isobutanol qualifies under the Renewable Fuel Standard in the United States for a renewable identification number (RIN). Accordingly, the main target of innovative fermentation technology is to position isobutanol as a fuel, either for automotive combustion engines or as a precursor for aviation fuel. The conversion of alcohols into jet fuel components proceeds via dehydration to remove the oxygen – which would hamper the combustion process in the turbine – an oligomerization to compounds with higher molecular weight, and finally hydrogenation to remove the remaining double bonds (Moiseev 2016). The fully saturated status of the linear and branched alkanes ensures the high energy density of the fuel (Díaz-Péres and Serrano-Ruiz 2020).

In the context of the endeavor to develop "green" polyethylene terephthalate (PET) for soft drink bottles, plastic containers, fibers, and films and, accordingly, to replace PET based on petrochemical resources, it was also planned to convert isobutanol via isobutene into p-xylene. PET is a linear ester made of ethylene glycol as the bifunctional alcohol component and terephthalic acid (TA) as the bifunctional acid component. TA might also be obtained from renewable raw materials, and the US-based fermentation company Gevo has developed a route based on fermentative isobutanol. Hereby, isobutanol would be subjected to dehydration (digitalrefining 2022) using a homo- or heterogeneous catalyst, and the obtained isobutylene would be dimerized to diisobutylene. The C8-adduct would then be rearranged using alumina-based or silica-based heterogenous catalysts to obtain p-xylene and hydrogen. p-Xylene would then be converted into TA using the established oxidation technology and the existing chemical plants (Collias et al. 2014; Wagemann 2014; Figure 9.1).

Pathway Design for Industrial Fermentation, First Edition. Walter Koch.

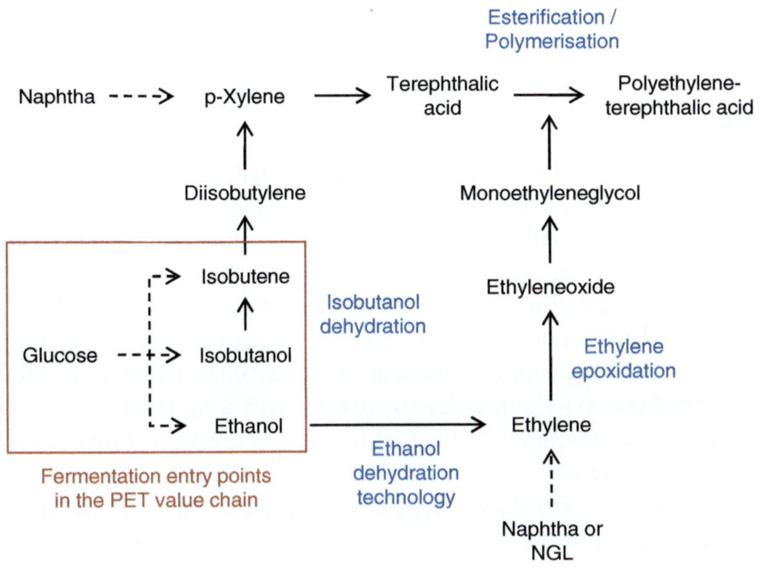

Figure 9.1 PET value chain and potential entry points via fermentation products. Source: Walter Koch.

The initiative for a biobased PET was supported by e.g. Gevo whose early business plans intended to supply isobutanol for p-xylene to serve a "green" PET deal with Coca-Cola (Chemical week 2013). However, it seems that the major brands at least partially decided to prefer recycled PET (Tournier et al. 2023), and Nestlѐ and Coco-Cola announced to increase the share of recycled PET in the bottle products and invest in recycling technology (Coca-Cola 2021).

9.2 Conventional Synthesis of Isobutanol

Isobutanol is primarily obtained as a by-product in the synthesis of butanol, based on the hydroformylation of propylene with carbon monoxide and hydrogen. The reaction obtains the linear butyraldehyde (NBA) and the branched isobutyraldehyde (IBA) in a molar ratio of about 80/20. About 70% of the global capacity applies the low-pressure Davy technology:

$$CH_3\text{–}CH=CH_2 + CO + H_2 \rightarrow CH_3\text{–}CH_2\text{–}CH_2\text{–}CHO$$

$$CH_3\text{–}CH=CH_2 + CO + H_2 \rightarrow CH_3\text{–}CH\left(CH_3\right)\text{–}CHO$$

The selectivity of the hydroformylation depends on the applied catalyst. Improved catalysts suppress the formation of the iso-intermediate to a greater extent. Butyraldehyde and isobutyraldehyde are then separated. Butyraldehyde is usually hydrogenated to obtain butanol. Isobutyraldehyde may be further processed under reaction of the aldehyde function toward various downstream products such as ethyl-hexanol or be hydrogenated to isobutanol (ECN 2018):

$$CH_3\text{–}CH\left(CH_3\right)\text{–}CHO + H_2 \rightarrow CH_3\text{–}CH\left(CH_3\right)\text{–}CH_2\text{–}OH$$

9.3 Metabolic Engineering

9.3.1 Isobutanol via 2-Oxoisovalerate in *Escherichia coli* with Balanced Reduction Equivalents

The probably most relevant pathway has originally been published by Atsumi and colleagues in the landmark publication on fermentative isobutanol (Atsumi et al. 2008, 2009). It partially exploits the established reactions in the synthesis of valine, which proceeds from pyruvate toward 2-acetolactate as intermediate. The keto function of 2-acetolactate is reduced with NAD(P)H to yield 2,3-dihydroxyisovalerate as intermediate, which is dehydrated to 2-oxoisovalerate (= 2-ketoisovalerate, 2-KIV). 2-Oxoisovalerate may either be converted into valine via a transamination with glutamate or – in the artificial pathway toward isobutanol – be decarboxylated to isobutyraldehyde. Isobutyraldehyde would then be hydrogenated with NAD(P)H to obtain isobutanol. In the first instance, it is assumed that pyruvate is provided via glycolysis (Table 9.1).

Since the acetolactate synthase and the alcohol dehydrogenase usually apply NADPH as a cofactor, the reduction equivalents of the reaction are not balanced, although two reduction equivalents are consumed and two reduction equivalents are generated. Thus, the metabolic engineering strategy needs to include respective measures like the introduction of a transhydrogenase or a cofactor swap by replacement of a NADPH-consuming isoenzyme by a NADH-consuming enzyme to achieve a balance of reduction equivalents. The overall pathway could then be stated in the following net equation:

$$\text{Glucose} + 2\ \text{ADP} + 2\ \text{Pi} \rightarrow \text{Isobutanol} + 2\ \text{ATP} + 2\ \text{CO}_2 + 3\ \text{H}_2\text{O}$$

A net impact of the recombinant isobutanol pathway on the redox balance would negatively interfere with the overall metabolism of the host. The generation of reduced reduction equivalents would likely enforce to pursue an anaerobic fermentation, which is associated with a lower carbon yield due to the high availability of ATP to sustain growth. The additional two mol of ATP would support growth and be available for maintenance metabolism. They might also feed a transporter system for the secretion of isobutanol reentering the cell or allow the production of membrane constituents to strengthen the product tolerance of the host.

Table 9.1 Metabolic pathway toward isobutanol via 2-oxoisovalerate.

Reaction	Enzyme
Glucose + 2 NAD^+ + 2 ADP + 2 Pi → 2 Pyruvate + 2 NADH + 2 H^+ + 2 ATP + 2 H_2O	
2 Pyruvate → 2-Acetolactate + CO_2	Acetolactate synthase
2-Acetolactate + NAD(P)H + H^+ → 2,3-Dihydroxyisovalerate + $NAD(P)^+$	Ketoacid-reducto-isomerase
2,3-Dihydroxyisovalerate → 2-Oxoisovalerate + H_2O	Dihydroxyacid-dehydratase
2-Oxoisovalerate → Isobutyraldehyde + CO_2	2-Oxoacid-decarboxylase
Isobutyraldehyde + NAD(P)H + H^+ → Isobutanol + $NAD(P)^+$	Alcohol dehydrogenase
Glucose + 2 ADP + 2Pi + 2 NAD(P)H + 2 NAD^+ → Isobutanol + 2 ATP + 2 $NAD(P)^+$ + 2 NADH + 2 CO_2 + 3 H_2O	

The pathway toward isobutanol via 2-ketoisovalerate has successfully been constructed in *E. coli*, which is not a natural producer of isobutanol. The strain carrying five overexpressed genes delivered 6.8 g/l in 24 h with a productivity of 0.28 g/lxh and 20 g/l in 112 h with a productivity of 0.18 g/lxh. The highest productivity of 7.0 g/lxd was found when four genes were overexpressed, two native genes of *E. coli* and two heterologous genes from *Bacillus subtilis* and *Lactococcus lactis*. The tolerance of *E. coli* toward isobutanol is limited, and it has been reported that *E. coli* is not able to grow at concentrations >8 g/l (Brynildsen and Liao 2009) though this does not imply that the existing biomass is no longer able to generate product (Figure 9.2):

The measured isobutanol titer without expression of the *kdc* gene of *L. lactis* revealed that the acetolactate synthase of *B. subtilis* is also capable to convert 2-ketoisovalerate into isobutyraldehyde. Initially, the hypothesis was pursued that the decarboxylation of 2-ketoisovalerate is to some extent facilitated by the pyruvate dehydrogenase or the 2-ketoglutarate dehydrogenase, though this assumption could be ruled out with knock-out experiments. It could be shown that the substrate specificity of the acetolactate synthase enzyme can be modeled if the amino acid at position 487 is varied. The experiments were pursued with M9 medium, cells were grown at 37 °C, and induction was carried out with 0.1 M IPTG. For production, the temperature was reduced to 30 °C, and a rotary shaker (250 rpm) with microaerobic conditions was applied. The oxygen demand is not immediately plausible and is not explained. It might indicate a slight unbalance of reduction equivalents, which can only be resolved if oxygen as an electron acceptor is offered. Though it might also be that the significant engineering measures cause metabolic stress, which requires an efficient ATP yield via respiratory chain phosphorylation.

The balance of the redox status has specifically been addressed in a separate study in *E. coli* based on a genome-scale metabolic model (GSMM) for reduction equivalents

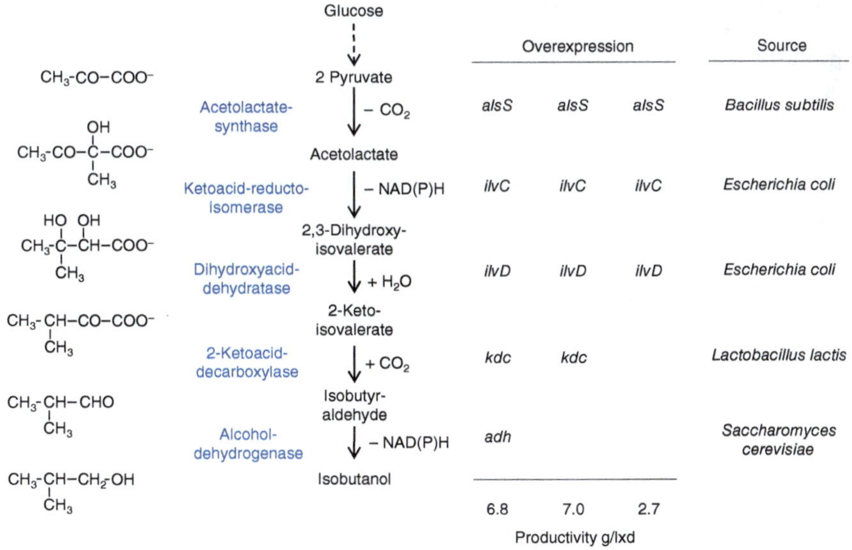

Figure 9.2 Metabolic pathway toward isobutanol in *E. coli* with overexpression of native and heterologous genes. The highest titer was obtained with 5 overexpressed genes. The *kdc* gene may also be referred to as *kivd* gene. Source: Adapted from Atsumi et al. (2009).

(Liu et al. 2015). The work used *E. coli* strain LA02 as a starting point, which displayed an isobutanol titer of 2.7 g/l based on the standard recombinant pathway via 2-oxoisovalerate (i.e. *alsS* from *B. subtilis*, *ilvC, ilvD,* and from *E. coli*, *kivd* from *L. lactis*, and *yqhd* [encoding an alcohol dehydrogenase] from *E. coli*). The theoretical model operated with over 300 redox reactions using NADH and over 100 redox reactions using NADPH. It was observed that the surplus of NADH generated in the glycolysis triggered the production of ethanol with 1.6 g/l and lactic acid with 12.6 g/l. It was additionally concluded that the metabolic engineering strategy should not rely on one specific measure, like the introduction and overexpression of a transhydrogenase, but various parameters are to be adjusted such that the interrelated impacts on the metabolic network are considered. It was finally revealed that the application of a NADPH-dependent glyceraldehyde-3-phosphate dehydrogenase provides the biggest lever to solve the redox imbalance. An isoenzyme with NADPH as cofactor was identified in *Clostridium acetobutylicum*, and the gapN gene was overexpressed in *E. coli*. The redox modulations resulted in an enhanced titer of 8.68 g/l isobutanol combined with a decreased generation of ethanol and lactate (Liu et al. 2015).

9.3.2 Carbon Yield with Balanced Reduction Equivalents

The pathway reconstruction via 2-oxoisovalerate reveals that one mol of glucose provides one mol of isobutanol, reflecting a maximum carbon yield of 0.41 t isobutanol/t glucose (74.1 g/mol/180.1 g/mol; Liang et al. 2018). This translates into a theoretical glucose input factor of 2.44 t glucose/t isobutanol. Since isobutanol or a derivative of isobutanol (e.g. via hydrogenation, dimerization) might be used as fuel additive, the energy efficiency of the reaction is of importance. The heating value of isobutanol is estimated using the following formula:

$$\text{Isobutanol} + 6\,O_2 \rightarrow 4\,CO_2 + 5\,H_2O\ \text{(liquid)}$$

The enthalpy of the reaction is estimated according to Hess's law:

$$\Delta H^{\circ}_{R} = \Sigma\Delta H^{\circ}_{P} - \Sigma\Delta H^{\circ}_{E} = (4 \times -393.5\ \text{KJ/mol} + 5 \times -285.8\ \text{KJ/mol}) \\ - -334.6\ \text{KJ/mol} = -2{,}668.6\ \text{KJ/mol given}$$

$$\Delta H_f^{\circ} CO_2\ (g) = -393.3\ \text{KJ/mol},$$

$$\Delta H_f^{\circ} H_2O\ (l) = -285.8\ \text{KJ/mol, and}$$

$$\Delta H_f^{\circ}\text{isobutanol}\ (l) = -334.6\ \text{KJ/mol}$$

This translates into an estimated heating value of 36.0 GJ/t (HHV) for isobutanol, which corresponds to the published data in literature (Fu et al. 2021). The fermentative production of isobutanol based on glucose would provide a significant "energy concentration" given a heating value of 15.6 GJ/t (HHV) for glucose. The energy efficiency of the theoretical reaction equation amounts to 95%, which would indicate a very efficient conversion.

$$\text{Energy efficiency (theory)} = 0.41\ \text{t isobutanol/t glucose} \\ = 0.41 \times 36.0\ \text{GJ/t}/15.6\ \text{GJ/t} = 95\%$$

It should be indicated that the realistic energy efficiency is lower since the process itself and especially the product isolation and purification will incur an energy investment in the form of power and steam. The theoretical maximum carbon yield would translate into a realistic carbon yield of 0.37 t isobutanol/t glucose in case it is assumed that 10% of the feedstock is consumed for biomass and maintenance metabolism. It should be indicated that about 90–92% of the maximum carbon yield is actually achieved in the fermentative production of ethanol and lactic acid based on glucose (lactic acid) or corn (ethanol), though both processes have been highly optimized in the past decades, and it is not to be expected that such a high value can be achieved in the comparably early technology development as in the case of isobutanol. The actual carbon yield achieved with the technology applied by Gevo and Butamax, the main players in the production of fermentative isobutanol, is not published. Nevertheless, the carbon yield can be derived based on an early company presentation of Gevo (Gevo 2011):

2.1 gallons isobutanol/bushel corn (Gevo 2011), derive
6.37 kg isobutanol/bushel corn (3.79 l/gallon, isobutanol density 0.8 kg/l), derive
0.25 t isobutanol/t corn (1 bushel = 25.4 kg), derive
0.35 t isobutanol/t glucose (starch in corn 65%, 1.1 t glucose/t starch) as realistic carbon yield.

The result indicates that about 85.5% of the theoretical carbon yield of 0.41 t isobutanol/t glucose is achieved. This figure does likely also include the loss in downstream processing although it is not explicitly mentioned in the Gevo company presentation. Since the feed preparation or the extraction of the fermentable carbohydrate out of the corn starch for ethanol and isobutanol are used for both processes in the Gevo "side-by-side" technology a high efficiency seems plausible. For comparison, in academia a carbon yield with 86% of the theoretical yield was reported based on a titer of 22 g/l after 112 h (Atsumi et al. 2008).

9.3.3 Isobutanol via 2-Oxoisovalerate in *E. coli* Without Direct Balance of Reduction Equivalents

The 2-ketoisovalerate reducto-isomerase or ketoacid reducto-isomerase ("KARI", *ilvC* in *E. coli*) does mostly apply NADPH as cofactor. Additionally, the alcohol dehydrogenase converting isobutyraldehyde into isobutanol might use $NADP^+$ as cofactor and generate a second mole of NADPH in the pathway. In case a transhydrogenase cannot balance the mismatch of reduced and oxidized reduction equivalents, the following reaction equation would be obtained:

$$\text{Glucose} + 2\,\text{ADP} + 2\,\text{Pi} + 2\,\text{NADPH} + 2\,\text{NAD}^+ \rightarrow \text{Isobutanol} + 2\,\text{ATP} + 2\,\text{CO}_2 + 2\,\text{NADH} + 2\,\text{NADP}^+ + 3\,\text{H}_2\text{O}$$

One approach of the metabolic engineering strategy might be to identify isoenzymes displaying a sufficient enzymatic efficiency with NADH as cofactor for both NADPH generating steps so that an unbalance of reduction equivalents is avoided. However, it might not be easy to implement this idea, especially with respect to the ketoacid reducto-isomerase. As a digression, this circumstance is likely the background of the dispute between Gevo and Butamax on the question whether the patents of Butamax rely on NADPH as cofactor for the "KARI" enzyme (Coons 2013). The patent dispute between Gevo and Butamax could be resumed after Gevo acquired the isobutanol-related patents of Butamax in 2021. The asset purchase agreement provides Gevo the full ownership over the isobutanol-related patent portfolio of Butamax (globesnewswire 2021).

Given the unbalance of reduction equivalents, two salvage measures might be chosen. The first addresses the surplus of NADH, which might be balanced in an aerobic process where oxygen is available as electron acceptor. This would imply that one mol of oxygen is needed to convert two mol of reduced NADH into two mol of oxidized NAD^+:

$$2\,NADH + 2\,H^+ + O_2 \rightarrow 2\,NAD^+ + 2\,H_2O$$

The second salvage measure addresses NADPH, which might be replenished via the pentose phosphate pathway and imply an additional consumption of glucose. The initial activation of glucose toward glucose-6-phosphate can be fed with the surplus of ATP generated in the pathway toward isobutanol:

$$1/6\,\text{Glucose} + 1/6\,ATP \rightarrow 1/6\,G6P + 1/6\,ADP$$

$$1/6\,G6P + 2\,NADP^+ + 7/6\,H_2O \rightarrow 2\,NADPH + 2\,H^+ + CO_2 + 1/6\,Pi$$

This would translate into an overall reaction equation as following (Stryer 1988):

$$1/6\,\text{Glucose} + 2\,NADP^+ + 1/6\,ATP + 7/6\,H_2O \rightarrow CO_2 + 2\,NADPH + 2\,H^+ + 1/6\,ADP + 1/6\,Pi$$

The conversion is plausible since glucose carries 24 electrons, and it can be concluded that 1/6 mol of glucose (i.e. four electrons) is sufficient to generate two mol of NADPH. The execution of the two salvage measures decreased the maximum molar yield from one mol isobutanol per mol glucose down to 0.86 mol isobutanol per mol glucose (1/1.166). Accordingly, the maximum carbon yield would decrease from 0.41 ton isobutanol per ton glucose down to 0.35 ton isobutanol per ton glucose ($74.1/180.2 \times 1.166$).

The strategy to channel glucose into the pentose phosphate pathway – via inactivation of the glucose-6-phosphate isomerase – has been pursued in academia, though was not successful (Smith et al. 2010). Several options have been evaluated to achieve a balance of reduction equivalents without hampering carbon yield. Experiments were pursued with *Shimwellia blattae* as host, which displays a sufficient tolerance toward isobutanol. The NAD-dependent alcohol dehydrogenase (the *adhA* gene) from *L. lactis* was overexpressed to avoid NADPH consumption in the final step of the pathway. Additionally, the pntAB transhydrogenase from *E. coli* was introduced to replenish the NADPH consumed in the production of 2,3-dihydroxyisovalerate. The aerobic fermentation process applied a M92X medium supplemented with 1.5 g/l yeast extract and 36 g/l glucose. Induction was pursued with 0.5 mM IPTG, and 250 ml flasks were incubated at 37 °C at 250 rpm. Finally, an isobutanol titer of 6 g/l was achieved. The productivity achieved was 0.19 g/lxh, with a specific productivity of 0.08 g/g biomass x h (Acedos et al. 2021).

An alternative approach to ensure balance of reduction equivalents relies on the classic Entner–Dourodoff (ED) pathway and the generation of NADPH by the glucose-6-phosphate dehydrogenase (Spaans et al. 2015; Ahmed et al. 2005):

$$G6P + NADP^+ \rightarrow \text{6-phosphoglucono-}\delta\text{-lactone} + NADPH + H^+ \quad \Delta G^{\circ\prime} = -17.6\,KJ/mol$$

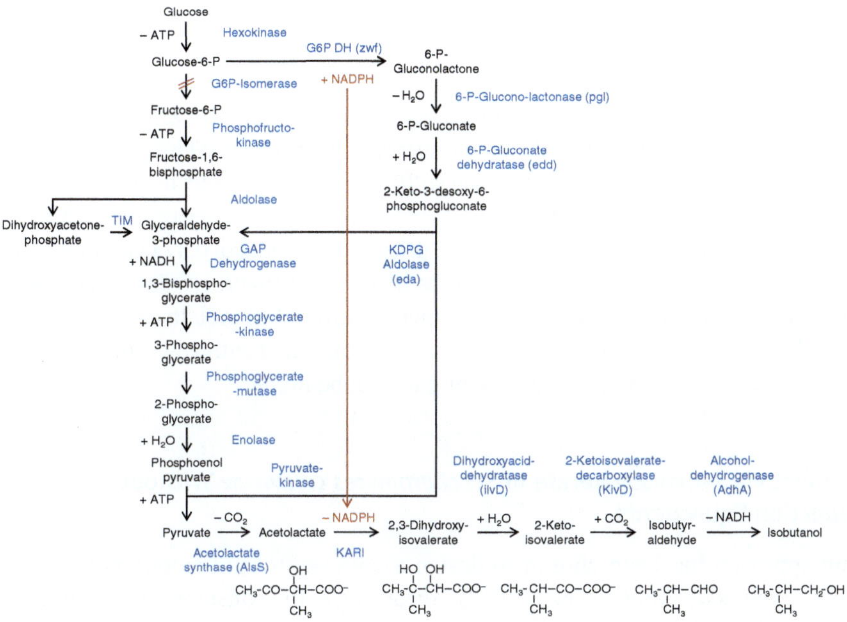

Figure 9.3 Recombinant pathway in *E. coli* for the synthesis of isobutanol. Flux via glycolysis is silenced on the level of the glucose-6-phosphate isomerase to channel glucose-6-phosphate into the ED pathway (Noda et al. 2019). KARI: NADPH-dependent Ketoacid reducto-isomerase (ilvC). Source: Walter Koch.

The recombinant metabolic pathway in *E. coli* foresees to combine parts of the ED pathway with a section of glycolysis to provide pyruvate with a surplus of NADPH, which is then available for the reduction of 2-acetolactate (Figure 9.3) (Voet et al. 2016):

The pathway for the conversion of pyruvate into isobutanol was established with the following enzymes from the indicated hosts (Table 9.2):

The cofactor demand and supply of the combined pathways are balanced since the ED pathway produces one mol of NADH and one mol of NADPH and the conversion of pyruvate demands one mol of NADH and one mol of NADPH. The pathway does not, in theory, require oxygen as an electron acceptor, though the experiments were pursued aerobically for 18 h after batch start and then run as an anaerobic process (Noda et al. 2019). The oxygen supply triggered the formation of biomass, which slowed down after 18 h, and the

Table 9.2 Enzyme portfolio for the generation of isobutanol (Noda et al. 2019).

Enzyme	Origin	Abbreviation	Cofactor
Acetolactate synthase	*B. subtilis*	AlsS	
Ketoacid reducto-isomerase (KARI)	*E. coli*	ilvC	NADPH
Dihydroxyacid-dehydratase	*E. coli*	ilvD	
2-Ketoisovalerate decarboxylase	*L. lactis*	KivD	
Alcohol dehydrogenase	*L. lactis*	AdhA	NADH

formation of isobutanol was not hampered after 18 h in anaerobic conditions. The best performing strain contained the heterologous enzymes glucose-6-phosphate dehydrogenase (*zwf*), 6-phosphoglucono-δ-lactonase (*pgl*), phosphogluconate dehydratase (*edd*), and the KDPG aldolase (*eda*). The glucose-6-phosphate isomerase (*pgi*) was knocked out to block the Embden–Meyerhof–Parnas pathway (conventional glycolysis). The lactate dehydrogenase (*ldhA*) and the pyruvate formate lyase (*pflB*) were knocked out to suppress the formation of acids and to protect selectivity.

The engineered *E. coli* strain achieved a titer of 15 g/l, a carbon yield of 0.37 t isobutanol/t glucose, and a productivity of 0.16 g/lxh. It should be pointed out that these data might not reveal the full potential given the limited host tolerance toward isobutanol. In case the pathway design with balanced reduction equivalents would be combined with an *in situ*-product recovery, the (calculated) titer and productivity might even be higher.

9.3.4 Isobutanol via 2-oxoisovalerate in *Saccharomyces cerevisiae* Without Balanced Reduction Equivalents

An interesting approach has been chosen to design a pathway toward isobutanol in the eukaryotic host *S. cerevisiae* (Matsuda et al. 2013). According to the considerations outlined above, the in principle feasible pathway via 2-oxoisovalerate has been pursued. The pathway displays an efficient conversion of the C6 precursor glucose into the C4 product isobutanol and two mol of carbon dioxide. The reduction equivalents are balanced but often do not allow a direct implementation given the non-fitting co-factor requirements of the involved enzymes. Thus, a surplus of NADH and a gap of NADPH are obtained:

$$\text{Glucose} + 2\,\text{ADP} + 2\,\text{Pi} + 2\,\text{NADPH} + 2\,\text{NAD}^+ \rightarrow \text{isobutanol} + 2\,\text{ATP} + 2\,\text{NADP}^+ + 2\,\text{NADH} + 2\,\text{CO}_2 + 3\,\text{H}_2\text{O}$$

This drawback has been solved with two measures. In the first instance, the pyridine nucleotide transhydrogenase has been introduced by metabolic engineering. This enzyme is involved in regulating the cellular redox state though *S. cerevisiae* does not possess a gene encoding this enzyme:

$$\text{NADH} + \text{NADP}^+ \rightarrow \text{NAD}^+ + \text{NADPH} \qquad \text{Transhydrogenase}$$

Furthermore, an ATP-driven cycle has been added, which proceeds via the pyruvate carboxylase and malic enzyme. The decisive reaction is the oxidative decarboxylation of malate toward pyruvate and carbon dioxide, which generates one mol of NADPH per mol of pyruvate. This step is prepared by the preceding ATP-driven carboxylation of pyruvate to oxaloacetate, which is then converted into malate. In case this cycle is proceeded twice, two mol of NADPH are generated (Table 9.3):

If now the overall general statement of the pathway and the NADPH generation cycle are combined, the imbalance of NADH and NADPH has indeed been solved and a straightforward equation of the reaction is obtained. Hereby, the surplus of ATP in the general statement of the pathway is invested in the recycling of NADPH to facilitate an efficient execution of the reduction steps:

$$\text{Glucose} \rightarrow \text{isobutanol} + 2\,\text{CO}_2 + \text{H}_2\text{O}$$

Table 9.3 ATP-driven conversion of NADH into NADPH.

2 Pyruvate + 2 ATP + 2 H_2O + 2 CO_2 → 2 oxalacetate + 2 ADP + 2 Pi	Pyruvate carboxylase
2 Oxalacetate + 2 NADH + 2 H^+ → 2 malate + 2 NAD^+	Malate dehydrogenase
2 Malate + 2 $NADP^+$ → 2 pyruvate + 2 NADPH + 2 H^+ + 2 CO_2	Malic enzyme
2 $NADP^+$ + 2 NADH + 2 ATP + 2 H_2O → 2 NADPH + 2 NAD^+ + 2 ADP + 2 Pi	

The valine pathway in *S. cerevisiae* is located in the mitochondrion, and each of the acetolactate synthase (ILV2), the ketoacid-reducto-isomerase (ILV5), and the dihydroxyacid-dehydratase (ILV3) possess a mitochondrial transit signal. This pathway is not disturbed to ensure a sufficient supply of valine. In addition, these enzymes without the transit signal were also expressed in the cytosol, in addition to the 2-oxoacid-decarboxylase (KIVD) and the alcohol dehydrogenase (ADH6). Furthermore, a truncated version of the malic enzyme (sMAE1), which is usually expressed in the mitochondrion, was expressed in the cytosol. The flask-scale fermentation was aerobically cultured with 10 g/l yeast extract, 20 g/l peptone, and 100 g/l glucose, with an initial cell concentration of 6.7 g/l dry weight cells. It cannot be excluded that the metabolic engineering measures caused a valine auxotrophy since also amino acids were added to the medium. An isobutanol titer of 1.6 g/l after 24 h, corresponding to a productivity of 0.06 g/lxh, could be achieved (Matsuda et al. 2013).

9.3.5 Isobutanol Based on Glycine and Butyryl-CoA

An alternative pathway toward isobutanol and butanol has been proposed based on glycine and butyryl-CoA (Branduardi et al. 2013). The recombinant pathway has been designed in *S. cerevisiae*. Glycine would, in the first step, be deaminated and oxidized to glyoxylate via a monooxygenase, and hydrogen peroxide would be obtained as by-product. Inasmuch as the generation of hydrogen peroxide requires a separate compartment – likely with a catalase for immediate degradation and detoxification – has not explicitly been discussed, though it should be kept in mind in case a commercial host is developed. Since a glycine oxidase has not been reported in *S. cerevisiae*, the respective gene of *B. subtilis* has constitutively been expressed. The second step relies on the substrate flexibility of the malate synthase, which facilitates the condensation of glyoxylate and acetyl-CoA in native systems but also accepts butyryl-CoA as reaction partner (Figure 9.4).

The analogue condensation of glyoxylate and butyryl-CoA derives ß-ethylmalate, which is then decarboxylated and dehydrogenated by the ß-isopropylmalate-dehydrogenase. This enzyme converts ß-isopropylmalate as its natural substrate to generate ß-ketoisocaproate, though can also generate α-ketovalerate. α-Ketovalerate is then isomerized toward α-ketoisovalerate, which is in the following steps subject to a decarboxylation and a dehydrogenation to finally obtain isobutanol. The last two steps are according to the conversion of pyruvate into ethanol and are carried out by the pyruvate decarboxylase and the alcohol dehydrogenase. The metabolic engineering of yeast metabolism was a success, and proof of concept has been achieved with a maximum isobutanol titer of 58 mg/l after 58 h, corresponding to a productivity of 0.001 g/lxh (Branduardi et al. 2013). However, the pathway has not been further developed. The overall carbon yield is low, and more than two mol of carbon dioxide molecules would be obtained per mol of isobutanol since

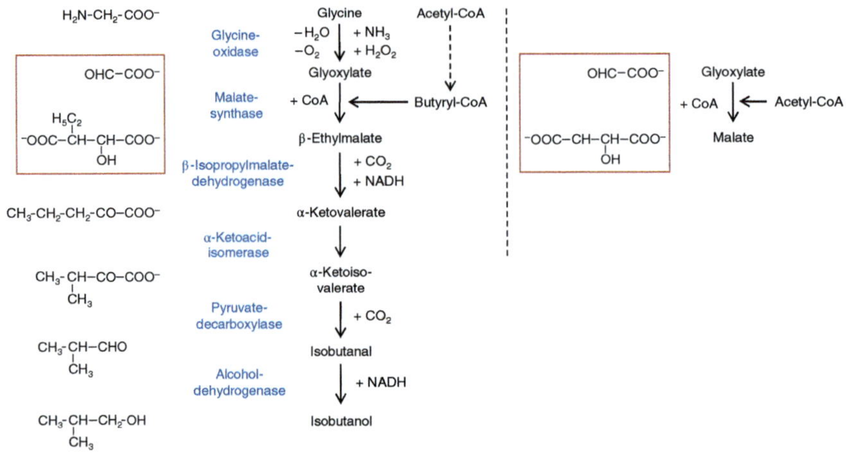

Figure 9.4 Recombinant pathway toward isobutanol based on glycine and butyryl-CoA. Source: Walter Koch.

the generation of acetyl-CoA and butyryl-CoA would imply a decarboxylation step in conventional metabolism.

9.3.6 Isobutanol Based on CO and CO_2

It has also been pursued to generate isobutanol with acetogenic bacteria based on carbon dioxide and carbon monoxide as carbon source and carbon monoxide and hydrogen as energy source. The pathway may be designed with different molar ratios of carbon dioxide, carbon monoxide, and hydrogen pending on the decision to either apply carbon monoxide or hydrogen as energy source and thermodynamic limitations. In cases where hydrogen is applied as an energy source, either carbon dioxide or carbon monoxide may function as carbon sources:

$$4\,CO_2 + 12\,H_2 \rightarrow \text{Isobutanol} + 7\,H_2O$$
$$4\,CO + 8\,H_2 \rightarrow \text{Isobutanol} + 3\,H_2O$$

The second reaction with the more reduced carbon monoxide as feedstock has a more negative $\Delta G°'$ value and is thermodynamically favored. However, the thermodynamics of the isobutanol formation might imply that a share of carbon monoxide is oxidized to carbon dioxide to facilitate the formation of reduced ferredoxin:

$$4\,CO_2 + 2\,CO + 10\,H_2 \rightarrow \text{Isobutanol} + 2\,CO_2 + 5\,H_2O$$

The Wood–Ljungdahl pathway generates acetyl-CoA, which is usually converted into acetate and ATP to replenish the ATP investment needed in the generation of formyl-THF. In the case where isobutanol is obtained, the acetyl-CoA pool would need to be channeled into the formation of pyruvate via reductive carboxylation. Two mol of pyruvate would then be converted into acetolactate, according to the landmark *Nature* publication on fermentative isobutanol as of 2008 (Atsumi et al. 2008). Hereby, the former metabolic route has been constituted in acetogenic bacteria, and a novel route via isobutyryl-CoA has been tested (Figure 9.5):

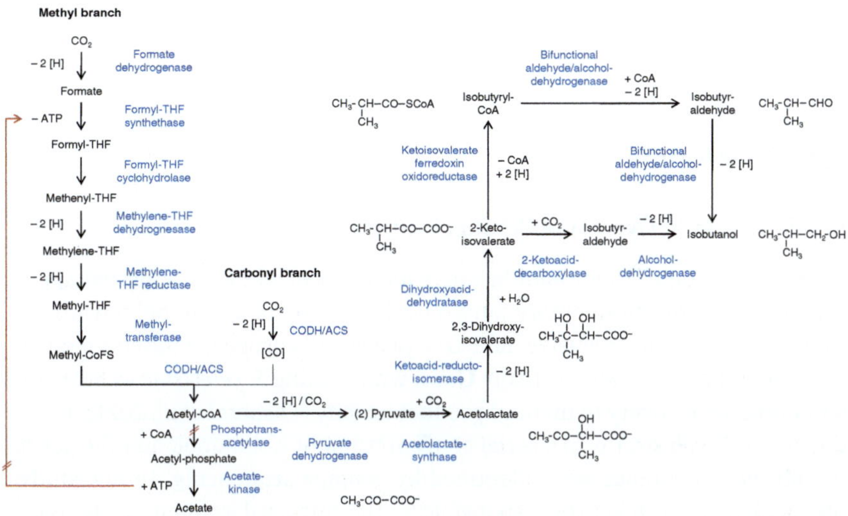

Figure 9.5 Recombinant pathway toward isobutanol based on carbon monoxide and hydrogen as energy source. Source: Walter Koch.

The novel route applies the 2-ketoisovalerate ferredoxin oxidoreductase from *Clostridium thermocellum* and the bifunctional aldehyde/alcohol dehydrogenase from *C. acetobutylicum* (Weitz et al. 2021). This route bears the specific circumstance that reduced ferredoxin as a critical reduction equivalent shall be generated during the formation of isobutyryl-CoA driven by the decarboxylation of 2-ketoisovalerate. Reduced ferredoxin can then be used to generate the proton or sodium gradient feeding the ATPase. The route via isobutyryl-CoA was tested in the acetogenic model organisms *Acetobacterium woodii* and *Clostridium ljungdahlii*. In *C. ljungdahlii*, no isobutanol could be detected unless additional 2-ketoisovalerate was fed under autotrophic conditions, revealing that the added enzymes are in principle working, though the host is likely unable to co-generate sufficient ATP along with the formation of isobutanol. In transgenic *A. woodii*, traces of isobutanol could be detected. The level of isobutanol could be raised after external 2-ketoisovalerate had been added (Weitz et al. 2021).

The second pathway proceeds via isobutyraldehyde. In order to fuel the 2-ketoisovalerate intermediate pool genes from *C. lungdahlii* encoding the preceding pathway enzymes acetolactate synthase, ketoacid-reducto-isomerase, and the dihydroxyacid-dehydratase, they were subcloned downstream of the alcohol dehydrogenase. *A. woodii* was able to provide 0.2 mM isobutanol under heterotrophic conditions, which could be increased to 0.4 mM isobutanol after additional 2-ketoisovalerate had been added. Under autotrophic conditions, no isobutanol was detected. The addition of 2-ketoisovalerate allowed the generation of 1.5 mM isobutanol under autotrophic conditions. With respect to *C. ljungdahlii*, no isobutanol was formed under heterotrophic conditions and 0.1 mM isobutanol under autotrophic conditions (Weitz et al. 2021). The results demonstrate a general proof of concept, though it remains to be seen if the thermodynamic constraints likely hampering the cogeneration of ATP can be overcome. The work with *C. ljungdahlii* has been extended, and the supply of 2-ketoisovalerate and pyruvate were identified as the major pathway bottlenecks.

The isobutanol titer could be raised to 0.13 g/l after the competing pathway to valine had been blocked (Hermann et al. 2021).

9.4 Process Development

9.4.1 Strain Adaptation to Overcome Product Toxicity

The fermentative production of isobutanol has in principle been shown. Industrial application is still hampered by low productivity caused by the toxicity of isobutanol toward the standard fermentation hosts. The exclusive industrial player still engaged in the commercial fermentative production of isobutanol is likely Gevo, who is using *S. cerevisiae* as host and pursues an *in situ* product recovery to manage product toxicity (Lakshmi et al. 2021). It has been pursued to adapt *E. coli* strains to a serial concentration of isobutanol, and the genes responsible for advanced tolerance were identified by genome sequencing. In one study, five gene knockouts were identified to be responsible for the advanced isobutanol tolerance, namely *arcA*, *gatY*, *tnaA*, *yhbJ*, and *marCRAB*. It is assumed that the genes *arcC* and *marC* encode transporters involved in the secretion of isobutanol (Atsumi et al. 2010). Another academic working group working on the same topic identified the genes *marC*, *arcA*, and *gatY* with a comparable approach (Minty et al. 2011). However, in both attempts, even the *E. coli* strains with the most improved isobutanol tolerance did not achieve a higher production of isobutanol (Wong et al. 2017). Adaptive laboratory evolution covering a growth duration of 60 days and more than 250 generations led to the isolation of *L. lactis* strains with improved tolerance up to 40 g/l, though the complex transcriptome analysis did not yet provide a suitable chassis concept for industrial application (Gupta et al. 2020).

9.4.2 *In Situ* Recovery of Isobutanol to Raise Productivity

In cases where the host tolerance cannot be improved, an *in situ* product recovery approach might provide a solution (Freeman et al. 1993). Several approaches have been discussed in the literature, which may rely on pure theoretical considerations or refer to empirical studies in academia or the industry. A comprehensive survey covers vacuum evaporation, adsorption, pervaporation, gas stripping, solvent extraction, salting out, and salting-out extraction (Fu et al. 2021). Solvent extraction might turn out to be a reasonable approach in case a solvent with high discriminating portioning coefficient can be identified. An organic solvent would be added to the fermentation broth, and a share of the broth and solvent mixture would be channeled into a separate reactor to isolate isobutanol and to recycle the other broth components. Alternatively, a part of the broth (without solvent) might be channeled into a separate reactor to add a solvent and to pursue the extraction. Both approaches would rely on an efficient extraction coefficient between the water and the organic phase. The broth would then be recycled into the fermentation process, and the isobutanol be isolated from the applied organic extractant.

Empirical data for a gas stripping process with *in situ* product recovery have been published (Baez et al. 2011). An air stream is bubbled through the broth with a flow rate of 0.5 vvm for the first three hours of isobutanol production, which is increased to 1.2 vvm afterwards. The air is stripping isobutanol out of the broth. Isobutanol evaporates, and the fermenter off-gas is captured and removed out of the reactor head space. The off-gas stream is then channeled into a receiver containing water, which is cooled with ice. The isobutanol condenses and is trapped in the water reservoir for later isolation. This condensation step is pursued twice. The pressure level in the fermenter headspace and the condensation section is not indicated, and it seems that ambient pressure is applied. The experiments were performed with an engineered *E. coli* host pursuing the metabolic pathway via pyruvate and 2-ketoisovalerate. Using this *in situ* recovery, accumulated titers up to 50 g/l isobutanol were obtained (Baez et al. 2011; Table 9.4):

The accumulated titer is not the actual titer in the broth, though obtained via recalculation. The isobutanol content in the broth and the extracted isobutanol derived in the air stripping process are added up. The actual titer in the broth between 6.4 g/l and 11 g/l is according to the usual tolerance level of *E. coli*. The data prove that calculated productivities above 0.7 g/lxh can be obtained. Nevertheless, the overall extraction efforts need to be kept in mind with respect to a potential commercial application. There is an energy demand for the stirrer and the air circulation; for industrial applications, the air stream might be sufficient and the stirrer might be avoided. The more cost-driving factor might be the product isolation since a mixture of water and isobutanol is obtained in the two consecutive condensation steps. Finally, the water needs to be removed, which requires steam. The isobutanol condensation should be designed to maximize the recovery yield out of the off-gas and limit the water content in the obtained mixture.

An *in situ* isobutanol recovery based on vacuum evaporation has been implemented by Gevo in its plant in Luverne called GIFT ("Gevo Integrated Fermentation Technology") process. Parts of the fermentation broth are channeled into the GIFT section, in which butanol and water are separated via distillation. The efficiency of this step is not straightforward since isobutanol has an evaporation point 8 °C higher than water under standard conditions. Accordingly, it may be assumed that mainly water would be evaporated and removed via distillation. However, Gevo is reporting to pursue a low-pressure evaporation in which isobutanol flashes out of the broth, "resulting in a vapor concentration nearly twenty times greater than what was in the fermenter" (gevo.com 2021). This step of the separation protocol exploits the azeotropic properties of isobutanol-water (Ryan 2019). The biomass is recycled and remains in the fermenter-GIFT system (gevo.com 2021). The corn is processed according to the dry milling technology, and after liquefaction and saccharification, the mash is fed into the main reactors. Of the four reactors available in Luverne, three are used for ethanol and one for isobutanol. The process may be reconstructed as following (Figure 9.6).

The vapor is condensed and has an isobutanol concentration 20-fold higher than that in the broth. The average isobutanol concentration in the fermenter is not indicated. It may be assumed that the broth concentration is in the range of 10 g/l isobutanol, which would

Table 9.4 Performance data with different strains.

Strain	Tolerance	Accumulated titer	Productivity	Actual titer
JCL260	Non-tolerant	50.8 g/l in 72 h	0.71 g/lxh	11.0 g/l after 22 h
JCL16	Non-tolerant	41.4 g/l in 72 h	0.58 g/lxh	11.0 g/l after 22 h
SA481	Tolerant	23.0 g/l in 68 h	0.34 g/lxh	6.4 g/l after 22 h

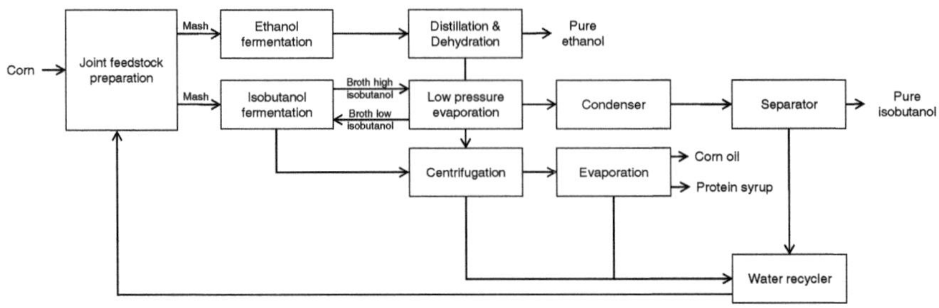

Figure 9.6 Fermentative production of isobutene as operated by Gevo in Luverne. Source: Walter Koch.

imply that the isobutanol concentration in the vapor amounts to 200 g/l given the claimed concentration factor of 20. The vapor is condensed, and a two-phase system of water and isobutanol is formed since the solubility threshold of isobutanol (85 g/l in water at 20 °C) is exceeded. After phase separation by decantation, the isobutanol is purified. The water phase is subject of a stripping distillation to remove the residual isobutanol, and the water is recycled into the process (Ryan 2019).

9.4.3 Recovery of Value Components into By-products

The ethanol industry in the United States has improved the fractionation of the value components in corn and often extracts the kernel oil fraction. Following this trend to access additional sales potential, Gevo retrofitted its plant in Luverne in 2018 with the respective equipment from Shockwave LLC to derive corn oil for food quality (ICIS News 2018). Gevo is additionally extracting the corn protein content and the protein generated during the fermentation process. It is indicated that for each million gallon of isobutanol (about 3 kt isobutanol), 4 kt protein animal feed or 1 kt protein are generated (Ryan 2019). This relation corresponds to a by-product factor of 0.33 t protein/t isobutanol, which is according to the estimated input factor of 4.0 t corn/t isobutene (0.25 t isobutanol/t corn, section 9.3.2) in the case where the protein content of corn is estimated at about 8.5%:

4 t corn/t isobutanol, and
0.085 t protein/t corn, derive
0.33 t protein/t isobutanol.

9.4.4 Cascade Biocatalysis to Overcome Product Toxicity

The very low product tolerance of the established fermentation hosts initiated the search for alternative options to derive isobutanol with biochemical means. An interesting attempt to overcome the critical product tolerance of the living cell is the cell-free approach, appealing to the historical breakthrough observation of Buchner that ethanol production is possible with yeast extracts (Dudley et al. 2015). This approach foresees to simply mix the needed carbon source with selected enzymes to accomplish a biochemical pathway outside living cells. Since the cofactors ATP and NAD(P)H would be too expensive if required in a molar

Table 9.5 Biocatalytic pathway toward isobutanol without ATP as cofactor.

Reaction	Enzyme
Glucose + NAD^+ → gluconate + NADH + H^+	Glucose dehydrogenase (GDH)
Gluconate → 2-keto-3-desoxygluconate + H_2O	Dihydroxy acid dehydratase (DHAD)
2-keto-3-desoxygluconate → pyruvate + glyceraldehyde	2-Keto-3-desoxygluconate aldolase (KDGA)
Glyceraldehyde + NAD^+ → glycerate + NADH + H^+	Glyceraldehyde dehydrogenase (AlDH)
Glycerate + H_2O → pyruvate	Dihydroxy acid dehydratase (DHAD)
2 Pyruvate → 2-acetolactate + CO_2	Acetolactate synthase (ALS)
2-Acetolactate + NAD(P)H + H^+ → 2,3-dihydroxyisovalerate + $NAD(P)^+$	Keto-acid-reducto-isomerase (KARI)
2,3-Dihydroxyisovalerate → 2-oxoisovalerate + H_2O	Dihydroacid-dehydratase (DHAD)
2-Oxoisovalerate → isobutyraldehyde + CO_2	2-Oxoacid-decarboxylase (KDC)
Isobutyraldehyde + NAD(P)H + H^+ → isobutanol + $NAD(P)^+$	Alcohol dehydrogenase (ADH)
Glucose → isobutanol + 2 CO_2 + H_2O	

ratio, a cofactor balance of the whole pathway is needed. Such a pathway toward isobutanol has successfully been designed, and proof of concept been achieved (Guterl et al. 2012; Table 9.5):

An analogue biocatalytic concept has been developed for butanol (Krutsakorn et al. 2013). With respect to the biocatalytic route toward isobutanol, the conversion of glucose into pyruvate follows a modified non-phosphorylative Entner–Dourodoff pathway (ED) found in hyperthermophilic archaea (Spaans et al. 2015; Ahmed et al. 2005). This unusual route for the conversion of glucose into pyruvate avoids any phosphorylation or dephosphorylation, which would commercially not be feasible in a cell-free system. In the first step, the aldehyde function of glucose is oxidized by the glucose dehydrogenase to the carboxylic acid function, and gluconate is obtained, which is converted into 2-keto-3-desoxygluconate via dehydration catalyzed by the dihydroxyacid dehydratase (DHAD) (Figure 9.7):

The dihydroxy acid dehydratase (DHAD) displays substrate promiscuity and is also able to pursue the analogue reaction with glycerate as substrate. Glycerate is in the (non-modified) non-phosphorylative ED pathway and converted into pyruvate via 2-phosphoglycerate and phosphoenolpyruvate, and the related investment and regeneration would define a hurdle for a biocatalytic system. Thereby, glucose is converted into two mol of pyruvate, and the standard pathway toward isobutanol via 2-oxoisovalerate can be followed. Finally, glucose is converted into isobutanol and two mol of carbon dioxide without any involvement of ADP/ATP and a balanced system of reduction equivalents. The glyceraldehyde dehydrogenase (AlDH) has a high substrate specificity and does not accept other aldehydes. As a prerequisite, the AlDH was rendered with directed evolution to work exclusively with NADH as a cofactor to limit the complexity of the system. The overall reaction equation does not provide a surplus of energy in the form of ATP. It would not be needed anyway since there is no biomass to be generated and no maintenance metabolism to be sustained. The cell-free system was able to convert a 19.1 mM glucose solution into a 10.3 mM isobutanol

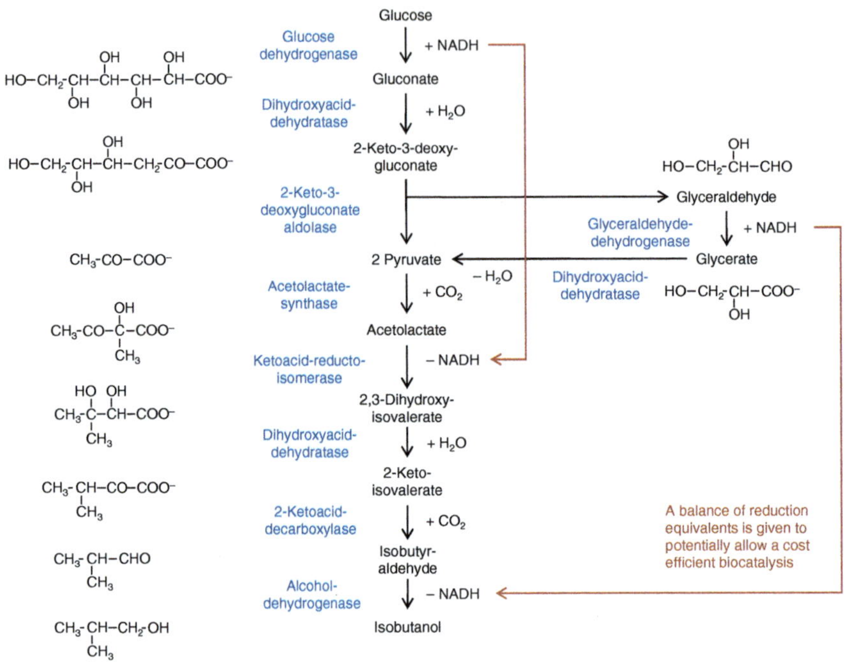

Figure 9.7 Biocatalytic conversion of glucose into isobutanol via the modified non-phosphorylative ED pathway. The intermediate glucono-1,5-lactone derived in the conversion of glucose, which undergoes a dehydration to obtain gluconate has been omitted. Source: Walter Koch.

solution, corresponding to a yield above 50%. Improvements like the immobilization of enzymes and a higher substrate concentration such as 230 g/l glucose are discussed though were so far not implemented (Kraus et al. 2015; Guterl et al. 2012).

However, the system still needs improvement to meet the threshold of industrial maturity. The substrate promiscuity of the dehydratase allows limiting the number of enzymes, though the enzyme is not ideally equipped to catalyze each step, and the activity of the trifunctional dehydratase has recently been identified as a bottleneck in the pathway. The applied dihydroxyacid dehydratase of *Sulfolobus solfataricus* displays an activity of 0.66 U/mg with gluconate, 0.011 U/mg with glycerate, and 0.38 U/mg with 2,3-dihydroxyisovalerate (Guterl et al. 2012). The k_{cat} values for the dehydration of gluconate and 2,3-dihydroxyisovalerate amount to 0.56 s^{-1} and 0.32 s^{-1}, respectively, and were identified as the slowest rates in the entire cascade (Bayaraa et al. 2022). It has also been found out that the activity of the dehydratase gets partially lost during the purification procedure due to the instability of the Fe-S cluster (Bayaraa et al. 2022). Enzyme evolution is pursued to improve the reaction kinetics and the stability of the enzyme. It might also turn out that a specific dehydratase for each dehydration step is recommended. In general, cell-free systems are a very helpful tool for exploring the boundaries of biochemical pathways, though their commercial application is often still restricted to high-value products or reactions that are not feasible with organic chemistry (Karim et al. 2017).

9.5 Industrial Application

The attempts to establish isobutanol as a biobased high-volume intermediate either as a fuel additive, a precursor for a green "PET" for soft drink bottles, or a building block for aviation fuel were successful to some extent, though business achievements are significantly below original marketing plans. Major R&D and marketing efforts were pursued by Gevo, which has been founded in 2005 and went public in 2010. The origin of this company is closely related to the work of Frances Arnold on "directed enzyme evolution" (Dougherty and Arnold 2009). Arnold won the Nobel Prize for chemistry in 2018 and co-founded Gevo. The biotech start-up constructed in the first step a pilot plant for demonstration scale in Silsbee (Texas) to prepare the technology beyond lab scale. Commercial production of fermentative isobutanol was started in Luverne (Minnesota) in 2012, when Gevo bought an existing idled plant originally designed for the fermentative production of ethanol based on corn. Due to the technical fallbacks of the retrofitted plant, the output was only 50,000 gallons in 2014. Only one of the four fermenters had been used. The original plan was to produce 18 million gallons per year (CEN 2015). Gevo had closed an additional public offering in 2013 and raised $ 22.6 million to finance a technical upgrade and completion of the isobutanol plant in Luverne. The efforts to suppress contamination had been underestimated, and the process was reported to suffer from non-sterile conditions at the beginning. The issues could be solved after steam capabilities were added and measures for operational discipline were implemented (Coons and Frost 2013; Chemical week 2013). The plant in Luverne has later on again been used for the production of ethanol based on corn and renewable gasoline blends. Gevo designed and implemented a "side-by-side" strategy for running the plant either to make ethanol or to make isobutanol, opportunistically decided pending on fluctuating demand. Both processes share the feedstock handling, the section on animal feed operation – processing of the yeast biomass for marketing into the feed industry – and the water recycling system.

Ethanol is sold into the market. For isobutanol, several markets are under development. It may be sold as specialty chemical and solvent or gasoline blendstock for marine and off-road. The market perspective for biobased isobutanol was strongly supported by the decision of the EPA to approve isobutanol at a 16% blend level in gasoline for on-road use in automobiles; the previous threshold was limited to 12.5%. Gevo has also partnered up with BP to address conventional automotive combustion engines (American City Business Journals 2018). Additionally, the isobutanol may be transferred to the Gevo plant in Silsbee (Texas), which has been equipped to convert isobutanol into isobutene, isooctene, isooctane, and finally jet fuel (isoparaffinic kerosene). It is operated in collaboration with South Hampton Resources (Ryan 2019). Jet fuel has been produced since 2011 and been used in certification tests. For the marketing of the products, Gevo has partnered up with several airlines, such as Alaska, United, and the U.S. Air Force. In 2019, an additional collaboration was started with the airline Virgin Australia, who had tested the jet fuel produced by Gevo (Bioenergy International 2019).

The production and marketing of fermentative isobutanol have also been pursued by Butamax Advanced Biofuels, a 50/50 joint venture of BP and Dupont. The company acquired an existing ethanol plant in Scandia (Kansas) in 2017 to prepare the fermentative production of isobutanol. The plant shall prove the commercial maturity of the technology and confirm strain improvement of the biocatalyst. The strategy is to license the technology (Vanguard 2017).

References

Acedos, M.G., de la Torre, I., Santos, V.E. et al. (2021). Modulating redox metabolism to improve isobutanol production in *Shimwellia blattae*. *Biotechnology for Biofuels* 14: 8. https://doi.org/10.1186/s13068-020-01862-1.

Ahmed, H., Ettema, T.J.G., Tjaden, B. et al. (2005). The semi-phosphorylative Entner-Dourodoff pathway in hyperthermophilic archaea: a re-evaluation. *Journal of Biochemistry* 390: 529–540.

American City Business Journals (2018). Nobel price winner co-founded this Colorado company, 3 October 2018.

Atsumi, S., Hanai, T., and Liao, J.C. (2008). Non-fermentative pathways for synthesis of branched-chain higher alcohols as biofuels. *Nature* 451: 86–89.

Atsumi, S., Li, Z., and Liao, J.C. (2009). Acetolactate synthase from *Bacillus subtilis* serves as a 2-ketoisovalerate decarboxylase for isobutanol biosynthesis in *Escherichia coli*. *Applied and Environmental Microbiology* 6306–6311.

Atsumi, S., Wu, T.-Y., Machado, I. et al. (2010). Evolution, genomic analysis, and reconstruction of isobutanol tolerance in *Escherichia coli*. *Molecular Systems Biology* 6: article number 449, https://doi.org/10.1038/msb.2010.98.

Baez, A., Cho, K.-M., and Liao, J.C. (2011). High-flux isobutanol production using engineered *Escherichia coli*: a bioreactor study with in situ product removal. *Applied Microbial Biotechnology* 90: 1681–1690.

Bayaraa, T., Lonhienne, T., Sutiono, S. et al. (2022). Structural and functional insight into the mechanism of the Fe-S cluster-dependent dehydratase from *Paralcaligenes ureilyticus*. *Chemistry A European Journal* https://doi.org/10.1002/chem.202203140.

Bioenergy International (2019). Virgin Australia passes 1 million kilometre milestone on Gevo's biojet fuel, 19 June 2019. https://bioenergyinternational.com/virgin-australia-passes-1-million-kilometre-milestone-on-gevos-biojet-fuel/ (accessed 3 August 2023).

Branduardi, P., Longo, V., Berterame, N.M. et al. (2013). A novel pathway to produce butanol and isobutanol in *Saccharomyces cerevisiae*. *Biotechnology for Biofuels* 6: 68, www.biotechnologyforbiofuels.com/content/6/1/68.

Brynildsen, M.P. and Liao, J.C. (2009). An integrated network approach identifies the isobutanol response network of *Escherichia coli*. *Molecular Systems Biology* 5: 277.

CEN (2015). Biobased chemicals: Butanol firms chart diverging paths in the Midwest, 26 January 2015, p. 7.

Chemical week (2013). Gevo raises $ 22.6 million to complete start-up at Luverne plant, 23./30 December 2013, p. 11.

Coca-Cola (2021). Coca-Cola progresses in 100% recycled or renewable journey in Germany. https://www.cocacolaep.com/media/news/2021/100-renewable-recycled-germany/ (accessed 29 August 2021).

Collias, D.I., Harris, A.M., Nagpal, V. et al. (2014). Biobased terephthalic acid technologies: A literature review. *Industrial Biotechnology* 10 (2): 91–105.

Coons, R. (2013). Gevo wins round in patent fray; Butamax plans appeal, IHS. *Chemical Week*, March 25/April 1, p. 16.

Coons, R. and Frost, L. (2013). Chemical week, 17./24.6.2013, Gevo resumes isobutanol production in Luverne, p. 10.

Díaz-Péres, M.A. and Serrano-Ruiz, J.C. (2020). Catalytic production of jet fuels from biomass. *Molecules* 25 (802): https://doi.org/10.3390/molecules25040802.

digitalrefining (2022). Gevo investigates proprietary dehydration technology at the hte. www.digitalrefining.com (accessed 4 March 2022).

Dougherty, J.M. and Arnold, F.H. (2009). Directed evolution – new parts and optimized function. *Current Opinion in Biotechnology* 20 (4): 486–491.

Dudley, Q.M., Karim, A.S., and Jewett, M.C. (2015). Cell-free metabolic engineering: Biomanufacturing beyond the cell. *Biotechnology Journal* 10: 69–82.

ECN (2018). Chemical profile: Europe oxo-alcohols, 17 May 2018.

Freeman, A., Woodley, J.M., and Lilly, M.D. (1993). In situ product removal as a tool for bioprocessing. *Nature* 11: 1007–1012. The paper is the first publication on in situ product recovery in fermentation processes.

Fu, C., Li, Z., Jia, C. et al. (2021). Recent advances on bio-based isobutanol separation. *Energy Conversion and Management: X* 100059: https://doi.org/10.1016/j.ecmx.2020.100059.

Gevo (2011). Gateway to Renewable Chemicals and Hydrocarbon Fuels, Corporate update, July 2011, p. 24.

Gevo.com (2021). Homepage. https://gevo.com/about-gevo/our-plants/luverne/ (accessed 14 June 2021).

Globesnewswire (2021). Gevo acquires Butamax patent estate, 23 September 2021. www.wallstreet-online.de/nachricht/14443733-gevo-acquires-butamax-patent-estate (accessed 24 September 2021).

Gupta, J.A., Thapa, S., Verma, M. et al. (2020). Genomics and transcriptomics analysis reveals the mechanism of isobutanol tolerance of a laboratory evolved *Lactococcus lactis* strain. *Scientific Reports* 10: 10850. https://doi.org/10.1038/s41589-020-67635-w.

Guterl, J.-K., Garbe, D., Carsten, J. et al. (2012). Cell-free metabolic engineering: Production of chemicals by minimized reaction cascades. *ChemSusChem* 5: 2165–2172.

Hermann, M., Teleki, A., Weitz, S. et al. (2021). Identifying and engineering bottlenecks of autotrophic isobutanol formation in recombinant *C. ljungdahlii* by systemic analysis. *Frontiers in Bioengineering and Biotechnology* 9: article 647853, https://doi.org/10.3389/fbioe.2021.647853.

ICIS News (2018). US Gevo to install new process at Minnesota isobutanol-ethanol plant. www.icis.com (14 August 2018).

Karim, A.S., Dudley, Q.M., and Jewett, M.C. (2017). Cell-free systems for metabolic engineering and biosynthetic pathway prototyping. In: *Industrial Biotechnology. Microorganisms* (ed. C. Wittmann and J.C. Liao), 125–148. Wiley.

Kraus, M., Koltermann, A., Kettling, U. et al. (2015). Cell-free and minimized metabolic reaction cascade for the production of chemicals, US2015/0218594.

Krutsakorn, B., Honda, K., Ye, X. et al. (2013). In vitro production of n-butanol from glucose. *Metabolic Engineering* 20: 84–91.

Lakshmi, N.M., Binod, P., Sindhu, R. et al. (2021). Microbial engineering for the production of isobutanol: current status and future directions. *Bioengineered* 12 (2): 12308–12321. https://doi.org/10.1080/21655979.2021.1978189.

Liang, S., Chen, H., Liu, J., and Wen, J. (2018). Rational design of a synthetic Entner-Dourodoff pathway for enhancing glucose transformation to isobutanol in *Escherichia coli*. *Journal of Industrial Microbial Biotechnology* 45: 187–199.

Liu, J., Qi, H., Wang, C., and Wen, J. (2015). Model-driven intracellular redox status modulation for increasing isobutanol production in *Escherichia coli*. *Biotechnology for Biofuels* 8: 108. https://doi.org/10.1186/s13068-015-0291-2.

Matsuda, F., Ishli, J., Kondo, T. et al. (2013). Increased isobutanol production in Saccharomyces cerevisiae by eliminating competing pathways and resolving cofactor imbalance. *Microbial Cell Factories* 12: 119. www.microbialcellfactories.com/content/12/1/119.

Minty, J.J., Lesnefsky, A.A., Lin, F. et al. (2011). Evolution combined with genomic study elucidates genetic bases of isobutanol tolerance in *Escherichia coli*. *Microbial Cell Factories* 10: 18. www.microbialcellfactories.com/content/10/1/18.

Moiseev, I.I. (2016). Biotechnology is storming the heights of petrochemistry. *Kinetics and Catalysis* 57 (4): 405–421.

Noda, S., Mori, Y., Oyama, S. et al. (2019). Reconstruction of metabolic pathway for isobutanol production in *Escherichia coli*. *Microbial Cell Factories* 18: 124. https://doi.org/10.1186/s12934-019-1171-4.

Ryan, C. (2019). An overview of Gevo's biobased isobutanol production process. www.gevo.com (accessed 1 January 2021).

Smith, K.M., Cho, K.-M., and Liao, J.C. (2010). Engineering *Corynebacterium glutamicum* for isobutanol production. *Applied Microbiology and Biotechnology* 87 (3): 1045–1055.

Spaans, S.K., Weusthuis, R.A., van der Oost, J., and Kengen, S.W.M. (2015). NAPDH-generating systems in bacteria and archaea. *Frontiers in Microbiology* 6: 742. https://doi.org/10.3389/fmicb.2015.00742.

Stryer, L. (1988). *Biochemistry*, 3e, 433. New York: W. H. Freeman and Company.

Tournier, V., Duquesne, S., Guillamot, F. et al. (2023). Enzymes' power for plastics degradation. *Chemical Reviews* https://doi.org/10.1021/acs.chemrev.2c00644.

Vanguard (2017). BP joint venture acquires ethanol facility in Kansas, US, 5.4.2017.

Voet, D., Voet, J.G., and Pratt, C.W. (2016). *Principles of Biochemistry*, 518. Wiley.

Wagemann, K. (2014). Herstellung von Grundchemikalien auf Basis nachwachsender Rohstoffe als Alternative zur Petrochemie. *Chemie Ingenieur Technik* 86 (12): 2115–2134.

Weitz, S., Hermann, M., Linder, S. et al. (2021). Isobutanol production by autotrophic acetogenic bacteria. *Frontiers in Bioengineering and Biotechnology* 9: article 657253, https://doi.org/10.3389/fbioe.2021.657253.

Wong, S.S., Mi, L., and Liao, J.C. (2017). Microbial production of butanols. In: *Industrial Biotechnology. Products and Processes*, vol. 4 (ed. C. Wittmann and J.C. Liao), 573–595. Wiley-VCH.

10

Isobutene

10.1 Application

Isobutene or isobutylene is a gaseous hydrocarbon with the chemical formula C_4H_8 and a molecular weight of 56.1 g/mol. It is applied for the synthesis of tertiary butanol via hydration (Brzeski and Skurski 2019), the fuel additive MTBE (methyl tertiary butylether) via hydration and etherification, as monomer for polymer production via polymerization or raw material for other downstream chemicals. The global production volume is estimated at 15 Mio. *t* corresponding to a value of 20 bn USD (Grand View Research 2016). In recent years, isobutene has gained attractiveness as a potential building block and intermediate in the endeavor to produce a green polyethylene terephthalic acid (PET) based on renewable resources, i.e. it was foreseen to convert isobutene into isooctane by dimerization followed by a dehydro-aromatization to obtain para-xylene. The bio-based para-xylene would then be oxidized to terephthalic acid and be applied for producing "green" PET based on the conventional technology using ethylene glycol and terephthalic acid. These approaches were supported by the major PET users who are seeking to improve their sustainability profile and to fulfill their carbon reduction targets (Collias et al. 2014; Wagemann 2014).

10.2 Conventional Synthesis

Isobutene is so far mainly be produced from petrochemical resources. It is conventionally obtained in the purification process of the C4 stream generated as a by-product in the cracking of naphtha to derive ethylene and propylene up to 50% of the ethylene amount. The C4 stream mainly consists of isobutene and 1,3-butadiene, which are separated and then channeled into the respective value chains (Arpe 2007). The global production volume is estimated to exceed 10 Mio. t pa (Saaret et al. 2021).

Alternatively, isobutene may be obtained in the back-cracking of MTBE which is derived as a by-product in the synthesis of PO (propylene oxide) according to the MTBE-PO technology. The technology reacts isobutane with oxygen to obtain TBA (tertiary butanol) and TBHP (tertiary butanol-hydroperoxide) in the first step. The TBHP as organic peroxide is then reacted with propylene to transfer the reactive oxygen into propylene to obtain PO. Hereby TBHP is also converted into TBA and in total about 2,5 t TBA/t PO are obtained. TBA may then be reacted and etherified with methanol to MTBE (Huntsman technology).

Pathway Design for Industrial Fermentation, First Edition. Walter Koch.

Depending on local markets and demand, the MTBE may be applied as fuel additive or channeled into the back-cracking into isobutene and methanol. The MTBE-PO technology may also dehydrate the by-product TBA into isobutene (LyondellBasell technology), which may then pending on local demand be sold into the market or be converted into MTBE, ETBE (ethyl tertiary butylether), iso-octene, and iso-octane.

The synthesis of isobutene based on fermentative isobutanol and subsequent chemical dehydration has been evaluated, though has not yet reached commercial maturity. This is partially due to the still limited productivity of the fermentative production of isobutanol given the toxicity of the compound toward the most relevant fermentation hosts. Furthermore, the gas phase dehydration of isobutanol to isobutene has in principle been shown with catalysts based on alumina, silica-alumina, zeolite, and solid phosphoric acid, though still suffers from low selectivity and high catalyst consumption (Suzuki 2018).

10.3 Pathway Design Toward Isobutene

Since isobutene is not a natural product and an "isobutene synthase" is not reported as a natural enzyme, all proposed biochemical pathways are based on metabolic engineering and especially try to exploit substrate promiscuity of natural enzymes for the final reaction. Several biochemical pathways to isobutene have been proposed (Van Leeuwen et al. 2012; Wilson et al. 2018). Three pathways proceed via 2-oxoisovalerate (α-ketoisovalerate). It is obtained via pyruvate which is then converted into the C5-intermediate 2-oxoisovalerate. Four additional pathways proceed via acetoacetyl-CoA; one proceeds via acetone and 3-hydroxyisovalerate and three additional routes proceed via 3-methylcrotonyl-CoA (insert 22) (Figure 10.1).

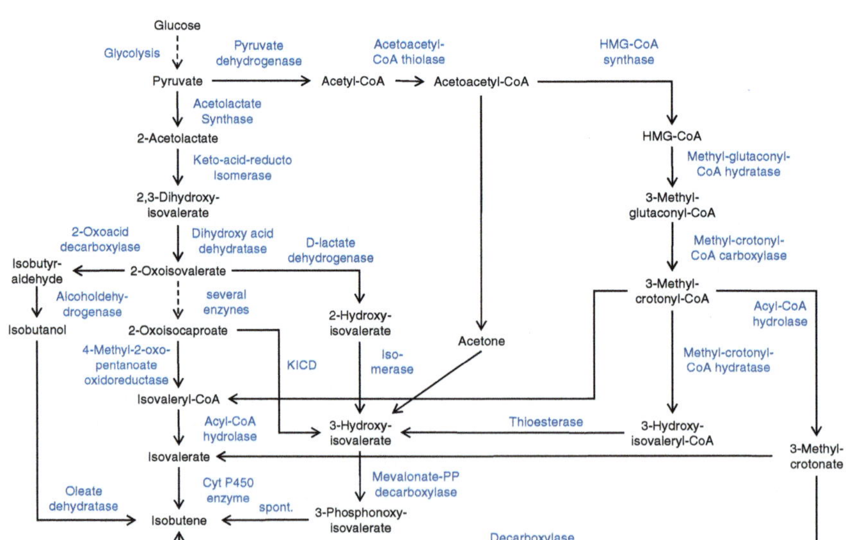

Figure 10.1 Proposed biochemical pathways toward isobutene. KICD: α-ketoisocaproate dioxygenase. Source: Walter Koch.

10.3.1 Isobutene via 2-oxoisovalerate and 2-hydroxyisovalerate

The first pathway via 2-oxoisovalerate proceeds via 2-hydroxyisovalerate in the first step via reduction of the keto-function. 2-hydroxyisovalerate is then isomerized to 3-hydroxyisovalerate. The decisive and interesting step facilitates combined dehydration and decarboxylation to obtain isobutene (Table 10.1).

Thus, one C6-precursor would be converted into one mol of isobutene and 2 mol of carbon dioxide. In case a transhydrogenase is assumed, which is able to convert NADPH into NADH, the reduction equivalents would be balanced. One additional mol of ATP would be derived. The research group of Global Bioenergies (Marlière 2010a, 2010b, 2011a, 2011b) and Iowa State University have developed a fermentation technology applying this route which uses the mevalonate-diphosphate decarboxylase (MVD) for the catalysis of the final step converting 3-hydroxyisovalerate into isobutene (Gogerty and Bobik 2010) (Figure 10.2).

This enzyme has the native function to convert mevalonate diphosphate into isopentenyl-pyrophosphate (IPP) in the mevalonate pathway. Given its natural promiscuity, it can also be employed to catalyze the chemically analog reaction of 3-hydroxyisovalerate to isobutene, which also converts a 3-hydroxy-3-methyl-acid via transient phosphorylation of the hydroxyl-group and decarboxylation. The isoenzyme of *Picrophilus torridus*, an obligate aerobic archaeon, displays the highest activity if transformed into *Escherichia coli* compared to the MVD isoenzyme of other organisms (Marlière 2010a, 2010b, 2011a, 2011b).

It has later been elucidated that the MVD enzyme of *Picrophilus torridus* is in the first instance a kinase that converts mevalonate into mevalonate-3-phosphate. Given its broader substrate specificity, it also phosphorylates 3-hydroxyisovalerate (and 3-hydroxybutyrate and 3-hydroxypropionate) into an unstable intermediate which then decarboxylates and releases one inorganic phosphate to obtain isobutene. The hydrolysis of a phosphoric acid anhydride bond in the conversion of ATP into ADP thermodynamically favors the formation of isobutene. Isobutene formation was detected using whole cells, crude extracts, and pure enzyme. The reported production rates amount to 507 pmol per min and per g cells using *E. coli* cells expressing the enzyme and 2,880 pmol per min and per mg protein

Table 10.1 Metabolic pathway toward isobutene.

Reaction	Enzyme
Glucose + 2 NAD^+ + 2 ADP + 2 Pi → 2 Pyruvate + 2 NADH + 2 H^+ + 2 ATP + 2 H_2O	Glycolysis
2 Pyruvate → 2-Acetolactate + CO_2	Acetolactate synthase
2-Acetolactate + NADPH + H^+ → 2,3-Dihydroxyisovalerate + $NADP^+$	Keto-acid-reducto-isomerase
2,3-Dihydroxyisovalerate → 2-Oxoisovalerate + H_2O	Dihydroxyacid-dehydratase
2-Oxoisovalerate + NADH + H^+ → 2-Hydroxyisovalerate + NAD^+	D-Lactate dehydrogenase
2-Hydroxyisovalerate → 3-Hydroxyisovalerate	Hypothetical isomerase
3-Hydroxyisovalerate + ATP → Isobutene + ADP + Pi + CO_2	Mevalonate-PP-decarboxylase
Glucose + NAD^+ + NADPH + ADP + Pi → Isobutene + NADH + $NADP^+$ + ATP + 2 CO_2 + 3 H_2O	

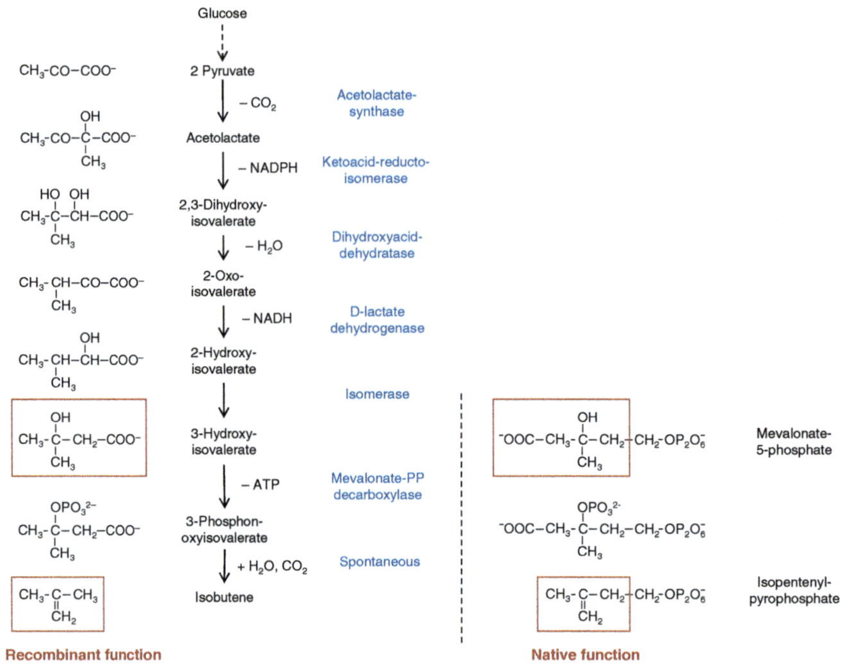

Figure 10.2 Biochemical pathway for the conversion of pyruvate into isobutene via 2-oxoisovalerate. Source: Adapted from Gogerty and Bobik (2010).

with the purified histidine-tagged enzyme (Rossoni et al. 2015). These data are sufficient to provide a general proof of concept, though are by far not sufficient to allow a commercial application. The overall conversion was improved by combining two separate enzymes with the first addressing the phosphorylation and activation step and the second providing isobutene via dehydration and decarboxylation (Mazaleyrat et al. 2018). In particular, the first step might be pursued with the 3-hydroxyisovalerate kinase of *Thermoplasma acidophilum*, whose activity could be further enhanced by genetic modifications, and the second step might be pursued with the mevalonate diphosphate decarboxylase of *Thermoplasma acidophilum*.

10.3.2 Isobutene via 2-oxoisovalerate and Isobutanol

An alternative route based on 2-oxoisovalerate as intermediate proceeds via isobutyraldehyde and isobutanol (Atsumi et al. 2009). Isobutene would be obtained by biochemical dehydration of isobutanol (Table 10.2).

2-Oxoisovalerate would be subjected to a non-oxidative decarboxylation to obtain isobutyraldehyde analog to the conversion of pyruvate to acetaldehyde in the ethanol fermentation process. The keto function at C2 is converted into an aldehyde and in the next step reduced to the respective alcohol. The alcohol would then be dehydrated to isobutene. This technology follows a joint path with a route for the fermentative production of isobutanol with the difference that in the last step, a biochemical dehydration is added (Figure 10.3).

Table 10.2 Metabolic pathway toward isobutene via 2-oxoisovalerate and isobutanol.

Glucose + 2 NAD^+ + 2 ADP + 2 Pi → 2 Pyruvate + 2 NADH + 2 H^+ + 2 ATP + 2 H_2O	Glycolysis
2 Pyruvate → 2-Acetolactate + CO_2	Acetolactate synthase
2-Acetolactate + NADPH + H^+ → 2,3-Dihydroxyisovalerate + $NADP^+$	Keto-acid-reducto-isomerase
2,3-Dihydroxyisovalerate → 2-Oxoisovalerate + H_2O	Dihydroxyacid-dehydratase
2-Oxoisovalerate → Isobutyraldehyde + CO_2	2-Oxoacid-decarboxylase
Isobutyraldehyde + NADH + H^+ → isobutanol + NAD^+	Alcohol dehydrogenase
Isobutanol → Isobutene + H_2O	Oleate hydratase
Glucose + NAD^+ + NADPH + 2 ADP → Isobutene + NADH + $NADP^+$ + 2 ATP + 2 CO_2 + 4 H_2O	

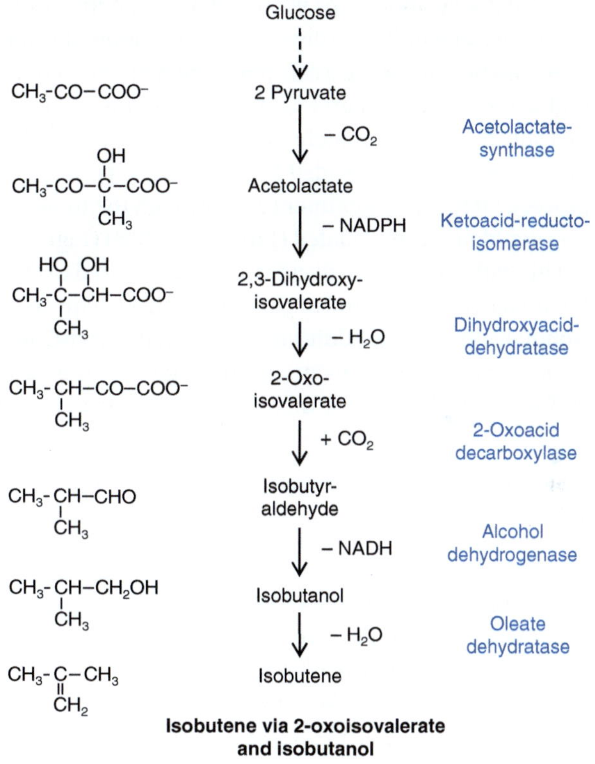

Figure 10.3 Biochemical pathway for the conversion of pyruvate into isobutene via 2-oxoisovalerate and isobutyraldehyde. For the last step, a promiscuous hydratase with different substrate specificity has been identified. Source: Walter Koch.

The direct biochemical dehydration followed by the secretion of isobutene would be the preferred approach since the gaseous isobutene can easily be isolated from the fermentation off-gas, whereas the liquid isobutanol would need to be purified from the fermentation broth via distillation and extraction. Additionally, isobutanol is reported to be toxic to most fermentation hosts and an *in situ* recovery needs to be developed to arrange a commercially robust process productivity. This pathway requires a dehydratase that allows the removal of water from isobutanol without adjacent ß-carbonyl-function like in fatty acid metabolism. Such an enzyme has been isolated from Elizabethkingia meningoseptica (Bevery et al. 2009). The natural function of the oleate hydratase (EC 4.2.1.53) is to catalyze the reaction of oleic acid to (R)-10-hydroxystearate. After several homologies of the gene were screened, an enzyme could be identified, which is in principle capable to convert isobutanol into isobutene (Marlière 2011a, 2011b).

10.3.3 Isobutene via 2-oxoisovalerate via Isovalerate

Finally, an interesting pathway via 2-oxoisovalerate has been proposed, which would then proceed with the condensation of 2-oxoisovalerate with acetyl-CoA to derive 2-oxoisocaproate via three biochemical steps (Van Leeuwen et al. 2012). The first step is known from the biosynthesis of the branched amino acid leucine via 2-acetolactate and 2-oxoisovalerate. This pathway proceeds via 2-oxosisocaproate which is formed via the condensation of 2-oxoisovalerate with acetyl-CoA. 2-Oxoisocaproate is then subjected to two decarboxylation steps via isovalerate to derive isobutene (Table 10.3).

The last step of this pathway – the formation of isobutene from isovalerate – has already been pursued in 1988 in a cell-free system of *Rhodotorula minuta*. The experiments revealed that isovalerate, isovaleryl-CoA, and α-keto-isocaproate (2-oxo-isocaproate) could be used as a substrate which matches the reconstruction of the pathway. The maximum activity of the enzyme extract provided 17 nl isobutene per mg cell extract per h. The reaction did not proceed after a heat treatment step, though the molecular mechanism behind this observation could not be elucidated (Fujii et al. 1988) (Figure 10.4).

This pathway has two disadvantages compared to the other routes via 2-oxoisovalerate. In the first instance, 1.5 mol of glucose would be required instead of 1.0 mol of glucose per mol isobutene in case the additionally required acetyl-CoA would be provided via the oxidative decarboxylation of pyruvate. Accordingly, the theoretical molar yield would be limited to 0.67 mol of isobutene per mol glucose. This would be about one-third below the two other

Table 10.3 Metabolic pathway toward isobutene via 2-oxoisovalerate and isovalerate.

Reaction	Enzyme
Glucose + 2 NAD^+ + 2 ADP + 2 Pi → 2 Pyruvate + 2 NADH + 2 H^+ + 2 ATP + 2 H_2O	Glycolysis
2 Pyruvate → 2-Acetolactate + CO_2	Acetolactate synthase
2-Acetolactate + NADPH + H^+ → 2,3-Dihydroxyisovalerate + $NADP^+$	Keto-acid-reducto-isomerase
2,3-Dihydroxyisovalerate → 2-Oxoisovalerate + H_2O	Dihydroxyacid-dehydratase
2-Oxoisovalerate + acetyl-CoA + H_2O + NAD^+	
→ 2-oxoisocaproate + CoA + CO_2 + NADH + H^+	(Several enzymes)
2-Oxoisocaproate + CoA + NAD^+ → Isovaleryl-CoA + CO_2 + NADH	4-Methyl-2-oxopentanoate oxido-RD
Isovaleryl-CoA + H_2O → Isovalerate + CoA	Acyl-CoA hydrolase
Isovalerate + NADPH + H^+ + O_2 → Isobutene + CO_2 + 2 H_2O + $NADP^+$	Cyt. P450-enzyme
Glucose + acetyl-CoA + 2 ADP + 2 Pi + 2 $NAD(P)^+$ + O_2 → Isobutene + 4 CO_2 + 2 ATP + 2 NAD(P)H + 2 H^+ + 3 H_2O + CoA	

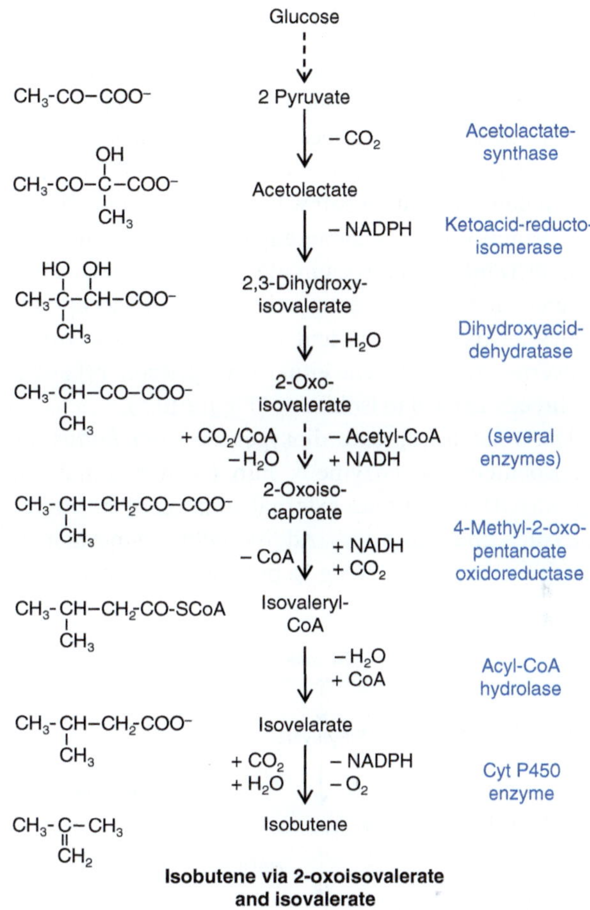

Figure 10.4 Biochemical pathway for the conversion of pyruvate into isobutene via 2-oxoisovalerate and isovalerate. Source: Walter Koch.

pathways via 2-oxoisovalerate. Additionally, the carbon footprint would be hampered, and the carbon dioxide emission would amount to four mol of carbon dioxide instead of two mol of carbon dioxide per ton of isobutene. The theoretical carbon dioxide by-product factor would increase from 1.6 t CO_2/t isobutene ($2\times 44/56.1$) to 3.2 t CO_2/t isobutene ($4\times 44/56.1$). It should be pointed out that the theoretical molar yield can be improved in case the additional acetyl-CoA would be generated via the non-oxidative glycolysis. This artificial pathway bypasses the oxidative decarboxylation of pyruvate into acetyl-CoA since acetyl-CoA is obtained via the cleavage of xylulose-5-P into glycerine-aldehyde-3-phosphate and acetyl-phosphate to be converted into acetyl-CoA. The overall equation of non-oxidative glycolysis can be stated as follows (Bogorad et al. 2013; Lin et al. 2018):

$$\text{Glucose} + \text{ATP} + 3\,\text{CoA} \rightarrow 3\,\text{Acetyl} - \text{CoA} + \text{ADP} + \text{Pi} + 2\,\text{H}_2\text{O}$$

Thus, three mol of acetyl-CoA can be generated from one mol of glucose. Then only 1.33 mol of glucose would be needed to supply one mol of acetyl-CoA and two mol of pyruvate and the theoretical molar yield would increase from 0.67 to 0.75 mol of isobutene/mol glucose. The required energy input in the form of ATP would be provided in the provision of the two mol of pyruvate, or in other words, the overall reaction equation of the route generates an ATP surplus which could be channeled into the non-oxidative glycolysis.

10.3.4 Isobutene via 2-oxoisocaproate and Isovalerate

An interesting pathway has been proposed in Synechocystis sp. which belongs to the group of photosynthetic microorganisms that convert CO_2 via the Calvin cyclus to various metabolites. The α-ketoisocaproate dioxygenase from *Rattus norvegicus* has been cloned in the host and small amounts of isobutene could be detected (Mustila et al. 2021). The pathway follows the native pathway toward L-valine, though 2-oxoisovalerate is not converted to L-valine via the enzyme encoded by ilvE, though 2-oxoisovalerate is converted with leuABCD into α-ketoisocaproate. This intermediate is then reacted to 3-hydroxyisovalerate whereby a dioxygenase, namely, the α-ketoisocaproate dioxygenase (KICD), which integrates an oxygen atom and facilitates a decarboxylation. 3-Hydroxyisovalerate is then converted into isobutene either via a phosphorylated intermediate as shown in the insert or directly reacted to isobutene (Figure 10.5).

The α-ketoisocaproate dioxygenase from *Rattus norvegicus* (RnKICD) achieves 104.6 ng isobutene/mg enzyme x min (with external addition of 2-oxoisocaproate) or 91 ng/lxhxOD750 (without external addition of substrate, Mustila et al. 2021). In both cases, the batch time was 7 days. Additionally, the mevalonate kinase was tested in a biocatalytic enzyme assay and 1.71 ng isobutene/mg enzyme x min could be detected. Alternatively,

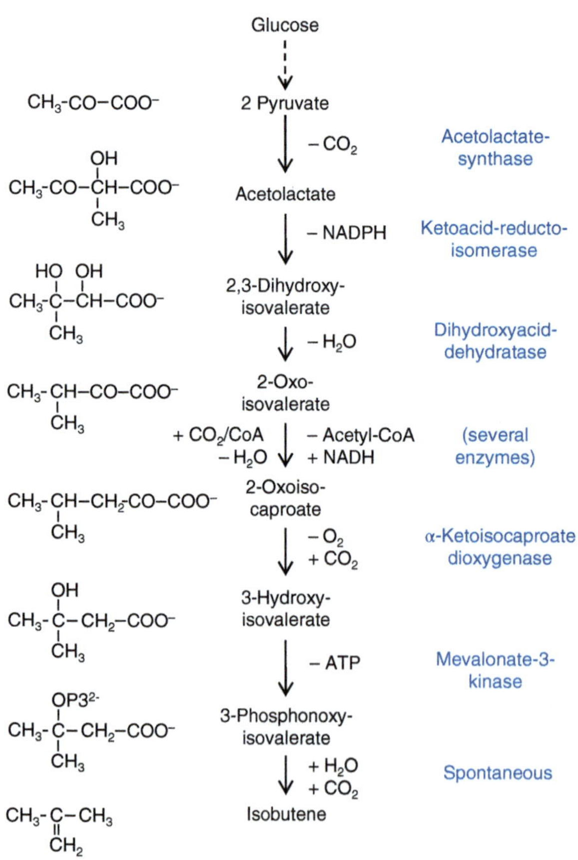

Figure 10.5 Biochemical pathway for the conversion of pyruvate into isobutene via 2-oxoisovalerate and 3-hydroxyisovalerate. Source: Adapted from Mustila et al. (2021).

the mevalonate kinase was integrated into *Synechocystis* and 79 ng isobutene/lxhxOD750 could be detected. These experiments indicate general feasibility, though are far away from commercial application.

10.3.5 Isobutene via acetoacetyl-CoA and 3-methylcrotonate

This route provides acetyl-CoA in the first instance and would then proceed via the mevalonate pathway to derive HMG-CoA. HMG-CoA is converted into 3-methylglutaconyl-CoA (3-MG-CoA) and 3-methylcrotonyl-CoA (3-MCT-CoA) (Table 10.4).

This pathway might proceed in two versions. The version outlined above pursues the reduction and hydrogenation of the double bond in 3-methylcrotonyl-CoA in the first step to derive isovaleryl-CoA (3-methylbutyric acid). Then the thioester of isovaleryl-CoA would be hydrolyzed to obtain isovalerate and CoA. Alternatively, 3-methylcrotonyl-CoA might in the first step be converted into 3-methylcrotonate via hydrolysis of the thioester bond, and then be converted into isovalerate in the second step (Marlière and Allard 2017). A prenylated FMN-dependent decarboxylase for the conversion of 3-methylcrotonate into isobutene is not known in nature. A ferulic acid decarboxylase from *Trichoderma atroviride* which usually decarboxylates aromatic acids or acrylic acids with extended conjugation has been modified via directed evolution to accept 3-methylcrotonate as substrate. The biocatalytic *in vitro* assay was performed with 10 mM 3-methylcrotonate and 0.3 mg/ml enzyme. Isobutene could be detected, though only in minor amounts (Saaret et al. 2021). For comparison, a modified ferulic acid decarboxylase has also been applied to generate 1,3-butadiene in *E. coli* based on glucose via cis, cis-muconic acid as intermediate (Mori et al. 2021). The pathway implies the need of 1.5 mol of glucose for one mol of isobutene and thereby achieves a theoretical molar yield on only 0.67 mol of isobutene per mol glucose. The carbon footprint would accordingly also be high since the surplus of carbon feed is fully oxidized to carbon dioxide. Also in this case the acetyl-CoA pool might be provided via the non-oxidative glycolysis and the theoretical molar yield would increase to 1.0 mol of isobutene per mol glucose (Table 10.5).

The net demand of ATP would require some glucose amount and the theoretical molar yield would be about 0.97 mol of isobutene per mol glucose. It is hereby assumed that one

Table 10.4 Metabolic pathway toward isobutene via acetoacetyl-CoA and 3-methylcrotonate.

Reaction	Enzyme
1.5 Glucose + 3 CoA + 3 ADP + 3 Pi + 6 NAD^+ → 3 Acetyl-CoA + 3 CO_2 + 3 ATP + 6 NADH + 6 H^+ + 3 H_2O	
2 Acetyl-CoA → Acetoacetyl-CoA + CoA	Acetyl-CoA thiolase
Acetoacetyl-CoA + Acetyl-CoA + H_2O → HMG-CoA + CoA	HMG-CoA synthase
HMG-CoA → 3-MG-CoA + H_2O	Methylglutaconyl-CoA hydratase
3-MG-CoA → 3-MCT-CoA + CO_2	3-MCT-CoA carboxylase
3-MCT-CoA + H_2O → 3-Methylcrotonate + CoA	Acyl-CoA hydratase
3-Methylcrotonate → Isobutene + CO_2	Decarboxylase
1.5 Glucose + 3 ADP + 3 Pi + 4 NAD^+ → Isobutene + 5 CO_2 + 3 ATP + 6 NADH + 6 H^+ + 2 H_2O	

Table 10.5 Metabolic pathway toward isobutene via non-oxidative glycolysis.

Glucose + ATP + 3 CoA → 3 Acetyl-CoA + ADP + Pi + 2 H_2O	Non-oxidative glycolysis
2 Acetyl-CoA → Acetoacetyl-CoA + CoA	Acetyl-CoA thiolase
Acetoacetyl-CoA + Acetyl-CoA + H_2O → HMG-CoA + CoA	HMG-CoA synthase
HMG-CoA → 3-MG-CoA + H_2O	Methylglutaconyl-CoA hydratase
3-MG-CoA → 3-MCT-CoA + CO_2	3-MCT-CoA carboxylase
3-MCT-CoA + H_2O → 3-Methylcrotonate + CoA	Acyl-CoA hydratase
3-Methylcrotonate → Isobutene + CO_2	Decarboxylase
Glucose + ATP → Isobutene + 2 CO_2 + ADP + Pi + H_2O	

mol glucose provides about 30 mol of ATP in respiratory chain phosphorylation. The two reduction equivalents would be balanced.

10.3.6 Isobutene via acetoacetyl-CoA and 3-Hydroxyisovaleryl-CoA

This route provides acetyl-CoA in the first instance and would then proceed via the mevalonate pathway to derive HMG-CoA. HMG-CoA is converted into 3-methylglutaconyl-CoA (3-MG-CoA) and 3-methylcrotonyl-CoA (3-MCT-CoA). After hydration of the double bond to derive 3-hydroxyisovaleryl-CoA, the thioester is cleaved to obtain 3-hydroxyisovalerate, which is then dehydrated and decarboxylated to finally yield isobutene (Mazaleyrat et al. 2018) (Table 10.6).

Though this pathway would also imply the need for 1.5 mol of glucose for one mol isobutene and thereby achieve a theoretical molar yield of only 0.67 mol of isobutene/mol glucose. The carbon footprint would accordingly also be high since the surplus of carbon

Table 10.6 Metabolic pathway toward isobutene via acetoacetyl-CoA and 3-hydroxyisovaleryl-CoA.

1.5 Glucose + 3 CoA + 3 ADP + 3 Pi + 6 NAD^+ → 3 Acetyl-CoA + 3 CO_2 + 3 ATP + 6 NADH + 6 H^+ + 3 H_2O	
2 Acetyl-CoA → Acetoacetyl-CoA + CoA	Acetyl-CoA thiolase
Acetoacetyl-CoA + Acetyl-CoA + H_2O → HMG-CoA + CoA	HMG-CoA synthase
HMG-CoA → 3-MG-CoA + H_2O	Methylglutaconyl-CoA hydratase
3-MG-CoA → 3-MCT-CoA + CO_2	3-MCT-CoA carboxylase
3-MCT-CoA + H_2O → 3-hydroxyisovaleryl-CoA	Methylcrotonyl-CoA hydratase
3-Hydroxyisovaleryl-CoA + H_2O → 3-Hydroxyisovalerate + CoA	Thioesterase
3-Hydroxyisovalerate + ATP → Isobutene + ADP + Pi + CO_2	Mevalonate-PP-decarboxylase
1.5 Glucose + 2 ADP + 2Pi + 6 NAD^+ → Isobutene + 5 CO_2 + 2 ATP + 6 NADH + 6 H^+ + H_2O	

feed is fully oxidized to carbon dioxide. Also in this case, the acetyl-CoA pool might be provided via the non-oxidative glycolysis and the theoretical molar yield might even increase up to 1.0 mol of isobutene per mol glucose:

$$\text{Glucose} + \text{ATP} \rightarrow \text{Isobutene} + 2\ CO_2 + \text{ADP} + \text{Pi} + H_2O$$

Though the net demand of ATP would require some glucose amount for the provision of reduction equivalents and the theoretical molar yield would be about 0.97 mol of isobutene/mol glucose. It is hereby assumed that one mol of glucose provides about 32 mol of ATP in respiratory chain phosphorylation.

10.3.7 Isobutene via Acetoacetyl-CoA and Acetone and 3-hydroxyisovalerate

This route provides acetyl-CoA in the first instance to generate acetoacetyl-CoA and would then proceed to isobutene via acetone and 3-hydroxyisovelarate (Table 10.7).

The important pathway intermediate 3-hydroxyisovalerate is obtained via the reaction of acetone and acetyl-CoA (Marlière 2010a, 2010b). This pathway would also consume 1.5 mol of glucose per mol isobutene. In case the acetyl-CoA is provided via the non-oxidative glycolysis, an improved molar yield could be achieved (Table 10.8).

The negative ATP balance would need to be accounted for which would allow a maximum molar yield of about 0.94 mol of isobutene per mol glucose in case it is assumed that 1 mol of glucose can generate about 32 ATP.

The results of the pathway analysis can be summarized (Table 10.9).

Table 10.7 Metabolic pathway toward isobutene via acetoacetyl-CoA and acetone and 3-hydroxyisovalerate.

Reaction	Enzyme
1.5 Glucose + 3 CoA + 3 ADP + 3 Pi + 6 NAD^+ → 3 Acetyl-CoA + 3 CO_2 + 3 ATP + 6 NADH + 6 H^+ + 3 H_2O	
2 Acetyl-CoA → Acetoacetyl-CoA + CoA	Acetyl-CoA thiolase
Acetoacetyl-CoA + H_2O → Acetone + CO_2 + CoA	Acetoacetate decarboxylase
Acetone + acetyl-CoA + H_2O → 3-hydroxyisovalerate + CoA	PksG protein
3-Hydroxyisovalerate + ATP → Isobutene + ADP + Pi + CO_2	Mevalonate-PP-decarboxylase
1.5 Glucose + 2 ADP + 2 Pi + 6 NAD^+ → Isobutene + 5 CO_2 + 2 ATP + 6 NADH + 6 H^+ + H_2O	

Table 10.8 Metabolic pathway toward isobutene via acetoacetyl-CoA and non-oxidative glycolysis.

Reaction	Enzyme
Glucose + ATP + 3 CoA → 3 acetyl-CoA + ADP + Pi + 2 H_2O	
2 Acetyl-CoA → Acetoacetyl-CoA + CoA	Acetyl-CoA thiolase
Acetoacetyl-CoA + H_2O → Acetone + CO_2 + CoA	Acetoacetate decarboxylase
Acetone + acetyl-CoA + H_2O → 3-hydroxyisovalerate + CoA	PksG protein
3-Hydroxyisovalerate + ATP → Isobutene + ADP + Pi + CO_2	Mevalonate-PP-decarboxylase
Glucose + 2 ATP → Isobutene + 2 CO_2 + 2 ADP + Pi	

Table 10.9 TMY and carbon yield of potential pathways toward isobutene.

Pathway	TMY mol IB/mol Glc	Carbon yield t IB/t Glc	Comment
IB via 2-Oxoisovalerate/2-hydroxyisovalerate	1.0	0.31	
IB via 2-Oxoisovalerate/isobutanol	1.0	0.31	
IB via 2-Oxoisovalerate/2-isovalerate	0.67	0.21	Oxidative glycolysis
	0.75	0.23	Non-oxidative glycolysis
IB via 2-Oxoisovalerate/ 3-methylcrotonate	0.67	0.21	Oxidative glycolysis
	0.97	0.30	Non-oxidative glycolysis
IB via Acetoacetyl-CoA/HMG-CoA	0.67	0.21	Oxidative glycolysis
& 3-hydroxyisovaleryl-CoA	0.97	0.30	Non-oxidative glycolysis
IB via Acetoacetyl-CoA/acetone	0.67	0.21	Oxidative glycolysis
& 3-hydroxyisovalerate	0.94	0.29	Non-oxidative glycolysis

10.4 Carbon Yield and Carbon Footprint

In case the pathway via 2-oxoisovalerate would be pursued, a theoretical molar yield of 1.0 mol of isobutene per mol glucose would be derived. This would translate into a maximum carbon yield of 0.31 t isobutene/t glucose. Based on the attractiveness of the pathways and the patent landscape, it may be assumed that Global Bioenergies pursues the route via 2-oxoisovalerate and 3-hydroxyisovalerate. The company revealed information on the yield of their technology in the laboratory which may be interpreted such that 87% of the theoretical maximum molar yield is obtained (Chemie.de 2018b). This would translate into a carbon yield of 0.27 t isobutene/t glucose ($56 \times 0.87/180.2$). The remaining glucose would be required for the generation of the biocatalyst (i.e. biomass) and carbon dioxide generated in cell maintenance. Thus, about 3.7 t glucose would be needed for the production of per ton isobutene in case a selectivity of 87% would be achieved and no isobutene is lost in the downstream and purification process. Additionally, the scale-up of the technology might also cause a lower selectivity.

10.5 Industrial Fermentation and Capacity

The fermentative production of isobutene is pursued by Global Bioenergies (founded 2008) based on their own patent-protected technology of Philipp Marlière and colleagues (Marlière 2010a, 2010b, 2011a, 2011b), the co-founder and at that time president of the Global Bioenergies' Scientific Advisory Board. Global Bioenergies has announced in 2013 to start with the construction of a pilot scale biorefinery plant in Leuna comprising two fermenters with 5 m^3 each and the respective downstream process funded by the German Ministry for Education and Research with 5.7 Mio. €. The target was to produce about 100 t p.a. to supply interested partners with application tests (transcript.de 2013). In parallel a fermentation pilot plant in Pomacle-Bazancourt has been constructed and funded with 5.2 Mio. € by the French government's Investissements d'Avenir Programme (SpecialChem 2016). It was then reported that isobutene samples were handed over to Audi, Arkema, Arlanxeo, CFBP, Clariant, Crystal Union, and L'Oréal which were interested in the new technology (Global Bioenergies 2016).

The various industry partners pursue different application areas for fermentative isobutene. L'Oréal is interested in fermentative isobutene since petrochemically derived isobutene is the precursor of several derivatives applied in cosmetics. The switch to an alternative raw material source would support the sustainability targets of L'Oréal (Chemie.de 2017). On the other hand, Audi is interested in biobased fuel additives, e.g. isooctane, ETBE, and isododecane (Chemie.de 2018a). Clariant has developed a new bio-based polymer that may be applied as a rheology modifier to improve viscosity in formulations of creams and lotions (ICIS 2018).

In order to reduce raw material costs, Global Bioenergies is also evaluating to apply a wheat straw hydrolysate as raw material instead of the conventional fermentation feed glucose. The wheat straw residues remain after the wheat grains for food application – the main value component – have been isolated and the raw material would thereby not compete with the limited agricultural land needed for food supply. The project was started in 2017 and was funded by the Bio-Based Industry Undertaking (BBI-JU) in the frame of the H2020-program with 9.8 Mio. €. The wheat straw hydrolysate provided by Clariant based on their proprietary Sun-Liquid technology shall be converted into isobutene, which will then be further converted into oligomers and polymers by Ineos (process.vogel.de 2019).

The strategy for ramping up the production capacity has been changed in the past. It has been communicated in 2019 by Global Bioenergies that the construction of a plant with a capacity of 30 kt p.a. for fermentative isobutene is planned in cooperation with Technip Engineering with a presumed investment of about 140 Mio. €. The investment would be pursued by IBN-One, a JV of Global Bioenergies and Crystal Union, and comprises the fermentation plant and capacity for derivatives as well. The marketing focus was broadened from fuel additives to additionally cosmetic ingredients to address a higher value perspective which will require further engineering efforts to meet the advanced quality requirements (Global Bioenergies 2018). Global Bioenergies has created an own make-up brand LAST® to better position its products in the market. A production capacity of around 100 t in Pomacle was announced in 2021. Commercial startup was in 2022 and the first batches of fermentative isobutene were delivered (Global Bioenergies 2022). The product shall allow customers to increase the naturalness of their make-up products. A larger capacity with a capacity of some thousands of tons shall be started in 2024 to serve the market for skincare and haircare application (Sapp 2021). The scale-up plan was specified in 2022 and foresees startup of a plant with a capacity of thousand tons pa to provide ingredients for the skin and hair care markets in 2024–2025 and startup of a plant with a capacity of ten thousand tons pa to provide a building block for sustainable aviation fuel after 2026 (Global Bioenergies 2022).

References

Arpe, H.-J. (2007). *Industrielle Organische Chemie. Bedeutende Vor- und Zwischenprodukte*, 72f. Wiley.

Atsumi, S., Li, Z., and Liao, J.C. (2009). Acetolactate synthase from Bacillus subtilis serves as a 2-ketoisovalerate decarboxylase for isobutanol biosynthesis in *Escherichia coli*. *Applied and Environmental Microbiology* 75 (19): 6306–6311.

Bevery, L.E., Pinske, M.W.H., Verhaert, P.D.E.M., and Hagen, W.R. (2009). Oleate hydratase catalyzes the hydration of a non-activated carbon-carbon bond. *Journal of Bacteriology* 191 (p): 5010–5012.

Bogorad, I.W., Lin, T.-S., and Liao, J.C.A. (2013). Synthetic non-oxidative glycolysis enables complete carbon conservation. *Nature* 502: 693–698.

Brzeski, J. and Skurski, P. (2019). The mechanisms of isobutene hydration yielding *tert*-butanol catalyzed by a strong mineral acid (H_2SO_4) and Lewis-Bronsted superacid (HF/SbF_5). *Heliyon* https://doi.org/10.1016/j.heliyon.2019.e 02133.

Chemie.de (2017). Global Bioenergies liefert erstes Bio-Isobuten an L'Oréal, 29 November 2017.

Chemie.de (2018a). Global Bioenergies liefert Biokraftstoff an Audi, 18 March 2018.

Chemie.de (2018b). Global Bioenergies erreicht 87% Ausbeute im Isobutenprozess, 12 October 2018.

Collias, D.I., Harris, A.M., Nagpal, V. et al. (2014). Biobased terephthalic acid technologies: a literature review. *Industrial Biotechnology* 10 (2): 91–105.

Fujii, T., Ogawa, T., and Fukuda, H. (1988). Preparation of a cell-free, isobutene-forming system from *Rhodotorola minuta*. *Applied and Environmental Microbiology* 583–584.

Global Bioenergies (2016). Press release 27 June 2016.

Global Bioenergies (2018). Global Bioenergies receives letters of intent for purchases covering the capacity of its first plant project, IBN-One, a joint venture with Christal Union. https://www.global-bioenergies.com/global-bioenergies-receives-letters-of-intent-for-purchases-covering-the-capacity-of-its-first-plant-project-ibn-one-a-joint-venture-with-cristal-union/?lang=en (accessed 3 August 2023).

Global Bioenergies (2022). Press release, Start of commercial production in Pomacle, 24 March 2022. https://www.global-bioenergies.com/start-of-commercial-production-in-pomacle/?lang=en (accessed 20 July 2022).

Gogerty, D.S. and Bobik, T.A. (2010). Formation of isobutene from 3-Hydroxy-3-Methylbutyrate by Diphospho-mevalonate Decarboxylase. *Applied and Environmental Microbiology* 76 (24): 8004–8010.

Grand View Research (2016). Isobutene Market Size, Share & Trends Analysis Report by Product (Methyl Tert-Butyl Ether, Ethyl Tert-Butyl Ether), by Application, by Region, and Segment Forecasts 2016–2024. https://www.grand viewresearch.com/industry-analysis/isobutene-market. Report ID: GVR-1-68038-174-0.

ICIS (2018). Clariant, partner develop new bio-based polymer, 2 January 2018.

Lin, P.P., Jaeger, A.J., Wu, T.-Y. et al. (2018). Construction and evolution of an *Escherichia coli* strain relying on nonoxidative glycolysis for sugar catabolism. *PNAS* 115 (14): 3538–3546.

Marlière, P. (2010a). Method for the enzymatic production of 3-hydroxy-3-methylbutyric acid from acetone and butyryl-CoA, WO 2010/032934.

Marlière, P. (2010b). Production of alkenes by enzymatic decarboxylation of 3-hydroxyalkanoic acids, WO 2010/001078.

Marlière, P. (2011a). Method for the enzymatic production of 3-hydroxy-3-methylbutyric acid from acetone and acetyl-CoA, EP 2295593.

Marlière, P. (2011b). Method for producing an alkene comprising step of converting an alcohol by an enzymatic dehydration, WO 2011/076691.

Marlière, P. and Allard, M. (2017). Method for producing isobutene from 3-methylcrotonyl-CoA, US 2017/0260548.

Mazaleyrat, S., Barbaroux, J.-B., Delcourt, M. et al. (2018). HIV kinase variants, US 2018/0223266.

Mori, Y., Noda, S., Shirai, T., and Kondo, A. (2021). Direct 1,3-butadiene biosynthesis in *Escherichia coli* via a tailored ferulic acid decarboxylase mutant. *Nature Communications* 12: 2195. https://doi.org/10.1038/s41467-021-22504-6.

Mustila, H., Kugler, A., and Stensjö, K. (2021). Isobutene production in *Synechocystis* sp. PCC 6803 by introducing α-ketoisocaproate dioxygenase from *Rattus norvegicus*. *Metabolic Engineering Communications* https://doi.org/10.1016/j.mec.2021.e00163.

process.vogel.de (2019). Demonstrationsanlage: Erfolgreicher Start der vorkommerziellen Produktion von Isobuten aus Weizenstroh, 10 February 2019.

Rossoni, L., Hall, S.J., Eastham, G. et al. (2015). The putative mevalonate diphosphate decarboxylase from *Picrophilus torridus* Is in reality a mevalonate-3-kinase with high potential for bioproduction of isobutene. *Applied and Environmental Microbiology* 81 (7): 2625–2634.

Saaret, A., Villiers, B., Stricher, F. et al. (2021). Directed evolution of prenylated FMN-dependent Fdc supports efficient in vivo isobutene production. *Nature Communications* https://doi.org/10.1038/s41467-021-25598-0.

Sapp, M. (2021). Global Bioenergies makes advances on bio-based isobutene, 20 October 2021. www.biofuelsdigest.com (accessed 22 October 2021).

SpecialChem (2016). Global Bioenergies confirms compatibility of bio-sourced isobutene with chemical oxidation, 14 June 2016, SpecialChem4coatings.com

Suzuki, T. (2018). Catalyst for isobutylene production and method for producing isobutylene, USPTO 2018/0072636.

transcript.de (2013). Leuna: Neubau von Bioraffinerie. Der französisch-deutsche Konzern Global Bioenergies baue eine neue Bioraffinerie-Pilotanlage in Leuna, also published in transcript 12/2013 published 28 November 2013.

Van Leeuwen, B.N.M., van der Wulp, A.M., Duijnstee, I. et al. (2012). Fermentative production of isobutene. *Applied Microbial Biotechnology* 93: 1277–1387.

Wagemann, K. (2014). Herstellung von Grundchemikalien auf Basis nachwachsender Rohstoffe als Alternative zur Petrochemie. *Chemie Ingenieur Technik* 86 (12): 2115–2134.

Wilson, J., Gering, S., Pinard, J. et al. (2018). Bio-production of gaseous alkenes: ethylene, isoprene, isobutene. *Biotechnology for Biofuels* 11: 234. https://doi.org/10.1186/s13068-018-1230-9.

11 1,4-Butanediol

11.1 Application

1,4-Butanediol is an organic alcohol with the formula $C_4H_{10}O_2$ and a molecular weight of 90.1 g/mol. The terminal hydroxyl functions favor the use of polymerization. Accordingly, the demand in the market is mainly addressing polymer applications in automotive plastics, electronics, and spandex for apparel. The biggest share with about 45% of global 1,4-butanediol production is converted into tetrahydrofurane (THF) via cyclodehydration (Rosales-Calderon and Arantes 2019), which is then polymerized to Poly-THF or PTMEG (Polytetramethyleneetherglycol) in the subsequent step:

$$\text{1,4-Butanediol} \rightarrow \text{Tetrahydrofurane} + H_2O$$

$$n\text{-Tetrahydrofurane} \rightarrow -\left(CH_2{-}CH_2{-}CH_2{-}CH_2{-}O\right)_n-$$

The preceding and separated synthesis of Poly-THF based on the ring-opening polymerization allows efficient removal of water in the first step compared to the direct polymerization of 1,4-butanediol. The release of water during the direct polymerization of 1,4-butanediol would imply a limited yield and cause a rather high residual monomer content since the water molecules drive the backward reaction. The obtained polyether Poly-THF may then be reacted with isocyanates to obtain elastomeric polyurethanes like Spandex.

1,4-Butanediol might also be reacted with diacids to obtain polyesters like terephthalic acid as a partnering monomer to derive polybutylene-terephthalate (PBT). About 25% of global 1,4-butanediol production is allocated to PBT (Rosales-Calderon and Arantes 2019). Polybutylene succinate (PBS) defines a further emerging application area because it can be produced based on two biobased monomers – 1,4-butanediol and succinic acid both produced via fermentation – and thereby satisfy the demand for sustainable plastics (Rosenboom et al. 2022). The biodegradability of PBS is a distinctive attribute compared to other polyesters and supports marketing for single-use food packaging (Rafiqah et al. 2021). 1,4-Butanediol is also the starting material for the "BDO-value chain" covering the intermediates γ-butyrolactone, 2-pyrrolidone (2-Pyr), N-vinylpyrrolidone (NVP), and polyvinylpyrrolidone (PVP) (Fischer and Bauer 2009). PVP is an important excipient for the formulation of solid dosage forms in the pharmaceutical industry (Bühler 2008).

Summarizing, a broad range of downstream technology and application areas have been developed, which have triggered the permanent growth of 1,4-butanediol production in the last decades. The global demand for 1,4-butanediol is estimated about 2 Mio t pa

Pathway Design for Industrial Fermentation, First Edition. Walter Koch.

(Coons 2019). The average market price of 1,4-butanediol in the years 2010–2014 amounts to 2.660 USD/t (Rosales-Calderon and Arantes 2019).

11.2 Conventional Synthesis of 1,4-Butanediol

Several organic chemistry technologies to produce 1,4-butanediol are pursued in the industry (Arpe 2007; Wagemann 2014). The historically first synthesis of 1,4-butanediol starts with acetylene and is part of the acetylene-based "Reppe chemistry" (Trotuş et al. 2014). Acetylene is reacted with formaldehyde at 100–110 °C and 5–20 bar (Reppe 1950). The primary intermediate 2-butine-1,4-diol is at 70–100 °C and 250–300 bar hydrogenated to 1,4-butanediol. The addition of two mol of hydrogen may be pursued in one step or two steps. A two-step process with 1,4-butanediol as intermediate allows the reduction of the pressure down to 25 bar:

$$\text{HC} \equiv \text{CH} + 2\,\text{HCHO} \rightarrow \text{HO–CH}_2\text{–C} \equiv \text{C–CH}_2\text{–OH}$$

$$\text{HO–CH}_2\text{–C} \equiv \text{C–CH}_2\text{–OH} + \text{H}_2 \rightarrow \text{HO–CH}_2\text{–CH} = \text{CH–CH}_2\text{–OH}$$

$$\text{HO–CH}_2\text{–CH} = \text{CH–CH}_2\text{–OH} + \text{H}_2 \rightarrow \text{HO–CH}_2\text{–CH}_2\text{–CH}_2\text{–CH}_2\text{–OH}$$

The Mitsubishi technology foresees the reaction of butadiene with acetic acid to derive 1,4-diacetoxy-2-butene (acetoxylation). The reaction is pursued at 70 °C and 70 bar and a Pd/C catalyst is used (Arpe 2007). This intermediate is then saturated or hydrogenated, respectively, at 60 °C and 50 bar with a conventional hydrogenation catalyst to derive 1,4-diacetoxybutane. Hydrolysis and cleavage of the ester bonds provide the free acetic acid and 1,4-butanediol:

$$\begin{aligned} &\text{H}_2\text{C} = \text{CH–CH} = \text{CH}_2 + 2\,\text{CH}_3\text{COOH} + 0.5\,\text{O}_2 \\ &\quad \rightarrow \text{CH}_3\text{–CO–O–CH}_2\text{–CH} = \text{CH–CH}_2\text{–O–CO–CH}_3 + \text{H}_2\text{O} \\ &\text{CH}_3\text{–CO–O–CH}_2\text{–CH} = \text{CH–CH}_2\text{–O–CO–CH}_3 + \text{H}_2 \\ &\quad \rightarrow \text{CH}_3\text{–CO–O–CH}_2\text{–CH}_2\text{–CH}_2\text{–CH}_2\text{–O–CO–CH}_3 \\ &\text{CH}_3\text{–CO–O–CH}_2\text{–CH}_2\text{–CH}_2\text{–CH}_2\text{–O–CO–CH}_3 + 2\,\text{H}_2\text{O} \\ &\quad \rightarrow \text{HO–CH}_2\text{–CH}_2\text{–CH}_2\text{–CH}_2\text{–OH} + 2\,\text{CH}_3\text{COOH} \end{aligned}$$

These technologies avoid the direct generation of carbon dioxide and are in this respect advantaged. The 1,4-butanediol technology starting with benzene is hampered by its low carbon efficiency since the main intermediate maleic acid anhydride generates two mol of carbon dioxide per mol maleic acid anhydride. The reaction is pursued at 400–450 °C and 2–5 bar:

$$\text{Benzene} + 4{,}5\,\text{O}_2 \rightarrow \text{maleic acid anhydride} + 2\,\text{CO}_2 + 2\,\text{H}_2\text{O}$$

As an alternative pathway, maleic acid anhydride may also be derived via the selective oxidation of butane. The intermediate maleic acid anhydride is then subjected to various technologies, which hydrolyze the anhydride bond, saturate the double bond of the backbone, hydrogenate the terminal carboxyl functions, and derive γ-butyrolactone or 1,4-butanediol.

1,4-Butanediol may additionally be produced with propylene oxide as starting material which is isomerized to allyl alcohol (2-propene-1-ol). The allyl alcohol is then subjected to a hydroformylation to derive 3-hydroxybutyraldehyde as an intermediate. A rhodium phosphine catalyst is applied which is efficiently recycled since the reaction is pursued in a two-phase system. The catalyst is solved in toluene, which separates from the polar product phase. After conventional chemical hydrogenation of the aldehyde function, the product 1,4-butanediol is obtained:

$$HO{-}CH_2{-}CH = CH_2 + CO + H_2 \rightarrow HO{-}CH_2{-}CH_2{-}CH_2{-}CHO$$

$$HO{-}CH_2{-}CH_2{-}CH_2{-}CHO + H_2 \rightarrow HO{-}CH_2{-}CH_2{-}CH_2{-}CH_2{-}OH$$

This reaction belongs to the overall group of olefin hydroformylations with hydrogen and carbon monoxide toward aldehydes as intermediates, which defines a landmark of organic chemistry applying homogeneous catalysis (Röper 2006). The technology is applied by LyondellBasell and Covestro (formerly Bayer Material Science). The biggest producers of petrochemically derived 1,4-butanediol are BASF, Dairen Chemical, LyondellBasell, Ashland (formerly ISP), Mitsubishi, and Xinjiang Markor Chemical Industry (Rosales-Calderon and Arantes 2019).

11.3 Pathway Design

11.3.1 Oxidative Citric Acid Cycle Until Succinyl-CoA

1,4-Butanediol is not a product found in nature. However, it could be shown in the 80th that strains of *Pseudomonas* are able to convert butanol into 1,4-butanediol in a whole-cell biocatalysis (Stieglitz and Weimer 1985). For the fermentative production of 1,4-butanediol, a recombinant pathway had to be designed. Genomatica, the leading technology provider for fermentative 1,4-butanediol, has therefore developed a proprietary bioinformatics tool called "SimPheny" to define, evaluate, and select recombinant pathways. The program initially identified 10 000 pathways that might in principle be suitable for the synthesis of 1,4-butanediol. The biotech company selected with their bioinformatics tools succinate and α-ketoglutarate as suitable precursors since both C4 intermediates of the citric acid cycle are highly favored for a reductive conversion into 1,4-butanediol (Yim et al. 2011). Accordingly, suitable enzymes were identified in biochemical databases, *Escherichia coli* was chosen as target host and plasmids engineered combined with an IPTG activation module. In particular, the lac-based P_{A01} promoter has been used, which is IPTG-inducible in strains containing *lacI*^{Q} such as those used in the experiments. For the industrial-scale commercial production later on, the genes were chromosomally integrated and IPTG-induction phased out since the cellular efforts for expressing plasmid-based genes are regarded as high. Thereby the inducible promoters were replaced by constitutive promoters (Barton et al. 2015). In the first instance, smaller sections of the overall pathway were pursued and tested in small-scale experiments (Figure 11.1).

Exclusively two steps were pursued with native enzymes of *E. coli* (SucCD and ADH). For the other steps, genes of other organisms were selected due to general feasibility or improved efficiency. The bioinformatics tool also suggested to knockout the (native) alcohol dehydrogenase (*adhE*), the pyruvate formate lyase (*pfl*), the lactate dehydrogenase (*ldh*), and the malate dehydrogenase (*mdh*) in order to suppress the generation of the fermentation products ethanol, formate, lactate, and succinate. The available reduction equivalents are saved for the target metabolic pathway. Additionally, since the pyruvate dehydrogenase

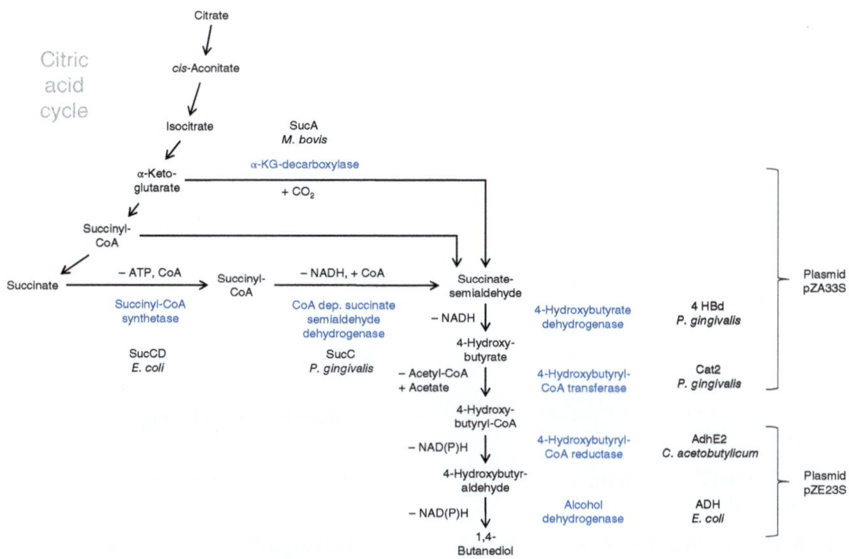

Figure 11.1 Biochemical pathway design toward 1,4-butanediol starting with succinate and α-ketoglutarate as intermediates of the citric acid cycle. Source: Adapted from Yim et al. (2011).

(*lpdA*) in *E. coli* is reported to have a low activity at high NADH levels, the native *E. coli lpdA* gene was replaced with the *lpdA* gene from *Klebsiella pneumonia* (Kim et al. 2008). Nicotinamide adenine dinucleotide (NADH) and nicotinamide adenine dinucleotide phosphate (NADPH) are employed as reduction equivalents. The combination of both vectors provided an initial titer of 18 g/l.

The pathway intermediate succinate semialdehyde might be obtained via different pathways. In the first instance, ketoglutarate might simply be decarboxylated to directly obtain succinate semialdehyde. This route seems to be predominant in the early R&D work since based on ^{13}C labeling experiments, it was elucidated that >95% of the pathway flux was via the oxidative citric acid cycle and included the decarboxylation of α-ketoglutarate (Yim et al. 2011). Additionally, succinate semialdehyde might be obtained with succinate as a starting, point which would then be activated to succinyl-CoA. This route is probably referred in the first publication of Genomatica describing the development of their technology: "The first route to 4HB starts from the tricarboxylic acid (TCA)-cycle intermediate succinate, which is activated as succinyl-CoA by the native *E. coli* enzyme succinyl-CoA synthetase" (Yim et al. 2011). However, the rationale of this route is not obvious since succinyl-CoA is also an intermediate of the citric acid cycle, and the generation of succinate and the immediate reactivation seems to be an idle cycle (Figure 11.1). The surplus of reduction equivalents is determined by the number of electrons carried by glucose and 1,4-butanediol and is thereby independent of the pathway. Also, the energy balance is not impacted since one mol of GTP is gained in the generation of succinate as reaction of the citric acid cycle which is, however, reinvested in the activation of succinate toward succinyl-CoA. In the later publications of Genomatica, a pathway is described that starts with succinyl-CoA as the intermediate of the citric acid cycle which is converted into succinate semialdehyde in the next step (Culler 2016; Burgard et al. 2016; Barton et al. 2015). The route via succinyl-CoA derived in the citric

Table 11.1 Pathway toward butanediol with succinyl-CoA as interlink of the CAC and the reduction steps toward BDO.

Reaction	Enzyme
Glucose + 2 NAD^+ → 2 PEP + 2 NADH + 2 H^+ + 2 H_2O	Glycolysis until PEP
PEP + CO_2 + H_2O → Oxalacetate + Pi	PEP carboxylase
PEP + ADP → Pyruvate + ATP	Pyruvate kinase
Pyruvate + CoA + NAD^+ → Acetyl-CoA + CO_2 + NADH + H^+	Pyruvate dehydrogenase
Oxalacetate + Acetyl-CoA + H_2O → Citrate + CoA	Citrate synthase
Citrate → Isocitrate	Aconitase
Isocitrate + $NADP^+$ → α-Ketoglutarate + CO_2 + NADPH + H^+	Isocitrate dehydrogenase
α-Ketoglutarate + NAD^+ → Succinyl-CoA + CO_2 + NADH + H^+	α-Ketoglutarate DH complex
Succinyl-CoA + NADH + H^+ → Succinate semialdehyde + CoA + NAD^+	CoA-dep. succinate semialdehyde DH
Succinate semialdehyde + NADH + H^+ → 4-Hydroxybutyrate NAD^+	4-Hydroxybutyrate dehydrogenase
4-Hydroxybutyrate + Acetyl-CoA → 4-Hydroxybutyryl-CoA + acetate	CoA-acyl transferase
4-Hydroxybyturyl-CoA + NAD(P)H + H^+ → 4-Hydroxybutanal + CoA	CoA-dep. aldehyde dehydrogenase
4-Hydroxybutanal + NADPH + H^+ → 1,4-Butanediol + $NADP^+$ + H^+	Alcohol dehydrogenase
Acetate + ATP → Acetyl-phosphate + ADP	Acetate kinase
Acetyl-phosphate + CoA → Acetyl-CoA + Pi	Phosphotransacetylase
Glucose + $NAD(P)^+$ → 1,4-Butanediol + 2 CO_2 + NAD(P)H + H^+	

acid cycle is chosen in this context to analyze the whole pathway. Carbon yield considerations propose that a share of the carbon dioxide derived in the formation of acetyl-CoA is recycled. Two mol of PEP (phosphoenolpyruvate) need to be generated in the first step to prepare the formation of citrate and oxalacetate (Burgard et al. 2016; Table 11.1).

The oxidative section of the citric acid cycle until succinyl-CoA together with reduction equivalents obtained in the generation of citrate is followed to reduce the two carboxylic acid functions of succinyl-CoA consuming four reduction equivalents. The net equation of the reaction reveals that the C6-skeleton of glucose is split into the C4-compound 1,4-butanediol and two mol of carbon dioxide. In the early phase of development, the by-product spectrum was not yet ideal and needed further improvement since after 120 h also 80 mM acetate, pyruvate, and 4-hydroxybutyrate were detected. ^{13}C-pulse labeling experiments using $^{13}C_6$-glucose as feed were performed which confirmed the rather high formation of 4-hydroxybutyryl-CoA compared to low levels of 1,4-butanediol and thus revealed the bottleneck of the pathway. Enzyme development was refocused on the identified bottleneck (Barton et al. 2015). The adenosine triphosphate (ATP) pool derived in substrate-level phosphorylation is balanced since the ATP generated in the synthesis of pyruvate is consumed later on in the regeneration of acetyl-CoA.

11.3.2 Reduction Equivalents and Oxygen Demand

The reduction equivalents are not balanced. One mol of NAD(P)H is generated in the overall course considering the gain of five reduction equivalents until succinyl-CoA and the four reduction equivalents needed for the reduction of succinyl-CoA to 1,4-butanediol. Glucose carries 24 electrons in total (without counting reduced oxygen) as opposed to only 22 electrons carried by 1,4-butanediol (without counting reduced oxygen). Accordingly, irrespective of the chosen metabolic pathway two electrons or one reduction equivalent will remain. The surplus of reduction equivalents would accumulate and hamper productivity. Several options might be taken.

Since a surplus of carbon dioxide and reduction equivalents is given one approach to recycle $NAD(P)^+$ might be the addition of the formate dehydrogenase with NAD(P)H as cofactor opposed to a formate hydrogenase with molecular hydrogen as cofactor, e.g. found in *Acetobacterium woodii* (Müller 2019). This is possible since the respective reduction potentials E_0 at pH 7.0 are in the same range. Alternatively, an aerobe fermentation process with oxygen as final electron acceptor would be straightforward option to recycle NAD^+:

$$2\,NADH + 2\,H^+ + O_2 \rightarrow 2\,NAD^+ + 2\,H_2O$$

Then the overall equation of the reaction is derived as follows (Burgard et al. 2016):

$$\text{Glucose} + 0.5\,O_2 \rightarrow 1,4\text{-Butanediol} + 2\,CO_2 + H_2O$$

A small ATP gain is achieved for cellular growth and cell maintenance since the oxidation of NADH allows the production of ATP in the respiratory chain phosphorylation. Formally 0.18 t oxygen is needed per t 1,4-butanediol (0.5 × 32 g/mol/90 g/mol) and, accordingly, Genomatica designed the fermentation as an aerobe process. It is interesting to note that this unit consumption rate is chosen in media aeration if the published quantitative process data are referred to. Cells were grown aerobically to a A_{600} density of about 10 and then the genes for the recombinant pathway are induced with 0.2 mM IPTG. One hour after induction the airflow was reduced to 0.02 l per minute feeding a 1 l culture volume in a 2 l bioreactor (Yim et al. 2011). This calculation assumes that this oxygen feed would allow maximum productivity of 2 g/lxh, which is exactly matching the threshold published by Genomatica as the commercially relevant target (Burgard et al. 2016; Table 11.2).

The reported productivity in the earlier publication states that the best-performing strains achieved a titer of 18 g/l with a productivity of 0.15 g/lxh (18 g/l in 5 d) (Yim et al. 2011), which is below the commercial target published afterward (Burgard et al. 2016).

Table 11.2 Alignment of oxygen transfer rate and productivity.

0.02 l air/l × min, derive
0.004 l oxygen/l × min, derive
5.71 mg oxygen/l × min, derive
0.343 g oxygen/l × h, which allow about
2 g 1,4-BDO/l × h according to the oxygen unit consumption rate of 0.18 t oxygen/t 1,4-BDO

11.3.3 Theoretical and Practically Achieved Yield

The maximum theoretical yield amounts to one mol of BDO per mol glucose according to the reaction equation as indicated above. The theoretical maximum carbon yield would be derived with 0.5 g 1,4-BDO/g glucose (90.1 g/mol/180.1 g/mol) in case the formation of biomass is excluded. Due to the formation of biomass, by-products, and product loss in the downstream and isolation process, the actually achieved carbon yield is lower. In the publication of Genomatica, a carbon yield of 0.37 g 1,4-BDO/g glucose calculated for an anaerobic process is mentioned (Yim et al. 2011). However, the technology is commercially pursued under microaerobic conditions and the theoretical maximum carbon yield under anaerobe conditions might deviate: "However, with microaeration these cells grew well and produced much more BDO than the wild-type host. We chose microaerobic conditions such that sufficient NAD(P)H is available to drive the BDO pathway, but any excess beyond the capacity of the pathway is respired to generate ATP that may be necessary for cell growth, maintenance, and product transport" (Yim et al. 2011). In case one would assume that the published 0.37 g 1,4-BDO/g glucose (Yim et al. 2011) also applies to the aerobe process, a glucose consumption rate of 2.7 t glucose/t 1,4-BDO would be derived. This figure is exactly according to the figure which can be recalculated from the carbon footprint estimation published for the bio-BDO grade "Qira" offered by the JV Qore of Cargill and Helm (http://myqira.com 2022; Table 11.3).

In the first three lines, the carbon dioxide emission of 1,4-butanediol synthesized with acetylene or maleic anhydride is estimated. With respect to fermentative 1,4-butanediol, the biogenic emission of carbon dioxide (scope 1) amounts to 2.0 t CO_2/t BDO given as the difference between the two values for the Qira product grades. The fermentative production of 1,4-butanediol generates a carbon dioxide emission (scope 1) due to the disproportionation of glucose. The glucose unit consumption rate can be derived if the difference between the reported scope 1 emission under real conditions and the scope 1 emission obtained under theoretical conditions are compared:

$$\text{Scope 1 emission of } 0.98\ \text{t CO}_2/\text{t 1,4-BDO}\ \left(\text{yield } 100\%\ \text{yield}, 2 \times 44\ \text{g/mol}/90\ \text{g/mol}\right), \text{and}$$

$$\text{Scope 1 emission of } 2.00\ \text{t CO}_2/\text{t 1,4-BDO}\ \ (\text{reflect real yield}), \text{derive}$$

$$\text{Scope 1 emission of } 1.02\ \text{t CO}_2/\text{t 1,4-BDO due to yield loss under real conditions.}$$

Table 11.3 Carbon footprint of 1,4-butanediol pending on synthesis route (http://myqira.com 2022).

Feedstock BDO technology	GHG emissions
Acetylene with hard coal	19.3 t CO_2/t BDO
Acetylene with natural gas	4.2
Maleic anhydride	3.9
Qira bio-BDO with biogenic carbon	2.7
Qira bio-BDO with sequestered CO_2	0.7

Source: Adapted from http://myqira.com 2022.

The additional scope 1 emission can be recalculated into the respective glucose consumption and the total glucose consumption can be estimated:

$$1.02\ \text{t}\ CO_2/\text{t}\ 1{,}4\text{-BDO derive}$$

$$0.278\ \text{t C/t}\ 1{,}4\text{-BDO, derive}$$

$$0.695\ \text{t glucose/t}\ 1{,}4\text{-BDO}\ \ (\text{given a C-content glucose of } 40\%).$$

Thus, the overall glucose consumption rate can be derived with 2.69 t glucose/t 1,4-BDO. This consumption rate corresponds to the published carbon yield of 0.375 g 1,4-BDO/t glucose (Yim et al. 2011).

11.3.4 Thermodynamically Challenging Reduction of Carboxylic Acid Functions

The pathway needs to reduce two carboxyl functions via the aldehyde-level down to the alcohol level and these conversions are thermodynamically not easy to achieve given the reduction potential of $E_0 = -580$ mV for the electron transfer on organic acids toward aldehydes. This may be achieved with reduced ferredoxin as in *Clostridium autoethanogenum* for the reduction of acetate toward acetaldehyde (Mock et al. 2015). However, reduced ferredoxin does not play a role in the metabolic engineering of *E. coli* for the production of 1,4-butanediol and, instead, the metabolic engineering strategy operates with activated thioesters in the form of succinyl-CoA and 4-hydroxybutyryl-CoA (Figure 11.2).

All modifications finally provided a recombinant host that delivered a titer >140 g/l and a productivity of 3.5 g/lxh (Burgard et al. 2016; Culler 2016). The carbon yield achieved values above 0.4 t 1,4-butanediol/t glucose which is >80% of the theoretical maximum carbon yield (Burgard et al. 2016). It is interesting to relate the published lab-scale data to the feed demand of a new plant with a disclosed capacity. The raw material demand of 30 000 bushels of corn per day to supply a plant in Iowa with a 1,4-butanediol capacity of 65 kt pa is somehow corresponding to this carbon yield, though a precise calculation is not possible given missing information on annual operation time and yield in downstream processing (Lane 2022). Summarizing, the pioneering work of Genomatica managed to run the oxidative citric acid cycle with predominant channeling of the gained reduction equivalents into the reductive synthesis of the product with a minor loss into respiratory chain phosphorylation.

11.4 Process Design for Fermentative 1,4-Butanediol Based on Glucose

The 1,4-butanediol technology of Genomatica is operated by Novamont who obtained a license. The exact performance data of the Genomatica technology (i.e. titer and productivity) industrially applied by Novamont are not disclosed. It may be assumed that the process parameters of the commercial technology are close to the R&D process as described in the publications of Genomatica. However, based on the qualitative process description, some potential constraints in the scale-up and commercial application can properly be described.

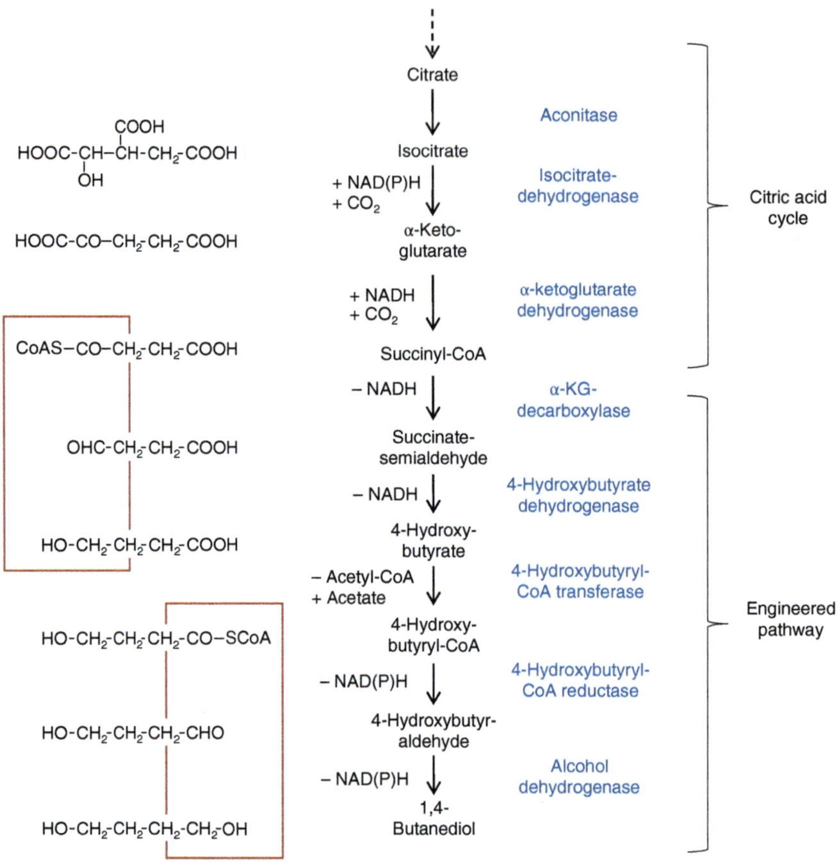

Figure 11.2 Biochemical pathway design toward 1,4-butanediol with α-ketoglutarate as starting point as intermediate of the citric acid cycle. Source: Adapted from Yim et al. (2011).

The host needs to endure specific conditions in the commercial reactor which it is not facing in lab-scale or mini-plant operation. This includes varying levels of oxygen due to the heterogeneous fluid dynamics in the reactor despite efficient stirring. Since titer and productivity strongly depend on oxygen supply, which should be proportional to the achieved oxygen transfer rate in the commercial reactor, high efforts are needed to scale-up the process and to keep the lab-scale performance in the commercial setup (Pooth et al. 2020). Additionally, the hydrostatic pressure in the bioreactor with a height of 25 m is enormous, and the generated carbon dioxide causes acidity via the formation of hydrogen carbonate which might not directly be buffered in the remote areas of the bioreactor. All effects might be detrimental to bacterial viability and cell growth. Accordingly, final strain selection was performed based on simulated upscale conditions to avoid a productivity gap in commercial production (Burgard et al. 2016).

After fermentation, the broth is channeled through a heated sterilization coil to kill the cells which are separated by centrifugation or microfiltration. Large biomolecules are then removed by nanofiltration and ultrafiltration membranes. The remaining ions are removed by ion exchange and raw 1,4-butanediol is obtained after the water has been evaporated. This step might imply rather a high cost since pending on the final titer at the batch end about 5–6 t steam might be needed per ton 1,4-butanediol. It is hereby assumed that in the first instance, one-ton steam is needed to evaporate one ton of water unless energy recycling measures like vapor compression are implemented to ensure a commercially acceptable energy efficiency. It might alternatively be considered to avoid the removal of water and instead pursue the next step in the downstream value chain as a liquid phase reaction in water. Accordingly, it has been tested to convert an aqueous 1,4-butanediol feed directly into THF applying a heterogeneous ZrO_2–Al_2O_3 metal oxide catalyst at temperatures of 200–240 °C and respective pressure (Li and Chen 2019). If successful, the steam consumption as part of the processing cost in the fermentative production of 1,4-butanediol might be avoided or at least be reduced.

11.5 1,4-Butanediol Derived by Chemical Hydrogenation of Succinic Acid

An alternative route toward 1,4-butanediol based on fermentation has been taken into consideration by some producers of succinic acid, a C4-intermediate for which robust fermentation technology has been developed. R&D and commercial production of succinic acid, respectively, has been pursued by Bioamber, Myriant, Reverdia, and Succinity. The diacid had been identified as a suitable platform molecule that might also provide biobased access into the 1,4-butanediol value chain (Bozell and Petersen 2010). Since initial market development of fermentative succinic acid did not match strategic planning, 1,4-butanediol was considered as alternative market outlet for (hydrogenated) succinic acid. The low greenhouse gas emissions of fermentative succinic acid may translate into greenhouse gas emissions of 1,4-butanediol which are lower than those of petrochemically derived 1,4-butanediol (Adom et al. 2014). The chemical conversion is basically a standard hydrogenation with molecular hydrogen as often applied in industrial chemistry. Succinic acid is catalytically hydrogenated in the liquid phase converting the terminal carboxylic acid functions into alcohol functions to obtain 1,4-butanediol (Hara and Inagaki 1991):

$$\mathrm{HOOC{-}CH_2{-}CH_2{-}COOH + 4\,H_2 \rightarrow HO{-}CH_2{-}CH_2{-}CH_2{-}CH_2{-}OH + 2\,H_2O}$$

A transient esterification and the hydrogenation of the respective ester may be favored – as also pursued in the conversion of fatty acids to fatty alcohols (Behr and Seidensticker 2018) – given an improved selectivity and yield:

$$\mathrm{ROOC{-}CH_2{-}CH_2{-}COOR + 4\,H_2 \rightarrow HO{-}CH_2{-}CH_2{-}CH_2{-}CH_2{-}OH + 2\,ROH}$$

These plans have in particular been pursued by BioAmber and Myriant who both were cooperating with catalyst companies to develop this step (Rosales-Calderon and Arantes 2019). BioAmber operated a mini-plant for the chemical hydrogenation of succinic acid under the license of a catalyst from Dupont and in fact obtained multi-ton product of 1,4-butanediol, THF, and GBL (Morden 2019; BioAmber 2012). Going one more step along the value chain, the joint venture AmberWorks was founded by BioAmber and NatureWorks in 2012 to produce PBS based on renewable BDO and renewable succinic acid (Adkins et al. 2012). It is also reported that Dupont applies 1,4-butanediol derived from succinic acid for the production of PBT (Chemical Week 2018).

11.6 Alternative Carbon and Energy Source for Fermentation

11.6.1 D-Xylose as Carbon Source via 2,5-Dioxopentanoate

The established and commercial technology for the fermentative production of 1,4-butanediol relies on glucose as carbon and energy source. Already Genomatica in their early development efforts tried to apply alternative sugars and could indeed show that the process in *E. coli* does also work with the disaccharide sucrose or mixed sugar streams derived from pretreated hydrolyzed biomass (Yim et al. 2011). An innovative and potentially cheaper alternative feed concept has been shown which foresees to convert lignocellulosic biomass and alternative side-streams into the intermediate 2,5-dioxopentanoate. This rather unusual intermediate shall bundle distinct feed streams and might be derived from D-xylose, L-arabinose (both ex lignocellulosic biomass), and D-galacturonate (ex citrus peel and sugar beet pulp) as well. All three precursors and alternative feed streams, respectively, are subjected to a nonphosphorylative metabolism leading to one common C5 intermediate (Figure 11.3).

The nonphosphorylative pathway for the conversion of the C5 sugars was originally identified in *Caulobacter crescentus*, *Pseudomonas fragi*, and *Azospirillum brasilense*. In this work, gene mining accessed putative gene clusters from various microorganisms. Once established, the 2,5-dioxopentanoate intermediate might be channeled into the citric acid cycle via oxidation of the aldehyde function to obtain α-ketoglutarate or directly be converted into 1,4-butanediol. Therefore, several decarboxylases were screened and their natural promiscuity was exploited to accept 2,5-dioxypentoate as substrate. Finally, the 2-ketoacid decarboxylase (Kivd) of *Lactobacillus lactis* was revealed as the most efficient decarboxylase. With the addition of an alcohol dehydrogenase (ADH), the

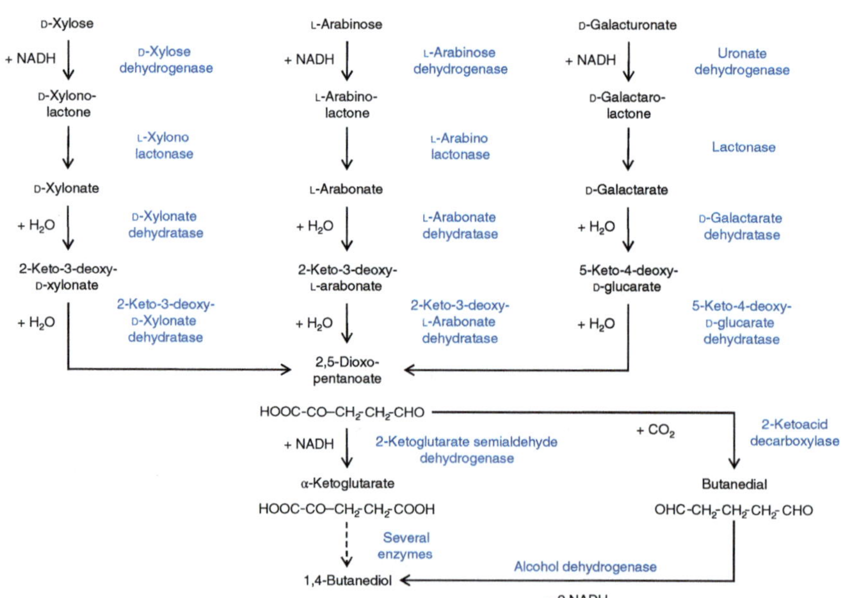

Figure 11.3 Alternative feedstock approach to derive α-ketoglutarate from lignocellulosic biomass and citrus peel/sugar beet pulp. α-Ketoglutarate as an intermediate of the citric acid cycle might be further converted into various target products. Source: Adapted from Tal et al. (2016).

Table 11.4 Performance data for the use of alternative feedstocks.

Carbon source	Titer (g/l)	Batch (h)	Productivity (g/lxh)
D-Xylose	12.0	30	0.40
L-Arabinose	15.6	75	0.21
D-Galacturonate	16.5	90	0.18

generation of 1,4-butanediol could be shown. The recombinant pathway proceeds via butanediol obtained in the decarboxylation of 2,5-dioxypentanoate (Figure 11.3). The dial-intermediate is then hydrogenated at both ends to yield 1,4-butanediol:

$HOOC{-}CO{-}CH_2{-}CH_2{-}CHO \rightarrow OHC{-}CH_2{-}CH_2{-}CHO + CO_2$	2-Ketoacid decarboxylase
$OHC{-}CH_2{-}CH_2{-}CHO + 2\,NADH + H^+ \rightarrow HO{-}CH_2{-}CH_2{-}CH_2{-}CH_2{-}OH + 2\,NAD^+$	Alcohol dehydrogenase

1,2,4-Butanetriol appeared as a disturbing by-product whose generation could be suppressed, though not fully be avoided. 0.37 g 1,4-butanediol/g D-xylose was achieved which is about 62% of the theoretical maximum carbon yield. The following titers and productivities for 1,4-butanediol were achieved (Tal et al. 2016; Table 11.4).

In these feeding trials, often pure model substances are applied. The chemical and physical digestion of lignocellulosic biomass often generates by-products that negatively interfere with the host and hamper the fermentation process. Future experiments need to confirm inasmuch the C5-sugars derived from lignocellulosic biomass have a purity that suffices the quality requirements of a fermentation process. Otherwise, costly purification protocols need to be developed which compromise the cost attractiveness of the alternative sugar source.

11.6.2 D-Xylose as Carbon Source via Xylonic Acid

The suitability of C5 precursors derived from lignocellulosic biomass for providing 1,4-butanediol has triggered the search for alternative pathways. A novel route toward 1,4-butanediol covering six enzymatic steps has been proposed. It initially proceeds via an acid intermediate and converts D-xylose into D-xylonic acid facilitated by the D-xylose dehydrogenase (*xdh*) from *C. crescentus* (Liu and Lu 2015). D-Xylonic acid is then subjected to a dehydration catalyzed by the D-xylonic acid dehydratase (*yjhG* and *yagF*) to an enol-intermediate with immediate rearrangement to obtain 2-dehdro-3-deoxy-D-xylonate. This intermediate is processed into α-ketoglutaric semialdehyde by the 2-dehydro-3-deoxy-D-xylonate dehydratase (*xylX* and *hvo*) from *C. crescentus* and *H. volcanii*. The terminal aldehyde function is subjected to a reduction and 5-hydroxy-α-ketoglutarate is derived. This step is catalyzed by an alcohol dehydrogenase (*sadh*). After decarboxylation via the ketoacid decarboxylase (*mdlC* and *kivD*) from *L. lactis* and *Pseudomonas putida,* the C4-skeleton is final and the remaining aldehyde function is reduced such that the target product 1,4-butanediol is obtained (Figure 11.4).

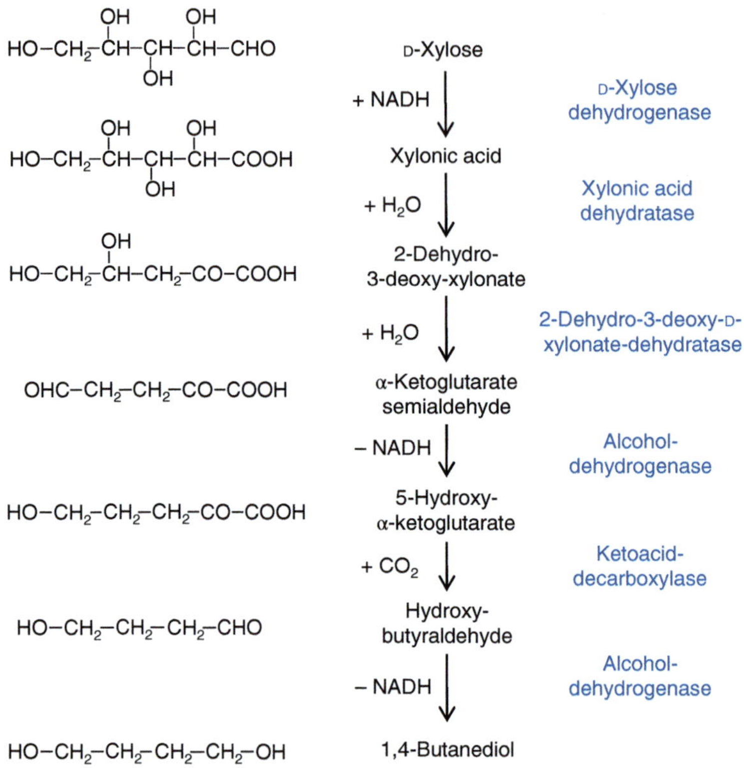

Figure 11.4 Alternative pathway toward 1,4-butanediol based on D-xylose (Liu and Lu 2015). An additional route from D-xylose to 1,4-butanediol proceeds via D-xylonic acid and 3,4-dihydroxybutanal. Source: Adapted from Guo et al. (2022).

The overall pathway starting with D-xylose can be stated as follows:

$$\begin{aligned}&HO\text{–}CH_2\text{–}CHOH\text{–}CHOH\text{–}CHOH\text{–}CHO + NADH + H^+\\ &\rightarrow HO\text{–}CH_2\text{–}CH_2\text{–}CH_2\text{–}CH_2\text{–}OH + CO_2 + H_2O + NAD^+\end{aligned}$$

The 20 electrons carried by D-xylose (without counting reduced oxygen) are transferred to 1,4-butanediol which carries 22 electrons (without counting reduced oxygen) and 2 additional electrons are missing. The net demand of one reduction equivalent favors fermentation technology in living cells such that the electron gap can be supplied by parallel auxiliary oxidation in the cell. The generated NAD^+ would be recycled within the cell and the costly external supply of NADH would be avoided. Fermentation experiments were pursued in 100 ml in 250 ml shaking flasks employing LB medium and appropriate antibiotics to avoid contamination. The incubation was pursued at 37 °C with 150 rpm and induction was initiated at OD_{600}. The pathway installed in recombinant *E. coli* cells achieved a titer of 0.44 g/l and a yield of 0.07 mol of 1,4-butanediol/mol D-xylose (Liu and Lu 2015).

11.7 Industrial Application and Capacity

The biotech company Genomatica started in 2000 with the focus on bioinformatics and the modeling of biochemical pathways. It pursues the strategy to develop innovative fermentation technology and to license the technology to industry partners, which carry the investment, take over plant operation, and apply the product for their own value chain or sell it into the market. Genomatica's income is based on royalties from industrial partners. In 2011, Genomatica entered cooperations with Mitsubishi, Tate & Lyle, and Mossi & Ghisolfi to further support the commercialization of its technology (Coons 2011). Mitsubishi is one of the major BDO producers based on petrochemistry and could facilitate the incorporation of a biobased product grade in brand concepts downstream in the value chain. The equivalence of the biobased products for current applications with petrochemically derived 1,4-butanediol is to be expected, though nevertheless needs to be shown before high-volume production. Genomatica provided material to Toray Industries which successfully produced PBT (Chemical Week 2013). The product of Genomatica had also been tested by DSM who confirmed technical equivalence in their Arnitel copolyesters (DSM 2013).

The R&D activities of Genomatica required a period of seven years to raise the titer from <20 g/l in 2009 to greater than 140 g/l in 2016 to allow a first small-scale commercial production in late 2012 (Culler 2016). The technology of Genomatica was finally launched on an industrial scale in 2016 with Novamont, a polymer company, starting the first commercial 1,4-butanediol plant in Italy applying the Genomatica technology (Coons 2019). Novamont had bought a former Ajinomoto biomanufacturing site in Italy to prepare production. The initial capacity is indicated with 30 kt pa (ilbioeconomista 2016).

In 2021, it has been announced by Cargill and Helm that an additional plant for fermentative 1,4-butanediol is planned to be constructed in Iowa. The plant will equally be owned by Cargill and Helm which both will hold a 50% share in the joint venture Qore. Both partners have different responsibilities: Cargill with a strong position in trading corn and glucose will provide the feedstock for the plant and provide its knowledge in fermentation production technology. Helm as a chemical distributor more closely to the customer side will market the introduced brand "QIRA" and try to persuade customers in the value chain to incorporate the brand into their product concepts. Especially, the carbon footprint analysis of the fermentative grade shall allow a differentiation toward the petrochemically derived products. The greenhouse gas emissions are claimed to be up to 93% lower compared to petrochemical products. In order to prepare the marketing of the biobased product, a comprehensive life cycle analysis was performed (Forte et al. 2016). The plant shall start production in 2024 and require an investment of 300 Mio. USD for a capacity of 65 kt pa which translates into a specific investment of about 4.600 USD/t (Coons 2021). The raw material demand amounts to 30 000 bushels of corn per day (Lane 2022).

References

Adkins, J., Pugh, S., McKenna, R., and Nielsen, D.R. (2012). Engineering microbial chemical factories to produce renewable "biomonomers". *Frontiers in Microbiology* 3: 313. https://doi.org/10.3389/fmicb.2012.00313.

Adom, F., Dunn, J.B., Han, J., and Sather, N. (2014). Life-cycle fossil energy consumption and greenhouse gas emissions of bioderived chemicals and their conventional counterparts. *Environmental Science & Technology* 48: 14624–14631. https://doi.org/10.1012/es503766e.

Arpe, H.-J. (2007). *Industrielle Organische Chemie. Bedeutende Vor- und Zwischenprodukte*. Wiley p. 107ff, p. 412ff.

Barton, N.R., Burgard, A.P., Burk, M.J. et al. (2015). An integrated biotechnology platform for developing sustainable chemical processes. *Journal of Microbial Biotechnology* 42: 349–360. https://doi.org/10.1007/s10295-014-1541-1.

Behr, A. and Seidensticker, T. (2018). *Einführung in die Chemie nachwachsender Rohstoffe.* Springer 41ff.

BioAmber (2012). BioAmber produces bio-based 1,4-butanediol (1,4-BDO). Source: https://www.bio-amber.com/bioamber/en/news/2012/biomaber-produces-bio-based-14-butanediol-14-bdo/516 (accessed 20 June 2017).

Bozell, J.J. and Petersen, G.R. (2010). Technology development for the production of biobased products from biorefinery carbohydrates – the US department of energy's "Top 10" revisited. *Green Chemistry* 12: 539–554.

Bühler, V. (2008). *Kollidon. Polyvinylpyrrolidone Excipients for the Pharmaceutical Industry.* Ludwigshafen: BASF SE.

Burgard, A., Burk, M.J., Osterhout, R. et al. (2016). Development of a commercial scale process for production of 1,4-butanediol from sugar. *ScienceDirect* 42: 118–125. https://doi.org/10.1016/j.copbio.2016.04.016.

Chemical Week (2013). BASF licenses Genomatica's renewable BDO technology, 13./20.5.2013, p. 35.

Chemical Week (2018) Cars drive thermoplastic copolyesters, 15./22.1.2018, p. 31.

Coons, R. (2011). Genomatica. Partnerships spree includes industry heavyweights, Chemical Week, 18./25.7.2011, p. 39.

Coons, R. (2019). Genomatica turns the corner, Chemical Week, 11./18.2.2019, p. 27.

Coons, R. (2021). Cargill, Helm plan $300-million renewable BDO facility in Iowa, Chemical Week, 14./21.6.2021, p. 10.

Culler, S. (2016). *A Bioengineering Platform to Industrialize Biotechnology*, 42–51. CEP, www.aiche.org.cep.

DSM (2013). Achieving higher bio-based content in DSM Arnitel copolyerster: results from testing 1,4-butanediol (BDO) made with Genomatica's process, Technical paper, 8.10.2013.

Fischer, F. and Bauer, S. (2009). Polyvinylpyrrolidon. Ein Tausendsassa der Chemie. *Chemie in unserer Zeit* 43: 376–383.

Forte, A., Zucaro, A., Basosi, R., and Fierro, A. (2016). LCA of 1,4-butanediol produced via direct fermentation of sugars from wheat straw feedstock within a territorial biorefinery. *Materials* 9 (563): https://doi.org/10.3390/ma9070563.

Guo, H., Liu, H., Jin, Y. et al. (2022). Advances in research of the bio-production of 1,4-butanediol by the engineered microbes. *Biochemical Engineering Journal* 185: https://doi.org/10.1016/j.bej.2022.108478.

Hara, Y., Inagaki, H. (1991). Method for producing 1,4-butanediol. Mitsubishi patent US 5077442.

http://myqira.com (2022). Qira offers fuel free replacement and a reduced carbon footprint. https://www.myqira.com/qira-sustainability (accessed 19 October 2022).

ilbioeconomista (2016). Novamont's bio-based butanediol plant will be on stream by the next September. source: https://ilbioeconomista.com/2016/06/20/novamonts-bio-based-butanediol-plant-in-northern-italy-will-be-on-stream-by-the-next-september (accessed 10 October 2021).

Kim, Y., Ingram, L.O., and Shanmugan, K.T. (2008). Dihydrolipoamide dehydrogenase mutation alters the NADH sensitivity of pyruvate dehydrogenase complex of *Escherichia coli* K-12. *Journal of Bacteriology* 190: 3851–3858.

Lane, J. (2022). Beyond ethanol: How 3 international companies are finding success in Iowa, www.biofuelsdigest.com (14 June 2022), https://www.biofuelsdigest.com/bdigest/2022/06/14/beyond-ethanol-how-3-international-companies-are-finding-success-in-iowa (accessed 16 June 2022).

Li, K.-T. and Chen, K.-W. (2019). Cyclodehydration of 1,4-butanediol over Zr-Al catalysts: Effect of reaction medium. *Materials* 12: https://doi.org/10.3390/ma12132092.

Liu, H. and Lu, T. (2015). Autonomous production of 1,4-butanediol via a *de novo* biosynthesis pathway in engineered Escherichia coli. *Metabolic Engineering* 29: 135–141. https://doi.org/10.1016/j.ymben.2015.03.009.

Mock, J., Zheng, Y., Mueller, A.P. et al. (2015). Energy conservation associated with ethanol formation from H_2 and CO_2 in *Clostridium autoethanogenum* involving electron bifurcation. *Journal of Bacteriology* 197 (18): 2965–2980.

Morden, P. (2019). Benefuel secures land at Sarnia bio-industrial park. Source: http://www.sarnialambton.on.ca/benefuel-secures-land-at-sarnia-bioindustrial-park (accessed 17 May 2019).

Müller, V. (2019). New horizons in acetogenic conversion on one-carbon substrates and biological hydrogen storage. *Trends in Biotechnology* 37 (12): 1344–1354. https://doi.org/10.1016/j.tibtech.2019.05.008.

Pooth, V., van Gaalen, K., Trenkamp, S. et al. (2020). Comprehensive analysis of metabolic sensitivity of 1,4-butanediol producing *Escherichia coli* toward substrate and oxygen availability. *Biotechnology Progress* https://doi.org/10.1002/btpr.2917.

Rafiqah, S.A., Khalina, A., Harmaen, A.S. et al. (2021). A review on properties and application of bio-based poly(butylene succinate). *Polymers* 13: 1436. https://doi.org/10.3390/polym13091436.

Reppe, W. (1950). Chemie und Technik der Acetylen-Druck-Reaktionen. *Chemie Ingenieur Technik* 23–34: 527–539.

Röper, M. (2006). Selektivität, Aktivität und Standzeit: Homogene Katalyse in der chemischen Industrie. *Chemie in unserer Zeit* 40: 126–135. https://doi.org/10.1002/ciuz.200600373.

Rosales-Calderon, O. and Arantes, V. (2019). A review on commercial-scale high-value products that can be produced alongside cellulosic ethanol. *Biotechnology for Biofuels* 12: 240. https://doi.org/10.1186/s13068-019-1529-1.

Rosenboom, J.-G., Langer, R., and Traverso, G. (2022). Bioplastics for a circular economy. *Plastics* 7: 117–137. https://doi.org/10.1038/s41578-021-00407-8.

Stieglitz, B. and Weimer, P.J. (1985). Novel microbial screen for detection of 1,4-butanediol, ethylene glycol, and adipic acid. *Applied and Environmental Microbiology* 49 (3): 593–598.

TaI, Y.-S., Xiong, M., Jambunathan, P. et al. (2016). Engineering nonphosphory-lative metabolism to generate lignocellulose-derived products. *Nature Chemical Biology* 12: 247–253. https://doi.org/10.1038/NCHEMBIO.2020.

Trotuş, I.-T., Zimmermann, T., and Schüth, F. (2014). Catalytic reactions of acetylene: a feedstock for the chemical industry revisited. *Chemical Reviews* 114: 1761–1782. https://doi.org/10.1021/cr400357r.

Wagemann, K. (2014). Herstellung von Grundchemikalien auf Basis nachwachsender Rohstoffe als Alternative zur Petrochemie. *Chemie Ingenieur Technik* 86 (12): 2115–2134.

Yim, H., Haselbeck, R., Niu, W. et al. (2011). Metabolic engineering of Escherichia coli for direct production of 1,4-butanediol. *Nature Chemical Biology* https://doi.org/10.1038/NCHEMBIO.580.

12

Succinic Acid

12.1 Application

Succinic acid or 1,2-ethanedicarboxylic acid is an organic diacid with the formula $C_4H_6O_4$ and a molecular weight of 118.1 g/mol. It is a medium-volume chemical intermediate that is used in food, medicine, fine chemicals, and as monomer for polyesters (Nghiem et al. 2017). The C4 diacid molecule profits from the rising demand in biobased polymers (McKinlay et al. 2007; Erickson et al. 2012) and has gained high attention in academia and the chemical industry as building block for polyesters based on renewable resources (Lomelí-Rodríguez et al. 2018) with improved greenhouse gas emissions compared to the alternative petrochemically derived product (Adom et al. 2014). As an example of marketing for "green" products, Clariant announced in 2019 the availability of the organic pigment quinacridone derived from biobased succinic acid to replace inorganic pigments containing heavy metals (Ravenscroft 2019). Covestro pursued the reduction of the carbon footprint in polyurethane coatings and developed the PU dispersion Impranil® eco in which the functionally comparable petrochemically derived adipic acid is replaced with biobased succinic acid (Goldbach 2018).

Polybutylene succinate (PBS) is an emerging polyester and is developed for various application areas. It can be produced from the two biobased monomers 1,4-butanediol and succinic acid and satisfies the interest in sustainable plastics completely derived from renewables (Rosenboom et al. 2022; Mehlhase 2018; Bomgardner 2014). Biodegradability of PBS is a distinctive attribute compared to other polyesters and supports marketing for single-use food packaging (Rafiqah et al. 2021).

Succinic acid was at the top of the list of biobased products that might be obtained in biorefineries based on lignocellulosic biomass (US Department of Energy (USDE) 2004). The production of fossil-based succinic acid reached a volume of 15 kt in 1999 (McKinlay et al. 2007) and 40 kt/pa in 2012 and was originally forecasted to increase to more than 700 kt/pa in 2020. However, USDA's list of biobased target molecules was reviewed, and it was stated that the prospected market development for succinic acid did not materialize (Bozell and Petersen 2010) and the market volume of 700 kt succinic acid in 2020 was not met because of high production costs (Market Watch 2021; Maximize Market Research 2019–2027). Biobased succinic acid is reported to achieve a price premium of about 10% given a price of around 2500 USD/t for fossil-based products and approximately 2.860 USD/t

Pathway Design for Industrial Fermentation, First Edition. Walter Koch.

for fermentative products in 2019 (Rosales-Calderon and Arantes 2019). The future market perspective is pending on cost reduction and the long-term price premium of "green" products.

12.2 Conventional Synthesis of Succinic Acid

The chemical synthesis of succinic acid relies mainly on the C4 intermediate maleic acid anhydride, the desaturated acid anhydride of succinic acid. This C4 building block may be converted into succinic acid anhydride, γ-butyrolactone, tetrahydrofurane (THF), maleic acid, or 1,4-butanediol. It can also be reacted with molecular hydrogen to obtain succinic acid.

$$\text{Maleic acid anhydride} + H_2 + H_2O \rightarrow HOOC{-}CH_2{-}CH_2{-}COOH$$

Maleic acid anhydride was originally a by-product in the synthesis of phthalic anhydride generated from coke oven-based naphthalene (Spitz 2019) until on-purpose technology became available. Maleic acid anhydride was afterwards derived from either benzene, butene, or butane (Arpe 2007). Benzene was the raw material of choice from about 1930–1975 (Hagen 2015). The benzene technology applies air at 2–5 bar and catalytically converts benzene at a temperature of 400–450 °C. The turnover achieves 85–95% and the weak selectivity of 60–65% in the early stage of technology development could in the meantime be increased up to 80%:

$$\text{Benzene} + 4.5\,O_2 \rightarrow \text{Maleic acid anhydride} + 2\,CO_2 + 2\,H_2O$$

The process suffers from the high generation of carbon dioxide since only four C-atoms of the raw material remain in the product and two C-atoms are completely oxidized. The replacement of benzene as feedstock and the switch to a C4-precursor define an advantage since the backbone of butene and butane is conserved and converted into the backbone of succinic acid, resulting in better carbon selectivity. The gas phase reaction of butene with oxygen is characterized by a selectivity of about 45–50% and several by-products are obtained:

$$\text{Butene} + 3\,O_2 \rightarrow \text{Maleic acid anhydride} + 3\,H_2O$$

Optimization of the process economics was facilitated with the application of butane as raw material, which is contained in natural gas. The reaction is pursued with a low turnover of about 80% to protect the selectivity, which amounts to 50–60%. The reaction is often carried out in a fixed-bed reactor with vanadium-phosphorous oxide as catalyst and the temperature is about 400–450 °C. Efficient removal of heat is facilitated by heat exchangers with a molten mixture of alkali nitrates (Hagen 2015):

$$\text{Butane} + 3.5\,O_2 \rightarrow \text{Maleic acid anhydride} + 4\,H_2O\ \Delta H_R{}^\circ = -1260\,\text{KJ/mol}$$

The switch to fermentative production of succinic could especially improve the carbon footprint of the product and thereby drive demand based on sustainability marketing toward the final consumer. The petrochemical routes based on benzene or natural gas incur a scope 3 carbon footprint in the production of the raw material as opposed to biobased glucose, which is derived by photosynthesis and has a negative carbon footprint (Cok et al. 2014).

12.3 Pathway Design and Metabolic Engineering

12.3.1 Pathway via Glycolysis, PEP Carboxylation, and Oxidative Citric Acid Cycle

Since succinic acid is a major biochemical pathway intermediate and part of the citric acid cycle, the design of biochemical pathways for metabolic engineering is rather easy. A variation in proposals for the best approach in metabolic engineering refers to the most promising direction of the citric acid cycle and the generation of precursors. The pathway toward succinic acid may travel along the citric acid cycle in the usual oxidative direction or in the unusual reductive direction.

In case the citric acid cycle would be pursued in the conventional oxidative direction, the two substrates of the citrate synthase, acetyl-CoA and oxalacetate, need to be generated. Acetyl-CoA would be generated via the pyruvate dehydrogenase and oxalacetate would be provided via the carboxylation of phosphoenolpyruvate (PEP) (Table 12.1).

One mol glucose would be converted into one mol succinic acid and two mol carbon dioxide. The 24 electrons of glucose would be rearranged to 14 electrons carried by succinic acid (without counting the electrons of the reduced oxygen) and five surplus reduction equivalents. The metabolic engineering concept would need to consider the native pathway network and its regulation. In case *Escherichia coli* as natural producer of succinic acid under anaerobe conditions shall be applied, the activity of the pyruvate dehydrogenase needs to be increased. Due to the inhibition by NADH, the activity of the enzyme is hardly detectable (Snoep et al. 1993). The inefficiency of this pathway is obvious since only one mol succinic acid is derived from one mol glucose. A maximum theoretical molar yield of 1.0 translates into a maximum carbon yield of 0.65 t succinic acid per ton glucose (118.1/180.1). Additionally, a high surplus of ATP and reduction equivalents is generated, which would have been better invested in restructuring the carbon backbone with higher C-atom efficiency. Accordingly, the reductive direction of the citric acid cycle is favored.

Table 12.1 Metabolic pathway toward succinic acid via PEP carboxylation.

Reaction	Enzyme
Glucose + 2 NAD^+ + 2 Pi → 2 PEP + 2 NADH + 2 H^+ + 2 H_2O	Glycolysis until PEP
PEP + ADP + CO_2 → Oxalacetate + ATP	PEP carboxykinase
PEP + ADP → Pyruvate + ATP	Pyruvate kinase
Pyruvate + CoA + NAD^+ → Acetyl-CoA + NADH + CO_2	Pyruvate dehydrogenase
Oxalacetate + Acetyl-CoA + H_2O → Citrate + CoA	Citrate synthase
Citrate → cis-aconitate + H_2O	cis-aconitase
Cis-aconitate + H_2O → Isocitrate	cis-aconitase
Isocitrate + NAD^+ → α-Ketoglutarate + NADH + CO_2	Isocitrate dehydrogenase
α-Ketoglutarate + CoA + NAD^+ → Succinyl-CoA + NADH + CO_2	α-Ketoglutarate dehydrogenase
Succinyl-CoA + GDP + Pi → Succinate + CoA + GTP	Succinyl-CoA synthetase
Glucose + 5 NAD^+ + 2 ADP + GDP + 3 Pi → Succinate + 2 CO_2 + 5 NADH + 5 H^+ + 2 ATP + GTP + H_2O	

12.3.2 Pathway via Glycolysis, PEP Carboxylation, and Reductive Citric Acid Cycle

The idea of proceeding the citric acid cycle in the reductive direction is an interesting concept of metabolic engineering. Unless thermodynamical hurdles render this option as non-realistic, a metabolic pathway may be followed in the reversed direction, which is pursued in the metabolic pathway toward adipic acid (reversed β-oxidation) or, in this context, in the metabolic pathway toward succinic acid (reductive citric acid cycle). The pathway toward succinic acid via the reductive citric acid cycle proceeds as follows (Table 12.2).

The respective chemical reaction can be stated as follows and the derived Gibbs free energy is sufficiently negative to drive the reaction in the right direction:

$$\text{Glucose} + 2\,HCO_3^- + 2\,H^+ + 2\,H_2 \rightarrow 2\,\text{Succinate} + 4\,H_2O \qquad \Delta G^{\circ\prime} = -317\,\text{KJ/mol}$$

The metabolic pathway and the potential carbon yield, respectively, profit from the structural integration of carbon dioxide, and therefore the value cannot directly be compared to other processes that derive carbon dioxide as by-product due to thermodynamical constraints, e.g. the fermentative production of ethanol and isobutene (Figure 12.1) (McKinlay et al. 2007).

The pathway via the reductive citric acid cycle exploits the surplus of reduction equivalents that would be obtained if the citric acid cycle were pursued in the conventional oxidative reaction. Hereby, the integration of carbon dioxide is possible, and glucose and carbon dioxide function as carbon donors. The repositioning of carbon dioxide as valuable chemical feedstock – opposed to its qualification as useless fuel waste stream of endothermic reactions or oxidation chemistry with low selectivity – supports the targets of sustainable chemistry. Structural integration of carbon dioxide qualifies succinic acid fermentation technology as a carbon capture and use technology (CCU) like the hydrogenation of carbon dioxide to methane by methanogenic bacteria or the production of urea in organic chemistry (Burkart et al. 2019). The biochemical usage of carbon dioxide may proceed via the reduction to formate catalyzed by the formate dehydrogenase or the coupling of carbon dioxide to another intermediate by formation of a new C—C bond ("carboxylation"). In the metabolic pathway toward succinic acid the CO_2-fixation might in principle be accomplished by four metabolic carboxylation reactions (Lee et al. 2006; Table 12.3).

The first two reactions profit from the high energy level of PEP, which renders the backward reaction as thermodynamically difficult. Both reactions should proceed efficiently if the carbon dioxide or bicarbonate concentration within the cell is sufficiently high. Metabolic engineering strategies may prefer the PEP carboxy kinase versus the

Table 12.2 Pathway toward succinic acid via reductive citric acid cycle.

Reaction	Enzyme
Glucose + 2 NAD^+ + 2 Pi → 2 PEP + 2 NADH + 2 H^+ + 2 H_2O	Glycolysis until PEP
2 PEP + 2 ADP + 2 CO_2 → 2 Oxalacetate + 2 ATP	PEP carboxykinase
2 Oxalacetate + 2 NADH + 2 H^+ → 2 Malate + 2 NAD^+	Malate dehydrogenase
2 Malate → 2 Fumarate + 2 H_2O	Fumarase
2 Fumarate + 2 NADH + 2 H^+ → 2 Succinate + 2 NAD^+	Succinate dehydrogenase
Glucose + 2 CO_2 + 2 NADH + 2 H^+ + 2 ADP + 2 Pi → 2 Succinate + 2 NAD^+ + 2 ATP + 4 H_2O	

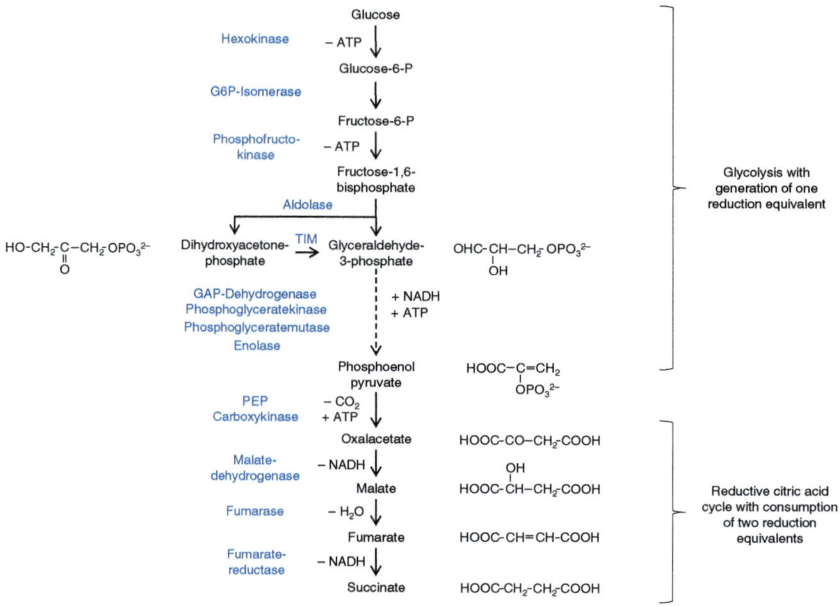

Figure 12.1 Metabolic pathway toward succinic acid with glucose as feed. The pathway combines section of glycolysis and citric acid cycle connected by the anaplerotic conversion of PEP into oxalacetate. Source: Walter Koch.

Table 12.3 Potential routes for the generation of the C4 backbone.

PEP + CO_2 → Oxalacetate + Pi	PEP carboxylase (*ppc*)
PEP + CO_2 + ADP → Oxalacetate + ATP	PEP carboxykinase (*pepck*)
Pyruvate + CO_2 + ATP → Oxalacetate + ADP + Pi	Pyruvate carboxylase (*pyc*)
Pyruvate + CO_2 + NADPH + H^+ → Malate + $NADP^+$	Malic enzyme

PEP carboxylase to allow a surplus of ATP in the pathway toward succinic acid required for growth and cell maintenance (Lee et al. 2006). Alternatively, the carboxylation of pyruvate to malate may be pursued. However, this reaction is kinetically not favored (Sauer and Eikmanns 2005). The following pathway reconstruction proceeds via the PEP carboxykinase.

On the first view reaction equation might indicate that two mol succinic acid may be derived from one mol glucose (a theoretical molar yield of 2.0 mol succinic acid per mol glucose) and two mol carbon dioxide, which would translate into a maximum theoretical carbon yield of 1.31 t succinic acid/t glucose (2 × 118.1/180.1). This selective view of the C-balance is misleading since it can also be deduced that one mol reduction equivalent is needed for the synthesis of one mol succinic acid. Since glucose carries 24 electrons (without counting reduced oxygen) and the two mol succinic acid carry 2 × 14 = 28 electrons it follows that two reduction equivalents are needed on the left side of the reaction equation. The electron gap needs to be accounted for in the estimation of the maximum theoretical yield. Several approaches are discussed to balance the gap in reduction equivalents. In the first instance, an anaerobe process via glycolysis may be chosen, which proceeds along the pathway scheme indicated above but requires a small amount of glucose to provide the two reduction equivalents. Since glycolysis (until pyruvate) provides two reduction equivalents per mol glucose a second mol glucose would be needed to replenish the gap in the reaction equation and altogether two mol glucose would be needed to obtain two mol succinic acid. The maximum theoretical yield would be 1.0 mol succinic acid per mol glucose and the maximum theoretical carbon yield would shrink to 0.65 t succinic acid/t glucose (118.1 g/mol/180.1 g/mol). This would obviously not be ideal, and a small addition of oxygen might be considered to partially allow the full oxidation of glucose. It might also be questioned if pyruvate, as the final product of glycolysis, would efficiently be exported and other more reduced by-products like lactic acid capture at least a share of the gained reduction equivalents.

In case it is assumed that one mol glucose generates 12 reduction equivalents under aerobe conditions, the generation of two mol reduction equivalents requires 0.083 mol glucose, which limits the maximum carbon yield to 1.21 t succinic acid/t glucose (2×118.1 g/mol/1.083×180.1 g/mol). The small surplus of energy in the form of ATP allows growth and cell maintenance.

12.3.3 Pathway via Glycolysis, PEP Carboxylation, and Reductive Citric Acid Cycle Combined with PPP

A balancing of reduction equivalents is possible in case where glucose is partially channeled through the pentose phosphate pathway (PPP). Following this pathway, only 5/6 of the carbon atoms of glucose would be converted into PEP and 1/6 be completely would be oxidized to carbon dioxide. The released electrons are transferred to NAD^+ and $NADP^+$ and the pathway can be reconstructed as follows (Table 12.4).

In this case, the 72 electrons provided by 3 mol glucose (3×24) would be transferred to 5 mol succinic acid, which carries 70 electrons (5×14). Accordingly, two electrons are left, which have been transferred to NAD^+ such that one mol NADH is gained. In the last step, both pathways would be combined in a ratio such that the surplus of reduction equivalents of the one pathway (i.e. PPP) is equalized by the gap of reduction equivalents of the other pathway (i.e. reductive citric acid cycle). It can be directly seen that both pathways need to be combined with a glucose consumption rate of 1 : 6 (i.e. 1 mol via glycolysis and 6 mol via PPP), and the overall reaction equation is obtained:

$$\text{Glucose} + 6\,CO_2 + 12\,\text{ADP} + 12\,\text{Pi} \rightarrow 12\,\text{Succinic acid} + 6\,H_2O + 12\,\text{ATP}$$

Seven mol glucose provide 12 mol succinic acid, and the theoretical maximum molar yield considering the restructuring of the carbon skeleton and electron flow amounts to 1.714 mol per mol glucose (Vemuri et al. 2002; McKinlay et al. 2007). The combination of pathways decreases the demand for exogenous carbon dioxide. In case where exclusively glycolysis and the reductive citric acid cycle would be pursued one mol exogenous carbon dioxide per mol succinic acid would be needed. For comparison, in case the reductive citric acid cycles and the PPP are combined, the carbon dioxide unit consumption rate decreases to 0.5 mol carbon dioxide per mol succinic acid (0.857/1.714) in case both pathways are mixed such that the reduction equivalents are fully balanced:

$$\text{Glucose} + 0.857\,CO_2 \rightarrow 1.714\,\text{Succinic acid} + 0.857\,H_2O$$

Table 12.4 Pathway toward succinate via reductive citric acid cycle and PPP.

Reaction	Enzyme
3 Glucose + 3 ATP → 3 G6P + 3 ADP	Hexokinase
3 G6P + 3 $NADP^+$ → 3 6PG-δ-lactone + 3 NADPH + 3 H^+	G6P dehydrogenase
3 6PG-δ-lactone + 3 H_2O → 3 6-Phosphogluconate	6-Phosphogluconolactonase
3 6-Phosphogluconate + 3 $NADP^+$ → 3 Ru5P + 3 NADPH + 3 CO_2	6-Phosphogluconate dehydrogenase
Ru5P → R5P	Ru5P isomerase
2 Ru5P → 2 Xu5P	Ru5P epimerase
R5P + Xu5P → S7P + GAP	Transketolase
S7P + GAP → F6P + E4P	Transaldolase
E4P + Xu5P → F6P + GAP	Transketolase
2 F6P + 2 ATP → 2 F-1,6-BP + 2 ADP	Phosphofructokinase
2 F-1,6-BP → 2 GAP + 2 DHAP	Aldolase
2 DHAP → 2 GAP	Triosephosphate isomerase
5 GAP + 5 NAD^+ → 5 1,3-BPG + 5 NADH + 5 H^+	GAP dehydrogenase
5 1,3-BPG + 5 ADP → 5 3-PG + 5 ATP	Phosphoglycerate kinase
5 3-PG → 5 2-PG	Phosphoglycerate mutase
5 2-PG → 5 PEP + 5 H_2O	Enolase
5 PEP + 5 ADP + 5 CO_2 → 5 Oxalacetate + 5 ATP	PEP carboxykinase
5 Oxalacetate + 5 NADH + 5 H^+ → 5 Malate + 5 NAD^+	Malate dehydrogenase
5 Malate → 5 Fumarate + 5 H_2O	Fumarase
5 Fumarate + 5 NADH + 5 H^+ → 5 Succinate + 5 NAD^+	Succinate dehydrogenase
6 NADPH + 6 NAD^+ → 6 $NADP^+$ + 6 NADH	Transhydrogenase
3 Glucose + 2 CO_2 + 5 ADP + 5 Pi + NAD^+ → 5 Succinate + 5 ATP + NADH + H^+ + 7 H_2O	

The respective chemical reaction can be stated as follows (Meng et al. 2016):

$$\text{Glucose} + 0.857\ \text{HCO}_3^- \rightarrow 1.714\ \text{Succinate} + 1.714\ \text{H}_2\text{O} + 2.57\ \text{H}^+$$
$$\Delta G°' = -173\ \text{kJ/mol}$$

The decreased carbon dioxide consumption rate is facilitated by the partial oxidation of glucose, which provides a surplus of reduction equivalents and carbon dioxide (McKinlay et al. 2007). This circumstance might provide a second advantage besides the balance of reduction equivalents in the mixed pathway strategy since the limited mass transfer rate of carbon dioxide or bicarbonate might hamper productivity. The Gibbs free energy is sufficiently negative to drive the reaction, and even a surplus of ATP is achieved. The net reaction equation of the biochemical pathway indicates that one mol ATP can be obtained per mol succinic acid, which is reasonable since the minimum Gibbs free energy to derive ATP under physiological conditions is estimated at 50–60 kJ/mol (Flamholz et al. 2012).

Table 12.5 Metabolic pathway toward succinic acid via glycolysis and glyoxylate cycle.

Glyoxylate + Acetyl-CoA + H_2O → Malate + CoA	Malate synthase
Malate + NAD^+ → Oxalacetate + NADH + H^+	Malate dehydrogenase
Oxalacetate + Acetyl-CoA + H_2O → Citrate + CoA	Citrate synthase
Citrate → Isocitrate	Aconitase
Isocitrate → Succinate + Glyoxylate	Isocitrate lyase
2 Acetyl-CoA + 2 H_2O + NAD^+ → Succinate + 2 CoA + NADH + H^+	

Table 12.6 Overall reaction toward succinic acid via the glyoxylate cycle.

Glucose + 2 CoA + 2 ADP + 2 Pi + 4 NAD^+ → 2 Acetyl-CoA + 2 H_2O + 2 CO_2 + 2 ATP + 4 NADH + 4 H^+
2 Acetyl-CoA + 2 H_2O + NAD^+ → Succinic acid + 2 CoA + NADH + H^+
Glucose + 5 NAD^+ + 2 ADP + 2 Pi → Succinic acid + 2 CO_2 + 5 NADH + 5 H^+ + 2 ATP

12.3.4 Alternative Pathway via Glycolysis and Glyoxylate Cycle

The glyoxylate cycle is also known to generate succinic acid via a different reaction than the citric acid cycle and might also be useful in the design of a metabolic pathway. The glyoxylate cycle is present in plants, bacteria, and fungi and possesses enzymes for the overall conversion of acetyl-CoA into oxalacetate. The net equation can be stated as follows:

$$\begin{aligned}&2\,\text{Acetyl–CoA} + 2\,\text{NAD}^+ + \text{FAD} + 3\,\text{H}_2\text{O} \rightarrow \text{Oxalacetate} + 2\,\text{CoA}\\&\quad + 2\,\text{NADH} + 2\,\text{H}^+ + \text{FADH}_2\end{aligned}$$

The specific reaction is catalyzed by the isocitrate lyase and splits isocitrate into succinate and glyoxylate. The by-product glyoxylate would be recycled via condensation with acetyl-CoA to finally obtain citrate. The net conversion of acetyl-CoA into succinate is derived as follows (Table 12.5).

The acetyl-CoA pool would be provided via glycolysis and the pyruvate dehydrogenase complex, and the overall reaction equation can be stated as follows:

$$\begin{aligned}&\text{Glucose} + 2\,\text{CoA} + 2\,\text{ADP} + 2\,\text{Pi} + 4\,\text{NAD}^+ \rightarrow 2\,\text{Acetyl-CoA} + 2\,\text{H}_2\text{O} + 2\,\text{CO}_2\\&\quad + 2\,\text{ATP} + 4\,\text{NADH} + 4\,\text{H}^+\end{aligned}$$

The overall reaction equation for the conversion of glucose into succinic acid according to the glyoxylate cycle can then be derived by combining the two reactions (Table 12.6).

The pathway does not provide an advantage compared to the combination of glycolysis and oxidative citric acid cycle, as shown above. The minor difference in the net reaction equation relates to the energy balance since the formation of succinic acid does not provide a GTP as the succinyl-CoA synthetase in the citric acid cycle (Cheng et al. 2013).

The pathway design and the carbon yield in the fermentative production of succinic acid can be improved if external carbon dioxide is applied as educt, which is facilitated with the reductive citric acid cycle as shown below. Hereby, PEP would be reacted with external carbon dioxide to obtain oxalacetate, which is then consumed in the generation of citrate. This step is not helpful in case the glyoxylate cycle is pursued since oxalacetate is derived in the recycling of glyoxylate ex malate (Figure 12.2).

12.3.5 Summary of TMY Pending on Pathway Design

The theoretical maximum yield (TMY) and the maximum carbon yield enabled by the different metabolic engineering strategies can be summarized as follows (Table 12.7).

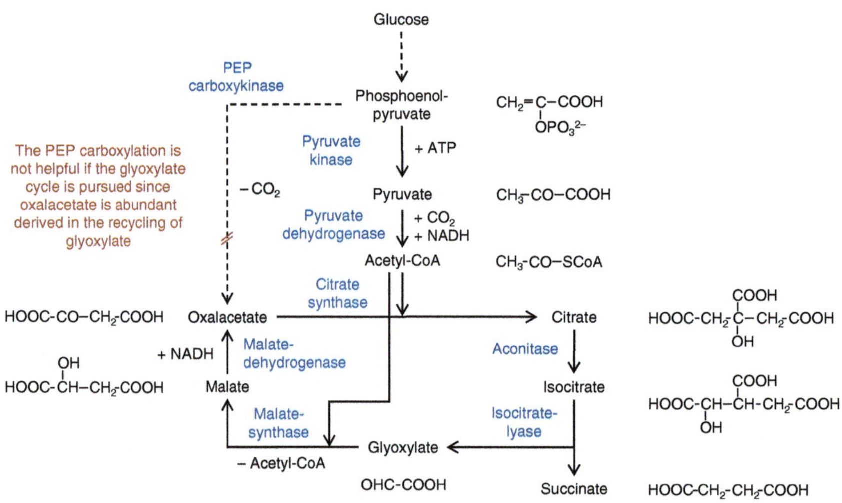

Figure 12.2 Glyoxylate cycle for the synthesis of succinate with recycling of glyoxylate toward oxalacetate. Source: Walter Koch.

Table 12.7 Theoretical molar yield and carbon yield for different routes.

Pathway	Theoretical molar yield Mol succinic acid/mol glucose	Maximum carbon yield (t/t)
Oxidative CAC	1.0	0.65 (118.1/180.1)
Reductive CAC	~~2.0~~	1.31 (2 × 118.1/180.1 g/mol)
with CO_2 feed	Not feasible since the reduction equivalents are not balanced	
Reductive CAC	1.0	0.65 (118.1/180.1)
with CO_2 feed	Anaerobe/one mol glucose needed to replenish two reduction equivalents	
Reductive CAC	1.84	1.21 (2 × 118.1/1.083 × 180.1)
with CO_2 feed	Aerobe/one mol glucose provides 12 reduction equivalents	
Reductive CAC & PPP	1.71	1.12 (1.714 × 118.1/180.1)
with CO_2 feed	Anaerobe production phase	

12.3.6 Case Study: Combination of Reductive Citric Acid Cycle and Pentose Phosphate Pathway

In one academic attempt, the metabolic engineering strategy was implemented in *E. coli* with a molar yield of 1.12 mol succinic acid per mol glucose as starting point. It could be increased up to 1.31 mol succinate per mol glucose after the inhibition of the pyruvate dehydrogenase by NADH was eliminated with mutations in the lpdA gene encoding the lipoamide dehydrogenase (Zhu et al. 2014). A boost of the PPP via enhanced activity of the transketolase and the introduction of a soluble transhydrogenase increased the molar yield from 1.12 to 1.33 mol succinic acid per mol glucose. The combination of both measures provided a molar yield of 1.5 mol succinic per mol glucose, which is 88% of the theoretical molar yield (Zhu et al. 2014).

In a second academic approach to combine glycolysis and PPP, an even higher yield of 1.54 mol succinate per mol glucose has been achieved, which represents about 90% of the TMY (Meng et al. 2016). Several measures were implemented such that the overall carbon flow can be visualized as follows (Figure 12.3).

The introduction of the glucose-6-phosphate dehydrogenase (*zwf243*) and the 6-phosphogluconate dehydrogenase (*gnd361*) from *Corynebacterium glutamicum* into *E. coli* provided an increase in the molar yield from 1.01 to 1.16 mol succinate per mol glucose. Overexpression of three native enzymes of the PPP, the 6-phosphogluconolactonase (pgl), the transketolase (*tktA*), and the transaldolase (*talB*), provided a further step toward 1.21 mol succinate per mol glucose. In order to support the formation of oxalacetate, the recombinant PEP carboxykinase (pepck) from *Actinobacillus succinogenes* was introduced.

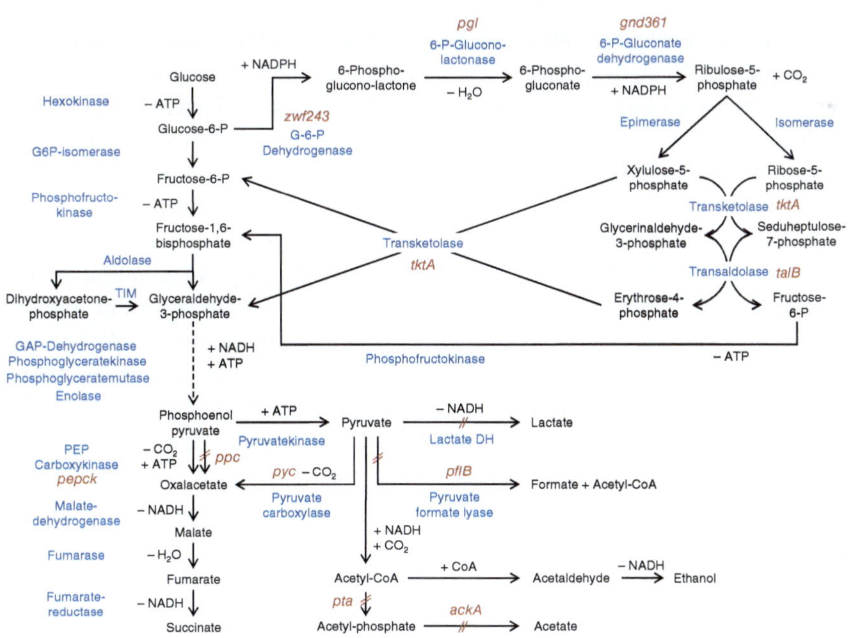

Figure 12.3 Combination of glycolysis and pentose phosphate pathways for the balance of reduction equivalents in the generation of glucose with glucose as starting material in *E. coli*. Source: Adapted from Meng et al. (2016).

Together with overexpression of the soluble transhydrogenase (*sthA*), the yield was further increased to 1.31 mol succinate per mol glucose. The next move in metabolic engineering addressed the formation of by-products and the phosphate acetyltransferase encoded by the *pta* gene and the acetate kinase encoded by the *ackA* gene toward acetate were knocked out. Combined with the recombinant expression of the pyruvate carboxylase (*pyc*) from *C. glutamicum* a yield of 1.4 mol succinate per mol glucose was achieved. Finally, suitable succinate exporters for efficient secretion into the medium encoded by the *dcuB* and *dcuC* genes provided a commercially interesting yield of 1.54 mol succinate per mol glucose, which represents 90% of the theoretical maximum molar yield of 1.714 mol succinic acid per mol glucose. In order to reduce consumption of PEP for competing reactions, the phosphotransferase (PTS) glucose uptake system has been knocked out (Meng et al. 2016).

Fermentation often pursues a disproportionation of glucose to obtain a target product with a lower oxidation level than glucose, which implies that a share of the glucose backbone needs to be fully oxidized to carbon dioxide. This kind of carbon restructuring applies typically if fuels are produced that have a high reduction level to bear a high heating value like ethanol, butanol, isobutanol, isobutene, or isoprene. In the pathway toward succinic acid, a mixed-substrate conversion would be applied, which is to be regarded as a comproportionation (synproportionation) since glucose is partially oxidized and carbon dioxide is reduced. The mixed-substrate approach allows to integrate carbon dioxide into the product though is restricted to molecules with medium oxidation level since otherwise additional reduction equivalents would need to be added (Steiger et al. 2017).

12.3.7 Glycerol as Carbon and Energy Source

The gap in reduction equivalents might also be closed if a carbon source with higher reduction level is offered. Whereas glucose carries 24 electrons and thereby 4 electrons per C-atom, glycerol represents a higher reduction level and carries 4.66 electrons per C-atom (without counting reduced oxygen; Nielsen et al. 2003). The decisive difference refers to the aldehyde function of glucose in the C1-position compared to the three hydroxyl functions of glycerol in the C1-, C2-, and C3-positions. The additional oxidation step provides one more reduction equivalent, and the 14 electrons carried by glycerol can be rearranged with the 14 electrons carried by succinic acid. The maximum molar yield amounts to one mol succinic acid per mol glycerol (Chen et al. 2010) in cases where sufficient carbon dioxide is offered. Hereby three possible pathways for initial glycerol dissimilation are distinguished (Blankschien et al. 2010). The native glycerol dissimilation of *E. coli* is initiated by the phosphorylation and activation of glycerol toward glycerol-3-phosphate by the glycerol-3-phosphate phosphatase (or glycerol kinase) encoded by the *glpK* gene. In the second step, the medium hydroxyl function of glycerol-3-phosphate is oxidized with flavin (Q) to obtain dihydroxyacetonephosphate. The dehydrogenase is encoded by the *glpD* gene or the *glpABC* gene in *E. coli* (Chen et al. 2010):

$$\text{Glycerol} + \text{ATP} \rightarrow \text{Glycerol-3-phosphate} + \text{ADP} \qquad \text{Glycerol-3-phosphate phosphatase}$$

$$\text{Glycerol-3-phosphate} + \text{Q} \rightarrow \text{Dihydroxyacetonephosphate} + \text{QH}_2 \qquad \text{Glycerol-3-phosphate dehydrogenase}$$

Since this route would not generate NADH through reduced flavin (QH_2), it is disregarded in metabolic engineering since the reductive steps from oxalacetate to succinic acid require NADH. Alternatively, glycerol might be converted into dihydroxyacetone via the glycerol dehydrogenase (gldA), which derives one mol NADH. Dihydroxyacetone would then be converted into dihydroxyacetone phosphate via the dihydroxyacetone kinase encoded by the *dhaKLM* gene. However, the native dhaKLM enzyme of *E. coli* uses PEP as donor of the phosphoric acid group and thereby consumes an intermediate, which is foreseen as substrate for the later carboxylation. Alternatively, an isoenzyme might employ ATP:

Glycerol + NAD^+ → Dihydroxyacetone + NADH + H^+	Glycerol dehydrogenase
Dihydroxyacetone + PEP (or ATP) → DHAP + Pyruvate (or ADP)	Dihydroxyacetonekinase

For the design of the pathway toward succinic acid, two options are considered. In the first option, it is assumed that the dhaKLM gene employs ATP as energy carrier and that PEP is the acceptor of the carboxylation reaction. The overall pathway can be stated as follows (Table 12.8).

In the second option, it is assumed that the formation of DHAP consumes PEP and that the decisive C3 to C4 conversion proceeds via the pyruvate carboxylase encoded by the *pyc* gene. Since *E. coli* lacks this enzyme, a heterologous enzyme might be introduced (Blankschien et al. 2010). The overall pathway can be stated as follows (Table 12.9).

Since no net ATP gain is accomplished, the pathway would not provide biochemical energy for growth and cell maintenance. This obstacle might be circumvented if the process is run microaerobically such that some reduction equivalents carry over their electrons on the offered oxygen and at least a minor ATP gain is possible. However, the TMY would be compromised and could not reach the 1.0 mol succinic acid per mol glycerol as proposed by the electron balance. Cell growth and production of succinic acid would be coupled since the pathway until fumarate replenishes the NADH pool. *E. coli* consumes glycerol under aerobic and anaerobic conditions. It is interesting to note that an elementary mode analysis – considering the complete metabolic system including biomass formation and mass

Table 12.8 Pathway toward succinic acid starting with glycerol.

Glycerol + NAD^+ → Dihydroxyacetone + NADH + H^+	Glycerol dehydrogenase
Dihydroxyacetone + ATP → DHAP + ADP	Dihydroxyacetonekinase
DHAP → GAP	Triosephosphate isomerase
GAP + NAD^+ → 1,3-BPG + NADH + H^+	GAP dehydrogenase
1,3-BPG + ADP → 3-Phosphoglycerate + ATP	Phosphoglycerate kinase
3-Phosphoglycerate → 2-Phosphoglycerate	Phosphoglycerate mutase
2-Phosphoglycerate → PEP + H_2O	Enolase
PEP + ADP + CO_2 → Oxalacetate + ATP	PEP carboxykinase
Oxalacetate + NADH + H^+ → Malate + NAD^+	Malate dehydrogenase
Malate → Fumarate + H_2O	Fumarase
Fumarate + NADH + H^+ → Succinic acid + NAD^+	Succinate dehydrogenase
Glycerol + CO_2 + ADP + 2 Pi → Succinic acid + ATP + 2 H_2O	

Table 12.9 Pathway toward succinic acid starting with glycerol via pyruvate carboxylase.

Glycerol + NAD^+ → Dihydroxyacetone + NADH + H^+	Glycerol dehydrogenase
Dihydroxyacetone + PEP → DHAP + pyruvate	Dihydroxyacetonekinase
DHAP → GAP	Triosephosphate isomerase
GAP + NAD^+ → 1,3-BPG + NADH + H^+	GAP dehydrogenase
1,3-BPG + ADP → 3-Phosphoglycerate + ATP	Phosphoglycerate kinase
3-Phosphoglycerate → 2-Phosphoglycerate	Phosphoglycerate mutase
2-Phosphoglycerate → PEP + H_2O	Enolase
Pyruvate + ATP + CO_2 → Oxalacetate + ADP + Pi	Pyruvate carboxylase
Oxalacetate + NADH + H^+ → Malate + NAD^+	Malate dehydrogenase
Malate → Fumarate + H_2O	Fumarase
Fumarate + NADH + H^+ → Succinic acid + NAD^+	Succinate dehydrogenase
Glycerol + CO_2 → Succinic acid + H_2O	

transfer rates between the medium and the host – predicts a higher succinic acid productivity in cases where the oxygen supply is limited to 0.65 mol oxygen per mol succinic acid. Overexpression of the PEP carboxylase or introduction of a heterologous pyruvate carboxylase is predicted to facilitate the highest succinic acid yield (Chen et al. 2010).

The pathway with PEP as donor of the phosphate group and substrate of the dihydroxyacetone kinase and a heterologous pyruvate carboxylase in *E. coli* has been experimentally verified. The *E. coli* biocatalyst achieved a specific productivity of 400 mg succinic acid per g cell weight per h and a carbon yield of 0.69 t succinate per ton glycerol. For comparison, the theoretical maximum carbon yield amounts to 1.28 t succinic acid per ton glycerol (118.1 g/mol/92.1 g/mol) without considering the minor ATP formation for cell growth and maintenance. A titer of 12.6 g/l succinic acid was reached after 60 and 72 h corresponding to a productivity of 0.21 g/lxh and 0.175 g/lxh (Blankschien et al. 2010). Experiments were pursued at 37 °C and ambient pressure cultured in MOPS medium supplemented with 10 g/l tryptone, 5 g/l yeast extract, and 10 g/l glycerol as feed. A rather high filling of the flask combined with foam plugs and shaking ensured aeration, though the oxygen transfer rate or the oxygen consumption was not measured. $CaCO_3$ was added at 55 g/l, which also stabilizes the pH with rising formation of the diacid product. Calcium succinate has a peak solubility of 12.6 g/l and precipitates above the threshold. Calcium succinate can be recovered by filtration, and after the addition of sulfuric acid fully the protonated succinic acid and calcium sulfate are obtained.

Glycerol might be conceptualized as the sole carbon source in the fermentative production of succinic acid or be used as a supplementary carbon source to boost the carbon yield. It has been reported that the carbon yield of *A. succinogenes* used for the fermentative production of succinic acid with a Napier grass hydrolysate as feedstock could be enhanced with the addition of glycerol. In particular, with a hydrolysate to glycerol ratio of 10 : 1 a carbon yield of 0.65 g/g was achieved, which could be optimized up to 0.88 g/g when the hydrolysate to glycerol ratio was shifted to 1 : 1 (Lee et al. 2022). Accordingly, glycerol might be an interesting carbon and energy source for the fermentative production of succinic acid

(Clomburg and Gonzalez 2013). This feedstock became available in the marketplace in high amounts and at comparably low prices with the rising production of biodiesel (fatty acid methylester), which derives glycerol as by-product. Though future availability might be difficult given the critical considerations for the production of biodiesel and the resulting deforestation in Southeast Asia. The research on a fermentative succinic acid with glycerol as feedstock and yeast as biocatalyst is continued in academia and funded by the German Research Foundation (Lempka 2022).

The idea to offer additional reduction equivalents has also been pursued in *E. coli* with a combination of glucose and the "sugar alcohol" sorbitol as feedstock and molecular hydrogen in the pressurized reactor headspace. Whereas the addition of sorbitol provided an advantage, the hydrogen supply in the reactor headspace was not successful (Hong and Lee 2002). The limitation of the second approach might be explained by the laboratory scale of the experiments, which were not really equipped for the efficient supply of molecular hydrogen. Additionally, the hydrogen uptake capacity of the host might be a challenge.

12.3.8 Oil Palm Frond Juice as Alternative Raw Material Source

The oil palm plantation industry in Indonesia and Malaysia produces about 83 Mio. t pa of oil palm frond (OPF) juice as by-product, which is valued at cost of only 22 USD/t (Tan et al. 2018), likely at producer economics and without cost for storage and transportation. The OPF juice contains in total up to 43 g/l of fermentable sugars covering glucose (28.6 g/l), sucrose (8.3 g/l), and fructose (6.0 g/l). Given the lower price of OPF juice, if calculated in glucose equivalents compared to conventional glucose, it is considered as cheap fermentation feedstock and has been tested for producing succinic acid. Carbon dioxide for the carboxylation of PEP is added at optimized sparging rate. The carbonate concentration in the medium was increased up to 500 mM, which facilitated the maximum ratio of succinic acid to acetic acid of 15 t succinic acid per t acetic acid, the most prominent by-product. This effect is rationalized by the competing pathway direction of PEP, which may either be carboxylated to oxalacetate or be converted into pyruvate, which may be further processed to obtain acetic acid. The pH of the medium is controlled with KOH to avoid acidification via the formation of bicarbonate. With *A. succinogenes* as host and OPF juice as feed, a titer of 29 g/l succinic acid after 40 h could be achieved, corresponding to a productivity of 0.725 g/lxg (Tan et al. 2018). The carbon yield achieved 0.71 t succinic acid per ton sugar mixture of OPF juice.

12.3.9 Succinic Acid Based on Glucose Combined with Microbial Electrosynthesis

The reduction equivalents for the conversion of glucose with carbon dioxide into succinic acid might partially be provided via microbial electrosynthesis (MES). In this specific case, the metabolic pathway via PEP carboxylation and the reductive citric acid cycle in the microbial host *E. coli* has been combined with the cathodic provision of additional electrons. It is known that the electroactive microbe *Shewnella oneidensis* MR-1 is able to express an electron transfer pathway that consists of a c-type outer membrane cytochrome (MtrC), a periplasmic c-type cytochrome (MtrA), a non-heme-outer-membrane ß-barrel protein (MtrB), and an inner membrane-associated quinol oxidase (CymA) (Ross et al. 2011). This pathway has been cloned in *E. coli* and indeed the electroactivity could be restored in the recombinant *E. coli* host (Wu et al. 2019). This success might open the door for further steps in the exploitation of recombinant microbial uptake of cathodic electrons because *E. coli* can easily be transformed, whereas the molecular biology toolbox for *S. oneidensis* is not mature. It is assumed that the electrons at the cathode are received by the added neutral red, which then interacts with the MtrC protein in the outer membrane of the recombinant *E. coli* host. The received electrons are then further channeled to the inner membrane and finally applied for the saturation of the double bond in fumarate to obtain succinic acid. The fermentation was pursued as strictly anaerobic process, and oxygen residues in the medium were removed by constant sparging with pure nitrogen. Glucose was added as the carbon and energy source for the generation of PEP and fumarate, respectively, and hydrogen carbonate was added as additional carbon source – as reaction partner for the PEP carboxylase and PEP carboxykinase – to allow a high carbon yield based on glucose. About 1.0 mol succinic acid per mol glucose could be achieved, which is a 1.6-fold improvement over the parent strain (Wu et al. 2019).

12.4 Production Host

12.4.1 Hosts Applied for Development of an Industrial Fermentation Process

The four industrial players, which were running plants for fermentative succinic acid, applied different production hosts. The "first movers" Bioamber and Myriant applied *E. coli* as host and Reverdia – supported by the yeast biocatalyst business of its JV partner DSM – chose the yeast *S. cerevisiae*. Succinity applied their proprietary host *Basfia succiniproducens,* which is almost identical to *Mannheimia succiniproducens* in genome sequence (Ahn et al. 2020). Host selection is driven by several criteria, and each host has advantages and disadvantages.

Since succinic acid is a native intermediate of the citric acid cycle, the required enzymes are in principle in all hosts available. However, the production of succinic acid under native conditions is different, and the starting baseline is distinguished. The established fermentation hosts *C. glutamicum*, *E. coli*, and *S. cerevisiae* are nonnatural producers and display low production rates, and often rely on dual-phase fermentation combining aerobic cultivation for the formation of biomass and anaerobic cultivation for production. *E .coli* proceeds with mixed-acid fermentation under anaerobe conditions and generates to some extent succinic acid besides acetate, formate, ethanol, carbon dioxide, and hydrogen (Clark 1989). Fermentation protocols were developed by Roquette, which foresaw an aerobe initial phase with glucose for the formation of biomass followed by an anaerobe secondary phase in the presence of carbon dioxide for the formation of succinic acid (Bernard and Segueilha 2010). The natural succinic acid producers *A. succinogenes* and *M. succiniproducens* are facultative anaerobic bacteria and display attractive production rates before any metabolic engineering program has been started (Lee et al. 2006). On the other hand, these rumen organisms are reported to require expensive nutrient sources and display an unstable performance in the fermentation process (Meng et al. 2016).

Fermentative development programs may favor a change of the host in case critical limits in metabolic engineering are identified. This refers to, e.g. the development program of Amyris for the production of artemisinic acid and, in this context, the decision of Bioamber to finally bet on a yeast host (i.e. *Pichia kudriavzevii*; Ahn et al. 2016) for commercial production. The former president and CEO of Bioamber, J.-F. Huc, referred to the pH sensitivity of *E. coli* toward the target diacid product, which necessitates neutralization during the fermentation and causes rather high efforts in the downstream process to isolate the protonated molecule. Accordingly, Bioamber exclusively licensed a pH-tolerant yeast host of Cargill originally developed for the production of lactic acid. It is reported that the superior performance of the yeast at lab scale could also be seen in the commercial process and that the obtained product had a higher quality than the beforehand used *E. coli* strain (ICIS Chemical Business 2016). The eukaryotic yeast host implies an additional complexity level compared to a bacterial host since the citric acid cycle takes place in the mitochondrion and the limited ability of some intermediates to pass the mitochondrial membrane needs to be considered. Since the inner mitochondrial membrane is impermeable to succinate, a specific transporter would need to be introduced. Alternatively, the metabolic pathway might install a cytoplasmic copy of the citric acid cycle and thereby bypass the permeability restrictions of the mitochondrion (Myriant 2015). In order to validate an efficient secretion of the product, the ratio of intracellular and extracellular succinic acid was determined and resulted in a ratio of 1 : 239 (Raab et al. 2010).

In academia, the bacterial host *C. glutamicum* has been tested, and an aerobic process has been proposed. For comparison, the succinic acid technology applying *E. coli*, yeast, or *B. succiniproducens* pursues the fermentation mainly without oxygen as an anaerobic process. The availability of oxygen might allow to generate 12 reduction equivalents per mol of glucose via glycolysis and the citric acid cycle and would thereby help to provide the needed reduction equivalents with minimum additional glucose consumption. However, the aerobic process does not yet achieve the performance level of the anaerobic process (Becker and Wittmann 2015). Additionally, fermentation hosts suitable for coproduction of succinic acid and other products have been explored in academia. *Enterobacter cloacae* has been enabled to provide 41 g/l 2,3-butanediol and 22 g/l succinic acid from xylose (Wu et al. 2017). Carbon flux could be redirected to obtain 40 g/l acetoin and 20 g/l succinic acid after 72 h facilitated by deletion of the lactate dehydrogenase encoded by *ldh* and 2,3-butanediol dehydrogenase encoded by *budC* (Su et al. 2021).

Promising results with *M. succiniproducens* were achieved in the early stages of technology development. The metabolically engineered strain LPK7 provided a titer of 13.4 g/l from 20 g/l glucose and achieved a molar yield of 0.97 mol succinic acid per mol glucose in a simple batch approach. The corresponding maximum productivity is 1.59 g/lxh. A fed-batch culture of the same host achieved a titer of 52.4 g/l accompanied by a molar yield of 1.16 mol succinic acid per mol glucose and a productivity of 1.8 g/lxh (Lee et al. 2006). The PEP carboxykinase was the most important enzyme for the carboxylation of oxalacetate. In order to limit the generation of by-products the lactate dehydrogenase (*ldh*), the pyruvate formate lyase (*pflB*), the phosphate acetyltransferase (*pta*), and the acetate kinase (*ackA*) were knocked out. *Since M. succiniproducens* does not use a PTS system for glucose uptake, which consumes PEP to uptake glucose and thereby generates glucose-6-phosphate, a switch of the glucose uptake system is not required (Lee et al. 2006).

12.4.2 Best Published Results on Technology Performance

An overview of the various natural producers and engineered hosts for the fermentative production of succinic acid reveals that quite promising performance data for natural microbes have also been observed, e.g. for *A. succinogenes* a titer of 105 g/l combined with a productivity of 1.36 g/lxh is reported (Liu et al. 2022). The best fermentation performance data were achieved via glycolysis until PEP, followed by the reductive citric acid cycle with the engineered host *M. succiniproducens* (strain PALKcgmdh). With a high-inoculum approach and dual carbon source combining glucose and glycerol, a titer of 134 g/l succinic acid after 6.3 h was obtained, which translates into a very high productivity of 21.3 g/lxh (Ahn et al. 2020). The molar yield was 1.25 mol succinic acid per mol glucose equivalent. The productivity is facilitated by the outstanding specific productivity of 1.19 g succinic acid per gram DCW (dry cell weight) per h (Ahn et al. 2020). In case the average glucose uptake rate is estimated with 10 mmol g per DCW and h a biomass concentration of 17.9 g DCW/l is obtained:

- A titer of 134 g/l and a productivity of 21.3 g/lxh imply a batch time of 6.3 h.
- A molar yield of 1.25 mol succinic acid/mol glucose equivalent implies a carbon yield of 0.625 g succinic acid/g glucose equivalent ($1.25 \times 90.1/180.1 = 0.625$).
- A specific productivity of 1.19 g succinic acid/g DCWxh and a productivity of 21.3 g/lxh imply a biomass concentration of 17.9 g DCW/l.
- A glucose uptake rate of 10 mmol/g DCWxh translates into 1.8 g glucose/g DCWxh and 32.2 g glucose/lxh (1.8×17.9).
- A glucose uptake and consumption rate of 32.2 g glucose/lxh allows a productivity of 20.1 g/lxh given a carbon yield of 0.625 t succinic acid/g glucose equivalent (32.2×0.625).

The metabolic engineering strategy focused on the specific activity of the malate dehydrogenase. Structural analysis of the crystallized enzyme revealed a susceptibility to substrate inhibition. After screening several enzyme candidates, the malate dehydrogenase of *C. glutamicum* was chosen given its low substrate inhibition. Malate dehydrogenases are phylogenetically classified into cytosolic and mitochondrial isoenzymes. Whereas the cytosolic malate dehydrogenases mainly participate in the malate–aspartate shuttle for NADH transfer and seem appropriate for the engineered pathway, the mitochondrial malate dehydrogenase acts in the citric acid cycle. The malate dehydrogenase isoenzymes applied so far for overproduction of succinic acid, the isoenzymes of *A. succinogenes*, *E. coli*, *M. succiniproducens*, *S. cerevisiae*, and *Yarrowia lypolytica*, belong to the mitochondrial malate dehydrogenase family. On the other hand, the malate dehydrogenase of *C. glutamicum* applied in this academic project belongs to the cytosolic enzyme family (Ahn et al. 2020).

Outstanding technology performance data were also published for the host *Vibrio natriegens*, a gram-negative δ-proteobacterium able to grow under aerobic and anaerobic conditions. The best-transformed host provided a titer of 60.4 g/l after seven hours, corresponding to a productivity of 8.6 g/lxh (Thoma et al. 2021). The succinate production rate achieved 1.33 g succinate per g DCW (dry cell weight) per h, which is a high value compared to other established production systems (Thoma et al. 2021). The metabolic engineering strategy activated the reductive citric acid cycle combined with the glyoxylate pathway. The experiments achieved a yield of 1.46 mol succinic acid per mol glucose.

12.5 Reactor Concepts

In most industrial fermentations, the suspended biocatalyst is rather equally distributed in the liquid phase of the reactor. Since the cells are subjected to sedimentation driven by the gravitational force, a stirring device is applied even if an anaerobe fermentation is pursued. The rotation of the medium might alternatively be achieved or supported by a pressurized gas stream like oxygen in an aerobe fermentation, hydrogen and carbon dioxide in the biological methanation, or hydrogen, carbon monoxide, and carbon dioxide in fermentations with acetogenic hosts to obtain ethanol and acetate. The application of fermenters with fixed-bed biofilm reactors, according to the fixed-bed reactor in organic chemistry, is rather unusual. An interesting and individual academic setup has been developed with a biofilm of the natural producer of succinic acid, *A. succinogenes*, in a packed-bed reactor (Ferone et al. 2018). The bioreactor is a lab-scale 166 ml glass bottle jacketed for heat removal and filled with a Tygon support to provide an area for the formation of the biofilm. No chemical agent had to be added to assist immobilization. The buildup of the biofilm required an incubation phase of 16 days with a glucose supply of 50 g/l.

On day 17, the production of succinic acid started again with 50 g/l glucose as feed. Additionally, carbon dioxide was sparged at the reactor bottom to provide the educt for the synthesis of oxalacetate as the first C4-intermediate of the pathway. The dilution rate is defined as the ratio of the volumetric flow rate of the medium in l/h to the (reaction) volume of the fixed-bed reactor and was set between 0.5 and 2.4 per h. The biofilm reactor's performance was evaluated after the system achieved a steady-state – defined as a period with constant concentration of energy feed and product – which takes about 2–4 days. The best results were obtained with the lowest dilution rate of 0.5 per h and the 50 g/l glucose was converted into 43 g/l succinic acid. This means that in one hour of incubation time, a glucose solution of 50 g/l with a volume of half the reaction volume was channeled through the fermenter. This implies a biofilm reactor productivity of about 22 g/lxh, which is high compared to the values published for a conventional fermentative production of succinic acid with suspended biocatalyst (Ferone et al. 2018). These performance figures seem attractive on first view, though the specific boundary conditions of the biofilm reactor need to be considered. The high productivity relies on a high cell density, which is acceptable if the biofilm has a long service life. The glucose consumption and the nonoperational reactor capacity increase with decreasing service life of the catalyst. Additionally, an efficient purification protocol with full recovery for the Tygon support is needed to avoid the consumption of an additional raw material.

12.6 Downstream Processing

After the fermentation process, the broth is often subjected to a filtration step to remove the biomass and potential residues of the sugar source. Pending on the titer, the succinic acid product would then be obtained as solution in the filtrated broth, given a solubility of 58 g/l at a temperature of 20 °C. The main task of the downstream process is the full protonation and the purification of succinic acid to obtain a polymer grade. The broth is neutralized during the neutralization to avoid the pH decreasing below the pH tolerance of the host. In case a pH-tolerant yeast is applied as catalyst a pH in the range of 5.5 up to 6.0 might be possible, though a value in the range of 7.0 is more realistic. Thus, a neutralization during fermentation is required, and the product is mainly deprotonated. One of the earliest protocols for the isolation of fermentative succinic acid proposes calcium oxide or calcium hydroxide for neutralization and insoluble calcium succinate is obtained, which can be isolated by precipitation (Lemme and Datta 1987):

$$\text{HOOC–CH}_2\text{–CH}_2\text{–COOH} + \text{Ca(OH)}_2 \rightarrow \text{Ca–Succinate} \downarrow + 2\,\text{H}_2\text{O}$$

The calcium succinate might be treated with sulfuric acid to obtain the protonated acid and, accordingly, calcium sulfate would be obtained as by-product. In case ammonia is applied as neutralization agent, ammonium-succinate would be obtained as the primary product, which might then be reconverted into the fully protonated acid by addition of sulfuric acid:

$$\text{HOOC-CH}_2\text{-CH}_2\text{-COOH} + 2\,\text{NH}_3 \rightarrow \text{NH}_4^{+\,-}\text{OOC-CH}_2\text{-CH}_2\text{-COO}^-\,\text{NH}_4^+ \quad \text{During the fermentation}$$

$$\text{NH}_4^{+\,-}\text{OOC-CH}_2\text{-CH}_2\text{-COO}^-\,\text{NH}_4^+ + \text{H}_2\text{SO}_4 \rightarrow \text{HOOC-CH}_2\text{–CH}_2\text{-COOH} + (\text{NH}_4)_2\text{SO}_4 \quad \text{After the fermentation}$$

The protonated succinic acid could then be further purified via crystallization, and ammonium sulfate (AS) is derived as by-product. In case it is assumed that a molar neutralization is needed, the AS by-product factor would be in the range of 1.12 t AS/t succinic acid (132.1 g/mol/118.1 g/mol). This is according to the process description given in the USDE comment on the fermentation technology of Myriant, which states that the plant of Myriant in Lake Providence has an output of 30 million lbs. succinic acid per year and not more than 40 million lbs. AS per year (US Department of Energy (USDE) 2010).

A comprehensive summary of the published downstream processes including crystallization, vacuum distillation, membrane separation, reactive extraction, bipolar membrane electrodialysis, and calcium precipitation indicates that the obtained purity ranges from 85% to 99% and the extraction yield may achieve up to 94% (Xiao et al. 2020). In all cases, succinic acid is obtained as product. An interesting purification strategy has been proposed that would derive succinimide as product, which could also be used in the downstream value chain as succinic acid equivalent. The fermentation broth would be filtrated, the pH set at 2.0, succinic acid be crystallized at 8 °C, and incubated for four hours. The crystals are removed, and the remaining mother liquor is subjected to a second crystallization at pH 2.0, 4 °C, and an incubation time of 12 h. Combined with the lower temperature to reduce the solubility of succinic acid, urea would be added, and urea-succinic acid cocrystals would be generated. The cocrystals are removed from the efficiently harvested mother liquor and united with the pure succinic acid crystals. The mixture of the succinic acid crystals and the urea-succinic acid crystals is then incubated at 195 °C for five hours to provide the heat for the endothermic formation of succinimide. The obtained succinimide is then recrystallized to obtain a pure succinimide (Xiao et al. 2020). The overall recovery yield of succinic acid in the downstream process is reported at 95% and thereof 80% is obtained as succinimide.

12.7 Industrial Capacity and Performance

The idea to produce succinic acid based on renewable resources attracted a lot of attention around 2005, driven by technological advancements in academia and the market interest for "green" products not derived from petrochemistry. Commercial production of biobased succinic acid started in the decade 2010–2020 (Fletcher et al. 2016). Four major players, or consortia, respectively, were initially engaged in the development of fermentative production of succinic acid (Nghiem et al. 2017):

Bioamber Inc. was founded in 2008 as a JV between ARD (Agroindustrie Recherches et Dévеloppments) and DNP Green Technology exclusively for the fermentative production of succinic acid. The company won the 2011 ICIS Innovation Award for its biobased succinic acid platform (Chemeurope.com 2011). The company went public in 2013 on the New York Stock Exchange. The production started with *E. coli* as production host and a pilot plant in Pomacle, France, with a capacity of 2 kt/pa. The reactor in Pomacle had a design volume of 350 m^3 (ICIS Chemical Business 2016). Construction of a second plant with a capacity of 34 kt/pa in Sarnia, Canada, was started including a capacity of 23 kt/pa for the downstream product 1,4-butanediol. Operation startup was envisioned for end of 2014, and the president and CEO of Bioamber, J.-F. Huc, explained that the production will be competitive as long as the oil price stays above $35 a barrel (Ravenscroft 2014; Il Bioeconomista 2013). Due to the switch of the biocatalyst from *E. coli* to yeast, some technical adjustments were required and startup was postponed to spring 2015 (ICIS Chemical Business 2016). Bioamber was also very active in the conceptual layout of the future succinic acid value chain. The business plan included the partial replacement of chemically produced succinic acid though also addressed new products profiting from the marketing of succinic acid as "green" product based on renewable resources. The senior vice president of Bioamber, Babette Pettersen, explained the strategy with focus on the downstream value chain: "Our business model is about partnering and open innovation. It is not just about being biobased but differentiating performance and, thus, the potential that these biobased materials have to add value. We have to work not just at the beginning of the value chain but all the way through – not just push our products to market but also to ensure that they get pull from the final applications" (Ravenscroft 2014). In this context, cooperations with Faurecia and Mitsubishi Chemical Corp. for production of biobased automotive plastics and with Lanxess on the development of biobased plasticizers were initiated (Stringer 2012). Furthermore, research on a suitable catalyst for the chemical hydrogenation of succinic acid toward 1,4-butanediol, THF, and γ-butyrolactone was started (Litsanov et al. 2014). The chemical conversion opens the business perspective for an additional chemical target derived from renewable resources, and it is reported that DowDupont uses 1,4-BDO derived from biosuccinic acid for the production of PBT (ChemicalWeek 2018). To improve production costs, a yeast microorganism developed by Cargill originally for the fermentative production of lactic acid was licensed, which displays an enhanced tolerance to low pH and is able to utilize lignocellulosic feedstocks (Rosales-Calderon and Arantes 2019). However, Bioamber filed for bankruptcy in May 2018 (Bioplastics 2021a).

Myriant Technologies LLC was founded in 2009 and was granted 50 USD m from the US Department of Energy for the construction of a biorefinery plant in Lake Providence, in the

United States, with a capacity of 13.6 kt/pa (30 million lbs. pa). Public funding was rationalized by a thorough analysis of the technology, which pointed out that fermentative production has a lower energy requirement per ton than a comparable petroleum-based process and that the consumption of carbon dioxide will reduce life-cycle greenhouse gas emissions (US Department of Energy (USDE) 2010). Myriant, like Bioamber, was interested in the production of 1,4-butanediol and started catalyst research with Johnson Matthey/Davy Process Technologies. Davy Technologies had developed a technology for the chemical production of 1,4-butanediol, and a combination of both technologies promised an improvement in carbon efficiency (Bioplastics 2021b). A marketing cooperation was started with PTT Chemical Group (Thailand) to facilitate penetration of the Asian market (Litsanov et al. 2014). Plans to expand the capacity up to 77 kt/pa were announced in 2015. Furthermore, the plan to construct a plant with a capacity of 100 kt/pa with the Chinese partner BlueStar was announced though did not yet materialize (Rosales-Calderon and Arantes 2019). The production base was restructured in 2017 and the focus shifted from production to R&D. The company name was officially changed to GC Innovation America in 2018 (Bioplastics 2021c). GC Innovation America is the US-based innovation arm of PTT Global Chemical Public Company Limited, the leading chemical company in Thailand.

Reverdia was a JV between DSM and Roquette. *S. cerevisiae* was chosen as production host to allow fermentation at a lower pH and to save cost for neutralization. The product was supposed to be sold under the brand name Biosuccinium™. A pilot plant was started in 2010 in Lestrem, France, to be followed by the first commercial plant with a capacity of 10 kt/pa in Cassano, Italy (Litsanov et al. 2014). In 2019, both parties decided to dissolve the JV, and the IP rights and obligations were transferred to Roquette. DSM as the original technology developer assumed the role of exclusive licensor and provided a nonexclusive license to Roquette. Operation in the plant in Cassano was continued (Barrett 2019). In 2022, DSM sold its Biosuccinium® technology and the respective IP rights and proprietary yeast strains to the Engineering & Technology company Technip (Technip 2022). Technip is thereby able to offer a technology package for producing succinic acid and the downstream polymer PBS (polybutylene succinate).

Succinity GmbH is a JV of the chemical company BASF and Purac, a subsidiary of CSM. The host has been isolated from bovine rumen and is a natural producer of succinic acid. The JV was founded in October 2012 (Bioplastics 2012) and originally planned to start production in 2013 with a capacity of 10 kt/pa applying the host *B. succiniproducens* as biocatalyst (Litsanov et al. 2014). The production was started in mid-2014 at the Corbion site in Montmeló, Spain, with a capacity of 10 kt/pa (BASF 2014).

Process performance data are published for the four major players that pursued commercial production of succinic acid (Ahn et al. 2016; Table 12.10).

Production costs for biobased succinic acid were estimated at 1.1–1.6 €/kg based on glucose for a plant with a capacity of 100 kt/pa (Hermann and Patel 2007). Another analysis with focus on the potential production host is estimating production costs for a titer of 100 g/l between 0.8 and 1.4 €/kg (Liebal et al. 2018). However, details on the considered downstream process are not disclosed. Since significant downstream efforts are required to render the neutralized dianion in the broth into the fully protonated acid, the assumed production costs might underestimate the full cost.

Table 12.10 Performance data for the four commercial technologies. These figures do not include the product loss in the isolation and downstream processing.

Company	Host	Titer (g/l)	Yield (mol/mol)	Yield (g/g)	Productivity (g/lxh)
Reverdia	*Saccharomyces cerevisiae*	43	n.a.	n.a.	0.45
Bioamber	*Pichia kudriavzevii*	48	0.69	0.45	0.97
Myriant	*Escherichia coli*	84	1.53	1.00	1.76
Succinity	*Basfia succiniproducens*	70	1.70	1.11	2.91

References

Adom, F., Dunn, J.B., Han, J., and Sather, N. (2014). Life-cycle fossil energy consumption and greenhouse gas emissions of bioderived chemicals and their conventional counterparts. *Environmental Science & Technology* 48: 14624–14631. https://doi.org/10.1012/es503766e.

Ahn, J.H., Jang, Y.-S., and Lee, S.Y. (2016). Production of succinic acid by metabolically engineered microorganisms. *Current Opinion in Biotechnology* 42: 54–56. https://doi.org/10.1016/j.copbio.2016.02.034.

Ahn, J.H., Seo, H., Park, W. et al. (2020). Enhanced succinic acid production by Mannheimia employing optimal malate dehydrogenase. *Nature Communication* 11: 1970. https://doi.org/10.1038/s41467-020-15839-z.

Arpe, H.-J. (2007). *Industrielle Organische Chemie. Bedeutende Vor- und Zwischenprodukte.* Wiley 407ff.

Barrett, A. (2019). DSM and Roquette take next steps in bio-based succinic acid. *Bioplastics News* https://bioplasticsnews.com/2019/02/22/dsm-and-roquette-take-next-step-in-bio-based-succinicacid/#:~:text=Effective%201%20April%202019%2C%20the%20joint%20venture%20Reverdia,plant%20and%20continue%20serving%20customers%20of%20Biosuccinium%20%C2%AE (accessed 29 October 2021).

BASF (2014). Press Release, Succinity produziert erste kommerzielle Mengen biobasierter Bernsteinsäure. https://www.basf.com/at/de/media/news-releases/2014/03/p-14-0303-ci.html (accessed 29 October 2021).

Becker, J. and Wittmann, C. (2015). Biotechnologie von morgen: Metabolisch optimierte Zellen für die bio-basierte Produktion von Chemikalien und Treibstoffen. *Materialien und Gesundheitsprodukten, Angewandte Chemie* 127: 3383–3407.

Bernard, C. and Segueilha, L. (2010). Process for producing succinic acid. Patent WO2010/092304 A2.

Bioplastics (2012). BASF and CSM establish 50/50 JV for biobased succinic acid, Bioplastics magazine. https://www.bioplasticsmagazine.com/en/news/meldungen/20121005_BASFCSM.php#:~:text=LUDWIGSHAFEN%2C%20GERMANY%2C%20and%20DIEMEN%2C%20THE%20NETHERLANDS%20%E2%80%93%20October,Succinity%20GmbH%20and%20will%20be%20operational%20in%202013 (accessed 3 August 2023).

Bioplastics (2021a). BioAmber. https://bioplasticsnews.com/bioamber (accessed 29 October 2021).

Bioplastics (2021b). Myriant. https://bioplasticsnews.com/myriant (accessed 29 October 2021).

Bioplastics (2021c). Myriant becomes GC Innovation America. A tactical move from PTT. https://bioplasticsnews.com/2018/08/21/myriant-becomes-gc-innovation-america-ptt-natureworks/

Blankschien, M.D., Clomburg, J.M., and Gonzalez, R. (2010). Metabolic engineering of *Escherichia coli* for the production of succinate from glycerol. *Metabolic Engineering* 12: 409–419. https://doi.org/10.1016/j.ymben.2010.06.002.

Bomgardner, M.M. (2014). Biobased polymers, CEN, pp. 10–12.

Bozell, J.J. and Petersen, G.R. (2010). Technology development for the production of biobased products from biorefinery carbohydrates – the US department of energy's "Top 10" revisited. *Green Chemistry* 12: 539–554.

Burkart, M.D., Hazari, N., Tway, C.L., and Zeitler, E.L. (2019). Opportunities and challenges for catalysis in carbon dioxide utilization. *ACS Catalysis* 9: 7937–7956. https://doi.org/10.1021/acscatal.9b02113.

Chemeurope.com (2011). Bioamber wins ICIS innovation award for best business innovation. https://www.chemeurope.com/en/news/134875/bioamber-wins-icis-innovation-award-for-best-business-innovation.html (accessed 29 October 2021).

ChemicalWeek (2018). Cars drive thermoplastic copolyesters. *Chemical Week* 31.

Chen, Z., Liu, H., Zhang, J., and Liu, D. (2010). Elementary mode analysis for the rational design of efficient succinate conversion of glycerol by *Escherichia coli*. *Journal of Biomedicine and Biotechnology* 2010: 518743. https://doi.org/10.1155/2010/518743.

Cheng, K.-K., Wang, G.-Y., Zeng, J., and Zhang, J.-A. (2013). Improved succinate production by metabolic engineering. *BioMed Research International* 2013: https://doi.org/10.1155/2013/538790.

Clark, D. and (1989). The fermentation pathway of *Escherichia coli*. *FEMS Microbiology Review* 5: 223–234. https://doi.org/10.1016/0168-6445(89)90033-8.

Clomburg, J.N. and Gonzalez, R. (2013). Anaerobic fermentation of glycerol: a platform of renewable fuels and chemicals. *Trends in Biotechnology* 31 (1): 20–28.

Cok, B., Tsiropoulos, I., Roes, A.L., and Patel, M.R. (2014). Succinic acid production derived from carbohydrates: an energy and greenhouse gas assessment of a platform chemical toward a bio-based economy. *Biofuels, Bioproducts and Biorefining* 8: 16–29. https://doi.org/10.1002/bbb.1427.

Erickson, B., Nelson, J.E., and Winters, P. (2012). Perspectives on opportunities in industrial biotechnology in renewable chemicals. *Biotechnology Journal* 7: 176–185. https://doi.org/10.1002/biot.201100069.

Ferone, M., Raganati, F., Ercole, A. et al. (2018). Continuous succinic acid fermentation by *Actinobacillus succinogenes* in a packed-bed biofilm reactor. *Biotechnology for Biofuels* 11: 138. https://doi.org/10.1186/s13068-018-1143-7.

Flamholz, A., Noor, E., Bar-Even, A., and Milo, R. (2012). eQuilibrator – the biochemical thermodynamics calculator. *Nucleic Acid Research* 40 (database issue): D770–D775.

Fletcher, M., Krivoruchko, A., and Nielsen, J. (2016). Industrial systems biology and its impact on synthetic biology of yeast cell factories. *Biotechnology and Bioengineering* 113 (6): 1164–1170. https://doi.org/10.1002/bit.25870/abstract.

Goldbach, V. (2018). Covestro presentation: alternative raw materials in a circular economy.CBPM Symposium, 18, https://bioplasticsnews.com/wp-content/uploads/2019/06/Alternative-Raw-Materials-in-a-Circular-Economy.pdf (accessed 20 October 2022).

Hagen, J. (2015). *Industrial Catalysis*, Weinheim, 278–279.

Hermann, B.G. and Patel, M. (2007). Today's and tomorrow's bio-based bulk chemicals from white biotechnology: a techno-economic analysis. *Applied Biochemistry and Biotechnology* 136: 361–388.

Hong, S.H. and Lee, S.Y. (2002). Importance of redox balance on the production of succinic acid by metabolically engineered *Escherichia coli*. *Applied Microbiology and Biotechnology* 58: 286–290. https://doi.org/10.1007/s00253-001-0899-y.

ICIS Chemical Business (2016). Bioamber Investment: Bio-Succinic Acid Makes the Big Time.

Il Bioeconomista (2013). Canada invests 12 million $ to support Bioamber Sarnia new bio-based succinic acid plant. https://ilbioeconomista.com/2013/09/23/canada-invests-12-million-to-support-bioamber-sarnia-new-bio-based-succinic-acid-plant/#:~:text=BioAmber%20Sarnia%E2%80%99s%20new%20bio-based%20succinic%20acid%20plant%2C%20the,Harper%20Government%20to%20support%20its%20construction%20in%20Sarnia

Lee, S.J., Song, H., and Lee, S.Y. (2006). Genome-based metabolic engineering of *Mannheimia succiniproducens* for succinic acid production. *Applied and Environmental Microbiology* 72 (3): 1939–1948. https://doi.org/10.1128/AEM.72.3.1939-1948.2006.

Lee, J.-S., Lin, C.-J., Lee, W.-C. et al. (2022). Production of succinic acid through the fermentation of *Actinobacillus succinogenes* on the hydrolysate of Napier grass. *Biotechnology for Biofuels and Bioproducts* 15: 9. https://doi.org/10.1186/s13068-022-02106-0.

Lemme, C.J. and Datta, R. (1987). Process for the production of succinic acid by anaerobic fermentation. Patent CA 1,339, 017.

Lempka, M. (2022). Research Success at Jacobs University: CO_2 Fixation in the Production of Succinic Acid. Press Release Jacobs University. https://idw-online.de/de/news802700 (accessed 14 October 2022).

Liebal, U.W., Blank, L.M., and Ebert, B.E. (2018). CO_2 to succinic acid – Estimating the potential of biocatalytic routes. *Metabolic Engineering Communications* 7: e00075. https://doi.org/10.1016/j.mec.2018.e00075.

Litsanov, B., Brocker, M., Oldiges, M., and Bott, M. (2014). Succinic acid. In: *Bioprocessing of Renewable Resources to Commodity Bioproducts* (ed. V.S. Bisaria and A. Kondo), 435–472. Wiley.

Liu, X., Zhao, G., Sun, S. et al. (2022). Biosynthetic pathway and metabolic engineering of succinic acid. *Frontiers in Bioengineering and Biotechnology* 10: 843887. https://doi.org/10.3389/fbioe.2022.843887.

Lomelí-Rodríguez, M., Corpas-Martínez, J.R., Willis, S. et al. (2018). Synthesis and characterization of renewable polyester coil coatings from biomass-derived isosorbide, FDCA, 1,5 pentanediol, succinic acid, and 1,3-propanediol. *Polymers* https://doi.org/10.3390/polym10060600.

Market Watch (2021). Succinic acid market trends future growth and revenue analysis forecast 2023.

Maximize Market Research, Global succinic acid market – industry analysis and forecast (2019–2027). https://www.maximizemarketresearch.com/market-report/global-succinic-acid-market/25839/ (accessed 29 October 2021).

McKinlay, J.B., Vielle, C., and Zeikus, J.G. (2007). Prospects for a bio-based succinate industry. *Applied Microbiology and Biotechnology* 76: 727–740. https://doi.org/10.1007/s00253-007-1057-y.

Mehlhase, S. (2018). Herstellung von Bio-Folien – Einblicke in die industrienahe. *Forschung* https://idw-online.de/de/news689767 (accessed 29 October 2021).

Meng, J., Wang, B., Liu, D. et al. (2016). High-yield anaerobic succinate production by strategically regulating multiple metabolic pathways based on stoichiometric maximum in *Escherichia coli*. *Microbial Cell Factories* 15: 141. https://doi.org/10.1186/s12934-016-0536-1.

Myriant (2015). *Production of organic acids by fermentation at low pH*. US Patent 2015/0240270, filed 14 Sep. 2012, Myriant Corporation, Quincy, MA, USA. Inventors: Yocum, R.R., Dole, S., and Pero, J.G.

Nghiem, N.P., Kleff, S., and Schwegmann, S. (2017). Succinic acid: technology and commercialization. *Fermentation* 3: 26. https://doi.org/10.3390/fermentation3020026.

Nielsen, J., Villadsen, J., and Liden, G. (2003). *Bioreaction Engineering Principles*, 104. Springer.

Raab, A.M., Gebhardt, G., Bolotina, N. et al. (2010). Metabolic engineering of Saccharomyces cerevisiae for the biotechnological production of succinic acid. *Metabolic Engineering* 518–525. https://doi.org/10.1016/j.ymben.2010.08.005.

Rafiqah, S.A., Khalina, A., Harmaen, A.S. et al. (2021). A review on properties and application of bio-based poly(butylene succinate). *Polymers* 13: 1436. https://doi.org/10.3390/polym13091436.

Ravenscroft, M. (2014). Bioplastics need to consider whole value chain, differentiate performance. European conference addresses industry challenges, IHS. *Chemical Week* 29–30.

Ravenscroft, M. (2019). Organic pigments: colors run to Asia. *Chemical Week* 23.

Rosales-Calderon, O. and Arantes, V. (2019). A review on commercial-scale high-value products that can be produced alongside cellulosic ethanol. *Biotechnology for Biofuels* 12: 240. http://doi.org/10.1186/s13068-019-1529-1.

Rosenboom, J.-G., Langer, R., and Traverso, G. (2022). Bioplastics for a circular economy. *Plastics* 7: 117–137. https://doi.org/10.1038/s41578-021-00407-8.

Ross, D.E., Flynn, J.M., Baron, D.B. et al. (2011). Towards electrosynthesis in *Shewanella*: energetics of reversing the Mtr pathway for reductive metabolism. *PLoS One* 6 (2): https://doi.org/10.1371/journal.pone.0016649.

Sauer, U. and Eikmanns, B.J. (2005). The PEP-pyruvate-oxalacetate node as the switch point for carbon flux distribution in bacteria. *FEMS Microbiology* 29: 209–226.

Snoep, J.L., de Graef, M.R., Westphal, A.H. et al. (1993). Differences in sensitivity to NADH of purified pyruvate dehydrogenase complexes of *Enterococcus faecalis*, *Lactobacillus lactis*, *Azetobacter vinelandii* and *Escherichia coli*: implication for their activity in vivo. *FEMS Microbiology Letters* 114: 279–283.

Spitz, P.H. (2019). Primed for success. In: *The Story of Scientific Design. How Chemical Engineers Created the Petrochemical Industry*. Springer 85ff.

Steiger, M.G., Mattanovich, D., and Sauer, M. (2017). Microbial organic acid production as carbon dioxide sink. *FEMS Microbiology Letters* 364 (21): https://doi.org/10.1093/femsle/fnx212.

Stringer, L. (2012). Germany's Lanxess to invest $10m in US-based bioamber. *ICIS* https://www.icis.com/explore/resources/news/2012/02/22/9535007/germany-s-lanxess-to-invest-10m-in-us-based-bioamber/ (accessed 29 October 2021).

Su, H.-Y., Li, H.-Y., Xie, C.-Y. et al. (2021). Co-production of acetoin and succinic acid by metabolically engineered *Enterobacter cloacae*. *Biotechnology for Biofuels* 14: 26. https://doi.org/10.1186/s13068-021-01878-1.

Tan, J.P., Luthfi, A.A.I., Manaf, S.F.A. et al. (2018). Incorporation of CO_2 during the production of succinic acid from sustainable oil palm fond juice. *Journal of CO_2 Utilization* 26: 595–601. https://doi.org/10.1016/j.jcou.2018.06.006.

Technip (2022). Press Release, Technip Energies acquires the Biosuccinium® technology from DSM for bio-sourced and fully biodegradable polymers production (accessed 1 September 2022).

Thoma, F., Schulze, C., Gutierrzez-Coto, C. et al. (2021). Metabolic engineering of *Vibrio natriegens* for anaerobic succinate production. *Microbial Biotechnology* 1–14. https://doi.org/10.1111/1751-7915.13983.

US Department of Energy (USDE) (2004). Top value added chemicals from biomass, Volume I: Results of screening for potential candidates from sugars and synthesis gas. http://www.osti.gov/bridge (accessed 13 February 2021).

US Department of Energy (USDE) (2010). Environmental assessment for the Myriant succinic acid biorefinery (MYSAB), Lake Provodence, Louisiana. https://www.energy.gov/sites/prod/files/nepapub/nepa_documents/RedDont/EA-1787-FEA-2010.pdf (accessed 29 October 2021).

Vemuri, G.N., Eitemann, M.A., and Altman, E. (2002). Effects of growth mode and pyruvate carboxylase on succinic acid production by metabolically engineered strains of *Escherichia coli*. *Applied and Environmental Microbiology* 68: 1715–1727.

Wu, J., Liu, H.J., Yan, X. et al. (2017). Co-production of 2,3-BDO and succinic acid using xylose by *Enterobacter cloacae*. *Journal of Chemical Technology and Biotechnology* 93: 1462–1467.

Wu, Z., Wang, J., Liu, J. et al. (2019). Engineering an electroactive Escherichia coli for the microbial electrosynthesis of succinate from glucose to CO_2. *Microbial Cell Factories* 18: 15. https://doi.org/10.1086/s12934-019-1067-3.

Xiao, Y., Zhang, Z., Wang, Y. et al. (2020). Two-stage crystallization combining direct succinimide synthesis for the recovery of succinic acid from fermentation broth. *Frontiers in Bioengineering and Biotechnology* https://doi.org/10.3389/fbioe.2019.00471.

Zhu, X., Tan, Z., Xu, H. et al. (2014). Metabolic evolution of two reducing equivalent-conserving pathways for high-yield succinate production in *Escherichia coli*. *Metabolic Engineering* 24: 87–96. https://doi.org/10.1016/j.ymben.2014.05.003.

13

Itaconic Acid

13.1 Application

Itaconic acid (methylenesuccinic acid) is an organic diacid with the formula $C_5H_6O_4$ and a molecular weight of 130.1 g/mol. The two acid functions with $pK_{a1} = 3.84$ and $pK_{a2} = 5.55$ at 25 °C explain the high acidity of the product (Cruz et al. 2018). Itaconic acid was on the list of the "Top value-added chemicals from Biomass" announced by the US Department of Energy in 2004 (US Department of Energy 2004; Bozell and Petersen 2010). The molecule was envisaged as "green" building block.

The two carboxylic acid functions define the most suited chemical functionality and, accordingly, itaconic acid is primarily applied to polymers. Copolymerization with a second or third monomer building block is presumably the main application. Pending on the chosen, e.g. bifunctional alcohols also defined polyesters can be generated. The application of itaconic acid copolymers covers various fields, including solvents, coatings, and binders (Magalhães et al. 2017). The saturation of the methylene group provides efficient access to an interesting methylated C4 building block whose carboxylic acid functions may also be hydrogenated to the aldehyde or the alcohol level. Additionally, the amination of the carboxylic acid function to either itaconic diamide or 2-methyl-1,4-butanediamine has been proposed.

13.2 Metabolic Engineering

13.2.1 Citrate Conversion via *cis*-Aconitate Toward Itaconate in *Aspergillus terreus* (*A. terreus*)

The standard route toward itaconate starts with citrate as the main intermediate of the citric acid cycle. Citrate can be efficiently converted into itaconate with two additional biochemical steps. These comprise the aconitase for the conversion of citrate into *cis*-aconitate via dehydration and the subsequent decarboxylation of *cis*-aconitate to itaconate facilitated by the *cis*-aconitate decarboxylase (Steiger et al. 2013). This enzyme is encoded by the *cadA* gene present in *A. terreus* (Table 13.1).

Pathway Design for Industrial Fermentation, First Edition. Walter Koch.

Table 13.1 Metabolic pathway toward itaconic acid.

Glucose + 2 ADP + 2 Pi + 2 NAD^+ → 2 Pyruvate + 2 ATP + 2 H_2O + 2 NADH + 2 H^+	Glycolysis
Pyruvate + CoA + NAD^+ → Acetyl-CoA + CO_2 + NADH + H^+	Pyruvate dehydrogenase
Pyruvate + CO_2 + ATP + H_2O → Oxalacetate + ADP + Pi	Pyruvate carboxylase
Oxalacetate + Acetyl-CoA + H_2O → Citrate + CoA	Citrate synthase
Citrate → *cis*-Aconitate + H_2O	*cis*-Aconitase
cis-Aconitase → Itaconic acid + CO_2	Aconitate decarboxylase
Glucose + ADP + Pi + 3 NAD^+ → Itaconic acid + CO_2 + ATP + H_2O + 3 NADH + 3 H^+	Net reaction

The theoretical maximum molar yield would amount to 1 mol itaconate per mol glucose. The C6 carbon and energy source is converted into the C5 product itaconate and 1 mol carbon dioxide. The pathway has a high C-atom efficiency since the carbon dioxide generated in the synthesis of acetyl-CoA is recycled by the pyruvate carboxylase. The release of carbon dioxide in the last step should pull the overall reaction cascade to the right side. This translates into a maximum carbon yield of 0.72 t itaconic acid/t glucose (130.1/180.2), which is according to the value indicated in literature (Kuenz and Krull 2018).

A higher theoretical maximum molar yield might be achieved since the overall reaction cascade provides a surplus of energy in the form of ATP and NADH. The 24 electrons carried by glucose (without counting reduced oxygen) are partially transferred to itaconic acid (18 electrons without counting reduced oxygen) and, accordingly, 3 reduction equivalents with 6 electrons remain. These byproducts might allow the conversion of carbon dioxide into formate, which might be rechanneled into the main metabolism and offer a higher theoretical molar yield. However, this would imply the implementation of an efficient carbon dioxide recycling pathway. Furthermore, a minor attrition rate of the theoretical maximum molar yield needs to be considered to address the formation of biomass and cellular maintenance.

Industrial fermentation toward itaconic acid employs *A. terreus* as eukaryotic fermentation host and is pursued as strict aerobic process (Klement and Büchs 2013). The oxygen demand is required to provide an electron acceptor for the surplus of reduction equivalents:

$$\text{NAD(P)H} + \text{H}^+ + 0.5\,\text{O}_2 \rightarrow \text{NAD(P)}^+ + \text{H}_2\text{O} \quad \Delta G^{\circ\prime} = -218\ \text{KJ/mol}$$

In case it is assumed that under realistic biochemical conditions about 2.5 mol ATP per mol NAD(P)H is obtained, the net reaction equation would be stated as follows (Voet et al. 2016):

$$\text{Glucose} + 8.5\,\text{ADP} + 8.5\,\text{Pi} + 1.5\,\text{O}_2 \rightarrow \text{Itaconic acid} + 8.5\,\text{ATP} + \text{CO}_2 + 11.5\,\text{H}_2\text{O}$$

Accordingly, 1.5 mol oxygen or 0.37 t oxygen per t itaconic acid needs to be supplied via the fermentation medium per mol itaconic acid. A stopping of aeration causes an interruption of itaconic acid production, which can only be restored after 24 h when aeration is resumed (Gyamerah 1995). It is plausible to assume that the oxygen supply is in the first instance needed as electron acceptor for the efficient regeneration of NAD^+ and especially to generate ATP for the maintenance of internal pH since the industrial fermentation process is pursued at an acidic pH between 2 and 3 (Gyamerah 1995). *A. terreus* is equipped with a native acid tolerance, which allows it to limit neutralization efforts and the generation of salt as by-product. However, since inhibitors of the plasma membrane ATPase do not negatively impact the synthesis of itaconic acid, it has been speculated that the plasma membrane is impermeable to H^+ and that the ATP demand might be related to the transportation of the product out of the cell (Gyamerah 1995).

The overall pathway toward itaconic acid proceeds in two biochemical compartments in the eukaryotic host *A. terreus*. Glucose metabolism including the synthesis of malate takes place in the cytosol. Pyruvate and malate as precursors for the citric acid cycle are then channeled into the mitochondrion, where *cis*-aconitate is generated as conventional intermediate of the citric acid cycle. Itaconate is then redirected into the cytosol and decarboxylated to itaconate. The product is secreted out of the cell into the fermentation broth (Figure 13.1).

13.2.2 Pathway Design: Citrate Conversion Toward *trans*-Itaconate in *Ustilago maydis* (*U. maydis*)

The standard fermentation host for the production of itaconate via citrate and *cis*-aconitate as pathway intermediates is *A. terreus*. A slightly different pathway toward itaconate has been elucidated in *U. maydis*. Pathway analysis based on the generation of fluorescent fusion proteins followed by fluorescence microscopy supports the model that *cis*-aconitate – the intermediate of the pathway in *Aspergillus niger* – is not directly decarboxylated to itaconate but isomerized to the thermodynamically more stable *trans*-aconitate via the cytosolic aconitate-Δ-isomerase encoded by the *adh1* gene (Geiser et al. 2016).

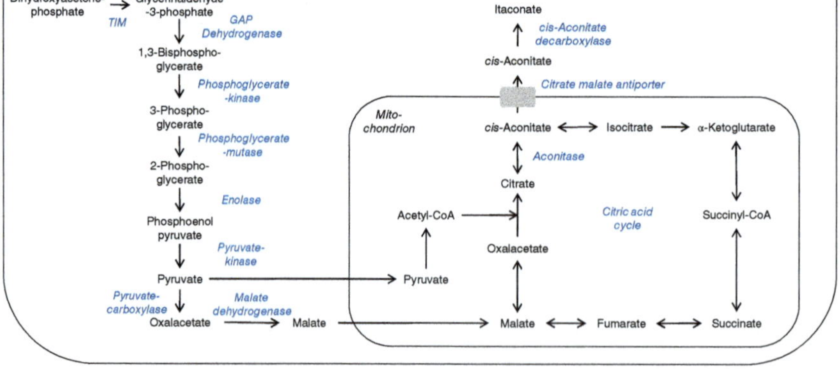

Figure 13.1 Metabolic pathway toward itaconic acid in *A. terreus* (Cruz et al. 2018; Kuenz and Krull 2018; Zhao et al. 2018). Source: Walter Koch.

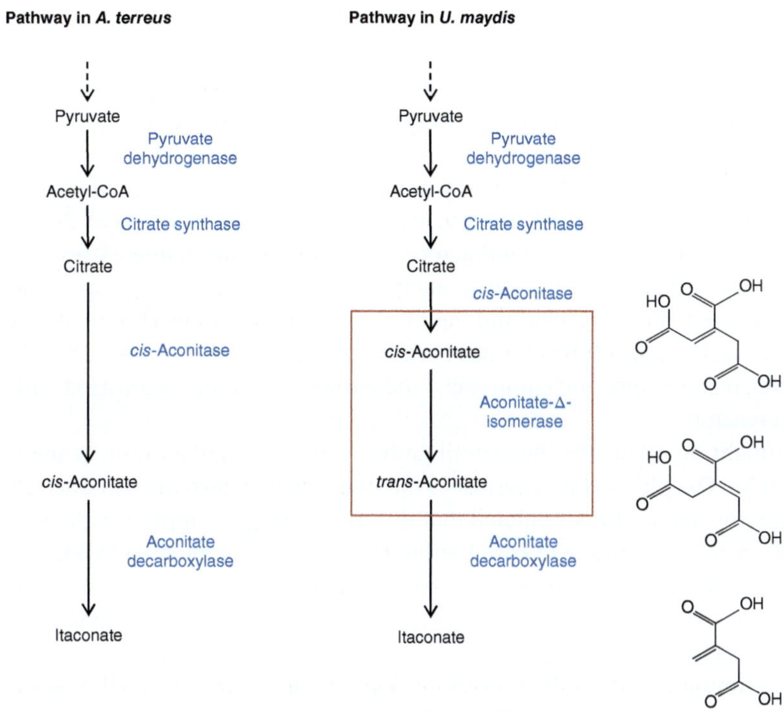

Figure 13.2 Conversion of *cis*-aconitate into itaconic acid (Geiser et al. 2016). Source: Walter Koch.

trans-Itaconate is then decarboxylated by the *trans*-itaconate decarboxylase encoded by the *tad1* gene. Itaconate is secreted into the fermentation broth (Figure 13.2).

These findings might give rise to more advanced metabolic engineering strategies in the future, though they do not yet offer an advantage over the conventional pathway.

13.3 Process Design

13.3.1 Fermentation with *A. terreus* as Host

From an industrial point of view, the fermentation technology for itaconic acid with *A. terreus* is established and can be regarded as rather mature. The best-published performance data indicate a titer of 129 g/l after a batch duration of 4.7 d, which translates into a productivity of 1.14 g/lxh (Hevekerl et al. 2014). The achieved carbon yield of 0.58 t itaconic acid/t glucose amounts to 80% of the maximum carbon yield of 0.72 t itaconic acid/t glucose (Hevekerl et al. 2014). It is to be expected that the fermentation performance values achieved in industrial practice – which are usually not published – are even higher. From this perspective, the search for alternative fermentation hosts seems less attractive.

However, there are still some reasons to evaluate alternatives. First of all, a very high titer above 200 g/l of citric acid is reported in the closely related *A. niger* host, which might propose a comparable threshold for itaconic acid in *A. terreus*. Additionally, the fungal morphology of *A. terreus* needs to be closely monitored. A short and swollen fungal hypha is likely ideal for the process. The reproducibility of fermentation runs is weak, and a high manganese ion concentration has a detrimental effect (Vuoristo et al. 2015). Another issue with *A. terreus* as host is the critical supply of oxygen. An interruption of oxygen supply in the aerobic process hampers production, and recovery of maximum productivity is difficult. A limited oxygen transfer rate (OTR) reflects the morphology of filamentous fungi. This might technically be overcome with a high energy input from the stirring device, though it might damage the cell morphology and distort the mycelia structure (Vuoristo et al. 2015). An efficient oxygen supply needs to be counterbalanced by the limited mechanical stress tolerance of the filamentous fungi *A. terreus*. A high energy input might negatively impact the cell integrity, partially destroy the mycelia, and thereby hamper productivity (Klement and Büchs 2013). Alternatively, an airlift reactor has been proposed (Okabe et al. 1993; Yahiro et al. 1997). However, it is not known if commercial industrial plants employ a stirred tank reactor or an airlift reactor.

These obstacles might be related to the overall pathway and its regulation or define a host-specific constraint. Only in the latter case does the switch to an alternative production host promises an improvement. It is speculated that the critical oxygen supply is explained by an inhibitory effect of *cis*-aconitate in case the intermediate is accumulated inside the cytosol (Cruz et al. 2018). In case of sufficient oxygen supply, *cis*-aconitate is efficiently transported into the cytosol via the Mtt transporter, which has an energy demand. In case of oxygen undersupply, the energy demand of this transporter might be impaired and *cis*-aconitate accumulates within the mitochondrion. There it is transformed into the thermodynamically more stable *trans*-aconitate, which has been described as an inhibitor of the enzymes aconitase (Laube et al. 1994) and fumarase (Rebholz and Northrop 1994). Thus, several studies with other hosts like *A. niger*, *Escherichia coli*, and *Corynebacterium glutamicum* are reported, though the indicated performance data are by far below the threshold of *A. terreus* (Steiger et al. 2017; Kuenz and Krull 2018; Cruz et al. 2018).

13.3.2 *U. maydis* as Alternative Production Host

Since the filamentous morphology of the standard host *A. terreus* for the fermentative production of itaconic acid presents obstacles in the production process and is sensitive toward mechanical stress, alternative hosts are evaluated. The basidiomycete fungal pathogen of maize, *U. maydis*, is one of the main candidates and promising results have been obtained. The pH sensitivity of the host causes a metabolic shift from acid production to polyol and glycolipids and a tight pH control is recommended (Demir et al. 2021). A marker-free chassis strain was obtained, which displayed a titer of 19.4 g/l itaconic acid with carbon feed of 50 g/l in the medium buffered with 100 mM MES. Further improvement was achieved with pH management and neutralization (Becker et al. 2019). The production of itaconic acid caused a pH decrease from 6.5 down to 3.7, which negatively impacts productivity since *U. maydis* is not tolerant to low pH and MES has been replaced by 66 g/l $CaCO_3$. The insoluble $CaCO_3$ provides a high buffer capacity and, additionally, an *in situ* product recovery because the solubility of calcium itaconate is limited to 14.5 g/l. The pH was kept between 6 and 7, and the glucose concentration was raised to 100 g/l. After full conversion of glucose, 43.7 g of itaconic acid/l was obtained after 194 h representing a productivity of 0.22 g/lxh (Becker et al. 2019). The potential use of *U. maydis* is still facing some hurdles given the generation of by-products. The R&D focus has shifted to strain engineering to channel metabolism into the target product and reduce the generation of by-products such as malic acid, triacylglyceroles, ustilagic acid, and mannosyl-erythritol lipids. Their abundance could be decreased using CRISPR/Cas9 and FLP/FRT (Becker et al. 2020).

13.3.3 *Pseudomonas putida* as Production Host with Pretreated Lignin as Feedstock

It is evaluated to apply pretreated lignin as fermentation feedstock given its high availability at a comparably low price. Lignin is often traded according to its energy value since it is still difficult to efficiently crack lignin into defined aromatic compounds with reasonable purity. An interesting lignin valorization approach foresees cracking lignin via chemical treatment into a mixture of aromatic compounds, which is fed to a fermentation host. The host is supposed to metabolize the various aromatic compounds into a few precursors, which can then be channeled into a related target product or into the citric acid cycle and thereby into the central metabolism of the host (Iram et al. 2021; Vardon et al. 2015). In this context, it has been tested to use the saprophytic bacterium *P. putida* as host for the whole-cell synthesis of itaconic acid with p-coumarate or mixed monomeric aromatics as feedstock. An itaconic acid titer of 1.3 and 1.4 g/l, respectively, after a batch duration of 48 h could be obtained (Elmore et al. 2021). These experiments might provide the path for the commercial usage of pretreated lignin fractions as feedstock for fermentation (Figure 13.3).

13.3.4 *A. terreus* as Production Host with Xylose as Feedstock

The application of a second-generation feedstock proposes to use pretreated lignocellulosic biomass as fermentation feedstock. However, the pretreatment step is known to generate byproducts like furfural, 5-hydroxymethylfurfural (HMF), and acetic acid, which might hamper the fermentation process. Especially xylose might be obtained as C5 sugar in the treatment of hemicellulose besides glucose derived in the hydrolysis of cellulose. Accordingly, fermentation efficiency in the presence of these contaminations was measured and revealed that a concentration of 188 mg/l acetic acid, 175 mg/l furfural, and 700 mg/l 5-hydroxymethylfurfural (HMF) led to a reduction of the itaconic acid titer. It was also found out that xylose can be exhaustively consumed, though the conversion into itaconic acid is lower than that achieved with glucose (Magalhães et al. 2020). It can be concluded that the conventional *A. terreus* host is in principle suited for the use of xylose and glucose mixtures derived from lignocellulosic biomass. The byproducts might need to be removed, and the xylose metabolism needs to be improved to ensure a productivity level achieved with glucose as feedstock.

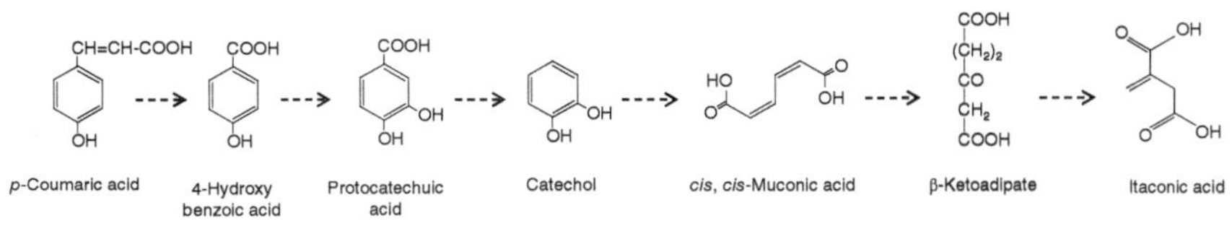

Figure 13.3 Pathway for the conversion of p-coumaric acid into itaconic acid. β-Ketoadipate would presumably be cleaved into succinyl-CoA and acetyl-CoA and both would be further metabolized via the citric acid cycle (Vardon et al. 2015; Beckham et al. 2016). Source: Walter Koch.

13.3.5 Overview of Achieved Fermentation Performance Data

An overview of the published technology performance data indicates that remarkable results have been achieved, represented by carbon yields of 0.58 g itaconic acid per g glucose and 0.64 g itaconic acid per g glucose, respectively. The solubility of itaconic acid in water has a maximum threshold of 83.1 g/l at 20 °C, which is exceeded in several cases in the reported titers at the batch end (Magalhães et al. 2017). This is explained by the high-temperature dependency of the solubility; the highest reported titer of 160 g/l was measured at a temperature of 35 °C, and the formation of crystals is not mentioned (Krull et al. 2017; Table 13.2).

13.3.6 Downstream Processing

Crystallization is the preferred technology for the isolation and purification of itaconic acid after the fermentation process. The broth is filtered at the batch end to remove the biocatalyst and other insoluble components in the media, and the filtered broth is then channeled into the crystallization unit. The formation of crystals may be achieved with cooling or evaporation crystallization. An industrial downstream process has been described (Lockwood 1975), which indicates that the product concentration is raised up to 350 g/l itaconic acid via evaporation and that crystal formation is triggered by a cooling step at 15 °C. The crystals are harvested and channeled into the next section of the process, and the mother liquor with a considerable product concentration is recycled (Figure 13.4).

Since byproducts like succinic acid, malic acid, and α-ketoglutarate are to some extent captured in the crystallization process and disturb the product quality a solubilization is needed, and altogether two crystallization steps (with intermediate solubilization) are pursued. The overall process yield has been reported at 80% for an industrial production including crystallization (Okabe et al. 2009). Alternative isolation technologies comprising precipitation, liquid–liquid extraction, membrane separation, and adsorption have been evaluated, though all suffer from limitations given the high affinity of the molecule in aqueous solution and the complex composition of the broth (Magalhães et al. 2017). Furthermore, an *in situ* product recovery (Freeman et al. 1993) has been tested using

Table 13.2 Fermentation performance data for itaconic acid.

Host	Titer (g/l)	Batch-time (d)	Productivity (g/lxh)	Carbon yield	Source
A. terreus	160	6.7	1.00	0.46 g/gglc	Krull et al. (2017)
A. terreus	129	4.7	1.14	0.58 g/gglc	Hevekerl et al. (2014)
A. terreus	95	11	0.36	0.64 g/gglc	Karaffa et al. (2015)
A. terreus	86	7	0.51	0.62 g/gglc	Kuenz et al. (2012)
U. maydis	113	12.7	0.37	0.38 g/gglc	Demir et al. (2021)
U. maydis	88.4	4.7	0.78	0.41 g/gglc	Demir et al. (2021)
U. maydis	45	6	0.31	0.24 g/gglc	Maassen et al. (2014)
U. maydis	20	3	0.28	n.a.	Klement et al. (2012)

Source: Adapted from Magalhães et al. (2017).

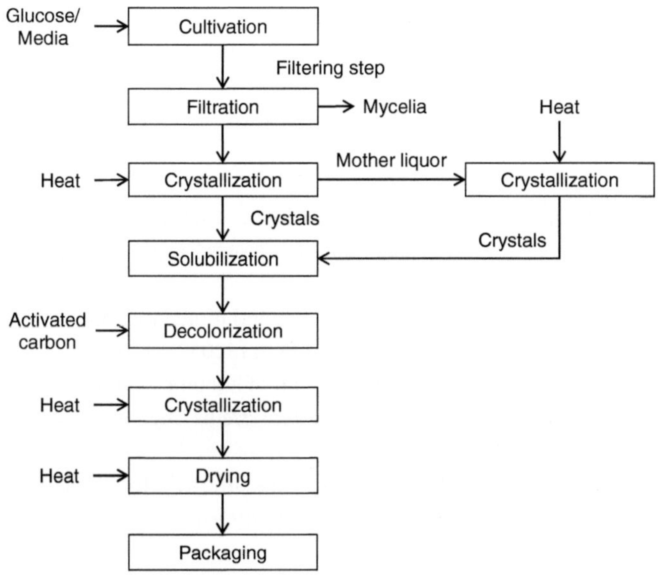

Figure 13.4 Downstream process for the purification of itaconic acid. Source: Adapted from Magalhães et al. (2017).

trioctylamine and isopropyl myristate as extractants, which are added to the broth during cultivation. The product titer was increased with the addition of an extractant from 70 up to 105 g/l, referring to the absolute amount of product divided by the total broth volume comprising the aqueous phase and the organic phase of the extractant. The productivity increased from 0.72 to 0.91 g/lxh (Kreyenschulte et al. 2018).

13.4 Industrial Application and Capacity

Itaconic acid was first produced by Pfizer in 1945, followed by Iwata Chemicals in 1970, Rhodia in 1995, and Cargill in 1996. Due to the enhanced market entry of Chinese producers, Cargill, Pfizer, and Rhodia stepped out of the business, and Chinese companies serve a major share of the market. Chinese producers include Qingdao Kehai Biochemistry, Chengdu Jindai Biology Engineering, Jinan Huaming Biochemistry, and Shandong Kaison Biochemistry (Cruz et al. 2018). The market size is estimated at 80 kt pa, and the average market price is estimated at 2 US$/kg itaconic acid (Kuenz and Krull 2018; Steiger et al. 2017). For comparison, the report of the US Department of Energy assumed in 2004 that the fermentative product needs to achieve a cost position of about 0.55 US$ per kg ($ 0.25/pound, US Department of Energy 2004), which is hardly possible if the maximum molar yield is taken into consideration.

References

Becker, J., Tehrani, H.H., Gauert, M. et al. (2019). An *Ustilago maydis* chassis for itaconic acid production without by-products. *Microbial Biotechnology* 13: 350–363.

Becker, J., Tehrani, H.H., Gauert, M. et al. (2020). An *Ustilago maydis* chassis for itaconic acid production without by-products. *Microbial Biotechnology* 13 (2): 350–362. https://doi.org/10.1111/1751-7915.13525.

Beckham, G.T., Johnson, C.W., Karp, E.M. et al. (2016). Opportunities and challenges in biological lignin valorization. *Current Opinion in Biotechnology* 42: 40–53. https://doi.org/10.1016/j.copbio.2016.02.030.

Bozell, J.J. and Petersen, G.R. (2010). Technology development for the production of biobased products from biorefinery carbohydrates – the US department of energy's "Top 10" revisited. *Green Chemistry* 12: 539–554.

Cruz, J.C., de Castro, A.M., and Sérvulo, E.F.C. (2018). World market and biotechnological production of itaconic acid. *3Biotech* 8: 138. https://doi.org/10.1007/s13205-018-1151-0.

Demir, H.T., Bezirci, E., Becker, J. et al. (2021). High level production of itaconic acid at low pH by *Ustilago maydis* with fed-batch fermentation. *Bioprocess and Biosystems Engineering* 44: 749–758. https://doi.org/10.1007/s00449-020-02483-6.

Elmore, J.R., Dexter, G.N., Salvachúa, D. et al. (2021). Production of itaconic acid from alkali pretreated lignin by dynamic two stage bioconversion. *Nature Communications* 12: 2261. https://doi.org/10.1038/s41467-021-22556-8.

Freeman, A., Woodley, J.M., and Lilly, M.D. (1993). *In situ* product removal as a tool for bioprocessing. *Nature* 11: 1007–1012. The paper is the first publication on *in situ* product recovery in fermentation processes.

Geiser, E., Pryzbilla, S.K., Friedrich, A. et al. (2016). *Ustilago maydis* produces itaconic acid via the unusual intermediate trans-aconitate. *Miciobial Biotechnology* 9 (1): 116–126.

Gyamerah, M.H. (1995). Oxygen requirement and energy relations of itaconic acid fermentation by *Aspergillus terreus* NRRL 1960. *Applied Microbiology and Biotechnology* 44: 20–26.

Hevekerl, A., Kuenz, A., and Vorlop, K.-D. (2014). Influence of the pH on the itaconic acid production with *Aspergillus terreus*. *Applied Microbiology and Biotechnology* 98: 10005–10012. https://doi.org/10.1007/s00253-014-6047-2.

Iram, A., Berenjian, A., and Demirci, A. (2021). A review on the utilization of lignin as a fermentation substrate to produce lignin-modifying enzymes and other value-added products. *Molecules* 26: 2960. https://doi.org/10.3390/molecules26102960.

Karaffa, L., Diaz, R., Papp, B. et al. (2015). A deficiency of manganese ions in the presence of high sugar concentrations is the critical parameter for achieving high yields of itaconic acid by *Aspergillus terreus*. *Applied Microbiology and Biotechnology* 99: 7937–7944. https://doi.org/10.1007/s00253-015-6735-6.

Klement, T. and Büchs, J. (2013). Itaconic acid – A biotechnological process in change. *Bioresource Technology* 135: 422–431.

Klement, T., Milker, S., Jäger, G. et al. (2012). Biomass pretreatment affects Ustilago maydis in producing itaconic acid. *Microbial Cell Factories* 11: 43.

Kreyenschulte, D., Heyman, B., Eggert, A. et al. (2018). In situ reactive extraction of itaconic acid during fermentation of *Aspergillus terreus*. *Biochemical Engineering Journal* 135: 133–141. https://doi.org/10.1016/j.bej.2018.04.014.

Krull, S., Hevekerl, A., Kuenz, A., and Prüße, U. (2017). Process development of itaconic acid production by a natural wild type strain of *Aspergillus terreus* to reach industrially relevant final titers. *Applied Microbiology and Biotechnology* 101: 4063–4072. https://doi.org/10.1007/s00253-017-8192-x.

Kuenz, A. and Krull, S. (2018). Biotechnological production of itaconic acid. *Applied Microbiology and Biotechnology* 102: 3901–3914.

Kuenz, A., Gallemüller, Y., Willke, T., and Vorlop, K.-D. (2012). Microbial platform of itaconic acid: developing a stable platform for high product concentrations. *Applied Microbiology and Biotechnology* 96: 1209–1216. https://doi.org/10.1007/s00253-012-4221-y.

Laube, H., Kennedy, M.C., and Beinert, H. (1994). Crystal structures of aconitate with trans-aconitate and nitrocitrate bound. *Journal of Molecular Biology* 237: 437–451.

Lockwood, L.B. (1975). Production of organic acids by fermentation. Part 3: Itaconic acid. *Microbial Technology* 367–373.

Maassen, N., Panakova, M., Wierckx, N. et al. (2014). Influence of carbon and nitrogen concentration on itaconic acid production by the smut fungus *Ustilago maydis*. *Engineering in Life Sciences* 14: 129–134. https://doi.org/10.1002/elsc.201300043.

Magalhães, A.I., de Carvalho, J.C., Medina, J.D.C., and Soccol, C.R. (2017). Downstream process development in biotechnological itaconic acid manufacturing. *Applied Microbiology and Biotechnology* 101: 1–12. https://doi.org/10.1007/s00253-016-7972-z.

Magalhães, A.I., de Carvalho, J.C., Thoms, J.F. et al. (2020). Second-generation itaconic acid: an alternative product for biorefineries? *Bioresource Technology* 308: 123319. https://doi.org/10.1016/j.biortech.2020.123319.

Okabe, M., Ohta, N., Park, Y., and (1993). Itaconic acid production in air-lift bioreactor using a modified draft tube. *Journal of Fermentation and Bioengineering* 76 (2): 117–122.

Okabe, M., Lies, D., Kanamasa, S., and Park, E.Y. (2009). Biotechnological production of itaconic acid and its biosynthesis in *Aspergillus terreus*. *Applied Microbiology and Biotechnology* 84: 597–606. https://doi.org/10.1007/s00253-009-2132-3.

Rebholz, K. and Northrop, D. (1994). Kinetics of enzymes with iso-mechanisms: dead-end inhibition of fumarase and carbonic anhydrase II. *Archives of Biochemistry and Biophysics* 312: 227–233.

Steiger, M.G., Blumhoff, M.L., Mattanovich, D., and Sauer, M. (2013). Biochemistry of microbial acid production. *Frontiers in Microbiology* 4: 23. https://doi.org/10.3389/fmicb.2013.00023.

Steiger, M., Wierckx, N., Blank, L.M. et al. (2017). Itaconic acid – an emerging building block, p. 465–466. In: *Industrial Biotechnology. Products and Processes* (ed. C. Wittmann and J.C. Liao), 453–472. Wiley.

US Department of Energy (2004). *Top Value Added Chemicals From Biomass, Volume I: Results of Screening for Potential Candidates from Sugars and Synthesis Gas*, 44. http://www.osti.gov/bridge (accessed 13 February 2021).

Vardon, D.R., Franden, M.A., Johnson, C.W. et al. (2015). Adipic acid production from lignin. *Energy & Environmental Science* 8: 617–628. 03230f, https://doi.org/10.1039/c4ee.

Voet, D., Voet, J.G., and Pratt, C.W. (2016). *Principles of Biochemistry*, 594. Wiley.

Vuoristo, K.S., Mars, A.E., Sangra, J.V. et al. (2015). Metabolic engineering of itaconate production in *Escherichia coli*. *Applied Microbial Biotechnology* 99: 221–228.

Yahiro, K., Takahama, T., Jai, S.-R. et al. (1997). Comparison or air-lift and stirred tank reactors fir itaconic acid production in *Aspergillus terreus*. *Biotechnology Letters* 19 (7): 619–621.

Zhao, M., Lu, X., Zong, H. et al. (2018). Itaconic acid production in microorganisms. *Biotechnology Letters* 40: 455–464. https://doi.org/10.1007/s10529-017-2500-5.

14

Glutamic Acid

14.1 Application

L-Glutamic acid (2-aminoglutaric acid) is a non-branched, aliphatic amino acid with the formula $C_5H_9NO_4$ and a molecular weight of 147.1 g/mol. It is one of the 20 natural amino acids present in proteins and enzymes. It has three functional groups, comprising an amino function with a pK_a value of 9.47 and two carboxylic acid functions with pK_a values of 2.10 and 4.07. L-Glutamic acid is especially used in Asia as food additive due to its ability as taste enhancer in the form of monosodium glutamate (MSG). The molecule is exclusively produced by fermentation, relying on the efficient generation and secretion by the bacterium *Corynebacterium glutamicum.* A chemical synthesis is possible though can be neglected. The global production of L-glutamic acid is estimated with 2.3 Mio t pa (Schmid 2016). Thereof, the market value of the biggest product grade MSG is expected to reach USD 5.8 bn in 2031 (PR Newswire 2022).

The synthesis of γ-PGA (poly-γ-glutamic acid) might develop as an interesting growth and application field for L-glutamate. It consists of L-glutamate and D-glutamate monomers coupled by peptide bonds and achieves a molecular weight of 10 kDa up to 2000 kDa. The peptide bonds are unusual since the α-amino group of one monomer reacts with the γ-carboxylic acid group of the next monomer instead of the α-carboxylic acid group. γ-PGAs were isolated for the first time isolated from the capsule of *Bacillus anthracis* and several established hosts are evaluated as potential production biocatalysts. The molecule might deserve more R&D attention in case where unique polymer properties in different industry fields counterbalance the non-optimized cost position (Li et al. 2022).

Furthermore, glutamate in the form of glutamic acid or monosodium glutamate might be applied as starting point for the biocatalytic production of γ-aminobutyric acid (GABA) (Dhakal et al. 2012). It is known that immobilized glutamic acid when incubated with the *Escherichia coli*-derived glutamate decarboxylase efficiently generates GABA via decarboxylation (Lee et al. 2013).

$$\text{Glutamic acid} \rightarrow \text{GABA} + CO_2$$

It would technically be easier and most cost-efficient if the fermentative production of GABA could be achieved directly from glucose in a one-step reaction. The glutamate decarboxylase has successfully been expressed in *C. glutamicum* and given a titer of 38.6 g/l and a productivity of 0.54 g/lxh commercially promising performance data were obtained (Choi et al. 2015).

Pathway Design for Industrial Fermentation, First Edition. Walter Koch.

14.2 Native Biochemical Pathway

14.2.1 Glutamic Acid via Reductive Amination of α-Ketoglutarate

The history of glutamate is linked to the birth of the Japanese company Ajinomoto, a major producer of glutamic acid. The chemist Kikunae Ikeda elucidated in 1908 that the decisive taste component in the traditional Japanese kombu dashi (kelp broth) is glutamic acid and managed to extract glutamate from kelp broth (Ajinomoto 2022). It has then in 1957 accidentally discovered that the bacterium *Micrococcus glutamicus* is able to produce and secrete glutamic acid (Kinoshita et al. 1957a, b). This host still defines the basis for the current industrial production of glutamate. The bacterium has in between been renamed into *C. glutamicum*. It is a rod-shaped, gram-positive, nonspore-forming bacterium, which grows aerobically and is characterized by a biotin auxotrophy. It is one of the leading workhorses in industrial biotechnology, and the product spectrum has been expanded from originally L-glutamate, L-lysine, L-arginine, and L-tryptophan to more than 30 value products (Wolf et al. 2021; Fahimitabar et al. 2021; Becker and Wittmann 2015). The host does not produce endotoxins, does not undergo phage lysis, and is generally recognized as safe (GRAS). The GRAS status has been granted by the United States Food and Drug Administration for the food and pharmaceutical industry (Wolf et al. 2021). L-Glutamate in the form of (protonated) L-glutamic acid and monosodium L-glutamate monohydrate produced by *C. glutamicum* is generally regarded as safe for feed and food applications (EFSA Panel 2022).

The fermentative production is following the natural ability of *C. glutamicum* to secrete glutamate and is using the native pathway. A metabolic engineering strategy was not required. The nitrogen supply is ensured via the addition of NH_3 into the medium. The cellular uptake is supported by a specific secondary active carrier in the form of NH_4^+ encoded by the *amt* gene (Siewe et al. 1996). Glucose is conventionally converted via glycolysis and the pyruvate dehydrogenase to feed the citric acid cycle with acetyl-CoA. The second substrate of the citrate synthase, oxalacetate, is partially provided via the anaplerotic carboxylation of pyruvate facilitated by the pyruvate carboxylase (Peters-Wendisch et al. 1998). Additionally, a share of the oxalacetate pool is provided via the PEP carboxylase (O'Regan et al. 1989) with PEP as starting molecule (Eggeling and Sahm 1999). Based on experiments with ^{13}C-labeled glucose fed under biotin limitation, it is believed that the major share of oxalacetate is provided via the PEP carboxylase (Hirasawa and Wachi 2017). The generation of α-ketoglutarate follows the citric acid cycle, and glutamate is finally derived via the reductive amination of the α-keto function (Table 14.1).

It has already been assumed by Japanese researchers in 1957 that *C. glutamicum* derives glutamate via the reductive amination of α-ketoglutarate (Chao and Foster 1959). The conversion of α-ketoglutarate into glutamate via the glutamate dehydrogenase displays a sufficiently exergonic Gibbs free energy of $\Delta G^{\circ\prime}$ −30 KJ/mol (Voet et al. 2016). The glutamate dehydrogenase in *C. glutamicum* uses NADPH as cofactor, which is provided by the preceding conversion of isocitrate into α-ketoglutarate (Tryfona and Bustard 2004). A transhydrogenation between NADH and NADPH in *C. glutamicum* is not known (Takaç et al. 1998). The mechanism of the reductive amination has been understood in detail. The keto function reacts with ammonium to form α-iminoglutarate as intermediate, which receives a hydride ion from NAD(P)H (McMurry and Begley 2006).

Table 14.1 Biochemical pathway toward glutamic acid.

Provision of acetyl-CoA	
Glucose + 2 ADP + 2 Pi + 2 NAD^+ → 2 Pyruvate + 2 ATP + 2 H_2O + 2 NADH+ 2 H^+	Glycolysis
Pyruvate + CoA + NAD^+ → Acetyl-CoA + CO_2 + NADH + H^+	Pyruvate dehydrogenase
Synthesis of citrate	
Pyruvate + CO_2 + ATP + H_2O → Oxalacetate + ADP + Pi	Pyruvate carboxylase
Oxalacetate + Acetyl-CoA + H_2O → Citric acid + CoA	Citrate synthase
Conversion of citrate into glutamate	
Citric acid → Isocitrate	Aconitase
Isocitrate + $NADP^+$ → α-Ketoglutarate + CO_2 + NADPH + H^+	Isocitrate dehydrogenase
α-Ketoglutarate + NH_3 + NADPH + H^+ → Glutamate + $NADP^+$ + H_2O	Glutamate dehydrogenase
Glucose + ADP + Pi +3 $NAD(P)^+$ + NH_3 → Glutamate + CO_2 + ATP + H_2O + 3 NAD(P)H + 3 H^+	

Table 14.2 Ammonia fixation via glutamine.

Glutamate + NH_3 + ATP + H_2O → Glutamine + ADP + P_i	Glutamine synthetase
α-Ketoglutarate + Glutamine + NAD(P)H + H^+ → 2 Glutamate + $NADP^+$ + H_2O	Glutamate synthase
α-Ketoglutarate + NH_3 + ATP + NAD(P)H + H^+ → Glutamate + $NADP^+$ + ADP	

The glutamate dehydrogenase has a low affinity for NH_4^+ and, accordingly, a high K_M value. At excess nitrogen supply (>1 mmol/l; Tryfona and Bustard 2004), ammonia fixation is mainly facilitated by the glutamate dehydrogenase (Tryfona and Bustard 2004). In case where the NH_4^+ concentration is low (<1 mmol/l; Tryfona and Bustard 2004), glutamate synthesis proceeds preferentially via the sequential reactions of the glutamine synthetase and the glutamate synthase (Table 14.2).

The glutamine synthesis requires an energy input and one mol ATP per fixed mol ammonium is consumed. Glutamine and α-ketoglutarate would then partner up in a transamination reaction and obtain two mol glutamate. The net generation of glutamate requires one reduction equivalent and one mol ATP. The glutamine synthetase has a high affinity for NH_4^+ and ensures an ammonia fixation at low concentrations.

The formation of glutamate competes with the conversion of α-ketoglutarate into succinate, facilitated by the α-ketoglutarate dehydrogenase, which would be the next usual step in the citric acid cycle. It has been found out that the suited production strains of *C. glutamicum* display a weakened activity of the α-ketoglutarate dehydrogenase. The activity of the α-ketoglutarate dehydrogenase (called ODHC in *C. glutamicum*) is subjected to a more complex regulation mechanism, which is not yet fully elucidated.

ODHC consists of a catalytic subunit E1o, a dihydrolipoamide acetyltransferase unit E2, and a dihydrolipoamide dehydrogenase unit E3. The ODHC (the E1o subunit) is inhibited by the unphosphorylated OdhI protein, which was discovered in *C. glutamicum* in 2006 (Niebisch et al. 2006). OdhI is phosphorylated by the *pknG* gene, which encodes a serine/threonine kinase. There are indications that the de novo synthesis of OdhI is needed for the penicillin-induced glutamate production by *C. glutamicum,* though the full regulation pattern has not yet been fully understood (Hirasawa and Wachi 2017). It is crucial that the citric acid cycle is still operating, though the main share of α-ketoglutarate is channeled into the synthesis of glutamate (Figure 14.1).

The reaction equation reveals that the maximum molar yield amounts to 1.0 mol glutamate per mol glucose. The C6 raw material glucose feedstock is converted into one mol C5 product, glutamate, and one mol carbon dioxide. Accordingly, the maximum carbon yield

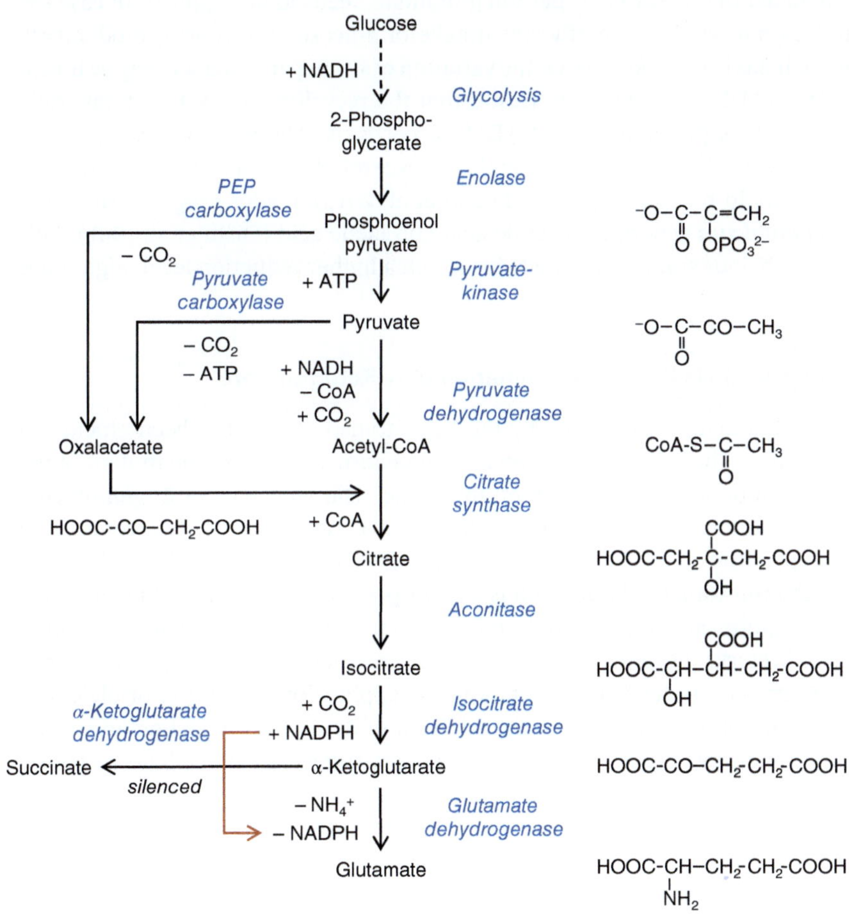

Figure 14.1 Biochemical pathway toward glutamic acid. The formation of oxalacetate might also be facilitated via the PEP carboxykinase, which conserves the energy content of PEP in the form of ATP. Source: Walter Koch.

is 0.82 ton glutamic acid per ton glucose (147.1/180). The biochemical pathway does also generate one mol ATP, which fuels cellular growth and cellular maintenance. Additionally, three reduction equivalents are formed since the 24 electrons of glucose (without counting the reduced oxygen) are allocated to one glutamate with 18 electrons (without counting the reduced oxygen and nitrogen) and 6 electrons in the biochemical form of 3 mol reduction equivalents remaining. The surplus of reduction equivalents implies that additional oxygen, besides glucose, needs to be offered as final electron acceptor (Voet et al. 2016).

$$\mathrm{NAD(P)H + H^+ + 0.5O_2 \rightarrow NAD(P)^+ + H_2O \Delta G^{\circ\prime} = -218KJ/mol}$$

The overall reaction equation can be stated as follows in case NAD^+ is fully recycled via the formation of water:

$$\mathrm{Glucose + 1.5\,O_2 + NH_3 \rightarrow Glutamate + CO_2 + 3\,H_2O}$$

It can be concluded that 0.32 ton O_2 per ton glutamate needs to be supplied. In case the reactor setup does not facilitate an efficient uptake of glucose and oxygen, productivity would be limited. It has been reported that the variation of strain performance may be linked to a lower NAD^+/NADH ratio, which indicates that the recycling of NAD^+ is temporally behind the formation of glutamate and NADH, respectively. The performance variation could be improved with the addition of sorbitol and glycerol, though a conclusive interpretation of the data could not yet be provided (Cao et al. 2015). Under oxygen-insufficient conditions, *C. glutamicum* generates succinic acid and lactic acid (Oishi et al. 1961). This observation is not obvious since by-products for which a higher reduction level might have been expected.

14.2.2 Glutamic Acid via Reductive Amination of α-Ketoglutarate

The secretion of glutamate is facilitated by the YggB-transporter. It has been elucidated that the commercial strains of *C. glutamicum* are characterized by a specific mutant of the transporter, which supports the secretion of glutamate. The wild type of *C. glutamicum* does not secrete glutamate (Eggeling and Sahm, 1999). Additionally, the secretion of glutamate is also pending on the cell permeability, which is impacted by the level of the cell membrane and the cell wall. *C. glutamicum* is a gram-positive bacterium that belongs to a well-separated subgroup characterized by a unique cell wall organization that is not found in Gram-positive bacteria (Eggeling and Sahm 2001). An efficient efflux of glutamate may be achieved with growth under biotin limitation, the application of glycerol or oleic acid auxotrophs, the addition of penicillin, or the addition of detergents (Eggeling and Sahm 1999). Since the execution of these measures does not trigger an immediate response, though the cells need to be incubated under the respective conditions, it was concluded that the secretion mechanism is impacted via an alteration of cell composition (Eggeling and Sahm 1999). The performance of the cell membrane is influenced by the availability of phospholipids, whose presence can be managed by the generation of malonyl-CoA and subsequently fatty acids. The robustness of the cell wall is weakened by penicillin, which may be added as media component to improve productivity. It is assumed that the biotin limitation in the media negatively impacts the activity of carboxylation enzymes in the synthesis of fatty acids and thereby interferes with fatty acid synthesis on a molecular level (Nampoothiri et al. 2002).

14.3 Metabolic Engineering

14.3.1 Finetuning of the Native Pathway

The first industrial strains of *C. glutamicum* applied for the synthesis of glutamate were not subject of a metabolic engineering project according to current standards. The respective molecular biology tools were not available in the first period of commercial production from the 1960 teeth onwards. The technology relied on the natural ability of *C. glutamicum* and metabolic engineering was restricted to random mutation and selection to achieve further improvement.

There are the first activities to establish the methylotrophic host *Bacillus methanolicus* as innovative production host for glutamate based on its natural ability to generate and secrete glutamate up to 59 g/l under fed-batch conditions (Frank et al. 2021; Heggeset et al. 2012). *B. methanolicus* is able to use methanol as sole carbon and energy source and might prepare the path to fermentation technology independent from glucose (Frank et al. 2021). The thermotolerant host grows at a temperature of 50 °C and pursues the ribulose monophosphate pathway (RuMP). Its development require a distinct metabolic engineering strategy, though it would offer access to methanol as feedstock. However, a commercial application has not yet been achieved.

14.3.2 Coproduction of Glutamate and 1,3-Propanediol

The surplus of three reduction equivalents in the synthesis of glutamate with glucose as starting material needs to be addressed in the metabolic engineering strategy to ensure a replenishment of NAD^+. The conventional fermentation process offers oxygen as final electron acceptor, and NAD^+ is recycled via the formation of water. Alternatively, the coproduction of glutamate and 1,3-propanediol has been proposed. In case glucose and glycerol are provided as feedstock in a molar ratio of 1 : 3 and the genes encoding the glycerol dehydratase and the 1,3-propanediol dehydrogenase are introduced, theoretically 3 mol 1,3-propanediol per mol glutamate might be obtained:

$$\text{Glucose} + 3\,\text{glycerol} + NH_3 \rightarrow \text{Glutamate} + CO_2 + 3\,1,3\text{–propanediol}$$

One mol glucose (24) and 3 mol glycerol (3 × 14) contribute in total 66 electrons (without counting reduced oxygen) on the left side of the reaction equation, which are allocated to one mol glutamate (18) and three mol 1,3-propanediol (3 × 16). The handover of electrons derived in the oxidative synthesis of glutamate for the reductive synthesis of 1,3-propanediol is indicated by the use of reduction equivalents (Figure 14.2).

In lab-scale experiments, the strain PY04 of *C. glutamicum* could be manufactured capable to generate 32.5 g/l glutamate with a molar yield of 0.5 mol glutamate per mol glucose (Huang et al. 2017). The strain managed in parallel to generate 14.4 g/l 1,3-propanediol with a molar yield of 0.89 mol 1,3-propanediol per mol glycerol. The glutamate yield was increased by 18.2% compared to the control strain without synthesis of 1,3-propanediol, which favors the interpretation that the production of 1,3-propanediol is beneficial for the glutamate yield due to a more balanced redox state. The amount of NADH consumed in the synthesis of 1,3-propanediol accounted for 28% of the NADH generated in the synthesis of glutamate. With respect to the downstream process, it is outlined that both products can

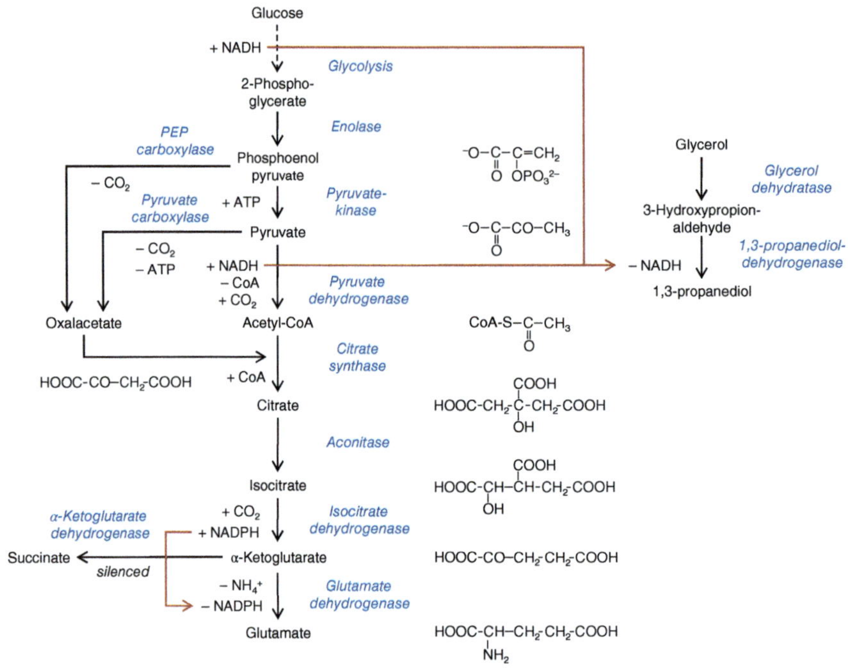

Figure 14.2 Redox balance in the joint synthesis of glutamate and 1,3-propanediol. Source: Walter Koch.

easily be separated and purified. The broth would undergo acidification and the protonated glutamic acid could be isolated via crystallization. 1,3-Propanediol would be obtained via distillation out of the mother liquor (Huang et al. 2017).

14.4 Process Development and Industrial Application

The process development in the first phase of commercial use, before the application of molecular biology and metabolic engineering tools, focused on host screening projects based on random strain variation and selection guided by commercial suitability. The media composition was also subjected to extended trial and error programs, and the ideal concentration of single additives was elucidated. These attempts comprise the limited addition of biotin, which is known to impact the efficiency of the process. *C. glutamicum* bears a biotin auxotrophy, and biotin needs to be added to some extent. A biotin limitation reduces the activity of the α-ketoglutarate dehydrogenase (Kimura et al. 1999) and the pyruvate carboxylase (Peters-Wendisch et al. 1997). Biotin availability also impacts phospholipid composition since the acyl-CoA carboxylase requires biotin. Furthermore, the early commercial processes used the addition of Tween 60 (Shiio et al. 1963) or dodecylammonium acetate to promote glutamate efflux (Duperray et al. 1992).

Since the current industrial glutamate production is a mature technology that achieves commercially acceptable performance data compared to other fermentation processes, the overall R&D efforts are limited. Activities refer to incremental process adjustments, the use of a 2nd generation carbon source, and the product isolation.

14.4.1 Adaption to a Carbon Source 2nd Generation

Glucose in the form of starch hydrolysate or fructose and sucrose might be used as carbon and energy sources. Molasses, which might provide a cost advantage compared to glucose, has been established as feedstock after a specific technical solution for biotin management was available given the natural excess biotin content of molasses in relation to the maximum biotin threshold in the broth (Hashimoto 2017). It is pursued for several fermentation products within the industry, with the major efforts likely allocated to ethanol, to switch from conventional glucose to a 2nd generation carbon source. In this context, the endeavor to use corn stover hydrolysate as feedstock for the fermentative production of glutamate is to be seen. The industrial strain *C. glutamicum* S9114 has been fed with a corn stover hydrolysate, and it was pursued to achieve commercially interesting fermentation performance data (Wen and Bao 2019). In the initial experiments, a weak productivity was observed, which is presumably due to the biotin content in the corn stover hydrolysate.

The maximum biotin concentration in the medium of an optimized commercial process amounts to about 2–5 μg/l, and at higher values, the host productivity is hampered (Wen and Bao 2019; Tryfona and Bustard 2004). For comparison, the fermentation medium displayed a biotin concentration of 60 μg/l after dilution of the corn stover hydrolysate with the other media components. It is known that a biotin concentration above the threshold of 2–5 μg/l can be managed, e.g. with the addition of β-lactam antibiotics, surfactants, or ethambutol, though these measures were excluded, also taking the other usual contaminations of 2nd generation feedstocks from lignocellulosic biomass like furfural, 5-hydroxymethyl-furfural, and phenolic aldehydes into account. Improvements in the use of corn stover hydrolysate were achieved after truncation of the C-terminal end of the glutamate secretion channel MscCG. The competing conversion of α-ketoglutarate into succinate was weakened by regulating the *odh* RBS sequence. The best strain was able to generate a glutamate titer of 65.2 g/l after 48 h corresponding to a productivity of 1.4 g/lxh. 149 g glucose was consumed and 93.9 g glutamate was produced, corresponding to a carbon yield of 0.63 t glutamate per ton glucose (Wen and Bao 2019).

14.4.2 Industrial Process Performance

It is reported that the first strains of *C. glutamicum* already achieved a molar yield of 0.25 mol glutamate per mol glucose, or 0.2 t glutamate per ton glucose (Chao and Foster 1959). The current industrial process achieves a carbon yield of 0.6–0.7 t glutamate per ton glucose (Sahm et al. 2013), which amounts to 73–85% of the maximum carbon yield of 0.82 t glutamate per ton glucose. The glutamate titer of the commercial process in the 500 m^3 fermenters may achieve up to 150 g/l after a batch duration of 60 h, corresponding to a productivity of 2.5 g/lxh (Schmid 2016). It is reported in academia that the product titer could be increased from 154 g/l up to 177 g/l after 36 h in a lab-scale process if with trypsin (0.5 g/l) a protease is added to the medium. Thereby, the secreted proteins will at least partially be decomposed into single amino acids and foaming could be suppressed. The released single amino acids can be used by the host as building blocks and are incorporated into other proteins. It is not known if this observation could be confirmed in the industrial process and if proteases are already considered in the media composition (Liu 2019).

The pH of the broth would decrease with increasing glutamate titer in the broth, and, accordingly, a pH management is needed. Ammonia is supplemented as nitrogen source since it serves as substrate of glutamate dehydrogenase and buffers the released protons:

$$^{-}OOC\text{–}CH(NH_3{}^{+})\text{–}CH_2\text{–}CH_2\text{–}COO^{-} + H^{+} + NH_3 \rightarrow {}^{-}OOC\text{–}CH(NH_3{}^{+})\text{–}CH_2\text{–}CH_2\text{–}COO^{-} + NH_4{}^{+}$$

Alternatively, NaOH might be used for neutralization of the medium, especially if sodium glutamate is obtained as final product. At batch end, the biomass is removed via ultrafiltration. The ammonium glutamate obtained in the broth is converted into sodium glutamate via an ion exchange process. Pure MSG is then derived via crystallization. Sodium glutamate may finally be converted into pure, protonated glutamic acid, pending the specification of the target application. Most commercial producers are located in Asia with close proximity to the major Asian markets as food ingredients.

References

Ajinomoto 2022 www.ajinomoto.com, https://www.ajinomoto.com/aboutus/history (accessed 18.7.2022).

Becker, J. and Wittmann, C. (2015). Biotechnologie von morgen: Metabolisch optimierte Zellen für die bio-basierte Produktion von Chemikalien und Treibstoffen. *Materialien und Gesundheitsprodukten, Angewandte Chemie* 127: 3383–3407.

Cao, Y., He, Z., Shi, Z., and Enock, M. (2015). Improving the stability of glutamate fermentation by *Corynebacterium glutamicum* via supplementing sorbitol or glycerol. *Bioresources and Bioprocessing* 2: 9. https://doi.org/10.1186/s40643-014-0032-6.

Chao, K.-C. and Foster, J.W. (1959). A glutamic acid-producing Bacillus. *Journal of Bacteriology* 715–725. https://doi.org/10.1128/jb.77.6.715-725.1959.

Choi, J.W., Yim, S.S., Lee, S.H. et al. (2015). Enhanced production of gamma-aminobutyrate (GABA) in recombinant *Corynebacterium glutamicum* by expressing glutamate decarboxylase in expanded pH range. *Microbial Cell Factories* 14: 21. https://doi.org/10.1186/s12934-015-0205-9.

Dhakal, R., Bajpai, V.K., and Baek, K.-H. (2012). Production of GABA (γ-aminobutyric acid) by microorganisms: a review. *Brazilian Journal of Microbiology* 1230–1241. 1517-8382.

Duperray, F., Jezequel, D., Ghazi, A. et al. (1992). Excretion of glutamate from *Corynebacterium glutamicum* triggered by surfactants. *Biochimica et Biophysica acta* 1103: 250–258.

EFSA Panel (2022). Safety and efficacy of the feed additives consisting of L-glutamic acid and monosodium L-glutamate monohydrate produced by *Corynebacterium glutamicum* NITE BP-01681 for all animal species. *EFSA Journal* https://doi.org/10.2903/j.efsa.2022.7156.

Eggeling, L. and Sahm, H. (1999). L-glutamate and L-lysine: traditional products with impetuous developments. *Applied Biology & Biotechnology* 52: 146–153.

Eggeling, L. and Sahm, H. (2001). The cell wall barrier of *Corynebacterium glutamicum* and amino acid efflux. *Journal of Bioscience and Bioengineering* 92 (3): 201–213.

Fahimitabar, A., Razavian, S.M.H., and Rezaei, S.A. (2021). Application of RSM for optimization of glutamic acid production by *Corynebacterium glutamicum* in bath culture. *Heliyon* https://doi.org/10.1016/j.heliyon.2021.e7359.

Frank, C., Hoffmann, T., Zelder, O. et al. (2021). Enhanced glutamate synthesis and export by the thermotolerant emerging industrial workhorse *Bacillus methanolicus* in response to high osmolarity. *Frontiers in Microbiology* 12: 640980. https://doi.org/10.3389/fmicb.2021.640980.

Hashimoto, S. (2017). Discovery and history of amino acid fermentation. *Advances in Biochemical Engineering/Biotechnology* 159: 15–34. https://doi.org/10.1007/10_2016_24.

Heggeset, T.M.B., Krog, A., Balzer, S. et al. (2012). Genome sequence of thermotolerant *Bacillus methanolicus*: features and regulation related to methylotrophy and production of L-lysine and L-glutamate from methanol. *Applied and Environmental Microbiology* 78 (15): 5170–5181. https://doi.org/10.1128/AEM.00703-12.

Hirasawa, T. and Wachi, M. (2017). Glutamate fermentation-2: mechanism of L-glutamate overproduction in *Corynebacterium glutamicum*. *Advances in Biochemical Engineering/Biotechnology* 159: 57–72. https://doi.org/10.1007/10_2016_26.

Huang, J., Wu, Y., Wu, W. et al. (2017). Cofactor recycling of co-production of 1,3-propanediol and glutamate by metabolically engineered *Corynebacterium glutamicum*. *Scientific Reports* 7: 42246. https://doi.org/10.1038/srep42246.

Kimura, E., Yagoshi, C., Kawahara, Y. et al. (1999). Glutamate overproduction in Corynebacterium glutamicum triggered by a decrease in the level of a complex comprising DtsR and a biotin-containing subunit. *Bioscience, Biotechnology and Biochemistry* 63: 1274–1278.

Kinoshita, S., Udaka, S., and Shimono, M. (1957a). Studies on the amino acid fermentation. I. Production of L-glutamate acid by various microorganisms. *Journal of General and Applied Microbiology* 3: 193–205.

Kinoshita, S., Tanaka, K., Udaka, S., and Akita, S. (1957b). Glutamic acid fermentation. In: *Paper No. 403, Symposium on Enzyme Chemistry, Japan*.

Lee, S., Ahn, J., Kim, Y.-G. et al. (2013). Gamma-aminobutyric acid using immobilized glutamate decarboxylase followed by downstream processing with cation exchange chromatography. *International Journal of Molecular Science* 14: 1728–1739. https://doi.org/10.3390/ijms14011728.

Li, D., Hou, L., Gao, Y. et al. (2022). Recent advances in microbial synthesis of poly-γ-glutamic acid: a review. *Foods* 11: 739. https://doi.org/10.3390/foods11050739.

Liu, X. (2019). Effects of proteases on L-glutamic acid fermentation. *Bioengineered* 10 (1): 646–658. https://doi.org/10.1080/21655979.2019.1688224.

McMurry, J. and Begley, T. (2006, Spektrum). *Organische Chemie der biologischen Stoffwechselwege*, 213–214. Elsevier Munich.

Nampoothiri, K.M., Hoischen, C., Bathe, B. et al. (2002). Expression of genes of lipid synthesis and altered lipid composition modulates L-glutamate efflux of *Corynebacterium glutamicum*. *Applied Microbiology & Biotechnology* 58: 89–96. https://doi.org/10.1007/s00253-001-0861-z.

Niebisch, A., Kabus, A., Schultz, C. et al. (2006). Corynebacterial protein kinase G controls 2-oxoglutarate dehydrogenase activity via the phosphorylation status of the OdhI protein. *Journal of Biological Chemistry* 281 (18): 12300–12307. https://doi.org/10.1074/jbc.M512515200.

Oishi, K., Aida, K., and Asai, T. (1961). Studies on amino acid fermentation. Part VIII. On the mechanism of conversion of L-glutamic acid fermentation to succinic acid fermentation. *Journal of General and Applied Microbiology* 7: 213–226.

O'Regan, M., Thierbach, G., Bachmann, B. et al. (1989). Cloning and nucleotide sequence of the phosphoenolpyruvate carboxylase-coding gene of *Corynebacterium glutamicum* ATCC13032. *Gene* 77 (2): 237–251.

Peters-Wendisch, P.G., Wendisch, V.F., Eikmanns, P.S., and Sahm, H. (1997). Pyruvate carboxylase as an anaplerotic enzyme in *Corynebacterium glutamicum*. *Microbiology* 143: 1095–1103.

Peters-Wendisch, P.G., Kreutzer, C., Kalinowski, J. et al. (1998). Pyruvate carboxylase from *Corynebacterium glutamicum*: characterization, expression and inactivation of the *pyc* gene. *Microbiology* 144 (pt 4): 915–927.

PR Newswire, 2022 Monosodium Glutamate (MSG) Market to Reach Value of US$ 5.8 Bn by 2031, https://www.prnewswire.com/news-releases/monosodium-glutamate-msg-market-to-reach-value-of-us-5-8-bn-by-2031--tmr-study-301646004.html (accessed 12 October 2022)

Sahm, H., Antranikian, G., Stahmann, K.-P., and Takors, R. (ed.) (2013). *Industrielle Mikrobiologie*, 117. Heidelberg: Springer Verlag.

Schmid, R.D. (2016). *Taschenatlas der Biotechnologie und Gentechnik*, 126. Wiley-VCH.

Shiio, I., Otsuka, S., and Katsuya, N. (1963). Cellular permeability and extracellular formation of glutamic acid in *Brevibacterium flavum*. *Journal of Biochemistry* 53: 333–340.

Siewe, R.M., Weil, B., Burkovski, A. et al. (1996). Functional and genetic characterization of the (methyl)ammonium uptake carrier system of *Corynebacterium glutamicum*. *Journal of Biological Chemistry* 271: 5398–5403.

Takaç, S., Çalik, G., Mavituna, F., and Dervakos, G. (1998). Metabolic flux distribution for the optimized production of L-glutamate. *Enzyme and Microbial Technology* 23: 286–300.

Tryfona, T. and Bustard, M.T. (2004). Mechanistic understanding of the fermentative L-glutamic acid overproduction by *Corynebacterium glutamicum* through combined metabolic flux profiling and transmembrane transport characteristics. *Journal of Chemical Technology and Biotechnology* 79: 1321–1339. https://doi.org/10.1002/jctb.1133.

Voet, D., Voet, J.G., and Pratt, C.W. (2016). *Principles of Biochemistry*, 728. Wiley.

Wen, J. and Bao, J. (2019). Engineering *Corynebacterium glutamicum* triggers glutamic acid accumulation in biotin-rich corn Stover hydrolysate. *Biotechnology for Biofuels* 12: 86. https://doi.org/10.1186/s13068-019-1428-5.

Wolf, S., Becker, J., Tsuge, Y. et al. (2021). Advances in metabolic engineering of *Corynebacterium glutamicum* to produce high-value active ingredients for food, feed, human health, and well-being. *Essays in Biochemistry* 65: 197–212. https://doi.org/10.1042/EBC20200134.

15

Isoprene

15.1 Application

Isoprene or 2-methyl-1,3-butadiene is a gaseous hydrocarbon with the chemical formula C_5H_8 and a molecular weight of 68.1 g/mol. Given its two double bonds as chemical functionality, the molecule is often applied for polymerization to obtain *cis*-polyisoprene or synthetic rubber. Synthetic rubber is a decisive component in tire manufacturing. Alternatively, isoprene may be copolymerized with other suitable monomers like isobutene and acrylonitrile to derive advanced isoprene polymers. Styrene-isoprene-styrene block copolymers are primarily used in hot-melt pressure-sensitive adhesive formulations (Whited et al. 2010). Isoprene is also suitable for the synthesis of terpene-related compounds based on its nonterminal methyl group.

15.2 Chemical Synthesis

Several routes have evolved in the last century for the synthesis of isoprene. A great share of supply is ensured by the cracking of petroleum or naphtha, which derives a complex mixture of C5 products and other hydrocarbons as by-products. This mixture is often used as feedstock for isoprene after it was discovered that polar solvents like acetonitrile, dimethylformamide, and N-methylpyrrolidone allow an efficient distillation and selective extraction of isoprene. However, since petroleum refineries tend to use a lighter feedstock or natural gas, the amount of isoprene as a by-product is diminishing. The closure of Shell's isoprene production unit in Pernis in 2009 was likely an additional trigger for the interest of the tire industry in the development of a fermentative isoprene technology besides the critical supply of natural rubber (Whited et al. 2010).

In the beginning, other chemical technologies will be presented. An early route has evolved in the age of acetylene chemistry (derived from calcium carbide) in Germany and reacts acetylene with acetone to derive methylbutinol and dimethylvinylcarbinol (DMVC) in the following steps. DMVC can then be dehydrated to introduce the second double bond and obtain isoprene (Figure 15.1).

The overall yield toward isoprene is estimated at 85% (Arpe 2007). Isoprene might also be obtained via the reaction of isobutene with two mol formaldehyde (Prins reaction). The generated cyclic C6 intermediate 4,4-dimethyl-1,3-dioxan can be converted into isoprene

Pathway Design for Industrial Fermentation, First Edition. Walter Koch.

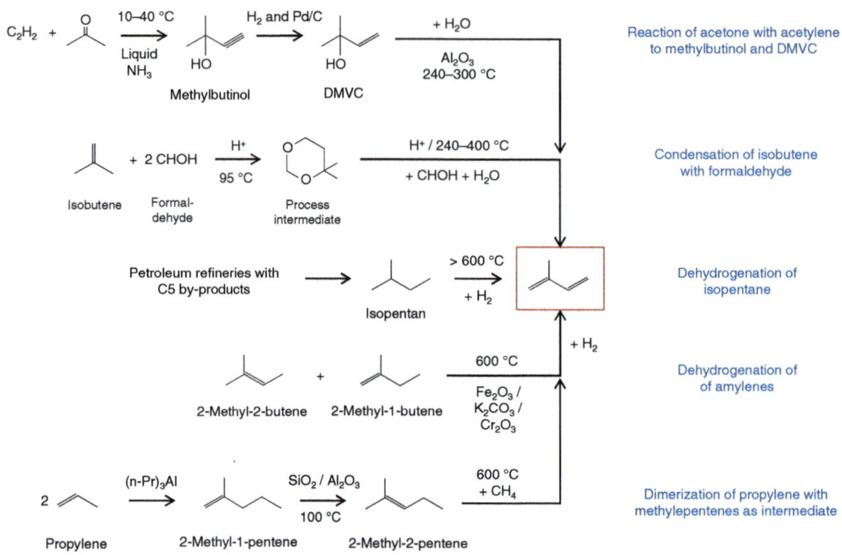

Figure 15.1 Routes for the chemical production of isoprene. Source: Adapted from Whited et al. (2010).

with formaldehyde and water as by-products. The overall yield toward isoprene is about 77% (Arpe 2007). A straightforward route to isoprene pursues a dual dehydrogenation of isopentane and obtains isoprene via the introduction of two double bonds. In cases where the C5 building blocks 2-methyl-2-butene and 2-methyl-1-butene are available, isoprene might simply be derived by a catalytic dehydrogenation. Finally, isoprene might be derived by the condensation of propylene to derive 2-methyl-1-pentene in the first step. Since the final generation of isoprene cannot be pursued with the terminal double bond, an isomerization step is needed to convert 2-methyl-1-pentene into 2-methyl-2-pentene. This technology was developed by Scientific Design and applied by Goodyear (Spitz 2019). The overall yield toward isoprene amounts to 50%, and the Goodyear plant applying this technology has been closed (Arpe 2007).

15.3 Pathway Design

15.3.1 Terpenes Derived from the Unique C5 Building Block IPP/DMAPP

Isoprene is a natural product made by various prokaryotic and eukaryotic organisms. It is known that *Bacillus subtilis* synthesizes and emits the volatile compound isoprene. It has been speculated that the emitted isoprene serves as a signal molecule or is to be regarded as an overflow metabolite (Julsing et al. 2007). Isoprene is especially synthesized in plant chloroplasts, which harbor the isoprene synthetase identified about 30 years ago (Silver and Fall 1991, 1995). Terpenes are released into the atmosphere by several plants and might impact the global climate. It has been estimated that plants collectively release about 600 Mio. t isoprene pa into the atmosphere (Guenther et al. 2006). This circumstance explains the high interest in the synthesis of isoprene in plant chloroplasts and its regulation pending on the atmospheric carbon dioxide concentration (Sharkey and Monson 2017).

Isoprene may be regarded as the simplest terpene or terpenoid, which comprises a huge repertoire of secondary metabolites and make up to more than 40 000 different compounds (Bohlmann and Keeling 2008; Breitmaier 2005). Terpenes are rather ubiquitous in biological systems and are found in virtually all organisms. The broadest number of terpenes and especially their advanced molecular functionality are found in plant secondary metabolism. The application field covers pharmaceuticals, flavors, fragrances, pesticides, disinfectants, and chemical intermediates. Several terpenes have successfully been introduced into the market including the anti-malarial drug precursor artemisinic acid, the anti-cancer intermediate taxadiene, and the vitamin E precursor farnesene. Terpenes are synthesized by either the mevalonate (MVA) or the deoxy-xylulose-5-phosphate (DXP) pathway also referred to as methyl-D-erythritol-4-phosphate (MEP) pathway. The MVA pathway is found in most eucaryotes and plant mitochondria. The DXP pathway is found in most eubacteria, cyanobacteria, green algae, and plant plastids.

The MVA and the DXP pathways also derive the shared isomeric C5-intermediates isopentenyl-pyrophosphate (IPP) and dimethylallyl-pyrophosphate (DMAPP), which define some of the most pervasive biochemical intermediates. They may be condensed to geranyl-diphosphate (GPP), farnesyl-pyrophosphate (FPP), geranylgeranyl-pyrophosphate (GGPP), or higher adducts (Figure 15.2).

As a digression, it should be mentioned that lepidopterans (i.e. butterflies and moths) possess a specific MVA pathway that condenses propionyl-CoA and acetyl-CoA in the first step instead of the classical two mol acetyl-CoA. Since the additional methyl function from propionyl-CoA is retained in the further pathway, a new class of terpene analogs is accessible. By using a heterologous propionyl-CoA ligase expressed in *Escherichia coli*, a first proof of principle for terpenes containing 16 carbon atoms could be shown, expanding the biochemical space for noncanonical terpenes (Eiben et al. 2019).

Terpenes are interesting targets for industrial fermentation and have attracted an enormous management attention in academia and industrial R&D after the first proof of concept for the synthesis of amorphadiene was shown (Martin et al. 2003).

15.3.2 Mevalonate (MVA) Pathway Without Pyruvate-Dehydrogenase Bypass

The MVA pathway starts with the formation of acetoacetyl-CoA based on two units of acetyl-CoA. The acetoacetyl-CoA thiolase transfers after release of CoA one acetyl-unit to the other acetyl-CoA reaction partner. In the next step, a third unit of acetyl-CoA is condensed with acetoacetyl-CoA via the HMG-CoA synthase to obtain HMG-CoA or 3-methyl-3-hydroxy-glutaric acid given as a thioester with CoA. The carboxylic acid function is in the next step reduced down to the alcohol level which is thermodynamically supported by its activated form as thioester. Hereby, two mol NADPH as reduction equivalents are consumed. The obtained terminal hydroxyl function is phosphorylated in two steps to obtain MVA-diphosphate. The two consecutive activations are pursued by different kinases and consume two mol ATP (Figure 15.3).

For the calculation of the theoretical molar yield of the MVA pathway, it needs to be distinguished how the required acetyl-CoA pool is supplied. The most straightforward pathway proceeds via the glycolysis and the oxidative pyruvate decarboxylation via the pyruvate

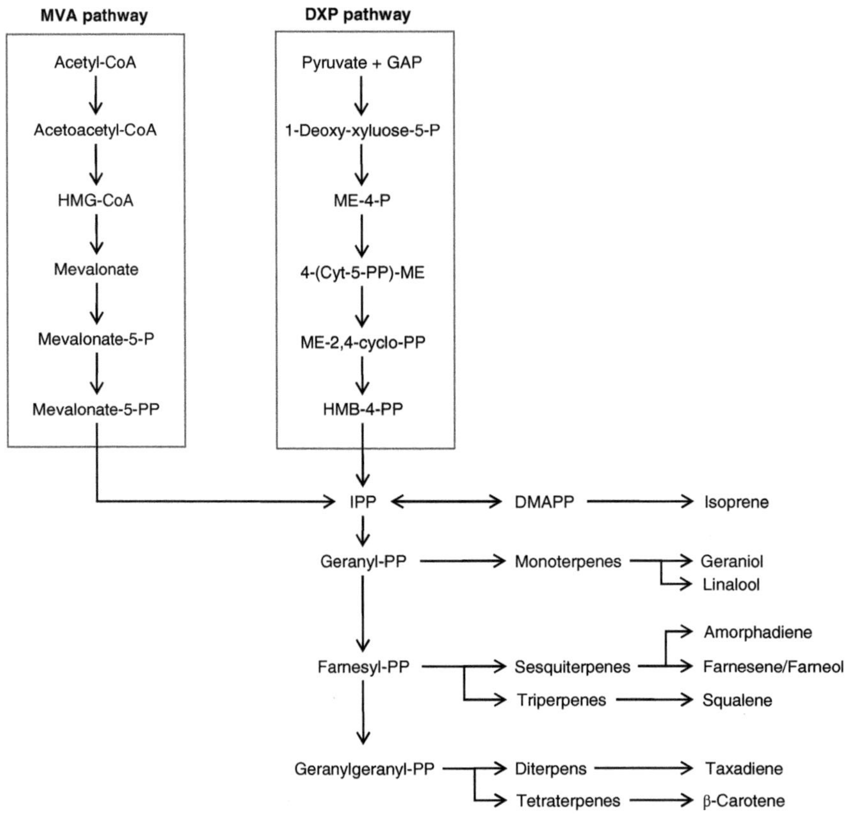

Figure 15.2 Simplified MVA and DXP pathways for the synthesis of IPP and DMAPP as precursors for higher terpenes. Terpenes are classified according to the number of integrated C5 isoprene building blocks into monoterpenes, sesquiterpenes, triterpenes, diterpenes, and tetraterpenes. On the right side, product examples are given. Source: Adapted from Hara et al. (2014).

dehydrogenase. Three mol acetyl-CoA need to be invested in the building of one mol isoprene (Table 15.1).

The equation reveals that 1.5 mol glucose is needed for the synthesis of 1 mol isoprene, which translates into a theoretical maximum molar yield of 0.66 mol isoprene per mol glucose and a maximum carbon yield of 0.25 t isoprene/t glucose (68.1/180.2 × 1.5). However, this value supposes that the NADPH pool for the reduction of HMG-CoA is efficiently replenished. This might be facilitated with a native or recombinant transhydrogenase or a specific HMG-CoA reductase that is able to consume NADH as well. Since the fermentation is pursued as an aerobic process to offer a final electron acceptor, the reaction can be stated as follows:

$$1{,}5\ \text{Glucose} + 2\ O_2 \rightarrow \text{isoprene} + 4\ CO_2 + 5\ H_2O$$

15.3.3 Mevalonate (MVA) Pathway with Pyruvate-Dehydrogenase Bypass

In the overall metabolic engineering strategy, it needs to be distinguished whether a prokaryotic or an eukaryotic host shall be employed. In case a eukaryotic yeast strain

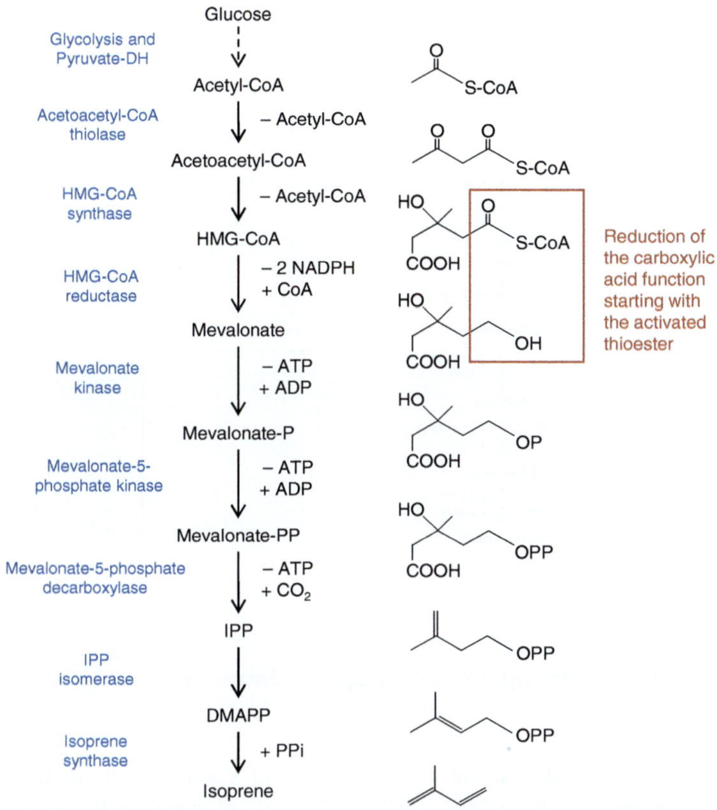

Figure 15.3 MVA pathway toward isopentenyl-PP and dimethylallylpyrophosphate. Source: Walter Koch.

Table 15.1 Metabolic pathway for the MVA pathway toward isoprene.

1,5 Glucose + 3 CoA + 3 ADP + 6 NAD^+ → 3 acetyl-CoA + 3 CO_2 + 3 ATP + 6 NADH + 6 H^+ + 3 H_2O
2 Acetyl-CoA → acetoacetyl-CoA + CoA
Acetoacetyl-CoA + acetyl-CoA + H_2O → HMG-CoA + CoA
HMG-CoA + 2 NADPH + 2 H^+ → mevalonate + CoA + 2 $NADP^+$
Mevalonate + ATP → mevalonate-5-P + ADP
Mevalonate-5-P + ATP → mevalonate-PP + ADP
Mevalonate-5-PP + ATP + H_2O → IPP + CO_2 + ADP + Pi
IPP → DMAPP
DMAPP → isoprene + 2 Pi
1,5 Glucose + 6 NAD^+ + 2 NADPH → isoprene + 4 CO_2 + H_2O + 2 $NADP^+$ + 6 NADH + 4 H^+

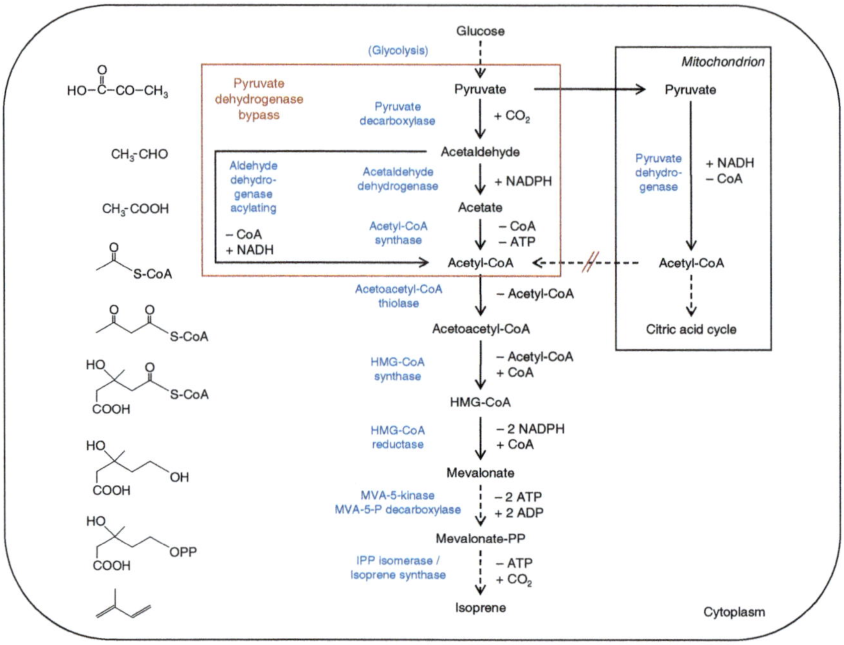

Figure 15.4 Schematic representation of the pyruvate-dehydrogenase bypass toward isoprene. Source: Walter Koch.

like *Saccharomyces cerevisiae* is applied, the native localization of specific steps in the mitochondrion as separate compartments or "reactors" needs to be addressed (Huttanus and Feng 2017). The MVA with acetyl-CoA as a precursor may be envisaged to occur in the cytosol, whereas the formation of acetyl-CoA via the pyruvate dehydrogenase takes place mainly in the mitochondrion to supply the citric acid cycle (TCA). On the other hand, the conversion of pyruvate in the cytosol proceeds via the pyruvate decarboxylase to yield acetaldehyde and potentially ethanol. The acetyl-CoA pool in the cytosol to supply the native MVA pathway is limited (Wegner et al. 2021), and a direct diffusion of mitochondrial acetyl-CoA into the cytosol is not reported. Therefore, genes for the cytosolic provision of acetyl-CoA have been overexpressed to bypass the mitochondrial pyruvate dehydrogenase (Figure 15.4).

In the context of the development of the fermentative production of amorphadiene or artemisinic acid, respectively, it has been managed to increase the cytosolic acetyl-CoA pool via overexpression of the acetaldehyde dehydrogenase and the acetyl-CoA synthase of *Salmonella enterica* in *S. cerevisiae* to enhance production of the terpene amorphadiene (Shiba et al. 2007). The net reaction equation for the supply of acetyl-CoA can be described as follows (Table 15.2).

It should be indicated that the conversion of acetaldehyde into acetyl-CoA might also be facilitated by the aldehyde dehydrogenase acylation (Figure 15.3; Amyris 2016). This route would save one mol ATP in the generation of acetyl-CoA compared to the two-step conversion of acetaldehyde via acetate. It has additionally been evaluated to express the bacterial dehydrogenase complex of *Enterococcus faecalis* in *S. cerevisiae* with cytosolic localization to facilitate a cytosolic, non-ATP-consuming conversion of pyruvate into

Table 15.2 Conversion of glucose into acetyl-CoA via the pyruvate-dehydrogenase bypass.

Reaction	Enzyme
Glucose + 2 ADP + 2 Pi + 2 NAD^+ → 2 Pyruvate + 2 ATP + 2 H_2O + 2 NADH + 2 H	Glycolysis
2 Pyruvate → 2 Acetaldehyde + 2 CO_2	Pyruvate decarboxylase
2 Acetaldehyde + 2 $NADP^+$ + 2 H_2O → 2 Acetate + 2 NADPH + 2 H^+	Acetaldehyde dehydrogenase
2 Acetate + 2 CoA + 2 ATP → 2 Acetyl-CoA + 2 ADP + 2 Pi	Acetyl-CoA Synthase
Glucose + 2 CoA + 2 $NADP^+$ + 2 NAD^+ → 2 Acetyl-CoA + 2 CO_2 + 2 NADPH + 2 NADH + 4 H	

Table 15.3 Conversion of glucose into acetyl-CoA via the pyruvate-dehydrogenase bypass. It is assumed that NADPH is the cofactor of the acetaldehyde dehydrogenase, according to the native enzyme in *S. cerevisiae.*

Reaction
1,5 Glucose + 3 CoA + 3 $NADP^+$ + 3 NAD^+ → 3 Acetyl-CoA + 3 CO_2 + 3 NADPH + 3 NADH + 6 H^+ + 3 H_2O
2 Acetyl-CoA → Acetoacetyl-CoA + CoA
Acetoacetyl-CoA + Acetyl-CoA + H_2O → HMG-CoA + CoA
HMG-CoA + 2 NADPH + 2 H^+ → Mevalonate + CoA + 2 $NADP^+$
Mevalonate + ATP → Mevalonate-5P + ADP
Mevalonate-5P + ATP → Mevalonate-PP + ADP
Mevalonate-5PP + ATP + H_2O → IPP + CO_2 + ADP + Pi
IPP → DMAPP
DMAPP → Isoprene + PPi
1,5 Glucose + 3 ATP + 3 NAD^+ + $NADP^+$ → Isoprene + 4 CO_2 + 3 ADP + H_2O + NADPH + 3 NADH + 4 H^+

Source: Adapted from Meadows et al. (2016).

acetyl-CoA (Kozak et al. 2014). In this context, the more ATP-consuming and thermodynamically favored pyruvate-dehydrogenase bypass via the acetyl-CoA synthase is chosen. The assumed cofactor of the acetaldehyde dehydrogenase is NADPH, according to the native enzyme in *S. cerevisiae* (Meadows et al. 2016). The conversion of pyruvate into acetaldehyde implies that two mol ATP need to be invested to finally convert acetate into acetyl-CoA such that the ATP surplus via substrate-level phosphorylation gained in the glycolysis is lost. If again three mol acetyl-CoA are supplied for the conversion of one mol isoprene, it turns out that the overall pathway displays an ATP gap (Table 15.3).

The 36 electrons of glucose (1.5 × 24) are transferred to isoprene carrying 28 electrons, and 4 reduction equivalents carrying 8 electrons do remain. The maximum carbon yield can be estimated at 0.25 t isoprene/t glucose (68.1/180.2 × 1.5) since the process is pursued as an aerobic process and the ATP gap can be supplied via respiratory chain respiration. This overall reaction scheme is likely to be applied in case a recombinant MVA pathway is pursued in the cytosol of an eukaryotic host like *S. cerevisiae*. The net demand of three mol ATP can likely be replenished with the surplus of reduction equivalents via respiratory chain phosphorylation in an aerobic process. It has been proposed to combine a glucose fermentation toward lactate with the biocatalytic whole cell conversion of MVA into isoprene in order to recycle the three mol ATP needed for the conversion of MVA into IPP (Cheng et al. 2021). Both reactions were facilitated by a recombinant *E. coli* strain under microaerobic conditions that achieved a lactic acid titer of 40 g/l combined with a (recalculated after gas capture in the reactor headspace) isoprene titer of 3.2 g/l after a batch incubation of 40 h. The coproduction is guided by the idea to leverage the surplus of two mol ATP generated per mol lactate, which shall support the energy demand in the biocatalytic conversion of mevalonate into isoprene (Cheng et al. 2021).

15.3.4 Mevalonate (MVA) Pathways Pursued in Archaea

It should be mentioned that noncanonical MVA pathways have been identified in archaea bacteria, which pursue a slightly adjusted route. In particular, two additional routes are distinguished besides the classical MVA pathway (Motoyama et al. 2019). The modified MVA pathway I (red line, Figure 15.3) has been found in halophilic bacteria and bacteria of the phylum *Chloroflexi* and proceeds via isopentenyl-phosphate derived by decarboxylation of MVA-phosphate. Isopentenyl-phosphate is then phosphorylated by the isopentenyl-P-kinase to obtain IPP as a common intermediate of all MVA pathways. The modified MVA pathway II (green line, Figure 15.3) proceeds via MVA-3-P and MVA-3,5-phosphate after two consecutive phosphorylation steps of MVA. After decarboxylation, isopentenyl-phosphate is obtained, which is phosphorylated to IPP according to the last step of the MVA pathway I. It is assumed that the MVA pathway II has specifically been invented by extreme acidophiles triggered by low pH (Vinokur et al. 2016; Figure 15.5).

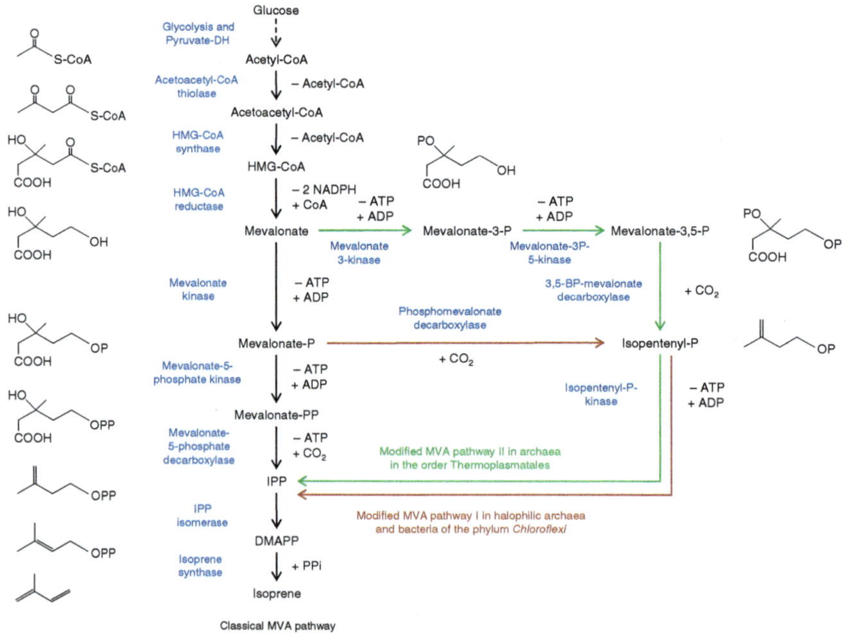

Figure 15.5 Classical MVA pathway compared with modified MVA pathway I (*red line*) and MVA pathway II (*green line*). Source: Adapted from Motoyama et al. (2019).

These specific versions of the MVA pathway are, so far, primarily of academic interest. They do not offer an improved C atom efficiency since in each case three mol acetyl-CoA are required for the synthesis of the common C5 building block IPP. The modified MVA pathway I has a slight advantage since the conversion of MVA to IPP does exclusively require two mol ATP compared to three mol ATP needed by the two other versions of the MVA pathway. This circumstance does nevertheless not allow for the improvement of theoretical maximum molar yield versus the classical MVA pathway since the net ATP demand of the classical MVA pathway can be addressed with the surplus of four reduction equivalents in an aerobic process.

15.3.5 Mevalonate (MVA) Pathway with Reduced ATP Consumption

Several innovative pathways that are closely related to or partially coincide with the classical MVA pathway have been designed (Wang et al. 2017; Bian et al. 2017). These pathways are possible since the biochemical reaction steps of decarboxylation, dehydration, and phosphorylation are rather independent from each other and may be executed in different orders in case the given substrate specificity is fully exploited or partially extended by rational evolution. Pathway modeling pursues different objectives comprising (i) the synthesis of related C5 products like prenol and isoprenol (Wang et al. 2019, 2022; Li et al. 2018; Zheng et al. 2013), (ii) the circumvention of the critical or at least partially toxic intermediate pathway intermediate IPP via the "IPP bypass" (Wang et al. 2022; Kang et al. 2016), (iii) the limitation of ATP consumption, and (iv) the circumvention of existing patents or, in other words, the provision of the mandatory freedom to operate for a commercial project.

The IPP bypass has been designed and implemented since IPP – besides HMG-CoA and the higher terpene building block FPP – is regarded as a critical intermediate whose concentration needs to be kept on a low level. Especially in the synthesis of isoprenol the IPP level might negatively impact host productivity and a high overexpression of the phosphatase NudB, which converts IPP into isoprenol, is recommended (George et al. 2015). The IPP toxicity mechanism is not yet fully elucidated, though it is assumed that a high IPP concentration inhibits cell growth and may trigger various stress responses. The stress mechanisms consume ATP and may compete with the rather high ATP demand of the MVA pathway, which requires three mol ATP in the conversion of MVA into IPP. A high level of IPP may also trigger the synthesis of GPP and FPP and thereby divert carbon into competing products. Accordingly, pathways that omit IPP have been designed to convert MVA directly into isoprenol or MVA-phosphate into isopentenyl-phosphate (Kang et al. 2016; Figure 15.6).

The innovative route via 3-hydroxy-3-methyl-pent-4-enoate has been explored by the polyamide producer Invista and has been patented (Invista 2015). The intention of the pathway design with isoprene as target product is likely to avoid the energy-intensive activation of MVA toward MVA-pyrophosphate requiring two mol ATP (Invista 2015).

The innovative route starts with HMG-CoA via 3-methylglutanconyl-CoA, 3-methyl-2-butenoyl-CoA, and 3-methyl-2-butenal toward prenol (Figure 15.6), and potentially higher terpenes like geraniol, limonene, farnesol, and lycopene were driven by a retro-biosynthetic approach (Clomburg et al. 2019). It defines a target product in the first instance and then looks for potential biochemical conversion steps to get to the target molecule instead of binding itself to the established pathways. The synthetic consecutive conversion steps are

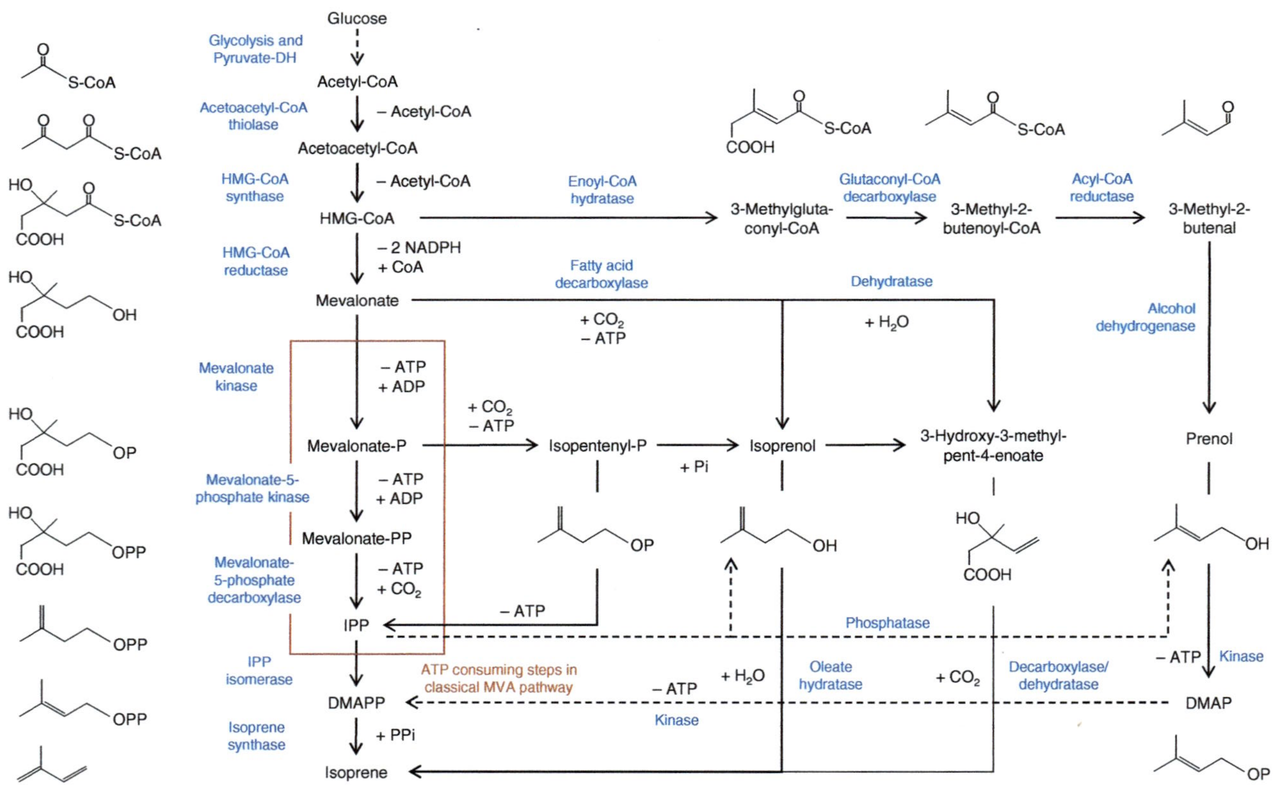

Figure 15.6 Advanced versions of the MVA pathway toward prenol, isoprenol, and isoprene (Clomburg et al. 2019; Li et al. 2018; Kang et al. 2016). Prenol and isoprenol might also be obtained in cases where IPP and DMAPP, respectively, are converted by a pyrophosphatase (George et al. 2015; Zheng et al. 2013). Source: Walter Koch.

referred to as the nonnatural isoprenoid alcohol pathway and might be especially interesting for the synthesis of prenol. Prenol might be then chemically dehydrated toward isoprene. The experiments were pursued in engineered *E. coli* strains with 20 g/l glycerol as carbon and energy source and were executed without and with 20% tributyl phosphate as organic layer to especially capture the product share, which would otherwise migrate into the reactor headspace. A titer of 2 g/l prenol after 48 h was obtained, corresponding to a productivity of 0.04 g/l×h (Clomburg et al. 2019). Alternatively, the IPA pathway might be extended, further enzymes might be included in the transformation vectors, and prenol might be further processed to derive isoprene or higher terpenes.

15.3.6 The Deoxy-Xylulose-Phosphate (DXP) Pathway

Alternatively, the DXP pathway may be applied in metabolic engineering. The DXP pathway is applied by most bacteria, and isoprene is a natural product. Since, in this case, not acetyl-CoA but pyruvate and glyceraldehyde-3-phosphate (GAP) are used as C3 building blocks for the C6 intermediate DXP, glycolysis is restated to provide pyruvate and GAP in equal shares. The DXP pathway starts with the condensation of pyruvate and GAP (Table 15.4).

One mol glucose is converted into one mol C5 intermediate, and only one mol carbon dioxide is generated. However, compared to the MVA pathway, the DXP pathway relies on three mol ATP and two reduction equivalents to achieve a carbon-efficient disproportionation of glucose. 1 mol glucose carries in total 24 electrons (without counting reduced oxygen). These are completely transferred to isoprene (carrying 28 electrons), which implies that a gap of four electrons or two reduction equivalents remains. This can be seen on the left side of the overall reaction equation, which indicates a demand of two NADH per mol isoprene. The pathway can be summarized as follows (Figure 15.7).

The reaction equation might propose that the maximum theoretical molar yield amounts to one mol isoprene per mol glucose. However, this value cannot be achieved because, for the calculation of the maximum yield, the energy and reduction equivalents also need

Table 15.4 DXP pathway toward isoprene.

Glucose + NAD^+ + Pi → Pyruvate + GAP + NADH + H^+ + 2 H_2O
Pyruvate + GAP → 1-Deoxy-xylulose-5-phosphate + CO_2
1-Deoxy-xylulose-5-phosphate + NADH → 2C-Methyl-erythritol-4-phosphate + NAD^+
2C-Methyl-erythritol-4-phosphate + CTP → 4-Diphosphocytidyl-2C-methyl-erythritol + PP_i
4-Diphosphocytidyl-2C-methyl-erythritol + ATP → 4-Diphosphocytidyl-2C-methyl-erythritol-2-phosphate + ADP
4-Diphosphocytidyl-2C-methyl-erythritol-2-phosphate → 2C-Methyl-erythritol-2,4-cyclo-PP + CMP
CMP + 2 ATP → CTP + 2 ADP
2C-Methyl-erythritol-2,4-cyclo-PP + NADH + H^+ → 1-Hydroxy-2-methyl-2-butenyl-4-PP + NAD^+ + H_2O
1-Hydroxy-2-methyl-2-butenyl-4-PP + NADH + H^+ → IPP + NAD^+ + H_2O
IPP → DMAPP
DMAPP → Isoprene + PPi
Glucose + 3 ATP + 2 NADH + 2 H^+ → Isoprene + CO_2 + 3 ADP + 3 Pi + 2 NAD^+ + 4 H_2O

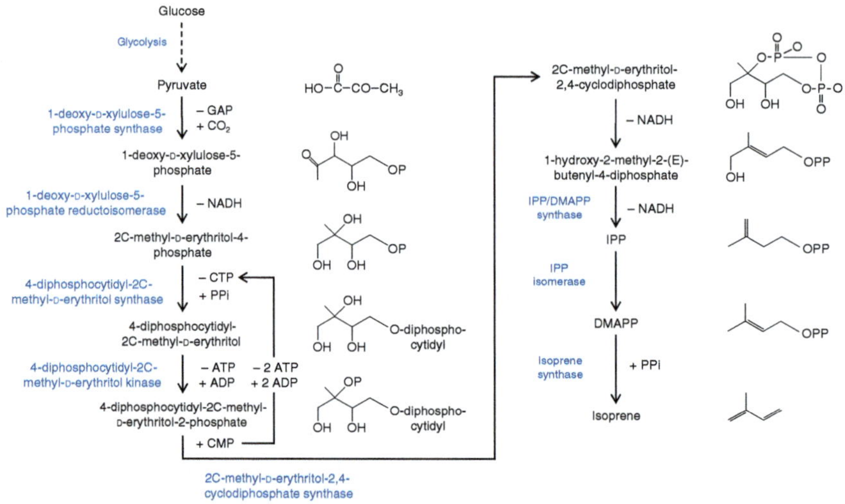

Figure 15.7 DXP pathway toward isoprene via IPP and DMAPP. Source: Walter Koch.

to be considered. The DXP pathway would be pursued as an aerobic pathway and the two reduction equivalents, or four electrons require 0.17 mol glucose (4/24), which also provides 0.67 ATP (0.17/4) in coproduction. The missing 2.33 mol ATP requires 0.07 mol glucose (2.33/32), such that in total an additional 0.24 mol glucose is needed. Accordingly, the maximum theoretical molar yield amounts to 0.806 mol isoprene per mol glucose (1/1.24), translating into a maximum carbon yield of 30.4% or 304 kg isoprene per ton glucose (68.1/1.24 × 180.2). The DXP or MEP pathway promises a higher yield than the MVA pathway but is disregarded in most of the published metabolic engineering programs. This is likely due to the lack of detailed knowledge on the regulatory control mechanisms (Yao et al. 2020). Since the DXP pathway would likely be implemented as an aerobic process, the overall reaction can be stated as follows, considering the additional 0.24 mol glucose per mol isoprene:

$$1.24\ \text{Glucose} + 0.44\ O_2 \rightarrow \text{Isoprene} + 2.44\ CO_2 + 3.44\ H_2O$$

15.4 Metabolic Engineering Toward Isoprene

The metabolic engineering attempts need to consider the specific boundary conditions of the pathway intermediates, the target product, and host characteristics. Since usually a high number of experiments with slightly changed enzymes (directed evolution) or pathway variations are required to finally derive a suitable isoenzyme or a robust pathway, an efficient high-throughput screening system is recommended. To some extent, the target reaction is linked to colored products, which can easily be measured and detected in the single wells of the microtiter plate. Thus, carotenoid-producing strains are preferred, which allow a direct colorimetric readout (Emmerstorfer-Augustin et al. 2016). The screening efficiency is significantly reduced and cheaper if the isolation of the product out of the small medium in the well can be avoided and instead a lean and direct colorimetric analysis is possible.

It is also to be considered that the typical host resistance toward IPP is critical, and a robust process will likely require an efficient conversion of IPP toward the next pathway intermediate or the final product. This circumstance has also triggered the search for biocatalytic approaches or the implementation of an IPP bypass (Kang et al. 2016). The establishment of a cytosolic MVA pathway in an eukaryotic host usually requires a pyruvate-dehydrogenase bypass to facilitate an efficient acetyl-CoA pool in the cytosol. Thus, the acetyl-CoA pool is addressed in most engineering approaches in case a eukaryotic host has been chosen (Lv et al. 2014). In the case where a more complex terpene than isoprene is addressed, which requires the efficient integration of a cytochrome P450-dependent terpene synthase, the boundary conditions for the host selection are rather tight. Functional expression of these enzymes is still difficult in prokaryotic hosts, and an eukaryotic host is recommended. Though this is not the case with isoprene as the target product. In the following section, different metabolic engineering approaches are presented.

15.4.1 Isoprene Production by a Metabolically Engineered *E. coli* Strain Under Aerobic Conditions

Promising fermentation data for the fermentative production of isoprene have been achieved within the research activities of the Goodyear Innovation Center and Genencor, which was formerly a division of Danisco until the acquisition of Danisco by Dupont (http://M.net 2012). The project bet on *E. coli* as a biocatalyst, which harbors its native DXP pathway for the synthesis of IPP and DMAPP. The pathway genes were arranged in two operons and either introduced as plasmids or integrated into the bacterial chromosome. The isoprene synthase is of plant origin either from *Populus alba* or *Kudzu* (Figure 15.8).

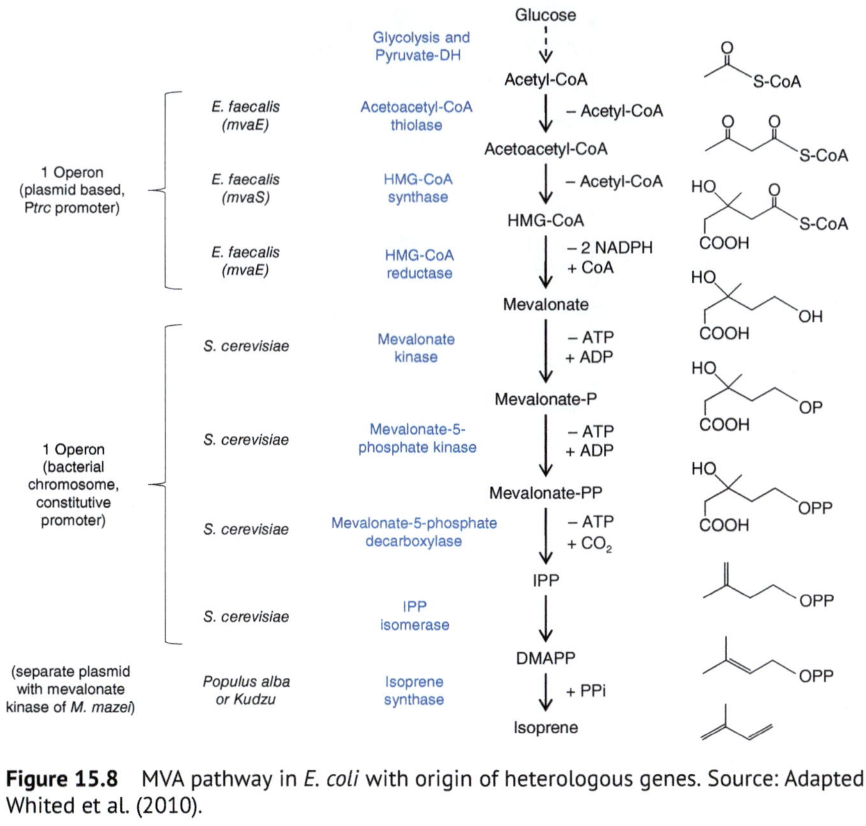

Figure 15.8 MVA pathway in *E. coli* with origin of heterologous genes. Source: Adapted from Whited et al. (2010).

The best-performing *E. coli* strain achieved a titer of 60 g/l in a 14 l fermentation reactor with a fed-batch feeding strategy. The titer was achieved after 60 h, which indicates a productivity of 1 g/l×h (Whited et al. 2010). The biobased isoprene has been polymerized to *cis*-1,4-polyisoprene following the standard polymerization protocol. The reaction was pursued with a heterogeneous catalyst containing titanium tetrachloride, anhydrous hexane, diphenyl ether, and triisobutyl aluminum at a temperature of 40 °C. An antioxidant is added to stabilize the polymer during transportation and storage. The obtained polymer displayed equivalent properties compared to the commercially produced Natsyn™ product grade of Goodyear (Whited et al. 2010). It could be shown that fermentative isoprene could from a polymer performance point of view be established as a technologically equivalent substitute to petrochemically derived isoprene.

A structurally equivalent approach has been pursued by an academic working group. For the development of an isoprene-producing *E. coli* strain, a recombinant MVA pathway with genes from *E. faecalis* and *Streptococcus pneumoniae* has been transferred combined with a codon-optimized isoprene synthase (*ispS*) from *Populus trichocarpa* (Kim et al. 2016). The MVA pathway genes were supported by a strong promoter and cloned on a high-copy plasmid. The formation of by-products was reduced with the deletion of nine genes, including the ethanol dehydrogenase (*adhE*) and the lactate dehydrogenase (*ldhA*) (Figure 15.9).

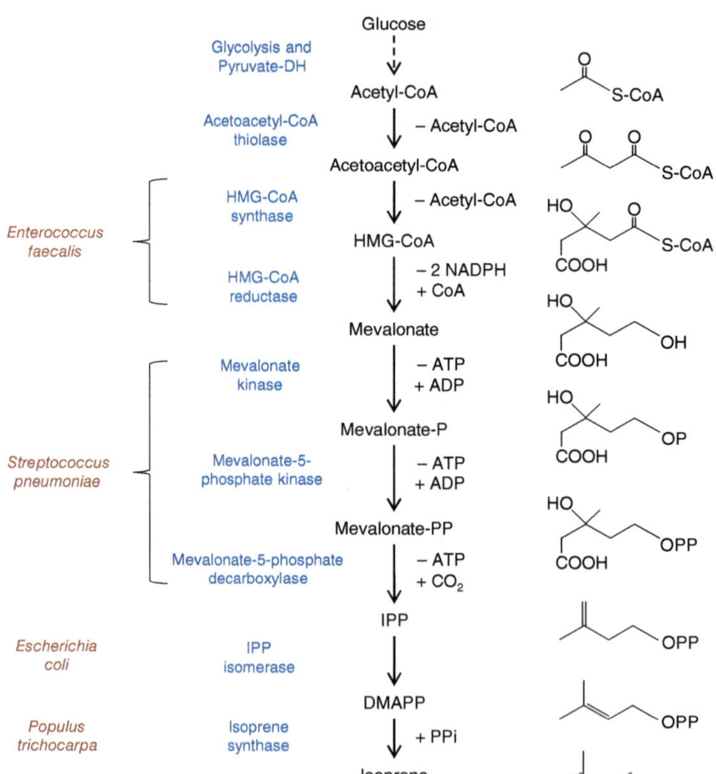

Figure 15.9 MVA pathway in *E. coli* with origin of heterologous genes. Source: Adapted from Kim et al. (2016).

The final strain achieved an isoprene titer of 1.8 g/l after a batch duration of 36 h (Kim et al. 2016). These values are according to a productivity of 0.05 g/l×h. The expression of the pathway genes was activated with the addition of 0.01 mM IPTG. The results are far below a commercially interesting level.

15.4.2 Coproduction of Isoprene and 1,3-Propanediol with Redox Recycling

The MVA pathway toward isoprene generates a surplus of reduction equivalents since 1.5 mol glucose bears more electrons than one mol isoprene. This surplus is usually removed via the generation of water in an aerobic process. However, it might be possible to exploit the surplus of reduction equivalents to supply a second pathway that displays a gap of reduction equivalents. This has indeed been analyzed for the coproduction of 1,3-propanediol with glycerol as the feedstock (Guo et al. 2019). The following net reaction equation can be derived (Table 15.5).

The electrons carried by glycerol (4 × 14) and glucose (1.5 × 24) are allocated to 1,3-propanediol (4 × 16) and isoprene (28, in all cases without reduced oxygen), and the reduction equivalents are balanced. The dual metabolic engineering approach was pursued in *E. coli,* and a recombinant NADH-dependent transhydrogenase encoded by the *pntAB* gene has been overexpressed. The most developed *E. coli* strain provided a titer of 665 mg/l isoprene and 2532 mg/l 1,3 propanediol under flask fermentation conditions (Guo et al. 2019). The achieved molar yields were 0.3 mol isoprene per mol glucose and 1.0 mol 1,3-propanediol per mol glycerol. The commercial viability of such a concept might be questioned given the different volume perspectives of both products and especially the limited volume perspective for 1,3-propanediol. In case a metabolic pathway generates a surplus of reduction equivalents, usually an aerobic process is pursued unless the host displays a critical oxygen sensitivity.

15.4.3 Isoprene via a Biocatalytic Conversion of Mevalonate

The reported toxicity of pathway intermediates, especially IPP, has triggered the search for an alternative process design to obtain isoprene. In the course of this approach, the setup of a biocatalytic conversion of MVA into isoprene with five added enzymes has been tested, which could provide a proof of concept (Cheng et al. 2017). The enzymes mevalonate kinase (MVK), phosphomevalonate kinase (PMK), diphosphomevalonate kinase (MVD), isopentenyl diphosphate isomerase (IDI), and isoprene synthase (ISPS) were purified and initially applied at a concentration of 0.5 μM each in a 2 ml in vitro system supplemented with the substrate MVA (2.5 mM) and ATP (12 mM) as cofactor. In the most optimized final setup

Table 15.5 Coproduction of 1,3-propanediol and isoprene.

4 Glycerol → 4 3-Propionaldehyde + 4 H_2O
4 3-Propionaldehyde + 4 NADPH + 4 H^+ → 4 1,3-Propanediol + 4 $NADP^+$
6 $NADP^+$ + 6 NADH → 6 NADPH + 6 NAD^+
1,5 Glucose + 6 NAD^+ + 2 NADPH → Isoprene + 4 CO_2 + H_2O + 2 $NADP^+$ + 6 NADH + 4 H^+
4 Glycerol + 1.5 glucose → 4 1,3-Propanediol + isoprene + 4 CO_2 + 5 H_2O

in a 50 ml reaction volume, the enzymes were applied in different concentrations – MVK, PMK, and MVD in 0.5 μM, IDI in 1.0 μM, and ISPS in 8.0 μM – and 302 mg/l isoprene were produced in 40 h (Cheng et al. 2017). The cost of the cofactor ATP, which is used with a concentration above the stoichiometric molar demand (3 mol ATP per mol isoprene) and which can likely not be recycled, will turn out to be the most critical cost factor in case the technology is commercially applied.

15.4.4 Mevalonate (MVA) Pathway with Reduced ATP Demand

The metabolic engineering trials positioning isoprenol as a pathway intermediate are driven by the objective to shorten the pathway by reducing enzymatic steps and to reduce the demand for ATP (Yang et al. 2016b). According to this strategy, a fatty acid decarboxylase from a *Jeotgalicoccus* species, which decarboxylates and dehydrates MVA to isoprenol, and an oelate hydratase from *Elizabethkingia meningoseptica*, which dehydrates isoprenol to isoprene, were introduced and overexpressed in *E. coli* (Figure 15.6). The net equation of the pathway – with acetyl-CoA generation by the pyruvate dehydrogenase in a prokaryotic host – would achieve a surplus of ATP and reduction equivalents (Table 15.6).

The experiments provided an isoprene titer of 620 mg/l after a batch time of 32 h, corresponding to a productivity of 19 mg/l×h (Yang et al. 2016b). The modest output may be explained by several hypotheses, which include the weak substrate specificity of the applied enzymes toward MVA and isoprenol and the thermodynamic efficiency of the pathway given the omitted phosphorylation steps of the common MVA pathway.

15.4.5 Isoprene via Mixed MVA and DXP Pathway in *E. coli*

It is obvious that synergies may be realized in case the MVA and the DXP pathways are combined and run in parallel such that the high carbon yield of the DXP pathway is fully exploited and the MVA pathway with weaker carbon yield is allowed a minor share, exactly matching the gap of energy and reduction equivalents of the DXP pathway. The net reactions of both pathways can be stated in the following simplified form (Table 15.7).

The gap of reduction equivalents of the DXP pathway would be resolved in case a molar ratio of 2 : 1 of the DXP pathway versus the MVA pathway could be implemented. Though

Table 15.6 Adjusted MVA pathway via isoprenol toward isoprene.

Reaction	Enzyme
1,5 Glucose + 3 CoA + 3 ADP + 3 Pi + 6 NAD^+ → 3 Acetyl-CoA + 3 CO_2 + 3 ATP + 6 NADH + 6 H^+ + 3 H_2O	
2 Acetyl-CoA → Acetoacetyl-CoA + CoA	
Acetoacetyl-CoA + Acetyl-CoA + H_2O → HMG-CoA + CoA	
HMG-CoA + 2 NADPH + 2 H^+ → Mevalonate + CoA + 2 $NADP^+$	
Mevalonate → Isoprenol + CO_2 + H_2O	Fatty acid decarboxylase
Isoprenol → Isoprene + H_2O	Oleate hydratase
1,5 Glucose + 3 ADP + 3 Pi + 6 NAD^+ + 2 NADPH → Isoprene + 4 CO_2 + 3 ATP + 4 H_2O + 2 $NADP^+$ + 6 NADH + 4 H^+	

Source: Adapted from Yang et al. (2016b).

Table 15.7 Simplified net reaction of the MVA and the DXP pathways toward isoprene.

MVA pathway: 1.5 Glucose + 4 $NAD(P)^+$ → Isoprene + 4 CO_2 + 4 NAD(P)H
DXP pathway: Glucose + 3 ATP + 2 NADH → Isoprene + CO_2 + 3 ADP + 2 NAD^+

Table 15.8 Simplified net reaction with matching combination of MVA and DXP pathway.

MVA pathway: 1.5 Glucose + 4 $NAD(P)^+$ → Isoprene + 4 CO_2 + 4 NAD(P)H
DXP pathway: 1.25 Glucose + 3.75 ATP + 2.5 NADH → 1.25 Isoprene + 1.25 CO_2 + 3.75 ADP + 2.5 NAD^+
2.75 glucose → 2.25 isoprene + 5.25 CO_2

there are still 3 mol ATP per mol isoprene generated via the DXP pathway to be addressed, which requires 1.2 mol reduction equivalents in case it is assumed that 1 mol NAD(P)H provides 2.5 mol ATP in an aerobic process. Given a pathway ratio of 2 : 1, the 1.2 mol reduction equivalents per mol isoprene (generated via the DXP pathway) imply that additionally, 0.6 mol isoprene needs to be generated via the MVA pathway, which derives an overall molar ratio of 1.25 for the DXP pathway versus the MVA pathway (2 : 1.6). The net reaction equation of both pathways in the matching ratio can be stated as follows (Table 15.8).

The carbon atoms are balanced (16.5 C atoms on both sites), and 66 electrons on the left side (24 mol electrons per mol glucose × 2.75) correspond to 63 electrons on the right site (28 mol electrons per mol isoprene × 2.25) since 1.5 mol NAD(P)H or 3 electrons are allocated to respiratory chain phosphorylation and are transferred to oxygen as the final acceptor. This estimation with a resulting maximum carbon yield of 0.309 kg isoprene per kg glucose (2.25 × 68.1 g/mol/2.75 × 180.2 g/mol) matches the published considerations (Yang et al. 2016a).

The synergistic metabolic engineering approach has been implemented in *E. coli,* and an impressive carbon yield of 0.267 g isoprene per kg glucose could be obtained. The titer was 24 g/l after a batch duration of 22 h, resulting in a productivity of 1.09 g/l×h (Yang et al. 2016a).

15.4.6 Isoprene via MVA Pathway in *Clostridium ljungdahlii* with CO/H_2 and Fructose as Feedstock

The emerging syngas fermentation technology might also provide a chance to use carbon monoxide and hydrogen as carbon and energy sources for the synthesis of isoprene. The central R&D department of Dupont has evaluated the suitability of syngas fermentation and equipped the acetogenic host *C. ljungdahlii* with a recombinant MVA pathway (Diner et al. 2018). *C. ljungdahlii* is gram-positive, strictly anaerobe, and possesses the autotrophic Wood–Ljungdahl pathway to convert carbon dioxide, carbon monoxide, and hydrogen into acetyl-CoA though is also able to grow heterotrophically on organic substrates like hexoses and pentoses. The genes were put on plasmids and could be successfully introduced into the host via electroporation. The synthesis relies on an efficient supply of acetyl-CoA, which is either provided by the existing Wood–Ljungdahl pathway or glycolysis followed by the pyruvate dehydrogenase. The host has been fed with syngas – 35% carbon monoxide, 36% hydrogen, 18% carbon dioxide, and 11% argon – and fructose as heterotrophic feedstock as well (Figure 15.10).

The overall intention of the mixotrophic pathway is to partially recapture the carbon dioxide released in the conversion of pyruvate into acetyl-CoA via the Wood–Ljungdahl pathway, given that sufficient reduction equivalents are available to provide the needed electrons. The decisive criterion to derive a judgment on the suitability of the overall concept is the ATP supply for the conversion of MVA into IPP. The MVA pathway pursued in an aerobic system is able to generate a surplus of ATP since the surplus of reduction equivalents can be bargained into ATP via respiratory chain phosphorylation. However, this is not possible in an anaerobic system required for the growth of a strictly anaerobic bacterium like *C. ljungdahlii*. Accordingly, the surplus of reduction equivalents of the MVA pathway is foreseen to support the Wood–Ljungdahl pathway in the generation of reduced ferredoxin such that a proton motive force is enabled to generate ATP via the Rnf complex and the related ATP synthase. A net calculation of the reaction – to describe the share and molar interaction of the Wood–Ljungdahl pathway and glycolysis – is difficult since it is not exactly known how efficiently electron bifurcation and the related generation of reduced ferredoxin can be facilitated in the respective recombinant system. It has been proposed that a related acetogenic bacterium, *Acetobacterium woodii*, is not able to run the Wood–Ljungdahl pathway toward isoprene with a surplus of ATP, but that the insufficient ATP generation would render the concept as not feasible (Bertsch and Mueller 2015). A more detailed analysis specified to pathway characteristics of *C. ljungdahlii* also taking maintenance ATP demand into consideration is required to derive a conclusive evaluation of the concept.

15.5 Metabolic Engineering Toward Mevalonate

15.5.1 Mevalonate via an Engineered MVA Pathway in *S. cerevisiae*

A proof of concept for the fermentative production of MVA in *S. cerevisiae* has been reported (Wegner et al. 2021). The pathway proceeds basically via the classical MVA pathway, though several improvements were implemented. It is generally acknowledged that the supply of cytosolic acetyl-CoA defines a critical point in the metabolic engineering strategy. Accordingly, the pyruvate-dehydrogenase bypass was implemented combined with the introduction of non-oxidative glycolysis to facilitate an efficient provision of acetyl-CoA (Bogorad et al. 2013). The transacetylase plays a central role in this pathway design since the enzyme channels acetyl-CoA (derived without C-loss) into the acetyl-CoA pool feeding the MVA pathway (Amyris 2016). The competing conversion of acetaldehyde into ethanol via the ADH1 enzyme and the conversion of glyceraldehyde-3-phosphate into glycerol via the GPD1 enzyme were blocked. Additionally, the availability of coenzyme A (CoA) was supported by the overexpression of the pantothenate kinase (encoded by *cab1*), and pantothenate as major building block was supplemented to the medium (Figure 15.11).

The overexpression of the acetyl-CoA synthetase encoded by the *acs* gene combined with the overexpression of the acetaldehyde dehydrogenase encoded by the *ald6* gene defines also critical modules of the metabolic engineering strategy. The use of the acetyl-CoA synthetase from *S. enterica* (Se-acs^{L641P}), which is feedback-insensitive, has been shown to

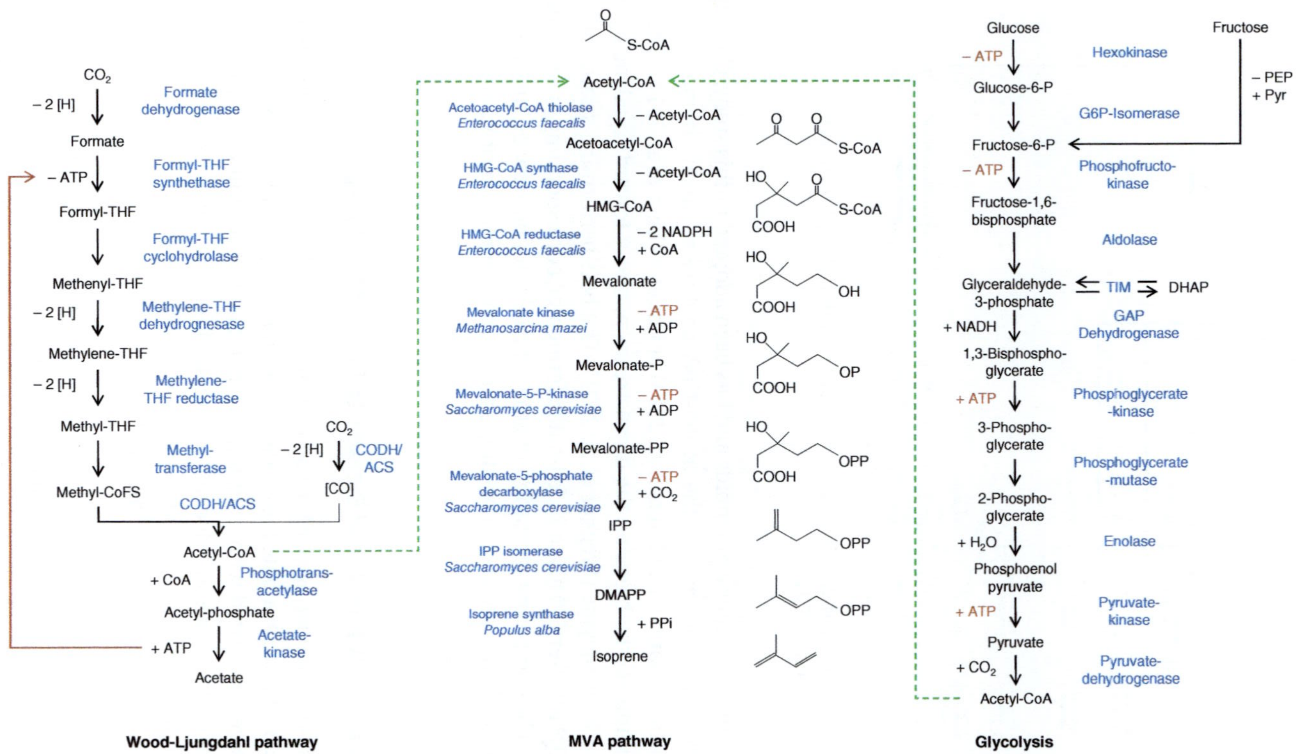

Figure 15.10 Mixotrophic pathway for the synthesis of isoprene in an acetogenic bacterium. Source: Adapted from Diner et al. (2018).

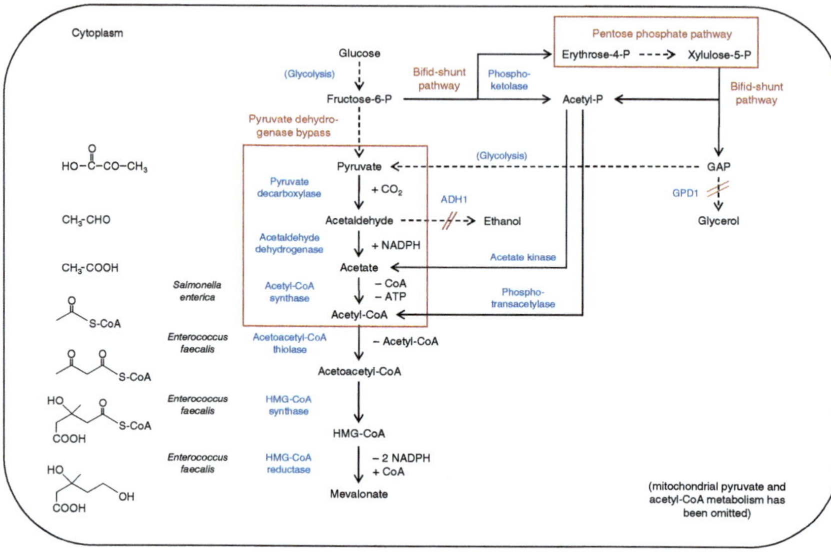

Figure 15.11 Engineered pathway for the synthesis of mevalonate in *S. cerevisiae*. Source: Adapted from Wegner et al. (2021).

support the synthesis of other isoprenoids and has been adopted in this study. Corresponding engineering studies addressing the acetyl-CoA pool were already pursued for other target products like α-santalene (Chen et al. 2013) or butanol (Lian et al. 2014). The results could also be improved by blocking the downstream pathway toward squalene by downregulating the ERG9 enzyme (Wegner et al. 2021). The best-performing strain achieved a MVA titer of 13.3 g/l after 160 h in a 2 l fed-batch fermenter, corresponding to a productivity of 0.083 g/l×h. The competing formation of ethanol could not entirely be suppressed, and a titer of about 6 g/l was detected. The supporting supplementation of pantothenate was 15 μM (Wegner et al. 2021).

15.5.2 Mixing of Cell-Free *E. coli* Extracts to Derive Mevalonate

Cell-free metabolic engineering (CFME) defines the approach to mix and incubate isolated enzymes and cofactors to evaluate the efficiency of a metabolic pathway. This approach has several advantages compared with traditional fermentation since unknown impacts of the cytoplasmic environment are avoided and several build-to-test iterations can easily be pursued without transformation of a production host. Especially in case, a rather long pathway shall be pursued a step-by-step experimental setup can be chosen, adding only one additional step after the efficiency of the preceding pathway section has been ensured (Dudley et al. 2015). This approach has also been applied to the MVA pathway. Hereby crude lysates were applied instead of completely purified and isolated enzymes and cofactors to limit the efforts for isolation and purification (Dudley et al. 2019). The experiment was designed such that glycolysis and pyruvate dehydrogenase, which convert glucose into acetyl-CoA were handled by the cell lysate, and three further steps were initiated with the addition of the respective enzyme lysates (Table 15.9).

Table 15.9 Conversion of glucose into acetyl-CoA via the pyruvate-dehydrogenase bypass.

1.5 Glucose + 3 CoA + 3 ADP + 3 Pi + 6 NAD^+	
→ 3 Acetyl-CoA + 3 CO_2 + 3 ATP + 6 NADH + 6 H^+ + 3 H_2O	
2 Acetyl-CoA → Acetoacetyl-CoA + CoA	Acetyl-CoA acetyltransferase
Acetoacetyl-CoA + Acetyl-CoA + H_2O → HMG-CoA + CoA	HMG-CoA synthase
HMG-CoA + 2 NADPH + 2 H^+ → Mevalonate + CoA + 2 $NADP^+$	HMG-CoA reductase
1.5 Glucose + 2 NADPH + 6 NAD^+ + 3 ADP + 3 Pi +	
→ Mevalonate + 3 CO_2 + 2 $NADP^+$ + 6 NADH + 4 H^+ + 3 ATP + 2 H_2O	

It can be seen that the mixture of cell lysates derives a surplus of energy equivalents. Additionally, the enzyme extract includes microsome vesicles that produce ATP via the proton motive force. On the other hand, the supply of reduction equivalents for the HMG-CoA reductase requires management attention since the enzyme requires NADPH instead of NADH derived from glycolysis. Several isoenzymes of the HMG-CoA reductase have been tested. The isoenzyme of *S. cerevisiae* exclusively relies on NADPH as a cofactor and provides as expected the lowest MVA titer. However, the low titer might also be explained by the circumstance that the enzyme is truncated and might not display its full activity in the biocatalytic system. It is interesting to note that a molar calculation revealed that the cell lysate was able to generate NADPH, which might be enabled via the pentose phosphate pathway or the soluble transhydrogenase (encoded by *sthA*) or the membrane-bound transhydrogenase (encoded by *PntBA*, Dudley et al. 2019). The titer could be increased threefold with the supplementation of the HMG-CoA reductase from *E. faecalis,* which is also NADPH-dependent but is a bifunctional enzyme linked to the acetyl-CoA acetyltransferase. The highest titer was achieved with the HMG-CoA reductase from *S. aureus,* which is able to use NADPH and NADH as well.

The R&D attempt achieved a general proof of concept. Given several optimization rounds, an in vitro titer of 17.6 g/l MVA after 20 h corresponding to a productivity of 0.88 g/l×h could be obtained. The carbon yield – unusually indicated per mol C atom – was 0.35 mol carbon per mol carbon substrate (Dudley et al. 2019). Given these impressive results, it remains to be seen if the original R&D method might finally also be applied as production technology.

15.5.3 Isoprene via Mevalonate in a Two-Step Process

The pathway toward isoprene via MVA might be separated into two separate fermentation procedures. This approach has been chosen to identify the most suitable process conditions for each single step that might profit from a dedicated steering. The conversion of glucose into MVA achieved a MVA titer of 84 g/l combined with a productivity of 1.8 g/l×h and *E. coli* as host (Liu et al. 2017). Afterwards, the biomass was removed via centrifugation, and the intermediate product MVA contained in the remaining medium was converted into mevalonolactone initiated with 3 HCl at pH 2.0, 45 °C and an incubation for 1 h. Mevalonolactone was isolated with ethylacetate and purified via vacuum rotary evaporation. After neutralization with NaOH, mevalonolactone was fed together with glucose into the second fermentation reactor, which achieved an isoprene titer of 11 g/l combined with a productivity of 0.23 g/l×h also with *E. coli* as the host.

The separation of the process into two steps allowed a high productivity in the first fermentation. However, usually the carbon yield is a decisive criterion to prepare a commercial process. It is claimed that the separation into two steps improved the carbon yield from 0.23 g MVA/g glucose up to 0.368 g MVA/g glucose (Liu et al. 2017). For comparison, the maximum carbon yield can be estimated with 0.54 g MVA/g glucose in case it is assumed that 1.5 mol glucose is needed for the synthesis of 1 mol MVA and that the consumed NADPH reduction equivalents can be recycled with the surplus of NADH.

15.5.4 Isoprene via Chemical Decarboxylation of Mevalonolactone

An interesting semi-synthetical approach combining fermentation and organic chemistry has been pursued within the multi-partner project "BioCatPolymers," which was funded by the EU BioCatPolymers initiative (Whited et al. 2010). The research project was a three-year activity supported by the European Union's Horizon 2020 research and innovation program. It pursued the fermentative production of mevalonolactone, the correlating cyclic anhydride of MVA, based on a recombinant *E. coli* host with biomass as feedstock. After the fermentation, the cells were removed and the broth titer of mevalonolactone was concentrated up to 500 g/l by vacuum evaporation at 80 °C and 0.5 bar. The intermediate was then catalytically converted into either 3-methyl-pentanediol or isoprene. In this context, especially the chemical conversion of mevalonolactone toward isoprene is of interest (Figure 15.12).

The reaction is pursued at temperatures of 225–250 °C and is facilitated with an amorphous SiO_2/Al_2O_3 catalyst, which avoids a secondary oligomerization of the generated isoprene. The project activities focused to a great extent on the optimization of the catalyst, and the reaction conditions were optimized. Finally, a decarboxylation yield of 60% of the theoretical maximum was achieved (Heracleous et al. 2020).

15.5.5 Overview of Reported Fermentation Performance Data

A selective overview of the published performance data for the fermentative production of isoprene and MVA indicates that already remarkable titers and productivities have been achieved (Table 15.10). The so far highest isoprene titer has been reported for the research endeavors of Genencor (acquired by Danisco) and The Goodyear Tire & Rubber Company (Whited et al. 2010 as publication; Cervin et al. 2009 as patent).

It should be kept in mind that the conventional fermentation titers are indicated in g/l (broth) since the product usually stays in the liquid phase and accumulates in the broth.

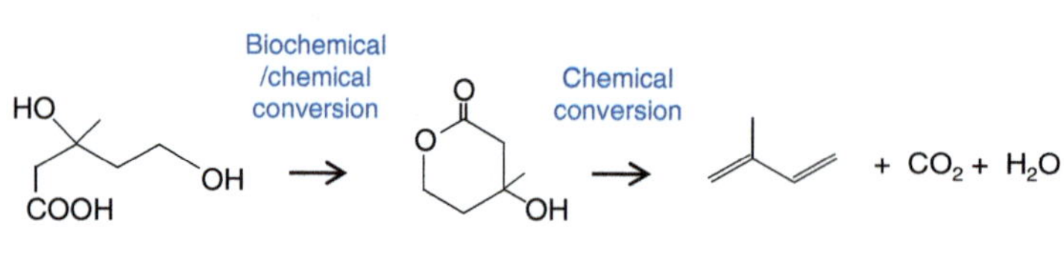

Figure 15.12 Conversion of mevalonate into mevalonolactone and isoprene. Source: Walter Koch.

Table 15.10 Selection of published performance data for the fermentative production of isoprene. The results of Genencor & Goodyear achieved a cell productivity of 0.85 g isoprene per g dry cell weight (Whited et al. 2010). For a more comprehensive overview, other publications may be referred to by Isar et al. (2022).

Host	Product	Isoprene synthase	Titer	Batch-time	Productivity	Carbon yield	Source
E. coli	Isoprene	*P. trichocarpa*	25 g/l	72 h	0.35 g/l×h	n.a.	Lee et al. (2020)
E. coli	Isoprene	*P. trichocarpa*	1.8 g/l	36 h	0.05 g/l×h	n.a.	Kim et al. (2016)
E. coli	Isoprene		24 g/l	22 h	1.09 g/l×h	267 g/kg glc	Yang et al. (2016a)
E. coli	Isoprene	*Populus alba*	6.3 g/l	40 h	0.16 g/l×h	70 g/kg glc	Yang et al. (2012)
S. cerevisiae	Isoprene	*Kudzu vine*	37 mg/l	70 h	0.53 mg/l×h	n.a.	Lv et al. (2014)
E. coli	Isoprene	*n.a.*	60 g/l	30 h	2.00 g/l×h	110 g/kg glc	Whited et al. (2010)

Host	Product	HMGR	Titer	Batch-time	Productivity	Carbon yield	Source
S. cerevisiae	MVA	*E. faecalis*	13.3 g/l	160 h	0.08 g/l×h	n.a.	Wegner et al. (2021)
(Cell lysate)	MVA	*Staphylococcus aureus*	17.6 g/l	20 h	0.88 g/l×h	n.a.	Dudley et al. (2019)
E. coli	MVA	n.a.	84 g/l	47 h	1.80 g/l×h	368 g/kg glc	Liu et al. (2017)
E. coli	MVA	*E. faecalis*	47 g/l	50 h	0.94 g/l×h	n.a.	Tabata and Hashimoto (2004)

However, isoprene has a high vapor pressure and migrates into the reactor headspace. The generated isoprene is measured in the reactor headspace and nevertheless reported as recalculated broth titer: "Bioisoprene titer was measured from the off-gas using a mass spectrometer and is presented as g/l fermentation broth" (Whited et al. 2010).

15.6 Downstream Processing

The isolation and purification of isoprene profits from the high volatility of the gaseous product, which after secretion, migrates into the reactor headspace due to physical reasons. The isoprene molecule has a vapor pressure of 400 mmHg at 20 °C. The culture medium is substantially free of the product (Whited et al. 2010). The physically solved isoprene share remaining in the broth is often not indicated, though it may be concluded that the share can be neglected. In case a significant share is measured, specific measures are recommended to maximize the carbon yield of the overall process. Dodecane might be added to the broth, and a second phase is formed at the batch end, which can be separated from the broth by decantation (Kim et al. 2016). Dodecane is used in several protocols developed for the fermentative production of terpenoids, does not display a negative impact on the host's performance, and can readily be recycled after the lipophilic product has been extracted (Rodriguez et al. 2014; Vickers et al. 2017). Alternatively, the broth might be channeled into a vacuum unit, and the remaining isoprene gas would be pulled into the gaseous phase.

Accordingly, the product by itself provides an in situ product recovery (Freeman et al. 1993) and can be captured as gas. A specific step for the separation of biomass or by-products is not needed. The product in the reactor headspace is given as a gaseous mixture with mainly nitrogen, oxygen, and carbon dioxide, which can be separated according to their different condensation points. It has been evaluated to coproduce lactic acid and isoprene. The highest titers amount to 16.1 g/l lactic acid combined with 0.1 g/l isoprene after a batch duration of 40 h (Cheng et al. 2021). Lactic acid would be extracted from the broth, and isoprene would be isolated from the gas mixture in the reactor headspace. However, two distinct processes for product extraction and isolation are required, and it remains to be seen inasmuch this concept is further developed to commercial maturity.

From an EHS point of view, the high flammability of isoprene and the related explosion risk need to be addressed in the downstream process, especially because fermentation protocols using the MVA pathway are pursued as an aerobe process. A direct further processing of the gas mixture in the reactor headspace including the immediate separation of isoprene and oxygen might be recommended.

15.7 Industrial Application and Capacity

The fermentative production of isoprene was pursued by several industry consortia, which were formed after 2010, when the United States shale gas revolution triggered the substitution of naphtha by ethane as alternative feedstock. This feedstock shift caused a reduction of the isoprene by-product generated by the cracker industry. Accordingly, the tire industry initiated a research initiative and searched for biotech companies as cooperation partners: Goodyear partnered with Dupont (Whited et al. 2010), Michelin with Amyris, and

Bridgestone with Ajinomoto. Furthermore, respective R&D activities were announced by GlycosBio, Aemetis, and Lanzatech (Liew et al. 2016).

However, a fermentative industrial isoprene technology has so far not been commercialized, although the published technology achievements of Goodyear and Genencor (later Danisco and Dupont) in *E. coli* with a titer of 60 g/l combined with a reasonable productivity should define a good starting point for a commercial technology (Whited et al. 2010). Goodyear is a leading producer of isoprene in the United States with a production capacity of about 120 kt pa in Beaumont, Texas (Whited et al. 2010). The discontinuation of R&D might be explained commercially in case the formerly petrochemical volume limitation did not occur. Alternatively, the failed commercialization might indicate technical hurdles of fermentative isoprene that could not be overcome.

References

Amyris (2016). Use of phosphoketolase and phosphotransacetylase for production of acetyl-coenzyme A derived compounds, Invertors: Hawkins, K.M., Mahatdejkul-Meadows, T.T., Meadows, A.L., Pickens, L.B., Tai, A., Tsong, A.E., US patent 9,410,214 B2.

Arpe, H.-J. (2007). *Industrielle Organische Chemie. Bedeutende Vor- und Zwischenprodukte*, 126–131. Wiley.

Bertsch, J. and Mueller, V. (2015). Bioenergetic constraints for conversion of syngas to biofuels in acetogenic bacteria. *Biotechnology for Biofuels* 8: 210. https://doi.org/10.1186/s13068-015-0393-x.

Bian, G., Deng, Z., and Liu, T. (2017). Strategies for terpenoid overproduction and new terpenoid discovery. *Current Opinion in Biotechnology* 48: 234–241. https://doi.org/10.1016/j.copbio.2017.07.002.

Bogorad, I.W., Lin, T.-S., and Liao, J.C.A. (2013). Synthetic non-oxidative glycolysis enables complete carbon conservation. *Nature* 502: 693–698.

Bohlmann, J. and Keeling, C.I. (2008). Terpenoid biomaterials. *The Plant Journal* 54: 656–669.

Breitmaier, E. (2005). *Terpene.* Aromen, Düfte, Pharmaka, Pheromone: Wiley-VCH.

Cervin, M.A., Whited, G.M., Chotani, G.K. et al. (2009). As patent inventors. Compositions and methods for producing isoprene, US 8709785.

Chen, Y., Daviet, L., Schalk, M. et al. (2013). Establishing a platform cell factory through engineering of yeast acetyl-CoA metabolism. *Metabolic Engineering* 15: 48–54. https://doi.org/10.1016/j.ymben.2012.11.002.

Cheng, T., Liu, H., Zou, H. et al. (2017). Enzymatic process optimization for the *in vitro* production of isoprene from mevalonate. *Microbial Cell Factories* 16: 8. https://doi.org/10.1186/s12934-016-0622-4.

Cheng, T., Liang, X., Wang, Y. et al. (2021). Co-production of isoprene and lactate by engineered *Escherichia coli* in microaerobic conditions. *Molecules* 26: https://doi.org/10.3390/molecules26237173.

Clomburg, J.M., Qian, S., Tan, Z. et al. (2019). The isoprenoid alcohol pathway, a synthetic route for isoprenoid biosynthesis. *PNAS* 116 (26): 12810–12815. https://doi.org/10.1073/pnas.1821004116.

Diner, B.A., Miles, J.F., Scotcher, M.C. et al. (2018). Synthesis of heterologous mevalonic acid pathway enzymes in *Clostridium ljungdahlii* for the conversion of fructose and of syngas to mevalonate and isoprene. *Applied and Environmental Microbiology* 84 (1): e01723–e01717. https://doi.org/10.1128/AEM.01723-17.

Dudley, Q.M., Karim, A.S., and Jewett, M.C. (2015). Cell-free metabolic engineering: biomanufacturing beyond the cell. *Biotechnology Journal* 10: 69–82.

Dudley, Q.M., Anderson, K.C., and Jewett, M.C. (2019). Cell-free mixing of *Escherichia coli* crude extracts to prototype and rationally engineer high-titer mevalonate synthesis. *ACS Synthetic Biology* 5 (12): 1578–1588. https://doi.org/10.1021/acssynbio.6b00154.

Eiben, C.B., de Rond, T., Bloszies, C. et al. (2019). Mevalonate pathway promiscuity enables noncanonical terpene production. *Synthetic Biology* 8: 2238–2247. https://doi.org/10.1021/acssynbio.9b00230.

Emmerstorfer-Augustin, A., Moser, S., and Pichler, H. (2016). Screening for improved isoprenoid biosynthesis in microorganisms. *Journal of Biotechnology* 235: 112–120. https://doi.org/10.1016/j.jbiotec.2016.03.051.

Freeman, A., Woodley, J.M., and Lilly, M.D. (1993). In situ product removal as a tool for bioprocessing. *Nature* 11: 1007–1012. The paper is the first publication on *in situ* product recovery in fermentation processes.

George, K.W., Thompson, M.G., Kang, A. et al. (2015). Metabolic engineering for the high-yield production of isoprenoid-based C5 alcohols in *E. coli*. *Scientific Reports* 5: 11128. https://doi.org/10.1038/srep11128.

Guenther, A., Karl, T., Harley, P. et al. (2006). Estimates of global terrestrial isoprene emissions using MEGAN (Model of Emissions of Gases and Aerosols from Nature). *Atmospheric Chemistry and Physics* 6: 3181–3210.

Guo, J., Cao, Y., Liu, H. et al. (2019). Improving the production of isoprene and 1,3-propanediol by metabolically engineered *Escherichia coli* through recycling redox cofactor between the dual pathways. *Applied Microbiology and Biotechnology* 103 (2597): 2597–2608. https://doi.org/10.1007/s00253-018-09578-x.

Hara, K.Y., Araki, M., Okai, N. et al. (2014). Development of bio-based fine chemical production through synthetic bioengineering. *Microbial Cell Factories* 13: 173. https://doi.org/10.1186/s12934-014-0173-5.

Heracleous, E., Pachatouridou, E., Louie, L. et al. (2020). Efficient route for the production of isoprene via decarboxylation of bioderived mevalonolactone. *ACS Catalysis* 10: 9649–9661. https://doi.org/10.1021/acsatal.0c01438.

http://M.net (2012). Dupont to acquire Danisco for $6.3 billion. https://www.manufacturing.net/operations/news/13087221/dupont-to-acquire-danisco-for-63-billion (accessed 20 December 2022).

Huttanus, H.M. and Feng, X. (2017). Compartmentalized metabolic engineering for biochemical and biofuel production. *Biotechnology Journal* 12: https://doi.org/10.1002/biot.201700052.

Invista (2015). Methods for biosynthesis of isoprene. Inventors: Botes, A.L., Conradie, A.V.E. patent WO2015/021045.

Isar, J., Jain, D., Joshi, H. et al. (2022). Microbial isoprene production: an overview. *World Journal of Microbiology and Biotechnology* 38: 122. https://doi.org/10.1007/s11274-022-03306-4.

Julsing, M.K., Rijpkema, M., Woerdenbag, H.J. et al. (2007). Functional analysis of genes involved in the biosynthesis of isoprene in *Bacillus subtilis*. *Applied Genetics and Molecular Biotechnology* 75: 1377–1384. https://doi.org/10.1007/s00253-007-0953-5.

Kang, A., George, K.W., Wang, G. et al. (2016). Isopentenyl diphosphate (IPP)-bypass mevalonate pathways for isopentenol production. *Metabolic Engineering* 34: 25–35. https://doi.org/10.1016/j.ymben.2015.12.002.

Kim, J.-H., Wang, C., Jang, H.-J. et al. (2016). Isoprene production by *Escherichia coli* through the exogenous mevalonate pathway with reduced formation of fermentation byproducts. *Microbial Cell Factories* 15: 214. https://doi.org/10.1186/s12934-016-0612-6.

Kozak, B.U., van Rossum, H.M., Luttik, M.A.H. et al. (2014). Engineering acetyl coenzyme A supply: functional expression of a bacterial pyruvate dehydrogenase complex in the cytosol of *Saccharomyces cerevisiae*. *mBio* 5 (5): e01696-14.

Lee, H.-W., Park, J.-H., Kim, W.-K. et al. (2020). Engineered *Escherichia coli* strains as platforms for biological production of isoprene. *FEBS Open Bio* 10: 780–788. https://doi.org/10.1002/2211-5463.12829.

Li, M., Nian, R., Xian, M., and Zhang, H. (2018). Metabolic engineering for the production of isoprene and isopentenol by *Escherichia coli*. *Applied Microbiology and Biotechnology* 102: 7725–7738. https://doi.org/10.1007/s00253-018-9200-5.

Lian, J., Si, T., Nair, N.U., and Zhao, H. (2014). Design and construction of acetyl-CoA overproducing *Saccharomyces cerevisiae* strain. *Metabolic Engineering* 24: 139–149. https://doi.org/10.1016/j.ymben.2014.05.010.

Liew, F.-M., Martin, M.E., Tappel, R.C. et al. (2016). Gas fermentation – a flexible platform for commercial scale production of low-carbon-fuels and chemicals from waste and renewable feedstocks. *Frontiers in Microbiology* 7: 694. https://doi.org/10.3389/fmicb.2016.00694.

Liu, H., Cheng, T., Zou, H. et al. (2017). High titer mevalonate fermentation and its feeding as a building block for isoprenoids (isoprene and sabinene) production in engineered *Escherichia coli*. *Process Biochemistry* 62: 1–9. https://doi.org/10.1016/j.procbio.2017.07.021.

Lv, X., Xie, W., Lu, W. et al. (2014). Enhanced isoprene biosynthesis in *Saccharomyces cerevisiae* by engineering of the native acetyl-CoA and mevalonic acid pathways with a push-pull-restrain strategy. *Journal of Biotechnology* 186: 128–136. https://doi.org/10.1016/j.jbiotec.2014.06.024.

Martin, V.J.J., Pitera, D.J., Withers, S.T. et al. (2003). Engineering a mevalonate pathway in *Escherichia coli* for production of terpenoids. *Nature Biotechnology* 21 (7): 796–802.

Meadows, A.L., Hawkins, K.M., Tsegaye, Y. et al. (2016). Rewriting yeast central carbon metabolism for industrial isoprenoid production. *Nature* 537: 694–697. https://doi.org/10.1038/nature19769.

Motoyama, K., Sobue, F., Kawaide, H. et al. (2019). Conversion of mevalonate 3-kinase into 5-phosphomevalonate 3-kinase by single amino acid mutations. *Applied and Environmental Microbiology* 85 (9): https://doi.org/10.1128/AEM.00256-19.

Rodriguez, S., Kirby, J., Denby, C.M., and Keasling, J.D. (2014). Production and quantification of sesquiterpenes in *Saccharomyces cerevisiae,* including extraction, detection and quantification of terpene products and key related metabolites. *Nature Protocols* 9 (8): 1980–1996. https://doi.org/10.1038/nprot.2014.132.

Sharkey, T.D. and Monson, R.K. (2017). Isoprene research – 60 years later, the biology is still enigmatic. *Plant, Cell and Environment* 40: 1671–1678. https://doi.org/10.1111/pce.12930.

Shiba, Y., Paradise, E.M., Kirby, J. et al. (2007). Engineering of the pyruvate dehydrogenase bypass in Saccharomyces cerevisiae for high-level production of isoprenoids. *Metabolic Engineering* 9: 160–168.

Silver, G.M. and Fall, R. (1991). Enzymatic synthesis of isoprene from dimethylallyldiphosphate in aspen leaf extracts. *Plant Physiology* 97: 1588–1591.

Silver, G.M. and Fall, R. (1995). Characterization of aspen isoprene synthase, an enzyme responsible for leaf isoprene emission to the atmosphere. *Journal of Biological Chemistry* 270: 13010–13016.

Spitz, P.H. (2019). *Primed for Success. The Story of Scientific Design. How Chemical Engineers Created the Petrochemical Industry*, 118–120. Springer.

Tabata, K. and Hashimoto, S.-I. (2004). Production of mevalonate by a metabolically-engineered *Escherichia coli*. *Biotechnology Letters* 26: 1487–1491.

Vickers, C.E., Williams, T.C., Peng, B., and Cherry, J. (2017). Recent advances in synthetic biology for engineering isoprenoid production in yeast. *Current Opinion in Chemical Biology* 40: 47–56. https://doi.org/10.1016/j.cbpa.2017.05.017.

Vinokur, J.M., Cummins, M.C., Kormann, T.P., and Bowie, J.U. (2016). An adaptation to life in acid through a novel mevalonate pathway. *Scientific Reports* 6: 39737. https://doi.org/10.1038/srep39737.

Wang, C., Zada, B., Wei, G., and Kim, S.-W. (2017). Metabolic engineering and synthetic biology approaches driving isoprenoid production in *Escherichia coli*. *Bioresource Technology* 241: 430–438. https://doi.org/10.1016/j.biortech.2017.05.168.

Wang, S., Cheng, G., Dong, J. et al. (2019). NaCl enhances *Escherichia coli* growth and isoprenol production in the presence of imidazolium ionic liquids. *Bioresource Technology Reports* 6: 1–5. https://doi.org/10.1016/j.biteb.2019.01.021.

Wang, X., Baidoo, E.E.K., Kakumanu, R. et al. (2022). Engineering of isoprenoids production in metabolically versatile microbial host *Pseudomonas putida*. *Biotechnology for Biofuels and Bioproducts* 15: 137. https://doi.org/10.1186/s13068-022-02235-6.

Wegner, S.A., Chen, J.-M., Ip, S.S. et al. (2021). Engineering acetyl-CoA supply and *ERG9* repression to enhance mevalonate production in *Saccharomyces cerevisiae*. *Journal of Industrial Microbiology and Biotechnology* 48: https://doi.org/10.1093/jimb/kuab050.

Whited, G.M., Feher, F.J., Benko, D.A. et al. (2010). Development of a gas-phase bioprocess for isoprene-monomer production using metabolic pathway engineering. *Industrial Biotechnology* 6 (3): 152–163. www.biocatpolymers.eu (accessed 15 December 2022).

Yang, J., Xian, M., Su, S. et al. (2012). Enhancing production of bio-isoprene using hybrid MVA pathway and isoprene synthase in *E. coli*. *PLoS One* 7 (4): https://doi.org/10.1371/journal.pone.0033509.

Yang, C., Gao, X., Jiang, Y. et al. (2016a). Synergy between methylerythritol phosphate pathway and mevalonate pathway for isoprene production in *Escherichia coli*. *Metabolic Engineering* 37: 79–91. https://doi.org/10.1016/j.ymben.2016.05.003.

Yang, J., Nie, Q., Liu, H. et al. (2016b). A novel MVA-mediated pathway for isoprene production in engineered *E. coli*. *BMC Biotechnology* 16: 5. https://doi.org/10.1186/s12896-016-0236-2.

Yao, P., You, S., Qi, W. et al. (2020). Investigation of fermentation conditions of biodiesel by-products for high production of β-farnesene by an engineered *Escherichia coli*. *Environmental Science and Pollution research* 27: 22758–22769. https://doi.org/10.1007/s11356-020-08893-z.

Zheng, Y., Liu, Q., Li, L. et al. (2013). Metabolic engineering of *Escherichia coli* for high-specificity production of isoprenol and prenol as next generation of biofuels. *Biotechnology for Biofuels* 6: 57. http://www.biotechnologyforbiofuels.com/content/6/1/57 (accessed 16 January 2023).

16

Pentamethylenediamine

16.1 Application

Pentamethylenediamine (PMD) is a linear diamine with a molecular weight of 102.2 g/mol. The molecule is also known as 1,5-diaminopentane or cadaverine. Its two terminal amino functions favor the application as a monomer for polyamides. Whereas conventional nylon PA-66 consists of adipic acid (ADA) and hexamethylenediamine (HMD) the innovative polyamide PA-56 consisting of PMD and ADA has been produced and introduced into the market (Xue et al. 2021):

$$
\begin{aligned}
& n\,H_2N\text{–}[CH_2]_5\text{–}NH_2 + n\,HOOC\text{–}[CH_2]_4\text{–}COOH \\
& \rightarrow [\text{–}CO\text{–}(CH_2)_4\text{–}CO\text{–}NH\text{–}(CH_2)_5\text{–}NH\text{–}]_n + 2_{n-1}\,H_2O
\end{aligned}
$$

The fabricated polyamide PA-56 has a melting point of 250 °C, a crystallization point of 220 °C, and a degradation temperature of 410 °C (Xue et al. 2021). Additionally, the innovative polyamides PA-54 and PA-510 have been synthesized to explore polymer functionality not provided by mainstream polyamides PA-66 and PA-6 (Chung et al. 2015). These polymers might completely be derived from renewable resources in case fermentative succinic acid is used for the production of PA-54 and biobased sebacic acid made from castor oil is used for the production of PA-510. In order to further extend the toolbox of biobased polyamides, also the fermentative production of monomers with different chain lengths is under evaluation including 1,3-diaminopropane and 1,4-diaminobutane (Wang et al. 2020; Tsuge et al. 2016).

Polyamides made from carbohydrate via fermentation exhibit similar or identical physical and chemical properties compared to polyamides derived from petrochemical precursors. They are considered as monomers for green plastics and market penetration profits from the low carbon footprints even in case the production costs are higher than petrochemistry. The different number of methylene functions shall provide additional functionality and offer in some applications excellent properties of mechanical strength, moisture absorption, air permeability, and flame retardancy. Respective polymers have been developed for application in apparels, food packaging, automobiles, electronics, and the textile industry (Kim et al. 2019).

PMD may also serve as starting point for the biocatalytic synthesis of 5-aminovalerate or the PMD chassis strain may be leveraged for the fermentative production of 5-aminovalerate

Pathway Design for Industrial Fermentation, First Edition. Walter Koch.

(Jorge et al. 2017). The molecule is additionally evaluated for producing the aliphatic isocyanate pentamethylene diisocyanate (PDI), which is used for polyurethane coatings in various fields (Wang et al. 2021).

16.2 Chemical Synthesis

A chemical synthesis of PMD from petrochemical precursors is possible, and several approaches are under evaluation. The hydrogenation of the dinitrile glutaronitrile as starting point with 5-aminovaleronitrile as an intermediate is according to the synthesis of hexamethylene-diamine based on the hydrogenation of adiponitrile (Huang et al. 2021):

$$NC\text{–}(CH_2)_3\text{–}CN + 2\,H_2 \rightarrow NC\text{–}(CH_2)_3\text{–}CH_2\text{–}NH_2$$
$$NC\text{–}(CH_2)_3\text{–}CH_2\text{–}NH_2 + 2\,H_2 \rightarrow H_2N\text{–}CH_2\text{–}(CH_2)_3\text{–}CH_2\text{–}NH_2$$

The reaction is facilitated with an amorphous Ni alloy catalyst, and the intermediate 5-aminovaleronitrile tends to a spontaneous cyclization toward piperidine. Alternatively, PMD may be produced via semi-synthesis comprising a chemical decarboxylation of fermentative and biobased L-lysine, which is readily available in the market. This reaction may be pursued in a one-step reaction by using 2-cyclohehexen-1-one as a homogeneous catalyst (Huang et al. 2021):

$$H_2N\text{–}[CH_2]_4\text{–}CH(NH_2)\text{–}COOH \rightarrow H_2N\text{–}CH_2\text{–}(CH_2)_3\text{–}CH_2\text{–}NH_2 + CO_2$$

A cadaverine yield of 87.8% has been reported. An alternative two-step decarboxylation of L-lysine (not shown) converts L-lysine with *N*-bromosuccinimide in the first step toward 5-aminovaleronitrile whose nitrile function is then hydrogenated toward PMD. This reaction achieves a PMD yield of 77% (Huang et al. 2021). Nevertheless, the chemical synthesis of PMD can be neglected from a commercial point of view, and the volumes available in the market are derived via fermentation or a biocatalytic decarboxylation of L-lysine. It is also to be considered that marketing activities for the innovative polyamide appeal to its biobased character which is likely not given in case a chemical production is pursued.

16.3 Pathway Design

16.3.1 Glucose as Feedstock via Lysine Decarboxylation

PMD is the decarboxylation product of lysine. Since lysine is efficiently produced with a mature technology from renewable resources with *C. glutamicum* as a host, it is readily available in the market and has replaced the former chemical technologies. Thus, it is recommended to leverage the existing platform and use *C. glutamicum* also for producing PMD:

$$H_2N\text{–}(CH_2)_4\text{–}CH(NH_2)\text{–}COOH \rightarrow H_2N\text{–}(CH_2)_5\text{–}NH_2 + CO_2$$

This reaction is not invented by metabolic engineering, though occurs in natural pathways. It is reported that lysine catabolism in *Streptomyces* spp. proceeds via the decarboxylation of lysine toward PMD which is then via 1-piperidine and δ-aminovalerate further degraded to finally carbon dioxide, water, and ammonia (Madduri et al. 1989). The most straightforward approach would be to overexpress a lysine decarboxylase into commercial strains of *C. glutamicum* with resolved feedback inhibition of the aspartate kinase to fully leverage the existing technology platform. Alternatively, lysine might be applied as feedstock for whole-cell biocatalysis with *E. coli* equipped with an efficient lysine decarboxylase. With respect to the latter option purified lysine or lysine-hydrochloride might be used or an integration of both technologies – reducing efforts for the isolation and purification of lysine – might be pursued. The considerations on the maximum theoretical molar yield and the maximum carbon yield are adopted from lysine. Accordingly, the maximum molar yield can be assumed with 0.82 mol PMD/mol glucose according to the respective figure for lysine (Wittmann and Becker 2007). This translates into a maximum theoretical carbon yield of 0.46 t PMD/t glucose ($0.82 \times 102.2/180.2$).

16.3.2 Methanol as Feedstock via Lysine Decarboxylation

Methanol is one of the major volume chemicals with an overall global production volume of about 56 Mio. t in 2015 (Tullo 2014), 70 Mio. t in 2015, and 80 Mio. t in 2020 (Chemical Week 2020). It is used as chemical building block for producing formaldehyde (26%), the methanol-to-olefins and methanol-to-propylene technology (MTO/MTP, 19%), MTBE/TAME (methyl-tertiary-butyl-ether/tertiary-amyl methyl-ether, 11%), acetic acid (8%), and other application (Chemical Week 2020). Additionally, it is applied as chemical energy storage and partially burnt as fuel in gasoline blending and as a methyl-component in biodiesel. Methanol is usually produced via syngas derived from natural gas with mature technologies (Arpe 2007). It would be available at competitive prices in sufficient quantity in all regions (Methanex company homepage 2022). A main criterion for the decision on a fermentation feedstock is the "energy price" in €/GJ especially in case a fuel shall be produced. With respect to this criterion methanol is clearly favored versus glucose (Table 16.1).

This cost advantage per energy unit is partially compromised by the comparably high carbon footprint of methanol derived from syngas and natural gas, respectively, whereas glucose is derived from renewable resources and has a negative carbon footprint. The carbon footprint of methanol depends on the production technology. Whereas conventional methanol technology is exclusively optimized with respect to production cost alternative technology is under development to reduce the carbon footprint (IRENA and Methanol Institute 2021; Hoppe et al. 2017). These emerging technologies also cover the approach to capture carbon dioxide from industrial processes to be reacted with green hydrogen generated via the electrolysis of water supplied with green power. Different methanol grades with either low cost and high carbon footprint and especially higher cost and low carbon

Table 16.1 Comparison of methanol and glucose.

Feedstock	Average price	Energy content	Energy price
Methanol	350 €/t (IRENA 2021)	22.7 GJ/t HHV	15.4 €/GJ
Glucose	350 €/t (Kölbl 2021)	15.6 GJ/t HHV	22.4 €/GJ

footprint will be available in the market (Tountas et al. 2021; Tountas et al. 2019; Marpani et al. 2017; Goeppert et al. 2014).

16.3.3 Pathway Design: The Ribulose Monophosphate Pathway in *Bacillus methanolicus*

Methylotrophy describes the ability of methylotrophic bacteria to grow on reduced C1 compounds as their sole carbon and energy source. Obligate methylotrophs can exclusively metabolize C1 compounds as carbon and energy source, while facultative methylotrophs may also use carbohydrates. These species have been characterized with respect to their physiology, metabolism, and genomes in the last two decades (Irla et al. 2014) and the genetic and metabolic engineering tools are evolving. The prokaryotic model organisms for methylotrophy are *Bacillus methanolicus* (Hansen et al. 1996; Brautaset et al. 2004; Brautaset et al. 2007) and *Methylobacterium extorquens* (Müller et al. 2015; Ochsner et al. 2015). Most interesting, the ability of *B. methanolicus* to grow on methanol has been shown to be dependent on plasmid pBM19, which gets lost if the cells are grown for more than 140 generations in a minimal medium with mannitol (Jakobsen et al. 2006).

Methanol metabolism in *B. methanolicus* is following the ribulose monophosphate pathway (RuMP). The RuMP cycle is pursued by gram-negative obligate and facultative methylotrophs and gram-positive facultative methylotrophs (Pfeifenschneider et al. 2020). This pathway is described with two versions, namely, the seduheptulose-1,7-bisphosphatase (SBPase variant) and the transaldolase variant (TA variant). It has formerly been thought that *B. methanolicus* uses exclusively the SBPase variant, but it has recently been shown that also the TA variant is operated (Pfeifenschneider et al. 2020). The pathway may be split into three sections fixation, cleavage, and rearrangement. Both pathway versions coincide in their approach for the fixation of methanol and the cleavage section though differ in the rearrangement section for the recycling of the C1 receiving intermediate of the first step (Brautaset et al. 2007; Stolzenberger et al. 2013).

In the SBPase variant, methanol is initially subjected to dehydrogenation with NAD^+ as an electron acceptor to derive formaldehyde. This reaction is catalyzed by the methanol dehydrogenase. The rather nonconventional biochemical C1 building block is then reacted with ribulose-5-phosphate by the hexulosephosphate synthase (Hps) to hexulose-6-phosphate. Hexulose-6-phosphate is isomerized by the phosphohexulose isomerase (Phi) to fructose-6-phosphate, which is channeled into central metabolism. In the first step of the cleavage section, fructose-6-phosphate is phosphorylated to obtain fructose-1,6-bisphosphate, which is split into glyceraldehyde-3-phosphate (GAP) and dihydroxyacetone-phosphate (DHAP). GAP is then reacted with fructose-6-phosphate to obtain xylulose-5-phosphate and erythrose-4-phosphate catalyzed by a transketolase. Erythrose-4-phosphate is then reacted with DHAP to derive seduheptulose-1,7-bisphosphate, which is converted into seduheptulose-7-phosphate facilitated by the fructose-bisphosphatase. Seduheptulose-7-phosphate together with GAP is then reacted with a transketolase to derive the two C5 intermediates ribose-5-phosphate and xylulose-5-phosphate, which can be isomerized to the C1 receiving unit ribulose-5-phosphate. The overall reaction schema displays the advanced rebuilding of the C-backbone with partial overlap of the pentose monophosphate pathway. The crossing lines represent the catalytic transfer of C2-units via two ketolases (Figure 16.1).

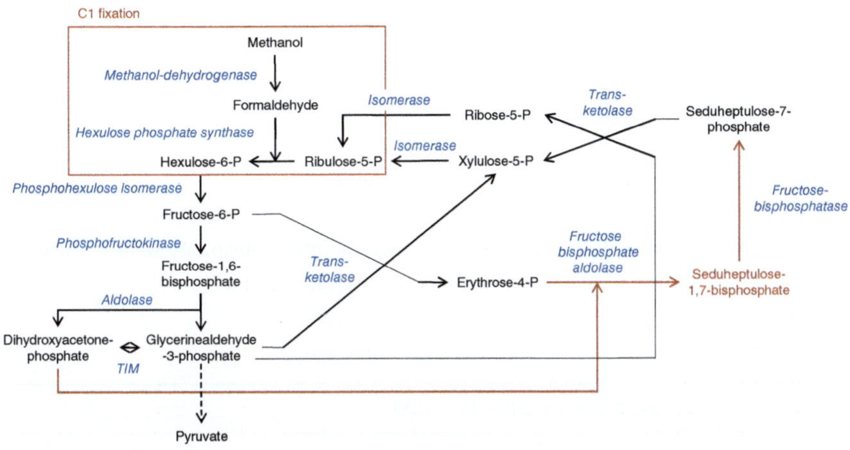

Figure 16.1 Ribulose monophosphate pathway (SBPase variant) in *B. methanolicus* (Pfeifenschneider et al. 2020). Source: Walter Koch.

The fixation of methanol with ribulose-5-phosphate as receiving group can be summarized as follows (Table 16.2).

The enzymes Hps and Phi are described as two different enzymes. In some organisms, both enzymes are linked to one polypeptide with dual function as shown for the archaeon *Pyrococcus horikoshii*. The combination of both enzymes displays a higher conversion than the single enzymes and it is believed that the combination enhances the thermostability of the Phi enzyme (Orita et al. 2005). The subsequent reactions of the ribulose-5-phosphate pathway convert fructose-6-phosphate into ribulose-5-phosphate to prepare the receiving unit for formaldehyde and to generate one mol of glycerinealdehyde-3-phosphate (GAP) for central metabolism ultimately derived from three mol of methanol (Table 16.3).

The overall two reaction cascades can then be combined into a net equation:

$$3\ \text{MeOH} + 2\ \text{ATP} + 3\ \text{NAD}^+ + \text{H}_2\text{O} \rightarrow \text{GAP} + 2\ \text{ADP} + \text{Pi} + 3\ \text{NADH} + 3\ \text{H}^+$$

The gained mol GAP is converted into pyruvate for comparability with glycolysis to obtain the following reaction equation:

$$3\ \text{MeOH} + 4\ \text{NAD}^+ \rightarrow \text{Pyruvate} + 4\ \text{NADH} + 4\ \text{H}^+$$

Alternatively, the RuMP pathway may process via the transaldolase variant which differs in the formation of seduheptulose-7-phosphate, which is derived in the reaction of fructose-6-phosphate with erythrose-4-phosphate to derive seduheptulose-7-phosphate and GAP catalyzed by a transaldolase (Figure 16.2).

Table 16.2 Fixation of methanol into fructose-6-phosphate.

Reaction	Enzyme
$3\ \text{MeOH} + 3\ \text{NAD}^+ \rightarrow 3\ \text{HCHO} + 3\ \text{NADH} + 3\ \text{H}^+$	MeOH dehydrogenase
$3\ \text{HCHO} + 3\ \text{Ribulose-5-P} \rightarrow 3\ \text{Hexulose-6-P}$	Hexulosephosphate synthase (Hps)
$3\ \text{Hexulose-6-P} \rightarrow 3\ \text{F6P}$	Phosphohexulose isomerase (Phi)
$3\ \text{MeOH} + 3\ \text{NAD}^+ + 3\ \text{Ribulose-5-P} \rightarrow 3\ \text{F-6-P} + 3\ \text{NADH} + 3\ \text{H}^+$	

Table 16.3 Generation of ribulose-5-phosphate as receiving group for formaldehyde.

2 F6P + 2 ATP → 2 Fructose-1,6-BP + 2 ADP	Phosphofructokinase (PFK)
2 Fructose-1,6-BP → 2 GAP + 2 DHAP	Aldolase (FBA)
DHAP → GAP	Isomerase
GAP + F6P → Xylulose-5-P + Erythrose-4-P	Transketolase (TKT)
Erythrose-4-P + DHAP → Seduheptulose-BP	(FBA)
Seduheptulose-BP + H_2O → Seduheptulose-7-P+ Pi	Seduheptulose-bisphosphatase (SBPase)
Seduheptulose-7-P + GAP → Ribose-5-P + Xylulose-5-P	Transketolase (TKT)
Ribose-5-P → Ribulose-5-P	Isomerase
2 Xylulose-5-P → 2 Ribulose-5-P	Isomerase
3-F6P + 2 ATP + H_2O→ 3 Ribulose-5-P + 2 ADP + Pi + GAP	

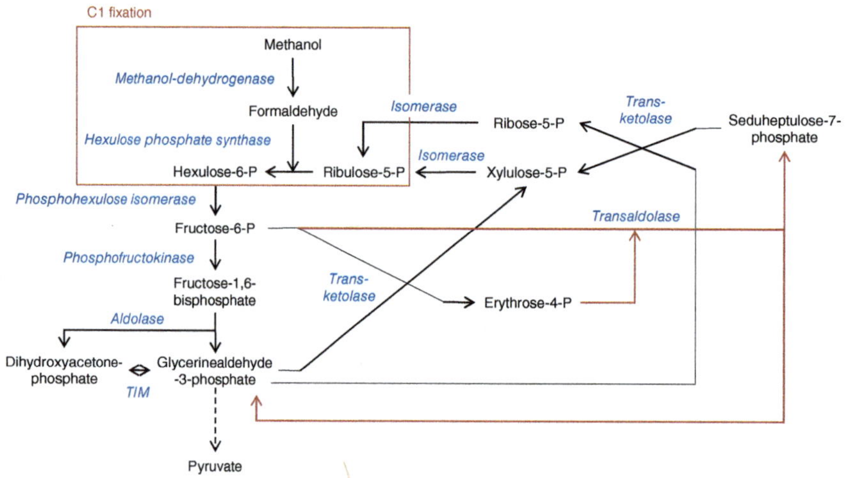

Figure 16.2 Ribulose monophosphate pathway (aldolase variant) in *B. methanolicus* (Pfeifenschneider et al. 2020). Source: Walter Koch.

The net reaction equation of the pathway can be derived as follows (Table 16.4):

Table 16.4 Generation of GAP ex MeOH according to the RuMP pathway (transaldolase variant).

3 MeOH + 3 NAD^+ → 3 HCHO + 3 NADH + 3 H^+	MeOH Dehydrogenase
3 HCHO + 3 Ribulose-5-P → 3 Hexulose-6-P	Hexulosephosphate synthase (Hps)
3 Hexulose-6-P → 3 F6P	Phosphohexulose isomerase (Phi)
F6P + ATP → Fructose-1,6-BP + ADP	Phosphofructokinase (PFK)
Fructose-1,6-BP → GAP + DHAP	Aldolase (FBA)
DHAP → GAP	Isomerase
GAP + F6P → Xylulose-5-P + Erythrose-4-P	Transketolase
Erythrose-4-P + F6P → Seduheptulose-7-P + GAP	Transaldolase
Seduheptulose-7-P + GAP → Ribose-5-P + Xylulose-5-P	Transketolase
Ribose-5-P → Ribulose-5-P	Isomerase
2 Xylulose-5-P → 2 Ribulose-5-P	Isomerase
3 MeOH + ATP + 3 NAD^+ → GAP + ADP + 3 NADH + 3 H^+	

The overall reaction cascade derives an energy investment of only one mol of ATP per mol GAP. The two versions of the RuMP differ in their net ATP demand (Woolston et al. 2018) since the aldolase variant consumes less fructose-1,6-bisphosphate – derived from glucose-6-phosphate – than the SBPase variant.

The gained mol GAP is converted into pyruvate for comparability with glycolysis to obtain the following reaction equation:

$$3\ \text{MeOH} + 4\ \text{NAD}^+ + \text{ADP} + \text{Pi} \rightarrow \text{Pyruvate} + 4\ \text{NADH} + 4\ \text{H}^+ + \text{ATP}$$

The application of methanol as the sole carbon and energy source provides one mol of pyruvate and four reduction equivalents derived from 3 mol of methanol. The three-mol of methanol contains in total of 18 electrons without considering the reduced oxygen atom of the hydroxyl group. One mol of pyruvate contains 10 electrons (without counting the electrons allocated to the reduced oxygen atoms) plus 8 electrons absorbed by the four mol of NADH add up to 18 electrons altogether. The reaction displays an aligned electron balance. On the other hand, conventional glycolysis can be stated as follows:

$$\text{Glucose} + 2\ \text{NAD}^+ + 2\ \text{ADP} + 2\ \text{Pi} \rightarrow 2\ \text{Pyruvate} + 2\ \text{NADH} + 2\ \text{ATP} + 2\ \text{H}_2\text{O} + 2\ \text{H}^+$$

The different ratio of reduction equivalents per mol pyruvate impacts the suitability of methanol feed and the RuMP for defined metabolic engineering pathways. In case the product has a higher reduction level than glucose – and a share of glucose would need to be invested for the provision of reduction equivalents – methanol or glycerol might be favored feedstocks given their higher reduction level and higher electron density (Table 16.5).

The conventional *B. methanolicus* strains seem to be suited for the development of an industrial technology for glutamate, lysine, and PMD (the decarboxylation product of lysine) given their natural metabolic profile. *B. methanolicus* is a natural overproducer of glutamate and in a methanol fed-batch cultivation, a titer of 60 g/l has been shown. With respect to lysine, mutants of *B. methanolicus* have been developed which show a lysine titer of 11 g/l and 65 g/l lysine (Müller et al. 2015). These observations suggest that these strains can efficiently transform methanol into intermediates of the citric acid cycle since glutamate is derived from α-ketoglutarate and lysine is derived from oxalacetate and aspartate. Hereby the enzymatic portfolio of anaplerosis needs to be kept in mind in case *B. methanolicus* shall be positioned as an alternative host versus *C. glutamicum,* which is usually applied for the fermentative production of glutamate and lysine. Whereas *C. glutamicum* possesses a pyruvate carboxylase (PC) and a PEP carboxylase (PEPC) as well, *Bacilli* usually lack the PEP carboxylase (Müller et al. 2015).

Accordingly, the metabolic engineering activities addressing methylotrophy are partially focusing on these products. The demand of 4 mol of NADPH in the metabolic pathway toward lysine and PMD does not fit with the circumstance that the methanol dehydrogenase of *B. methanolicus* is specific to NADH as cofactor (Arfman et al. 1989; Krog et al. n.d.).

Table 16.5 Ratio of reducing equivalents per building block.

Feedstock	Electrons	Pathways	C-building block	NADH/pyruvate	ATP/pyruvate
Methanol	6 e/mol	RuMP/SBPase	Pyruvate	4 mol/mol	0 mol/mol
Methanol	6 e/mol	RuMP/aldolase	Pyruvate	4 mol/mol	1 mol/mol
Glucose	24 e/mol	Glycolysis	Pyruvate	2 mol/mol	2 mol/mol

However, it has been shown that the replacement of some amino acids accomplished a mixed cofactor acceptance including $NADP^+$, which might be useful in future R&D projects (Ochsner et al. 2014).

16.3.4 Kinetics and Thermodynamics of the Methanol Dehydrogenase

The oxidation of methanol with NAD^+ toward formaldehyde and NADH catalyzed by the methanol dehydrogenase (Mdh) is one of the key enzymes in the RuMP. This step has an estimated positive free Gibbs energy:

$$CH_3OH + NAD^+ \rightarrow CH_2O + NADH + H^+ \; \Delta G^{\circ\prime} = 34.2\,\text{KJ/mol}$$

The net reaction can be divided into two half-reactions (Woolston et al. 2018). The standard reduction potentials reveal that the exergonic oxidation of methanol is not strong enough to deliver the needed reduction potential for the hydrogenation of NAD^+ under standard conditions:

$$CH_3OH \rightarrow CH_2O + 2\,H^+ + 2\,e \qquad E^{\circ\prime} = 0.143\,\text{V (based on recalculation)}$$

$$NAD^+ + H^+ + 2\,e \rightarrow NADH \qquad E^{\circ\prime} = -0.320\,\text{V (Voet, 2016)}$$

$$CH_3OH + NAD^+ \rightarrow CH_2O + NADH + H^+ \quad \Delta E^{\circ\prime} = -0.177\,\text{V}$$

The net overall potential of the two combined reactions derives the Gibbs free energy:

$$\Delta G^{\circ\prime} = -n\,\text{F}\,\Delta E^{\circ\prime} = -2 \times 96.485\,\text{J/V} \times \text{mol} \times -0.177\,\text{V} = 34.2\,\text{KJ/mol},$$

whereby the Faraday constant F is given as 96.485 C/mol or 96.485 J/V and $n = 2$ since two electrons are transferred. Accordingly, the reaction is thermodynamically unfavored and the equilibrium under standard conditions is mainly on the left side. $\Delta G^{\circ\prime}$ is the Gibbs free energy under standard conditions defined with 1 M concentration for all reactants except protons, which are expressed as physiologic pH of 7.0. The real Gibbs free energy of a reaction depends on the Gibbs free energy $\Delta G^{\circ\prime}$ under standard conditions adjusted by the concentration of the reactants:

$$\Delta G = \Delta G^{\circ\prime} + \text{RT} \ln K_{eq} \text{ with } K_{eq} = c\,[\text{products}]/c\,[\text{educts}]$$
$$= c\,[CH_2O]\,c\,[NADH]/c\,[CH_3OH]\,c\,[NAD^+]$$

Since $\Delta G = 0$ at equilibrium, the product-to-educt ratio can be estimated as follows:

$$\Delta G^{\circ\prime} = -\text{RT} \ln \text{K}_{eq}$$

The gas constant R is given as 8.315 J/mol × K the following equation is derived:

$$34.2\,\text{KJ/mol} = -8.315\,\text{J/mol K} \times 310\,\text{K} \times \ln\,K_{eq}, \text{ derives}$$

$$34.2\,\text{KJ/mol} = -2.577\,\text{J/mol} \times \ln\,K_{eq}, \text{ derives}$$

$$-13.27 = \ln K, \text{ derives}$$
$$e^{-13.27} = K_{eq}, \text{ derives}$$
$$K_{eq} = 1.72 \times 10^{-6}$$

The usual ratio of the reduced and oxidated form of NADH (i.e. nicotine-adenosine-dinucleotide) is important for the efficiency of methanol conversion. In standard eukaryotic cells – which may not be representative for the respective host – it may be assumed that the ratio of NAD^+/NADH is about 1,000 to favor metabolic oxidation (Voet et al. 2016). This ratio ensures that a sufficient quantity of oxidized reduction equivalent is available to receive electrons from feedstocks to be oxidized. The equilibrium constant reveals that even if a NAD^+/NADH ratio in the range of 1,000 is assumed only about one mol of formaldehyde is present in a pool of 1,000 mol of methanol. This circumstance might imply a bottleneck of the overall pathway and needs to be considered in the metabolic engineering strategy.

16.4 Metabolic Engineering

As mentioned above, PMD might be generated via a one-step fermentation of glucose or a whole-cell biocatalytic decarboxylation of lysine. The two distinct options are discussed below in more detail.

16.4.1 Glucose as Carbon and Energy Source with *C. glutamicum* as Host

It is straightforward to use *C. glutamicum* for the fermentative production of PMD since highly mature strains for the production of lysine are available which exclusively miss the terminal decarboxylase to constitute as a promising prototype. Although it is not known that this host does generate PMD naturally, it was able to grow in the presence of a concentration of 0.3 M (Minitsuka et al. n.d.) and a sufficient product tolerance may be assumed. Accordingly, the lysine decarboxylase *ldcC* was overexpressed in the strain *C. glutamicum* LYS-12, which is a high-performing lysine producer (Kind et al. 2014). The *lysE* gene encoding a lysine exporter was deleted to avoid loss of the final pathway intermediate into the medium. The aerobe fermentation was pursued at 30 °C and the pH was controlled in the neutral range with a 25% ammonia solution. The feed was adjusted such that the glucose concentration is kept at 10 g/l. The final strain *C. glutamicum* DAP-16 achieved a titer of 88 g/l PMD within 50 h representing a productivity of 1.76 g/l × h. The product yield was 0.29 t PMD per t glucose which equals 63% of the maximum carbon yield (Kind et al. 2014).

The fermentative production of PMD in *C. glutamicum* may generate *N*-acetyl-diaminopentane as an undesired by-product, which is formed by acetylation with acetyl-CoA as a donor (Kind et al. 2010). It was observed that up to 20% of the PMD pool appears in the acetylated form. After identification of potential genes with acetyl-transferase activity, the gene NCgl1469 was identified to cause the respective acetylation, and the corresponding deletion strain DAP-4 provided secreted product without contamination with the acetylated by-product. The PMD yield increased by 11% and the deletion strain was not negatively impacted by its growth behavior and glucose uptake rate (Kind et al. 2010).

A wild-type strain of the facultative anaerobic bacterium *C. glutamicum* produces lactate and succinate under anaerobic conditions and the carbon flux is directed toward glycolysis with less than 5% of the flux channeled into the pentose phosphate pathway. The production of lysine usually proceeds under aerobe conditions and a lysine producer with attenuated feedback inhibition by lysine under aerobic conditions channels about 53% of the carbon flux through glycolysis and about 47% into the pentose phosphate pathway. The flux via the pentose phosphate pathway is essential since the production of one mol of lysine (or PMD) requires 4 mol of NADPH, which are difficult to generate via a transhydrogenase or an engineered GAPDH with changed cofactor specificity (Kobayashi et al. 2020). It might be helpful to create a metabolic switch to regulate flux between the two pathways and two products under different oxygen regimes. Since a chemical induction bears additional cost and might add complexity in the purification process of the product, an interesting oxygen-dependent system has been installed in *C. glutamicum*. For this purpose, the originating *C. glutamicum* strain ATCC 13032 was metabolically engineered to provide PMD and succinate under aerobic and anaerobic conditions. In particular, the following measures were implemented in the resulting strain YT12:

- Deletion of the lactate dehydrogenase gene (*ldhA*) to block the formation of lactate,
- Deletion of the menaquinone oxidoreductase gene (*pqo*), the acetyl-CoA transferase (*cat*), the acetate kinase (*ack*) and the phosphotransacetylase gene (*pta*) to suppress the formation of acetate,
- Mutation of the *lysC* gene encoding the aspartate kinase to deregulate the native feedback inhibition by lysine,
- Mutation of the isocitrate dehydrogenase gene (*icd*) and the pyruvate decarboxylase gene (*pyc*) to enhance the formation of oxalacetate,
- Overexpression of the *pyc* gene by exchanging the native promoter by the *tuf* promoter, and
- Introduction of the lysine decarboxylase (*ldcC*) to generate PMD.

The strain YT12 provided under aerobic conditions a PMD titer of 0.22 g/l within 14 h corresponding to a productivity of 0.015 g/l × h combined with a carbon yield of 0.023 g PMD/g glucose. No lysine and succinate were detected. Under anaerobic conditions, a succinate titer of 18.2 g/l was observed with a batch time of 10 h corresponding to a productivity of 1.8 g/l × h and a related carbon yield of 0.53 g product/g glucose (Kobayashi et al. 2020). The formation of PMD was not observed.

The strain YT12 has further been modified by replacing the wild-type promoter of the glucose-6-phosphate isomerase gene (*pgi*) with an anaerobic-specific promoter of the lactate dehydrogenase gene (*ldhA*). This measure of the metabolism engineering strategy attenuates glycolytic flux under aerobic conditions and allows a high glycolytic flux under anaerobic conditions. The obtained strain YT13 displayed a PMD titer of 1.01 g/l with a productivity of 0.039 g/l × h under aerobic conditions. The carbon yield was 0.1 g PMD/g glucose (Kobayashi et al. 2020). The changed promoter of the glucose-6-phosphate isomerase facilitated under aerobic conditions a partial shift of metabolic flux into the pentose phosphate pathway. The improved availability of NADPH explains the raised formation of PMD (Figures 16.3 and 16.4).

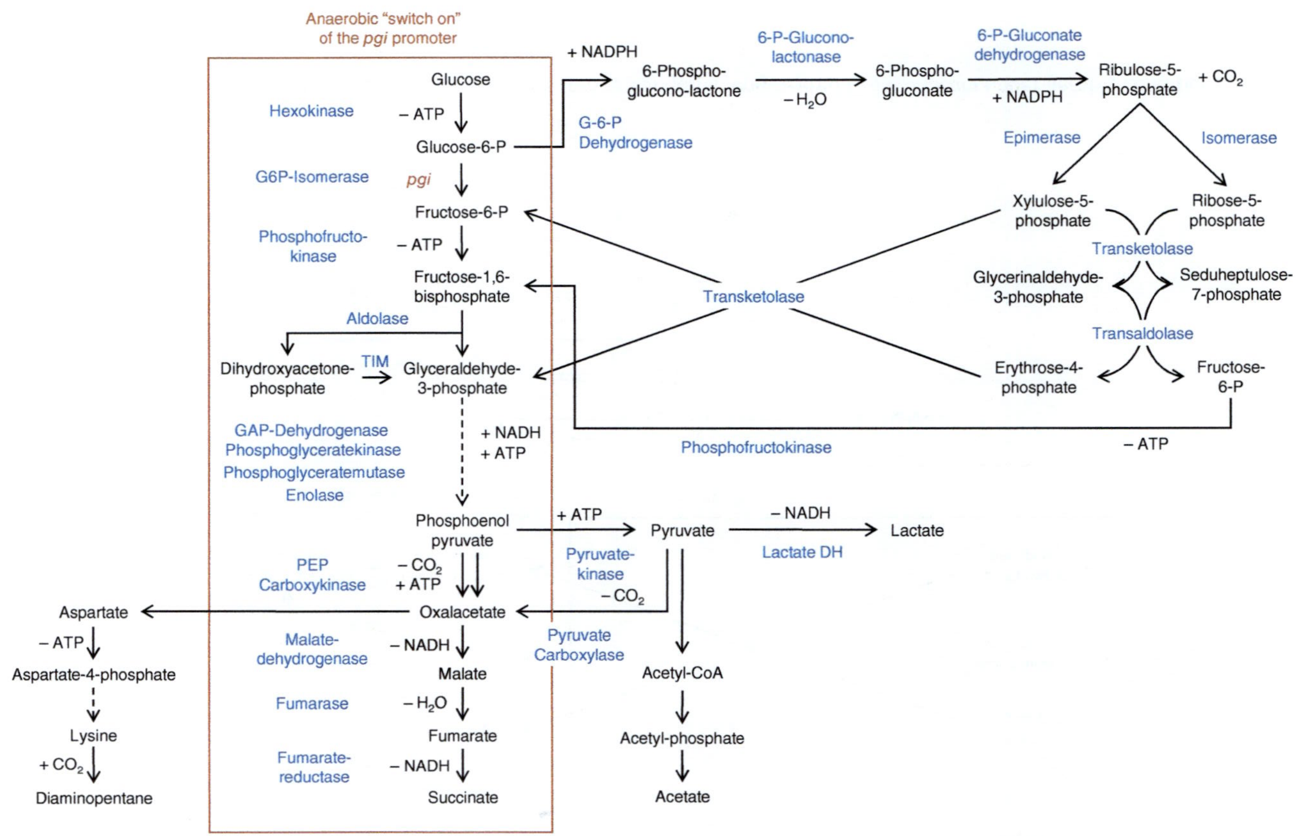

Figure 16.3 Metabolic flux via glycolysis toward succinate (red rectangle) in engineered *C. glutamicum* under anaerobic conditions (Kobayashi et al. 2020). Source: Walter Koch.

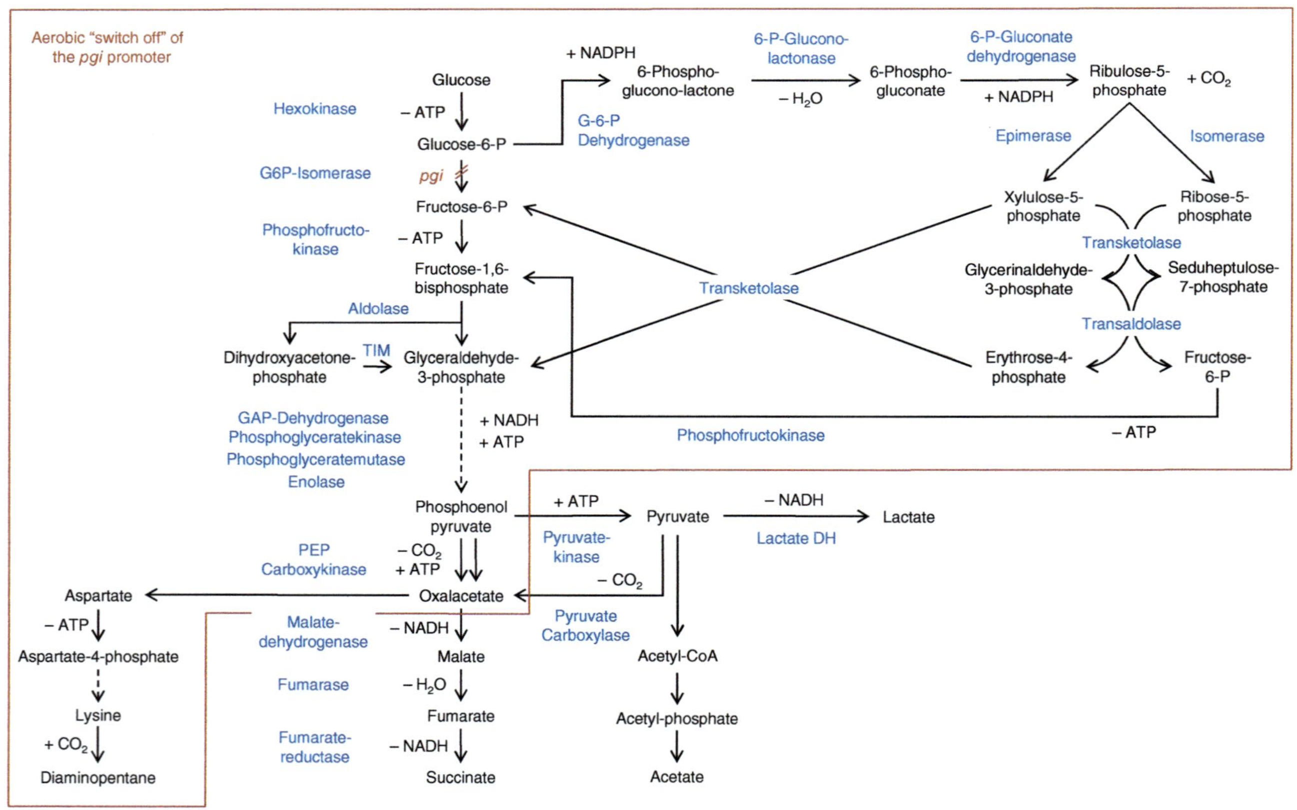

Figure 16.4 Metabolic flux via PPP toward pentamethylenediamine (red rectangle) in engineered *C. glutamicum* under aerobic conditions. Source: Walter Koch.

The concept could successfully be implemented in *C. glutamicum* and titer, yield and production rate of PMD could be increased by factor 4.6, 4.4, and 2.6, respectively (Kobayashi et al. 2020). However, the strain is still far beyond a commercially interesting performance.

16.4.2 Xylose as Carbon and Energy Source with *C. glutamicum* as Host

It has also been evaluated to feed the C5 carbohydrate xylose as a carbon and energy source instead of glucose (Buschke et al. 2013). This would prepare the use of lignocellulosic biomass as sustainable feedstock, which is not in competition with acreage occupied for food production. Glucose turned out to be the preferred carbon source. The carbohydrate uptake rate q_s was both for glucose and xylose with about 4 mmol/g DCW × h in the same range (Buschke et al. 2013). However, it turned out that the assimilation of xylose could not be translated into an efficient biosynthesis. Flux analysis revealed that the citric acid circle showed enhanced activity and a significant share of xylose was converted into carbon dioxide. The modified strain DAP-Xyl12 was equipped with an overexpressed 1,6-bisphosphatase (*fbp*) and transketolase (*tkt*). The genes encoding the diaminopentane *N*-acetyl transferase (*cg1722*) and the lysine exporter (*lysE*) were deleted. The strain achieved a titer of 103 g/l PMD from xylose with a molar yield of 0.32 mol product/mol xylose (Buschke et al. 2013).

16.4.3 Glucose as Carbon and Energy Source with *E. coli* as Host

The fermentative production of PMD could also be shown in *E. coli* cells. An early technology step has been made with a reasonable metabolic engineering design of the native lysine pathway (Qian et al. 2011). The conventional pathway employed by *C. glutamicum* for the fermentative production of lysine has been followed combined with several modifications. The genes encoding the aspartate kinase, the dihydrodipicolinate synthase (*dapA*), the dihydrodipicolinate reductase (*dapB*), the meso-diaminopimelate dehydrogenase (*ddh*), and the diaminopimelate decarboxylase (*lysA*) have been inducible lysine decarboxylase (*cadA*) overexpressed. PMD might be further converted into undesired by-product before being secreted. Thus, the genes encoding enzymes for aminopropylcadaverine (*speE*), N-acetyl-cadaverine (*speG*), glutamylcadaverine (*puuA*), and 5-aminopentanal (*ygjG*). The gene encoding the putrescine importer (pup) was knocked out. The final strain achieved a product titer of 9.6 g/l after a batch duration of 30 h corresponding to productivity of 0.32 g/l × h (Qian et al. 2011).

Since it had been reported that the growth of *E. coli* cells decreased by 35% after a PMD titer of 20.4 g/l has been exceeded, specifically the product tolerance was subject to improvement efforts. *E. coli* cells were treated with radiation and ARTP mutagenesis to create random genetic diversity and the strains with the highest PMD tolerance were isolated. The obtained chassis cell was genetically prepared for the conversion of glucose into PMD and used in the experiments. Finally, a product titer of 58.7 g combined with a carbon yield of 39.6% could be achieved. The batch time is reported at 60 h translating into a productivity of 0.98 g/l × h (Wang et al. 2021).

Results of academic attempts for a one-step fermentation of glucose are summarized (Table 16.6).

Table 16.6 Selected performance data for the fermentation of glucose toward PMD.

Host	Decarboxylase origin	Titer	Yield	Batch	Productivity	Source
C. glutamicum	*E. coli (LdcC)*	88 g/l	0.29 t/t	50 h	1.76 g/l × h	Kind et al. (2014)
C. glutamicum	*H. alvei (LdcC)*	125 g/l	0.34 t/t	70 h	1.79 g/l × h	Kim et al. (2020)
E. coli	*E. coli*	9.6 g/l	0.10 t/t	30 h	0.32 g/l × h	Qian et al. (2011)
E. coli	*E. coli*	58.7 g/l	0.40 t/t	60 h	0.98 g/l × h	Wang et al. (2021)

16.4.4 Whole-Cell Biocatalysis as Production Technology

PMD may be produced via a conventional fermentation with carbohydrate as carbon and energy source or whole-cell biocatalysis with L-lysine as feedstock. The latter approach is advantageous since lysine is readily available in the market and offered at low prices facilitated by the efficient and mature fermentative production of lysine. Lysine decarboxylation in plants is catalyzed by the lysine/ornithine decarboxylase L/ornithine decarboxylase (ODC), which is the branching point for the biosynthesis of lysine-derived alkaloids (Shimizu et al. 2019). The experiments reported in academia mainly use the *E. coli*-derived inducible L-lysine decarboxylases CadA or the constitutive L-lysine decarboxylase LdcC (Krithika et al. 2010; Leong et al. 2020). The experimental setup foresees to express a suitable decarboxylase in *E. coli*, which is not secreted, and to incubate the substrate L-lysine in the suspended *E. coli* cells.

$$\text{L-lysine} \rightarrow \text{Pentamethylenediamine} + CO_2$$

A systematic analysis of the reaction conditions was pursued with *E. coli* as a biocatalyst, and the optimum reaction conditions were described (Leong et al. 2020). The applied biocatalyst was produced with *E. coli* overexpressing the *cadA* gene and induction with 0.1 mM IPTG at the beginning of the production phase of 6 h at 32 °C (Leong et al. 2020). pH management is critical since a high substrate concentration of L-lysine might cause a high pH, which might negatively impact the activity of the lysine decarboxylase. Tests were pursued with varied concentration of a phosphate buffer from 0 to 500 mM (adjusted at pH 5.8) and revealed that the production of PMD was identical. Thus, the technology may be pursued without buffer system to reduce production costs. Also, the dependency on the pH has been evaluated and experiments were pursued with a biocatalyst concentration of 0.66 g DCW/l. It could be observed that the highest enzyme activity of the L-lysine decarboxylase was from pH 4 to 7 with a L-lysine consumption of 96% and a PMD production of 93% (Leong et al. 2020). Since pyridoxal-5-phosphate (PLP) acts as a cofactor in the decarboxylation of lysine small amounts need to be added to the reaction medium. The best performance data were achieved with a biocatalyst concentration of 1.98 g DCW/l which provided a PMD titer of 145 g/l with a molar yield of 95%. The batch time is indicated with 2 h corresponding to productivity of 72.5 g/l × h (Leong et al. 2020).

Similar experiments were pursued with *E. coli* by applying the lysine decarboxylase of *Hafnia alvai* (Kim et al. 2019). The lysine decarboxylase displayed with 1825 U/mg protein, a higher activity than the lysine decarboxylase of *E. coli* with an activity of 1003 U/mg protein. The laboratory setup achieved a PMD titer of 136 g/l given a lysine concentration of 200 g/l which is corresponding to a molar yield of 97%. An induction with IPTG was not required and especially the addition of pyridoxal-5-phosphate could be omitted (Kim et al. 2019). The isolated product was used for producing PA-56, which has similar physical and chemical properties then PA-56 made from petrochemical precursors (Kim et al. 2019). However, the productivity with 1.13 g/l × h is rather low given a batch duration of 120 h. This element of the technology would be a great disadvantage in case a commercial setup is planned. Other recombinant *E. coli* strains carrying the lysine decarboxylase of *Hafnia alvai* achieved productivities in the range of 10 g/l × h though connected to titers.

The potential bottleneck of the lysine uptake and the product export or secretion out of the biocatalyst cell, respectively, was addressed in a structurally similar experiment with a specific permeabilization step (Rui et al. n.d.). *E. coli* cells were constructed such that the *cadA* gene can be induced with L-lysine instead of IPTG. For the permeabilization of the biocatalyst, the *E. coli* cells were harvested, centrifuged, and the precipitate resuspended with 35% (v/v) ethanol. The permeabilization was finalized after an incubation of 10 min at room temperature. The whole-cell biocatalysis was pursued at 37 °C with 100 mM phosphate buffer at a pH of 5.6. Pyridoxal phosphate with a concentration of 0.1 mM was added to ensure a high functionality of the lysine decarboxylase. The 2 l batch setup provided a titer of 205 g/l PMD with a yield of 92%. The batch time of 20 h translates into a productivity of 10.3 g/l × h (Rui et al. n.d.).

Results of academic attempts with decarboxylases of different origins are summarized (Table 16.7).

The results summarized above were achieved with pure L-lysine added to the cell suspension of the *E. coli* biocatalyst with overexpressed lysine decarboxylase. It has also been tried to couple a conventional production of lysine with the conversion into PMD to omit the costly lysine isolation and purification process. A whole-cell biocatalyst *E. coli* with an overexpressed cadA gene has been applied. The experiments revealed that a direct integration of the decarboxylation step with the synthesis of lysine is in principle possible, and a solution of 1 M lysine was converted to 80% after a batch time of 2 h. The alkaline pH after the production of lysine was adjusted with the addition of HCl down to 6 before the whole-cell biocatalysis has been started (Kim et al. 2015).

A cascade reaction in one fermenter has been proposed in academia (Wang et al. 2018). The concept foresees to establish in the first step a fermentative conversion of glucose into lysine with *E. coli* NT1004 as a catalyst. This host would then be cocultured with another *E. coli* CAD03 strain equipped for the biocatalytic conversion of lysine into PMD. In order to

Table 16.7 Performance data for the biocatalytic decarboxylation of lysine toward PMD.

Host	Decarboxylase origin	Titer	Yield	Batch	Productivity	Source
E. coli	*K. oxytoca*	112 g/l	100%	1.4 h	80.3 g/l × h	Li (2016), Naiqiang (2016)
E. coli	*H. alvai* (LdcC)	136 g/l	97%	120 h	1.13 g/l × h	Kim et al. (2019)
E. coli	*E. coli* (CadA)	145 g/l	95%	2.0 h	72.5 g/l × h	Leong et al. (2020)
E. coli	*E. coli* (CadA)	221 g/l	92%	16 h	13.8 g/l × h	Kim et al. (2015)
E. coli	*E. coli* (CadA)	205 g/l	92%	20 h	10.3 g/l × h	Rui et al. (n.d.)
E. coli	*E. coli* (LdcC)	102 g/l	83%	9.0 h	11.3 g/l × h	Shin et al. (2018)

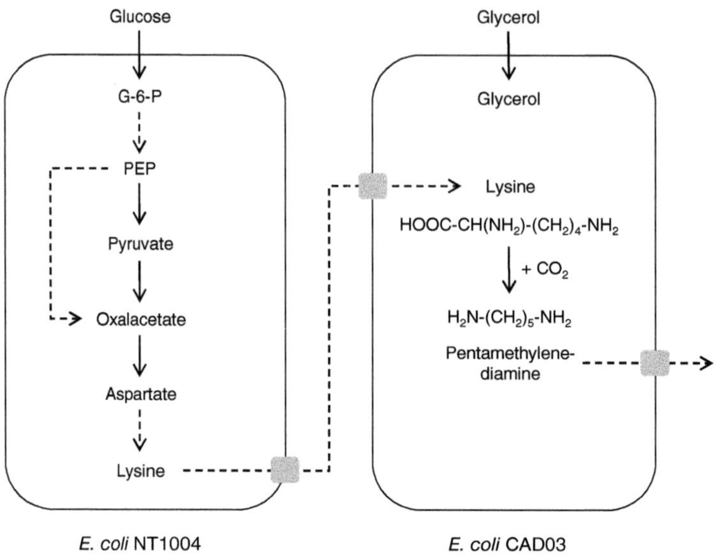

Figure 16.5 Consortium of two *E. coli* strain to synthesize PMD. Source: Adapted from Wang et al. (2018).

avoid a competition for glucose uptake the phosphotransferase system (PTS) of the second *E. coli* strain is disabled due to a CRISPR-Cas9 modification of the respective gene. Instead, the strain CAD03 consumes exclusively glycerol for cell growth and maintenance which is also added to the medium (Figure 16.5).

The final product is secreted by the CAD03 strain and can be harvested. A titer of 28.5 g/l PMD combined with a molar yield of 0.209 t/t glucose was obtained (Wang et al. 2018). The consortium approach with two *E. coli* strains bears several aspects, which need to be watched closely in case commercial use is envisaged. In the first instance, it is to be questioned if the *E. coli* NT1004 strain provides the same yield and productivity as the most-developed commercial *C. glutamicum* strains. Additionally, a full conversion of glucose and lysine is recommended before batch end to save raw material costs and to avoid separation efforts to isolate lysine and PMD.

16.4.5 Methanol as Carbon and Energy Source with *Bacillus methanolicus* as Host

First attempts to accomplish a methylotrophic host for high-volume production were pursued with *B. methanolicus* and addressed other target products than PMD. The *B. methanolicus* strain MGA3 is able to produce 55 g/l glutamate using the RuMP for carbon assimilation. The obtained yield is 0.36 t glutamate per ton methanol, which is according to 86% of the theoretical maximum yield (Brautaset et al. 2003; Lee et al. 1996). It is also reported that a methylotrophic Bacillus achieved a lysine titer of 19 g/l at cell densities of 20 g/l and phosphate limitation after a cultivation time of 32 h (Schendel et al. 1990). The developing toolbox of metabolic engineering allows more rational metabolic engineering approaches and several genes were cloned and overexpressed comprising the aspartate semialdehyde dehydrogenase (*asd*), the dihydrodipicolinate synthase (*dapA*), and the aspartate kinase (*dapG*, *yclM*). The strain displayed a lysine titer of 11 g/l (Jakobsen et al. 2006). The growth optimum of *B. methanolicus* is at 50–55 °C and fermentation is usually pursued in that range (Müller et al. 2015). For the introduction of heterologous genes, the temperature sensitivity of the enzyme needs to be ensured which might be weak at a rather high temperature of the biochemical system. A remarkable result has been achieved after the genes *cadA* and *ldcC* from *E. coli* encoding two different lysine decarboxylased were expressed in *B. methanolicus*. Both genes were shown to be functional in the host cultivated at 50 °C and a PMD titer of 300 mg/l under shake flask conditions was obtained. The product titer could be increased up to volume corrected 6.5 g/l in a fed-batch fermentation and high-cell density conditions (Nærdal et al. n.d.).

Natural methylotrophy as given in *B. methanolicus* is in competition with synthetic methylotrophy which is tested in *C. glutamicum* after the methanol dehydrogenase from *B. methanolicus* has been integrated by transformation. The genes *ald* and *fadH* encoding the endogenous aldehyde dehydrogenase were deleted to avoid a competing conversion of methanol. Since growth with methanol as the sole carbon source could not be observed, the incorporation of ^{13}C-methanol supplemented as co-substrate to sugars was monitored and indeed ^{13}C label from methanol could be detected in the secreted product PMD (Leßmeier et al. 2015).

16.5 Downstream Processing

The downstream processing activities have focused so far on the process with glucose as feedstock. Several isolation and purification protocols for PMD out of the reaction medium have been developed. In case PMD is derived by biocatalytic decarboxylation of lysine, the reaction medium would be neutralized with HCl to protect the activity of the permeabilized *E. coli* cells (Huang et al. 2021). After the conversion has been completed the permeabilized cells are removed and PMD is deprotonated triggered by the addition of NaOH. PMD would then efficiently be extracted with n-butanol which has been shown to possess a suitable discriminating partition coefficient (Kind et al. 2014). A phase separation between the water and the organic phase is formed and the n-butanol/PMD phase is captured and channeled into a rectification column. Butanol is removed via the column head and PMD can be isolated (Liu et al. 2020). Alternatively, it is also reported that PMD was purified with chloroform extraction (Kim et al. 2020).

R&D activities with methylotrophic hosts focus primarily on the optimization of the general fermentation performance criteria including titer, productivity, and molar yield. The isolation and purification protocols may partially be adopted and customized after the methylotrophic hosts have entered the stage of commercial application.

16.6 Industrial Application

The commercialization of fermentative PMD has been initiated by Cathay Industrial Biotech, which has strategically pursued and implemented the production of biobased polyamides. Cathay Biotech has tested its proprietary technology in a pilot plant and achieved a breakthrough in 2014. Quality testing in automotive surface paint was successful and the product won the ECS Innovation Award (Cathay, Company homepage

Table 16.8 Carbon footprint of polyamides.

PA-6	PA-66	PA-56
9.91 t CO2e/t	8.69 t CO2e/t	4.38 t CO2e/t

Source: Adapted from Cathay, Company homepage (2022).

2022). The company received 135 $ m additional funding in 2015 to increase its production capacities (ICIS 2016). It is currently running two production sites. In the county Jinxiang (Shandong, China), Cathay is operating a PMD pilot plant with a capacity of 1 kt pa and a commercial plant with a capacity of 40 kt biobased long-chain diacids (e.g. dodecanedioic acid). In the city of Wusu (Xinjiang, China), the company is operating a plant with a PMD capacity of 50 kt pa combined with a capacity of 30 kt biobased long-chain diacids and 100 kt polyamides (PR Newswire 2022). Cathay is marketing the biobased polyamide grade with PMD as an amino component in the textile industry under the brand name TERRYL®, which is used in clothing, bags, carpets, work suits, tents, and other textile downstream fields. Products for the engineering materials industry are marketed under the brand ECOPENT® for application in automotives, electronics and electricals, cable ties, film, and other engineering materials (Cathay, Company homepage 2022). The reduced greenhouse warming potential or the optimized carbon footprint, respectively, of PA-56 is positioned as a marketing tool. The PA-56 offered by Cathay has a significantly reduced emission of carbon dioxide than the conventional mainstream polyamides PA-6 and PA-66 (Table 16.8).

The exact technology applied by Cathay is not disclosed except the circumstance that the product is based on renewable plant materials. It may be assumed that either a direct fermentation ex glucose or a biocatalytic conversion with *E. coli* as a host of lysine generated with *C. glutamicum* is applied.

Mitsui Chemicals is offering the poly-isocyanate Stabio™ in the market which is partially made of biobased PMD. However, it is not clear inasmuch Mitsui Chemicals is sourcing the biobased educt in the market or operates its own plant for fermentative PMD (Mitsui, Company homepage 2022).

References

Arfman, N., Watling, E.M., Clement, W. et al. (1989). Methanol metabolism in thermotolerant methylotrophic *Bacillus* strains involving a novel catalytic NAD-dependent methanol dehydrogenase as a key enzyme. *Archives of Microbiology* 152: 280–288.

Arpe, H.-J. (2007). *Industrielle Organische Chemie. Bedeutende Vor- und Zwischenprodukte.* Wiley.

Brautaset, T., Williams, M.D., Dillingham, R.D. et al. (2003). Role of the *Bacillus methanolicus* citrate synthase II gene, *citY*, in regulating the secretion of glutamate in L-lysine-secreting mutants. *Applied and Environmental Microbiology* 69 (7): 3986–3995.

Brautaset, T., Jakobsen, T.Ø.M., Flickinger, M.C. et al. (2004). Plasmid-dependent methylotrophy in thermotolerant *Bacillus methanolicus*. *Journal of Bacteriology* 186: 1229–1238.

Brautaset, T., Jakobsen, T.Ø.M., Josefsen, K.D. et al. (2007). *Bacillus methanolicus*: a candidate for industrial production of amino acids from methanol at 50 °C. *Applied Microbiology and Biotechnology* 74: 22–34.

Buschke, N., Becker, J., Schäfer, R. et al. (2013). Systems metabolic engineering of xylose-utilizing *Corynebacterium glutamicum* for production of 1,5-diaminopentane. *Biotechnology Journal* 8: 557–570. https://doi.org/10.10027biot.201200367.

Cathay, Company homepage, (2022). www.cathaybiotech.com/en/projxags/.aspx#content, (accessed 20 Oct 2022).

Chemical Week, Methanol demand to lag GDP growth, (2020). 30.11.2020/7.12.2020, p. 22

Chung, H., Yang, J.E., Ha, J.Y. et al. (2015). Bio-based production of monomers and polymers by metabolically engineered microorganisms. *Current Opinion in Biotechnology* 36: 73–84. https://doi.org/10.1016/j.copbio.2015.07.007.

Goeppert, A., Czaun, M., Jones, J.-P. et al. (2014). *Recycling of carbon dioxide to methanol and derived products – closing the loop*. Royal Society of Chemistry. https://doi.org/10.1039/c4cs00122b.

Hansen, R.S., Dillingham, R., and Olson, P. (1996). Production of L-lysine and some other amino acids by mutants of *Bacillus methanolicus*. In: *Proceedings of the 8th International Symposium on Microbial Growth on C1 compounds* (ed. M.E. Lidstrom, F. Tabita, and F. Robert). Dordrecht: Kluwer.

Hoppe, W., Thonemann, N., and Bringezu, S. (2017). Life cycle assessment of carbon dioxide-based production of methane and methanol and derived polymers. *Journal of Industrial Ecology* 22 (2): 327–340. https://doi.org/10.1111/jiec.12583.

Huang, Y., Ji, X., Ma, X. et al. (2021). Green chemical and biological synthesis of cadaverine: recent development and challenges. *RSC Advances* 11: 23922–23942. https://doi.org/10.1039/d1ra02764f.

ICIS, (2016). Special Report Green and Biobased Materials: Biobased polymers keep textiles green, 27.6.-3.7.2016, p. 27

IRENA (2021). Innovation Outlook. *Renewable Methanol* 15.

IRENA and Methanol Institute, 2021, Innovation Outlook: Renewable Methanol, www.irena.org/publications (accessed 26 July 2023)

Irla, M., Neshat, A., Winkler, A. et al. (2014). Complete genome sequence of *Bacillus methanolicus* MGA3, a thermotolerant amino acid producing methylotroph. *Journal of Biotechnology* 188: 110–111. https://doi.org/10.1016/j.biotec.2014.08.013.

Jakobsen, M., Benichou, A., Flickinger, M.C. et al. (2006). Upregulated transcription of plasmid and chromosomal ribulose monophosphate pathway genes is critical for methanol assimilation rate and methanol tolerance in the methylotrophic bacterium *Bacillus methanolicus*. *Journal of Bacteriology* 188: 3063–3072.

Jorge, J.M.P., Pérez-García, F., and Wendisch, V.F. (2017). A new metabolic route for the fermentative production of 5-aminovalerate from glucose and alternative carbon sources. *Bioresource Technology* 245: 1701–1709. https://doi.org/10.1016/j.biortech. 2017.04.108.

Kim, H.J., Kim, Y.H., Shin, J.-H. et al. (2015). Optimization of direct lysine decarboxylase biotransformation for cadaverine production with whole-cell biocatalysts at high lysine concentration. *Journal of Microbiology and Biotechnology* 25 (7): 1108–1113. https://doi.org/10.4014/jmb.1412.12052.

Kim, H.T., Baritugo, K.-A., Oh, Y.H. et al. (2019). High-level conversion of L-lysine into cadaverine by *Escherichia coli* whole cell biocatalyst expressing *Hafnia alvai* L-lysine decarboxylase. *Polymers* 11: 1184. https://doi.org/10.3390/polym11071184.

Kim, H.T., Baritugo, K.-A., Hyun, S.M. et al. (2020). Development of metabolically engineered *Corynebacterium glutamicum* for enhanced production of cadaverine and its use for the synthesis of bio-polyamide 510. *ACS Sustainable Chemistry & Engineering* 8: 129–138. https://doi.org/10.1021/acssuschemeng.9b04693.

Kind, S., Jeong, W.K., Schröder, H. et al. (2010). Identification and elimination of the competing N-acetyldiamino-pentane pathway for improved production of diaminopentane by *Corynebacterium glutamicum*. *Applied and Environmental Microbiology* 76 (15): 5175–5180. https://doi.org/10.1128/AEM.00834-10.

Kind, S., Neubauer, S., Becker, J. et al. (2014). From zero to hero – production of bio-based nylon from renewable resources using engineered *Corynebacterium glutamicum*. *Metabolic Engineering* 25: 113–125. https://doi.org/10.1016/j.ymben.2014.05.007.

Kobayashi, S., Kawaguchi, H., Shirai, T. et al. (2020). Automatic redirection of carbon flux between glycolysis and pentose phosphate pathway using an oxygen-responsive metabolic switch in *Corynebacterium glutamicum*. *ACS Synthetic Biology* 9: 814–826.

Kölbl, T., German Corporate Conference, UniCredit/Kepler Cheuvreux, Südzucker company presentation, 202118.-20.1.2021, p. 61, www.suedzucker.de/de/investor-relations/publikationen, (accessed 2 March 2021).

Krithika, G., Arunachalam, J., Preyinka, H., and Indulekha, K. (2010). Two forms of lysine decarboxylase; kinetics and effect of expression in relation to acid tolerance response in *E. coli*. *Journal of Experimental Sciences* 1: 10–21.

Krog, A., Heggeset, T.M., Muller, J.E. et al. Methylotrophic *Bacillus methanolicus* encodes two chromosomal and one plasmid born NAD+ dependent methanol dehydrogenase paralogs with different catalytic and biochemical properties. *Plos One* 8: e59188. https://doi.org/10.1371/journal.pone.0059188.

Lee, G.H., Hur, W., Bremmon, C.E., and Flickinger, M.C. (1996). Lysine production from methanol at 50 °C using *Bacillus methanolicus*: modeling volume control, lysine concentration, and productivity using a three-phase continuous simulation. *Biotechnology and Bioengineering* 49: 639–653.

Leong, Y.K., Chen, C.-H., Huang, S.-F. et al. (2020). High-level L-lysine bioconversion into cadaverine with enhanced productivity using engineered *Escherichia coli* whole-cell biocatalyst. *Biochemical Engineering Journal* 157: https://doi.org/10.1016/j.bej.2020.107547.

Leßmeier, L., Pfeifenschneider, J., Carnicer, M. et al. (2015). Production of carbon-13-labeled cadaverine by engineered *Corynebacterium glutamicum* using carbon-13-labeled methanol as co-substrate. *Applied Microbiology and Biotechnology* 99: 10163–10176. https://doi.org/10.1007/s00253-015-6906-5.

Li, N., Chou, H., Xu, Y., (2016). Improved cadaverine production from mutant Klebsiella oxytoca lysine decarboxylase, Engineering in Life Sciences, 16: 299–305, doi: 10.1002/elsc.201500037.

Liu, Y., Zheng, Y., Wu, H. et al. (2020). Development of an integrated process for the production of high-purity cadaverine from lysine-decarboxylase. *Journal of Chemical Technology and Biotechnology* 95: 1542–1549. https://doi.org/10.1002/jctb.6348.

Madduri, K., Stuttard, C., Vining, L. C., Lysine catabolism in *Streptomyces* spp. is primarily through cadaverine: ß-Lactam producers also make α-aminoadipate, 1989, Vol. 171, No. 1, p. 299–302.

Marpani, F., Pinelo, M., and Meyer, A.S. (2017). Enzymatic conversion of CO_2 to CH_3OH via reverse dehydrogenase cascade biocatalysis: Quantitative comparison of efficiencies of immobilized enzyme systems. *Biochemical Engineering Journal* 127: 217–228.

Methanex company homepage, (2022) Current posted prices, https://www.methanex.com/our-business/pricing, (accessed 31 Oct 2022).

Minitsuka, T., Sawai, H., Hatsu, M., and Yamada, K. Metabolic engineering of *Corynebacterium glutamicum* for cadaverine fermentation. *Bioscience, Biotechnology and Biochemistry* 71: 2130–2135.

Mitsui, Company homepage, (2022). product list, https://jp.mitsuichemicals.com/en/service/product/stabio.htm, (accessed 20 Oct 2022).

Müller, J.E.N., Heggeset, T.M.B., Wendisch, V.F. et al. (2015). Methylotrophy in the thermophilic *Bacillus methanolicus*, basic insights and application for commodity production from methanol. *Applied Microbiology and Biotechnology* 99: 535–551. https://doi.org/10.1007/s00253-014-6224-3.

Nærdal, I., Pfeifenschneider, J., Brautaset, T., and Wendisch, V.F. Methanol-based cadaverine production by genetically engineered *Bacillus methanolicus* strains. *Microbial Biotechnology* 8: 342–350. The product titer was later on corrected and reduced from 11.3 to 6.5 g/l in Microbial Biotechnology, 2019 Vol. 12, p. 182–183.

Naiqiang, L., Chou, H., and Xu, Y. (2016). Improved cadaverine production from mutant *Klebsiella oxytoca* lysine decarboxylase. *Engineering in Life Sciences* 16: 299–305. https://doi.org/10.1002/elsc.201500037.

Ochsner, A.M., Müller, J.E., Mora, C.A., and Vorholt, J.A. (2014). In vitro activation of NAD-dependent alcohol dehydrogenase by Nudix hydrolases in more widespread than assumed. *FEBS Letters* 588: 2993–2999. https://doi.org/10.1016/j.febslet.2014.06.008.

Ochsner, A.M., Sonntag, F., Buchhaupt, M. et al. (2015). *Methylobacterium extorquens*: methylotrophy and biotechnological applications. *Applied Microbiology and Biotechnology* 99: 517–534. https://doi.org/10.1007/s00253-014-6240-3.

Orita, I.H., Yurimoto, R., Hirai, R. et al. (2005). The archaeon *Pyrococcus horikoshii* possesses a bifunctional enzyme for formaldehyde fixation via the ribulose monophosphate pathway. *Journal of Bacteriology* 187: 3636–3642.

Pfeifenschneider, J., Markert, B., Stolzenberger, J. et al. (2020). Transaldolase in *Bacillus methanolicus*: biochemical characterization and biological role in ribulose monophosphate cycle. *BMC Microbiology* 20: 63. https://doi.org/10.1186/s12866-020-01750-6.

PR Newswire, (2022). Cathay Industrial Biotech invests an additional USD 500 Million to expand its production capacity for new bio-based materials, https://www.prnewswire.com/news-releases/cathay-industrial-biotech-invests-an-additional-usd-500-million-to-expand-its-production-capacity-for-new-bio-based-materials-300763246.html, (accessed 20 Oct 2022).

Qian, Z.G., Xia, X.-X., and Lee, S.Y. (2011). Metabolic engineering of *Escherichia coli* for the production of cadaverine: a five carbon diamine. *Biotechnology and Bioengineering* 108 (1): https://doi.org/10.1002/bit.22918.

Rui, J., You, S., Zheng, Y. et al. High-efficiency and low-cost production of cadaverine from a permeabilized-cell bioconversion by a lysine-induced engineered *Escherichia coli*. *Bioresource Technology* 302: https://doi.org/10.1016/j.biortech.2020.122844.

Schendel, F.J., Bremmon, C.E., Flickinger, M.C. et al. (1990). L-lysine production at 50 °C by mutants of a newly isolated and characterized methylotrophic *Bacillus* sp. *Applied and Environmental Microbiology* 56: 963–970.

Shimizu, Y., Rai, A., Okawa, Y. et al. (2019). Metabolic diversification of nitrogen-containing metabolites by the expression of a heterologous lysine decarboxylase gene in *Arabidopsis*. *The Plant Journal* 100: 505–521. https://doi.org/10.1111/tpj.14454.

Shin, J., Joo, J.C., Lee, E. et al. (2018). Characterization of a whole-cell biotransformation using a constitutive lysine decarboxylation from *Escherichia coli* for the high-level production of cadaverine from industrial grade L-lysine. *Applied Biochemistry and Biotechnology* 185: 909–924. https://doi.org/10.1007/s12010-018-2696-4.

Stolzenberger, J., Lindner, S.N., Persicke, M. et al. (2013). Characterization of fructose-1,6-bisphosphatase and seduheptulose 1,7-bisphosphatase from the facultative ribulose monophosphate cycle methylotroph *Bacillus methanolicus*. *Journal of Bacteriology* 195 (22): 5112–5122. https://doi.org/10.1128/JB.00672-13.

Tountas, A.A., Peng, X., Tavasoli, A.V. et al. (2019). Towards solar methanol: past, present, and future. *Advanced Science* 6 (1801903): 1–52.

Tountas, A.A., Ozin, G.A., and Sain, M.M. (2021). Solar methanol energy storage. *Nature Catalysis* 4: 934–942.

Tsuge, Y., Kawaguchi, H., Sasaki, K., and Kondo, A. (2016). Engineering cell factories for producing building block chemicals for bio-polymer synthesis. *Microbial Cell Factories* 15: 19. https://doi.org/10.1186/s12934-016-0411-0.

Tullo, A.H. (2014). *Methanol for Sale*, 12–13. CEN.

Voet, D., Voet, J.G., and Pratt, C.W. (2016). *Principles of Biochemistry*. Wiley.

Wang, J., Lu, X., Ying, H. et al. (2018). A novel process for cadaverine bio-production using a consortium of two engineered *Escherichia coli*. *Frontiers in Microbiology* 9: article 1312. https://doi.org/10.3389/fmicb.2018.01312.

Wang, L., Li, G., and Deng, Y. (2020). Diamine biosynthesis: research progress and application prospects. *Applied and Environmental Microbiology* 86 (23): https://doi.org/10.1128/AEM.01972-20.

Wang, X., Guo, X., Wang, J. et al. (2021). Ameliorating end-product inhibition to improve cadaverine production in engineered *Escherichia coli* and its application in the synthesis of bio-based diisocyanates. *Synthetic and Systems Biotechnology* 6: 243–253. https://doi.org/10.1016/j.synbio.2021.09.004.

Wittmann, C. and Becker, J. (2007). The L-lysine story: from metabolic pathways to industrial production. In: *Amino Acid Biosynthesis* (ed. V.F. Wendisch), 40–70. Springer Heidelberg https://doi.org/10.1007/7171_2006_089.

Woolston, B.M., King, J.R., Reiter, M. et al. (2018). Improving formaldehyde consumption drives methanol assimilation in engineered *E. coli*. *Nature Communications* 9: 2387. https://doi.org/10.1038/s41467-018-04795-4.

Xue, C., Hsu, K.-M., Chiu, C.-Y. et al. (2021). Fabrication of bio-based polyamide 56 and antibacterial nanofiber membrane from cadaverine. *Chemosphere* 266: https://doi.org/10.1016/j.chemosphere.2020.128967.

17
Lysine

17.1 Application

Lysine is a natural amino acid with the chemical formula $C_6H_{14}N_2O_2$ and a molecular weight of 146.2 g/mol. It bears an amino and a carboxylic acid function as the common backbone of all proteogenic amino acids and additionally a terminal amino function in the side chain. The second carbon atom is chiral and lysine may be given as a racemic mixture of either the L-lysine form or the D-lysine form. In this context, exclusively L-lysine is referred to. Lysine belongs to the high-volume fermentation products with a global capacity above 2.5 Mio. t pa in 2016 and a growth rate of 7% pa (Cheng et al. 2018). The market price is fluctuating, though is oscillating around a value of 1.5 $/kg (Cheng et al. 2018; Ikeda 2017) up to 2.0 $/kg (Qian et al. 2011).

Lysine is mainly applied as a feed additive especially for swine and other animals to supplement the natural amino acid composition of other feedstocks. Since the protein fraction of conventional animal feed such as grain, soybean meal, or agricultural by-products does often display a share of single amino acids below the ideal threshold some single amino acids are added in the feed mixing production process. Lysine is regarded as the first limiting amino acid in pig diets based on maize and soybean meal and therefore it is applied as a feed additive (Smiricky et al. 2004). It is also used as a supplement or formulation aid in the food, beverage, and pharmaceutical industry.

It is interesting to note that the high-volume availability of lysine for a comparable low price has prepared the path of a downstream chemistry that pursues further chemical steps to derive advanced target products (Cheng et al. 2018). These products comprise cadaverine or pentamethylenediamine which can be obtained via a decarboxylation of lysine:

$$\mathrm{HOOC{-}CH(NH_2){-}(CH_2)_4{-}NH_2 \rightarrow H_2N{-}(CH_2)_5{-}NH_2 + CO_2}$$

Pentamethylenediamine might be obtained via biocatalytic or chemical decarboxylation of lysine (Leong et al. 2020) or might directly be obtained in a one-step fermentation of glucose. Further potential lysine downstream products including L-pipecolic acid and 5-aminovalerate (Cheng et al. 2018) and 1,5-pentanediol (Cho et al. 2022) are evaluated in academia and it remains to be seen inasmuch these concepts gain commercial relevance.

Pathway Design for Industrial Fermentation, First Edition. Walter Koch.

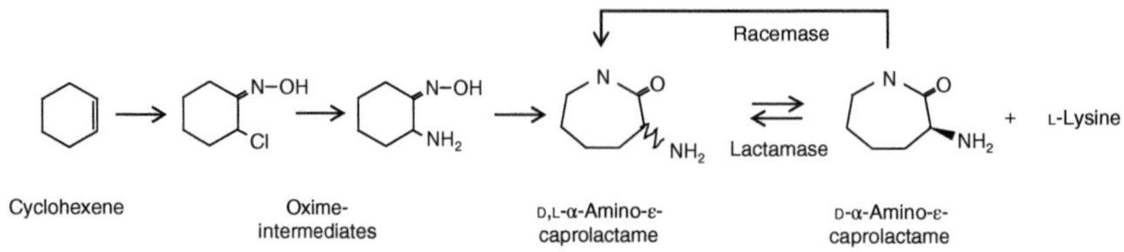

Figure 17.1 Biocatalytic synthesis of L-lysine based on cyclohexene as starting material. Source: Walter Koch.

17.2 Chemical Production

Lysine has originally been produced via natural extraction. It is reported that already in 1889 casein was used as protein-rich raw material and the protein fraction was hydrolyzed to split the peptide bonds to obtain single amino acids (Ikeda 2003). Protein hydrolysis was the common lysine technology until 1960 and organic chemical companies sold the product for 3–5$ per pound (Connor 2008). This price level is too high for the use by feed manufacturers on a regular basis and the hydrolysis technology did not allow the significant market development enabled by fermentation technology starting in the 60th by Ajinomoto and Kyowa Hakko. The hydrolysis of vegetable protein can be neglected in the current analysis of supply though may still play a role in the production of pharmaceutical-grade lysine (Connor 2008).

Additionally, lysine has also been produced by chemical means with cyclohexene as C6 starting material. The reaction with nitrosyl chloride provides an oxime intermediate whose chlorine function is replaced by an amino function. Thereof the D,L-racemate of α-amino-ε-caprolactam is available via a rearrangement reaction (Grunwald 2018; Arpe 2007). The L-lysine-1,6-lactam hydrolase from *Cryptococcus laurentii* selectively converts L-α-amino-ε-caprolactam into L-lysine, whereas the D-α-amino-ε-caprolactam remains unconverted. The left D-intermediate is racemized with the aid of the α-amino-ε-caprolactam racemase from *Achrobacter obae* into the D,L-racemate of α-amino-ε-caprolactam for recycling and finally a high selectivity can be achieved (Figure 17.1).

Since the fermentation production technology is rather efficient and has been further optimized in the last decades, the chemical production of lysine has been phased out. It is also to be kept in mind that the chemical synthesis provides a racemic mixture of L-lysine and D-lysine in the first instance and a costly downstream process is needed to obtain pure L-lysine.

17.3 Metabolic Pathway via DAP and Metabolic Engineering

17.3.1 Pathway via Aspartate and Diaminopimelate (DAP)

Lysine is the second amino acid after glutamic acid, which has been produced via fermentation on an industrial level. The process has initially been commercialized by the Japanese company Koywa Hakko Bio Co. in 1958 after the company had observed that a homoserine-auxotrophic mutant of *Corynebacterium glutamicum* secretes high amounts of lysine (Cheng et al. 2018; Ikeda 2017). The predominant host for the fermentative production of lysine (and glutamic acid) applied in academia and the industry is still the gram-positive, facultative anaerobic soil bacterium *C. glutamicum*. The product spectrum of this mainstream biocatalyst in industrial biotechnology besides, e.g. *Escherichia coli* and *S. cerevisiae*, has been extended to more than 30 chemical intermediates (Becker and Wittmann 2015). Highly optimized lysine-producing strains of the species *C. glutamicum* are preferentially applied for pathway extension projects. The idea is to simply add some genes via molecular biology and to keep the high carbon yield and productivity of the lysine technology for producing suitable lysine downstream products. For example, the fermentative production of glutaric acid has been achieved with a titer of 105.3 g/l after 69 h. The recombinant pathway is adding sequential steps to the established route toward lysine and can thereby leverage the existing production chassis (Han et al. 2020; Rohles et al. 2016).

Strains of *C. glutamicum*, which natively generate lysine, were optimized in the lab and subsequently in the fermentation plant based on mutant generation and selection without detailed flux analysis. This has changed after the more advanced toolbox of molecular biology allowed a more defined analysis and rational manipulation of the metabolic pathway. The first strains for the fermentative production of lysine used the native and efficiently functioning pathway which is still – with minor adaptions – followed today. A deregulation of the natural and common feedback mechanisms is a prerequisite for the overproduction of lysine. Especially, the aspartate kinase encoded by the *lysC* gene is of high importance and is subjected to feedback inhibition by lysine and threonine. Point mutations in the ß-subunit of the *lysC* have been shown to overcome the feedback mechanism (Tsuge and Yamaguchi 2021; Wittmann and Becker 2007). Several routes to lysine have been described and four proceed via aspartate. Aspartate is generated via a transamination with oxalacetate as substrate with the keto function as receiving center for the transferred amino function. The overall pathway for the generation of aspartate can be summarized as follows (Table 17.1).

The pathway assumes glucose as starting point and the conventional activation toward glucose-6-phosphate with ATP facilitated by the hexokinase. Pending on the used host and the respective glucose uptake system glucose might be activated with PEP during the uptake via the phosphotransferase system (PTS) system:

$$\text{Glucose (outside)} + \text{PEP} \rightarrow \text{Glucose-6-phosphate (inside)} + \text{pyruvate}$$

Table 17.1 First section of the metabolic pathway toward lysine until aspartate. The carbon dioxide is not externally applied though generated via decarboxylation in the second section of the pathway.

Glucose + 2 ADP + 2 Pi + 2 NAD^+ → 2 Pyruvate + 2 ATP + 2 H_2O + 2 NADH + 2 H^+	Glycolysis
2 Pyruvate + 2 ATP + + 2 H_2O + 2 CO_2 → 2 Oxalacetate + 2 ADP + 2 Pi	Pyruvate carboxylase
2 α-Ketoglutarate + 2 NH_3 + 2 NAD(P)H + 2 H^+ → 2 Glutamate + 2 $NAD(P)^+$ + 2 H_2O	Glutamate synthase
2 Oxalacetate + 2 Glutamate → 2 Aspartate + 2 α-Ketoglutarate	Aminotransferase
Glucose + 2 CO_2 + 2 NH_3 → 2 Aspartate + 2 H_2O	

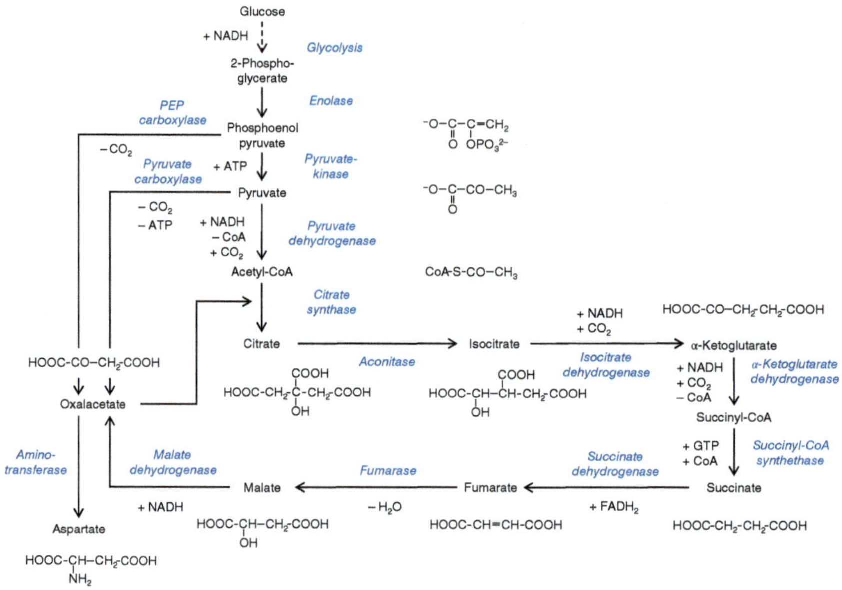

Figure 17.2 Pathway covering the section glucose to aspartate (section 1; Wittmann 2010; Tadepally 2019). Source: Walter Koch.

This uptake avoids the energy investment of the hexokinase. The impact on the energy balance of the pathway is nevertheless very limited since the ATP gain in the usual conversion of PEP into pyruvate via the pyruvate kinase is omitted. The reduction equivalents of the first section of the overall pathway are balanced since the 24 electrons carried by glucose (without counting reduced oxygen) are equally allocated to 2 mol of aspartate each carrying 12 electrons (without counting reduced oxygen and nitrogen). The formation of oxalacetate may proceed via the pyruvate carboxylase as exemplified above, the PEP carboxylase or the PEP carboxykinase as alternative anaplerotic reactions. The provision of the amino function may rely on glutamate or aspartate as an amine-donor and the equation of the reaction operates with an abstract keto-acid and amino acid (Figure 17.2).

It should be pointed out that this pathway section is heuristically chosen to split the overall lysine pathway into reasonable sections. It does not suit as metabolic pathway design for the fermentative synthesis of aspartate with the appealing theoretical molar yield of two mol of aspartate per mol glucose facilitated by the co-utilization of carbon dioxide. The fermentative production of aspartate with glucose and carbon dioxide as carbon sources is not yet working on a commercially interesting level (Piao et al. 2019). Aspartate is commercially be produced via the biocatalytic amination of fumarate (Appleton and Rosentrater 2021).

In the second section of the pathway, aspartate is converted into lysine. Aspartate is initially be transformed into L-tetrahydrodipicolinate (L-THP). This intermediate is then converted into meso-diaminopimelate via four potential routes comprising the direct dehydrogenase pathway and the aminotransferase pathway on the one hand and the more complex acetylase and succinylase pathway on the other hand. The two last routes proceed via a secondary amine with either L-N-Succinyl-2-amino-6-ketopimelate (SAKP) or L-N-acetyl-2-amino-6-ketopimelate (AAKP), respectively, as intermediate. Both are further converted into N-Succinyl-L,L-2,6-diaminopimelate (SDAP) and

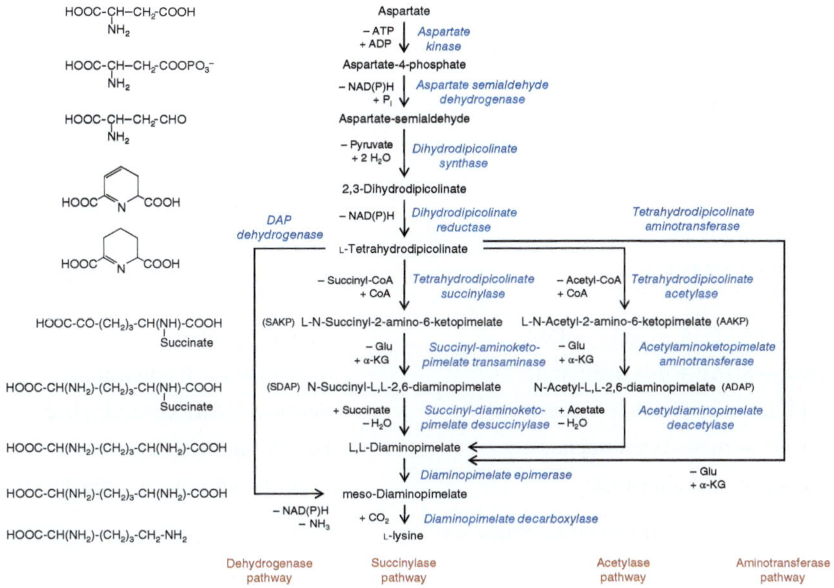

Figure 17.3 Pathway toward lysine covering the section aspartate to lysine (section 2; Wittmann 2010; Tadepally 2019). L-Tetrahydrodipicolinate is identical to L-Piperidin-2,6-dicarboxylate. Source: Walter Koch.

N-Acetyl-L,L-2,6-diaminopimelate (ADAP) via a transamination with glutamate as a donor of the amine function (McCoy et al. 2006; Figure 17.3).

In the initial step of the second pathway section, aspartate is activated with ATP to obtain aspartate-4-phosphate as energized intermediate. The activated carboxylic acid function can then be reduced to the aldehyde level with NADH as a reducing agent to aspartate semialdehyde. In the following steps, the C4 intermediate aspartate semialdehyde is subject to a chain elongation with the addition of pyruvate to provide the cyclic C7 intermediate 2,3-dihydropicolinate. It should be mentioned that the two succinyl-intermediates of the succinylase pathway (second column) and the two acetyl-intermediates of the acetylase pathway (third column) are secondary amines and have the same molecular structure except that the succinate-moiety is replaced by an acetate-moiety. The reaction equation of the second pathway section can be stated as follows (Table 17.2).

The overall lysine pathway combining the first and the second section can be summarized as follows:

$$\text{Glucose} + 2\,\text{NH}_3 + \text{ATP} + 2\,\text{NAD(P)H} + 2\,\text{H}^+$$
$$\rightarrow \text{Lysine} + \text{ADP} + \text{Pi} + 2\,\text{NAD(P)}^+ + 3\,\text{H}_2\text{O}$$

It is remarkable that *C. glutamicum* is indeed using the dehydrogenase and the succinylase pathways in parallel. The repertoire of two pathways allows an adaption to low and abundant ammonia supply and reflects the circumstance that diaminopimelate is a critical building block also for the essential generation of peptidoglycan. The dehydrogenase pathway relies on an abundant ammonia supply and is saving energy since the generation

Table 17.2 Biochemical pathway toward lysine (section 2; Wittmann 2010).

Generation of lysine	
Aspartate + ATP → Aspartate-4-P + ADP	Aspartate kinase
Aspartate-4-P + NAD(P)H + H^+ → Aspartate semialdehyde + Pi + $NAD(P)^+$	Aspartate semialdehyde DH
Aspartate semialdehyde + pyruvate → 2,3-dihydropicolinate + 2 H_2O	Dihydrodipicolinate synthase
2,3-dihydropicolinate + NAD(P)H + H^+ → L-THP + $NAD(P)^+$	Dihydrodipicolinate reductase
L-THP + H_2O + Succinyl-CoA → SAKP + CoA	THP-succinylase
SAKP + glutamate → SDAP + α-ketoglutarate	Succinyl-DAKP transaminase
SDAP + H_2O → LL-Diaminopimelate + succinate	Succinyl-DAKP desuccinylase
LL-Diaminopimelate → meso-Diaminopimelate	Diaminopimelate epimerase
meso-Diaminopimelate → Lysine + CO_2	Diaminopimelate decarboxylase
Recycling of pyruvate, α-ketoglutarate and succinyl-CoA	
0.5 Glucose + ADP + Pi + NAD^+ → Pyruvate + ATP + H_2O + NADH+ H^+	Glycolysis
Succinate + CoA + ATP → Succinyl-CoA + ADP + Pi	Succinyl-CoA synthetase
α-Ketoglutarate + NH_3 + NAD(P)H + H^+ → Glutamate + $NAD(P)^+$ + H_2O	Glutamate dehydrogenase
0.5 Glucose + Aspartate + NH_3 + 2 NAD(P)H + 2 H^+ + ATP → Lysine + CO_2 + ADP + Pi + 2 $NAD(P)^+$ + 2 H_2O	

Source: Adapted from Wittmann (2010).

of succinyl-CoA in the alternative pathway is consuming ATP. On the other hand, the succinylase pathway ensures the synthesis of diaminopimelate in periods with low ammonia supply and thereby justifies the investment of ATP (Eggeling and Sahm 1999).

17.3.2 Theoretical Maximum Molar Yield of the DAP Pathway via Succinylase

For the estimation of the maximum theoretical molar yield, the reduction equivalents and the energy balance need to be considered. In the first instance, the pathway view is disregarded and the electron balance is referred to. The net reaction equation indicates that 2 mol of reduction equivalents are needed per mol lysine since the 24 electrons of glucose are not sufficient to deliver the 28 electrons carried in the lysine molecule without considering the reduced oxygen and nitrogen. Thus, a minor share of glucose needs to be oxidized to balance the reduction equivalents and it can be deduced that 0.166 additional mol of glucose are needed per mol lysine because the complete oxidation of glucose would in total provide 24 electrons (4/24). These 0.166 mol of glucose would derive 0.664 mol ATP in coproduction and the remaining 0.336 mol of ATP still needs to be accounted for. Since the lysine production is an aerobe process, it may be assumed that 1 mol of glucose provides 32 mol of ATP and the overall glucose demand to supply the 2 reduction equivalents and the 1 energy equivalent is estimated at 0.176 mol (0.166 + 0.336/32). Accordingly, the maximum theoretical molar yield would amount to 0.85 (1/1.176) mol of lysine per mol glucose, and the maximum theoretical carbon yield would amount to 0.69 t lysine per ton glucose (0.85 × 146.2/180.2).

The estimation outlined above refers exclusively to the electron flow and stipulates one joint pool of reduction equivalents with interchangeable sub-pools of NADH and NADPH. This assumption might be realized in case a transhydrogenase that converts NADH into NADPH is present (Spaans et al. 2015):

$$\text{NADH} + \text{NADP}^+ \rightarrow \text{NAD}^+ + \text{NADPH} \quad \Delta G^{\circ\prime} = 1.0\ \text{KJ/mol}$$

It is not known that *C. glutamicum* possesses a transhydrogenase and the heterologous expression of genes from other organisms was tried. The pntAB genes of *E. coli* are encoding a membrane-bound transhydrogenase that catalyzes the oxidation of NADH to NAD^+ coupled to the reduction of $NADP^+$ to NADPH. The heterologous expression of this gene in *C. glutamicum* could in fact enhance the generation of lysine (Kabus et al. 2007). The effect depends on the carbon source and the lysine concentration was increased by 10% with glucose feed, by 70% with fructose feed, by 50% with glucose and fructose feed, and by even 300% with sucrose feed. However, a full interchange of reduction equivalents might not be achieved under commercial physiological conditions and a more defined pathway-related and thereby host-specific view is recommended.

It is known that the succinylase pathway in *C. glutamicum* bears a demand of four mol of NADPH per mol lysine (Xu et al. 2020; Vassilev et al. 2018; Takeno et al. 2010) and delivers two mol of NADH per mol lysine. In case it is assumed in the first round that the two mol of pyruvate are provided via glycolysis, the following reaction equation would be derived:

$$\text{Glucose} + 2\,\text{NH}_3 + \text{ATP} + 4\,\text{NADPH} + 4\,\text{H}^+ + 2\,\text{NAD}^+$$
$$\rightarrow \text{Lysine} + \text{ADP} + \text{Pi} + 4\,\text{NADP}^+ + 2\,\text{NADH} + 2\,\text{H}^+ + 3\,\text{H}_2\text{O}$$

The reaction could not yet proceed and the supply of the four mol of NADPH needs to be accounted for. This might in principle be facilitated with different approaches. The simplest approach might be to assume that the 4 mol of NADPH are exclusively be provided via the pentose phosphate pathway, which generates up to 12 mol of NADPH per mol glucose-6-phosphate based on the oxidative steps catalyzed by the glucose-6-phosphate dehydrogenase and the 6-phosphogluconate dehydrogenase:

$$\text{Glucose-6-phosphate} + 12\,\text{NADP}^+ + 7\,\text{H}_2\text{O} \rightarrow 12\,\text{NADPH} + 12\,\text{H}^+ + 6\,\text{CO}_2 + \text{Pi}$$

The supply of the four mol of NADPH per mol lysine via the pentose phosphate pathway defines a major target of metabolic engineering activities. Theoretical flux models revealed that indeed the NADPH pool is a limiting factor for the synthesis of lysine and that the pentose phosphate pathway is the main supply channel (Wittmann and Becker 2007). Accordingly, the concentration of the respective enzymes has been increased and the overexpression of the glucose-6-phosphate dehydrogenase, the transketolase, and the transaldolase triggered a higher flux through the pentose phosphate pathway (Becker et al. 2005, 2007, 2011). After a sufficient flux via the pentose phosphate pathway is ensured, it should be avoided that the two mol of NADH generated via the glyceraldehyde dehydrogenase cannot be used to replenish the pool of NADPH. In case the 4 mol of NADPH would exclusively be provided via the pentose phosphate pathway already 0.33 mol of glucose per mol lysine (8/24) would be needed for providing the reduction equivalents and 2 mol of

NADH would either be spent for ATP generation or trigger the generation of by-products. The maximum theoretical molar yield would then shrink to 0.750 mol of lysine per mol glucose (1/1.33) and the maximum carbon yield would amount to 0.61 t lysine per ton glucose ($0.750 \times 146.2/180.2$).

The applied approaches in metabolic engineering try to avoid this drop in the theoretical maximum yield and implement alternative measures for the generation of NADPH without detrimental impact on the carbon yield. Enzymes with different cofactor specificity may be introduced (Takeno et al. 2010). In this context, the most prominent target is the glyceraldehyde dehydrogenase (GAPDH) which usually generates NADH though might also generate NADPH in the oxidation of glyceraldehyde-3-phosphate. Comparable efforts addressing the cofactor specificity of other enzymes (e.g. formate dehydrogenase, mannitol-2-dehydrogeanse, and keto-acid reductoisomerase) focused on residues surrounding the cofactor binding site and achieved some cofactor flexibility. Since most of the residues are involved in cofactor specificity and catalytic efficiency, a complete reversal was not accomplished (Bommareddy et al. 2014).

However, an isoenzyme of the GAPDH of *C. glutamicum* with elevated catalytic efficiency and similar preference toward NADH and NADPH has been constructed based on the modification of four amino acid residues adjacent to the coenzyme binding site. The recombinant enzyme provided a 60% increase in lysine production (Bommareddy et al. 2014). In the course of this idea, the amino acid sequence of the GAPDH of *E. coli* has been changed at positions 34, 188, and 189 and a dual cofactor specificity with recruitment of NAD^+ and $NADP^+$ as well (Slivinskaya et al. 2022). The overexpression of the engineered GAPDH led to a significant increase in the media concentration of lysine, threonine, and proline in the respective *E. coli* production strains. In particular, the lysine concentration was raised 1.6 and 3.0 times for glucose and xylose as carbon sources (Slivinskaya et al. 2022).

In the ideal case the GAPDH would generate two mol of NADPH and a futile oxidation of NADH is avoided, though in reality only a mixed cofactor specificity might be accomplished. In general, the theoretical yield consideration derives different results pending on the assumed interchangeability of reduction equivalents. It is a common view in the literature that the pentose phosphate pathway is at least partially be pursued to replenish the pool of $NADP^+$ in the metabolic pathway toward lysine in *C. glutamicum* (Wittmann 2010). A rechanneling of flux into the pentose phosphate pathway may be achieved by silencing the phosphoglucose isomerase or overexpression of the glucose-6-phosphate dehydrogenase. The commercial metabolic engineering strategies likely realize a mixture of measures which in sum achieve that a share of the 2 mol of NADH in the theoretical model are either avoided via a GAPDH enzyme with dual cofactor specificity or are converted into NADPH via a transhydrogenase (Figure 17.4).

It may also be the case that the isocitrate dehydrogenase, which consumes $NADP^+$ in bacterial systems, and malic enzymes are involved (Spaans et al. 2015). The real metabolism of the commercial lysine-producing *C. glutamicum* strains is likely combining glycolysis and the pentose phosphate pathway, an engineered GAPDH cofactor specificity and transferred transhydrogenase activity to approach a net demand of only four reduction equivalents and thereby to limit an electron drain into respiratory chain phosphorylation or reduced by-products. The theoretical maximum yield may be given as a window between the two described scenarios (Table 17.3).

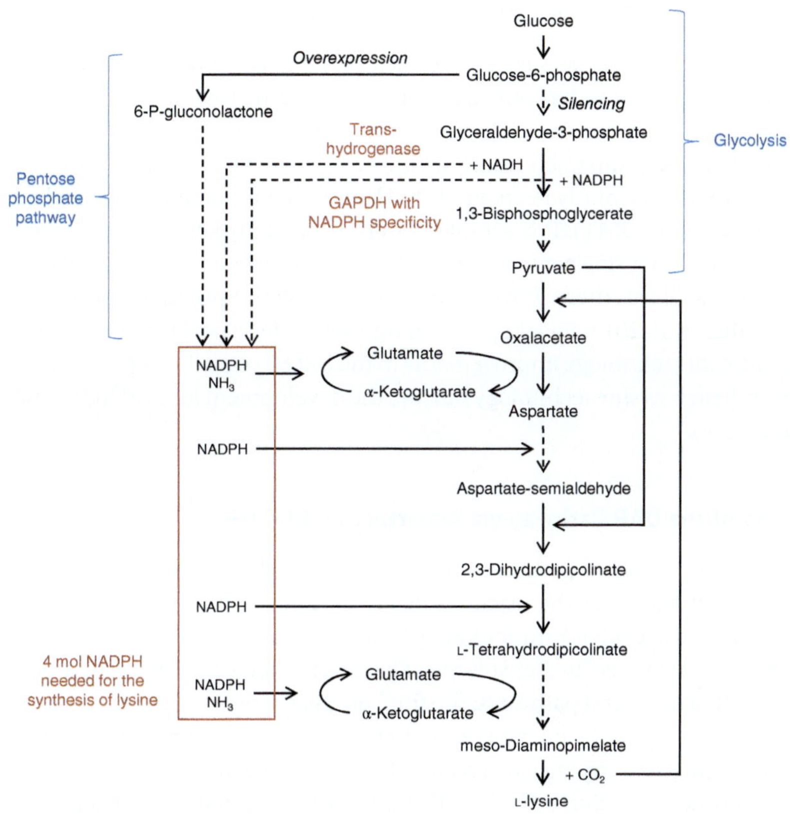

Figure 17.4 Simplified pathway model with focus on reduction equivalents. Source: Walter Koch.

Table 17.3 Technology development of lysine via fermentation. The maximum carbon yield of 0.81 t lysine/t glucose (146.2/180.2) is not plausible since the higher reduction level of lysine compared to glucose is not considered.

Supply of reduction equivalents	Theoretical molar yield	Maximum carbon yield
Pathway considerations on yield		
Net demand of 4 red. Equivalents (only 2 NADPH via PPP)	0.850 mol/mol (1/1.176)	0.69 t lysine/t glucose
4 mol NADPH via PPP (and 2 NADH lost in by-products)	0.750 mol/mol (1/1.333)	0.61 t lysine/t glucose
Yield figures published in literature		
	0.82 mol/mol (Wittmann and Becker 2007)	
		0.66 t lysine/t glucose (Liu et al. 2022)
		0.81 t lysine/t glucose (Xu et al. 2020)

17.3.3 Achieved Carbon Yield of the Pathway via Aspartate and L-THP

The carbon yield achieved in the industry is not disclosed by the major producers. A carbon yield of 0.54 t lysine per ton glucose has already been published in 1999 by an academic group funded with grants of the commercial lysine producer Evonik (formerly Degussa) (Eggeling and Sahm 1999). A comparable value with 0.55 t lysine per ton glucose for the *C. glutamicum* strain LYS-12 has jointly been published by another academic group and an industrial partner (Becker et al. 2011). It is estimated that the actually achieved values are substantially higher than 0.55 t lysine per ton glucose (Sahm et al. 2013). The industrial process is highly mature and achieves titers above 170 g/l combined with a productivity of up to 4 g/lxh (Ikeda 2017; Sahm et al. 2013; Eggeling and Sahm 1999). These performance figures demonstrate the significant technology improvements in the last 60 years if compared to the starting point of the industry. Lysine technology mirrors the development leaps of industrial biotechnology (Table 17.4).

17.3.4 Engineering of the DAP Pathway via Aspartate and L-THF

There are several aspects of the main pathway which are addressed in metabolic engineering activities. One element addresses the usual feedback mechanism. It is known that the aspartate kinase is subject to a feedback inhibition by lysine and threonine. The investment of ATP is exclusively justified in case further volume of lysine or other downstream products are needed. As exemplified in several pathways, the final product avoids its surplus production by suppressing the activity of the enzyme initiating the generation of precursor intermediates. Accordingly, mutants carrying isoenzymes of the aspartate kinase were searched for which display a silenced inhibition by lysine. Mutants with deregulated feedback inhibition were originally generated by random mutagenesis and the optimized performance in production could often not be related to a defined gene and enzyme, respectively. The more advanced engineering approaches manage to integrate rational modifications into the wild-type host (Chen et al. 2011a, b) and avoid the accumulation of undesired mutations, which remain undetected though tend to compromise the production efficiency.

One further element of metabolic engineering uses the overexpression of enzymes to enhance the catalytic efficiency of defined pathway steps. In particular, the copy number of the genes *dapA* (dihydrodipicolinate synthase), *dapB* (dihydrodipicolinate reductase), *dapC* (succinyl-aminoketo-pimelate transaminase), and *dapF* (diaminopimelate epimerase) has

Table 17.4 Technology development of lysine via fermentation. The performance date indicated at the bottom of the table represents the highest lysine titer and carbon yield achieved. The results as of 2018 rely on the addition of biotin with 2.4 mg/l and are related to a specific productivity of 0.73 g/g DCWxh corresponding to a biomass concentration of 5.2 g/l.

Technology status	Titer (g/l)	Carbon yield	Productivity (g/lxh)	Source
1950/1960	40–60	25%	≪4.0	Ikeda (2017)
1970/1980	100	40–50%	<4.0	Ikeda (2017)
2000/2010	170	>55%	4.0	Sahm et al. (2013)
2018	182	64.6%	3.8	Xu et al. (2018)/lab scale

Source: Adapted from Xu et al. (2018).

been elevated and provided increased levels of lysine (Anastassiadis 2007). High efforts have also been allocated in flux direction toward the pentose phosphate pathway to ensure the availability of the 4 mol of NADPH per mol lysine. It should be pointed out that the carbon metabolism in wild-type strains of *C. glutamicum* does more or less exclusively proceed via glycolysis and less than 5% of the flux is directed via the pentose phosphate pathway. For comparison, 47% of the flux is channeled via the pentose phosphate pathway in strains with feedback resistance to lysine under aerobic conditions as needed for commercial application. Accordingly, the genes of the pentose phosphate pathway have been overexpressed (Kobayashi et al. 2020).

17.4 Metabolic Pathway via α-Aminoadipate in Fungi

It is known that fungi pursue a different pathway toward lysine, which is referred to as the α-aminoadipate (AAA) pathway according to its central intermediate metabolically differentiated versus the diaminopimelate (DAP) pathway (Torruella et al. 2009; Xu et al. 2006; Velasco et al. 2002). The pathway is commercially not used for the fermentative production of α-lysine though has gained interest since aminoadipate is also a precursor for penicillin. The pathway starts with α-ketoglutarate as an available intermediate of the citric acid cycle and starts with the condensation of α-ketoglutarate and acetyl-CoA toward homocitrate. The pathway proceeds via homoisocitrate to reach α-ketoadipate (Figure 17.5).

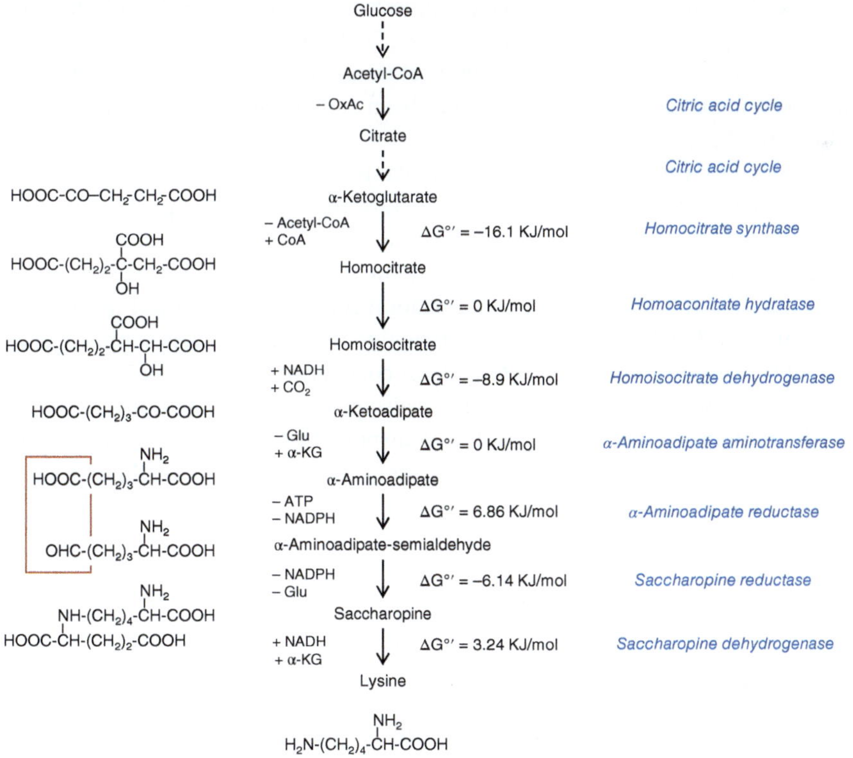

Figure 17.5 Synthetic lysine pathway via aminoadipate in fungi. Source: Adapted from Pissarra and Nielsen (1997).

The conversion of homocitrate into homoisocitrate is shown as a one-step reaction in the simplified pathway scheme above though indeed two enzymes might be involved with homoaconitate with a double bond between the two carboxylic acid functions as intermediate (Fazius et al. 2012). The conversion of α-aminoadipate into α-aminoadipate-semialdehyde defines a crucial step of the pathway since it comprises the biochemical reduction of a carboxylic acid function into an aldehyde function. The challenging thermodynamics of this reaction type is overcome by the investment of one mol of ATP per mol α-aminoadipate. It is assumed that the reaction proceeds in a two-step reaction covering a adenylation with δ-5-adenylyl-L-α-aminoadipate as an intermediate and a reduction reaction (Xu et al. 2006). The reaction is nevertheless estimated to have a positive Gibbs free energy, though the pathway should proceed with sufficient efficiency in case the proceeding steps provide a critical pull (Pissarra and Nielsen 1997).

17.5 Secretion of Lysine

The understanding of export systems for small products defines a landmark in the exploration of host physiology and fermentation technology. The secretion of products relies on membrane proteins and may be facilitated by general exporters with a rather broad metabolite spectrum or specialized and selective exporters (Marin and Krämer 2007). The secretion of lysine in *C. glutamicum* is managed by the LysE protein which exports the amino acids lysine and arginine. This mechanism could be demonstrated by overexpression of the *lysE* gene, which resulted in a fivefold increase in secretion (Burkovski and Krämer 2002; Vrljić et al. 1996). It relies on H^+ antiport or OH^- symport as an energy source. The secretion protein can also be transferred to other hosts as has been shown for *Methylophilus methylotrophicus* (Gunji and Yasueda 2006).

The lysine exporter has a natural function that explains its survival in the evolutionary mechanism of energy saving. Bacteria bear the ability to absorb extracellular peptides which can be broken in parts with internal hydrolyzing enzymes to provide cheap building blocks for protein synthesis. Since *C. glutamicum* has no lysine-degrading activity, an option to remove lysine is required if the metabolite concentration is exceeding a critical threshold. In experiments with a *lysE* deletion mutant, the dipeptide lysyl-alanine was supplied as a media component besides glucose and an intracellular concentration of up to 1 M could be measured (Eggeling and Sahm 1999). *C. glutamicum* also possesses a lysine uptake system LysP. It is a carrier protein that belongs to the APC transport superfamily and alternatively transports lysine or alanine. LysP ensures the intracellular and the extracellular lysine concentration are balanced (Liu et al. 2022).

17.6 Process Development

17.6.1 Aerobic Process and Oxygen Demand

The fermentative production of lysine is pursued as an aerobic process and a stirrer facilitates an efficient dispersion of air (i.e. oxygen) into the medium (Locher 2022). The power input of the stirrer causes a significant share of the overall power demand and increases the cooling efforts (Marinussen and Kool 2010). The availability of dissolved oxygen needs to be monitored carefully and an efficient oxygen transfer rate (OTR) needs to be maintained. The OTR can be enhanced by increasing the stirring rate, by applying pure oxygen for the aeration, or by pressurizing the reactor (Liu et al. 2022).

The oxygen demand is not straightforward given the basic reaction equation for the fermentative production of lysine, which indicates a gap of two reduction equivalents per mol lysine. Accordingly, the oxygen supplementation does not function as a final electron acceptor to remove a surplus of reduction equivalents. The reaction of NADH or NADPH with oxygen would limit the pool of reduced reduction equivalents and require the additional oxidation of glucose. Instead, *C. glutamicum* relies on oxygen as electron acceptor to sustain growth. Alternatively, nitrate may be offered as final electron acceptor though the accumulation of nitrite would be getting toxic. The anaerobic fermentation with *C. glutamicum* is possible to some extent and mainly provides lactate, succinate, and acetate as products (Tsuge and Yamaguchi 2021; Vassilev et al. 2018; Okino et al. 2008).

17.6.2 Higher Flexibility of Carbon and Energy Source

The industrial production of lysine is mainly pursued with starch hydrolysate from corn and cassava (i.e. glucose and dextrose) often combined with cane molasses and beet molasses (Ikeda 2017). In line with the general endeavor to establish a second-generation carbon source for fermentation technology, alternative feedstocks have been tested. These attempts including lactose, starch, arabinose, xylose, glycerol, and cellulose could provide a proof of concept (Liu et al. 2022). In parallel to academic work with a focus on single carbon sources, there are industry-driven initiatives to better understand the consumption of molasses (including biological active substances) and mixed sugars (Xu et al. 2020). It remains to be seen inasmuch as second-generation feedstocks will in the future be able to replace conventional first-generation carbohydrates.

17.6.3 Product Isolation and Purification

The industrial process tries to enhance the reactor productivity and the switch to a semi-continuous operation with interim harvest during the production allowed a productivity increase of up to 50% (Locher 2022). The temperature during the fermentation is kept at about 30–34 °C via cooling. The pH is kept around 7.0–7.5 (Liu et al. 2022). In case no pH management system is installed, the pH of the broth would increase during the fermentation since lysine is carrying two amino functions and one carboxylic acid function. In case ammonia or ammonia water is used as a nitrogen source, hydrochloric acid or sulfuric acid might be added during the fermentation to neutralize the broth. The protonated lysine would then be converted into lysine hydrochloride or lysine sulfate:

$$
\begin{aligned}
&{}^{-}\mathrm{OOC{-}CH(NH_3)^{+}{-}(CH_2)_4{-}NH_3^{+} + OH^{-} + HCl} \\
&\quad \rightarrow {}^{-}\mathrm{OOC{-}CH(NH_3)^{+}{-}(CH_2)_4{-}NH_3^{+} + Cl^{-} + H_2O}
\end{aligned}
$$

It is recommended to combine the supplementation of nitrogen with the neutralization system and often ammonium sulfate is added as preferred nitrogen source. Lysine is then given as lysine sulfate and the sulfate anions can be recovered in the further downstream process.

The isolation and purification of lysine are often following a simple path in case the product shall be applied as a feed additive which is the biggest market outlet. Since the industrial host *C. glutamicum* is generally recognized as safe, it qualifies as feed and protein source and a removal of the biomass is not required for the use as feed ingredient. The biomass is inactivated at the batch end and the water is removed by evaporation or spray-drying, respectively (Eggeling and Sahm 1999). The obtained broth components including lysine are subject to a granulation step and finally dried. Evonik (formerly Degussa-Hüls) is offering a Biolys® product grade in the market which contains L-lysine sulfate with the formula $[C_6H_{14}N_2O_2]_2 \times H_2SO_4$ and a molecular weight of 390.4 g/mol (Evonik 2015). The lysine content is indicated with a minimum of 54.6% of the sulfate salt.

Alternatively, a lysine grade without media residues might be needed for non-feed application and an appropriate downstream process is installed. Often lysine hydrochloride is the preferred product given its low hygroscopicity compared to pure lysine. A standard process is to neutralize the medium with sulfuric acid – i.e. ammonium sulfate as nitrogen source – during the fermentation such that lysine in the broth is given as lysine sulfate at the batch end. The biomass is inactivated at the batch end and removed via filtration. In an ion exchange step, the lysine sulfate is converted into free lysine and ammonium sulfate is obtained as by-product. The free lysine is concentrated by evaporation and hydrochloric acid is added before the subsequent crystallization step. The lysine hydrochloride crystals are harvested, dried, and packaged containing 98.5% lysine hydrochloride (Eggeling and Sahm 1999). The remaining mother liquor is recycled. The yield in the downstream process to obtain pure lysine hydrochloride is estimated at 90% compared to 95% for the downstream process toward Biolys with 95% (Sahm et al. 2013; Figure 17.6).

Since the concentration and evaporation of water consumes energy also liquid lysine grades with lower costs are offered in the market like the Liquid Lysine 60 feed grade from Ajinomoto. It contains free lysine and lysine hydrochloride at a concentration of 60% (Ikeda 2017). The energy demand of the overall lysine fermentation process – with the final product in the form of lysine hydrochloride – has been estimated with 3.94 MWh/t lysine × HCl and 5.8 t steam/t lysine HCl (Marinussen and Kool 2010). The utility consumption seems

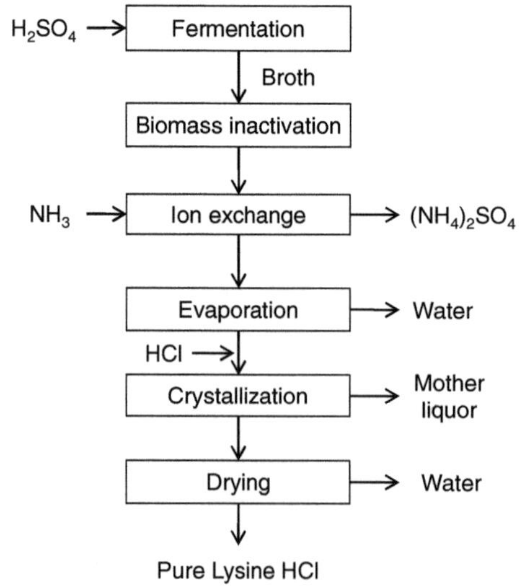

Figure 17.6 Downstream process for the generation of pure lysine hydrochloride (Sahm et al. 2013; Mishra et al. 2016). Source: Walter Koch.

rather high, though also reflects the high energy input for the stirrer to facilitate an efficient aeration, which is estimated at 3.75 MWh/t lysine HCl (Marinussen and Kool 2010). The actually achieved energy input is likely lower reflecting incremental process optimization.

Fermentation studies were also pursued with immobilized cells of *C. glutamicum* MH 20–22 B to explore the potential of further process improvements. The impact of batch duration, pH, temperature, glucose concentration, airflow rate, and aeration rate was optimized for both the conventional submers fermentation and the alternative approach with immobilized cells. Under the chosen process conditions, the approach with immobilized cells displays a slightly higher titer at 31.6 g/l compared to 26.3 g/l with suspended cells (Razak and Viswanath 2015a, b). However, the titer is significantly below the achieved commercial titers and, accordingly, the minor titer upside may not be confirmed on industrial scale.

17.7 Industrial Application

The fermentative lysine technology was invented by the Japanese company Kyowa Hakko who started to produce monosodium glutamate in 1956 and followed already in 1958 with lysine as the second fermentation product. Both processes offer synergies since *C. glutamicum* is employed as host and molasses is used as feedstock. Ajinomoto as the second Japanese producer entered the market in 1960. The high market growth as feed supplement attracted other companies to enter the market and with Cargill and ADM two major US starch companies leveraged their own supply with glucose as feedstock and gained market share with their vertical integration strategy. A detailed analysis on lysine capacity and producer has been published for the year 2000 (Connor 2008). The global production capacity was estimated at 848 kt pa covered by several producers (Table 17.5).

The global demand was estimated at 445 kt pa translating into a global use of capacity of 52% (Connor 2008). The global capacity has increased and is estimated with above 2.5 Mio. t pa in 2016 and a growth rate of 7% pa (Cheng et al. 2018). The supply structure has seen significant changes and the current major producers are CJ CheilJedang (South Korea), Global Bio-Chem Technology Group (China), Ajinomoto (Japan), ADM (United States), Evonik (formerly Degussa, Germany) and among others (Ikeda 2017). The major commercial

Table 17.5 Global lysine capacity based on fermentation, selection of major players.

Producer	Capacity (kt pa)
ADM	250
Ajinomoto	205
Cargill/Degussa	75
Cheil Sugar	60
Kyowa Hakko	97
Sewon/Miwon	135

Source: Adapted from Connor (2008).

plants are operated in China, the United States, Brazil, Indonesia, and Russia. About 65% of the global production capacity is located in China (Kennedy 2021).

An updated and comprehensive view of capacities and plants is not available. The companies disclosed in former press releases capacity expansion steps which, however, do not allow a reconstruction of their overall production capacity. The South Korean biotech company CJ CheilJedang claims that a market share of 25% has been served in 2019 compared to 19% in 2011 (CJ Company Information 2022). Evonik has a JV with Shandong Cathay Lineng Biotechnology with a former capacity of 40 kt pa, which was expanded to 120 kt pa in 2007/2008 (Cabi 2005). Evonik also expanded its capacity at its major site in Blair (Nebraska) from 140 kt pa up to 280 kt pa in 2012 (Evonik Company Information 2012). ADM announced in 2011 plans to expand its lysine and threonine capacity in Decatur (Illinois) up to 340 kt pa in 2013. Additionally, the company declared to expand its lysine capacity in Clinton (Iowa) up to 70 kt pa (Food Ingredients 2011).

References

Anastassiadis, S. (2007). L-lysine fermentation. *Recent Patents on Biotechnology* 1: 11–24.

Appleton, H. and Rosentrater, K.A. (2021). Sweet dreams (are made of this): a review and perspectives on aspartic acid production. *Fermentation* 7: 49. https://doi.org/10.3390/fermentation7020049.

Arpe, H.-J. (2007). *Industrielle Organische Chemie. Bedeutende Vor- und Zwischenprodukte*, 288–289. Wiley.

Becker, J. and Wittmann, C. (2015). Biotechnologie von Morgen: metabolisch optimierte Zellen für die bio-basierte Produktion von Chemikalien und Treibstoffen. *Materialien und Gesundheitsprodukten, Angewandte Chemie* 127: 3383–3407. https://doi.org/10.1002/ange.201409033.

Becker, J., Klopprogge, C., Zelder, O. et al. (2005). Amplified expression of fructose 1,6-bisphosphatase in *Corynebacterium glutamicum* increase in vivo flux through the pentose phosphate pathway and lysine production on different carbon sources. *Applied and Environmental Microbiology* 71 (12): 8587–8596. https://doi.org/10.1128/AEM.71.12.8587-8596.2005.

Becker, J., Klopprogge, C., Herold, A. et al. (2007). Metabolic flux engineering of L-lysine production in *Corynebacterium glutamicum* – overexpression and modification of G6P dehydrogenase. *Journal of Biotechnology* 132: 99–109.

Becker, J., Zelder, O., Häfner, S. et al. (2011). From zero to hero – design-based systems metabolic engineering of *Corynebacterium glutamicum* for L-lysine production. *Metabolic Engineering* 13: 159–168. https://doi.org/10.1016/j.ymben.2011.01.003.

Bommareddy, R.R., Chen, Z., Rappert, S., and Zeng, A.-P. (2014). A *de novo* NADPH generation pathway for lysine production of *Corynebacterium glutamicum* by rational design of the coenzyme specificity of glyceraldehyde-3-phosphate dehydrogenase. *Metabolic Engineering* 25: 30–37. https://doi.org/10.1016/j.ymben.2014.06.005.

Burkovski, A. and Krämer, R. (2002). Bacterial amino acid transport proteins: occurrence, functions, and significance for biotechnological applications. *Applied Microbiology and Biotechnology* 58: 265–274.

Cabi (2005). *New Lysine Capacity Set for China*. Cabi. https://www.cabi.org/animalscience/news/13532 (accessed 10 September 2022).

Chen, Z., Meyer, W., Rappert, S., and Sun, J. (2011a). Coevolutionary analysis enabled rational deregulation of allosteric enzyme inhibition in *Corynebacterium glutamicum* for lysine production. *Applied and Environmental Microbiology* 77: 4352–4360.

Chen, Z., Rappert, S., Sun, J., and Zeng, A.-P. (2011b). Integrating molecular dynamics and co-evolutionary analysis for reliable target prediction and deregulation of the allosteric inhibition of aspartokinase for amino acid production. *Journal of Biotechnology* 154: 248–254.

Cheng, J., Chen, P., Song, A. et al. (2018). Expanding lysine industry: industrial biomanufacturing of lysine and its derivatives. *Journal of Industrial Microbiology & Biotechnology* 45: 719–734. https://doi.org/10.1007/s10295-018-2030-8.

Cho, J.S., Kim, G.B., Eun, H. et al. (2022). Designing microbial cell factories for the production of chemicals. *JACS* 2: 1781–1799. https://doi.org/10.1021/jacsau.2c00344.

CJ Company Information (2022). www.cj.co.kr/en/about/business/bio (accessed 10 September 2022)

Connor, J.M. (2008). Evolution of the global lysine industry 1960-2000. *SSRN Electronic Journal* https://doi.org/10.2139/ssrn.1188512.

Eggeling, L. and Sahm, H. (1999). L-glutamate and L-lysine: traditional products with impetuous developments. *Applied Biology & Biotechnology* 52: 146–153.

Evonik (2015). *Product Information Biolys®*, 1–2. Evonik Nutrition & Care GmbH.

Evonik Company Information (2012). U.S. expansion of L-lysine capacity expanded, https://corporate.evonik.com/en/investor-relations/us-expansion-of-l-lysine-capacity-completed-107026.html (accessed 10 September 2022)

Fazius, F., Shelest, E., Gebhardt, P., and Brock, M. (2012). The fungal α-aminoadipate pathway for lysine biosynthesis requires two enzymes of the aconitase family for the isomerization of homocitrate to homoisocitrate. *Molecular Microbiology* 86 (6): 1508–1530. https://doi.org/10.1111/mmi-12076.

Food Ingredients (2011). https://www.foodingredientsfirst.com/news/ADM-to-Expand-Lysine-and-Threonine-Production.html (accessed 10 September 2022)

Grunwald, P. (2018). Biocatalysis. In: *Biochemical Fundamentals and Applications*, 755. World Scientific Publishing.

Gunji, Y. and Yasueda, H. (2006). Enhancement of L-lysine production in methylotroph *Methylophilus methylotrophus* by introducing a mutant LysE exporter. *Journal of Biotechnology* 127: 1–13.

Han, T., Kim, G.B., and Lee, S.Y. (2020). Glutaric acid production by systems metabolic engineering of an L-lysine overproducing *Corynebacterium glutamicum*. *PNAS* 117 (48): 30328–30334. https://doi.org/10.1073/pnas.2017483117.

Ikeda, M. (2003). Amino acid production processes. *Advances in Biochemical Engineering/Biotechnology* 79: 1–35. https://doi.org/10.1007/3-540-45989-8.

Ikeda, M. (2017). Lysine fermentation: history and genome breeding. In: Amino Acid Fermentation, Advances in Biochemical Engineering/Biotechnology, vol. 159 (ed. A. Yokata and M. Ikeda), 73–102. Springer. https://doi.org/10.1007/10_2016_27.

Kabus, A., Georgi, T., Wendisch, V.F., and Bott, M. (2007). Expression of the *Escherichia coli* pntAB genes encoding a membrane-bound transhydrogenase in *Corynebacterium glutamicum* improves L-lysine formation. *Applied Microbiology and Biotechnology* 75 (1): 47–53. https://doi.org/10.1007/s00253-006-0804-9.

Kennedy, M. (2021). Cash market moves, 22.11.2021. *Progressive Farmer*. https://www.dtnpf.com/agriculture/web/AG/columns/cash-market-moves/article/2021/11/22/fuss-lysine-shortage (accessed 10 September 2021).

Kobayashi, S., Kawaguchi, H., Shirai, T. et al. (2020). Automatic redirection of carbon flux between glycolysis and pentose phosphate pathway using an oxygen-responsive metabolic switch in *Corynebacterium glutamicum*. *ACS Synthetic Biology* 9: 814–816. https://doi.org/10.1021/acssynbio.9b00493.

Leong, Y.K., Chen, C.-H., Huang, S.-F. et al. (2020). High-level L-lysine bioconversion into cadaverine with enhanced productivity using engineered *Escherichia coli* whole-cell biocatalyst. *Biochemical Engineering Journal* 157: https://doi.org/10.1016/j.bej.2020.107547.

Liu, J., Xu, J.-Z., Rao, Z.-M., and Zhang, W.-G. (2022). Industrial production of L-lysine in *Corynebacterium glutamicum*: progress and prospects. *Microbiological Research* 262: https://doi.org/10.1016/j.micres.2022.127101.

Locher, A., *Toward Climate Neutrality, BioLys Process Evonik* 2022, Company presentation, https://elements.evonik.de/forschung-und-innovation/die-kraft-der-mikro-organismen (accessed 2 September 2022)

Marin, K. and Krämer, R. (2007). Amino acid transport systems in biotechnologically relevant bacteria. In: *Amino Acid Biosynthesis* (ed. V. Wendisch), 289–317. Heidelberg: Springer.

Marinussen, M. and Kool, A. (2010). *Environmental Impacts of Synthetic Amino Acid Production*. Blonk Milieu Advies BV. https://website-production-s3bucket-1nevfd7531z8u.s3.eu-west-1.amazonaws.com/public/website/download/f0e6feeb-d39c-45d2-af25-f94d1de9e074/amino-acids.pdf (accessed 3 September 2022).

McCoy, A.J., Adams, N.E., Hudson, A.O. et al. (2006). L-diaminopimelate aminotransferase, a trans-kingdom enzyme shared by *Chlamydia* and plants for synthesis of diaminopimelate/lysine. *PNAS* 103 (47): 17909–17914.

Mishra, H., Dangwal, J., Sharma, K., Soman, D., 2016, Production of amino acids (lysine) by fermentation technology, http://www.slideshare.net/kskuldeep1995/lysine-industrial-production (accessed 13 October 2022)

Okino, S., Suda, M., Fujikura, K. et al. (2008). Production of D-lactic acid by *Corynebacterium glutamicum* under oxygen deprivation. *Applied Microbiology and Biotechnology* 78: 449–454. https://doi.org/10.1007/s00253-007-1336-7.

Piao, X., Wang, L., Lin, B. et al. (2019). Metabolic engineering of *Escherichia coli* for production of L-aspartate and its derivative ß-alanine with solid stoichiometric yield. *Metabolic Engineering* 54: 244–254. https://doi.org/10.1016/j.ymben.2019.04.012.

Pissarra, P.D. and Nielsen, J.B. (1997). Thermodynamics of metabolic pathways for penicillin production: analysis of thermodynamics feasibility and free energy changes during fed-batch cultivation. *Biotechnology Progress* 13: 156–165.

Qian, Z.G., Xia, X.-X., and Lee, S.Y. (2011). Metabolic engineering of *Escherichia coli* for the production of cadaverine: a five carbon diamine. *Biotechnology and Bioengineering* 108 (1): https://doi.org/10.1002/bit.22918.

Razak, M.A. and Viswanath, B. (2015a). Optimization of fermentation upstream parameters and immobilization of *Corynebacterium glutamicum* MH 20-22 B cells to enhance the production of L-lysine, 3. *Biotech* 5: 531–540. https://doi.org/10.1007/s13205-014-0252-7.

Razak, M.A. and Viswanath, B. (2015b). Comparative studies for the biotechnological production of L-lysine by immobilized cells of wild-type *Corynebacterium glutamicum* ATCC

13032 and mutant MH 20-22 B. *3Biotech* 5: 765–774. https://doi.org/10.1007/s13205-015-0275-8.

Rohles, C.M., Gießelmann, G., Kohlstedt, M. et al. (2016). Systems metabolic engineering of *Corynebacterium glutamicum* for the production of the carbon-5 platform chemicals 5-aminovalerate and glutarate. *Microbial Cell Factories* 15: 154. https://doi.org/10.1186/s12934-016-0553-0.

Sahm, H., Antranikian, G., Stahmann, K.-P., and Takors, R. (ed.) (2013). *Industrielle Mikrobiologie*, 121. Springer Verlag, Heidelberg.

Slivinskaya, E.A., Plekhanova, N.S., Altman, I.B., and Yampolskaya, T.A. (2022). Engineering of Escherichia coli glyceraldehyde-3-phosphate dehydrogenase with dual NAD+/NADP+ cofactor specificity for improving amino acid production. *Microorganisms* 10: 976. https://doi.org/10.3390/microorganisms10050976.

Smiricky, M.R., Mavromichalis, I., Albin, D.M. et al. (2004). Bioefficacy of L-lysine sulfate compared to feed grade lysine-HCl in young pigs. *Journal of Animal Science*. https://doi.org/10.2527/2004.8292610x.

Spaans, S.K., Weusthuis, R.A., van der Oost, J., and Kengen, S.W.M. (2015). NADPH-generating systems in bacteria and archaea. *Frontiers in Microbiology* 6: 742. https://doi.org/10.3389/fmicb.2015.00742.

Tadepally, H.D. (2019, (Chapter 8)). Recent advances in the industrial production of L-lysine by bacteria. In: *Recent Developments in Applied Microbiology and Biochemistry*, 97–106. https://doi.org/10.1016/B978-0-12-816328-3.00008-8.

Takeno, S., Murata, R., Kobayashi, R. et al. (2010). Engineering of *Corynebacterium glutamicum* with an NADPH-generating glycolytic pathway for L-lysine production. *Applied and Environmental Microbiology* 76 (21): 7154–7160. https://doi.org/10.1128/AEM.01464-10.

Torruella, G., Suga, H., Riutort, M. et al. (2009). The evolutionary history of lysine biosynthesis pathways within eukaryotes. *Journal of Molecular Evolution* 69: 240–248. https://doi.org/10.1007/s00239-009-9266-x.

Tsuge, Y. and Yamaguchi, A. (2021). Physiological characteristics of *Corynebacterium glutamicum* as a cell factory under anaerobic conditions. *Applied Microbiology and Biotechnology* 105: 6173–6181. https://doi.org/10.1007/s00253-021-11474-w.

Vassilev, I., Gießelmann, G., Schwechheimer, S.K. et al. (2018). Anodic electro-fermentation: anaerobic production of L-lysine by recombinant *Corynebacterium glutamicum*. *Biotechnology and Bioengineering* 115: 1499–1508. https://doi.org/10.1002/bit.26562.

Velasco, A.M., Leguina, J.I., and Lazcano, A. (2002). Molecular evolution of the lysine biosynthetic pathways. *Journal of Molecular Evolution* 55: 445–459. https://doi.org/10.1007/s00239-002-2340-2.

Vrljić, M., Sahm, H., and Eggeling, L. (1996). A new type of transporter with a new type of cellular function: L-lysine export from *Corynebacterium glutamicum*. *Molecular Microbiology* 22: 815–826.

Wittmann, C. (2010). Entwicklung und Einsatz neuer Tools zur metabolischen Netzwerkanalyse des industriellen Aminosäure-Produzenten *Corynebacterium glutamicum*. In: *Schriftenreihe des Instituts für Bioverfahrenstechnik der Technischen Universität Braunschweig*, vol. No. 50. Göttingen: Cuvillier Verlag.

Wittmann, C. and Becker, J. (2007). The L-lysine story: from metabolic pathways to industrial production. In: *Amino Acid Biosynthesis* (ed. V.F. Wendisch), 40–70. Springer Heidelberg: https://doi.org/10.1007/7171_2006_089.

Xu, H., Andi, B., Qiam, J. et al. (2006). The α-aminoadipate pathway for lysine biosynthesis in fungi. *Cell Biochemistry and Biophysics* 46: 43–64.

Xu, J.-Z., Wu, Z.-H., Gao, S.-J., and Zhang, W. (2018). Rational modification of tricarboxylic acid cycle for improving L-lysine production in *Corynebacterium glutamicum*. *Microbial Cell Factories* 17: 105. https://doi.org/10.1186/s12934-018-0958-z.

Xu, J.-Z., Ruan, H.-Z., Yu, H.-B. et al. (2020). Metabolic engineering of carbohydrate metabolism systems in *Corynebacterium glutamicum* for improving the efficiency of L-lysine production from mixed sugars. *Microbial Cell Factories* 19: 39. https://doi.org/10.1186/s12934-020-1294-7.

18

Citric Acid

18.1 Application

Citric acid (2-hydroxy,1,2,3-propanetricarboxylic acid) is an organic acid with the formula $C_6H_8O_7$ and a molecular weight of 192.1 g/mol. It bears three carboxylic acid functions with the corresponding pK_a values of 3.1, 4.7, and 6.4 (Max et al. 2010). The global production of citric acid was reported at 0.5 Mio. t in 1997 (Arpe 2007) and increased to 1.6 Mio. t in 2013 (Angumeental and Venkappayya 2013). The market price in 2013 is reported at 1 €/kg (Sahm et al. 2013). Citric acid is mainly used in the food and beverage industry for acidification and taste adjustment. It is also used in the formulation of pharmaceuticals for the rapid dissolution of active ingredients (Show et al. 2015; Schmidt and Lang 2013). In the dishwashing industry, it is used to reduce the water hardness due to its capability as a chelating agent to trap metal cations. In the chemical industry, it is used as an anti-foaming agent and plasticizer. The product may be offered as pure acid in the form of crystalline citric acid anhydrous or citric acid monohydrate or alternatively as salt in the form of trisodium dihydrate or trisodium anhydrous (www.jungbunzlauer.com 2022).

18.2 Chemical Production and Natural Extraction

The rather high molecular functionality of citric acid with branched C-backbone and four functional groups favors fermentation technology versus organic chemistry. Fermentation technology allows to bundle several chemical steps in one process, without the otherwise required isolation and purification of intermediates, and this advantage is leveraged with increasing molecular functionality. Accordingly, the fermentation industry contributes more than 90% of the global production (Ye et al. 2022). The chemical production of citric acid is minor and can be neglected.

Citric acid might alternatively be derived by natural extraction. Several fruits have a natural content of citric acid which reaches a level of 70 g/l in lime and 56 g/l in lemon (Ciriminna et al. 2017). The natural citric acid in fruits can be isolated and the first commercial extraction was started in England in 1826 using imported lemons as feedstock. Citric acid can be precipitated as calcium citrate with the addition of calcium oxide.

$$CaO + H_2O \rightarrow Ca(OH)_2$$

$$2\ C_6H_8O_7 + 3\ ca(OH)_2 \rightarrow Ca_3(C_6H_5O_7)_2 \downarrow + 6\ H_2O$$

Pathway Design for Industrial Fermentation, First Edition. Walter Koch.

The calcium citrate is captured by filtration and then reconverted into the fully protonated citric acid with the addition of sulfuric acid. The finding in 1893 that *Citromyces* spp. is able to convert carbohydrate into citric acid was not used as an industrial practice due to contamination issues and the long fermentation process time (Show et al. 2015). The breakthrough research was conducted in 1916 by James Currie who found out that *Aspergillus niger* is able to generate citric acid and can grow at a pH of about 2.5–3.5 (Show et al. 2015). The production volume of citric acid via raw material extraction peaked in 1915–1916 with 18 kt pa and declined afterward because fermentation technology was gaining market share (di San n.d.). The extraction was still pursued in 2010 in smaller plants in Mexico and Africa, though is not relevant anymore with respect to the total production capacity within the industry (Max et al. 2010).

18.3 Biochemical Pathway

18.3.1 Provisioning of the Two Main Precursors and Pathway Split on Two Compartments

Citric acid is a natural and common intermediate in biochemical pathways and is generated in the first step of the citric acid cycle, namely, in the condensation of oxalacetate and acetyl-CoA. The formerly proposed generation of citric acid via the reductive condensation of malic acid and acetic acid, the condensation of oxalacetate with acetic acid, or the oxidation of gluconic acid could not be confirmed (Abou-Zeid and Ashy 1984). A specific pathway development was not needed. The first steps toward industrial technology have already been elucidated in the nineteenth century and relied on natural producers. Specific metabolic engineering programs were technically not possible at that point in time due to limited analytical methods and a missing toolbox for molecular engineering. The observed high productivity of *A. niger* in carbohydrate metabolism and the secretion of citric acid could directly be translated into a production process that focused on media optimization and management of host morphology.

The synthesis of citrate requires acetyl-CoA and oxalacetate according to the first reaction of the citric acid cycle. The natural metabolic pathway is straightforward and proceeds via the cleavage of the glucose C6 backbone into 2 mol of pyruvate. The 2 mol of pyruvate are then further converted into acetyl-CoA and oxalacetate to provide the two substrates for the citrate synthase (Table 18.1).

It is to be kept in mind that *A. niger* is an eukaryotic host and the involved pathways glycolysis and citric acid cycle are taking place in two different compartments. The pathway dissection into two separate compartments implies an additional level of complexity. The 2 mol of pyruvate derived in the glycolysis are initially located in the cytosol. One mol of pyruvate is then transferred into the mitochondrion for the generation of acetyl-CoA via the pyruvate dehydrogenase. Thereby, the first substrate of the citrate synthase has been provided (Figure 18.1).

The second substrate of the citrate synthase, oxalacetate, is derived from malate and is generated in the mitochondrion via the mitochondrial malate dehydrogenase. Malate is transported into the mitochondrion via the citrate malate carrier. The cytosolic malate pool needed for the citrate malate carrier is replenished by the cytosolic pyruvate carboxylase

Table 18.1 Biochemical pathway toward citric acid.

Provision of acetyl-CoA	
Glucose + 2 ADP + 2 Pi + 2 NAD^+ → 2 Pyruvate + 2 ATP + 2 H_2O + 2 NADH + 2 H^+	Glycolysis
Pyruvate + CoA + NAD^+ → Acetyl-CoA + CO_2 + NADH + H^+	Pyruvate dehydrogenase
Synthesis of citrate	
Pyruvate + CO_2 + ATP + H_2O → Oxalacetate + ADP + Pi	Pyruvate carboxylase
Oxalacetate + Acetyl-CoA + H_2O → Citric acid + CoA	Citrate synthase
Glucose + ADP + Pi + 3 NAD^+ → Citric acid + ATP + 3 NADH + 3 H^+	Net reaction

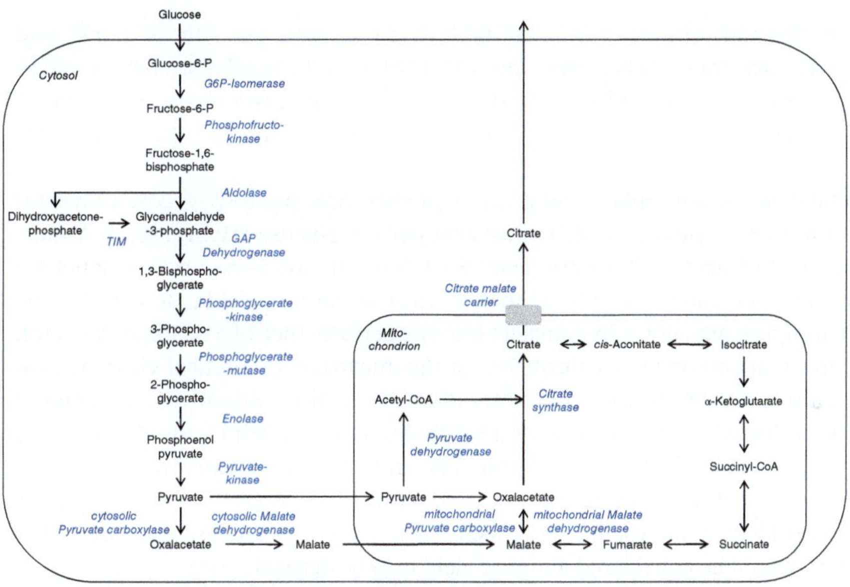

Figure 18.1 Metabolic pathway toward citric acid. Source: Walter Koch.

and the cytosolic malate dehydrogenase (Tong et al. 2019; Pel et al. 2007). The citric acid cycle is in parallel proceeding its natural reactions since the pathway intermediates are needed as starting molecules for essential biomolecules. Thus, a minor share of citrate is not secreted, though converted into *cis*-aconitate and isocitrate, respectively. The pool of reduction equivalents in the mitochondrion is not balanced. Two reactions in the mitochondrion, the conversion of pyruvate into acetyl-CoA and the conversion of malate into oxalacetate generate NADH which is not metabolically consumed but oxidized in the respiratory chain phosphorylation.

18.3.2 Surplus or Reduction Equivalents and Oxygen Demand

The net reaction equation reveals that in theory 1 mol of citric acid can be derived from 1 mol of glucose. The process also yields a surplus of ATP to allow cellular maintenance and some growth during the production phase. Additionally, three surplus reduction equivalents are obtained which reflect the electron density of the feedstock and the product (Nielsen et al. 2003). Whereas glucose bears 24 electrons (without counting the reduced oxygen) and citric acid bears 18 electrons (without counting the reduced oxygen). The surplus of six electrons implies that three reduction equivalents appear on the right side of the net reaction equation. In order to recycle the oxidized reduction equivalents for the additional processing of glucose, the presence of molecular oxygen as final electron acceptor is required and an aerobic fermentation protocol is needed:

$$3\,\text{NADH} + 3\,\text{H}^+ + 1.5\,\text{O}_2 \rightarrow 3\,\text{NAD}^+ + 3\,\text{H}_2\text{O}$$

The net oxygen demand of the reaction can be derived if both equations are combined (the internally consumed ATP has been omitted):

$$\text{Glucose} + 1.5\,\text{O}_2 \rightarrow \text{Citric acid} + 2\,\text{H}_2\text{O}$$

The oxygen demand can be estimated with 0.25 t oxygen per ton of citric acid (3 × 16/192.1). The aeration rate is in the range of 0.1–1.0 vvm during the production phase to ensure a dissolved oxygen level of at least 20–25% of the saturation concentration. *A. niger* is very sensitive toward brief interruptions of the oxygen supply (Max et al. 2010; Soccol et al. 2006).

The theoretical maximum molar yield of 1 mol of citric acid per mol glucose translates into a maximum carbon yield of 1.067 t citric acid per ton glucose (192.1/180.1). A maximum carbon yield of up to 1.03 t citric acid per ton of glucose is reported in lab-scale experiments, which indicates that 96% of the theoretical maximum yield is achieved (Tong et al. 2019). This figure may not fully translate into the performance of a commercial plant, though indicates that carbon yields above 90% of the maximum theoretical yield are possible. It is in some instances claimed that >80% of the theoretical carbon yield is achieved combined with a titer of about 200 g/l after a batch duration of 150 h (Schmid 2016). The gap is mainly explained by the formation of biomass and efforts for cellular maintenance.

The estimation of the theoretical maximum yield outlined above (i.e. 1 mol of citrate per mol glucose) is guided by the established biochemical pathways. In case exclusively the electron balance would be considered a higher yield may be derived:

$$0.75\,\text{Glucose} + 1.5\,\text{CO}_2 \rightarrow \text{Citric acid} + 0.5\,\text{H}_2\text{O} \quad \Delta G^{\circ\prime} = 144\,\text{KJ/mol}$$

The 18 electrons carried by 0.75 mol of glucose might simply be rearranged together with the carbon backbone to derive 1 mol of citrate carrying 18 electrons (without carrying reduced oxygen) (van Rossum et al. 2016). The theoretical maximum molar yield would increase to 1.33 mol of citrate per mol glucose (van Rossum et al. 2016). This deduction assumes that the electron surplus of glucose carrying 24 electrons per mol compared to 18 electrons carried per mol citric acid (without counting reduced oxygen) can be used to integrate carbon dioxide as a second carbon source, which is reduced with the surplus of electrons. The fixation of carbon dioxide may be facilitated with the pyruvate carboxylase, the PEP carboxylase, the PEP carboxykinase, or malic enzyme. A respective biochemical pathway might in the future be designed and implemented supported by synthetic biology, though it would define a high hurdle for a metabolic engineering project. Furthermore, the pathway would thermodynamically not be feasible. Accordingly, for the estimation of the realistic maximum molar yield, the equimolar ratio of educt and product should be preferred to apply an attrition rate for the formation of byproducts, biomass, and cellular maintenance. The realistic maximum molar yield amounts then to 0.9 mol of citric acid per mol glucose.

18.4 Process Development

Since an adjustment of the metabolic pathway and a finetuning of the metabolic engineering strategy were initially not required due to the natural abilities of the *A. niger* host the early process development could focus on other topics. Management attention of R&D projects in the "pre-genomic era" was directed to host selection, feedstocks, metal ions in the broth, the type of fermentation technology, and product isolation. After the first sequencing and the publication of the *A. niger*, genome (Pel et al. 2007) projects were initiated to develop metabolic models to better understand limitations and to reveal levers for further improvement (Tong et al. 2019; Upton et al. 2017).

18.4.1 Host Selection

The fermentative production of citric acid is mainly pursued with *A. niger*, which is a natural producer and has a high acid tolerance which allows to pursue the fermentation at a low pH (Angumeental and Venkappayya 2013). *A. niger* is a filamentous fungus widely exploited in the industry for the production of organic acids (Yang et al. 2017; Pel et al. 2007) and enzymes (Li et al. 2022; Park et al. 2017). It had already been discovered about 100 years ago by James Currie that *A. niger* is able to grow in a medium containing sugar and salts at an initial pH of 2.5–3.5 (Currie 1917). The native properties of *A. niger* with respect to product formation and acid tolerance are rather ideal conditions for a technology development program that can directly focus on process finetuning. Often natural and non-genetically modified hosts are applied which might provide an advantage in the product marketing toward the food industry. The fermentation starts from fungal spores whose germination requires a pH above 5.0. The pH in the production phase (submerged technology) is in the range of 2.5–3.5 (Max et al. 2010; Soccol et al. 2006) and in some references, it is claimed that a pH below 2 is required to sufficiently repress the competing formation of gluconic acid (Kubicek et al. 1985). The formation of gluconic acid is presumably caused by the expression of a partially mycelium-bound glucose oxidase converting glucose into gluconic acid. The glucose oxidase is induced once a pH of 4 has been exceeded (Kubicek et al. 1985). A pH of 2 in the broth with fungal mycelium as host defines an exception for fermentation technology, which usually is pursued at neutral pH in the case single-cell hosts like *Escherichia coli* or *Saccharomyces cerevisiae* are applied. The contamination risk is low, though nevertheless in most processes a sterilization step seems to be included. Yeast hosts like *Candida* sp. and *Yarrowia* sp. were evaluated as alternative host, though could finally not provide an advantage compared to *A. niger*. The generation of the by-product iso-citric acid by yeast is a disadvantage (Tong et al. 2019; Show et al. 2015).

18.4.2 Feedstock

The fermentative production of citric acid according to the submerged technology applies glucose, sucrose, or starch hydrolysates as carbon and energy source. *A. niger* strains secrete a broad variety of enzymes in their habitat to release sugars from carbohydrates, which allows the application of disaccharides of polysaccharides as feedstock (Tong et al. 2019; Pel et al. 2007). Accordingly, it is reported already in the very early publications that *A. niger* is able to produce citric acid from oligosaccharides (3–12 units) and that the maximum yield is achieved from sucrose (Doelger and Prescott 1934). The use of cane or beet molasses is recommended for reducing production costs. Cane molasses is preferred due to its low price, though bears the advantage that the contained metal ions may have a negative impact on productivity. Especially, the iron content may need to be reduced via the dilution of cane molasses with other feedstocks or an ion exchange purification step (Abou-Zeid and Ashy 1984).

In the 1970s, the use of hydrocarbons has attracted a high management attention and several R&D programs were pursued to establish hydrocarbons as cheap and efficient feedstock. Hydrocarbons as a by-product of the petrochemistry were regarded as cheap feedstock allowing a carbon yield of 140–150% (w/w), a productivity of 1.4 g/l × h, and a titer of 225 g/l (Kubicek et al. 1985). In case hydrocarbons are fed, the pH must be kept above 5 since otherwise the polyols mannitol and erythritol accumulate at the expense of citric acid and thiamin must be added to avoid the formation of α-keto acids. The citric acid can be captured in the organic phase of the remaining hydrocarbons after interruption of the agitation (Anastassiadis et al. 2001). However, the R&D attempts were finally not a success, and hydrocarbons were not established as feedstock for commercial production.

18.4.3 Metal Ions in the Broth

An efficient fermentation process toward citric acid implies a thorough management of trace elements or metals, respectively, contained in the fermentation broth. In general, small concentrations of Zn, Mn, Fe, and Cu should be available, though defined thresholds are to be obeyed (Max et al. 2010). The use of chelating ion exchange to remove polyvalent metals from sugar has been established as a useful tool (Noguchi and Johnson 1961) in the early stages of technology development. Since the fermentation is pursued at a rather low pH of about 3.0, the inner (metal) reactor surface may be covered by a specific plastic layer to avoid corrosion and the extraction of metal ions.

Out of the portfolio of trace elements, especially manganese plays a crucial role. The critical manganese concentration is related to the host morphology. The liquid phase submerged fermentation was in the first technology generations characterized by a biocatalyst growth phase (trophophase) of about 20–40 hours in which glucose is mainly converted into biomass. This growth phase was followed by a production phase (idiophase) of about four to six days in which the citric acid is formed (Max et al. 2010). The overall batch time had a duration of about 170 hours. The later technology generations managed to improve productivity by shortening the batch time. The host morphology formed in the growth phase is decisive for productivity in the production phase. The most suited mycelium structure is described by pellets with a diameter > 0.5 mm, which is achieved by a strict manganese limitation (Takahashi et al. 1958). The manganese limitation hampers the formation of chitin and thereby triggers the development of an unusual host morphology ideal for the production period. The unusual mycelium structure has a higher permeability for oxygen and avoids a potential interim oxygen limitation, which would negatively impact the productivity.

18.4.4 Type of Fermentation Technology

Solid-state fermentation technology (Mitchell and Krieger 2020) was originally pursued with surface cultures ("surface process"). One type is the "Koji" process, which is the simplest fermentation production technology for citric acid. It was developed in Japan and applies fruit wastes, sweet potato fibrous residues, and rice and wheat bran (Grewal and Karla n.d.). The feedstock is moistened to about 70% moisture and placed in trays that are inoculated with *A. niger*. The pH is in the range of 4.5–6.0 and the temperature is in the range of 28–30 °C. The fermentation is pursued batchwise and stopped after 4–5 days. The processing is non-sterile and in most of the cases done manually (Vazhacharickal et al. 2018). A disadvantage is the presence of trace elements in the feedstock which cannot be removed by standard methods (Soccol et al. 2006; Grewal and Karla 1995).

The "Koji" process has been replaced by a more elaborated type of surface technology, which has already been started in 1919 in Belgium where the first industrial plant with *A. niger* as host has been constructed. The fermentation takes place in fermentation chambers with a great number of trays arranged on shelves. An air circulation is installed to supply oxygen and control humidity in the chamber. Heat removal is facilitated via evaporative cooling. The process is pursued batchwise and takes up to 12 days. Afterward the grown mycelial mats are harvested and washed to remove the impregnated citric acid (Soccol et al. 2006; Grewal and Karla 1995).

The liquid phase or submerged fermentation technology was for the first time applied in the 1940s and was then more and more applied in the 1950s and replaced the surface fermentation. The technology achieves the lowest production costs and provides the highest productivity (Soccol et al. 2006; Grewal and Karla n.d.). It achieves a titer of about 140 g/l (Sahm et al. 2013). This figure is corresponding to different sources, which indicate a titer of 110–140 g/l citric acid and a biomass titer of 10–15 g/l (Kubicek 2001; Marzona 1996). In small-scale experiments or mini plants operated in academia, a maximum titer up to 185.7 g/l citric acid combined with a carbon yield of 1.03 tons citric acid per ton glucose and a productivity of 2.58 g/l × h is reported (Tong et al. 2019). The actual figures realized in industrial plants are not published, though it may be assumed that titers >150 g/l combined with a productivity >2.5 g/l × h are achieved. Most of the industrial production is done batchwise in continuously stirred tank reactors or airlift reactors with a design volume of 120–250 m^3 up to 900 m^3 (Max et al. 2010). Also, a continuous fermentation with yeast has been described (Anastassiadis et al. 2001).

18.4.5 Product Isolation

The standard protocol for the downstream processing of citric acid after the submerged fermentation foresees that the biomass is in the first step removed via filtration. $Ca(OH)_2$ or CaO is added at pH 7 and a temperature of 90 °C to trigger the formation of insoluble calcium citrate, which precipitates and is captured with a filtration device. This step profits from the rather low solubility of calcium citrate and allows an efficient removal of citrate out of the filtrated broth:

$$2\,C_6H_8O_7 + 3\,ca(OH)_2 \rightarrow Ca_3(C_6H_5O_7)_2 \downarrow + 6\,H_2O$$

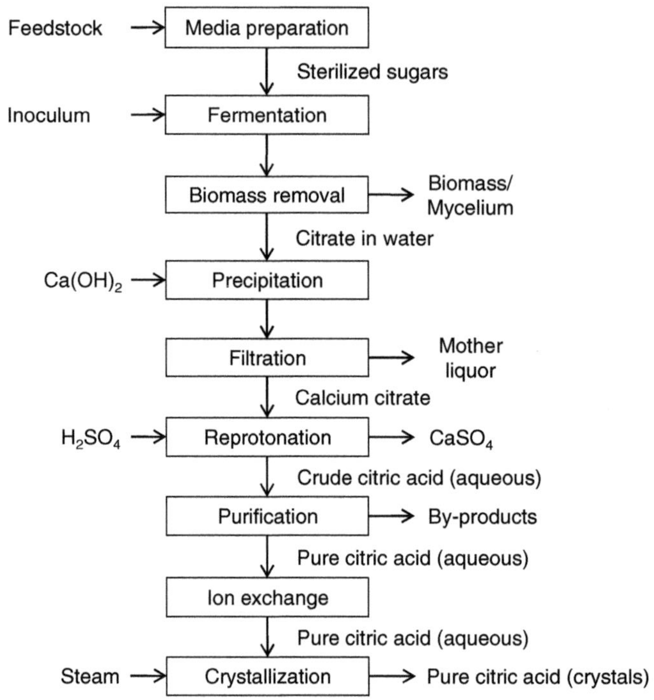

Figure 18.2 Process flow for the fermentative production and isolation of citric acid. Source: Walter Koch.

The free acid is then obtained with the addition of H_2SO_4 at 50 °C and calcium sulfate is obtained as by-product:

$$Ca_3(C_6H_5O_7)_2 + 3\,H_2SO_4 \rightarrow 2\,C_6H_8O_7 + 3\,CaSO_4 \downarrow$$

The citric acid is then, pending on the application and the required specification, purified with active carbon and ion exchange. The water content is evaporated with steam and a pure crystallized citric acid is derived (Stanbury et al. 2017). This step may be performed with vacuum crystallizers at a temperature of 20–25 °C (Kubicek et al. 1985). The overall process for the fermentative production of citric acid is summarized as follows in Figure 18.2 (Stanbury et al. 2017; Sahm et al. 2013; Soccol et al. 2006):

This isolation technology has the disadvantage that with 1.06 t $CaSO_4$ per ton citric acid (136.1 × 3/192.1 × 2) a high volume of byproduct is obtained. The gypsum may be sold into the market or, pending on regional demand, be deposited as landfill. Since the wastewater generation derived in the removal of the biomass, the filtration of calcium citrate and the crystallization of citric acid may sum up to 50–60 t per ton citric acid a specific downstream process has been worked out to limit the discharge. The wastewater including the biomass (i.e. the fungal mycelium) needs further processing and may be used as feedstock for an anaerobic fermentation to obtain biomethane which may be supplemented with an aerobic digestion (Xu et al. 2015; Xu et al. 2014).

It is reported that about 30%–40% of the overall production cost is caused by the precipitation of citric acid with $Ca(OH)_2$ and the subsequent reprotonation with H_2SO_4 (Ye et al. 2022). Thus, alternative isolation methods have been evaluated including extraction, adsorption, and membrane filtration. An interesting approach has been proposed in academia which foresees to convert calcium citrate with CO_2 or H_2CO_3, respectively, to obtain citric acid and $CaCO_3$. The $CaCO_3$ has also a low solubility and could efficiently be removed by precipitation (Ye et al. 2022):

$$CO_2 + H_2O \rightarrow H_2CO_3$$
$$Ca_3\left(C_6H_5O_7\right)_2 + 3\,H_2CO_3 \rightarrow 2\,C_6H_8O_7 + 3\,CaCO_3 \downarrow$$

The reaction of calcium citrate with CO_2 or H_2CO_3, respectively, achieved a conversion of 94.7% at 65 °C and 1.8 bar after 4 hours of incubation. The calcium citrate in the aqueous starting solution was 2 g/l (Ye et al. 2022). This reaction scheme has the advantage that the byproduct $CaCO_3$ is not a waste stream, though might be converted into $Ca(OH)_2$ and be recycled for the synthesis of calcium citrate. The regeneration of $CaCO_3$ toward $Ca(OH)_2$ proceeds as follows (Bertau et al. 2013):

$$CaCO_3 \rightarrow 3\,CaO + CO_2$$
$$CaO + H_2O \rightarrow Ca(OH)_2$$

The generation of Ca_2SO_4 is indeed avoided, though the synthesis of CaO with $CaCO_3$ as feedstock implies a complete different endothermic chemical step with high energy input. It remains to be seen if this recycling idea for the byproduct of citrate precipitation and reprotonation gains market share in the future. Alternatively, citric acid may be converted into a complex with trilaurylamine and then be extracted with a mixture of alkanes and 1-octanol. This downstream protocol also avoids the formation of $CaSO_4$ as by-product (Schmid 2016).

18.5 Industrial Production

The production of citric acid is mainly based on fermentation with *A. niger* and the small volumes still derived by natural extraction serve exclusively market niches. The producers with the biggest production capacity comprise Gadot Biochemical Industries (Israel), Weifang Ensign Industry (China), Huangshi Xinghua Biochemical (China), RZBC (China), Anhui COFCO Biochemical (China), Cargill (United States), ADM (United States), Citrique Belge (Belgium), Jungbunzlauer (Switzerland), and Tate & Lyle (UK) (Ciriminna et al. 2017). A large extent of the production capacity has been moved to China and the production capacity in Europa and the United States decreased. Nine commercial plants in Europe, the United States, and India were closed until 2010 with an overall production capacity of 190 kt. Dutch States Mines (DSM) closed its plant in Wuxi (China) with a capacity of 40 kt in 2009 (Ciriminna et al. 2017).

The Chinese production around 2010 achieved a titer of about 147 g/l citric acid combined with a productivity of 2.7 g/l × h (Hu et al. 2014). These thresholds were regarded as inferior compared to the technology performance of the best Western plants and triggered the start of research activities. Host development programs generated a mutant collection with X-ray irradiation and the performance of the resulting mutants was screened. A mutant H42002 was fed with a corn meal hydrolysate and displayed a titer of 178–196 g/l based on a sugar concentration of 200–236 g/l. The related productivity achieved 3.0–3.3 g/l × h (Hu et al. 2014), which closed the performance gap. R&D projects focus also on the use of cheap raw material sources and corn meals were hydrolyzed with α-amylase at 60–72 °C for 30 minutes and 95–95 °C for 10 minutes to provide mono- and disaccharides as feedstock (Hu et al. 2014).

References

Abou-Zeid, A. and Ashy, M.A. (1984). Production of citric acid: a review. *Agricultural Wastes* 9: 51–76.

Anastassiadis, S., Aivasidis, A., Wandrey, C. (investors) (2001). Process for the continuous production of citric acid by fermentation, US Patent 6,303,351B1.

Angumeental, A.R. and Venkappayya, D. (2013). *An Overview of Citric Acid Production*, vol. 50, 367–370. LWT-Food Science and Technology.

Arpe, H.-J. (2007). *Industrielle Organische Chemie. Bedeutende Vor- und Zwischenprodukte*, 415. Weinheim: Wiley-VCH.

Bertau, M., Müller, A., Fröhlich, P., and Katzberg, M. (2013). *Industrielle Organische Chemie*, 233–234. Weinheim: Wiley-VCH.

Ciriminna, R., Meneguzzo, F., Delisi, R., and Pagliaro, M. (2017). Citric acid: emerging applications of key biotechnology induistrial product. *Chemistry Central Journal* 11: 22. https://doi.org/10.1186/s13065-017-0251-y.

Currie, J.N. (1917). The citric acid fermentation of *A. niger*. *Journal of Biological Chemistry* 31: 5.

Doelger, W.P. and Prescott, S.C. (1934). Citric acid fermentation. *Industrial and Engineering Chemistry* 26 (11): 1142–1149.

Grewal, H.S. and Karla, K.L. (1995). Fungal production of citric acid. *Biotechnology Advances* 13 (2): 209–234.

Hu, W., Liu, J., Chen, J.-H. et al. (2014). A mutation of *Aspergillus niger* for hyper-production of citric acid from corn meal hydrolysate in a bioreactor. *Biomedicine & Biotechnology* 15 (11): 1006–1010. https://doi.org/10.1631/jzus.B1400132.

Kubicek, C.P. (2001). Organic acids, Chapter 14. In: *Basic Biotechnology*, 2e (ed. C. Ratledge and B. Kristiansen), 305–324. Cambridge, UK: Cambridge University Press.

Kubicek, C.P., Röhr, M., and Rehm, H.J. (1985). Citric acid fermentation. *Critical Reviews in Biotechnology* 3 (4): 331–373. https://doi.org/10.3109/07388558509150788.

Li, Q., Lu, J., Zhang, G. et al. (2022). Recent advances in the development of *Aspergillus* for protein production. *Bioresource Technology* 348: https://doi.org/10.1016/j.biortech.2022.126768.

Marzona, M. (1996). *Chimica della Fermentazioni & Microbiologica Industriale*, 2e. Padua, Italy: Piccin.

Max, B., Salgado, J.M., Rodríguez, N. et al. (2010). Biotechnological production of citric acid. *Brazilian Journal of Microbiology* 41: 862–875.

Mitchell, D.A. and Krieger, N. (2020). Solid-state cultivation bioreactors. In: *Essentials in Fermentation Technology* (ed. A. Berenjian), 105–133. Springer.

Nielsen, J., Villadsen, J., and Liden, G. (2003). *Bioreaction Engineering Principles*, 104. Springer.

Noguchi, Y. and Johnson, M.J. (1961). Citric acid fermentation of sugars purified with chelating agents. *Journal of Bacteriology* 538–541.

Park, H.-S., Jun, S.-C., Han, K.-H. et al. (2017). Diversity, application, and synthetic biology of industrially important *Aspergillus* fungi. *Advances in Applied Microbiology* 100: 161–186. https://doi.org/10.1016/bs.aambs. 2017.03.001.

Pel, H.J., de Winde, J.H., Archer, D.B. et al. (2007). Genome sequencing and analysis of the versatile cell factory *Aspergillus niger* CBS 513.88. *Nature Biotechnology* 25 (2): 221–231. https://doi.org/10.1038/nbt1282.

van Rossum, H.M., Kozak, B.U., Pronk, J.T., and van Maris, A.J.A. (2016). Engineering cytosolic acetyl-CoA supply in *Saccharomyces cerevisiae*: pathway stoichiometry, free-energy conservation and redox-factor balancing. *Metabolic Engineering* 36: 99–115. https://doi.org/10.1016/j.ymben.2016.03.006.

Sahm, H., Antranikian, G., Stahmann, K.-P., and Takors, R. (ed.) (2013). *Industrielle Mikrobiologie*, 99–103. Heidelberg: Springer Verlag.

di San, A.F.P. (1998). The Italian chemical industry from 1861 to 1918. In: *The Chemical Industry in Europe, 1850–1914. Industrial Growth, Pollution, and Professionalization* (ed. E. Homburg, A.S. Travis, and H.G. Schröter). Dordrecht: Springer Science + Business Media.

Schmid, R.D. (2016). *Taschenatlas der Biotechnologie und Gentechnik*, 146–147. Wiley-VCH.

Schmidt, P.C. and Lang, S. (2013). Pharmazeutische Hilfsstoffe. In: *Eigenschaften, Anwendung und Handelsprodukte*, 328–330. Eschborn: Govi-Verlag Pharmazeutischer Verlag.

Show, P.L., Oladele, K.O., Siew, Q.Y. et al. (2015). Overview of citric acid production from *Aspergillus niger*. *Frontiers in Life Science* 8 (3): 271–283. https://doi.org/10.1080/21553769.2015.1033653.

Soccol, C.R., Vandenberghe, L.P.S., Rodrigues, C., and Pandey, A. (2006). New perspectives for citric acid production and application. *Food Technology and Biotechnology* 44 (2): 141–149.

Stanbury, P.F., Whitaker, A., and Hall, S.J. (2017). *Principles of Fermentation Technology*, 3e. Elsevier, section on sterilization p. 673.

Takahashi, J., Yamada, K., and Aasai, T. (1958). Studies on the effect of some physical conditions on the submerged mold culture: part I: the process of pellet formation of *aspergillus Niger* under the shaking culture, and the effect of the inoculum size on the shape of the pellet. *Journal of Agricultural Science* 32: 501–506.

Tong, Z., Zheng, X., Tong, Y. et al. (2019). Systems metabolic engineering for citric acid production by *Aspergillus niger* in the post-genomic area. *Microbial Cell Factories* 18: 28. https://doi.org/10.1186/s12934-019-1064-6.

Upton, D.J., McQueen-Mason, S.J., and Wood, A.J. (2017). An accurate description of *Aspergillus niger* organic acid batch fermentation though dynamic metabolic modelling. *Biotechnology for Biofuels* 10: 258. https://doi.org/10.1186/s13068-017-0950-6.

Vazhacharickal, P.J., Mathew, J.J., and Sajeshkumar, N.K. (2018). *Industrial Biotechnology: An Introduction*, 153. Independent Publishing. http://www.jungbunzlauer.com/en/products/citrics.html (accessed 4 April 2022).

Xu, J., Chen, Y.-Q., Zhang, H.-J. et al. (2014). Production of citric acid using its extraction wastewater treated by anaerobic digestion and ion exchange in an integrated citric acid-methane fermentation process. *Bioprocess and Biosystems Engineering* 37: 1659–1668. https://doi.org/10.1007/s00449-014-1138-0.

Xu, J., Su, X.-F., Bao, J.-W. et al. (2015). Cleaner production of citric acid by recycling its extraction wastewater treated with anaerobic digestion and electrodialysis in an integrated citric acid-methane production process. *Bioresource Technology* 189: 186–194. https://doi.org/10.1016/j.biortech.2015.04.016.

Yang, L., Lübeck, M., and Lübeck, P.S. (2017). *Aspergillus* as a versatile cell factory for organic acid production. *Fungal Biology Reviews* 31: 33–49. https://doi.org/10.1016/j.fbr.2016.11.001.

Ye, R., Zhao, Z., Gao, R. et al. (2022). Conversion of calcium citrate to citric acid with compressed CO_2. *ACS Omega* 7: 683–687. https://doi.org/10.1021/acsomega.1c05316.

19

Adipic Acid

19.1 Application

Adipic Acid (ADA) is an aliphatic diacid with the formula $C_6H_{10}O_4$ and a molecular weight of 146.1 g/mol. An aqueous solution of ADA is rather acidic given pK_a values of 4.4 and 5,4, respectively, for the two carboxylic acid functions (Karlsson et al. 2017; Deng et al. 2016). ADA is mainly applied as one of the alternating building blocks besides Hexamethylenediamine (HMD) for the synthesis of the polyamide PA-66. In the first step, an aqueous solution of ADA and HMD is generated (called "AH salt"), which is then polymerized to PA-66. ADA is additionally used for producing polyester polyols for polyurethanes. A further application is the use as food additives for acidification. ADA also has applications in the form of sodium or potassium adipate. The global production volume of ADA was 2.6 Mio. t in 2010, with a share of 65% allocated for the synthesis of PA-66 (Kruyer and Peralta-Yahya 2017). The global production volume increased to 3.3 Mio. t in 2016 and is expected to grow by 3–3.5% pa (Raj et al. 2018).

19.2 Chemical Production of Adipic Acid

ADA is traditionally produced from benzene, which is hydrogenated to cyclohexane in the first step. This reaction proceeds with molecular hydrogen and achieves a high yield. Cyclohexane is subjected to oxidation, and a mixture of cyclohexanol and cyclohexanone is obtained. This difficult reaction is pursued with a very low turnover of about 10%–12% to keep the selectivity in the range of 80–85% (Arpe 2007; Polen et al. 2013; Deng and Mao 2015; Bart and Cavallaro 2015; Figure 19.1).

A higher turnover of the cyclohexane oxidation would be possible, though the selectivity would decline. In the next step, the mixture of cyclohexanol and cyclohexanone is oxidized with nitric acid to ADA. Pending on the specific technology, the reaction may be pursued at a temperature of 150 °C and a pressure of 10 bar (Arpe 2007).

The hydroformylation of butadiene with ADA or hexanediol as product is under development. Hereby, butadiene would be reacted with H_2/CO to obtain the respective C6 dialdehyde. Since the direct reduction of the aldehyde function shall be avoided, the aldehyde

Pathway Design for Industrial Fermentation, First Edition. Walter Koch.

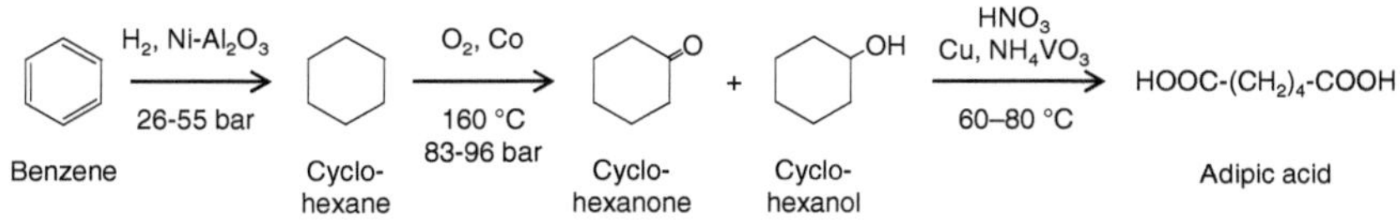

Figure 19.1 Chemical pathway for the synthesis of adipic acid ex benzene. Source: Walter Koch.

is *in situ* protected with glycol to obtain an acetal as an intermediate product. The acetal intermediate could then be oxidized to derive ADA (Mormul et al. 2016):

$$CH_2 = CH\text{–}CH = CH_2 + 2\,CO + 2\,H_2$$
$$\rightarrow OHC\text{–}CH_2\text{–}CH_2\text{–}CH_2\text{–}CH_2\text{–}CHO\ (\textit{in situ}\ \text{as acetal})$$
$$OHC\text{–}CH_2\text{–}CH_2\text{–}CH_2\text{–}CH_2\text{–}CHO + O_2 \rightarrow HOOC\text{–}CH_2\text{–}CH_2\text{–}CH_2\text{–}CH_2\text{–}COOH$$

The alternative reduction of the acetal intermediate would provide hexanediol. An innovative route toward ADA starting with glucose or fructose as biobased precursors has been evaluated and developed by Rennovia. An R&D program pursues the high-throughput screening of suitable catalysts to convert the —CHOH— functions of the sugar backbone into methylene groups to finally obtain a reduced backbone whose terminal C-atoms can then be oxidized to carboxylic acid functions. The biobased ADA technology is offered in the market, though Rennovia has also considered building its own plants (Coons 2017). Supply of ADA will also depend on the progress of recycling technology, especially of PA-66 plastics consisting of ADA and HMD (Damayanti et al. 2021; Vollmer et al. 2020; Duch and Allgeier 2006).

19.3 Metabolic Engineering for Fermentation

ADA is not an intermediate in biochemical pathways involved in the synthesis of cellular building blocks. However, ADA is an intermediate in the degradation of cyclohexane, cyclohexanol, cyclohexanone, and ε-caprolactam, which can be metabolized by some *Pseudomonas* strains (Cheng et al. 2002; Steffensen and Alexander 1995). It has also been elucidated that some species of *Pseudomonas* are able to degrade benzoic acid into *cis, cis*-muconic, which may accumulate or be converted into muconolactone in case the muconate cycloisomerase is present (Polen et al. 2013). Muconolactone is further processed via β-ketoadipate-enol-lactone or 2-oxo-2,3-dihydrofuran-5-acetate, respectively, into β-ketoadipate and finally acetyl-CoA and succinyl-CoA (Yoneda et al. 2016). Both activated acid thioesters are recycled into the citric acid cycle (Figure 19.2).

This benzoic acid dissimilation route has partially been used for the development of recombinant pathways toward ADA. The intermediate *cis, cis*-muconic acid may either chemically or biochemically be hydrogenated with an enoate reductase. Alternatively, the recycling pathway leading to succinyl-CoA and acetyl-CoA may be reversed and in the opposite direction be used to generate β-ketoadipyl-CoA, a suitable precursor for ADA. Both options have been pursued in academic and commercial attempts to derive a

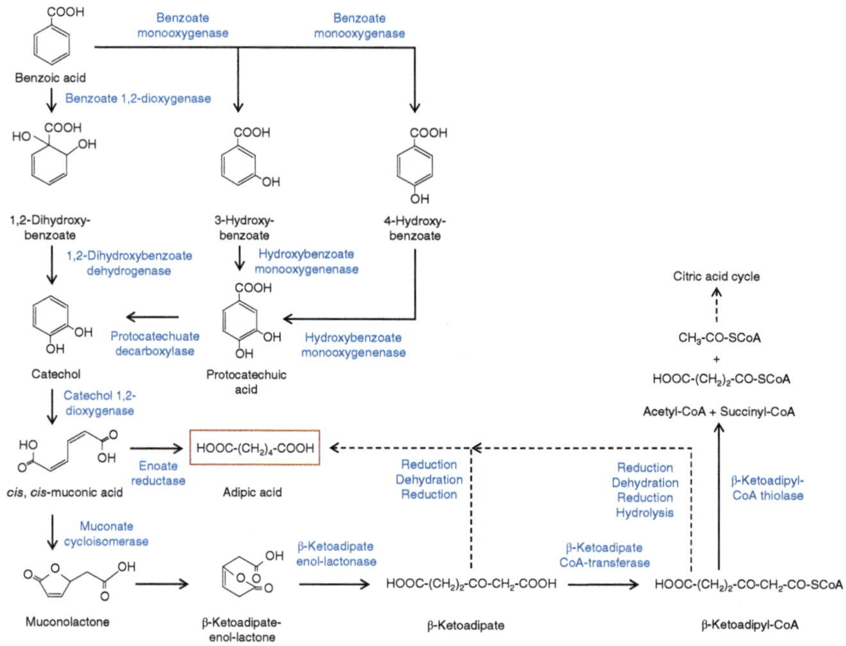

Figure 19.2 Natural degradation pathway for aromatic compounds. Source: Adapted from Polen et al. (2013).

fermentative ADA technology. The conversion of the lignin-derived aromatics protocatechuate and protocatechuate into acetyl-CoA and succinyl-CoA was also foreseen for other target products like triglycerides (Liu et al. 2018).

19.3.1 Adipic Acid via the Reverse Adipate Degradation Pathway

There are a high number of potential fermentative pathways toward ADA published in academic literature, which are comprehensively summarized in several publications (Raj et al. 2018). The most promising pathway, referred to as the reverse degradation pathway, proceeds via succinyl-CoA and acetyl-CoA (Genomatica patent n.d.). The pathway has been discovered in the microorganism *Thermobifida fusca,* which is reported to convert acetyl-CoA and succinyl-CoA into ADA (Deng and Mao 2015). The main precursors succinyl-CoA and acetyl-CoA are provided via glycolysis, followed by the pyruvate dehydrogenase and the reductive citric acid cycle, and their condensation obtains 3-oxoadipyl-CoA as the first linear C6 building block. The keto function would then be removed via the conventional biochemical conversion steps according to the degradation of fatty acids (Table 19.1).

This route has the advantage that the formation of the C6 backbone does not imply the generation of carbon dioxide since 1 mol carbon dioxide generated in the formation of acetyl-CoA is recycled in the synthesis of oxalacetate. The recycling is facilitated by the carboxylation of pyruvate to oxalacetate, catalyzed by the pyruvate carboxylase. Alternatively, the carboxylation of PEP to oxalacetate may be chosen. The energy may be regarded as balanced since the consumption of GTP is compensated by the generation of

Table 19.1 Biochemical steps for the conversion of glucose into adipate.

Provision of pyruvate	
Glucose + 2 ADP + 2 Pi + 2 NAD^+ → 2 Pyruvate + 2 ATP + 2 H_2O + 2 NADH + 2 H^+	
Provision of acetyl-CoA	
Pyruvate + CoA + NAD^+ → Acetyl-CoA + CO_2 + NADH + H^+	Pyruvate dehydrogenase
Provision of Succinyl-CoA	
Pyruvate + CO_2 + ATP + H_2O → Oxalacetate + ADP + Pi	Pyruvate carboxylase
Oxalacetate + NAD(P)H + H^+ → Malate + NAD^+	Malate dehydrogenase
Malate → Fumarate + H_2O	Fumarase
Fumarate + $FADH_2$ → Succinate + FAD	Succinate dehydrogenase
Succinate + CoA + GTP → Succinyl-CoA + GDP + Pi	Succinyl-CoA synthase
Generation of ADA	
Succinyl-CoA + Acetyl-CoA → 3-Oxoadipyl-CoA + CoA	Succinyl-CoA-acetyl-CoA transferase
3-Oxoadipyl-CoA + NADH + H^+ → 3-Hydroxyadipyl-CoA + NAD^+	3-Hydroxyacyl-CoA dehydrogenase
3-Hydroxyadipyl-CoA → 5-Carboxy-2-pentenoyl-CoA + H_2O	3-Hydroxyadipyl-CoA dehydratase
5-Carboxy-2-pentenoyl-CoA + NADH + H^+ → Adipyl-CoA + NAD^+	5-Carboxy-2-pentenoyl-CoA reductase
Adipyl-CoA + H_2O → Adipate + CoA	Adipyl-CoA thioesterase
Glucose + GTP + ADP + $FADH_2$ → Adipate + GDP + ATP + FAD + 2 H_2O	

ATP. The thermodynamics of the pathway have been analyzed and reveal that the route is sufficiently exergonic starting from succinyl-CoA and acetyl-CoA (Kallscheuer et al. 2017; Table 19.2).

The Gibbs free energy value for the synthesis of adipyl-CoA is calculated with FAD as cofactor, though this difference, i.e. NADH or $FADH_2$, may be neglected. The original source refers to 2,3-dehydroadipyl-CoA as the third pathway intermediate, which is identical with the alternative name 5-carboxy-2-pentenoyl-CoA (Kallscheuer et al. 2017). The pathway from succinyl-CoA to adipate may proceed via the CoA route or the acid

Table 19.2 Thermodynamics for the conversion of succinyl-CoA and acetyl-CoA into adipate.

Succinyl-CoA + Acetyl-CoA → 3-Oxoadipyl-CoA + CoA	$\Delta G^{\circ\prime} = +26.8$ KJ/Mol
3-Oxoadipyl-CoA + NADH + H^+ → 3-Hydroxyadipyl-CoA + NAD^+	$\Delta G^{\circ\prime} = -40.2$ KJ/Mol
3-Hydroxyadipyl-CoA → 5-Carboxy-2-pentenoyl-CoA + H_2O	$\Delta G^{\circ\prime} = +2.9$ KJ/Mol
5-Carboxy-2-pentenoyl-CoA + $FADH_2$ → Adipyl-CoA + FAD	$\Delta G^{\circ\prime} = -71.2$ KJ/Mol
Adipyl-CoA + H_2O → Adipate + CoA	$\Delta G^{\circ\prime} = +3.3$ KJ/Mol
Succinyl-CoA + Acetyl-CoA + NADH + H^+ + $FADH_2$ → Adipate + NAD^+ + FAD + 2 CoA	$\Delta G^{\circ\prime} = -78.4$ KJ/Mol

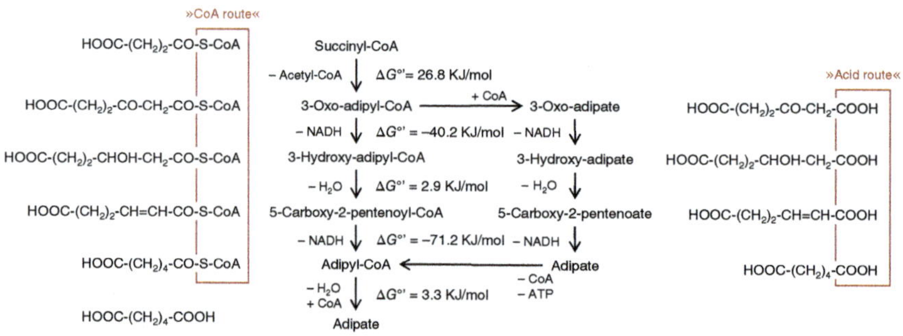

Figure 19.3 Metabolic pathway toward adipic acid with succinyl-CoA and acetyl-CoA as starting point. Source: Walter Koch.

route. The C6 building block subject of the metabolic reactions may be given as a CoA thioester or as free acid (Figure 19.3).

The reduction equivalents are not balanced because ADA bears 26 electrons (without counting the electrons of the reduced oxygen) compared to 24 electrons provided by 1 mol glucose. Accordingly, one reduction equivalent needs to be provided by the oxidation of glucose, and the glucose unit consumption rate is increased. Since glucose carries in total 24 electrons, about 0.083 mol glucose (2/24) needs to be oxidized to carbon dioxide and water, and the theoretical equation of the reaction can be stated as follows:

$$1.083\ \text{Glucose} \rightarrow 1.0\ \text{ADA} + 0.498\ CO_2 + 1.498\ H_2O$$

The maximum theoretical yield for the reverse degradation pathway can be estimated at 0.923 mol ADA per mol glucose, corresponding to a maximum carbon yield of 0.75 t ADA per ton glucose. This calculated value corresponds to the maximum carbon yield for the reverse degradation pathway and the β-ketoadipate pathway published in literature (Raj et al. 2018).

The bacterium *Thermobifida fusca* possesses a native reverse adipate degradation pathway and has been used as fermentation host to make ADA (Deng and Mao 2015). The native pathway comprises the β-ketothiolase (Tfu_0875), the 3-hydroxyacyl-CoA dehydrogenase (Tfu_2399), the 3-hydroxyadipyl-CoA dehydrogenase (Tfu_0067), the 5-carboxy-2-pentenoyl-CoA reductase (Tfu_1647), and the adipyl-CoA synthetase (Tfu_2576-7) and provided an ADA titer of 2.23 g/l combined with a productivity of 0.034 g/l × h starting from 50 g/l glucose (Deng and Mao 2015). Improvements were blocked by the limited molecular engineering toolbox for *Thermobifida fusca*.

The technology could be improved with a recombinant reconstruction of the pathway in *E. coli*. The expression of the 3-oxoadipyl-CoA thiolase (PaaJ), the 3-hydroxyadipyl-CoA dehydrogenase (PaaH), and the 2,3-dihydroxyadipyl-CoA hydratase (PaaF) provided an ADA titer of 12 μg/l (Babu et al. 2015) and 0.6 mg/l (Yu et al. 2014), respectively. The last two reactions, the reduction of 2,3-dehydroxyadipyl-CoA and the cleavage of adipyl-CoA, were identified as the rate-limiting steps, and suitable enzymes were identified. With the best candidates applied, an ADA titer of 36 mg/l after 30 h of cultivation was obtained (Kallscheuer et al. 2017). Further progress could be made by another academic working group pursuing a similar metabolic engineering strategy (Zhao et al. 2018). The temperature in the production phase after the addition of IPTG was set at 30 °C, and the pH was controlled with NaOH at 7.0. Glycerol was offered as major substrate, which supports the reductive conversion given the higher electron density than glucose (i.e. 4.0 e/C in glucose and 4.6 e/C in glycerol). Pending on the dissolved oxygen (DO) concentration, it varied with DO 0% (non-control), DO 5%, and DO 10%. ADA titers of 68 g/l, 42.3 g/l, and 36.1 g/l were achieved. The maximum titer was achieved after a batch duration of 64 h, corresponding to a productivity of 1.06 g/l × h (Zhao et al. 2018). The performance data were supported by a partial inhibition of the succinate production from succinyl-CoA, a deletion of the *atoB* gene to suppress the formation of butyric acid, and a deletion of the *ldhA* gene to eliminate lactic acid production.

19.3.2 *cis, cis*-Muconic Acid Ex Glucose via the Partial Shikimic Acid Pathway

The fermentative production of *cis, cis*-muconic acid as suitable intermediate to obtain biobased ADA has already been shown about 30 years ago (Draths and Frost 1994). An *E. coli* strain lacking a shikimate dehydrogenase to convert 3-dehydroshikimic acid (DHS) into shikimic acid has been selected as starting point. A 3-deoxy-D-arabinoheptulosonic acid (DAHP) synthase and a 3-dehydroquinic acid (DHQ) synthase together with a transketolase were overexpressed with plasmid pKD136. An increased synthesis of DHS was achieved with plasmid AB2834. The genes *aroZ* and *aro Y* encoding the DHS dehydratase and the protocatechuate decarboxylase were carried in plasmid pKD8.243A. Finally, the ring-opening conversion of catechol into *cis, cis*-muconic acid was facilitated by the catechol 1,2-dioxygenase encoded by the *catA* gene from *Acinetobacter calcoaceticus* on plasmid pKD8.292. In a 1 l scale laboratory shake flask, a feedstock of 56 mM glucose provided 16.8 mM *cis, cis*-muconic acid. The *cis, cis*-muconic acid was chemically converted with molecular hydrogen and a platinum catalyst on carbon with a yield of 90% (Draths and Frost 1994). Also, the partial saturation of *cis, cis*-muconic acid via electrodialysis to obtain an unsaturated PA-6,6 has been shown (Suastegui et al. 2015). The semi-synthesis of ADA via *cis, cis*-muconic acid is shown in Figure 19.4.

These results were confirmed and further enhanced later on, achieving a titer of 36.8 g/l *cis, cis*-muconic acid after 48 h, corresponding to a productivity of about 0.8 g/l × h (Niu et al. 2002). The optimization was facilitated by the chromosomal integration of two copies of the *aroZ* gene encoding the DHS dehydratase from *Klebsiella pneumoniae*. Additionally, the *aroY* gene encoding the PCA decarboxylase from *Klebsiella pneumoniae* and the *catA* gene encoding the 1,2-dioxygenase from *Acinetobacter calcoaceticus* were expressed from an extrachromosomal plasmid (Niu et al. 2002). The *cis, cis*-muconic acid technology developed by Draths has later been acquired by the renewable fuel biotech company Amyris, though it is not known whether the technology is actually applied (Greencarcongress.com 2011).

The same metabolic pathway was also established in the yeast host, *S. cerevisiae* (Weber et al. 2012). The higher acid tolerance of the eukaryotic host shall reduce neutralization efforts during the fermentation required to avoid a too-low pH triggered by the increasing product titer. Three heterologous genes were overexpressed: the 3-dehydroshikimate synthase, the protocatechuic acid decarboxylase, and the catechol 1,2-dioxygenase. The fermentation was pursued under microaerobic conditions with 10% DO air saturation sufficient to provide oxygen for the catechol 1,2-dioxygenase, though obviously not critical for the AroY enzyme, which is reported to display an oxygen sensitivity. A *cis, cis*-muconic acid titer of 1.56 mg/l could be achieved (Weber et al. 2012).

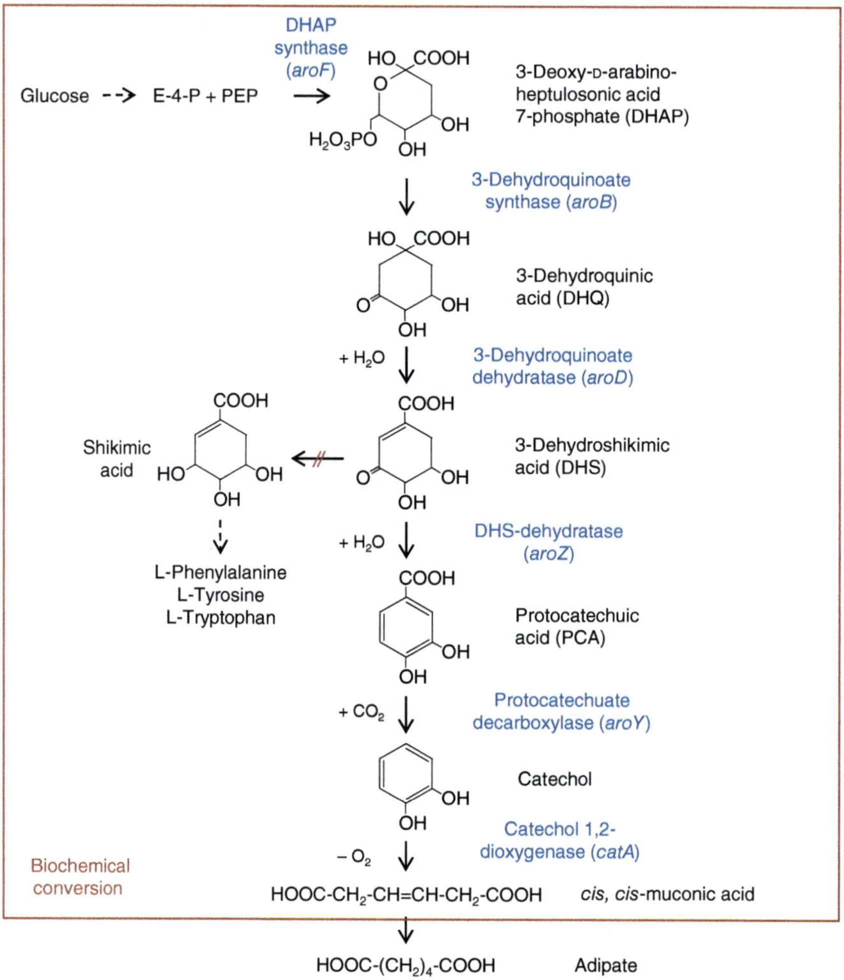

Figure 19.4 Metabolic pathway toward *cis, cis*-muconic acid based on glucose. Source: Walter Koch.

With an independent academic R&D project, a titer of 141 mg/l *cis, cis*-muconic acid could be achieved in *S. cerevisiae* (Curran et al. 2013). Three non-native enzymes were introduced: the dehydroshikimate dehydratase from *Podospora anserina*, the protocatechuic acid decarboxylase from *Enterobacter cloacae,* and the catechol 1,2-dioxygenase from *Candida albicans*. The transketolase responsible for providing erythrose-4-phosphate was overexpressed. The *aro3* gene was knocked out and a feedback-resistant mutant aro4 introduced to overcome feedback inhibition in the shikimate pathway (Curran et al. 2013). The titer of *cis, cis*-muconic acid in *S. cerevisiae* could be further increased up to 20.8 g/l with a productivity of 139 mg/l × h (Wang et al. 2020). The yield achieved was 66.2 mg *cis, cis*-muconic acid per g glucose. For the high-throughput screening of strains, a specific GFP protein was used whose expression is triggered by *cis, cis*-muconic acid and which allows an UV efficient screening of strains (Wang et al. 2020). The low titer may be explained by the complex regulation of aromatic acid biosynthesis in *S. cerevisiae,* which needs more management attention in the metabolic engineering toward ADA (Kruyer and Peralta-Yahya 2017). A comprehensive overview of the achieved titer and productivity for *cis, cis*-muconic acid in *E. coli*, *S. cerevisiae,* and *P. putida* with different feedstocks reflects the high interest in this topic in academia (Choi et al. 2020).

19.3.3 *cis, cis*-Muconic Acid via the Partial Shikimic Acid Pathway in Specific Organelles

The fermentative production of *cis, cis*-muconic acid has also been shown in specific organelles in *S. cerevisiae*. The approach refers to plant storage vesicles derived by separation from the endoplasmic reticulum. These vesicles usually store proteins, which serve as amino acid reservoirs for the developing germ. The modified N-terminal sequence of the usually stored γ-zein protein in maize is sufficient to initiate the formation of the vesicles. Accordingly, the genes encoding the DHA dehydratase, the protocatechuate decarboxylase, and the catechol-1,2-dioxygenase were fused with the signal peptide of the γ-zein storage protein and expressed in *S. cerevisiae.* A vesicle formation was shown with the aid of a fluorescence marker that was fused with the γ-zein protein. 3-Dehydroshikimic acid entered the vesicle and *cis, cis*-muconic acid was released. The host was fed with glucose and after 150 h a *cis, cis*-muconic acid titer of 62 mg/l could be obtained. The alternative execution of the three pathway steps in the cytosol could be excluded since the three enzymes with partially deleted signal peptides displayed a very low activity. However, the pathway design within partial localization in a specific vesicle did not yet provide an advantage compared to the approach to pursue the complete pathway in the cytosol, which provided a titer of 159 mg/l after 150 h (Tripp et al. 2022).

19.3.4 Adipic Acid Ex Glucose via the Partial Shikimic Acid Pathway

The semi-synthesis of ADA, which combines the fermentative conversion of glucose into *cis, cis*-muconic acid followed by the chemical hydrogenation of *cis, cis*-muconic acid into ADA, might be simplified if the last step could be integrated into the fermentation process. Suitable enzymes for this step were not known until two families of flavoenzymes were tested for the biocatalytic hydrogenation of alkenes: the so-called old yellow enzymes (OYEs, EC: 1.6.99.1) and enoate reductases (ERs, EC 1.3.1.31). Hereby, four enoate reductases from *Clostridia* and *Bacillus coagulans* were found capable of hydrogenating 2-hexenedioic acid and *cis, cis*-muconic acid *in vitro* and *in vivo.* The substrate conversion rate of 94% after 24 h was shown *in vivo* (Joo et al. 2017). A proof of concept for this idea could be achieved with the expression of the enoate reductase of *Bacillus coagulans* in *S. cerevisiae* in *E. coli* (Raj et al. 2018). The yeast strain MuA12, which had been shown to generate the highest titer of *cis, cis*-muconic acid, has been transformed with three plasmids carrying a transketolase for the provision of erythrose-4-phosphate, a 3-dehydroshikimate dehydratase, a catechol dioxygenase, and a protocatechuate decarboxylase from *Enterobacter cloacae* (Figure 19.5).

The host produced 2.59 mg/l ADA after 48 h corresponding to a productivity of 0.054 mg/l × h. The *cis, cis*-muconic acid titer amounted to >284 mg/l indicating that only a minor share of *cis, cis*-muconic acid was converted (Raj et al. 2018). For producing the biocatalyst, an interesting three-stage process has been chosen. Initially, biomass has been

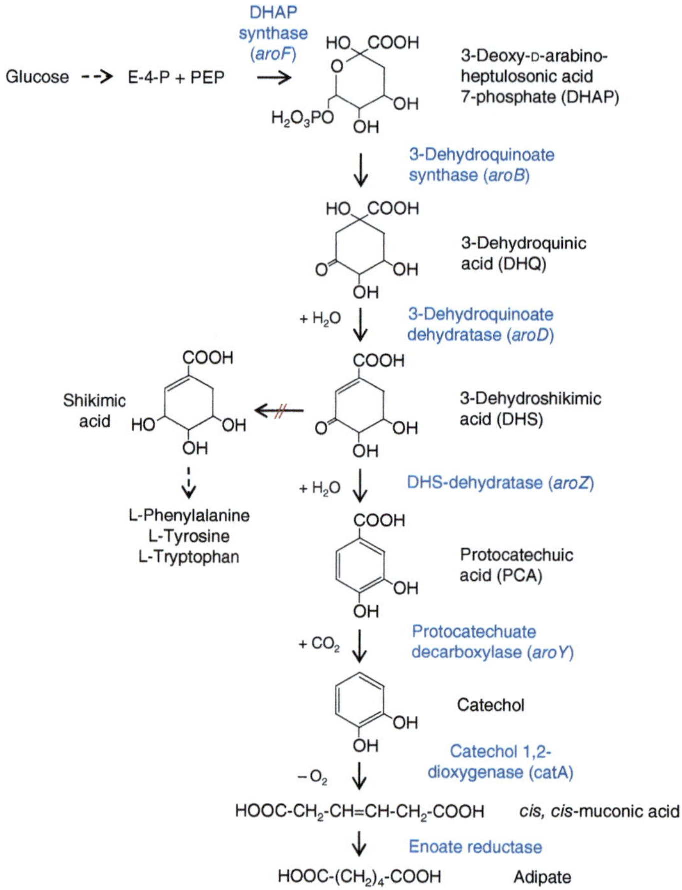

Figure 19.5 Metabolic pathway toward adipic acid based on glucose. Source: Walter Koch.

formed under aerobic conditions to provide sufficient ATP via respiratory chain respiration for the buildup of biomass (48 h). In the next period, the synthesis of a functional enoate reductase was ensured under anaerobic conditions (48 h). It is known that the maturation of an active enoate reductase requires anaerobic conditions for the synthesis and integration of an iron–sulfur cluster. The final production phase has been pursued under aerobic conditions since the catechol dioxygenase requires molecular oxygen as feedstock (48 h).

19.3.5 Coculturing of Two Hosts with 3-Dehydroshikimate as Handover

An interesting coculture approach combining two *E. coli* strains for producing *cis, cis*-muconic acid has been proposed in academia (Zhang et al. 2015). The two hosts in a mixed culture imply a higher complexity than in the traditional single-host fermentation. The idea is to install a system in which glucose and xylose are efficiently consumed. In case carbohydrate from lignocellulosic biomass is applied as feedstock, a mixture of C5 and C6 sugars, e.g. xylose and glucose, might be offered. A separation would in principle be possible, though it would increase the production costs. If the sugar mixture is applied, the host would usually display a priority of sugar digestion and would at first consume glucose and not start digestion of xylose before the glucose supply is exhausted. This observation is called carbon catabolite repression (Kremling et al. 2015; Brückner and Titgemeyer 2002) and might cause an inefficient use of the C5 sugars. The coculture of two *E. coli* hosts is designed such that the first host exclusively consumes xylose and the second *E. coli* host exclusively consumes glucose. In particular, the first strain is designed for xylose metabolism, and the gene encoding the PTS (phosphotransferase system) system for glucose uptake has been knocked out. It converts xylose into 3-dehydroshikimic acid via the shikimic acid pathway. The competing conversion of 3-dehydroshikimic acid toward the aromatic amino acids has been blocked. 3-dehydroshikimic acid would then be secreted and taken up by the second strain equipped with the DHS transporter encoded by the *shlA* gene. The second strain consumes exclusively glucose and pursues a whole-cell biocatalysis to convert 3-dehydroshikimic acid into *cis, cis*-muconic acid. The second host is not able to consume xylose, and the technology allows a selective feeding strategy for both strains (Figure 19.6).

In the first experiments with induced expression of *catA* (*Acinetobacter calcoaceticus*), *aroY* (*Klebsiella pneumoniae*), and *aroZ* (*Klebsiella pneumoniae*) and induction with IPTG (isopropyl-β-thiogalacto-pyranoside) and the inducible T7 promoter about 400 mg/l 3-dehydroshikimic acid were observed. With the constitutive expression of the three pathway genes and the constitutive *Zymomonas mobilis* pyruvate decarboxylase promoter, the titer could be increased and 850 mg/l 3-dehydroshikimic were measured. The parallel experiment with the monoculture approach did not provide relevant titers of *cis, cis*-muconic acid, or pathway intermediates. The broth concentration of alternative products, i.e. non-converted 3-dehydroshikimic acid and the pathway intermediates catechol and protocatechuic acid, were also analyzed and revealed for each a value below 200 mg/l. The medium of the coculture contained 3.3 g/l xylose and 6.6 g/l glucose. The coculture system had a high capacity for handling changing carbon feeds and achieved a titer of 682 mg/l *cis, cis*-muconic acid with a reversed medium containing 6.6 g/l xylose and 3.3 g/l glucose. Further improvements in a 1.3 l batch-mode reactor provided 4 g/l *cis, cis*-muconic acid produced given a feed of 6.6 g/l xylose and 13.2 g/l glucose in the medium. The titer could be further increased up to 4.7 g/l *cis, cis*-muconic acid with

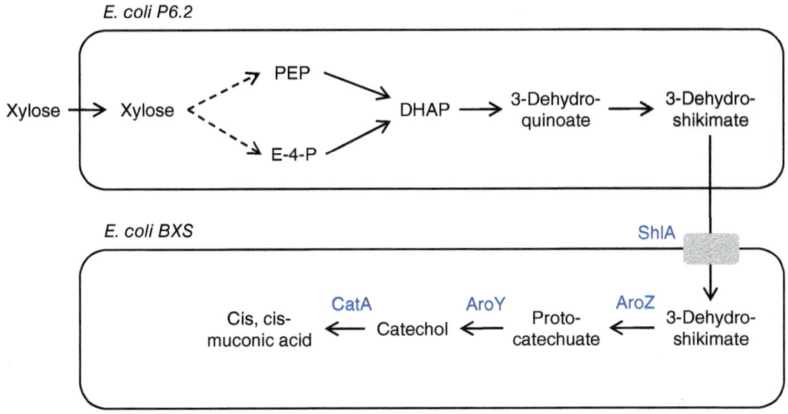

Figure 19.6 Allocation of pathway sections toward *cis, cis*-muconic acid in an *E. coli* coculture. Source: Adapted from Zhang et al. (2015).

fed-batch reactor mode, and a carbon yield of 0.35 g/g was achieved (Zhang et al. 2015). Similar results were reported by another academic working group, which reported a titer of 4.09 g/l *cis, cis*-muconic acid based on glucose with parallel feed of xylose consumed for energy synthesis and the generation of biomass constituents (Fujiwara et al. 2020).

19.3.6 *cis, cis*-Muconic Acid Derived from a Lignin Fraction via "Biological Funneling"

Lignin is a complex natural polymer integrated into plant cell walls for defense and mechanical stability purposes. It is a by-product of biorefineries in the processing of biomass. Whereas the cellulose and hemicellulose fractions of biomass may be converted into biobased fuels, lignin is often burned to exploit its energy value. The complex structure of lignin and its inherent inhomogeneity and recalcitrance define a high technical hurdle for a chemical use like lignosulfonates or vanillin precursors. A cost-competitive and industrially applied extraction of defined aromatics has so far not been established (Sun et al. 2018; Upton and Kasko 2015). In order to allow a chemical use of lignin with higher value perspective than its use as fuel, several valorization concepts are under development (Beckham et al. 2016). One idea is to start with a lignin breakdown and to use the obtained heterogeneous molecule mixture as feedstock for a biocatalytic conversion with *Pseudomonas putida* KT2440. The advantage of this system would be that *Pseudomonas* is able to convert several precursors into one defined molecule, i.e. catechol, which can then be converted with a ring-opening step catalyzed by a dioxygenase into *cis, cis*-muconic acid. The "biological funneling" might reduce the complexity, which otherwise would result if the mixture of aromatic compounds was channeled through a fractionation process (Vardon et al. 2015). The *cis, cis*-muconic acid might then be further converted with an enoate reductase or isolated and chemically hydrogenated to obtain ADA (Figure 19.7):

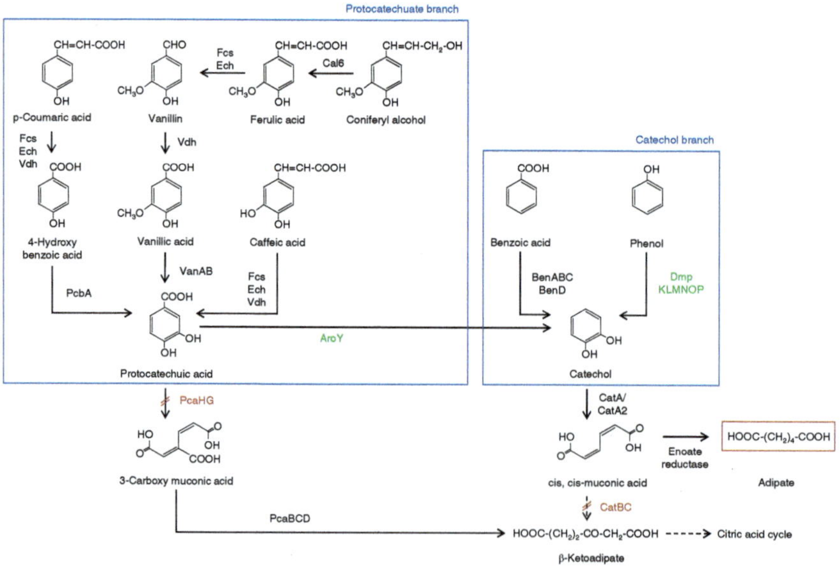

Figure 19.7 Conversion of lignin derivatives into *cis, cis*-muconic acid. Source: Adapted from Vardon et al. (2015).

In the first step, the concept was evaluated with the use of a model lignin-derived compound. Hereby, a titer of *cis, cis*-muconic acid of 13.5 g/l could be obtained in a fed-batch cultivation after 78.5 h (Vardon et al. 2015). The molecule was isolated with a purity of >97% and a yield of 74% after treatment with activated carbon and a crystallization at pH 2 and 5 °C. The chemical conversion into ADA was facilitated with a Pd/C catalyst and achieved a conversion of >97% and a selectivity of >97% (Vardon et al. 2015). These results are promising, though the process might suffer from difficulties if a real lignin mixture is applied. An independent academic research group pursuing a similar concept for lignin valorization achieved a titer of 64.2 g/l *cis, cis*-muconic acid in a fed-batch process with *Pseudomonas putida* KT2440 (Kohlstedt et al. 2018). After a hydrothermal depolymerization of softwood lignin to catechol, phenol, and cresols, a titer of 13 g/l *cis, cis*-muconic acid could be obtained. The conversion was able to convert o-cresol, m-cresol, and p-cresol into 3-methylcatechol with the enzyme DmpKLMOP, which provided 2-methyl-*cis, cis*-muconate as additional product (Kohlstedt et al. 2018).

The biocatalytic conversion of aromatic compounds might suffer from the toxicity of the feedstocks toward the biocatalyst. Since the host *Amycolatopsis* is known to tolerate a high level of vanillin (Feige et al. 2016; Feige et al. 2013), the strain *Amycolatopsis* ATCC 39116 has been tried for the conversion of lignin-based aromatics, i.e. catechol, guaiacol, phenol, toluene, p-coumarate, and benzoate into *cis and cis*-muconic acid. Based on guaiacol as model compound, a titer of 3.1 g/l *cis, cis*-muconic acid could be obtained within 24 h with a conversion yield of 96%. The mutant was also capable of converting a guaiacol-rich lignin hydrolysate into *cis, cis*-muconic acid. Additionally, the conversion of o-cresol into 2-methyl-*cis, cis*-muconate was shown (Barton et al. 2018).

Further experiments with *Corynebacterium glutamicum* as biocatalyst also provided promising results. A mutant with knocked-out muconate cycloisomerase (*catB*), to suppress the conversion of *cis, cis*-muconate into β-ketoadipate, showed an efficient conversion with 100% yield of catechol, phenol, and benzoic acid into *cis, cis*-muconic acid. The overexpression of catechol 1,2-dioxygenase (*catA*) led a further optimization. The best-performing mutant provided a titer of 85 g/l *cis, cis*-muconic acid from catechol after an incubation of 60 h and achieved a maximum productivity of 2.4 g/l × h (Becker et al. 2018).

Summarizing, these attempts might in the future allow a chemical access to defined and biobased chemical compounds, e.g. lignin. The production cost of biobased ADA might be higher than the production cost of petrochemically derived ADA, though the carbon footprint of the biobased product is expected to be lower. A life cycle assessment revealed that the greenhouse gas emissions of biobased ADA amount to 4.87 kg CO_2/kg ADA, which would provide a reduction of 62%–78% compared to conventional ADA (Corona et al. 2018). The technology would basically be a biocatalytic conversion of aromatic precursors, and glucose will need to be added as growth substrate unless at least a share of the compounds is channeled into central host metabolism.

19.4 Digression: Metabolic Engineering for C6+ Diacids

19.4.1 Long-Chain Carboxylic Acids via Biotransformation

The chemical synthesis of long-chain dicarboxylic fatty acids according to the formula HOOC-$(CH_2)_n$-COOH with $n > 12$ is challenging and thereby defines an attractive area for biotransformation and fermentation. Dicarboxylic fatty acids (DCA) with a chain length of C16, C18, C20, or longer are applied as monomer building blocks for high-grade polymers in several applications. The market for C18 dicarboxylic acid was valued at 6.6 USD bn in 2016 and is expected to grow up to 9.4 USD bn in 2024 (Bauwelinck et al. 2021). It is known that some organisms are able to convert the terminal methyl functions of alkanes or monocarboxylic fatty acids into carboxylic acid functions, and thereby dicarboxylic fatty acids are in principle accessible. Alkanes or monocarboxylic acids may be applied as substrates:

$$CH_3{-}(CH_2)_n{-}CH_3 \rightarrow\rightarrow\rightarrow HOOC{-}(CH_2)_n{-}COOH$$

$$HOOC{-}(CH_2)_n{-}CH_3 \rightarrow\rightarrow\rightarrow HOOC{-}(CH_2)_n{-}COOH$$

19.4.2 Generation of C6+ Diacids with *Candida tropicalis* Based on Waste Streams

Feedstock costs are often a critical factor in the overall cost position of a fermentative technology, which might hamper a broader commercial application. In order to reduce raw material costs, a waste stream or by-product stream has been explored as potential feedstock. In particular, a high-fat-containing side stream of the chocolate industry and sludge after treatment with dissolved air floatation (DAF) have been used as starting material for a bioconversion toward α,ω-dicarboxylic acids with a genetically modified strain of *Candida tropicalis* (Bauwelinck et al. 2021). The addition of both complex feedstocks in reasonable amounts did not show a negative impact on growth and dicarboxylic acid production. The fermentation was pursued under aerobic conditions with an aeration rate of 1.5 l/min (1.5 vvm) to provide an electron acceptor for the generated reduction equivalents. The pH of the potentially more acidic medium was kept at 8.2 using 4 M KOH as neutralization agent. However, it is known that *C. tropicalis* is able to tolerate and survive in acidic conditions when sufficient nutrients are present and – pending on the final titer of dicarboxylic acids – neutralization efforts may be limited. The strain was incubated at 30 °C and shaken at 200 rpm. In order to prevent that the strain used the monocarboxylic acids as feedstock, the first gene of the ß-oxidation pathway and the POX genes encoding the acyl-CoA-oxidase were knocked out. This step ensures that the feedstock is completely available for product formation. The following DCA titers and productivities were achieved (Bauwelinck et al. 2021; Table 19.3).

The host was not able to consume the sucrose available in the waste stream of the chocolate industry although *C. tropicalis* mutants are described, which are able to grow on sucrose. It might be the case that the respective α-glucosidase is not available in the

Table 19.3 Performance data for the conversion of fats into DCAs.

Feedstock	Titer	Batch duration	Productivity	Selectivity
Waste stream chocolate industry	5.8 g/l	120 h	0.048 g/l × h	47.5%
Sludge after DAF	4.9 g/l	72 h	0.068 g/l × h	37.5%

Source: Adapted from Bauwelinck et al. (2021).

used strain. However, since the use of sucrose as energy source was not possible and the β-oxidation of fatty acids has been blocked an energy source had to be provided and glucose was dosed at a rate of 0.4 g/l × h.

19.5 Process Development

Process development needs to address pH management in the first instance. ADA bears two carboxylic acid functions with pk_a values of 4.4 and 5.4, respectively, which inevitably acidify the fermentation medium beyond the tolerance level of the fermentation host (Karlsson et al. 2017). The resulting pH shall be estimated by simplified addition of the proton concentration caused by the two carboxylic acid functions. Hereby, it is assumed that a titer of 100 g/l ADA is achieved, which translates into a concentration of 0.684 mol/l (Table 19.4).

Accordingly, the overall proton concentration amounts to about 0.008 mol/l and a pH of 2.1 would be derived without taking the buffering effect of the media components into consideration. The most straightforward approach to neutralize the broth and to protect productivity would be the addition of a base. The product could be re-protonated with H_2SO_4 or HCl in the downstream process, and a salt would be derived as by-product. The salt might not represent a valuable product, though it might imply an environmental fee for landfilling, especially in case a higher volume of ADA shall be produced. Accordingly, *in situ* product removal approaches were tested, which might allow to limit the neutralization efforts (Coons 2011).

It has been speculated that the low titers and productivities achieved in R&D projects might also be due to a poor cellular tolerance to ADA itself (Karlsson et al. 2017). The experiments were pursued with an ADA concentration of 50–100 g/l since 100 g/l would be a reasonable titer for a commercial production process. The pH was adjusted to 7.0, and it was found out that all bacterial hosts, including *E. coli* lost their ability to grow. This observation might not yet exclude the option to apply a bacterial host since a resting culture might be used and the culture growth might be done before the generation of ADA is induced. On the other hand, the yeast *A. niger* was able to grow on glucose in the presence of 50–100 g/l ADA at pH 5.0. The yeast species *C. viswanathi* exhibited the highest tolerance toward ADA and was almost unaffected at the highest concentration of 100 g/l (Karlsson et al. 2017). The difference in growth supports the hypothesis that a higher inflow of ADA into bacterial cells than into yeast cells does occur.

Table 19.4 Estimation of the pH given an adipic acid titer of 100 g/l without buffering.

$pK_{a1} = 4.4$ and $K_{a1} = 3.98 \times 10^{-5}$, and

$$c[H_3O^+] = -K_{a1}/2 + (K_{a1}/4 + K_{a1} \times c[HA])^{0.5}$$

$$= -\,3.98 \times 10^{-5}/2 + (3.98 \times 10^{-5}/4 + 3.98 \times 10^{-5} \times 0.684)^{0.5} = 0.00607\ \text{mol/l, and}$$

$pK_{a2} = 5.4$ and $K_{a2} = 3.98 \times 10^{-6}$, and

$$c[H_3O^+] = -\,K_{a2}/2 + (K_{a2}/4 + K_{a2} \times c[HA])^{0.5}$$

$$= 3.98 \times 10^{-6}/2 + (3.98 \times 10^{-6}/4 + 3.98 \times 10^{-6} \times 0.684)^{0.5} = 0.00192\ \text{mol/l}$$

19.6 Industrial Application and Capacity

The industrial fermentation of ADA is under preparation, though a commercial high-volume supply is not yet established. Toray is a producer of PA-66 based on the two monomers HMD and ADA. The company announced in 2022 that it had successfully pursued the fermentative production of ADA from sugars and inedible biomass (Toray press release 2022). Toray has genetically modified a host and achieved an increase in the native concentration by a factor of 1000. The process shall especially profit from the use of reverse osmosis membranes, which concentrate the product after the fermentation step and thereby reduce the downstream efforts for conventional evaporation. It is pointed out by Toray that fermentation technology does not generate dinitrogen monoxide (N_2O, laughing gas) as by-product as opposed to the conventional petrochemical production route. The company is scaling up its production capacity in this area and is planning to commercialize fermentative ADA and the respective polyamide PA-66 grades around 2030 (Toray press release 2022). It remains to be seen inasmuch as fermentation technology is finally able to compete with mature petrochemistry and establish a biobased and "green" market share.

In general, the diacid has been addressed in several patents and a high number of academic publications, and more companies might offer an ADA grade based on renewables in the future. Several companies, including the Dutch food ingredient producer DSM and the R&D company Verdezyne, had announced plans to evaluate fermentative technology to produce ADA (Coons 2011). DSM formed an industrial biotechnology group in 2005 and explored synergies in the biobased production of succinic acid and ADA. For both products, a fermentation protocol for *in situ* product recovery had been considered to reduce the consumption of a base during the fermentation. DSM confirmed in 2013 that proof of concept for a proprietary technology has been developed but declared that "commercialization within five years cannot be confirmed at this stage" (Ravenscroft 2013).

References

Arpe, H.-J. (2007). *Industrielle Organische Chemie. Bedeutende Vor- und Zwischenprodukte.* Wiley p. 264ff.

Babu, T., Yun, E.J., Kim, D.H. et al. (2015). Engineering *Escherichia coli* for the production of adipic acid through the reversed ß-oxidation pathway. *Process Biochemistry* 50 (12): 2066–2071.

Bart, J.C., Cavallaro, J., and (2015). Transiting from adipic acid to bioadipic acid. 1, petroleum-based processes. *Industrial & Engineering Chemistry Research* 54: 1–46. https://doi.org/10.1021/ie5020734.

Barton, N., Horbal, L., Starck, S. et al. (2018). Enabling the valorization of guaiacol-based lignin: integrated chemical and biochemical production of *cis, cis*-muconic acid using metabolically engineered *Amycolatopsis* sp ATCC 39116. *Metabolic Engineering* 45: 200–210. https://doi.org/10.1016/j.ymben.2017.12.001.

Bauwelinck, J., Caluwe, M., Wijnants, M. et al. (2021). Chocolate industry side streams as a valuable feedstock for microbial long-chain dicarboxylic acid production. *Biochemical Engineering Journal* 167: http://doi.org/10.1016/j.bej.2020.107888.

Becker, J., Kuhl, M., Kohlstedt, M. et al. (2018). Metabolic engineering of *Corynebacterium glutamicum* for the production of *cis, cis*-muconic acid from lignin. *Microbial Cell Factories* 17: 115. https://doi.org/10.1186/s12934-018-0963-2.

Beckham, G.T., Johnson, C.W., Karp, E.M. et al. (2016). Opportunities and challenges in biological lignin valorization. *Current Opinion in Biotechnology* 42: 40–53. https://doi.org/10.1016/j.copbio. 2016.02.030.

Brückner, R. and Titgemeyer, F. (2002). Carbon catabolite repression in bacteria: choice of the carbon source and autoregulatory limitation of sugar utilization. *FEMS Microbiology Letters* 209: 141–148.

Cheng, Q., Thomas, S.M., and Rouviere, P. (2002). Biological conversion of cyclic alkanes and cyclic alcohols into dicarboxcylic acids: biochemical and molecular basis. *Applied Microbiology and Biotechnology* 58: 704–711.

Choi, S., Lee, H.-N., Park, E. et al. (2020). *Recent advances in microbial production of cis, cis-muconic acid, biomolecules* 10: 1238. https://doi.org/10.3390/biom10091238.

Coons, R. (2011). DSM adds adipic acid to biobased chemical efforts. *Chemical Week* October 17: 14.

Coons, R. (2017). Picking the right biobased targets. Funding still possible for companies that hit milestones, deliver value. *Chemical Week* October 2017 (23/30): 35.

Corona, A., Biddy, M.J., Vardon, D.R. et al. (2018). Life cycle assessment of adipic acid production from lignin. *Green Chemistry* 20: 3857–3866. https://doi.org/10.1039/c8gc00868j.

Curran, K.A., Leavitt, J.M., Karim, A.S., and Alper, H.S. (2013). Metabolic engineering of muconic acid production in *Saccharomyces cerevisiae*. *Metabolic Engineering* 15: 55–66. https://doi.org/10.1016/j.ymben.2012.10.003.

Damayanti, D., Wulandari, L.A., Bagaskoro, A., and Rianjanu, A. (2021). Possibility routes for textile recycling technology. *Polymers* 13: 3834. https://doi.org/10.3390/polym13213834.

Deng, Y. and Mao, Y. (2015). Production of adipic acid by the native occurring pathway in *Thermobifida fusca* B6. *Journal of Applied Microbiology* 119: 1057–1063.

Deng, Y., Ma, L., and Mao, Y. (2016). Biological production of adipic acid from renewable resources: current and future products. *Biochemical Engineering Journal* 105: 16–26. https://doi.org/10.1016/j.bej.2015.08.015.

Draths, K.M. and Frost, J.W. (1994). Environmentally compatible synthesis of adipic acid from D-glucose. *Journal of the American Chemical Society* 116: 399–400.

Duch, M.W. and Allgeier, A.M. (2006). Deactivation of nitrile hydrogenation cataysts: new mechanistic insight from a nylon recycle process. *Applied Catalysis* 318: 190–198. https://doi.org/10.1016/j.apcata.2006.11.003.

Feige, C., Hansen, G., and Steinbüchel, A. (2013). Investigation of the *Amycolatopsis* sp. strain ATCC 39116 vanillin dehydrogenase and its impact on the biotechnical production of vanillin. *Applied and Environmental Microbiology* 79 (1): 81–90. https://doi.org/10.1128/AEM.02358-12.

Feige, C., Meyer, F., and Steinbüchel, A. (2016). Metabolic engineering of the actinomycete *Amycolatopsis* sp. strain ATCC 39116 towards enhanced production of natural vanillin. *Applied and Environmental Microbiology* 82 (11): 3410–3419. https://doi.org/10.1128/AEM.00802-16.

Fujiwara, R., Noda, S., Tanaka, T., and Kondo, A. (2020). Metabolic engineering of *Escherichia coli* for shikimate pathway derivative production from glucose-xylose co-substrate. *Nature Communication* 11: 279. https://doi.org/10.1038/s41467-019-14024.

Genomatica patent, Inventors (n.d.): Burgard, A., Pharkya, P., Osterhout, R. E., Microorganisms for the production of adipic acid and other products, WO2009/151728 A2

http://Greencarcongress.com, Amyris to buy Draths, 2011, source: http://www.greencarcongress.com/2011/11/amyris-20111101.html.

Joo, J.C., Khusnutdinova, A.N., Flick, R. et al. (2017). Alkene hydrogenation activity of enoate reductases for an environmentally benign biosynthesis of adipic acid. *Chemical Science* 8: 1406. https://doi.org/10.1039/c6sc02842j.

Kallscheuer, N., Gätgens, J., Lübcke, M. et al. (2017). Improved production of adipate with *Escherichia coli* by reversal of ß-oxidation. *Applied Microbial Biotechnology* 101: 2371–2382. https://doi.org/10.1007/s00253-016-8033-3.

Karlsson, E., Mapelli, V., and Olsson, L. (2017). Adipic acid tolerance screening for potential adipic acid production hosts. *Microbial Cell Factories* 16: 20. https://doi.org/10.1186/s12934-017-0636-6.

Kohlstedt, M., Starck, S., Barton, N. et al. (2018). From lignin to nylon: cascased chemical and biochemical conversion using metabolically engineered *Pseudomonas putida*. *Metabolic Engineering* 47: 279–293. https://doi.org/10.1016/j.ymben.2018.03.003.

Kremling, A., Geiselmann, J., Ropers, D., and de Jong, H. (2015). Understanding carbon catabolite repression in *Escherichia coli* using quantitative models. *Trends in Microbiology* 23: 99–109.

Kruyer, N.S. and Peralta-Yahya, P. (2017). Metabolic engineering strategies to bio-adipic acid production. *Current Opinion in Biotechnology* 45: 136–143. https://doi.org/10.1016/j.copbio.2017.03.006.

Liu, Z.H., Xie, S., Lin, F. et al. (2018). Combinatorial pretreatment and fermentation optimization enabled a record yield on lignin bioconversion. *Biotechnology for Biofuels* 11: 21. https://doi.org/10.1186/s13068-018-1021-3.

Mormul, J., Breitenfeld, J., Trapp, O. et al. (2016). Synthesis of adipic acid, 1,6-hexanediamine, and 1,6-hexanediol via double-n-selective hydroformylation of 1,3-butadiene. *ACS Catalysis* 6: 2802–2810. https://doi.org/10.1021/acscatal.6b00189.

Niu, W., Draths, K.M., and Frost, J.W. (2002). Benzene-free synthesis of adipic acid. *Biotechnology Progress* 18: 201–211.

Polen, T., Spelberg, M., and Bott, M. (2013). *Journal of Biotechnology* 167: 75–84. https://doi.org/10.1016/j.biotec. 2012.07.008.

Raj, K., Partov, S., Correia, K. et al. (2018). Biocatalytic production of adipic acid from glucose using engineered *Saccharomyces cerevisiae*. *Metabolic Engineering Communications* 6: 28–32. https://doi.org/10.1016/j.meteno.2018.02.001.

Ravenscroft, M. (2013). Industrial biotechnology. Chemical building blocks from renewable resources, IHS. *Chemical Week* 5 (12): 19–22.

Steffensen, W.S. and Alexander, M. (1995). Role of competition for inorganic nutrients in the biodegradation of mixtures of substrates. *Applied and Environment Microbiology* 61: 2859–2862.

Suastegui, M., Matthiesen, J.E., Carraher, J.M. et al. (2015). Building a metabolic engineering and electrocatalysis pipeline for the production of polyamides from sugar. *Communication* https://doi.org/10.1002/ange.201509653.

Sun, Z., Fridrich, B., de Santi, A. et al. (2018). Bright side of lignin depolymerization: toward new platform chemicals. *Chemical Reviews* 118: 614–678. https://doi.org/10.1021/acs.chemrev.7b00558.

Toray press release, Toray invents 100% Bio-based adipic acid from sugars derived from inedible biomass, scaling up for application to eco-friendly nylon 66, 2022, https://www.toray.com/global/news/details/20220818103248.html.

Tripp, J., Oreb, M., and Boles, E. (2022). Neuartige, metabolisch aktive Organellen in eukaryotischen Zellen. *BIOspektrum* 28: 397–399. https://doi.org/10.1007/s12268-022-1777-7.

Upton, B.M. and Kasko, A.M. (2015). Strategies for the conversion of lignin to high-value polymeric materials: review and perspective. *Chemical Reviews* 116: 2275–2306. https://doi.org/10.1021/acs.chemrev.5b00345.

Vardon, D.R., Franden, M.A., Johnson, C.W. et al. (2015). Adipic acid production from lignin. *Energy & Environmental Science* 8: 617–628. https://doi.org/10.1039/c4ee 03230f.

Vollmer, I., Jenks, M.J.F., Roelands, M.C.P. et al. (2020). Beyond mechanical recycling: giving new life to plastic waste. *Angewandte Chemie* 59: 15402–15423. https://doi.org/10.1002/anie.201915651.

Wang, G., Øzmerih, S., Guerreiro, R. et al. (2020). Improvement of cis, cis-muconic acid production in *Saccharomyces cerevisiae* through biosensor-aided genome engineering. *ACS Synthetic Biology* 9: 634–646. https://doi.org/10.1021/acssynbio.9b00477.

Weber, C., Brückner, C., Weinreb, S. et al. (2012). Biosynthesis of *cis, cis*-muconic acid and its aromatic precursors, catechol and protocatechuic acid, from renewable feedstocks by *Saccharomyces cerevisiae*. *Applied and Environmental Microbiology* 78 (23): 8421–8430. https://doi.org/10.1128/AEM.01983-12.

Yoneda, A., Henson, W.R., Goldner, N.K. et al. (2016). Comparative transcriptomics elucidates adaptive phenol tolerance and utilization in lipid-accumulating *Rhodococcus opacus* PD630. *Nucleic Acid Research* 44 (5): 2240–2254. https://doi.org/10.1093/nar/gkw055.

Yu, J.L., Xia, X.X., Zhong, J.J., and Qian, Z.G. (2014). Direct biosynthesis of adipic acid from a synthetic pathway in recombinant *Escherichia coli*. *Biotechnology and Bioengineering* 111 (12): 2580–2586.

Zhang, H., Zhenjun, L., and Stephanopoulos, G. (2015). Engineering *Escherichia coli* coculture systems for the production of biochemical products. *Proceedings of the National Academy of Sciences of the United States* 112 (27): 8266–8271.

Zhao, M., Huang, D., Zhang, X. et al. (2018). Metabolic engineering of *Escherichia coli* for producing adipic acid through the reverse adipate-degradation pathway. *Metabolic Engineering* 47: 254–262.

20

Hexamethylenediamine

20.1 Application

Hexamethylenediamine (HMD) is an aliphatic diamine with the formula $C_6H_{16}N_2$ and a molecular weight of 116.2 g/mol. The global production volume of HMD is estimated at 2 Mio. t pa. An aqueous solution of HMD is basic given pK_b values of 10.24 and 11.02 for the two amino functions (Sargent et al. 2022; Bednarz et al. 2017). HMD is one of the two building blocks for the polyamide PA-66 ("Nylon") derived by polymerization of the two alternating building blocks adipic acid (ADA) and HMD:

$$n\,H_2N\text{–}[CH_2]_6\text{–}NH_2 + n\,HOOC\text{–}[CH_2]_4\text{–}COOH$$
$$\rightarrow [\text{–}CO\text{–}(CH_2)_4\text{–}CO\text{–}NH\text{–}(CH_2)_6\text{–}NH\text{–}]_n + 2_{n-1}\,H_2O$$

PA-66 is, besides PA-6, the main polyamide invented in the last century with versatile chemical and physical properties, which has paved its way in various consumer and industrial applications. The total global PA-66 capacity amounts to 3.1 Mio. t pa (Boswell 2022), corresponding to a value of the PA-66 market of 6.4 bn USD (Covestro press release 2022). The polymerization of ADA and HMD is highly mature and efficiently manages the removal of water generated in the formation of the amide bonds to pull the equilibrium of the reaction to the right side. The selectivity has achieved values of 98%–99%, and the art of the reaction is to achieve exactly the chain length and viscosity, respectively, of the target PA-66 grade. A low residual monomer content may be critical in some applications. The second major application of HMD comprises the phosgenation of the amino functions to obtain the aliphatic diisocyanate hexamethylene diisocyanate (HDI) or 1,6-diisocyanohexane. The amino functions are converted with the reactive phosgene as transfer unit for CO and form the diisocyanate:

$$H_2N\text{–}CH_2\text{–}CH_2\text{–}CH_2\text{–}CH_2\text{–}CH_2\text{–}CH_2\text{–}NH_2 + 2\,COCl_2$$
$$\rightarrow OCN\text{–}CH_2\text{–}CH_2\text{–}CH_2\text{–}CH_2\text{–}CH_2\text{–}CH_2\text{–}NCO + 4\,HCl$$

HDI is a major building block for urethanes, which are conventionally produced through the reaction of diisocyanates with polyols. Polyurethanes based on aliphatic diisocyanates are often used as coatings and elastomers.

Pathway Design for Industrial Fermentation, First Edition. Walter Koch.

20.2 Chemical Production of HMD

HMD is produced via the catalytic liquid-phase hydrogenation of adiponitrile (ADN), which is the dominant technology within the industry. The hydrogenation with molecular hydrogen was originally pursued at rather extreme conditions with 600–650 bar and a temperature of 100–135 °C, though the technology could be optimized and more moderate conditions are allowed. The advanced technology is pursued at a temperature of 75 °C and a pressure of 30 bar and achieves a yield of up to 99% (Arpe 2007):

$$N{\equiv}C\text{–}CH_2\text{–}CH_2\text{–}CH_2\text{–}CH_2\text{–}C{\equiv}N + 4\,H_2 \\ \rightarrow H_2N\text{–}CH_2\text{–}CH_2\text{–}CH_2\text{–}CH_2\text{–}CH_2\text{–}CH_2\text{–}NH_2$$

Hereby, the nitrile function is reduced and converted into a methylene function with terminal amino group. It is interesting to note that this industrial reaction indeed combines two different reactions since the mechanism likely proceeds via the mixed nitrile-amine intermediate. The different substrates of both steps would usually require specific reaction conditions, though it has been managed to pursue both steps in a combined one-pot reaction. The PA-66 synthesis with HMD as the main monomer profits from the economically attractive availability of its precursor ADN, which may be derived via two competing technologies: ADN may be produced by the dual hydrocyanation of butadiene ("C4 route"), a technology originally developed and commercially applied in 1971 by Dupont (Bart and Cavallaro 2015). The PA-66 business of Dupont is now owned by Invista (Hagen 2015; Arpe 2007):

$$CH_2{=}CH\text{–}CH{=}CH_2 + 2\,HCN \rightarrow N{\equiv}C\text{–}CH_2\text{–}CH_2\text{–}CH_2\text{–}CH_2\text{–}C{\equiv}N$$

Alternatively, ADN may be obtained via the electro-dimerization of acrylonitrile ("C3 route"):

$$2\,CH_2{=}CH\text{–}C{\equiv}N + 2\,H^+ + 2\,e^- \rightarrow N{\equiv}C\text{–}CH_2\text{–}CH_2\text{–}CH_2\text{–}CH_2\text{–}C{\equiv}N$$

The C3 technology is owned by Ascend (Boswell 2019). The technology was originally developed by Monsanto, which allocated the PA-66 business to Solutia. Solutia established Ascend as a separate business unit for PA-66, which was acquired by SK in 2009 (Valk 2019):

The first industrial route toward ADN used ADA as starting material, which is easily available via the oxidation of cyclohexanol. The original gas phase process was operated at temperatures of 300–350 °C with a high molar surplus of ammonia and in the presence of a catalyst (ICI 1964). Hereby, ammonium adipate is obtained as an intermediate, which is then twice dehydrated to obtain ADN:

$$HOOC\text{–}CH_2\text{–}CH_2\text{–}CH_2\text{–}CH_2\text{–}COOH + 2\,NH_3 \\ \rightarrow H_4NOOC\text{–}CH_2\text{–}CH_2\text{–}CH_2\text{–}CH_2\text{–}COONH_4$$

$$H_4NOOC\text{–}CH_2\text{–}CH_2\text{–}CH_2\text{–}CH_2\text{–}COONH_4 \\ \rightarrow H_2NOC\text{–}CH_2\text{–}CH_2\text{–}CH_2\text{–}CH_2\text{–}CONH_2 + 2\,H_2O$$

$$H_2NOC\text{–}CH_2\text{–}CH_2\text{–}CH_2\text{–}CH_2\text{–}CONH_2 \\ \rightarrow N{\equiv}C\text{–}CH_2\text{–}CH_2\text{–}CH_2\text{–}CH_2\text{–}C{\equiv}N + 2\,H_2O$$

The selectivity of the reaction could be improved from about 80% to up to 90%, though the cheap availability of butadiene ("C4 route") and propylene ("C3 route") has nevertheless replaced the route via ADA (Arpe 2007).

Alternatively, HMD is accessible via the amination of hexanediol, which can be derived via the reduction of ADA or 6-hydroxycaproic acid. The terminal hydroxyl functions may be aminated via the simple reaction with ammonia and generate water as by-product.

$$\begin{aligned} &HO{-}CH_2{-}CH_2{-}CH_2{-}CH_2{-}CH_2{-}CH_2{-}OH + 2\,NH_3 \\ &\quad \rightarrow H_2N{-}CH_2{-}CH_2{-}CH_2{-}CH_2{-}CH_2{-}CH_2{-}NH_2 + 2\,H_2O \end{aligned}$$

The amination reaction is pursued with a typical Raney Nickel catalyst and is operated at a temperature of about 200 °C and a pressure of 230 bar in the presence of ammonia. The yield is indicated at 90%, though the industrially achieved performance might achieve a value above 90%. The main by-products are hexamethylene-imine and 1,6-aminohexanol (Lee et al. 2019; Dias et al. 2014; Arpe 2007). The amination of 1,6-hexanediol might allow the application of a biobased raw material for the synthesis of HMD. The catalyst screening company Rennovia has proposed to convert biobased furfural into 1,6-hexanediol or 1,2,6-hexanetriol with molecular hydrogen and a Pt catalyst (Rennovia patent n.d.).

The recent development of alternative technology towards HMD does not primarily pursue further reducing the production cost because current petrochemistry is rather mature and allows an attractive cost level. The main driving force for biobased routes and fermentation is the attempt to reduce the carbon footprint of the products and, in the long term, to identify an alternative feedstock not relying on petrochemistry (Wang et al. 2021; Lee et al. 2019; Harmsen et al. 2013). Furthermore, growing volumes of recycled HMD may be available in the future in case chemical PA-66 depolymerization technology gains market share versus "fresh" HMD and accomplishes recycling a greater share of PA-66 waste streams (Zhou et al. 2021; Damayanti et al. n.d.; Vollmer et al. 2020; Duch and Allgeier 2006).

20.3 Metabolic Engineering for Fermentation Technology

20.3.1 Fermentation Based on Glucose via Adipyl-CoA

It is not known whether HMD is a natural intermediate in biochemical pathways. Accordingly, a recombinant pathway needs to be developed. This should be possible and technically realistic given the rather easy functionality of the molecule and the abundant amine function in the natural biochemical reservoir. The regularly observed promiscuity of enzymes proposes the generation of a linear C6-backbone with terminal carboxylic acid functions, which can then be reduced to the respective aldehyde. The aldehyde can then be converted into an amine via transamination. This rather straightforward idea describes the basic strategy of most of the prominent pathways originally described by Genomatica in several patent applications (Genomatica patents 2009, 2010, 2011, 2012). The metabolic engineering strategies and pathways towards HMD are often linked to the respective strategies for ADA and caprolactam (CPL) since all three aliphatic C6-intermediates share the same backbone, which can be built via several approaches. ADA is a suitable precursor for HMD, following twice the cascade of reduction and amination as outlined above. Furthermore, the fermentative production of CPL would likely proceed via 6-aminocaproic acid (6-ACA), which could also be converted into HMD following the cascade of reduction and amination. Accordingly, the original pathway design and several patents of Genomatica cover these three C6 target molecules.

The main approach for the synthesis of the C6-building block refers to the condensation of succinyl-CoA with acetyl-CoA. The initial steps of the pathway are designed to generate these two precursors. The formation of acetyl-CoA via pyruvate dehydrogenase generates carbon dioxide as by-product. This potential carbon loss is recycled since the precursor for succinyl-CoA in the citric acid cycle can be obtained via the carboxylation of pyruvate, whereby one ATP is invested. Carbon dioxide may also be recycled via the PEP carboxykinase, however, in this context, the pyruvate carboxylase is chosen. Succinyl-CoA and acetyl-CoA are condensed to 3-oxoadipyl-CoA, facilitated by the succinyl-CoA-acetyl-CoA transferase (compare WO 2014/099725 A1; Table 20.1):

In the following steps, the keto function of the obtained C6 building block, 3-oxoadipyl-CoA, is removed according to the conventional steps in the degradation or the synthesis of fatty acids. Hereby, adipyl-CoA appears as pathway intermediate, and the overall pathway is sometimes referred to as the adipyl-CoA route (Turk et al. 2016) or the reverse adipate degradation pathway (Genomatica patent 2012). The two carboxylic acid functions of adipyl-CoA need then to be reduced to the aldehyde level to prepare the transamination. The conversion of the 3-oxo-function of 3-oxo-adipyl-CoA may be pursued at the activated C6-thioester ("CoA route", left column) or the free acid ("acid route," right column). The acid route may provide an advantage in metabolic engineering, though it slightly diminishes the theoretical carbon yield since the energy loss in the hydrolysis of the thioester to convert 3-oxo-adipyl-CoA to 3-oxoadipate cannot be recovered. For the generation of adipyl-CoA from adipate, additional ATP is needed (Figure 20.1). The thermodynamics of the route indicate that the initial condensation of succinyl-CoA and acetyl-CoA is not favored and has a Gibbs free energy of 26.8 KJ/mol. However, the pathway should be possible if the subsequent steps toward 3-hydroxyadipyl-CoA and adipyl-CoA can efficiently be catalyzed to generate a sufficient pull (Kallscheuer et al. 2017; Figure 20.1):

A direct reduction of organic acids to the respective aldehyde with NADH as electron donor is difficult in biochemical systems given the reduction potentials of −580 mV for organic acids to aldehydes and −320 mV for NADH and NADPH. The biochemical conversion of the acid function to the aldehyde function defines a critical step in the pathway toward HMD. The pathway described by Genomatica proceeds via an activated acid thioester with CoA to facilitate the reduction to the aldehyde level. The first acid thioester is given after the condensation of succinyl-CoA and acetyl-CoA and allows the reduction in the conversion of adipyl-CoA to adipate semialdehyde (i.e. formylvaleric acid, FVA).

The second acid thioester is generated in the reaction of 6-aminocaproate to 6-aminocaproyl-CoA via the 6-aminocaproyl-CoA synthase according to the conversion of succinate to succinyl-CoA in the reductive citric acid cycle. One mol ATP needs to be invested to ensure an overall negative Gibbs free energy of the combined ATP hydrolysis to ADP and Pi. The introduced aldehyde function is used as preparation step for the amination. The integration of the amino function may operate with different amino donors applied in biochemical systems like glutamate or glutamine. The use of glutamate

Table 20.1 Biochemical conversion of glucose into HMD.

Provision of pyruvate	
Glucose + 2 ADP + 2 NAD^+ → 2 Pyruvate + 2 ATP + 2 NADH + 2 H^+ + 2 H_2O	Glycolysis
Provision of acetyl-CoA	
Pyruvate + CoA + NAD^+ → Acetyl-CoA + CO_2 + NADH + H^+	Pyruvate dehydrogenase
Provision of succinyl-CoA	*Reductive citric acid cycle*
Pyruvate + CO_2 + ATP + H_2O → Oxalacetate + ADP + Pi	Pyruvate carboxylase
Oxalacetate + NAD(P)H + H^+ → Malate + $NAD(P)^+$	Malate dehydrogenase
Malate → Fumarate + H_2O	Fumarase
Fumarate + $FADH_2$ → Succinate + FAD	Succinate dehydrogenase
Succinate + CoA + GTP → Succinyl-CoA + GDP + Pi	Succinyl-CoA synthase
Provision of glutamate	
2 α-KG + 2 NH_3 + 2 NADH + 2 H^+ → 2 Glutamate + 2 NAD^+ + 2 H_2O	Glutamate dehydrogenase
Generation of HMD	
Succinyl-CoA + Acetyl-CoA → 3-Oxoadipyl-CoA + CoA	Succinyl-CoA-acetyl-CoA transferase
3-Oxoadipyl-CoA + NADH + H^+ → 3-Hydroxyadipyl-CoA + NAD^+	3-Hydroxyacyl-CoA dehydrogenase
3-Hydroxyadipyl-CoA → 5-Carboxy-2-pentenoyl-CoA + H_2O	3-Hydroxyadipyl-CoA dehydratase
5-Carboxy-2-pentenoyl-CoA + NADH + H^+ → Adipyl-CoA + NAD^+	5-Carboxy-2-pentenoyl-CoA reductase
Adipyl-CoA + NADH + H^+ → Adipate semialdehyde + CoA + NAD^+	Aldehyde dehydrogenase
Adipate semialdehyde + glutamate → 6-Aminocaproate + α-KG	Transaminase
6-Aminocaproate + ATP + CoA → 6-Aminocaproyl-CoA + ADP + Pi	6-Aminocaproyl-CoA synthase
6-Aminocaproyl-CoA + NADH + H^+ → 6-Aminocaproate-SA + NAD^+ + CoA	6-Aminocaproyl-CoA reductase
6-Aminocaproate SA + Glutamate → HMD + α-KG	Transaminase
Glucose + GTP + 4 NADH + 4 H^+ + $FADH_2$ + 2 NH_3 → HMD + GDP + Pi + 4 NAD^+ + FAD + 5 H_2O	

is chosen in this pathway, which is in the transamination converted into α-ketoglutarate. In order to install an efficient recycling of the amino donor, the reductive amination of α-ketoglutarate with NADH and ammonia is foreseen. Nitrogen in the reduced form of ammonia as an additional educt – besides glucose as the carbon and energy source – is supplemented to the fermentation medium.

Since the reduction equivalents can be interchanged via a transhydrogenase, the NADH and the $FADH_2$ pools may be merged for simplification, and the overall reaction can be stated as follows:

$$\text{Glucose} + \text{GTP} + 5\,\text{NADH} + 5\,\text{H}^+ + 2\,\text{NH}_3 \rightarrow \text{HMD} + \text{GDP} + \text{Pi} + 5\,\text{NAD}^+ + 5\,\text{H}_2\text{O}$$

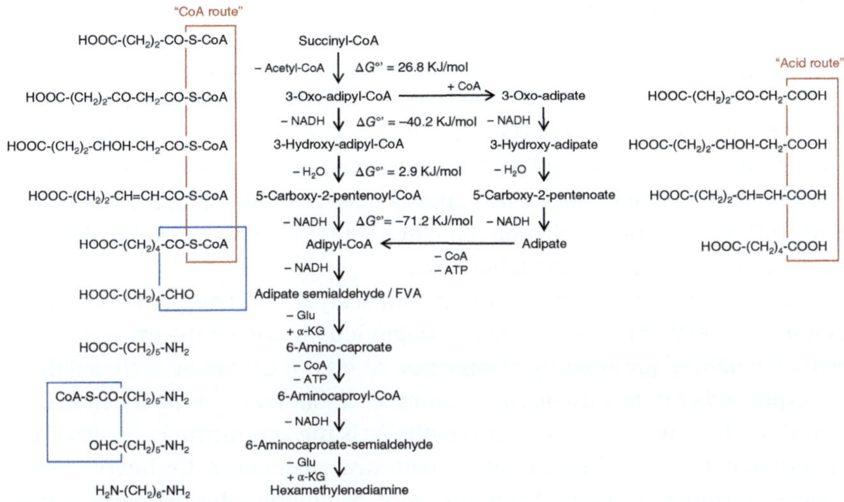

Figure 20.1 Reverse adipate pathway towards HMD based on the condensation of succinyl-CoA and acetyl-CoA. Source: Walter Koch.

It can be directly seen that the gap of reduction equivalents is decisive for the maximum theoretical yield of the reaction. The C6 backbone might indicate that one mol glucose derives one mol HMD. However, the much higher reduction level of HMD compared to glucose implies an additional input of glucose to provide the required electrons. Whereas glucose carries 24 electrons (without counting the electrons of the reduced oxygen), HMD carries altogether 34 electrons (without counting the reduced oxygen and nitrogen). Thus, in case the C6-backbone of glucose is remodeled, 10 additional electrons or 5 reduction equivalents, respectively, need to be supplied, which require 0.416 mol glucose (5/12). The complete oxidation of glucose to carbon dioxide and water provides 12 reduction equivalents. The additional consumption of one mol GTP (ATP) per mol HMD should not incur an additional consumption of glucose since sufficient ATP will be cogenerated via the recycling reduction equivalents in respiratory chain elongation. Thus, the overall reaction equation can be stated as follows:

$$1.42\,\text{Glucose} + 2.00\,\text{NH}_3 + 0.02\,\text{O}_2 \rightarrow \text{HMD} + 2.52\,\text{CO}_2 + 3.52\,\text{H}_2\text{O}$$

In order to balance the reaction equation, a small addition of oxygen is needed. The oxygen input factor can be estimated at about 0.006 t per ton HMD ($0.02 \times 32/116.2$), which is rather minor and should not define a limitation to achieve high productivity. However, oxygen needs to be supplemented carefully to avoid the surplus generation of ATP. The generated reduction equivalents provided in the oxidation of glucose feed need to be efficiently absorbed in the metabolic pathway, which can be supported by a strict limitation of oxygen.

Alternative fermentation routes will be outlined in the following sections: Since the investment of 1.43 mol glucose per mol HMD and the corresponding 0.7 mol HMD per mol glucose are regarded as the theoretical maximum molar yield, the yield of the alternative routes is not calculated in detail. Alternative routes display a higher complexity and likely incur a higher generation of carbon dioxide as by-product.

20.3.2 Fermentation Based on Glucose via 6-Hydroxyhexanoate

A further metabolic pathway has been proposed by Invista with acetyl-CoA as starting point. It initially foresees to generate a C4 intermediate, namely acetoacetyl-CoA, which is then converted to butanoyl-CoA (Invista n.d.). Butanoyl-CoA is then again reacted with acetyl-CoA to obtain 3-oxohexanoyl-CoA as linear C6 intermediate, which is then further processed to finally obtain HMD. The pathway does not imply a decarboxylation in the buildup of HMD ex acetyl-CoA, though it might imply the release of carbon dioxide pending on the generation of acetyl-CoA (Table 20.2):

The overall pathway comprises 14 steps, starting with acetyl-CoA. This pathway length requires high efforts in metabolic engineering and might tend to trigger the generation of by-products given the natural promiscuity of enzymes. At least it cannot be excluded that some of the overexpressed enzymes also partially catalyze similar reactions on functionally analogous substrates. The unique feature of this pathway is the specific functionalization of the second terminal C-atom. Whereas other pathways generate a C6-intermediate with functionalized terminal C-atoms from the very beginning, this pathway starts with a hexanoic acid derivative (i.e. 3-oxohexanoyl-CoA) with terminal methyl function (Figure 20.2):

The functionalization and activation of this methyl function are then foreseen at the level of hexanoate, which shall be hydroxylated with a P450-dependent monooxygenase. This reaction should in principle be possible, though it will further impact the metabolic burden of the pathway. Oxygen can be provided, though the functionally efficient integration of an efficient cytochrome P450 monooxygenase is not straightforward (Wang and Kim 2015; Bernhardt and Urlacher 2014).

Table 20.2 Biochemical conversion of glucose into HMD via 6-hydroxyhexanoate.

2 Acetyl-CoA → Acetoacetyl-CoA + CoA	Acetoacetyl-CoA thiolase
Acetoacetyl-CoA + NADPH + H^+ → 3-Hydroxybutanoyl-CoA + $NADP^+$	3-Acetoacetyl-CoA synthase
3-Hydroxybutanoyl-CoA → Crotonyl-CoA + H_2O	Enoyl-CoA synthase
Crotonyl-CoA + NADH + H^+ → Butanoyl-CoA + $NADP^+$	Trans-2-enoyl-CoA reductase
Butanoyl-CoA + Acetyl-CoA → 3-Oxohexanoyl-CoA + CoA	Acetyl-CoA C-acetyltransferase
3-Oxohexanoyl-CoA + NADPH + H^+ → 3-Hydroxyhexanoyl-CoA + $NADP^+$	3-Oxoacyl-CoA reductase
3-Hydroxyhexanoyl-CoA → Hex-2-enoyl-CoA + H_2O	Enoyl-CoA hydratase
Hex-2-enoyl-CoA + NADPH + H^+ → Hexanoyl-CoA + $NADP^+$	Trans-2-enoyl-CoA reductase
Hexanoyl-CoA + H_2O → Hexanoate + CoA	Thioesterase
Hexanoate + O_2 + NADPH + H^+ → 6-Hydroxyhexanoate + NADP + H_2O	Monooxygenase
6-Hydroxyhexanoate + NAD^+ → Adipate semialdehyde + NADH + H^+	Alcohol dehydrogenase
Adipate semialdehyde + Glutamate → 6-Aminohexanoate + α-Ketoglutarate	ω-Transaminase
6-Aminohexanoate + NADPH + H^+ → 6-Aminohexanal + $NADP^+$	Carboxylate reductase
6-Aminohexanal + Glutamate → HMD + α-Ketoglutarate	Diamine transferase

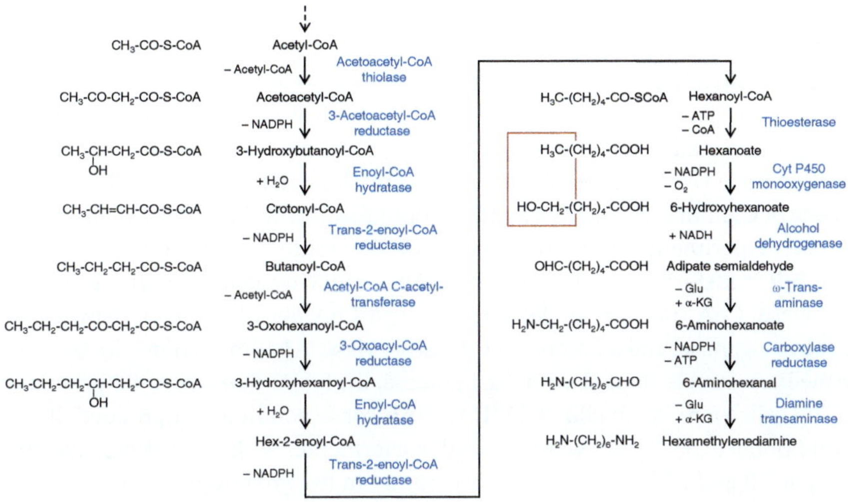

Figure 20.2 Biochemical pathway towards HMD via 6-hydroxyhexanoate. Source: Walter Koch.

20.3.3 Fermentation Based on Glucose via 2-Amino-6-Oxopimelate

An alternative route has been proposed by Bioamber which proceeds via the C7 precursor 2-amino-6-oxopimelate (Bioamber n.d.). The precursor for the backbone, a modified version of the C7 diacid pimelic acid, can in general be constructed in biochemical systems with acetyl-CoA as main starting point. The pathway pursues the removal of the keto-function at C6 via the usual hydrogenation-dehydration-saturation cycle and prepares the first amination via the reduction of the carboxyl function at C7 with a carboxylic acid reductase. It is interesting to note that the second terminal amino function is obtained via decarboxylation (Figure 20.3).

The pathway displays similarities with the routes proposed by Genomatica. The conversion of the keto function of 2-amino-6-oxopimelate into a reduced methylene function might be executed on a backbone with free acid function (left side) or an activated thioester with CoA (right side). The pathway from 2-amino-6-oxopimelate may be installed with

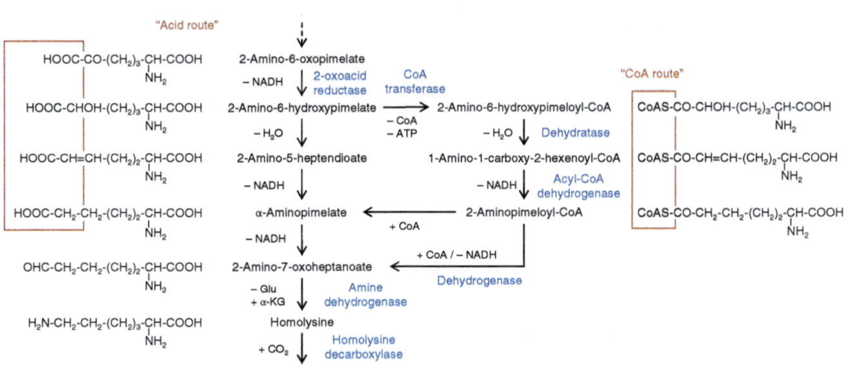

Figure 20.3 Biochemical pathway towards HMD based on 2-amino-6-oxopimelate as starting point. Source: Walter Koch.

different enzymes, and Bioamber is proposing several suitable enzymes for each step (Bioamber n.d.). The theoretical maximum molar yield of the pathway depends on the recycling of carbon dioxide during the buildup of 2-amino-6-oxopimelate. Since finally an α-amino acid is subject to decarboxylation, the theoretical maximum molar yield is compared to the reverse adipate pathway proposed by Genomatica. The pathway design might primarily pursue the objective of offering a technology that is not impeded by the patent landscape of Genomatica and less a further optimization of the carbon yield.

In shake flask experiments, the CoA-version of the route has been implemented in *Escherichia coli* transformed with plasmid pBA901 carrying genes for the enzymes 2-oxoacid reductase, CoA transferase, dehydratase, acyl-CoA dehydrogenase, dehydrogenase, amine dehydrogenase, and a homolysine decarboxylase. After induction with 0.2 mM IPTG in M9 medium supplemented with 2 g/l glucose, an HMD titer above 0.5 g/l could be detected. The cells were incubated at 37 °C for 48 h corresponding to a productivity of 0.01 g/l × h. Additional experiments were pursued in a fermenter with 2 l working capacity, pH maintained at 7.0 and 37 °C incubation temperature in the growth phase and 27 °C in the production phase. Oxygen supply was ensured with 20% of air saturation by adjusting the air feed and the agitation speed. The glucose concentration in the fermenter was maintained in the range of 5–20 g/l. An HMD titer of 0.5 μg/l could be detected after 48 h (Bioamber n.d.).

20.3.4 Fermentation Based on Glucose via α-Ketopimelic Acid

An alternative route towards HMD may be designed, which proceeds via α-ketopimelic acid. The main building block would be obtained starting from α-ketoglutarate, which is subject to two chain elongation cycles with acetyl-CoA as C2 transfer unit. The net result of each cycle accomplishes the integration of a methylene function, and α-ketoglutarate (C5) is thereby converted into α-ketoadipic acid (C6) and subsequently α-ketopimelic acid (C7). One of the two terminal aldehyde functions finally needed for the introduction of the two amino functions is introduced via the decarboxylation of an α-keto acid (Figure 20.4):

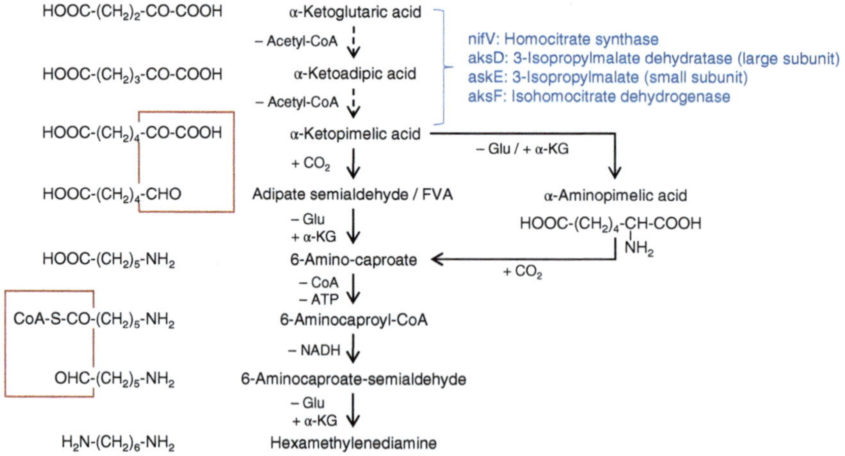

Figure 20.4 Biochemical pathway towards HMD via α-ketopimelic acid. Source: Walter Koch.

The generation of an aldehyde function via the decarboxylation of α-ketoadipic acid to obtain adipate semialdehyde is a known reaction type that is structurally analogous to the conversion of pyruvate into acetaldehyde in the fermentative production of ethanol. This reaction type may be advantaged from a thermodynamics point of view compared to the use of a carboxylic acid reductase or the pathway via an activated acid thioester. However, the conversion of a α-ketoacid into an aldehyde increases the carbon dioxide release concurrent with the biochemical pathway.

Empirical data on the pathway via pimelic acid intermediates toward HMD are not reported. The conversion of α-ketopimelic acid into adipate semialdehyde and α-aminopimelic acid into 6-aminocaproate is claimed in patents of Genomatica (Genomatica patent 2017) and DSM (DSM patent 2011). The overall strategic direction of the patents is likely more in the field of CPL than of HMD. Some steps of the pathway, especially the conversion of α-ketoglutaric acid into α-ketopimelic acid have been applied for the fermentative production of CPL (Chae et al. 2017; Sierra and Alper 2020). The genes encoding the homocitrate synthase from *Azotobacter vinelandii* (*nifV*), the 3-isopropylmalate dehydratase (*aksD*, *aksE*), and the isohomocitrate dehydrogenase (*aksF*) were cloned into *E. coli*, and a general proof of concept for the pathway could be shown given a CPL titer of about 80 μg/l (Chae et al. 2017).

20.3.5 Fermentation Based on Glucose via 6-amino-2-Hydroxy Hexanoic Acid

An alternative route towards HMD based on 6-amino-2-hydroxy hexanoic acid (6-AHEA) has been proposed by DSM (DSM patent 2005). The starting point 6-AHEA will likely also be generated via fermentation, though the patent focuses on the conversion of 6-AHEA to HMD. A suitable precursor would be lysine, whose 2-amino function can be converted into a hydroxyl group to obtain 6-AHEA. The proposed biochemical steps are reasonable given the typical dehydration of alcohols and reduction of double bonds in biochemical systems (Figure 20.5):

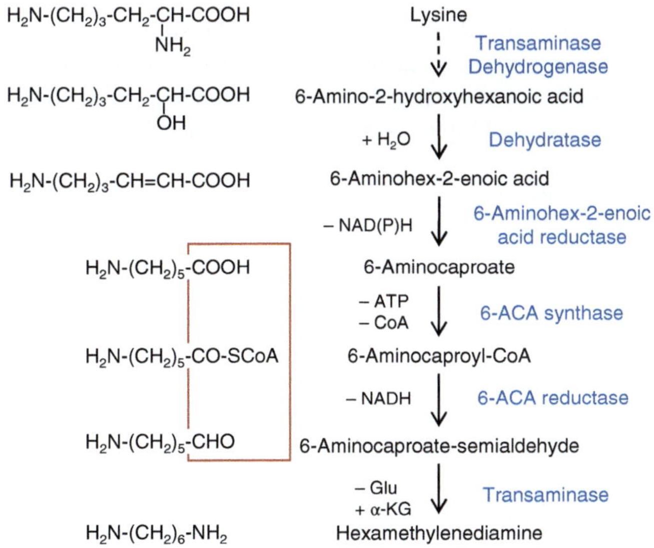

Figure 20.5 Biochemical pathway toward HMD based on 6-amino-2-hydroxy hexanoic acid. Source: Walter Koch.

The precursor 6-AHEA already possesses a terminal amine function. The second amine function is introduced via a conventional transamination with an aldehyde as starting point. The critical reduction of the carboxylic acid function is obtained via the reduction of the activated carboxylic acid thioester.

20.4 Biocatalytic Routes Towards HMD

20.4.1 Biocatalytic Conversion of ADA

HMD may also be accessed via the biocatalytic conversion of ADA. This pathway proceeds via adipic semialdehyde, and the thermodynamically challenging reduction step is not facilitated via the 6-aminocaproyl-CoA synthase – as outlined in the patents of Genomatica – though a carboxylic acid reductase. The mechanism of this enzyme class proceeds via the reaction of the carboxylic acid function with ATP to form with AMP-acyl a first intermediate under the release of pyrophosphate (Thomas et al. 2019). The AMP-acyl intermediate is converted into PPT-acyl thioester (i.e. phosphopantetheine-binding domain of the enzyme) according to the thioester in the activation of the carboxyl function with CoA. The acid thioester is then reduced with $NADP^+$ to obtain the aldehyde. Accordingly, this type of acid-to-aldehyde reduction consumes one mol ATP and generates one mol NADPH. A purified carboxylic acid reductase has also been used for the conversion of the aromatic carboxylic acid function of vanillic acid to vanillin (Li and Rosazza 2000) or may be used for the biocatalytic conversion of ADA to 1,6-hexanediol (Khusnutdinova et al. 2017). The pathway employing a carboxylic acid reductase for the generation of HMD proceeds as follows (Figure 20.6). The conversions at C1 and C6 seem to be independent from each other, and HMD might be equally obtained via the dialdehyde adipaldehyde or 6-ACA (Figure 20.6).

In the second step of the pathway, the aldehyde is converted into an amine via a transaminase, which operates with pyridoxal-5-phosphate as cofactor and catalyzes the transfer of an amino group from the amino donor to the aldehyde or ketone as the amino acceptor. The reaction provides a terminal amine in case an aldehyde is applied as educt or a non-terminal amine in case a ketone is applied as educt (e.g. in the conversion of pyruvate into L-alanine). Using an isolated and purified wild-type carboxylic acid reductase combined

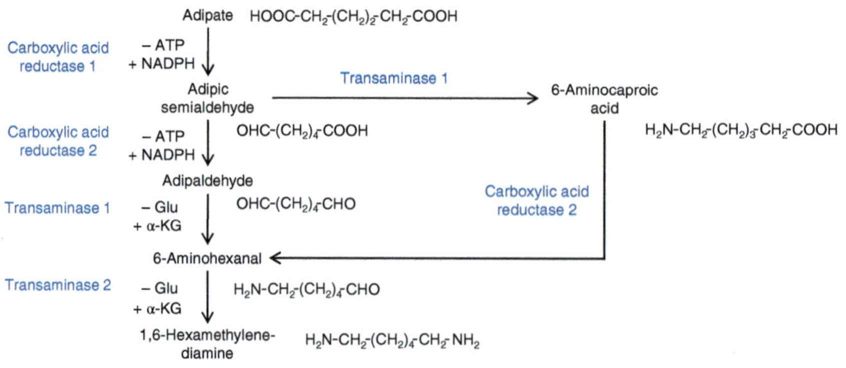

Figure 20.6 Biocatalytic routes to HMD based on ADA. The thioester intermediates appearing in the conversion catalyzed by the carboxylic acid reductase are not shown. Source: Walter Koch.

with an isolated and purified transaminase, a conversion of ADA to 6-ACA with a yield of 95% could be achieved (Fedorchuk et al. 2019). An efficient cofactor recycling system for the carboxylic acid reductase was used, which relies on polyphosphate for the regeneration of AMP to ATP and the oxidation of formate to carbon dioxide for the regeneration of NADPH according to the acid-to-alcohol pathway to 1,6-hexanediol (Khusnutdinova et al. 2017). The amino function transferred by the transaminase was donated by either alanine or glutamate, which were recycled with ammonia and the alanine dehydrogenase and the glutamate dehydrogenase, respectively, which both consume NADPH.

An alternative cofactor regeneration system has been described that is able to convert several aliphatic or aromatic carboxylic acids into the respective aldehydes. AMP is reconverted to ATP using the polyphosphate kinase from *Meiothermus ruber* and *Sinorhizobium melioti,* and NADPH is recycled with the aid of the glucose dehydrogenase oxidizing glucose to gluconic acid. The released pyrophosphate is hydrolyzed by a pyrophosphatase (Strohmeier et al. 2019). The second reduction step addresses the second carboxylic acid function and its implementation turned out to be more difficult. Several mutant carboxylic acid reductases were screened to finally achieve a 50%–75% conversion of 6-ACA into HMD with the CAR L342E protein, the most suitable mutant carboxylic acid reductase. The search for a suitable carboxylic acid reductase exploits the natural broad substrate specificity of carboxylic acid reductases (Kramer et al. 2018). Finally, both steps were combined, and the wild-type carboxylic acid reductase, the mutant reductase CAR L342E, and two different transaminases were incubated to achieve a 30% conversion of ADA into HMD and 70% into 6-ACA (Fedorchuk et al. 2019). An industrial application of this route mainly depends on the recycling of cofactors and the production cost and availability of ADA. Commercial use is not known.

20.4.2 Digression: Biocatalytic Synthesis of ADN via C6 Dialdehyde

HMD might also be accessed via a semi-synthesis combining the biocatalytic production of ADN coupled with the conventional hydrogenation of ADN to HMD. This interesting biotechnological access has been proposed by academia with 1,6-dialdehyde (or the bis-dimethyl acetal) as starting material. The 1,6-dialdehyde would be reacted with hydroxylamine in the presence of HCl and sodium carbonate to generate 1,6-dialdoxime, which is then in the second step biocatalytically converted into ADN (Hinzmann et al. 2021; Betke et al. 2018):

$$\begin{aligned}&\text{OHC–CH}_2\text{–CH}_2\text{–CH}_2\text{–CH}_2\text{–CHO} + 2\,\text{NH}_2\text{OH}\\&\quad\rightarrow \text{HO–N=CH–CH}_2\text{–CH}_2\text{–CH}_2\text{–CH}_2\text{–CH=N–OH} + 2\,\text{H}_2\text{O}\end{aligned}$$

$$\begin{aligned}&\text{HO–N=CH–CH}_2\text{–CH}_2\text{–CH}_2\text{–CH}_2\text{–CH=N–OH}\\&\quad\rightarrow \text{N}{\equiv}\text{C–CH}_2\text{–CH}_2\text{–CH}_2\text{–CH}_2\text{–C}{\equiv}\text{N} + 2\,\text{H}_2\text{O}\end{aligned}$$

Several aldoxime dehydratases were screened in *E. coli* as whole-cell biocatalyst, and especially the enzymes from *Pseudomonas chlororaphis* and *Bacillus sp. OxB-1* turned out to be useful. Starting from an educt concentration of 100 g/l 1,6-dialdoxime, a conversion of 80% and a high selectivity could be achieved. In laboratory experiments on liter-scale, an educt concentration of 50 g/l could be converted to ADN with a yield >99%. The best trials were performed with 50 g of dialdoxime in a 1 l reactor at 30 °C. A whole-cell catalyst with an activity of in total 6000 U managed to facilitate a nearly complete conversion in 27 h. A commercial implementation is likely pending on a cheap supply of the precursor 1,6-dialdehyde, which is proposed to be derived from 1-hexanol or 1,6-hexanediol (Betke et al. 2018).

20.5 Process Design

After a suitable biocatalyst with an efficient metabolic pathway has been constructed, the fermentation protocol needs to be developed. The main question will refer to the neutralization concept given the high alkaline character of HMD in aqueous solution. HMD has two amino functions with corresponding pK_b values of 10.24 and 11.02 (Bednarz et al. 2017). The HMD molecule in aqueous solution is twice protonated at neutral pH (Bednarz et al. 2017), which abstracts protons from water and causes an alkaline pH. A bacterial or an eucaryotic host might display some tolerance towards alkaline conditions, though above 8.0, a negative impact on productivity is difficult to avoid. Accordingly, the medium needs to be neutralized with an acid like HCl or H_2SO_4 to neutralize the hydroxide ions generated in the protonation of the amino functions of HMD:

$$H_2N\text{-}CH_2\text{-}CH_2\text{-}CH_2\text{-}CH_2\text{-}CH_2\text{-}CH_2\text{-}NH_2 + H_2SO_4$$
$$\rightarrow {}^{+}H_3N\text{-}CH_2\text{-}CH_2\text{-}CH_2\text{-}CH_2\text{-}CH_2\text{-}CH_2\text{-}NH_3{}^{+} + SO_4{}^{2-}$$

Since the neutral HMD molecule is finally targeted as a product, a base like KOH, NaOH, or $Ca(OH)_2$ needs to be added in the downstream process, which is rather costly. In case it is assumed that each amino function needs to be neutralized during the fermentation and re-protonated in the downstream process, about 0.69 t NaOH per ton HMD ($2 \times 40/116.2$) is required for neutralization management:

$${}^{+}H_3N\text{-}CH_2\text{-}CH_2\text{-}CH_2\text{-}CH_2\text{-}CH_2\text{-}CH_2\text{-}NH_3{}^{+} + SO_4{}^{2-} + Ca(OH)_2$$
$$\rightarrow H_2N\text{-}CH_2\text{-}CH_2\text{-}CH_2\text{-}CH_2\text{-}CH_2\text{-}CH_2\text{-}NH_2 + CaSO_4 + 2\,H_2O$$

H_2SO_4 and $Ca(OH)_2$ might be chosen as neutralization agents to obtain $CaSO_4$ or gypsum as by-products according to the conventional fermentative production of lactic acid. It can be deduced that about 1.48 t $CaSO_4$ per ton HMD would be obtained (172.1/116.2) and a suitable, local concept for dihydrate gypsum would be needed. Several measures might be conceived to limit neutralization efforts. The product might be removed *in situ* from the fermentation broth via ion exchange technology or electrodialysis. The removal of HMD from an aqueous solution with the aid of an electrically charged membrane has been investigated by Rhone-Poulenc though the application context seems to be recycling (Rhone-Poulenc 1998). Alternatively, HMD might be removed from the fermentation broth with the aid of a reactive extraction (Figure 20.7):

A small share of the broth might be permanently removed from the medium and channeled into a parallel extraction unit. In the separate extraction unit, a solvent would be applied to extract HMD out of the aqueous phase. Since HMD is positively charged at neutral pH, a conventional lipophilic extraction solvent will likely not suffice to pull the positively charged HMD ions out of the hydrophilic water phase. Full deprotonation of HMD would suppose raising the pH above 10, and the overall process would not be

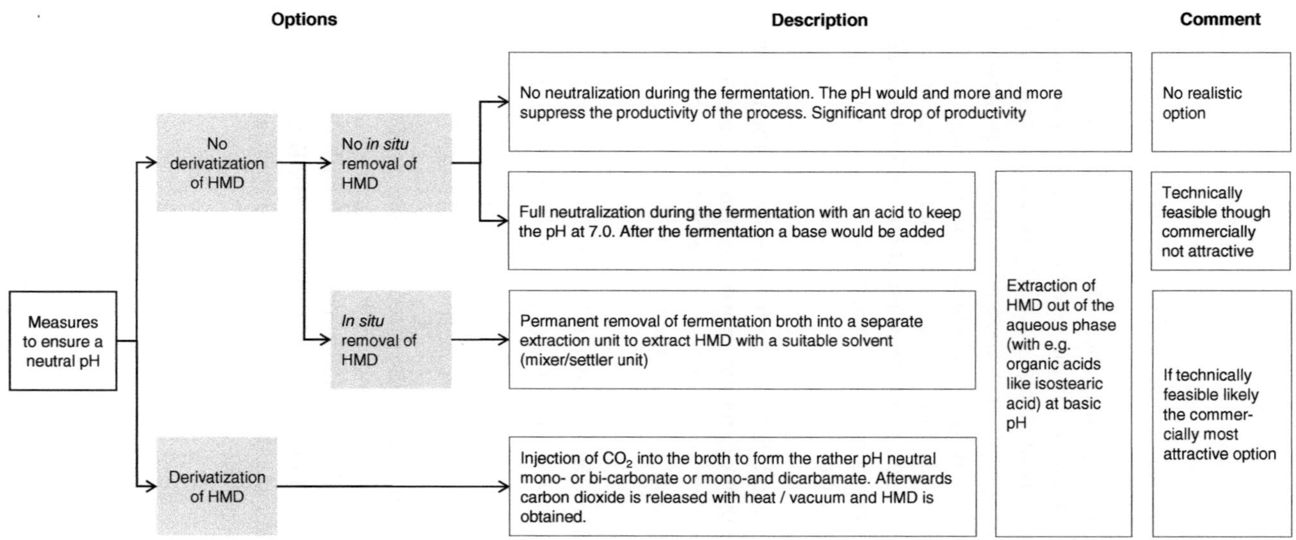

Figure 20.7 Options for pH management in the fermentative production of HMD. Source: Walter Koch.

simplified (Sargent et al. 2021). Accordingly, designed extraction aids have been developed, comprising isostearic acid and di-(2-ethylhexyl)phosphoric acid (D2EHPA), which bear an acid function and are nevertheless sufficiently hydrophobic to form a second phase (Bednarz et al. 2017). In case an efficient extraction is achieved, the HMD product and the specific extraction aids might then be separated by distillation. Alternatively, HMD might be extracted out of the aqueous phase with a solution of NaOH and salt or butanol supported by NaOH (Sargent et al. 2021; Sargent et al. 2022). The aqueous phase might be subjected to a second extraction step and then be rechanneled into the fermentation process.

Another concept would be product derivatization. Hereby, the product would be modified such that a pH neutrality in the fermentation broth would be achieved. This might be done with an additional biochemical step inside the cell, e.g. the transfer of an acetyl function according to the esterification of geraniol to geranyl-acetate (Zada et al. 2018; Liu et al. 2016). Alternatively, the released product might be chemically modified in the fermentation broth by addition of a suitable chemical agent. In case the free base is thereby converted into a neutral intermediate the neutralization efforts might not be needed or at least be significantly reduced. An innovative concept of product derivatization inside the broth has been proposed by Genomatica (Genomatica patent 2017). The basic idea is to inject a gaseous carbon dioxide feed into the medium during the fermentation. The solubility of carbon dioxide or its conversion into bicarbonate might be facilitated with the addition of the enzyme carbonic anhydrase. The released HMD molecule would then be converted into either the mono- or bi-carbonate or carbamate; the reaction equation shown below shows the formation of the respective mono-bicarbonate and mono-carbamate:

$$\begin{aligned}&H_2N\text{-}CH_2\text{-}CH_2\text{-}CH_2\text{-}CH_2\text{-}CH_2\text{-}CH_2\text{-}NH_2 + CO_2\\&\quad\rightarrow [H_2N\text{-}CH_2\text{-}CH_2\text{-}CH_2\text{-}CH_2\text{-}CH_2\text{-}CH_2\text{-}NH_3^+]\ HCO_3^-\end{aligned}$$

$$\begin{aligned}&H_2N\text{-}CH_2\text{-}CH_2\text{-}CH_2\text{-}CH_2\text{-}CH_2\text{-}CH_2\text{-}NH_2 + CO_2\\&\quad\rightarrow H_2N\text{-}CH_2\text{-}CH_2\text{-}CH_2\text{-}CH_2\text{-}CH_2\text{-}CH_2\text{-}NH\text{-}CO_2H\end{aligned}$$

The outlined process as described in the patent assumes a pH of 8–9 in the fermentation broth. At the batch end, the broth is channeled into a separate filtration unit for the removal of biomass. Afterwards, the carbon dioxide is stripped out with air, and the bicarbonate and the carbamate are reconverted into the free base HMD. HMD is then extracted out of the water-HMD mixture with a suitable solvent and further purified to meet the polyamide specification (Genomatica patent 2017).

20.6 Commercial Application

The fermentative production of HMD is evaluated by several industrial players, as indicated by the patent landscape of the last 20 years, including Genomatica, DSM, Invista, and Bioamber. With respect to pathway design and the identification of suitable enzymes, especially the conceptual progress of Genomatica is to be mentioned and reflected in several patents. Genomatica and Covestro announced already a collaboration in 2019 on the development of high-performance materials based on renewable feedstocks

(Genomatica press release 2019). In 2022, a breakthrough in the development of HMD was announced, and the first tons of the product were obtained in high purity. Covestro has the option to obtain a license from Genomatica in case the fermentation technology is finally commercialized (Covestro press release 2022).

References

Arpe, H.-J. (2007). *Industrielle Organische Chemie. Bedeutende Vor- und Zwischenprodukte*, 269–276. Wiley.

Bart, J.C.J. and Cavallaro, S. (2015). Transiting from adipic acid to bioadipic acid. 1, Petroleum-based processes. *Industrial & Engineering Chemistry Research* 54: 1–46. https://doi.org/10.1021/ie5020734.

Bednarz, A., Spieß, A., and Pfennig, A. (2017). Reactive and physical fermentation of bio-based diamines from fermentation media. *Journal of Chemical Technology & Biotechnology* 92: 1817–1824. https://doi.org/10.1002/jctb.5183.

Bernhardt, R. and Urlacher, V.B. (2014). Cytochromes P450 as promising catalysts for biotechnological application: chances and limitations. *Applied Microbiology and Biotechnology* 98: 6185–6203. https://doi.org/10.1007/s00253-014-5767-7.

Betke, T., Maier, M., Gruber-Wölfler, H., and Gröger, H. (2018). Biocatalytic production of adiponitrile and related aliphatic linear α,ω-dinitriles. *Nature Communication* 9: 5112. https://doi.org/10.1038/s41467-018-07434-0.

Bioamber Patent (n.d.), Recombinant bacterial cells producing (S)-2-amino-6-hydroxypimelate, Inventor: Lau, M., K., US 9,890,405 B2.

Boswell, C. (2019). Nylon engineering resin outlook mixed. Healthy demand, but overcapacity for nylon-6 and tight supply for nylon-6,6. *Chemcial Week* 3 (10): 24–25.

Boswell, C. (2022). Invista's Camden nylon-6,6 site to exit fiber, supply merchant resin. *Chemical Week*, 20/27 June 9.

Chae, T.U., Ko, Y.-S., Hwang, K.-S., and Lee, S.Y. (2017). Metabolic engineering of *Escherichia coli* for the production of four-, five-, and six-carbon lactams. *Metabolic Engineering* 43: 82–91. https://doi.org/10.1016/y.mben.2017.04. 001.

Covestro press release (2022). Covestro und Genomatica produzieren wichtigen Chemierohstoff mit Biotechnologie, https://www.covestro.com/press/de/covestro-und-genomatica-produzieren-wichtigen-chemierohstoff-mit-biotechnologie/ (accessed 11.2.2022).

Damayanti, D., Wulandari, L.A., Bagaskoro, A., and Rianjanu, A. Possibility routes for textile recycling technology. *Polymers* 13: 3834. https://doi.org/10.3390/polym13213834.

Dias, E. L., Shoemaker, J. A. W., Boussie, T. R., Murphy, V. J., Process for production of hexamethylenediamine from carbohydrate-containing materials and intermediates therefor, Rennovia Inc., 2014, US 8853458 B2.

DSM patent (2005). Biochemical synthesis of 6-amino caproic acid, Inventors. In: (ed. P.C. Raemakers-Franken, P.M.M. Nossin, P.M. Brandts, et al.). WO 2005/068643.

DSM patent (2011). Preparation of a compound comprising an amine group from an alpha-keto acid, Investors. In: (ed. P.C. Raemakers-Franken, M. Schürmann, M. Müller, et al.). WO 2011/031147.

Duch, M.W. and Allgeier, A.M. (2006). Deactivation of nitrile hydrogenation cataysts: new mechanistic insight from a nylon recycle process. *Applied Catalysis* 318: 190–198. https://doi.org/10.1016/j.apcata.2006.11.003.

Fedorchuk, T.P., Khusnutdinova, A.N., Evdokimova, E. et al. (2019). One-pot biocatalytic transformation of adipic acid to 6-aminocaproic acid and 1,6-hexamethylenediamine using carboxylic acid reductases and transaminases. *Journal of the American Chemical Society* 142: 1038–1048. https://doi.org/10.1021/jacs.9b11761.

Genomatica patent (2009). Microorganisms for the production of adipic acid and other compounds, Inventors: Burgard, A., Pharkya, P., Osterhout, R. E., WO 2009/151728.

Genomatica patent (2010). Microorganisms and methods for the biosynthesis of adipate, hexamethylenediamine and 6-caproic acid, Inventors: Burk, M. J., Burgard, A. P., Osterhout, R. E., Pharkya, P., WO 2010/129936.

Genomatica patent (2011). Methods for increasing product yields, Inventors: Burgard, A. P., Osterhout, R. E., Sun, J., Pharkya, P., US 2011/0207189.

Genomatica patent (2012). Microorganisms for producing 6-aminocaproic acid, Inventors: Burgard, A., Osterhout, R. E., Sun, J., Pharkya, P., WO 2012/177721 A1.

Genomatica patent (2017). Method of producing & processing diamines, Inventors: Suominen, L. H., Galleher, C. J., Japs, M., Burk, M. J., Tracewell, C., US 2017/0369913 A1.

Genomatica patent, Preparation of 6-aminocaproic acid from 5-formyl valeric acidInventors: Raemakers-Franken, P.C., Schurmann, M., Trefzer, A. C., De Wildeman, S. M. A., 2017, US 9,663,805.

Genomatica press release (2019). Covestro and Genomatica launch partnership for sustainable growth, https://www.genomatica.com/covestro-genomatica-partnership/ (accessed 11.2.2022).

Hagen, J. (2015). *Industrial catalysis. A practical approach*. Weinheim, 63.

Harmsen, P.F.H., Hackman, M.M., and Bos, H.L. (2013). Green building blocks for bio-based plastics. *Biofuels, Bioproducts & Biorefining* 8: 306–324. https://doi.org/10.1002/bbb.1468.

Hinzmann, A., Betke, T., Asano, Y., and Gröger, H. (2021). Synthetic processes towards nitriles without the use of cyanide: a water biocatalytic concept based in dehydration of aldoximes in water. *Chemistry Europe* 27: 5313–5321. https://doi.org/10.1002/chem.202001647.

ICI Patent (1964). Process for the manufacture of adiponitrile, Inventors: Mather, J. G., Williams, R. A., patent 3,325,531.

Invista Patent (n.d.). Recombinant bacterial cells producing (S)-2-amino-6-hydroxypimelate, Inventor: Lau, M. L., US 9,890,405 B2.

Kallscheuer, N., Gätgens, J., Lübcke, M. et al. (2017). Improved production of adipate with *Escherichia coli* by reversal of ß-oxidation. *Applied Microbial Biotechnology* 101: 2371–2382. https://doi.org/10.1007/s00253-016-8033-3.

Khusnutdinova, A.N., Flick, R., Popovic, A. et al. (2017). Exploring bacterial carboxylate reductases for the reduction of bifunctional carboxylic acid. *Biotechnology Journal* 12 (11): https://doi.org/10.1002/biot.201600751.

Kramer, L., Hankore, E.D., Liu, Y. et al. (2018). Characterization of carboxylic acid reductases for biocatalytic synthesis of industrial chemicals. *ChemBioChem* 19: 1452–1460. https://doi.org/10.1002/cbic. 201800157.

Lee, Y., Lin, K.-Y.A., Kwon, E.E., and Lee, J. (2019). Renewable routes to monomeric precursors of nylon 66 and nylon 6 from food waste. *Journal of Cleaner Production* 227: 624–633. https://doi.org/10.1016/j.clepro.2019.04.194.

Li, T. and Rosazza, J.P. (2000). Biocatalytic synthesis of vanillin. *Applied and Environmental Microbiology* 66 (2): 684–687.

Liu, W., Xu, X., Zhang, R. et al. (2016). Engineering *Escherichia coli* for high-yield geraniol production with biotransformation of geranyl acetate to geraniol under fed-batch culture. *Biotechnology for Biofuels* 9: 58. https://doi.org/10.1186/s13068-016-0466-5.

Rennovia patent (n.d). Process for production of hexamethylenediamine from 5-hydroxymethylfurfural, Inventors: Dias, E. L., Shoemaker, J. A. W., Boussie, T. R., Murphy, V. J., AU 2017/208286 A1.

Rhone-Poulenc Patent (1998). Method for extracting amine compounds from a liquid medium, Inventors: Canivenc, E., Horbez, D., WO 98/15341.

Sargent, M.J., Slater, N.K.H., Dennis, J.S. et al. (2021). The extraction of hexamethylenediamine from aqueous solution by pH control and salt addition. *Chemical Engineering Journal* 419: 129428. https://doi.org/10.1016/j.cej.2021.129428.

Sargent, M.J., Slater, N.K.H., Dennis, J.S. et al. (2022). The separation of hexamethylenediamine from aqueous solution by solvent extraction with butanol. *Separation and Purification Technology* 288: 120655. https://doi.org/10.1016/j.seppur.2022.120655.

Sierra, A.R.G. and Alper, H.S. (2020). Progress in the metabolic engineering of bio-based lactams and their omega ω-amino acids precursors. *Biotechnology Advances* 43: 107587. https://doi.org/10.1016/j.biotechadv.2020.107587.

Strohmeier, G.A., Eiteljörg, I.C., Schwarz, A., and Winkler, M. (2019). Enzymatic one-step reduction of carboxylates to aldehydes with cell-free regeneration of ATP and NADPH. *Chemistry. A European Journal* 25: 6119–6123. https://doi.org/10.1002/chem.201901147.

Thomas, A., Cutlan, R., Finnigan, W. et al. (2019). Highly thermostable carboxylic acid reductases generated by ancestral sequence reconstruction. *Communications Biology* 2: 249. https://doi.org/10.1038/s42003-019-0677-y.

Turk, S.C.H., Kloosterman, W.P., Ninaber, D.K. et al. (2016). Metabolic engineering toward sustainable production of Nylon-6. *Synthetic Biology* 5: 65–73. https://doi.org/10.1021/acssynbio.5b00129.

Valk, V. (2019). SK Capital focuses on operations. A private equity firm with an emphasis on chemicals and strategy. *Chemical Week* 17 (24): 20–21.

Vollmer, I., Jenks, M.J.F., Roelands, M.C.P. et al. (2020). Beyond mechanical recycling: giving new life to plastic waste. *Angewandte Chemie* 59: 15402–15423. https://doi.org/10.1002/anie.201915651.

Wang, C. and Kim, S.-W. (2015). Shaking up ancient scents: insights into santalol synthesis in engineered *Escherichia coli*. *Process Biochemistry* 50: 1177–1183. https://doi.org/10.1016/j.procbio.2015.04.018.

Wang, X., Gao, S., Wang, J. et al. (2021). The production of biobased amines from renewable carbon sources: current advances and perspectives. *Chinese Journal of Chemical Engineering* 30: 4–13. https://doi.org/10.1016/j.cjche.2020.12.009.

Zada, B., Wang, C., Park, J.-B. et al. (2018). Metabolic engineering of Escherichia coli for production of mixed isoprenoid alcohols and their derivatives. *Biotechnology for Biofuels* 11: 210. https://doi.org/10.1186/s13068-018-1210-0.

Zhou, J., Chen, Y., Zhang, Y. et al. (2021). *Biotransformation of nylon-6,6 hydrolysate to bacterial cellulose*, vol. 23, 7805. Green Chemistry. https://doi.org/10.1039/d1gc02129j.

21

Caprolactam and 6-Aminocaproic Acid

21.1 Application

Caprolactam (CPL) is a cyclic amide with the formula $C_6H_{11}NO$ and a molecular weight of 113.2 g/mol. It is more or less exclusively used for the production of the plastic PA-6 ("Perlon") via ring-opening chain polymerization catalyzed with water. PA-6, is besides PA-66, the second high-volume polyamide with a global production volume of about 6 Mio. t (Zong et al. 2017). The polymerization of CPL, the cyclic anhydride of 6-amino caproic acid (6-ACA), achieves a high conversion since exclusively a minor amount of water is needed to initiate the reaction, and no additional molecules of water are generated in the chain elongation process. PA-6 can be formed as a fiber and is applied in engineering plastics, carpets, and textiles. CPL may also be used for producing the amino acid lysine, though the chemical synthesis of lysine has been replaced by fermentation with *Corynebacterium glutamicum* (Sahm et al. 2013; Arpe 2007).

21.2 Chemical Production of CPL

The current production of CPL is, to a great extent, based on petrochemistry. The final step comprises the conversion of cyclohexanone-oxime into CPL, which is referred to as the "Beckmann reaction," reminding us that E. Beckmann discovered the reaction in 1886. Hereby, cyclohexanone-oxime is subjected to a liquid-phase cyclization at a temperature of 90–120 °C in the presence of highly concentrated sulfuric acid or oleum, respectively. The applied oleum is afterwards neutralized with ammonia and ammonium sulfate is obtained as by-product. The yield is in the range of 98% (Arpe 2007).

$$\text{Cyclohexanone} - \text{oxime} + H_2SO_4 + 2\,NH_3 \rightarrow \text{Caprolactam} + \left(NH_4\right)_2SO_4$$

The Beckmann reaction may be pursued as liquid-phase reaction with oleum or as gas-phase reaction without oleum. The mechanism of the liquid-phase reaction is not yet fully understood (Nguyen et al. 1997), and the reaction medium may be regarded as ionic liquid (Fábos et al. 2008). The gas-phase process has been developed to avoid ammonium sulfate as by-product. The conversion of the gas-phase process is higher than 99%, and the yield is reported at 96.5% (Zong et al. 2017). However, the market share of the gas-phase technology is very small, and disadvantages compared to the traditional

Pathway Design for Industrial Fermentation, First Edition. Walter Koch.

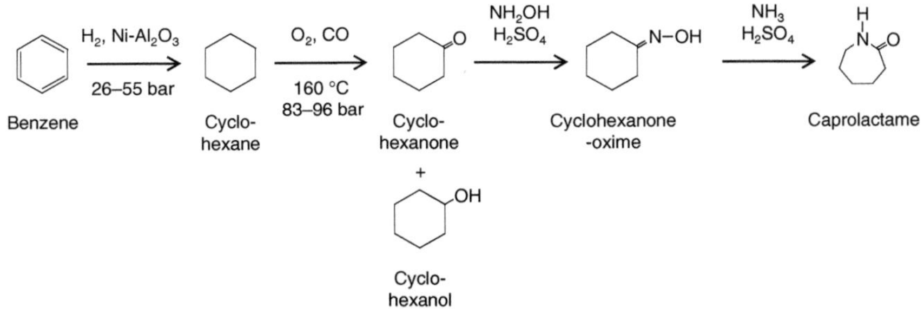

Figure 21.1 Chemical synthesis of CPL starting with benzene. Source: Walter Koch.

liquid-phase Beckmann seem to block a broader application (Zong et al. 2017). The precursor cyclohexanone-oxime is derived from cyclohexanone (Figure 21.1).

A mixture of cyclohexanone and cyclohexanol is derived from the oxidation process of cyclohexane with air under moderate conditions. Since exclusively cyclohexanone is suitable for the synthesis of CPL, cyclohexanol might be subject to catalytic dehydrogenation or be used for other purposes. Cyclohexane is obtained through the hydrogenation of benzene. Different technologies may be used for producing cyclohexanone-oxime and the introduction of the oxime function. A main route applies hydroxylammonium sulfate (Arpe 2007):

$$\text{Cyclohexanone} + \text{NH}_2\text{OH} \times \text{H}_2\text{SO}_4 \rightarrow \text{Cyclohexanone} - \text{oxime} + \text{H}_2\text{O} + \text{H}_2\text{SO}_4$$

A biobased synthesis of CPL had been pursued by the biotech company Draths, which proposed a thermal cyclization and intramolecular dehydration of lysine to obtain amino-caprolactam. The intermediate amino-caprolactam should then be reacted with hydrogen to obtain CPL and ammonia. Draths has been acquired by Amyris in 2011 (http://Greencarcongress.com 2011). It is not known whether the technology is commercially available.

Alternatively, the recycling of PA-6 has received significant management attention in the last few years since it would allow the avoidance of waste accumulation in the environment. Several biocatalytic (Tournier et al. 2023) and chemical technologies to depolymerize and convert PA-6 into CPL are under evaluation (Wursthorn et al. 2023; Kumar et al. 2020), and it remains to be seen inasmuch as recycling or fermentation will finally contribute toward the expected replacement of petrochemistry.

21.3 Metabolic Engineering for Fermentation Technology via Adipyl-CoA

21.3.1 Fermentation Toward CPL Based on Glucose via Adipyl-CoA

According to the metabolic pathway toward hexamethylenediamine (HMD), the most promising pathway toward CPL starts with the condensation of succinyl-CoA and acetyl-CoA to derive 3-oxoadipyl-CoA (Genomatica patents 2009, 2010, 2011, 2012). 3-oxoadipyl-CoA would then be converted into 3-hydroxyadipyl-CoA and 6-aminocaproic acid (6-ACA) (Table 21.1).

The attractiveness of this route relates to the recycling of carbon dioxide in the generation of succinyl-CoA. The C4 building block succinyl-CoA is derived from a carboxylation

Table 21.1 Biochemical pathway for the conversion of glucose into CPL.

Provision of pyruvate	
Glucose + 2 ADP + 2 Pi + 2 NAD^+ → 2 Pyruvate + 2 ATP + 2 NADH + 2 H^+ + 2 H_2O	
Provision of acetyl-CoA	
Pyruvate + CoA + NAD^+ → Acetyl-CoA + CO_2 + NADH	Oxidative pyruvate decarboxylase
Provision of Succinyl-CoA	
Pyruvate + CO_2 + ATP → Oxalacetate + ADP + Pi	Pyruvate carboxylase
Oxalacetate + NAD(P)H → Malate + NAD^+	Malate dehydrogenase
Malate → Fumarate + H_2O	Fumarase
Fumarate + $FADH_2$ → Succinate + FAD	Succinate dehydrogenase
Succinate + CoA + GTP → Succinyl-CoA + ADP + Pi	Succinyl-CoA Synthase
Provision of Glutamate	
α-KG + NH_4^+ + NADH → Glutamate + NAD^+	Glutamate dehydrogenase
Generation of HMD	
Succinyl-CoA + Acetyl-CoA → 3-Oxoadipyl-CoA + CoA	Succinyl-CoA-acetyl-CoA transferase
3-Oxoadipyl-CoA + NADH → 3-Hydroxyadipyl-CoA + NAD^+	3-hydroxyacyl-CoA dehydrogenase
3-Hydroxyadipyl-CoA → 5-Carboxy-2-pentenoyl-CoA + H_2O	3-hydroxyadipyl-CoA dehydratase
5-Carboxy-2-pentenoyl-CoA + NADH → Adipyl-CoA + NAD^+	5-carboxy-2-pentenoyl-CoA reductase
Adipyl-CoA + NADH + H^+ → Adipate semialdehyde + NAD^+	Aldehyde dehydrogenase
Adipate semialdehyde + glutamate → 6-Aminocaproate + α-KG	Transaminase
6-aminocaproate + ATP + CoA → 6-Aminocaproyl-CoA + ADP + Pi	6-Aminocaproyl-CoA synthase
6-aminocaproyl-CoA + H_2O → Caprolactam + CoA	Spontaneous cyclization
Glucose + GTP + 2 NADH + 2 H^+ + $FADH_2$ + NH_3 → CPL + GDP + Pi + 2 NAD^+ + FAD	

of pyruvate facilitated by the pyruvate carboxylase – alternatively by the carboxylation of PEP facilitated by the PEP carboxylase or the PEP carboxykinase – and thereby recycles a carbon dioxide molecule generated in the synthesis of acetyl-CoA. Additionally, the critical thermodynamics of the condensation of succinyl-CoA and acetyl-CoA with a Gibbs free energy of $\Delta G°' = 26.8$ KJ/mol can be overcome in the subsequent steps (Kallscheuer et al. 2017; Figure 21.2).

The conversion of glucose into CPL requires that per mol glucose, six additional electrons need to be provided since glucose bears 24 electrons and CPL bears 30 electrons (without counting the reduced oxygen and nitrogen). Accordingly, three reduction equivalents in the

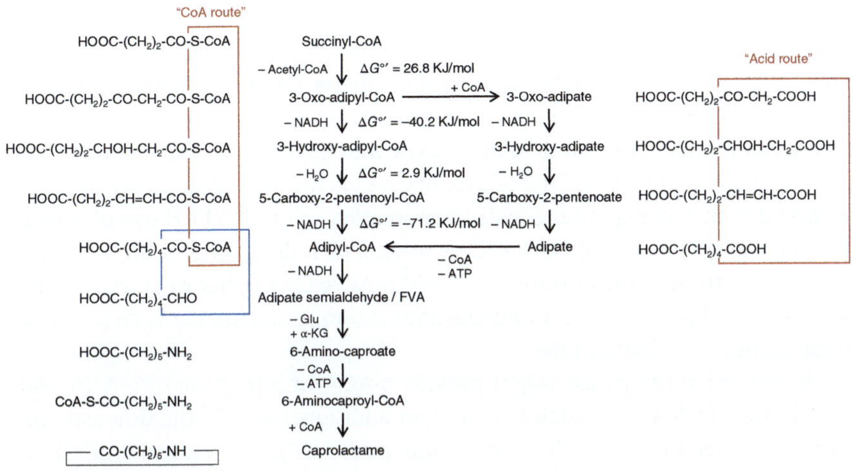

Figure 21.2 Reverse adipate pathway toward caprolactam. Source: Walter Koch.

form of NADH and $FADH_2$ appear on the left side of the biochemical net equation of the reaction. These three reduction equivalents imply that additional 0.25 mol glucose (3/12) per mol CPL needs to be oxidized into carbon dioxide and water to generate the reduction equivalents. Since it may be assumed that the breakdown of the 0.25 mol glucose also provides 1 mol energy equivalents, it can be concluded that in total 1.25 mol glucose is needed per mol CPL, and the theoretical maximum molar yield may be estimated at 0.8 mol CPL per mol glucose. The process may be designed as aerobic and a minor amount of glucose will also be needed to account for cellular maintenance and the formation of biomass.

21.3.2 Pathway Engineering Toward 6-ACA or CPL: The Cyclization

The metabolic pathway tries to obtain CPL as final product and would thereby allow a "drop-in" of fermentative CPL into the existing chemical value chain. It is known that bacteria are able to utilize CPL as their sole carbon source and form 6-ACA in the first step (Baxi et al. 2019). The reverse reaction, the spontaneous cyclization of 6-ACA, is reported to be endothermic and requires energy (Chae et al. 2017):

$$\text{6-ACA} \rightarrow \text{CPL} + H_2O \qquad \text{Endothermic reaction with 21.0 KJ/mol}$$

Thus, a direct biochemical conversion of 6-ACA into CPL via a potential "amidohydrolase" seems challenging (Genomatica patent 2011). It may be accomplished indirectly with the addition of an activation step of 6-ACA to derive 6-Aminocaproyl-CoA (6-ACA-CoA) and might be facilitated with the ≫investment≪ of ATP or acetyl-phosphate (acetyl-P), which provide a negative Gibbs free energy $\Delta G^{\circ\prime}$ of −30.5 KJ/mol or −43.1 KJ/mol, respectively (Voet et al. 2016).

$$\text{6-ACA} + \text{CoA} + \text{ATP} \rightarrow \text{6-ACA-CoA} + \text{ADP} + \text{Pi} + 2\,H_2O$$

$$\text{6-ACA} + \text{CoA} + \text{Acetyl-P} \rightarrow \text{6-ACA-CoA} + \text{Acetate} + \text{Pi}$$

An activation of 6-ACA is in principle possible, though the conversion defines a hurdle for a potential commercial application. There are several enzymes known to be able to convert

6-ACA into 6-ACA-CoA. These enzymes comprise the lipase B of *Candida antarctica*, the acyl-CoA ligase from *Streptomyces aizunensis*, the crotonobetaine CoA ligase of *Escherichia coli*, the β-alanine CoA transferase of *Clostridium propionicum*, and the 3-hydroxybutyrate dehydrogenase from *Citrobacter freundii* (Sierra and Alper 2020). For the first proof of concept to generate fermentative CPL, the β-alanine CoA transferase of *C. propionicum* has been applied. The formation of CPL and the release of CoA via the hydrolytic cleavage of the thioester bond would then occur spontaneously (Genomatica patent 2011). It was observed in experiments that the cyclization of 6-ACA-CoA toward CPL does occur with a conversion rate of 59% starting from a concentration of 1 mM 6-ACA-CoA (Chae et al. 2017). The conversion is likely not sufficient for a commercial application, and a full conversion might be achieved with a longer incubation time.

Otherwise, the fermentation process might provide 6-ACA or a mixture of 6-ACA and CPL as products in the broth at the batch end, and an additional step in the downstream processing might be needed to ensure a full conversion toward CPL. A chemical cyclization of 6-ACA is possible, and a protocol is described in a patent to derive 6-ACA via fermentative conversion of 6-aminohex-2-enoic acid (6-AHEA). The process proceeds with superheated steam at a temperature of 270–400 °C and a pressure of 5–20 bar in the absence of a catalyst (DSM patent 2005).

21.3.3 Fermentation Toward 6-ACA Based on Glucose via α-Ketopimelic Acid

The evaluation of fermentation technology toward CPL has been pursued by DSM, which formerly operated its own chemical CPL production, which later on was sold (EOG Asia 2018; Beerthuis et al. 2015; Figure 21.5). R&D activities were driven by the promise of reduced greenhouse gas emissions in case the petrochemically based production could be replaced by biobased and renewable precursors. Proof of concept could be achieved for the pathway via α-ketopimelate and 6-ACA (Turk et al. 2016). α-Ketopimelateα is an intermediate in the biosynthesis of coenzyme B in methanogenic bacteria. It is derived by two subsequent chain elongation cycles in which each an α-ketoacid with n methylene functions is converted into an α-ketoacid with $n + 1$ methylene functions with the investment of acetyl-CoA and the release of carbon dioxide in each cycle (Figure 21.3).

The pathway is implemented with basically four enzymes. The NifV enzyme functions as homocitrate synthase and condenses the α-ketoacid with acetyl-CoA to obtain an α-keto dicarboxylic acid as intermediate. This reaction is structurally similar to the condensation of oxalacetate and acetyl-CoA in the first step of the citric acid cycle. The enzymes AksD and AksE catalyze the isomerization of the intermediate via dehydration and rehydration. The final decarboxylation of the chain elongation cycle is facilitated by the enzyme AksF, which is homologs to the isocitrate dehydrogenase (Zhou et al. 2015). The chain elongation toward α-ketopimelic acid in Archaea is catalyzed by the enzymes AksA, AksD, AksE, and AksF. Since these enzymes do not have a selective substrate specificity but act uniquely on substrates with varying chain lengths, alternative enzymes were screened. The final constructs providing the best results contained the homocitrate synthase NifV from *Azotobacter vinelandii* and the *aks* genes to encode AksD, AksE, and AksF from *Methanococcus aeolicus*. The aminotransferase Vfl is derived from *Vibrio fluvialis* (Figure 21.4).

The conversion of α-ketopimelic acid into 6-ACA may proceed via adipate semialdehyde or α-aminopimelic acid. In the metabolic route via adipate semialdehyde, the aldehyde

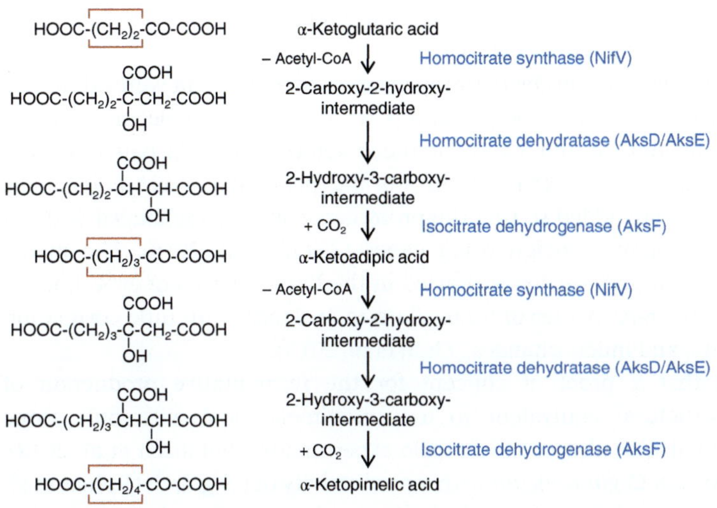

Figure 21.3 Chain elongation from α-ketoglutarate to α-ketopimelic acid. The dehydratase enzyme also has a hydratase activity and shifts the hydroxyl function within the molecule. The name of the dehydrogenase (AksF) is misleading since the enzyme catalyzes a non-oxidative decarboxylation. Source: Walter Koch.

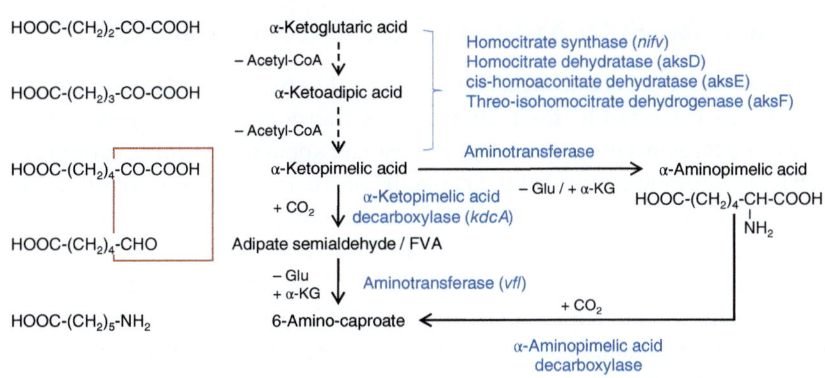

Figure 21.4 α-Ketopimelate pathwayα toward CPL. Source: Walter Koch.

function needed for the generation of an amino function via transamination is not obtained by the reduction of a carboxylic acid though the decarboxylation of an α-ketoacid. The order of the final catalytic steps toward 6-ACA, the decarboxylation, and the transamination, can be reversed. In the experiments, the route via adipate semialdehyde was executed. The selected decarboxylase KdcA originates from *Lactococcus lactis* and displays a K_M value of 53 Mm toward α-ketopimelic acid and 71 Mm toward α-ketoadipate. The host selection mainly relied on product tolerance. Whereas 100 g/l 6-ACA caused a growth rate reduction of 21% in *Saccharomyces cerevisiae*, no growth reduction was observed for *E. coli*. Accordingly, the genes were cloned into the host *E. coli*, which was fermented at 30 °C and a pH of 6.7. The oxygen concentration during the fermentation was controlled by adjusting the stirrer speed. After the target oxygen concentration was reached, the production phase was started with 0.1 mM IPTG. In lab-scale batch fermentation, a 6-ACA titer of 0.16 g/l was obtained. The total amount of 6-ACA and related compounds achieved a titer of 2 g/l in lab-scale fed-batch fermentations (Turk et al. 2016).

The high amounts of by-products in the pathway via α-ketopimelic acid were addressed in further experiments by another academic group with advanced molecular biology approach. The genes encoding the enzymes for the α-ketopimelic acid pathway were organized as monocistronic units under the control of independent T7 RNA polymerase promoters, and insulators were added so that the promoters could be exchanged without interfering with the expression of neighboring genes (Zhou et al. 2015). The media components Fe^{3+}, vitamin B_1 and Mg^{2+} were varied in the library design of experiments. The experiments provided a 6-ACA titer of 5.7 mg/l up to 46.8 mg/l with little correlation between titer, growth rate, and media changes (Zhou et al. 2015).

It should be added that a proof of concept for the fermentative production of 5-aminovalerate, the structural equivalent to α-aminopimelic acid with only three methylene groups, could also be shown in lab-scale experiments (Wendisch et al. 2018). Titers achieved up to 28 g/l in *C. glutamicum* with a productivity of 0.9 g/l × h (Rohles et al. 2016) and 40 g/l with a productivity of 0.54 g/l × h (Joo et al. 2017). 5-Aminovalerate is a lysine derivative, and the metabolic engineering strategy started with a high-producing lysine strain and added the respective genes to obtain the target product. There is no direct pathway similarity to the synthesis of α-aminocaproic acid via α-aminopimelic acid.

21.3.4 Fermentation Toward CPL Based on Glucose via α-Ketopimelic Acid

The pathway as outlined **α**above could be further extended. The 6-ACA needs to be activated to 6-aminocaproyl-CoA catalyzed by the β-alanine CoA transferase from *C. propionicum* (Chae et al. 2017). The fermentation used *E. coli* as host and applied glycerol as carbon and energy source. The medium was buffered to keep a pH of 7.0 and cultivated at 30 °C. The production phase was initiated with 0.2 mM IPTG, and the stirring velocity was reduced to 120 rpm for limited oxygen transfer. Cells were cultivated for 72 h and a titer of 80 μg/l CPL could be achieved. The productivity was 9.5 μg/l × h given a batch duration of about 8 h. The carbon yield was 3.88 μg CPL per g glycerol (Chae et al. 2017). These results are far below the threshold of a commercial application, though they show a first proof of principle for the fermentative production of CPL.

21.3.5 Fermentative Production of ADA and Chemical Conversion Toward CPL

A semi-synthesis of CPL starting with fermentative adipic acid (ADA) to be chemically converted into CPL has been proposed by Bioamber, a company mainly known for its endeavor to explore fermentative succinic acid technology. The route toward CPL starts with adipic acid, which already possesses the right C6 backbone for the synthesis of CPL (http://Greencarcongress.com 2011). The chemical step comprises the conversion of adipic acid into CPL in the presence of ammonia and hydrogen, which will likely proceed via 6-ACA as an intermediate. It is to be kept in mind that the fermentative production of adipic acid does not directly provide the pure and protonated acid but pending on the base a solution of an adipic acid salt (i.e. monoammonium adipate and diammonium adipate). In case the pH is kept around 7.0 with ammonia, the respective ammonium salt would be obtained as product. The innovative CPL technology might bear the advantage

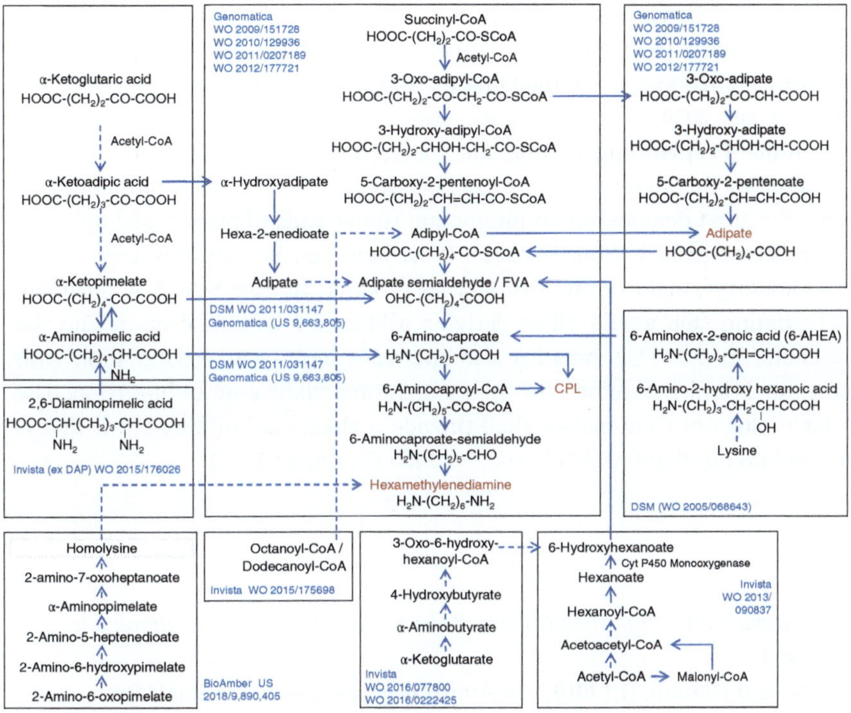

Figure 21.5 Proposed metabolic engineering routes toward ADA, CPL, and HMD.

that the adipic acid, obtained as ammonium adipate in the fermentation broth, would not need to be protonated and isolated as adipic acid, which would add additional steps and costs. Instead, it might be converted with hydrogen and a hydrogenation catalyst to yield CPL. Here a free line as spacer to separate the following sentences as a new section: In general, an extensive and impressing patent landscape has been developed in the last two decades which propose different pathways towards CPL, HMD and ADA as fermentative target products. The biggest contributions have probably been provided by the biotech company Genomatica, especially via their main idea to condense acetyl-CoA and succinyl-CoA. Though also Invista, DSM, BioAmber designed innovative routes toward linear C6 fermentation products. A non-comprehensive overview is given in Figure 21.5.

21.4 Industrial Application

Current CPL technology is based on petrochemistry and mainly proceeds via cyclohexanone and cyclohexanone-oxime, which are then reacted according to the Beckmann process to derive CPL. It has been announced by the biotechnology company Genomatica and several industrial players in the PA-6 value chain that a fermentative technology toward a biobased CPL is under development (globesnewswire 2018). Their activities are united in the project "Effective," which is supported by the EU's Horizon 2020 program. The project has the objective to "develop a sustainable nylon made from biobased CPL produced using renewable feedstocks". The partners of Genomatica comprise, the sugar producer Südzucker, the polymer specialist Aquafil, and the textile brands and retailers H&M, Carvico, and Vaude (biorrefineria-blogspot.com 2020). Genomatica and Aquafil teamed up in January 2020 to provide the "first ton of bio-nylon precursor" at pilot scale (http://Genomatica.com 2020). Genomatica focuses on the development of fermentation technology, and Aquafil is planning to build and operate a demonstration plant to further develop the technology for high-volume production. In July 2022, it was announced by Genomatica that the "first demonstration production run at scale of plant-based nylon-6" had successfully been finished (Shoaib 2022). It is, however, not entirely clear if the fermentation technology pursues 6-ACA or CPL as product. In case 6-ACA is obtained, a chemical dehydration step would allow deriving a biobased CPL as drop-in into the existing value chain. In 2021, Genomatica also entered a multi-year collaboration with the Canadian textile brand Lululemon for the joint commercialization of bio-nylon. The fermentation technology of Genomatica shall provide a chemical building block that is then incorporated into Lululemon's fabric supply chain (Kennedy 2021).

References

Arpe, H.-J. (2007). *Industrielle Organische Chemie. Bedeutende Vor- und Zwischenprodukte.* Wiley p. 278ff, 288ff.

Baxi, N.N., Patel, S., and Hansoti, D. (2019). An *Arthrobacter citreus* strain suitable for degrading ε-caprolactam in polyamide waste and accumulation of glutamic acid. *AMB Express* 9: 161. https://doi.org/10.1186/s13568-019-0887-1.

Beerthuis, R., Rothenberg, G., and Shiju, N.R. (2015). Catalytic routes towards acrylic acid, adipic acid and ε-caprolactam starting from biorenewables. *Green Chemistry* 17: 1341–1361. https://doi.org/10.1039/C4G02076F.

Biorefinery.com (2020). Genomatica produces the world's first ton of renewable nylon intermediate. https://biorrefineria.blogspot.com/2020/02/genomatica-produces-worlds-first-ton-of-renewable-nylon-intermediate-caprolactam.html (accessed 11 March 2022).

Chae, T.U., Ko, Y.-S., Hwang, K.-S., and Lee, S.Y. (2017). Metabolic engineering of *Escherichia coli* for the production of four-, five-, and six-carbon lactams. *Metabolic Engineering* 43: 82–91. https://doi.org/10.1016/y.mben.2017.04.001.

DSM patent (2005). Biochemical synthesis of 6-amino caproic acid, Inventors. In: (ed. P.C. Raemakers-Franken, P.M.M. Nossin, P.M. Brandts, et al.). WO 2005/068643.

EOG Asia, DSM sells Fibrant's caprolactam business to Highsun, 2018, source: https://www.eog-asia.com/dsm-sells-fibrants-caprolactam-business-to-highsun, (accessed 11 March 2022).

Fábos, V., Lantos, D., Bodor, A. et al. (2008). ε-Caprolactamium hydrogen sulfate: an ionic liquid used for decades in the large-scale production of ε-caprolactam. *ChemSusChem* 1: 189–192.

Genomatica patent (2009). Microorganisms for the production of adipic acid and other compounds, Inventors: Burgard, A., Pharkya, P., Osterhout, R. E., WO 2009/151728.

Genomatica patent (2010). Microorganisms and methods for the biosynthesis of adipate, hexamethylenediamine and 6-caproic acid, Inventors: Burk, M. J., Burgard, A. P., Osterhout, R. E., Pharkya, P., WO 2010/129936.

Genomatica patent (2011). Methods for increasing product yields, Inventors: Burgard, A. P., Osterhout, R. E., Sun, J., Pharkya, P., US 2011/0207189.

Genomatica patent (2012). Microorganisms for producing 6-aminocaproic acid, Inventors: Burgard, A., Osterhout, R. E., Sun, J., Pharkya, P., WO 2012/177721 A1.

Globesnewswire, Aquafil and Genomatica collaborate on project Effective to increase adoption of sustainable materials, 2018, https://www.globenewswire.com/news-release/2018/05/22/1509741/0/en/Aquafil-and-Geno-matica-Collaborate-on-Project-EFFECTIVE-to-Increase-Adoption-of-Sustainable-Materials.html, (accessed 02 March 2022).

Genomatica.com, Genomatica to scale Bio-nylon 50-fold with Aquafil to meet surging brand demand, 2020, https://www.genomatica.com/bio-nylon-scaling-50x-to-support-global-brands, (accessed 02 March 2022).

Greencarcongress.com, Amyris to buy Draths, 2011, source: http://www.greencarcongress.com/2011/11/amyris-20111101.html, (accessed 25 Feb 2020).

Joo, J.C., Oh, Y.H., Yu, J.H. et al. (2017). Production of 5-aminovaleric acid in recombinant *Corynebacterium glutamicum* strains from a *Miscanthus hydrolysate* solution prepared by a newly developed *Miscanthus* hydrolysis process. *Bioresource Technology* 245: 1692–1700. https://doi.org/10.1016/j.biortech.2017.05.031.

Kallscheuer, N., Gätgens, J., Lübcke, M. et al. (2017). Improved production of adipate with *Escherichia coli* by reversal of ß-oxidation. *Applied Microbial Biotechnology* 101: 2371–2382. https://doi.org/10.1007/s00253-016-8033-3.

Kennedy, H. T., lululemon + Genomatica = Bio-Nylon, 2021, www.biofuelsdigest.com, accessed 2.3.2021.

Kumar, A., von Wolff, N., Rauch, M. et al. (2020). Hydrogenative polymerization of nylons. *Journal of the American Chemical Society* 142: 14267–14275. https://doi.org/10.1021/jacs.0c05675.

Nguyen, M.T., Raspoet, G., and Vanquickenborne, L.G. (1997). Mechanism of the Beckmann rearrangement in sulfuric acid solution. *Journal of the Chemical Society, Perkin Transactions* 2: 821–825.

Rohles, C.M., Gießelmann, G., Kohlstedt, M. et al. (2016). Systems metabolic engineering of *Corynebacterium glutamicum* for the production of the carbon-5 platform chemicals 5-aminovalerate and glutarate. *Microbial Cell Factories* 15: 154. https://doi.org/10.1186/s12934-016-0553-0.

Sahm, H., Antranikian, G., Stahmann, K.-P., and Takors, R. (ed.) (2013). *Industrielle Mikrobiologie*. Heidelberg: Springer Verlag, p. 118ff.

Shoaib, M., Plant-based nylon: now it's fact, not fantasy, Genomatica Press Release, 2022, https://www.voguebusiness.com/sustainability/plant-based-nylon-now-its-fact-not-fantasy, accessed 21.7.2022.

Sierra, A.R.G. and Alper, H.S. (2020). Progress in the metabolic engineering of bio-based lactams and their ω-amino acids precursors. *Biotechnology Reviews* 107587, https://doi.org/10.1016/j.biotechadv.2020.107587.

Tournier, V., Duquesne, S., Guillamot, F. et al. (2023). Enzymes' power for plastics degradation. *Chemical Reviews*. https://doi.org/10.1021/acs.chemrev.2c00644.

Turk, S.C.H., Kloosterman, W.P., Ninaber, D.K. et al. (2016). Metabolic engineering toward sustainable production of Nylon-6. *Synthetic Biology* 5: 65–73. https://doi.org/10.1021/acssynbio.5b00129.

Voet, D., Voet, J.G., and Pratt, C.W. (2016). *Principles of Biochemistry*, 454. Wiley.

Wendisch, V. F., Mindt, M., Pérez-García, S., Biotechnological production of mono- and diamines using bacteria: recent progress, application, and perspectives, *Applied Microbiology and Biotechnology*, 2018, Vol. 102, p. 3583–3594, https://doi.org/10.1007/s00253-018-8890-z.

Wursthorn, L., Beckett, K., Rothbaum, J.O. et al. (2023). Selective lanthanide-organic catalyzed depolymerization of nylon-6 to ε-caprolactam. *Angewandte Chemie*. https://doi.org/10.1002/anie.202212543.

Zhou, H., Vonk, B., Roubos, J.A. et al. (2015). Algorithmic co-optimization of genetic constructs and growth conditions: application to 6-ACA, a potential nylon-6 precursor. *Nucleic Acid Research* 43 (21): 10560–10570.

Zong, B., Sun, B., Cheng, S. et al. (2017). Green production of the monomer of nylon-6: caprolactam. *Engineering* 3: 379–384. https://doi.org/10.1016/J.ENG.2017.03.003.

22

Anthranilic Acid and Aniline

22.1 Application

Anthranilic acid or 2-aminobenzoic acid is an aromatic compound with the chemical formula $C_7H_7NO_2$ and a molecular weight of 137.1 g/mol. Anthranilic acid is a low-volume chemical with application in dyes, perfumes, and pharmaceutical compounds. It has been discussed as building block for fungal peptidyl alkaloid multicyclic scaffolds, which may serve as fungicides in crop protection (Walsh et al. 2013). The demand might significantly increase in the future since Covestro, the former Material Science business of Bayer, is trying to develop a biobased entry point into the MDI (methylene diphenyl isocyanate) value chain. MDI is reacted with polyols to derive polyurethanes and rigid foams, which play a major role in, e.g. insulators and refrigerators. The MDI value chain is conventionally based on benzene derived from petrochemical sources and proceeds via several intermediates to finally obtain MDI. The global MDI production capacity was 5.4 Mio. t pa in 2010 and has been estimated to reach 8.5 Mio. t pa in 2020 (Covestro 2015) (Figure 22.1).

In the first step of the MDI value chain, benzene is subjected to a liquid-phase nitration with "nitrating acid," a mixture of nitric acid (HNO_3) and sulfuric acid (H_2SO_4). Thereby, the very active nitrosyl-cation, NO_2^+, is generated, which reacts with the π-electrons of the aromatic system to derive mono-nitrobenzene (MNB). The selectivity of the reaction is rather high. A second or third nitration of the aromatic ring is suppressed since the first nitro function limits the reactivity of the aromatic system. The original isothermal process has mostly been replaced by an adiabatic process, resulting in reduced energy consumption (Wong and Birkhoff 2005). The nitro function of MNB is then hydrogenated with molecular hydrogen to aniline. This reaction is catalyzed and pursued as either liquid or gas-phase reaction and achieves a yield of about 99% (Wong and Birkhoff 2005). The alternative ammonolysis of phenol to derive aniline has no industrial relevance. Since aniline has so far only one functional group, it is in the next step reacted with formaldehyde to derive 4,4-methylenediamine (MDA). Thereby, a methylene-"bridge" between two aromatic rings is formed, and the obtained molecule has two functional amino functions (Arpe 2007; Franck and Stadelhofer 1987).

The attempts to replace the non-sustainable use of petrochemical raw materials cover the fermentative production of anthranilic acid (Covestro 2020) and the application of biobased, green benzene (Covestro 2021). The most favored biobased entry into the MDI value chain would be the direct fermentative production of aniline, which could as a "drop in"-compound, directly replace aniline derived from petrochemical resources (Figure 22.2).

Pathway Design for Industrial Fermentation, First Edition. Walter Koch.

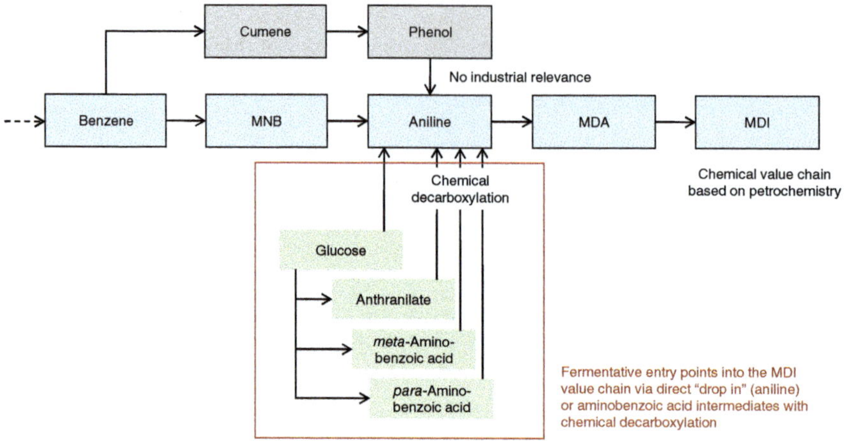

Figure 22.1 MDI value chain based on petrochemistry with alternative fermentation routes. Source: Walter Koch.

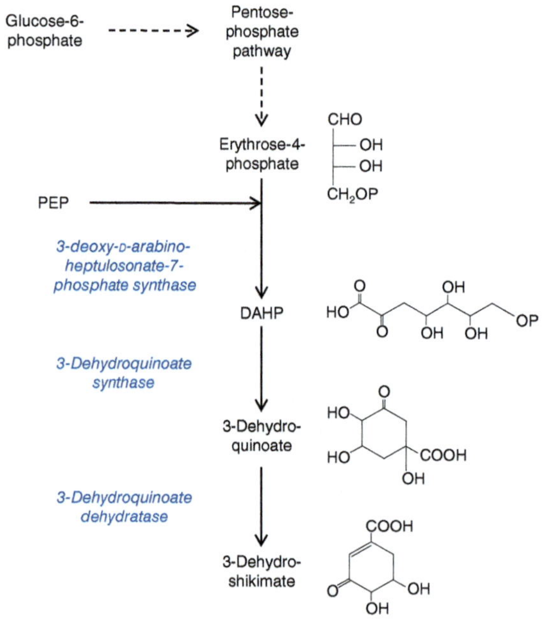

Figure 22.2 Initial pathway steps until 3-dehydroshikimate for the fermentative production of anthranilate. Source: Walter Koch.

The global production of aniline is estimated at 5 Mio. t pa and the demand is growing by 5% pa. Covestro has a production capacity of about 1 Mio. t and is one of the leading producers (Chemietechnik 2017). The fermentative production of aniline has originally been addressed in patent activities of Genomatica (Pharkya 2011, 2019), though it does not seem to be pursued. The feasibility of this route relies on the identification of a suitable enzyme catalyzing an efficient decarboxylation of anthranilate to obtain aniline. Since such an enzyme could not be identified Covestro has decided to focus on fermentative anthranilic acid with chemical decarboxylation in the following step: "A direct fermentation of sugar to aniline as a one-step conversion was thought to be most cost efficient if based on a biosynthesis pathway including an enzymatic, in vivo, decarboxylation of anthranilate to aniline as the final reaction step. Since an aminobenzoate decarboxylase could not successfully be identified or developed through protein engineering, the decarboxylation reaction of anthranilate to aniline could not be carried out by pure enzymatic means" (Covestro US 2017c/0066713). Covestro tries instead to produce anthranilic acid via fermentation, which is then in a second chemical step decarboxylated to aniline. Accordingly, the focus will be on the fermentative production of a suitable precursor for aniline. As an alternative toward anthranilate, *meta*-aminobenzoic acid (2-aminobenzoate) or *para*-aminobenzoic acid (3-aminobenzoate) might be produced via fermentation, which would then be chemically decarboxylated.

22.2 Pathway Design

The metabolic engineering strategies outlined below address either *ortho*-aminobenzoate (anthranilate), *meta*-aminobenzoate, or *para*-aminobenzoate as target products. For all three products, data with proof of concept are published. Irrespective of the relative position of the carboxylic acid function and the amino function on the benzene moiety, a chemical decarboxylation should be possible.

22.2.1 Shikimate Pathway via Chorismate Toward Anthranilate and Aniline

Anthranilate represents a rising variety of aromatic compounds that is gaining interest in the academic fermentation community. These comprise several pathway intermediates and derivatives along the shikimate pathway besides the aromatic amino acids L-tryptophan, L-phenylalanine, and L-tyrosine. Technology development is triggered by a robust understanding of the shikimate pathway and its regulation. The broad literature mirrors the portfolio of target products, and proof of concept has been achieved in several cases preferably by applying *Corynebacterium glutamicum*, *Escherichia coli*, and *Saccharomyces cerevisiae* as production hosts (Noda and Kondo 2017; Averesch and Krömer 2018; Shen et al. 2020). However, titers and yields are still not sufficient for an industrial application, which is mainly explained by the antimicrobial activity and toxicity of aromatics toward the host. Further tolerance management is needed or counterbalancing *in situ* recovery technology needs to be developed. Thus, commercial use is still limited to fine chemicals and pharma markets.

Anthranilate as one representative of this group is not an artificial target for metabolic engineering since it is a major biochemical intermediate in the synthesis of the amino acid L-tryptophan and has been identified as natural precursor in the, e.g. synthesis of 4-hydroxy-2-alkylquinolines of *Pseudomonas aeruginosa* (Bredenbruch et al. 2005). The conventional biochemical pathway toward anthranilic acid starts with phosphoenolpyruvate (PEP) and erythrose-4-phosphate (E-4-P), which are provided via glycolysis and the pentose phosphate pathway. The condensation of PEP and E-4-P derives the C7-intermediate 3-deoxy-D-arabino-heptulosonate-7-phosphate (DHAP), which is further converted into 3-dehydroquinoate and 3-dehydroshikimate (Figure 22.2).

3-Dehydroshikimate is phosphorylated to obtain shikimate-3-phosphate, which is condensed with PEP to 3-enoylpyruvylshikimate-5-phosphate (EPSP). EPSP is converted into chorismite facilitated by the chorismite synthase. Chorismate is the common precursor in the synthesis of the three aromatic amino acids tyrosine, phenylalanine, and tryptophan. The alternative routes toward anthranilate via *para*-aminobenzoate and *meta*-aminobenzoate are indicated (Figure 22.3).

The overall reaction scheme is derived by addition of the single reaction steps and elimination of the pathway intermediates. Hereby, PEP is derived via glycolysis and E-4-P via the pentose phosphate pathway (Table 22.1).

The net reaction reveals that a surplus of four reduction equivalents is obtained. The 36 electrons carried by 1.5 mol glucose (1.5 × 24) are transferred to 1 mol anthranilate, which carries 28 electrons without considering the reduced oxygen and nitrogen. The anthranilate synthase is usually referred to as one enzyme, though it catalyzes basically two distinct reactions with identified intermediate. The anthranilate synthase of *Sulfolobus solfataricus* has been crystallized without physiological ligands and its structure has been revealed based on x-ray crystallography. The enzyme is a heterotetramer combining the anthranilate synthase (TrpE) and the glutamine amidotransferase subunit (TrpG). It catalyzes the reversible reaction of chorismate with ammonia (provided via glutamine) to 2-amino-2-deoxyiso-chorismate (ADIC), followed by the irreversible elimination of pyruvate from ADIC (Knöchel et al. 1999; Figure 22.4).

The most efficient construction of the carbon backbone might suggest that 1.5 mol glucose is needed for the generation of 1 mol anthranilate. However, the net consumption of 4 mol ATP needs to be accounted for and the reduction equivalents are not balanced. The net reaction of the pathway provides a surplus of reduction equivalents since the feedstock carries 36 electrons (1.5 × 24 electrons per mol glucose without counting reduced oxygen), whereas anthranilate carries 28 electrons (without counting the reduced oxygen and nitrogen). Accordingly, a surplus of four reduction equivalents is caused, and the process would need to be pursued as an anaerobe fermentation with oxygen serving as electron acceptor:

$$\text{NAD(P)H} + \text{H}^+ + 0.5\,\text{O}_2 \rightarrow \text{NAD(P)}^+ + \text{H}_2\text{O} \qquad \Delta G^{\circ\prime} = -218\,\text{KJ/mol}$$

The oxidation of NAD(P)H might facilitate the synthesis of several mol ATP since the formation of an anhydride bond between two phosphoric acid moieties requires 30.5 kJ/mol ATP (Voet et al. 2016). However, since the theoretical thermodynamic maximum is never achieved in real biochemical systems, it is assumed that indeed 2.5 mol ATP per mol NAD(P)H is achieved (Voet et al. 2016). Accordingly, it can be assumed that the generated 4 mol NAD(P)H provides about 10 mol ATP in the respiratory chain and thereby balances the energy demand. Hereby also NAD(P)H would be recycled. The net reaction equation including energy and balanced reduction equivalents could be stated as follows:

$$1.5\,\text{Glucose} + \text{NH}_3 + 2\,\text{O}_2 + 6\,\text{ADP} + 6\,\text{Pi} \rightarrow \text{Anthranilate} + 2\,\text{CO}_2 + 6\,\text{ATP} + 13\,\text{H}_2\text{O}$$

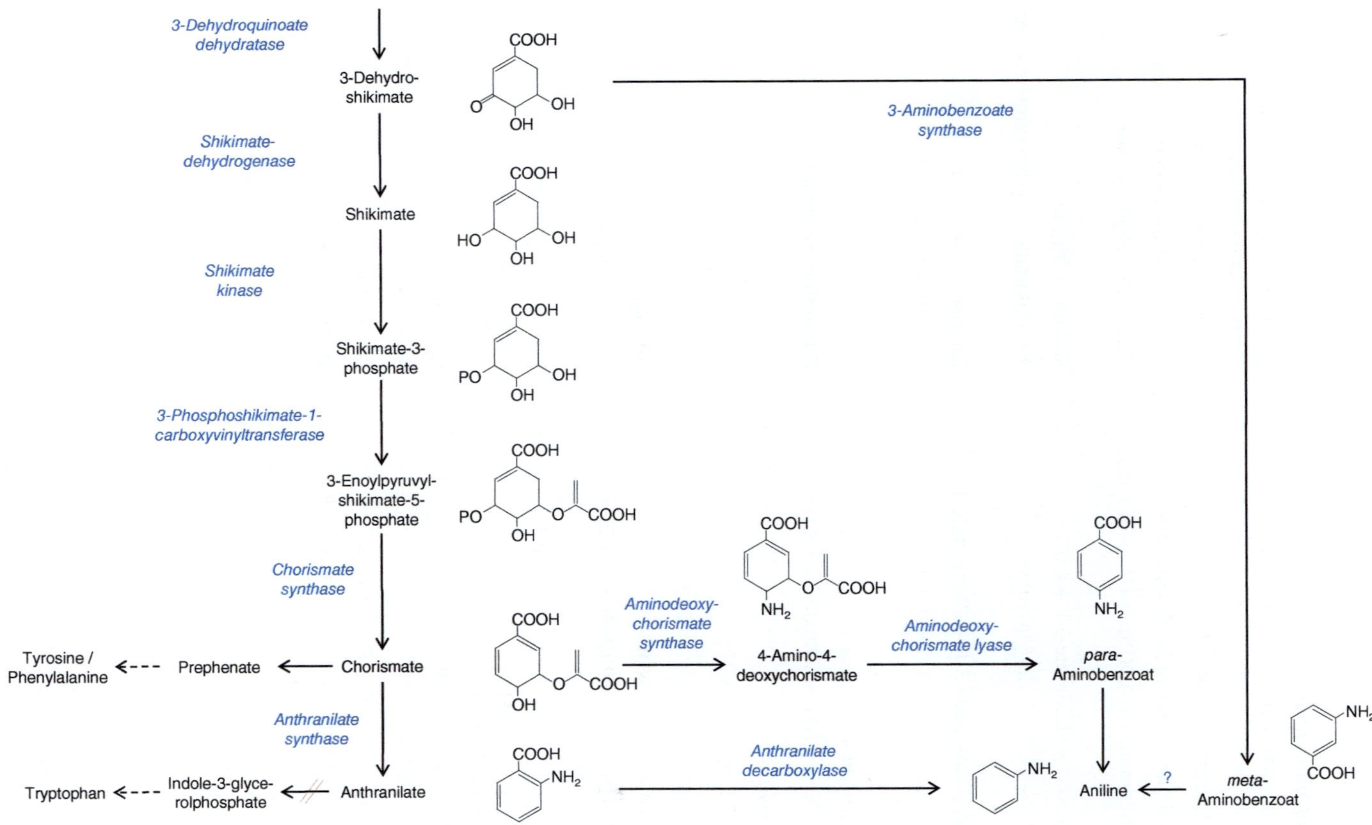

Figure 22.3 Metabolic network for the synthesis of anthranilate (Hirayama et al. 2013, 2015). Source: Walter Koch.

Table 22.1 Biochemical pathway to anthranilate.

Provision of PEP	
0.5 Glucose + Pi + NAD^+ → PEP + NADH + H^+ + H_2O	Glycolysis
Generation of anthranilate	
Erythrose-4-P + PEP + H_2O → DHAP + Pi	2-Keto-3-deoxy-D-arabino-heptulosonate -7-phosphate-synthase
DHAP → 3-Dehydroquinate + Pi	3-Dehydroquinoate synthase
3-Dehydroquinate → 3-Dehydroshikimate + H_2O	3-Dehydroquinoate dehydratase
3-Dehydroshikimate + NAD(P)H + H^+ → Shikimate + $NAD(P)^+$	Shikimate dehydrogenase
Shikimate + ATP → Shikimate-3-P + ADP	Shikimate kinase
Shikimate-3-P + PEP → 3-Enoylpyruvyl-shikimate-5-P + Pi	3-P-shikimate-1-carboxyvinyltransferase
3-Enoylpyruvyl-shikimate-5-P → Chorismate + Pi	Chorismate synthase
Chorismate + Glutamine → Anthranilate + Glutamate + Pyruvate	Anthranilate synthase
Provision of glutamine	
Glutamate + NH_3 + ATP → Glutamine + ADP + Pi	Glutamine synthetase
Recycling of pyruvate	
Pyruvate + ATP → PEP + ADP	PEP kinase
Provision of E-4-P	
Glucose + ATP → G-6-P + ADP	Hexokinase
2 G-6-P + 2 $NADP^+$ → 2x 6-P-Glucono-δ-lactone + 2 NADPH + 2 H^+	Dehydrogenase
2x 6-P-Glucono-δ-lactone + 2 H_2O → 2x 6-P-Gluconate	Lactonase/Hydratase
2x 6-P-Gluconate + 2 $NADP^+$ → 2 Ribu-5-P + 2 CO_2 + 2 NADPH + 2 H^+	6-Phosphogluconate dehydrogenase
Ribulose-5-P → Xylulose-5-P	Epimerase
Ribulose-5-P → Ribose-5-P	Isomerase
Ribose-5-P + Xylulose-5-P → SH-7-P + GAP	Transketolase
SH-7-P + GAP → Erythrose-4-P + Fructose-6-P	Transaldolase
Fructose-6-P → Glucose-6-P	G-6-P Isomerase
1.5 Glucose + NH_3 + 4 ATP + 4 $NADP^+$ + H_2O → Anthranilate + 2 CO_2 + 4 ADP + 4 Pi + 4 NADPH + 4 H^+	

It can also be concluded that about 1.33 mol oxygen per mol glucose needs to be provided, which allows regulating the oxygen supply device and the air volume to be dispersed. Given the net reaction equation, a maximum theoretical molar yield of 0.67 mol anthranilic acid per mol glucose (1/1.5) would be obtained, and the maximum carbon yield would amount to about 0.5 t anthranilate/t glucose. The energy surplus – 6 mol ATP per mol anthranilate – might be needed for the formation of biomass, cell maintenance, or even be employed to partially recycle carbon dioxide via formate dehydrogenase. The theoretical molar yield

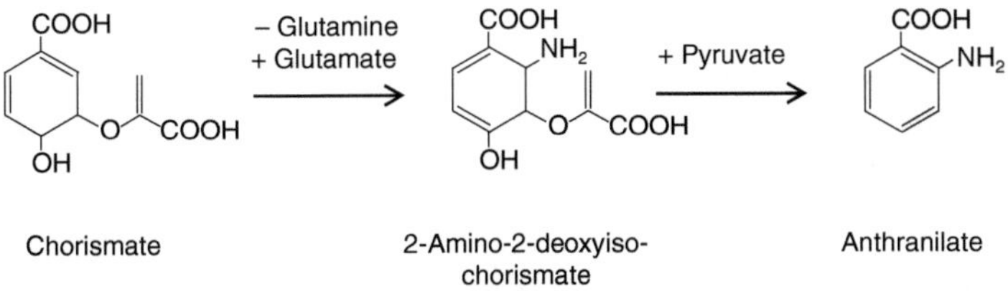

Figure 22.4 Mechanism of the anthranilate synthase with 2-amino-2-deoxyisochorismate (ADIC) as intermediate. Source: Walter Koch.

might then be slightly higher. However, this would suppose significant efforts in metabolic engineering to render the cell into an efficient producer of anthranilic acid and an efficient carbon dioxide recycling unit as well. It is claimed in the literature that the maximum carbon yield amounts to 0.435 t anthranilate/t glucose (Balderas-Hernández et al. 2009). For this estimation, the elementary mode flux analysis tool "META-TOOL" has been used (Schuster et al. 1999). This value is about 13% lower than the value derived from the simple pathway scheme. Explanation of the differences would require a thorough analysis of the inherent assumptions applied by the model, including host specification, glucose uptake mechanism, realistic mixture of pathways, ATP demand for maintenance, and formation of biomass.

22.2.2 Tryptophan Degradation Pathway Toward Anthranilate

An alternative pathway toward anthranilate has been described for *P. aeruginosa* and the generation of the *Pseudomonas* quinolone signal PQS. *P. aeruginosa* is a gram-negative bacterium whose kynurenine pathway genes are induced during growth on L-tryptophan. The biochemical studies derive the proposal that a second pathway toward anthranilate proceeds via the degradation of L-tryptophan: in the first step, L-tryptophan is transformed into formyl-kynurenine via the tryptophane-2,3-dioxygenase. The obtained formyl-kynurenine is then processed by the kynurenine formamidase to kynurenine, which is further converted into anthranilic acid via the kynureninase (Farrow and Pesci 2007). This pathway might indeed be pursued, though it is intrinsically disadvantaged compared to the direct synthesis of anthranilate via the shikimate pathway. The application of L-tryptophan as precursor for a potential biocatalytic whole-cell process would be too expensive.

22.3 Metabolic Engineering for Anthranilate as Fermentation Product

22.3.1 Anthranilate in *Bacillus subtilis*

An early work on the fermentative production of anthranilate is described in a BASF patent (Cooper et al. 1992). A strain of the species *B. subtilis* with L-tryptophan auxotrophy and resistance toward sulfaguanidine and flourotryptophan generated by mutagenesis displayed secretion of anthranilate. Two example cases are described: In the first example, an anthranilate titer of 1.5 g/l and an acetoin titer of 10 g/l were obtained after 48 h in a glucose medium with L-tryptophan supplementation though no addition of yeast extract.

In the second example, an anthranilate titer of 3.5 g/l and an acetoin titer of 25 g/l have been obtained after 60 h in a glucose medium with L-tryptophan supplementation and yeast extract (1 g/l). Given the low concentration of yeast extract, the increase in the anthranilate titer cannot be caused by the direct metabolism of aromatic precursors.

22.3.2 Anthranilate in *E. coli*

Basic principles of the metabolic engineering strategy have already been published in an early fermentation study in *E. coli* (Balderas-Hernández et al. 2009), primarily interested in the further elucidation of the metabolic pathway toward L-tryptophan. In *E. coli,* the first two steps in the synthesis of L-tryptophan, namely the enzymes for the conversion of chorismate into anthranilate and the conversion of anthranilate into N-(5′-phosphoribosyl)-anthranilate, are encoded by the enzyme complex anthranilate synthase-phosphoribosyl transferase (*trpE-trpD*) (Figure 22.5).

An *E. coli* mutant (W3310 trpD9923) and L-tryptophan auxotroph strain obtained via UV mutagenesis by UV light exposure displayed secretion of anthranilate and gave rise to the idea of developing a host for the fermentative production of anthranilate. 20 μg/ml

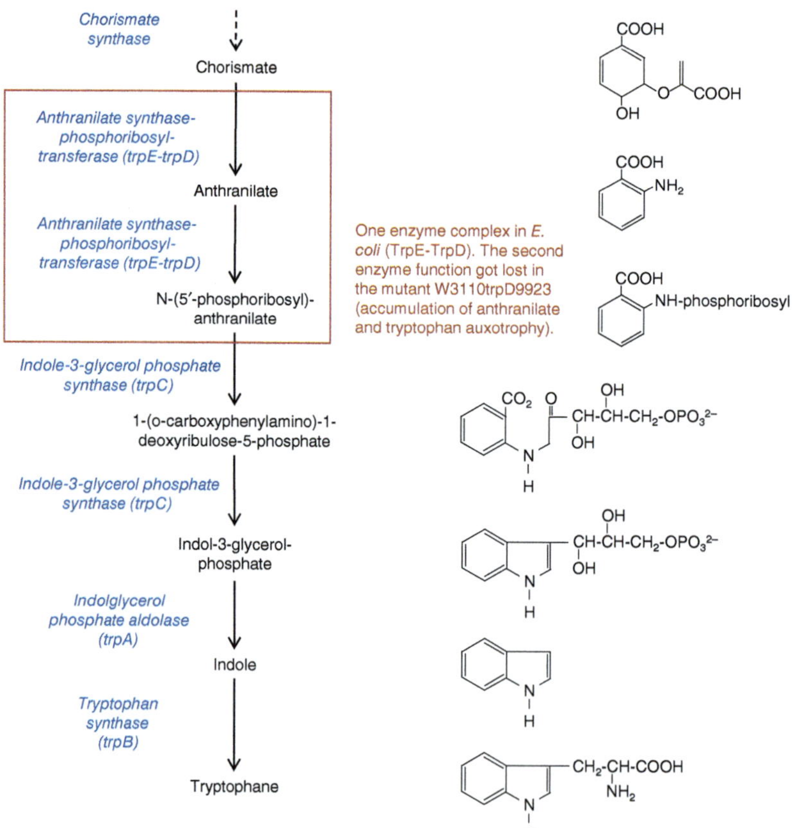

Figure 22.5 Biochemical conversion of chorismate into tryptophan including pathway intermediates. Source: Adapted from Rodriguez et al. (2014).

L-tryptophan is added to the medium unless a complex medium with yeast extract is applied. Based on sequence comparison, it was confirmed that the glutamine amidotransferase activity for the synthesis of anthranilate has been kept, though the anthranilate phosphoribosyl transferase activity got lost. In general, the organization of the genes *trpE* and *trpD* needs to be considered in the chosen metabolic engineering strategy and the attempt to silence or block the synthesis of tryptophan. In case where both functionalities are arranged in one enzyme complex exclusively, the glutamine amidotransferase activity needs to be conserved. The phosphoribosyl transferase activity needs to be silenced or might entirely be knocked out if tryptophan is added to the medium.

The feedback inhibition of the central enzyme DAHP synthase, the 3-deoxy-D-arabino-heptulosonate-7-phosphate synthase encoded by *aroG*, is not displayed by a feedback-resistant version of the strain. This approach has also been adopted in other projects for the metabolic engineering of the shikimate pathway. E.g. for the synthesis of hydroxybenzoate in *C. glutamicum,* the DAHP synthase, which is controlled by allosteric product inhibition, has been replaced by the isoenzymes AroH and AroF* from *E. coli.* AroH is exclusively feedback-inhibited by L-tryptophan, and AroF* is a truncated version of the L-tyrosine-sensitive DAHP synthase, which lost its feedback regulation due to the lack of 11 N-terminal amino acids (Kallscheuer and Marienhagen 2018).

The metabolic engineering strategy also addressed substrate availability for the central condensation of PEP and E-4-P toward DAHP. Altogether, two mol PEP are needed for the generation of one mol anthranilic acid, according to the biochemical pathway outlined above. Earlier studies on L-tryptophan suggested that PEP might be a limiting factor in the synthesis of aromatics and recommended that competing enzymes consuming PEP might be silenced or knocked out. One central consumer of PEP within the *E. coli* network is the PEP sugar phosphotransferase system (PTS) glucose uptake system, which converts glucose and PEP into glucose-6-phosphate and pyruvate:

$$\text{Glucose (outside)} + \text{PEP} \rightarrow \text{Glucose-6-phosphate (inside)} + \text{pyruvate}$$

Accordingly, the PTS operon has been knocked out. As could be expected, the respective strain displayed a reduced glucose uptake and a limited growth though the anthranilate accumulation in the broth was 2.2-fold higher than in experiments with the strain that possesses a functional PTS glucose uptake system. In order to overcome the severe growth limitation, the genes *glk* and *galP* were introduced encoding the glucokinase and the galactose permease. Additionally, the second substrate of the DAHP synthase has been addressed, and the transketolase (*tktA*) involved in the synthesis of E-4-P has been overexpressed. An extension of the growth phase was achieved in a fed-batch process with complex medium comprising 30 g/l yeast extract and 90 g/l glucose, and a titer of 14 g/l anthranilate after 34 h was measured. This corresponds to a productivity of 0.41 g/lxh. However, the authors of the study remark that these data cannot be compared with experiments exclusively applying glucose as carbon and energy source since precursors in the yeast extract might have been channeled in the synthesis of anthranilate.

22.3.3 Anthranilate in *Pseudomonas putida*

R&D results based on *P. putida* as host have been published by Bayer Technology GmbH (Kuepper et al. 2015). The choice of *P. putida* is rationalized by its high biomass yield and high growth rate. It has been shown to withstand chemical stress caused by a second phase of toluene, octanol, and styrene, which might be especially relevant in case an *in situ* recovery out of the fermentation broth is envisaged. It should be indicated that anthranilate is in general a toxic or at least growth-limiting compound for microorganisms, and the potential production host should be carefully selected with respect to product tolerance (Kuepper et al. 2020). In general, *P. putida* is gaining importance as an industrial fermentation host besides the most established chassis strains, *E. coli* and *S. cerevisiae* (Weimer et al. 2020). Glucose is applied at a concentration of 1 M as carbon and energy source in the anaerobic fermentation process with pH regulation to pH 7. The metabolic engineering strategy employs a feedback-insensitive DAHP synthase, and the corresponding *aroG* gene is overexpressed. Additionally, the anthranilate synthase *trpE* gene has been overexpressed as well. The competing pathway toward L-tryptophan has been blocked, and the *trpDC* operon encoding the anthranilate phosphoribosyl transferase (*trpD*) and the indole-3-glycerol phosphate synthase (*trpC*) has been deleted. L-Tryptophan is supplemented in a molar glucose to tryptophan ratio of 1000 : 1 or 2000 : 1 to ensure supply for the formation of biomass. The *pheA* gene encoding the bifunctional chorismate mutase and prephenate dehydratase for the conversion of chorismate toward phenylalanine has been knocked out. This knockout besides the blocking of the L-tryptophan synthesis supports the accumulation of anthranilate. The resulting L-phenylalanine and L-tyrosine auxotrophy needs to be addressed in the fermentation feed. *P. putida* is able to convert L-phenylalanine into L-tyrosine and exclusively phenylalanine needs to be supplemented to the fermentation medium. In particular, phenylpyruvate is added; the keto function of phenylpyruvate is then subjected to a transamination to derive L-phenylalanine according to the conversion of pyruvate into alanine. A titer of 1.5 g/l anthranilate has been achieved, which is by a factor of 10 lower than the previous results obtained in *E. coli* outlined above. The publication points out that exclusively glucose is applied as carbon and energy source, and the addition of the more costly yeast extract – which has been added to achieve the 14 g/l anthranilate titer in *E. coli* – could be avoided. It is speculated that the higher titer is pending on the addition of 30 g/l yeast extract, which provides anthranilate precursors and increases the stress tolerance of microorganisms. The data show the principal feasibility of the project though is by far not sufficient for industrial fermentation technology, which should be in the range of 100 g/l with a productivity in the range of 4 g/lxh.

22.3.4 Anthranilate in *S. cerevisiae*

Proof of concept has also been shown for the fermentative production of anthranilate in the eukaryotic host *S. cerevisiae*. The feedback inhibition of the DAHP synthase was overcome by genomic integration of a respective single-copy gene. The genes converting DAHP into EPSP are carried by the pentafunctional enzyme Aro1 in *S. cerevisiae*. The conversion of EPSP to chorismate is catalyzed by the Aro2 enzyme. The TRP4 enzyme required for the conversion of anthranilate into tryptophan was knocked out. After some strain optimizations, a titer of 200 mg/l after 72 h in a minimal medium (1% glucose and tryptophane supplementation) was achieved. Cultivation in a SCD medium (2% glucose and amino acids except for tyrosine and phenylalanine) provided a titer of 568 mg/l after 45 h (Kuivanen et al. 2021).

22.4 Derivatives of Anthranilate as Fermentation Product

22.4.1 Metabolic Engineering: N-Methylanthranilate in *C. glutamicum*

It is additionally pursued in academia to accomplish the fermentative production of N-methylanthranilate (Walter et al. 2020), a derivative of anthranilate in which the amino function is methylated to obtain the respective secondary amine. The experiments are carried out with the host *C. glutamicum*. The metabolic pathway proceeds via anthranilate, which is converted into N-methylanthranilate using the N-methylanthranilate transferase from *Ruta graveolens* with SAM as second substrate and methyl donor (Irla and Wendisch 2022). This work is interesting in this context since an anthranilate-secreting strain has also been generated, which achieved a maximum titer of 3.1 g/l after 48 h of shake flask cultivation. These data correspond to a productivity of 0.06 g/lxh. The metabolic engineering strategy covers several elements: according to the other approaches described in literature, a feedback-insensitive aroG gene encoding the 2-keto-3-deoxy-D-arabinoheptulosonate-7-phosphate synthase from *E. coli* has been used. The pyruvate dehydrogenase (*ldhA* gene) converting pyruvate into lactate, the PEP carboxylase (*ppc* gene) converting PEP into oxaloacetate, the shikimate dehydrogenase (*qsuD* gene) converting shikimate into 3-dehydroshikimate, and the 3-dehydroshikimate dehydratase (*qsuB* gene) converting 3-dehydroshikimate into protocatechuate have been knocked out.

In *C. glutamicum*, glucose supplied in the medium is transported into the cell with the support of the phosphotransferase system, according to the PTS in *E. coli*, which consumes one mol PEP per mol glucose uptake from the medium. In order to reserve the PEP pool for the synthesis of DAHP, the PTS in *C. glutamicum* is usually knocked out in case the shikimate pathway is pursued. The alternative myo-inositol symporter IolT1 may then be employed as transport system, offering a phosphotransferase-independent transport mechanism. Since the expression of the corresponding *iolT1* gene is repressed by the regulator IolR in the absence of myo-inositol, the lolR binding site of the *iolT1* gene has been changed. The removal of the IolR binding site blocks the binding of lolR and allows the constitutive expression of the alternative glucose transport system (Kallscheuer and Marienhagen 2018). The repressing effect of lolR has also been removed in the host constructs on N-methylanthranilate.

22.4.2 Methyl Anthranilate in *C. glutamicum*

An interesting approach toward anthranilic acid has been proposed by the working group of S. Y. Lee who especially considered the isolation of the product out of the fermentation broth as a central criterion for technology development (Luo et al. 2019). Hereby methyl anthranilate is chosen as the primary fermentation product given its low solubility in water or the fermentation broth, respectively. Methyl anthranilate is a natural flavor compound found in grapes. Its natural function may relate to pest resistance in plants based on the observation that an elevated level of methyl anthranilate in strawberries confers resistance against the pest fly *Drosophila suzukii* (Bräcker et al. 2020). Additionally, it has been identified as a precursor for secondary metabolites like the quinolone signal of *Pseudomonas* (Farrow and Pesci 2007). An organic phase using tributyrin as an extractant is applied to achieve an *in situ* recovery of methyl anthranilate out of the fermentation broth. The used

6.6 l fermenter contained 0.5 l tributyrin, which corresponds to a concentration of about 10% (v/v) in case a reasonable filling rate is assumed. The experiments were performed with the two industrial hosts *E. coli* and *C. glutamicum*. About 97% (*E. coli*) and 98% (*C. glutamicum*) of the methyl anthranilate was extracted into the tributyrin phase, showing the sufficiently discriminating partition coefficient, which may also be exploited in the downstream process to extract the remaining methyl anthranilate out of the broth.

The applied biochemical pathway proceeds as indicated above via the condensation of E-4-P and PEP to DAHP, which is then further converted into anthranilate. Anthranilate is the substrate for the formation of methyl anthranilate. Since the methyl anthranilate metabolism in the applied hosts is not yet elucidated, the biosynthesis in plants has been referred to as guiding model. In plants, two enzymes for the conversion of anthranilate into methyl anthranilate have been identified. One route is initiated by the CoA activation of anthranilate by an anthranilate-CoA ligase and an acyl-transfer in the second step catalyzed by the anthraniloyl-CoA methanol acyltransferase using CoA, ATP, and methanol as co-substrates. The second route proceeds directly via the *S*-adenosyl-L-methionine-dependent methyltransferase (SAM) (Pillet et al. 2017). This route has been chosen and the anthranilic acid methyltransferase 1 encoded by the aamt1 gene from *Zea mays* has been cloned into the two hosts. For both hosts, a feedback-resistant mutant of the DAHP synthase has been employed. Access to the aromatic amino acids L-phenylalanine and L-tyrosine is still given since their precursor chorismate is also a precursor of anthranilate. The pathway toward L-tryptophan with anthranilate as starting point has been blocked for flux optimization, and the trpD gene, which consumes anthranilate as substrate, has been knocked out. For compensation, L-tryptophan has been added to the fermentation medium at a concentration of 40 mg/l. In order to save these costs, the trpD gene would likely be silenced – opposed to a complete knockout – in an industrial production strain. On the other hand, the L-tryptophan auxotrophy of the host might allow to steer the product-to-biomass ratio if the supplementation is decreased so far that the ratio is optimized and the maximum share of the glucose supply is channeled into the product synthesis. The competing synthesis of glycerol in *C. glutamicum* has been blocked via the gene deletion of the hdpA gene, which converts DHAP into dihydroxyacetone (DHA).

The concept study indicated that the enzyme encoded by the *aamt1* gene might not fully convert anthranilate into methyl anthranilate and that the non-converted anthranilate might also be secreted by the host, and a coproduction of both products might be observed. In order to ensure that sufficient substrate for the methyltransferase is provided, an engineered anthranilate overproducer, namely the bacterial host *C. glutamicum* without the *aamt1* gene, has been tested. Hereby, a titer of 26.4 g/l anthranilate after 84 h – corresponding to a productivity of 0.31 g/lxh – in a glucose minimal media could be measured, which is the highest titer in microbial hosts reported so far. The reported titers of methyl anthranilate amount to 4.5 g/l (*E. coli*) and 5.7 g/l (*C. glutamicum*) when achieved with a minimal media (Luo et al. 2019). These data do not allow a final decision on the preferred production host though indicate the principal feasibility of a fermentative production of anthranilate or methyl anthranilate.

22.4.3 Methylanthranilate in *S. cerevisiae*

Proof of concept has also been shown for the fermentative production of methylanthranilate in the eukaryotic host *S. cerevisiae*. The feedback inhibition of the DAHP synthase was overcome by genomic integration of a respective single-copy gene. The genes converting DAHP into EPSP are carried by the pentafunctional enzyme Aro1 in *S. cerevisiae*. The conversion of EPSP to chorismate is catalyzed by the Aro2 enzyme. The TRP4 enzyme required for the conversion of anthranilate into tryptophan was knocked out. An anthranilic acid methyltransferase from *Medicago truncatula* was expressed in *S. cerevisiae* to extend the pathway toward methylanthranilate. A titer of 414 mg/l methylanthranilate was obtained after five days (Kuivanen et al. 2021).

22.5 Alternative Fermentation Precursors for Aniline

22.5.1 *meta*-Aminobenzoate (3-Aminobenzoate) via Dehydroshikimate

An alternative pathway toward 3-aminobenzoate (*meta*-aminobenzoate) and potentially toward aniline has been proposed by G. Stephanopoulos and colleagues (Zhang and Stephanopoulos 2016). The product of the fermentation process would be *meta*-aminobenzoate, which is derived from one-step conversion of 3-dehydroshikimate by the 3AB synthase. The 3AB synthase (*pctV* gene) from *Streptomyces pactum* belongs to the group of aminotransferases and catalyzes the multi-step reaction mechanism with PLP as cofactor and glutamine as donor of the amino function. The keto function of 3-dehydroshikimate is aminated and α-ketoglutarate is derived as by-product (Hirayama et al. 2013, 2015). This enzymatic reaction has been analyzed since it catalyzes a step in the biochemical synthesis of the antitumor antibiotic pactamycin.

The *pctV* gene has been cloned into *E. coli* (Hirayama et al. 2013, 2015). The inducible T7 promoter was utilized to express PctV. The mechanism of the 3AB synthase is not yet elucidated though it is known that 3-dehydroshikimate is the substrate and glutamine is applied as amino donor according to the amination of chorismate to anthranilate or p-aminobenzoate. The used LB medium contained 2% (w/v) glucose, yeast extract, and each 40 mg/ml L-tyrosine, L-phenylalanine, and L-tryptophan to compensate for potential auxotrophy caused by the engineered shikimate pathway. These supplements are reasonable from a methodology point of view though do not exclude the option that the measured product titer is at least partially generated via whole-cell biocatalysis. The obtained strain achieves a titer of 1.5 mg/l 3-amino-benzoate. In order to improve productivity, a coculture of *E. coli* strains has been developed: The conversion of glucose into 3-dehydroshikimate is pursued by the *E. coli* P5.2 host, and the secretion of the product is ensured by expression of the *shiA* gene encoding a transmembrane transporter. The product secreted by the first host of the consortium is then taken up by the second host *E. coli* XL10-VS to be converted into 3-aminobenzoate. The coculture of two *E. coli* hosts achieves a titer of 48 mg/l 3-aminobenzoate (Figure 22.6).

The R&D activities of Zhang and Stephanopoulos do not primarily pursue the fermentative production of aniline or respective precursors though are driven by the endeavor to evaluate host cocultures in charge of defined pathway modules as efficient screening tools for metabolic engineering (Zhang et al. 2015). The last step toward aniline would need to be

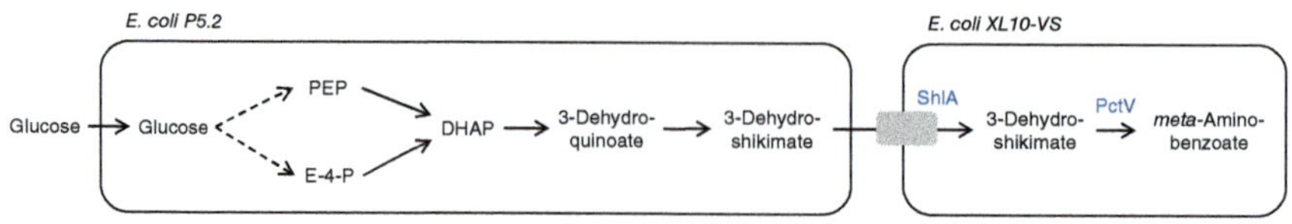

Figure 22.6 Coculture of *E. coli* strains for the fermentative production of *para*-aminobenzoate. Source: Adapted from Zhang and Stephanopoulos (2016).

added. This might be facilitated with a suitable decarboxylase – e.g. the anthranilate decarboxylase in case some substrate promiscuity is given – or a separate chemical step after the isolation of 3-aminobenzoic acid out of the fermentation broth.

22.5.2 *Para*-Aminobenzoic Acid via Shikimate

Besides *meta*-aminobenzoate acid, *para*-aminobenzoate might also be a suitable precursor for the synthesis of aniline via a chemical decarboxylation after the fermentation. Chorismate as central intermediate of the shikimate pathway is converted by the aminodeoxychorismate synthase, which consists of the two subunits. The first component encoded by *pabA* provides an amino function of glutamine and the second component encoded by *pabB* replaces the hydroxyl function of chorismate with an amino function. 4-Aminodeoxychorismate is then reacted by the aminodeoxychorismate lyase encoded by pabC, which releases a pyruvate and pursues an aromatization of the cyclohexedien-ring to obtain *para*-aminobenzoate (Figure 22.3).

The fermentative production of *para*-aminobenzoate has already been shown in several hosts. In the eukaryotic yeast host *S. cerevisiae* a hardly detectable titer of 0.03 g/l was achieved using glucose as the sole carbon source (Krömer et al. 2013). Based on glycerol and ethanol, a titer of 215 mg/l could be achieved in *S. cerevisiae* (Averesch et al. 2016). In *Lysobacter antibioticus*, a bacterium of the Xanthomodaceae family with application potential as biocontrol agent against bacterial and fungal diseases, a titer of 442 mg/l has been achieved after 24 h (Laborda et al. 2018). In bacteria higher titers were obtained with 4.8 g/l in *E. coli* based on glucose (Koma et al. 2014). The highest titer reported so far has been reported at 43 g/l in *C. glutamicum* based on glucose (Kubota et al. 2016). Besides the genes of the shikimate pathway, an *aroG* gene encoding a phenylalanine feedback-resistant mutant protein of *E. coli* was integrated into the bacterial chromosome of *C. glutamicum*. The *ldhA* gene encoding the lactate dehydrogenase was knocked out to limit the formation of lactate as by-product. Additionally, the *pabAB* gene of *Corynebacterium callunae* and the *pabC* gene of *Xenorhabdus bovienii* were introduced via plasmid into the host strain. It has been discovered that an undesired by-product was generated in the reaction of the terminal aldehyde function of glucose and the amino function of aminobenzoate. It is assumed that the by-product was N-glucopyranosyl-aminobenzoate since it could be reconverted into glucose and aminobenzoate with acid treatment though a direct isolation and identification were not feasible. The product yield based on glucose after the acid treatment was about 20%.

An interesting metabolic engineering R&D approach has been pursued with respect to *B. subtilis*. The two genes for the conversion of chorismate into *p*-aminobenzoate, the aminodeoxychorismate synthase (*pabAB* from *C. callunae*) and the aminodeoxychorismate lyase

Table 22.2 Titer and productivity for biochemical precursors for aniline.

Host	Product	Titer	Productivity	Source
S. cerevisiae	Anthranilate	568 mg/l	0.013 g/lxh	Kuivanen et al. (2021)
C. glutamicum	Anthranilate	26.4 g/l	0.31 g/lxh	Luo et al. (2019)
E. coli	Anthranilate	14 g/l0	0.41 g/lxh	Balderas-Hernández et al. (2009)
B. subtilis	Anthranilate	1.5 g/l	0.031 g/lxh	Cooper et al. (1992)
B. subtilis	Anthranilate	3.5 g/l	0.058 g/lxh	Cooper et al. (1992)
S. cerevisiae	Methylanthranilate	414 mg/l	0.003 g/lxh	Kuivanen et al. (2021)
E. coli	Methylanthranilate	4.5 g/	n.a.	Luo et al. (2019)
C. glutamicum	Methylanthranilate	5.7 g/l	n.a.	Luo et al. (2019)
C. glutamicum	N-Methylanthranilate	3.1 g/l	0.06 g/lxh	Walter et al. (2020)
E. coli	*meta*-Aminobenzoate	1.5 mg/l	n.a.	Zhang and Stephanopoulos (2016)
E. coli coculture	*meta*-Aminobenzoate	48 mg/l	n.a.	Zhang and Stephanopoulos (2016)
S. cerevisiae	*para*-Aminobenzoate	30 mg/l	n.a.	Krömer et al. (2013)
S. cerevisiae	*para*-Aminobenzoate	215 mg/l	n.a.	Averesch et al. (2016)
L. antibioticus	*para*-Aminobenzoate	442 mg/l	0.018 g/lxh	Laborda et al. (2018)
E. coli	*para*-Aminobenzoate	4.8 g/l	0.1 g/lxh	Koma et al. (2014)
C. glutamicum	*para*-Aminobenzoate	43 g/l	n.a.	Kubota et al. (2016)

(*pabC* from *X. bovienii*) were transformed separately and as a fusion protein spaced by a glycine-serine linker between the two enzymes (Averesch and Rothschild 2019). Such a fusion protein shall facilitate the direct conversion of the product of the first enzyme by the second enzyme without free diffusion of the intermediate into the cytosol. However, the fusion protein did not provide higher titers.

Summarizing, the following titers and productivities were achieved (Table 22.2).

22.6 Process Development with Focus on Product Isolation

22.6.1 Anthranilic Acid with *S. cerevisiae* as Host and *In Situ*-Product Recovery

An integrated approach for the fermentative production of anthranilic acid has been proposed by Covestro (2017a). The metabolic engineering strategy is not explained in detail. The conventional shikimate pathway starting with PEP and E-4-P is pursued, and the anthranilate phosphoribosyltransferase activity needs to be silenced or be knocked out, respectively. Accordingly, the strain would be auxotrophic, and L-tryptophan would be added to the fermentation medium unless a minimum remaining enzymatic activity could be sustained. The process shall employ a yeast species as host and several strains are mentioned like *Ashbya gossypii*, *Pichia pastoris*, *Hansenula polymorpha*, *Yarrowia lipolytica*, *Zygosaccharomyces bailii*, and *S. cerevisiae*. In particular, the eukaryotic species *S. cerevisiae* has been chosen as fermentation host since it displays a reasonable product formation in the stationary phase and decouples biomass growth and product formation.

Additionally, *S. cerevisiae* tolerates an acidic pH as opposed to the conventional prokaryotic bacterial hosts *E. coli* of *C. glutamicum*. The fermentation shall be pursued at a pH between 3.5 and 4.0. The host selection is supposed to allow the *in situ* recovery of anthranilic acid out of the fermentation broth because its isoelectric point pI has been estimated at 3.6. The pK_a values for the two functional groups of anthranilic acid have been estimated with $pK_{a1} = 2.2$ and $pK_{a2} = 4.9$ (Zapala et al. 2014), and an isoelectric point of $pI = 3.5$ is derived ($pI = 0.5$ ($pk_{a1} + pK_{a2}$; Voet et al. 2016). The solubility of anthranilic acid in water depends strongly on the pH, and the minimum solubility is achieved at its isoelectric point. In the patent application, the solubility is estimated at approximately 5 g/l anthranilate at a pH 3.6. The low product concentration in the broth during fermentation has three advantages. First of all, the known toxic effect of anthranilic acid on the host is very limited. Second, the low solubility facilitates an efficient removal of the product via crystallization, which is additionally supported by the application of seeding crystals of anthranilic acid. Third, the demand of base (e.g. NaOH, NH_3) can be limited. In case the fermentation would be pursued at pH 7 to enable a high productivity of, e.g. *E. coli* or *P. putida* and a titer, e.g. above 100 g/l a base would be needed to neutralize the acidity of the product. Afterwards, anthranilic acid would need to be reconverted in its fully protonated form and an acid would be needed. The resulting salt freight of the process can be handled though would increase cost and might hamper the sustainability profile of the product.

The oxygen supply shall be limited and follow an "anoxic" supply regime defined as 5–40% oxygen saturation compared to an "oxygen saturation which would be present in the fermentation broth under atmospheric oxygen partial pressure at the given temperature with oxygen consumption" (Covestro 2017a). The limited oxygen supply ensures the obligatory oxygen supply for some mono-oxygenases though suppresses the formation of biomass, which would increase the input factor of the carbon and energy source. The reactor might be a conventional stirring reactor or an airlift reactor, providing the required broth mixing via air stream. The stirring velocity shall be adjusted such that the crystallized anthranilic acid settles at the bottom of the reactor and the yeast cells are kept in suspension. The exact steering of the process relies on the density difference between biomass and crystals. The density of yeast cells is estimated in the range of 1.08–1.1 g/cm^3, and the density of the anthranilic acid crystals is estimated at 1.41 g/cm^3. The *in situ* product removal of crystallized anthranilic acid shall be possible without the removal of fermentation broth, thereby potentially enabling a continuous fermentation and harvest process.

In order to maximize the carbon yield of the process, it is also planned to recover the soluble share of anthranilic acid in the broth, which cannot be captured via crystallization. It is proposed to add dodecanol to the broth during the fermentation process, which binds anthranilic acid. This might completely avoid any toxicity of anthranilic acid toward the host since the product would stick to the hydrophobic solvent and hamper interaction of the product with the host surface. It might then be possible to permanently remove a minor share of the broth out of the reactor, to incubate the broth until a phase separation is visible, and then to separate the liquid broth including biomass from the organic phase containing dodecanol and anthranilic acid. Alternatively, the phase separation might be pursued at the batch end. In any case, the process concept envisaged in the patent supposes that the formed anthranilic acid can exhaustively be harvested, be it as *in situ* recovery via crystallization out of the broth or the remaining soluble fraction in the broth via solvent extraction. Finally, anthranilic acid shall be converted into aniline via a chemical, thermal decarboxylation with the aid of a heterogeneous catalyst (e.g. zeolite H—Y containing SiO_2/Al_2O_3 with a pore size of 0.7–0.8 nm). The thermal decarboxylation shall be pursued at a temperature of about 180 °C up to 200 °C with an incubation time of 0.5–3 h.

22.6.2 Anthranilic Acid with *C. glutamicum* as Host and *In Situ*-Product Recovery

In case the fermentation host cannot endure an acidic pH and an *in situ* product recovery via crystallization out of the broth *during* the fermentation is not possible, an alternative process might be developed with pH shift and crystallization of anthranilic acid *after* the fermentation. Such a process has been evaluated by Covestro as outlined in a patent application (Covestro 2017b). The prokaryotic bacterium *C. glutamicum* is pursued as host, and the fermentation run at a pH >6.5 up to 8.0. The fermentation might be pursued as batch, fed-batch, or continuous fermentation, and in the first step after the fermentation, the biomass would be removed via filtration or centrifugation. The obtained product solution in water might be decolorized to meet the final product specification by applying ion exchange technology. The pH would then be reduced to 3.5 with HCl to meet the isoelectric point of anthranilic acid and to minimize its solubility. The broth concentration of anthranilic acid after the fermentation is not explicitly mentioned in the patent, though it might be assumed to be in the range of about 80 g/l since the concentration in the example experiments is indicated as 91 g/l. Hereby, it is considered that the removal of biomass will cause a concentration step of about 10–20%. Crystallization might be supported by the addition of seeding crystals. After 1 h of incubation, including the addition of HCl, the crystallized product was isolated by decantation. The crystallization yield amounted to 85–93% pending on the chosen reaction conditions. The crystals of anthranilic acid are washed in water, and a resolving step with second crystallization to improve the purity is not reported. Anthranilic acid is finally chemically decarboxylated at 180–200 °C and a catalyst (e.g. zeolite) might be applied.

22.6.3 Anthranilic Acid Isolation via Organic Solvent Extraction After the Fermentation

Anthranilic acid might also be isolated based on an organic extraction protocol, as opposed to the crystallization approaches outlined above. A respective protocol has been described by Covestro (US2017/0066713) for *S. cerevisiae* or a bacterial host like *E. coli* with silenced functionality of the TrpD like the recombinant host W3110 trpD9923 (Balderas-Hernández et al. 2009). The technology protocol foresees to apply ammonia as N-donor and neutralizing agent to keep the pH neutral during the fermentation. At batch end, the biomass would be removed by, e.g. filtration, and the aqueous solution of ammonium-anthranilate be subjected to a chemical decarboxylation at 24 bar to obtain aniline. The carbon dioxide would be removed as off-gas, and the aniline be extracted with dodecanol. The organic and the water phase are separated by decantation, and the aniline fraction is isolated via distillation. A residual water phase would be evaporated, condensed, and rechanneled into the fermentation step (Figure 22.7).

The remaining water fraction obtained as second phase in the extraction of aniline is subjected to ammonia stripping, which splits $(NH_4)_2CO_3$, NH_4HCO_3, and $NH_4NH_2CO_2$ into

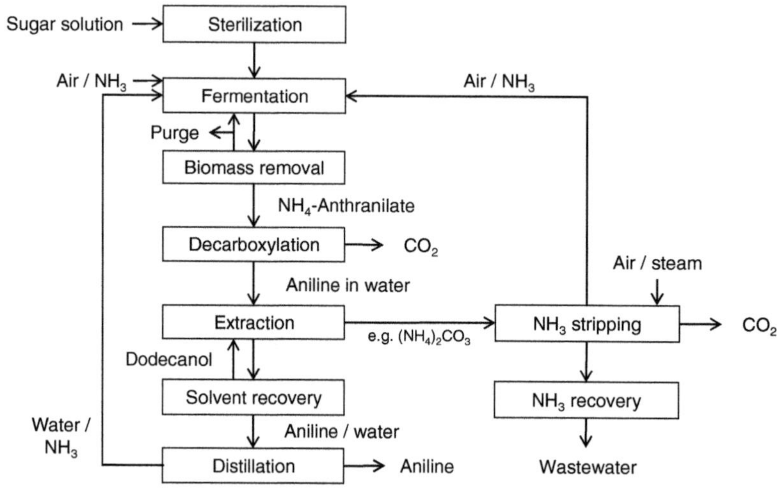

Figure 22.7 Process scheme for the fermentative production and isolation of aniline using solvent extraction with dodecanol (Covestro US 2017c/0066713). Source: Walter Koch.

ammonia and carbon dioxide. The ammonia gained is recycled and applied in the fermentation step. The efficiency and, especially, the yield of this downstream process relies on the portioning coefficient of aniline in dodecanol and water. A second extraction step might be recommended to limit production costs unless a high share of the aniline, presumably a threshold above 95%, is targeted.

22.7 Industrial Fermentation

R&D on the fermentative production of aniline or anthranilic acid, respectively, has been initiated by Covestro 2012 in collaboration with the University of Stuttgart, the CAT Catalytic Center at RWTH Aachen and Bayer (Tullo 2017). The fermentation product remains unnamed in the official announcements, though it may be assumed that anthranilic acid will be derived by fermentation, which is then chemically decarboxylated to obtain aniline. The first samples of the key chemical aniline from renewable raw materials were obtained in 2017 (Covestro 2020). The development of an industrially mature and robust process requires several years, and the bridge from lab-scale to pilot-scale and finally industrial scope – a world-scale aniline plant has a capacity above 100 kt/a – is ambitious. Covestro has repeatedly announced plans to further pursue activities, and an industrial plant might be started in the next few years.

References

Arpe, H.-J. (2007). *Industrielle Organische Chemie. Bedeutende Vor- und Zwischenprodukte*. Wiley.

Averesch, N.J.H. and Krömer, J.O. (2018). Metabolic engineering of the shikimate pathway for production of aromatics and derived compounds – present and future strain construction strategies. *Frontiers in Bioengineering and Biotechnology* 6: 32. https://doi.org/10.3389/fbioe.2018.00032.

Averesch, N.J.H. and Rothschild, L.J. (2019). Metabolic engineering of *Bacillus subtilis* for production of para-aminobenzoic acid – unexpected importance of carbon source is an advantage for space application. *Microbial Biotechnology* 12: 703–714.

Averesch, N.J.H., Winter, G., and Krömer, J.O. (2016). Production of para-aminobenzoic acid from different carbon-sources in engineered Saccharomyces cerevisiae. *Microbial Cell Factories* 15: 1–16.

Balderas-Hernández, V.E., Sabido-Ramos, A., Silva, P. et al. (2009). Metabolic engineering for improving anthranilate synthesis from glucose in *Escherichia coli. Microbial Cell Factories* 8: 19. https://doi.org/10.1186/1475-2859-8-19.

Bräcker, L.B., Gong, X., Schmid, C. et al. (2020). A strawberry accession with elevated methyl anthranilate fruit concentration is naturally resistant to the pest fly *Drosophila suzukii. PLoS One*. https://doi.org/10.1371/journal.pone.0234040.

Bredenbruch, F., Nimtz, M., Wray, V. et al. (2005). Biosynthetic pathway of *Pseudomonas aeruginosa* 4-hydroxy-2-alkylquinolines. *Journal of Bacteriology* 187 (11): 3630–3635. https://doi.org/10.1128/JB.187.11.3630-3635.2005.

Chemietechnik (2017). Covestro erzeugt Anilin aus Pflanzen. https://www.chemietechnik.de/markt/covestro-erzeugt-anilin-aus-pflanzen.html (accessed 1 February 2021).

Cooper, B., Meyer, J., and Euler, K. (1992). Production of anthranilic acid by a strain of *Bacillus subtilis* resistant to sulfaguanidine and fluorotryptophan. US patent 5,422,256.

Covestro (2015). Covestro investor presentation. https://www.covestro.com/en/investors/reports-and-presentations (accessed 2 February 2021).

Covestro (2017a). Process for the production of ortho-aminobenzoic acid and/or aniline by using recombinant yeast. Patent WO2017/102853.

Covestro (2017b). Method for producing aminobenzoic acid or an aminobenzoic acid derivative. Patent WO2017/085170.

Covestro (2017c). Production of aniline via anthranilate. Patent US2017/0066713.

Covestro (2020). Conference presentation: pushing boundaries in innovation and technology. https://investor.covestro.com/en/presentations (accessed 30 January 2021).

Covestro (2021). Covestro press release: Covestro received first delivery of sustainable benzene from Total. http://www.covestro.com/press/covestro-receives-first-delivery-of-sustainable-benzene-from-total/ (accessed 30 January 2021).

Farrow, J.M. and Pesci, E.C. (2007). Two distinct pathways supply anthranilate as a precursor of the *Pseudomonas* quinolone signal. *Journal of Bacteriology* 189 (9): 3425–3433. https://doi.org/10.1128/JB.00209-07.

Franck, H.-G. and Stadelhofer, J.W. (1987). *Industrielle Aromatenchemie*. Springer.

Hirayama, A., Eguchi, T., and Kudo, F. (2013). A single PLP-dependent enzyme PctV catalyzes the transformation of 3-dehydroshikimate into 3-aminobenzoate in the biosynthesis of pactamycin. *ChemBioChem* 14: 1198–1203.

Hirayama, A., Miyanaga, A., Kudo, F., and Eguchi, T. (2015). Mechanism-based trapping of the quinoid intermediate by using the K276R mutant of PLP-dependent 3-aminobenzoate synthase PctV in the biosynthesis of Pactamycin. *ChemBioChem* 16: 2484–2490.

Irla, M. and Wendisch, V.F. (2022). Efficient cell factories for the production of N-methylated amino acids of methanol-based amino acid production. *Microbial Biotechnology* 15 (8): 2145–2159. https://doi.org/10.1111/1751-7915.14067.

Kallscheuer, N. and Marienhagen, J. (2018). *Corynebacterium glutamicum* as platform for the production of hydroxybenzoic acid. *Microbial Cell Factories* 17: 70. https://doi.org/10.1186/s12934-018-0923-x.

Knöchel, T., Ivens, A., Hester, G. et al. (1999). The crystal structure of anthranilate synthase from *Sulfolobus solfataricus*: functional implications. *Proceedings of the National Academy of Science USA* 96: 9479–9484.

Koma, D., Yamanaka, H., Moriyoshi, K. et al. (2014). Production of para-aminobenzoic acid by metabolically engineered *Escherichia coli*. *Bioscience, Biotechnology and Biochemistry* 78: 350–357.

Krömer, J.O., Nunez-Bernal, D., Averesch, N.J.H. et al. (2013). Production of aromatics in *Saccharomyces cerevisiae* – a feasibility study. *Journal of Biotechnology* 163: 184–193.

Kubota, T., Watanabe, A., Suda, M. et al. (2016). Production of para-aminobenzoate by genetically engineered Corynebacterium glutamicum and non-biological formation of an N-glycosyl byproduct. *Metabolic Engineering* 38: 322–330.

Kuepper, J., Dickler, J., Biggel, M. et al. (2015). Metabolic engineering of *Pseudomonas putida* KT2240 to produce anthranilate from glucose. *Frontiers in Microbiology* 6: 1310. https://doi.org/10.3389/fmicb.2015.01310.

Kuepper, J., Otto, M., Dickler, J. et al. (2020). Adaptive laboratory evolution of *Pseudomonas putida* and *Corynebacterium glutamicum* to enhance anthranilate tolerance. *Microbiology* 166: 1025–1037. https://doi.org/10.1099/mic.0.000982.

Kuivanen, J., Kannisto, M., Mojzita, D. et al. (2021). Engineering of *Saccharomyces cerevisiae* for anthranilate and methyl anthranilate production. *Microbial Cell Factories* 20: 34. https://doi.org/10.1186/s12934-021-01532-3.

Laborda, P., Zhao, Y., Ling, J. et al. (2018). Production of antifungal p-aminobenzoic acid in *Lysobacter antibioticus* OH13. *Journal of Agricultural and Food Chemistry* 66: 630–636.

Luo, Z.W., Cho, J.S., and Lee, S.Y. (2019). Microbial production of methyl anthranilate, a grape flavor compound. *PNAS* 116 (22): 10749–10756.

Noda, S. and Kondo, A. (2017). Recent advances in microbial production of aromatic chemicals and derivatives. *Trends in Biotechnology* 35 (8): 785–796. https://doi.org/10.1016/t.tibtech.2017.05.006.

Pharkya, P. (2011). Microorganisms for the production of aniline. Patent US2011/0097767.

Pharkya, P. (2019). Microorganisms for the production of aniline. Patent US2019/0153454.

Pillet, J., Chambers, A.H., Barber, C. et al. (2017). Identification of a methyltransferase catalyzing the final step of methyl anthranilate synthesis in cultivated strawberry. *BMC Plant Biology* 17: 147. https://doi.org/10.1186/s12870-017-1088-1.

Rodriguez, A., Martínez, J.A., Flores, N. et al. (2014). Engineering *Escherichia coli* to overproduce aromatic amino acids and derived compounds. *Microbial Cell Factories* 13: 126, http://www.microbialcellfactories.com/content/13/1/126.

Schuster, S., Dandekar, T., and Fell, D.A. (1999). Detection of elementary flux modes in biochemical networks: a promoting tool for pathway analysis and metabolic engineering. *Trends in Biotechnology* 17: 53–60.

Shen, Y.-P., Niu, F.-X., Yan, Z.-B. et al. (2020). Recent advances in metabolically engineered microorganisms for the production of aromatic chemicals derived from aromatic amino acids. *Frontiers in Bioengineering and Biotechnology*. https://doi.org/10.3389/fbioe.2020.00407.

Tullo, A.H. (2017). Covestro pursues biobased aniline, c&en. https://cen.acs.org/articles/95/i23/Covestro-pursues-biobased-aniline.html (accessed 17 February 2023).

Voet, D., Voet, J.G., and Pratt, C.W. (2016). *Principles of Biochemistry*. Wiley.

Walsh, C.T., Haynes, S.W., Ames, B.D. et al. (2013). Short pathways to complexity generation: fungal peptidyl alkaloid multicyclic scaffolds from anthranilate building blocks. *ACS Chemical Biology* 8 (7): 1366–1382. https://doi.org/10.1021/cb4001684.

Walter, T., Medani, N.A., Burgardt, A. et al. (2020). Fermentative N-methylanthranilate production by engineered *Corynebacterium glutamicum*. *Microorganisms* 8: 866. https://doi.org/10.3390/microorganisms8060866.

Weimer, A., Kohlstedt, M., Volke, D.C. et al. (2020). Industrial biotechnology of *Pseudomonas putida*: advances and prospects. *Applied Microbiology and Biotechnology* 104: 7745–7766. https://doi.org/10.1007/s00253-020-10811-9.

Wong, E.W. and Birkhoff, R. (2005). DuPont/KBR aniline process. In: *Handbook of Petrochemicals Production Processes* (ed. R.A. Meyers). McGraw-Hill (Chapter 2.1).

Zapala, L., Woźnicka, E., and Kalembkiewicz, J. (2014). Tautomeric and microscopic protonation equilibria of anthranilic acid and its derivatives. *Journal of Solution Chemistry* 43: 1167–1183. https://doi.org/10.1007/s10953-014-0190-3.

Zhang, H. and Stephanopoulos, G. (2016). Co-culture engineering for microbial biosynthesis of 3-amino-benzoic acid in *Escherichia coli*. *Biotechnology Journal* 11: 981–987. https://doi.org/10.1002/biot.201600013.

Zhang, H., Zhenjun, L., and Stephanopoulos, G. (2015). Engineering *Escherichia coli* coculture systems for the production of biochemical products. *PNAS* 112 (27): 8266–8271.

23

Farnesene

23.1 Application

β-Farnesene or (6*E*)7,11-dimethyl-3-methylene-1,6,10-dodecatriene is a natural sesquiterpene with the formula $C_{15}H_{24}$ and a molecular weight of 204.4 g/mol. β-farnesene belongs to the broad group of terpenes (McGarvey and Croteau 1995) and bears an interesting chemical functionality given its two branched methyl functions and two terminal and two non-terminal double bonds. Farnesene exists in several stereoisomers since the double bonds of the molecule may appear in the *cis*- or the *trans*-form and one or two terminal double bonds may apply. β-farnesene bears two terminal double bonds and appears in two stereoisomers since the double bond in the C6/C7 position may be given in the *cis*- or *trans*-form (Figure 23.1):

The commercially relevant β-farnesene addressed in this context is shown on the left side (6E stereoisomer). β-farnesene needs to be distinguished from α-farnesene, which is characterized by one terminal and three non-terminal double bonds. α-Farnesene exists in four stereoisomers, of which one representative is exemplary shown above (Figure 23.1). The business interest in β-farnesene arose after the biotechnology company Amyris, together with the University of Berkeley led by Jay Keasling, successfully developed a terpene yeast strain that was originally designed to generate amorphadiene and artemisinic acid as precursors for artemisinin. The first results of this research initiative were published in 2003 in the journal Nature Biotechnology and demonstrated the fermentative synthesis of amorphadiene in *Escherichia coli*, facilitated by a recombinant MVA pathway of *Saccharomyces cerevisiae* (Martin et al. 2003). The later research strategy switched to *S. cerevisiae* as a host and managed to provide the more advanced artemisinic acid (Kung et al. 2018; Vickers et al. 2017).

Artemisinic acid can chemically be converted into artemisinin, which is used as an essential component in artemisinin-combination therapies (ACTs) against malaria (Arnaud 2016; Schäfer 2014). Since the natural extraction of artemisinin out of the leaves of the sweet wormwood plant *Artemisia annua* is very expensive and the generation of the raw material was critical (Hommel 2008), a high price fluctuation between 195 and 1000 USD/kg hampered a steady supply (Eisenmann 2012). The Bill and Melissa Gates Foundation supports the fight against malaria and financially supported the R&D efforts of Amyris with a total fund of 42.6 Mio. USD (Mackenzie 2013). The R&D program of Amyris, boosted with private funding, was successful, and the semi-synthetically derived artemisinin has successfully been introduced into the market and supports a cheaper

Pathway Design for Industrial Fermentation, First Edition. Walter Koch.

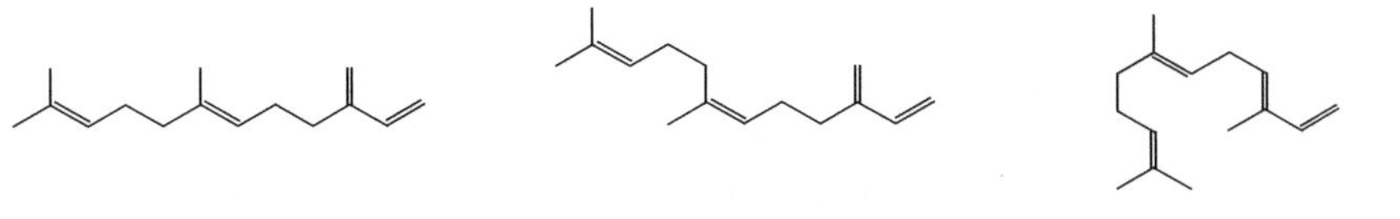

Figure 23.1 Structure of β-farnesene (left side) and α-farnesene (right side). Source: Walter Koch.

production of malaria drugs. In 2013 and 2014, 35 and 60 t artemisinin, respectively, were produced. The IP rights were given to Sanofi under the agreement of a "no profit, no loss" model. A chemical technology for the conversion of artemisinic acid into artemisinin had also been developed by Amyris, though Sanofi managed to design a more efficient route with higher overall yield (Osterath 2014; Turconi et al. 2013).

The original *S. cerevisiae* strain for artemisinic acid has then been leveraged by Amyris as a platform host for other terpenes (Carsanba et al. 2021) like bisabolene (Pang et al. 2018; Phelan et al. 2015; Peralta-Yahya et al. 2011), and especially high efforts were invested in the development of a β-farnesene fermentation protocol (Leavell et al. 2016). It is estimated that the total R&D investment in the period 2008–2012 amounts to >250 R&D person years (Karim et al. 2017). John Melo, the CEO of Amyris at that time, indicated that the R&D efforts for farnesene sum up to 200 Mio. USD (IHS Chemical Week 2017). Amyris entered several cooperations within the oil refining and aviation industry to finance the project and anticipated a high-volume business in case a cost-competitive technology could be established. Total invested 82 Mio. USD into Amyris to secure commercial application rights (IHS Chemical Week 2021; Ingredient networks 2021). β-Farnesene was supposed to be generated via fermentation and then subjected to chemical hydrogenation to saturate the double bonds, obtaining farnesane (Peralta-Yahya and Keasling 2010):

$$\text{Farnesene } (C_{15}H_{24}) + 4\,H_2 \rightarrow \text{Farnesane } (C_{15}H_{32})$$

The hydrogenation of β-farnesene relies on conventional chemical hydrogenation technology, which has been customized. It uses either Raney Nickel or a Pd catalyst such as 10% Pd/C (Amyris WO 2008/045555). β-Farnesene has been considered as a renewable jet fuel component that could partially replace petrochemically derived jet fuel and thereby reduce the carbon dioxide emissions of the aviation industry (Zhou et al. 2018; Liu et al. 2021). The commercialization has been enabled by the revised ASTM International fuel standard, which allows 10% blends of Amyris's sugar-derived farnesane (Bomgardner 2014). Amyris announced at the 5th Congress of the Brazilian Biotechnology Society that the farnesane based biofuel – marketed under the brand Biofene® – is designed as a cost-competitive, drop-in replacement for petroleum-derived fuels that is compatible with existing turbines. Sales of renewable diesel were initiated in the metropolitan areas of Brazil, and the industry certification for renewable jet fuel was pursued (Liebsch 2014). The production assets are located in Brazil to profit from the abundant sugarcane feedstock and the low cost of fermentable sugars.

According to this story line, Amyris celebrated its Initial Public Offering (IPO) in September 2010 with the following announcement: "Our first commercialization efforts have been focused on a molecule called farnesene, which forms the basis for a wide range of products varying from specialty chemical applications such as detergents, cosmetics, perfumes and industrial lubricants, to transportation fuels such as diesel. We have focused our research and development, business development and production operation on the use of Brazilian sugarcane as our primary feedstock for the foreseeable future, because it is abundant, low cost and relatively price stable" (Mackenzie 2013).

Given the R&D efforts and the commercial availability of β-farnesene at medium costs, a downstream chemistry has been developed or is under development. The molecule is used as educt for the chemical synthesis of advanced aroma chemicals. The attempts comprise biomuguet, homofarnesol, and ambrofix (Armamino et al. 2020; Lecourt and Antoniotti 2020). β-Farnesene is a natural alarm pheromone for many species of pest aphids and causes dispersal and repellency. It has also been considered to apply β-farnesene as an insect pheromone for the protection of agricultural crop plants. In this interesting approach, it is foreseen to apply a genetically modified wheat variety that is enabled to synthesize and presumably to some extent release β-farnesene. It could indeed be shown under laboratory conditions that the concept does in principle work, and three species of cereal aphids were repelled. However, the effect could not yet be repeated in the field (Bruce et al. 2015).

23.2 Chemical Production

It is not known that β-farnesene is commercially produced in significant volumes with organic chemistry. This is explained by the rather advanced functionality of the C15 molecule. The high synthesis efforts to introduce the two methyl functions and the four double bonds in the correct stereoisomeric form imply a multi-step process. It would cause a cost level that might be reasonable for agrochemical or pharmaceutical applications, though it would exclude the usage as a medium-volume chemical intermediate or fuel. Accordingly, the chemical synthesis can be neglected, and exclusively fermentation technology might provide a cost-efficient access to β-farnesene.

23.3 Biochemical Pathway

In case the conversion of glucose into farnesene would theoretically and pathway-independently be designed, one would derive a reaction equation that ideally requires only 3.5 mol of glucose. Farnesene carries 84 electrons, which can be provided by a minimum 3.5 mol of glucose carrying 24 electrons (3.5×24, without counting reduced oxygen) (van Rossum et al. 2016):

$$3{,}5\ \text{Glucose} \rightarrow \text{farnesene} + 6\ CO_2 + 9\ H_2O \qquad \Delta G^{\circ\prime} = -870\ \text{KJ/mol}$$

However, this reaction might be a useful guidance for synthetic biology, though does not reflect that inevitably a known pathway needs to be followed. There are basically two biochemical pathways for the synthesis of terpenes. Both pathways generate an equilibrium mixture of isopentenyl-pyrophosphate (IPP) and dimethylallyl-pyrophosphate (DMAPP), which is then further converted into terpenes via terpene synthases. The mevalonate (MVA) pathway and the deoxy-xylulose-phosphate (DXP) pathway define two alternative options to convert glucose into the two common terpenoid building blocks, IPP and DMAPP. The

MVA pathway is the common pathway in eukaryotic organisms like mammals, yeast, fungi, archaea, and the cytoplasm and the mitochondrion of plants (Liao et al. 2016). It should be indicated that recently a specific Lepidopteran MVA pathway has been identified that starts with the condensation of acetyl-CoA and propionyl-CoA and thereby allows the synthesis of novel terpenes carrying an additional methylene group in each C5 building block (Eiben et al. 2019). On the other hand, the DXP pathway is typically for prokaryotic organisms and is also found in green algae and plant plastids (Kemper et al. 2017; Kuzuyama and Seto 2012).

23.3.1 Mevalonate Pathway Toward IPP/DMAPP and β-Farnesene via Oxidative Glycolysis

The mevalonate pathway starts with two mol acetyl-CoA, which are condensed to acetoacetyl-CoA via the acetoacetyl-CoA thiolase in the first step (Kuzuyama and Seto 2012). The formation of 3-hydroxy-3-methyl-glutaryl-CoA (HMG-CoA) facilitated by the HMG-CoA) synthase requires a third mol of acetyl-CoA to provide the C6 pathway intermediate. The subsequent reduction of one of the carboxylic acid functions is thermodynamically supported since the function is given as CoA-thioester. The HMG-CoA reductase manages the transfer of four electrons to convert the carboxylic acid function into an alcohol function with NADPH as a cofactor and mevalonate – which serves for the naming of the pathway – is derived. The following steps, facilitated by the mevalonate kinase and the phosphomevalonate kinase, activate the terminal alcohol function at the expense of two mol of ATP to provide mevalonate-diphosphate. The two common C5 intermediates, IPP and DMAPP, are finally obtained by the decarboxylation of mevalonate-diphosphate at the expense of one mol of ATP (Figure 23.2).

The equilibrium of the C5 intermediates IPP and DMAPP is the basis for the formation of β-farnesene. According to the net equation of the MVA pathway in the synthesis of isoprene, it can be deduced that three mol of acetyl-CoA or 1.5 mol of glucose are needed for the synthesis of one mol of IPP. In this analysis, the acetyl-CoA building blocks are generated via glycolysis and oxidative pyruvate dehydrogenation. Additionally, it is assumed that the generation of acetyl-CoA does not proceed via malonyl-CoA as an activated intermediate, which would require an additional mol of ATP (Table 23.1).

Geranyl-PP as the C10 intermediate is formed by the reaction of one mol of IPP with DMAPP, whereby the double bond of IPP attacks the carbocation derived by the release of the negatively charged pyrophosphate function of DMAPP. The C15 building block farnesyl-PP is obtained after the addition of one additional mol of IPP. Finally, β-farnesene is synthesized by a specific terpene or terpenoid synthase, in this case the β-farnesene synthase. The pyrophosphate function is released in the first step of the mechanism, and a carbocation is formed. The high diversity of terpenes is explained by the remarkable ability of the terpene synthases, which manage to steer an electron rearrangement, a partial reshuffling of the double bonds, and finally a deprotonation at the right C-atom to create an individual acyclic (Gao et al. 2012; Yoshikuni et al. 2006; McGarvey and Croteau 1995) or cyclic terpene (Christianson 2017). The reconstruction of the overall reaction equation reveals that altogether nine mol of acetyl-CoA or 4.5 mol of glucose, respectively, are needed for the synthesis of β-farnesene:

$$\begin{aligned}&4{,}5\,\text{Glucose} + 18\,\text{NAD}^+ + 6\,\text{NADPH} \rightarrow \beta\text{-farnesene} + 12\,\text{CO}_2 + 6\,\text{H}_2\text{O}\\&\quad + 6\,\text{NADP}^+ + 18\,\text{NADH} + 12\,\text{H}^+\end{aligned}$$

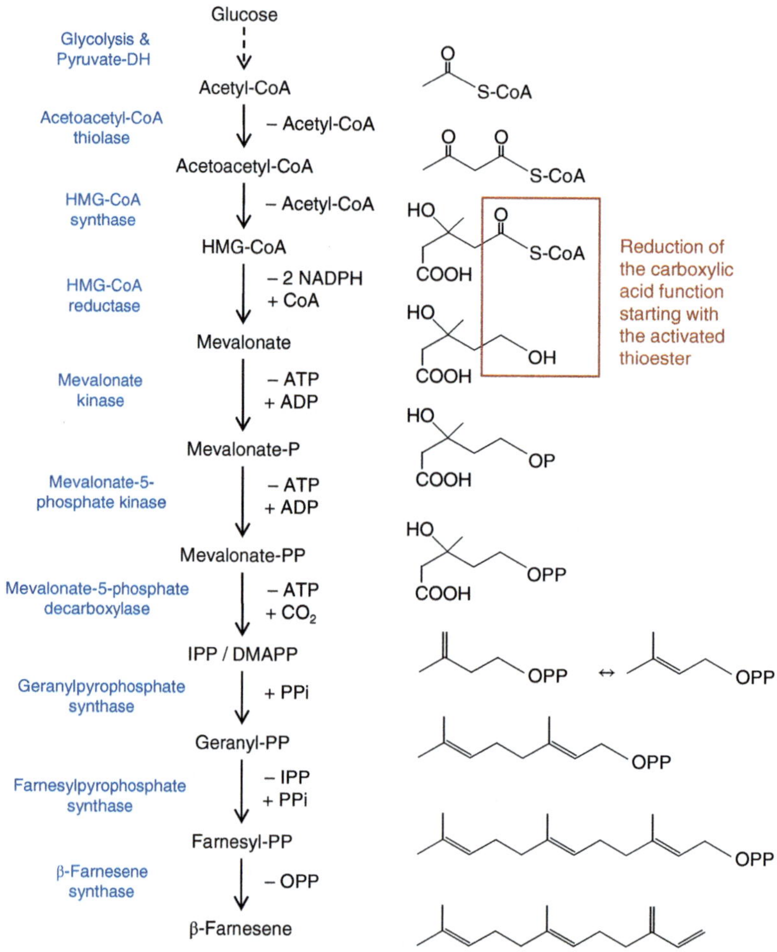

Figure 23.2 MVA pathway for the conversion of acetyl-CoA into IPP and DMAPP toward β-farnesene. Source: Walter Koch.

Table 23.1 Metabolic pathway for the MVA pathway toward IPP.

1,5 Glucose + 3 CoA + 3 ADP + 3 Pi + 6 NAD$^+$ → 3 Acetyl-CoA + 3 CO_2 + 3 ATP + 6 NADH + 6 H$^+$ + 3 H_2O
2 Acetyl-CoA → Acetoacetyl-CoA + CoA
Acetoacetyl-CoA + Acetyl-CoA + H_2O → HMG-CoA + CoA
HMG-CoA + 2 NADPH + 2 H$^+$ → Mevalonate + CoA + 2 NADP$^+$
Mevalonate + ATP → Mevalonate-5-P + ADP
Mevalonate-5-P + ATP → Mevalonate-PP + ADP
Mevalonate-5-PP + ATP → IPP + CO_2 + ADP + Pi
1,5 Glucose + 6 NAD$^+$ + 2 NADPH + 2 Pi → IPP + 4 CO_2 + 2 H_2O + 2 NADP$^+$ + 6 NADH + 4 H$^+$

The 4.5 mol of glucose carries a total of 108 electrons (4.5 × 24, without counting reduced oxygen). Thereof 84 are transferred to β-farnesene, which carries 84 electrons and, thus, a surplus of 24 electrons or 12 reduction equivalents is obtained. This can be seen on the right side of the overall reaction equation, which indicates a surplus of 18 NADH counter-balanced by the consumption of six NADPH, such that an overall surplus of 12 reduction equivalents is given. The reaction equation implies that the maximum theoretical molar yield amounts to 0.222 mol β-farnesene per mol glucose. The maximum carbon yield is 25%, or 0.25 t β-farnesene per t glucose (204.4/4.5 × 180.2 g/mol) in case by-products and the generation of biomass are excluded. This is according to the maximum carbon yield of isoprene in case the formation of acetyl-CoA is modeled according to conventional glycolysis.

It is to be kept in mind that the acetyl-CoA supply in eukaryotic hosts might precede via a cytosolic pyruvate dehydrogenase bypass since the mitochondrial or the peroxisomal acetyl-CoA pools are not accessible for a cytosolic MVA pathway (Shiba et al. 2007). In this case the overall reaction equation would need to consider an additional nine mol of ATP since the reactivation of one mol of acetate toward acetyl-CoA requires one mol of ATP (split into AMP and PPi, Figure 23.3). The overall reaction equation can be stated as follows in case it is assumed that for each mol of acetyl-CoA generated with the aid of the acetyl-CoA synthase, one mol of ATP needs to be invested:

$$\begin{aligned}4{,}5\ \text{Glucose} + 18\ \text{NAD}^+ + 6\ \text{NADPH} + 9\ \text{ATP} + 3\ \text{H}_2\text{O} \rightarrow\ & \beta\text{-farnesene} \\ + 12\ \text{CO}_2 + 6\ \text{NADP}^+ + 18\ \text{NADH} + 12\ \text{H}^+ + 9\ \text{AMP} + \text{PPi}\end{aligned}$$

The pyruvate dehydrogenase bypass seems to be the preferred solution to overcome the non-accessibility of the mitochondrial acetyl-CoA pool for recombinant cytosolic pathways, although it has been shown that the pyruvate dehydrogenase complex of *Enterococcus*

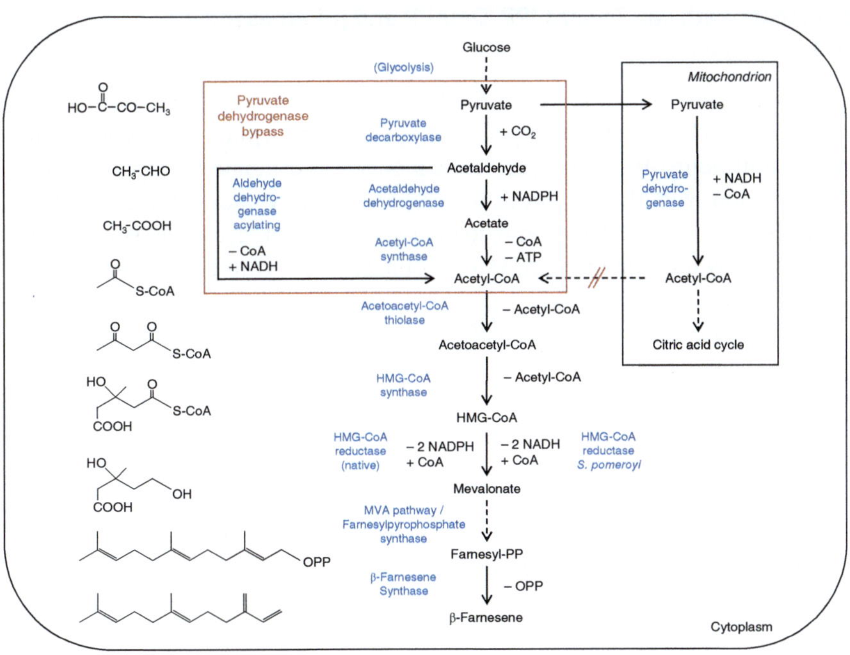

Figure 23.3 MVA pathway for the synthesis of β-farnesene with pyruvate dehydrogenase bypass. Source: Walter Koch.

faecalis, a complex enzyme with several subunits, can efficiently be expressed and assembled in the cytosol of *S. cerevisiae* (Kozak et al. 2014). The existing acetaldehyde dehydrogenase and the acetyl-CoA synthase in *S. cerevisiae* are useful, and the native formation of ethanol via a knock-out of the ethanol dehydrogenase can be blocked. It has also been evaluated to overexpress the ATP citrate lyase (Gruchattka and Kayser 2015), which converts citrate into acetyl-CoA and oxalacetate:

$$\text{Citrate} + \text{ATP} + \text{CoA} \rightarrow \text{Oxalacetate} + \text{acetyl} - \text{CoA} + \text{ADP} + \text{Pi}$$

The usefulness of this enzyme in the metabolic engineering strategy is not obvious. The expression and efficient functioning of this enzyme would need to ensure a sufficient cytosolic supply of citrate. The generation of oxalacetate would imply a high carbon burden and negatively impact the carbon yield.

The pyruvate dehydrogenase bypass should not imply a reduction of the theoretical maximum molar yield since the surplus of reduction equivalents should be able to compensate the ATP investment in case an aerobic process is assumed. Nevertheless, an elaborated *in silico* model has been programmed for *S. cerevisiae*, which derived the result that the maximum theoretical molar yield is indeed limited to 0.21 mol β-farnesene per mol glucose, which may be explained by the depletion of the NADPH pool (Meadows et al. 2016). The ATP investment in the synthesis of acetyl-CoA might be saved in case the aldehyde dehydrogenase acylating (ADA) is added and an isoenzyme of the HMG-CoA reductase from *Silicibacter pomeroyi* is introduced, which consumes NADH instead of NADPH (Figure 23.3):

The calculated maximum theoretical yield for this modeled pathway amounts exactly to 0.22 mol β-farnesene per mol glucose (Meadows et al. 2016).

23.3.2 Mevalonate Pathway Toward IPP/DMAPP and β-Farnesene via Non-oxidative Glycolysis

The yield of the MVA pathway suffers from the high unit consumption rate of the intermediate acetyl-CoA, which cannot be compensated by the high surplus of reduction equivalents or energy equivalents. It is therefore recommended to evaluate the option to provide acetyl-CoA via non-oxidative glycolysis which has a better C atom efficiency (Bogorad et al. 2013; Lin et al. 2018). The non-oxidative glycolysis, proceeds via several conversion steps initiated by the split of fructose-6-phosphate into erythrose-4-phosphate and acetyl-phosphate (Table 23.2):

The overall equation of non-oxidative glycolysis for the synthesis of β-farnesene can be stated as follows in (Figure 23.4):

The generation of acetyl-CoA via non-oxidative glycolysis does allow a very efficient generation of IPP, the main intermediate in the synthesis of terpenes like β-farnesene (Table 23.3):

The 24 electrons carried by glucose (without counting the reduced oxygen) are transferred to isopentenyl-pyrophosphate, which carries 28 electrons, which implies that two additional mol reduction equivalents are needed. Since one mol β-farnesene is derived from three mol of isopentenyl-pyrophosphate and the overall reaction equation can be stated as follows:

$$\begin{aligned} 3\,\text{Glucose} + 6\,\text{NADPH} + 6\,\text{H}^+ + 12\,\text{ATP} &\rightarrow \beta\text{-farnesene} + 3\,\text{CO}_2 + 3\,\text{H}_2\text{O} \\ &\quad + 6\,\text{NADP}^+ + 12\,\text{ADP} + 12\,\text{Pi} \end{aligned}$$

Table 23.2 Simplified metabolic pathway of non-oxidative glycolysis (Bogorad et al. 2013).

Glucose + ATP → Glucose-6-P + ADP
Glucose-6-P → Fructose-6-P
Fructose-6-P + GAP → Erythrose-4-P + Xylulose-5-P
Xylulose-5-P + Pi → GAP + Acetyl-P
Erythrose-4-P + Fructose-6-P → Seduheptulose-7-P + GAP
Seduheptulose-7-P + GAP → Xylulose-5-P + Ribose-5-P
Ribose-5-P → Ribulose-5-P
Ribulose-5-P → Xylulose-5-P
2-Xylulose-5-P + 2 Pi → 2 GAP + 2 Acetyl-P + 2 H_2O
3 Acetyl-P + 3 CoA → 3 Acetyl-CoA + 3 Pi
GAP → DHAP
GAP + DHAP → Fructose-1,6-BP
Fructose-1,6-BP + ADP → Fructose-6-P + ATP
Glucose + ATP + 3 CoA → 3 Acetyl-CoA + ADP + Pi + 2 H_2O

This carbon-conserving route is partially also referred to as bifid shunt (Wegner et al. 2021). GAP, glyceraldehyde-3-phosphate, DHAP, dihydroxyacetonephosphate.

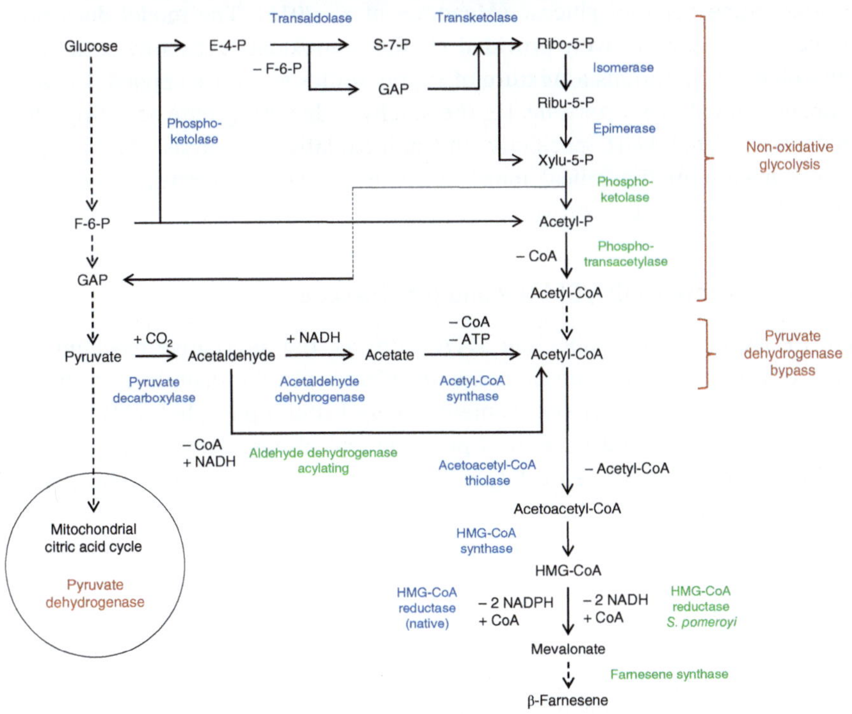

Figure 23.4 MVA pathway for the synthesis of β-farnesene with acetyl-CoA supply via a combined pyruvate dehydrogenase bypass and non-oxidative glycolysis. The MVA pathway takes place in the cytoplasm and the mitochondrial pyruvate dehydrogenase can be neglected. The native enzymes are shown in blue, recombinant enzymes in green (Meadows et al. 2016). F-6-P, fructose-6-phosphate, E-4-P, erythrose-4-phosphate; S-7-P, seduheptulose-7-phosphate; Ribo-5-P, ribose-5-phosphate; Ribu-5-P, ribulose-5-phosphate; Xylu-5-P, xylulose-5-phosphate. Source: Adapted from Meadows et al. (2016).

Table 23.3 Metabolic pathway for the MVA pathway toward IPP.

Glucose + ATP + 3 CoA → 3 Acetyl-CoA + ADP + Pi + 2 H_2O
2 Acetyl-CoA → Acetoacetyl-CoA + CoA
Acetoacetyl-CoA + Acetyl-CoA + H_2O → HMG-CoA + CoA
HMG-CoA + 2 NADPH + 2 H^+ → Mevalonate + CoA + 2 $NADP^+$
Mevalonate + ATP → Mevalonate-5-P + ADP
Mevalonate-5-P + ATP → Mevalonate-PP + ADP
Mevalonate-5-PP + ATP → IPP + CO_2 + ADP + Pi
Glucose + 2 NADPH + 2 H^+ + 4 ATP → IPP + CO_2 + 2 $NADP^+$ + 4 ADP + 2 Pi + H_2O

For the estimation of the maximum theoretical molar yield, the gap of 6 reduction equivalents needs to be considered and requires 0.5 mol glucose under aerobic conditions. Hereby, also two mol of ATP are generated as coproduction. The remaining gap of 10 mol ATP requires 0.312 mol glucose (10/32) and, in total 3.812 mol glucose is needed per mol β-farnesene. The maximum theoretical molar yield may be estimated at 0.262 mol glucose per mol β-farnesene. The elaborated *in silico* metabolic model derives the result that the introduction of a non-reductive glycolysis via the expression of a heterologous phosphoketolase and a phosphotransacetylase should allow a maximum theoretical yield of 0.241 mol β-farnesene per mol glucose (Meadows et al. 2016). The model does not suppose that acetyl-CoA is exclusively provided via the non-oxidative glycolysis, though under real physiological conditions a mixture of supply routes is always applied. In case all measures are combined in *S. cerevisiae*, i.e. the aldehyde dehydrogenase acylating, the HMG-CoA reductase with NADH specificity, and non-oxidative glycolysis, the in silico model predicts a maximum theoretical molar yield of 0.263 mol β-farnesene per mol glucose is obtained (Meadows et al. 2016).

23.3.3 DXP Pathway Toward IPP/DMAPP and β-Farnesene

The DXP or MEP pathway is, besides the MVA pathway, the alternative biochemical pathway toward terpenes and has been named with reference to the main intermediates 1-deoxy-D-xylulose-5-phosphate (DXP) or 2C-methyl-D-erythritol-4-phosphate (MEP). The pathway is initiated with the condensation of pyruvate and glyceraldehyde-3-phosphate (GAP) to obtain DXP in the first step catalyzed by the 1-deoxy-D-xylulose-5-phosphate synthase (Kuzuyama and Seto 2012). DXP is reduced and isomerized to obtain MEP in the second step, facilitated by the MEP reductoisomerase. The pathway intermediate MEP is then linked with a cytosine monophosphate (CMP) unit transferred by CTP to obtain 4-diphosphocytidyl-2C-methyl-D-erythritol-2-phosphate (CDP-ME), which is in the next steps phosphorylated, and a cytidyl unit (CMP) is released to yield 2C-methyl-D-erythritol-2,4-cyclodiphosphate (MEC). MEC is then reduced in the next step to provide the C5 intermediate 1-hydroxy-2-methyl-2-(E)-butenyl-4-diphosphate (HMBPP). After an additional reduction step, IPP is obtained, which is in equilibrium with DMAPP (Figure 23.5):

IPP and DMAPP are then converted into β-farnesene according to the former route toward β-farnesene. The three last steps are identical in both pathways. The overall pathway can be stated as follows, assuming that pyruvate and GAP are generated via conventional glycolysis (Table 23.4):

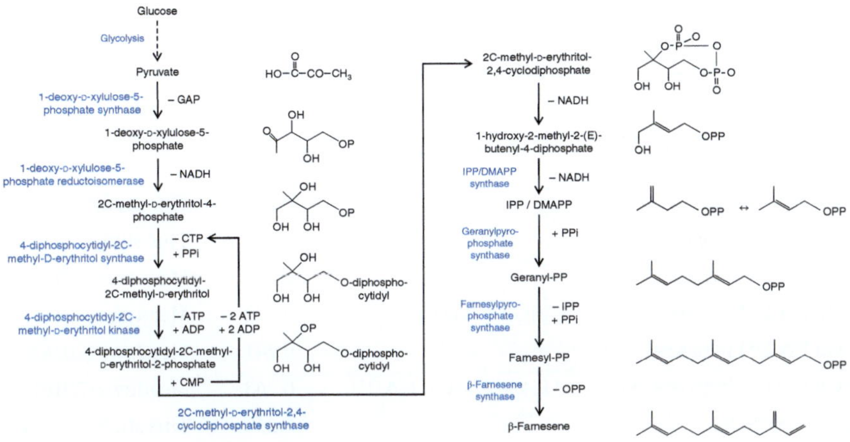

Figure 23.5 Simplified DXP pathway for the conversion of pyruvate and GAP into IPP and DMAPP toward β-farnesene. Source: Walter Koch.

Table 23.4 Metabolic pathway for the DXP pathway toward IPP.

Glucose + NAD^+ → Pyruvate + GAP + NADH + H^+
Pyruvate + GAP → 1-Deoxy-xylulose-5-phosphate + CO_2
1-Deoxy-xylulose-5-phosphate + NADH → 2C-Methyl-erythritol-4-phosphate + NAD^+
2C-Methyl-erythritol-4-phosphate + CTP → 4-Diphosphocytidyl-2C-methyl-erythritol + PP_i
4-Diphosphocytidyl-2C-methyl-erythritol + ATP → 4-Diphosphocytidyl-2C-methyl-erythritol-2-phosphate + ADP
4-Diphosphocytidyl-2C-methyl-erythritol-2-phosphate → 2C-Methyl-erythritol-2,4-cyclo-diphosphate + CMP
CMP + 2 ATP → CTP + 2 ADP
2C-Methyl-erythritol-2,4-cyclo-PP + NADH + H^+ → 1-Hydroxy-2-methyl-2-butenyl-4-PP + NAD^+ + H_2O
1-Hydroxy-2-methyl-2-butenyl-4-PP + NADH + H^+ → IPP + NAD^+ + H_2O
Glucose + 3 ATP + 2 NADH + 2 H^+ → IPP + CO_2 + 3 ADP + Pi + 2 NAD^+ + 2 H_2O

One mol of glucose is converted into one mol of C5 intermediate and only one mol of carbon dioxide is generated. However, compared to the MVA pathway the DXP pathway relies on three mol ATP and two reduction equivalents to achieve a carbon efficient disproportionation of glucose. The overall reaction equation toward β-farnesene can be stated as follows:

$$3\ \text{Glucose} + 9\ \text{ATP} + 6\ \text{NADH} + 6\ \text{H}^+ \rightarrow \beta\text{-farnesene} + 3\ \text{CO}_2 + 9\ \text{ADP} + 6\ \text{NAD}^+$$

The three mol of glucose carry a total of 72 electrons (3 × 24, without counting reduced oxygen). These are completely transferred to β-farnesene, which implies that a gap of 12 electrons or 6 reduction equivalents remains. This can be seen on the left side of the overall reaction equation, which indicates a demand of six NADH per mol β-farnesene.

Table 23.5 Summary of maximum theoretical molar yield (TMY) for several pathway designs in the fermentative synthesis of β-farnesene.

Pathway modules	TMY	Source
MVA pathway		
Acetyl-CoA via Pyr DH	0.222	This study
Acetyl-CoA via Pyr DH bypass	0.222	This study
Acetyl-CoA via Pyr DH bypass	0.210	Meadows (2016)
Acetyl-CoA via Pyr DH bypass + ADA + HMGR NADH	0.222	Meadows (2016)
Acetyl-CoA via Pyr DH bypass + NOG	0.241	Meadows (2016)
Acetyl-CoA via Pyr DH bypass + NOG + ADA + HMGR NADH	0.263	Meadows (2016)
Acetyl-CoA excl. via NOG	0.262	This study
DXP pathway	0.268	This study
Pathway independent	0.286	van Rossum (2016)

ADA: aldehyde dehydrogenase acylating, HMGR NADH: HMG-CoA reductase with NADH as cofactor, NOG: Non-oxidative glycolysis, Pyr DH: pyruvate dehydrogenase, Pyr DH bypass: generation of acetyl-CoA via acetaldehyde dehydrogenase and acetyl-CoA synthase. The values derived in this study refer to a theoretical flux model exclusively proceeding via one pathway; the values derived by Meadows et al. 2016, calculate a more realistic combination of pathways.

The reaction equation might propose that the maximum theoretical molar yield amounts to 0.33 mol β-farnesene per mol glucose (1/3). However, this value cannot be achieved because, for the calculation of the maximum carbon yield also the energy and reduction equivalents need to be considered. An aerobic process is assumed, and the six reduction equivalents require 0.5 mol of glucose, which also provides two ATP in coproduction. The missing seven mol of ATP require 0.22 mol of glucose (7/32) such that, in total, an additional 0.72 mol of glucose is needed. Accordingly, the maximum theoretical molar yield amounts to 0.268 mol β-farnesene per mol glucose (1/3.72), translating into a maximum carbon yield of 30% or 300 kg β-farnesene per ton glucose (204.4/3.72 × 180.2). The DXP or MEP pathway promises a higher yield than the MVA pathway but is disregarded in most of the published metabolic engineering programs. This is likely due to the lack of detailed knowledge on the regulatory control mechanisms (Yao et al. 2020).

The discussed maximum theoretical yields are summarized in Table 23.5:

23.4 Metabolic Engineering

23.4.1 Recombinant Reconstruction of Complete Pathways in Non-native Hosts

The availability of two equivalent pathways with comparable carbon yields in the range of 25% (MVA) and 30% (DXP), respectively, broadens the toolbox for metabolic engineering approaches. Since the native isoprenoid pathway in *E. coli* (DXP) or *S. cerevisiae* (MVA), the most common hosts in the fermentative production of terpenoids, has a natural function managed by a rather complex regulatory system, the introduction of a second pathway whose regulation is not controlled by a native regulation bears a great advantage (Martin et al. 2003). A recombinant second pathway bypasses the native regulation and should, at least in theory, allow a more fine-tuned steering of flux and selectivity compared to the approach that causes metabolic interference with the native pathway. This assumption is fueling the idea to add a second unregulated pathway for enhancing precursor availability (Milker and Holtmann 2021).

Accordingly, it has been pursued and shown that the bacterial isoprenoid pathway can be transferred to yeast and that the eukaryotic pathway can be transferred to *E. coli*. In the initial study, the eukaryotic MVA pathway, including a terpene synthase (i.e. the amorpha-4,11-diene synthase), was successfully engineered in *E. coli* for the production of amorpha-4,11-diene (Martin et al. 2003). Hereby, the MVA pathway has been expressed in two multi-gene vectors comprising the acetoacetyl-CoA thiolase, the HMG-CoA synthase, and the HMG-CoA reductase in the first plasmid and the mevalonate kinase, the phosphomevalonate kinase, the mevalonate-5-P decarboxylase, the IPP isomerase, and the FPP synthase in the second plasmid (Martin et al. 2003).

In the reverse direction, the bacterial isoprenoid pathway was successfully reconstructed in *S. cerevisiae* for the production of valencene, which could be detected by gas chromatography-mass spectrometry (GC-MS). The strain was able to produce up to 1 mg/l of valencene (Maury et al. 2008). Since the precursor of the valencene synthase might also be supplied by the native MVA pathway, the final proof needs to include an efficient blocking of the native pathway. Accordingly, the experiments with batch cultivation and galactose as carbon source were also pursued in the presence of lovastatin, an established inhibitor of the HMG-CoA reductase. The respective strain continued growth in the presence of 150 μg lovastatin, which proved the functionality of the recombinant DXP pathway (Maury et al. 2008). In case the native and recombinant pathways are not completely separated with respect to intermediates or regulatory elements and ≫cross talk≪ is observed, as appealed to in a patent of Amyris (Amyris WO 2008/045555), disturbing interference is to be expected.

Despite the potential regulatory interference, the implementation of an engineered MVA pathway in yeast may be regarded as the dominant pathway approach. This is due to several reason. The yeast *S. cerevisiae* allows an easier expression of plant P450 enzymes needed for the synthesis of several terpenes. The existing pyruvate dehydrogenase bypass can be fine-tuned to ensure an efficient cytosolic supply of acetyl-CoA, and often a truncated version of the HMG-CoA synthase with high solubility and less regulation is applied.

Another element to be addressed in metabolic engineering is the circumstance that some hosts, like yeast strains, suffer from a high internal accumulation of farnesyl-pyrophosphate. Accordingly, an efficient further conversion needs to be ensured. In case a critical cytosolic threshold is exceeded, a de-phosphorylation is observed, which results in the formation of farnesol and nerolidol, which are finally released and captured in the dodecane droplets usually given into the fermentation medium (Rodriguez et al. 2014).

23.4.2 Formation of Acetoacetyl-CoA via Malonyl-CoA

The fermentative synthesis of β-farnesene in *S. cerevisiae* via the native MVA pathway proceeds via acetoacetyl-CoA as pathway intermediate. It is known that the Claisen-type coupling of two mol of acetyl-CoA into acetoacetyl-CoA under the release of one mol of CoA via the acetoacetyl-CoA synthase is thermodynamically not favored because this step displays a positive Gibbs free energy (Tippmann et al. 2017; Lan and Liao 2012).

$$\text{Acetyl-CoA} + \text{acetyl-CoA} \rightarrow \text{Acetoacetyl-CoA} + \text{CoA} \quad \text{Acetyl-CoA thiolase}$$

$$\Delta G^{\circ\prime} \approx 7\ \text{kcal/mol}\ (= 29.3\ \text{kJ/mol})$$

In other publications, a slightly less positive value Gibbs free energy of this step with 26.1 kJ/mol is reported, though this deviation does not change the general picture (Vögeli et al. 2018). It is interesting to note that the acetoacetyl-CoA thiolase and the HMG-CoA synthase in some archaeal organisms are connected via a third protein with binding sites for both enzymes, which may help to overcome the thermodynamical hurdle by a fast conversion and accordingly, low concentration, of acetoacetyl-CoA (Vögeli et al. 2018). The alternative generation of acetoacetyl-CoA with acetyl-CoA and activated malonyl-CoA as substrates offers a more compelling thermodynamical profile and displays a negative Gibbs free energy:

$$\text{Acetyl-CoA} + CO_2 + \text{ATP} \rightarrow \text{Malonyl-CoA} + \text{ADP} + \text{Pi} \quad \text{Acetyl-CoA carboxylase}$$

$$\text{Malonyl-CoA} + \text{acetyl-CoA} \rightarrow \text{Acetoacetyl-CoA} + \text{CoA} + CO_2 \quad \text{Acetyl-CoA synthase}$$

$$\Delta G^{\circ\prime} \approx -0.9\ \text{kcal/mol}\ (= -3.8\ \text{KJ/mol}$$

The formation of malonyl-CoA via the carboxylation of acetyl-CoA is catalyzed by the acetyl-CoA carboxylase and demands the investment of one mol of ATP with a Gibbs free energy of 30.5 kJ/mol Tippmann et al. 2017). The *nphT7* gene encoding this enzyme has been found in a mevalonate gene cluster in *Streptomyces* sp. strain CL190, and its co-expression with the HMG-CoA synthase provided a 3.5-fold higher production of mevalonate (Okamura et al. 2010). The acetoacetyl-CoA synthase from *Streptomyces* sp. encoded by the *nphT7* gene has been cloned and a plasmid-based version expressed in *S. cerevisiae.* The plasmid is also harboring a farnesene synthase of *Citrus junos* and a truncated version of the HMG-CoA reductase, which reduces HMG-CoA to mevalonate but lacks the NH_2-terminal transmembrane domain (Tippmann et al. 2017). However, this strain with a dual supply of acetoacetyl-CoA did not provide an advanced titer compared to a strain not containing the *nphT7* gene with a farnesene titer in the range of 10 mg/l. Despite the theoretical thermodynamical advantage also a strain with an exclusive malonyl-CoA bypass and expression of a modified isoenzyme of the acetyl-CoA carboxylase, the ACC1 enzyme, did not show a titer improvement (Tippmann et al. 2017).

Since acetoacetyl-CoA is also a main intermediate in the metabolic engineering approaches for other target products, the related thermodynamic hurdle has already been addressed elsewhere. This refers especially to butanol and the respective metabolic engineering projects. Basically, two strategies were pursued, which comprise an advanced conversion of HMG-CoA, enabled by changed cofactor specificity (Nguyen et al. 2019), and the introduction of the malonyl-CoA bypass (Schadeweg 2016a,b).

23.4.3 β-Farnesene with Plant Oil as Feedstocks with Dual MVA Pathway in *Yarrowia lipolytica*

In order to overcome the dependency on glucose or sucrose as feedstock, alternative raw material sources are pursued. Since the MVA pathway proceeds with acetyl-CoA as central biochemical intermediate, which also serves as the standard building block in the deconstruction and breakdown of plant oils or fatty acids, plant oils may be regarded as a suitable feedstock. In particular, oleic acid and waste cooking oil have been used as feedstock for an engineered *Y. lipolytica* strain (Liu et al. 2022a). The native MVA pathway is taking place in the cytosol, and the β-oxidation starting from fatty acyl-CoA is taking place in the peroxisomes. Initially, the three cytosolic MVA pathway enzymes, acetoacetyl-CoA thiolase (AtoB), the HMG-CoA synthase (HMGS), and the HMG-CoA reductase (HMGR), were overexpressed, and the strain produced 1.96 g/l mevalonate using glucose as carbon source. With the switch to oleic acid as carbon source the titer dropped to 0.34 g/l, which was caused by the insufficient availability of acetyl-CoA in the cytoplasm. It is known that acetyl-CoA derived in the peroxisomes in the β-oxidation of fatty acids cannot directly be channeled into the cytosol, though usually the glyoxylate shunt or a shuttle is needed. Thus, the established peroxisomal signaling sequence ePTS1 was fused with the C-termini of the AtoB, HMGS, and HMGR enzymes, and the target localization of the enzymes was verified by fluorescence microscopy. Mevalonate can freely diffuse from the peroxisome to the cytoplasm. A heterologous β-farnesene synthase from *Artemisia annua* was selected and optimized by rational variation of amino acids around the farnesyl-pyrophosphate binding site. Additionally, the enzymes converting mevalonate into β-farnesene were overexpressed to enhance metabolic flux (Figure 23.6):

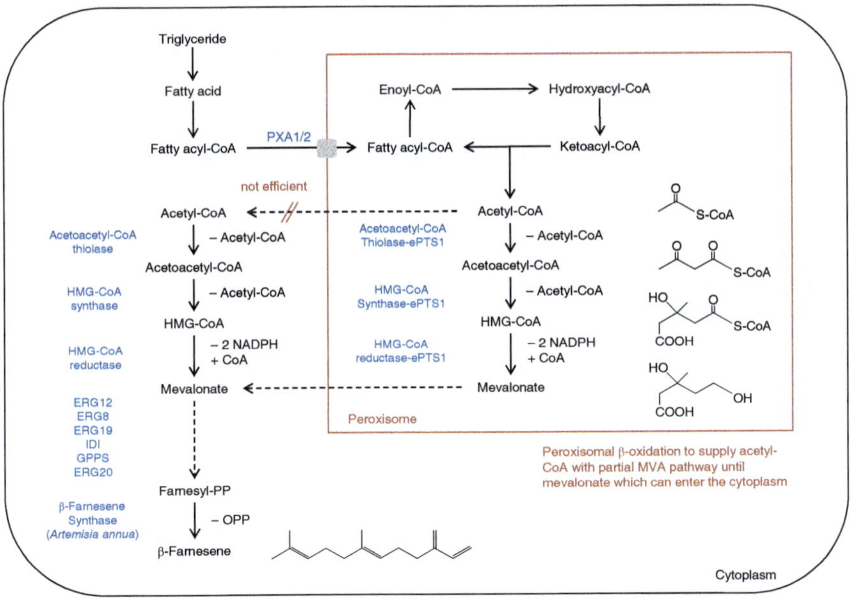

Figure 23.6 β-farnesene based on triglycerides/fatty acid with partial peroxisomal MVA pathway. Source: Adapted from Liu et al. (2022a).

The experiments were pursued in a 5 l bioreactor at pH 6.0 and 1.5 vvm airflow. The final β-farnesene titer was 35.2 g/l with oleic acid and 31.9 g/l with waste cooking oil as feedstock, achieved after a batch duration of 216 h. These results correspond to a productivity of 0.16 g/l × h and 0.15 g/l × h, respectively (Liu et al. 2022a). The results provide a proof of concept. A commercial application depends heavily on the availability and pricing of waste cooking oil. Whereas a minor production volume might be interesting from a business point of view, a high-volume production addressing the fuel market would likely face a feedstock limitation.

23.4.4 HMG-CoA Reductase with Changed Cofactor

The MVA pathway implies an unbalance of cofactors since the synthesis of one mol β-farnesene generates a surplus of 18 mol NADH combined with a gap of six mol NADPH. The unbalance is caused by the enzyme HMG-CoA reductase, which consumes two mol of NADPH for the reduction of the activated carboxylic acid function of HMG-CoA toward the reduced level of the alcohol function in mevalonate (Figure 23.4). Since three mol of mevalonate are needed for the synthesis of one mol of β-farnesene a gap of six mol of NADPH is caused. In order to address this unbalance, isoenzymes of the HMG-CoA reductase were searched, and the HMG-CoA reductase of *S. pomeroyi*, which is highly specific for NADH, was identified and cloned. This isoenzyme has successfully been employed in the design of a recombinant MVA pathway in *S. cerevisiae* (Meadows et al. 2016).

It is also reported that a HMG-CoA reductase with cofactor specificity for NADH has been employed for the fermentative generation of squalene. A metabolically engineered *S. cerevisiae* strain with overexpression of the enzymes that catalyze the conversion of acetyl-CoA to squalene serves as reference point. This strain provided a squalene titer of 978 mg/l from ethanol as carbon source in a fed-batch production. The titer increased up to 1086 mg/l after the HMG-CoA reductase from *S. pomeroyi* was introduced (Li et al. 2020). These data confirm a general proof of concept, though the overall impact of the changed cofactor specificity is low. Furthermore, the squalene product titer does not suffice for a commercially interesting technology.

23.4.5 Technology Development in *Cupriavidus necator* to Access Lithoautotrophic Feedstocks

The common fermentation hosts *E. coli* and *S. cerevisiae* rely on heterotrophic carbohydrates such as glucose, fructose, or glycerol as preferred feedstock. In order to adapt a potential high-volume production to the alternative feedstocks of carbon dioxide and hydrogen, experiments were also pursued with the alternative host *Cupriavidus necator*, which is able to grow on heterotrophic and lithoautotrophic feedstocks as well. After the β-farnesene synthase of *A. annua* was expressed, a titer of 5.4 mg/l (26.3 μM) could be detected using fructose in a fed-batch process (Milker and Holtmann 2021). This performance with the promising alternative host *C. necator* but still a heterotrophic feedstock is far below a commercially interesting threshold, but may prepare the path for alternative feedstocks or even microbial electrosynthesis in the future.

23.4.6 Digression: α-Farnesene with Fused Pathway Enzymes

The fermentative production of α-farnesene has been tested in *E. coli* with an engineered mevalonate pathway (Wang et al. 2011). The fermentation was pursued at 29 °C with 2YT medium with 16 g tryptone and 10 g yeast extract per 1 l medium. A recombinant α-farnesene synthase from wild-type apple was added, which provided some product generation after a codon-optimized gene was applied. Additionally, the farnesyl-pyrophosphate synthase gene (*ispA* gene) was overexpressed. Since the initial product titer was very low and did not exceed 60 mg/l, the farnesyl-pyrophosphate synthase and the α-farnesene synthase were fused to facilitate a more efficient substrate channeling of farnesyl-pyrophosphate to the active center of the α-farnesene synthase (Figure 23.7):

The experiments with the fused dual enzymes did in fact provide an α-farnesene titer of 380 mg/l (Wang et al. 2011). Enzyme fusion and metabolic channeling are discussed as means to increase selectivity and productivity since the unwanted redirection of pathway intermediates through diffusion, utilization by rival enzymes, or even degradation is avoided (Dueber et al. n.d.). Substrate channeling turned out to be a useful tool in the design of other fermentation pathways. In a structurally equivalent approach, a terpene synthase had been fused with the ERG20 enzyme, which catalyzes the conversion of IPP and DMAPP

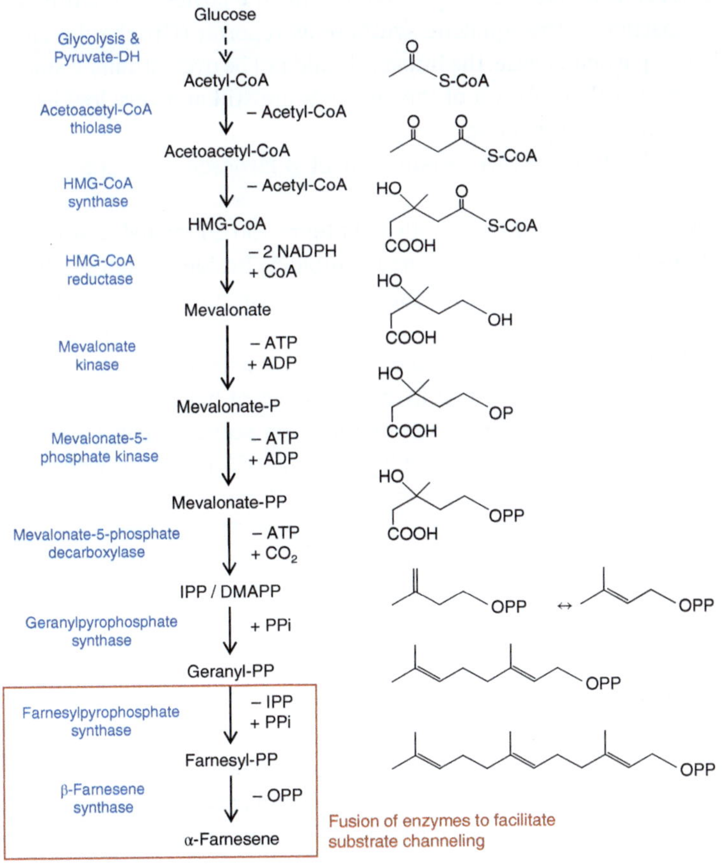

Figure 23.7 Pathway toward farnesene in *E. coli* with enzyme fusion. Source: Walter Koch.

Table 23.6 Selection of published fermentation performance data on β-farnesene. All R&D experiments choose the MVA pathway in their metabolic engineering strategy. The enzyme column refers to the origin of the β-farnesene synthase.

Host	Enzyme	Titer	Batch-time	Productivity	Carbon yield	Source
S. cerevisiae	(heterologous)	130 g/l	58 h	2.24 g/l × h	173 g/kg glc	Meadows (2016)
S. cerevisiae	*C. junos*	13.8 mg/l	18 h	0.77 mg/l × h	na (glc)	Tippmann (2017)
E. coli	*A. annua*	10.3 g/l	48 h	0.21 g/l × h	na (gly)	Yao (2020)
E. coli	*A. annua*	8.74 g/l	120 h	0.07 g/l × h	21.5 g/kg gly	You (2017)
E. coli	*A. annua*	4.06 g/l	48 h	0.08 g/l × h	na (glc)	You et al. (2017)
C. necator	*A. annua*	5.4 mg/l	290 h	0.02 mg/l × h	na (glc)	Milker and Holtmann (2021)

glc, glucose; gly, glycerol. A comprehensive overview is given elsewhere (Liu et al. 2022b).
Source: Adapted from Liu et al. (2022b).

into geranyl-pyrophosphate and farnesyl-pyrophosphate in the native yeast MVA pathway, and indeed a yield increase could be observed (Albertsen et al. 2011). Furthermore, a flux improvement could be achieved with a fusion enzyme combining the farnesyl-PP synthase and a terpene synthase, in particular the squalene synthase as reporter (Gruchattka and Kayser 2015). However, in this particular case, the impact should not be overestimated since the α-farnesene titers with and without fusion of the enzymes are still at a very low level and significantly below a commercial threshold.

The published results on the fermentative production of β-farnesene are compared (Table 23.6):

It is obvious that the *S. cerevisiae* strain with a published titer of 130 g/l related to Amyris (Meadows et al. 2016) defines the superior β-farnesene technology. Besides a high productivity, the carbon yield is also convincing given a theoretical maximum carbon yield of 250 kg per t glucose without including the generation of biomass. The selection of glycerol as carbon and energy source (You et al. 2017) seems to speculate on a cheap raw material price since glycerol is obtained as undesired by-product in the synthesis of biodiesel. However, this feedstock choice is not plausible from a pathway engineering point of view. Glycerol has a higher reduction level than glucose, which is not helpful since the MVA pathway generates a surplus of reduction equivalents. The R&D attempts in *E. coli* achieving a titer of 4.06 g/l pursue a separated hydrolysis of corncobs and fermentation. The acid treatment of the lignocellulosic biomass is reported to achieve a cellulose conversion rate into glucose of 96.83% with 2% sulfuric acid (You et al. 2019).

23.5 Process Design with Second Liquid Phase

Protocols for the fermentative production of terpenes have been published, summarizing the impressive technological development in the last 20 years (Vickers et al. 2016; Rodriguez et al. 2014). One protocol is focusing on the bicyclic sesquiterpene amorphadiene in *S. cerevisiae*, though the results are representative and can be adopted for the linear β-farnesene molecule as well (Rodriguez et al. 2014). A specific characteristic of sesquiterpenes is their high hydrophobicity and the resulting low solubility in water. β-farnesene is not soluble in water. Accordingly, with increased product secretion, an emulsion is emerging and small oil droplets are forming, which might not form a clearly defined second phase at batch end. The missing clear second phase might be an issue in the product isolation, and countermeasures were evaluated (Heeres et al. 2014). Since terpenes are volatile and a share of the product might disappear into the reactor head space – which is usually dismissed and not considered in the downstream product isolation process – the direct addition of a lipophilic agent into the fermentation broth is advised. For the fermentative production of sesquiterpenes often a lipophilic second phase in the reactor is preferred compared to the traditional liquid one-phase reaction medium. Dodecane can be regarded as the standard additive for terpenoid fermentation since it turned out to efficiently capture the secreted product. It is also applied for the fermentative production of bisabolene (Phelan et al. 2015). For producing limonene and bisabolene, it is reported that significantly higher product titers were achieved in case a dodecane layer was added (Davies et al. 2014).

The dodecane volume makes up to 10% (v/v) of the overall liquid phase (Rodriguez et al. 2014), though the share may be set to reach up to 20% (v/v) of the overall broth volume (Milker and Holtmann 2021). It avoids a potential toxic impact of the sesquiterpene on the host and blocks a partial evaporation of the volatile terpenoid into the reactor head space. Product found in the reactor head space could be isolated in the downstream process as well, though it is easier and more cost efficient if the isolation and purification section can exclusively focus on the fermenter broth. At batch end, the stirring of the broth is stopped, and after an incubation period, a dodecane layer is formed. It should be analyzed inasmuch as product residues remain in the water phase of the broth, which might be extracted in a second step. β-farnesene can then be extracted out of the dodecane fraction, and the dodecane is recycled (Figure 23.8):

The option to extract the product out of the aqueous broth avoids a laborious isolation of β-farnesene since the high boiling points of sesquiterpenes ranging from 250–280 °C might require the evaporation of water with a rather high energy input. No major issues are reported with the application of dodecane or decane as media supplementation. It might negatively interfere with the chromatography step to ensure the needed purity, and dodecane residues might cause a tailing of volatile isoprenoids. However, this obstacle can be solved with defined solvent mixtures (Vickers et al. 2016).

The volatile β-farnesene, which despite a liquid two-phase setup escaped into the reactor head space, can also be captured in case a critical threshold is exceeded. Amyris is describing a process in, which the fermentation off-gas in the reactor head space is recovered by venting the off-gas through a gas-washer containing heptanol. The β-farnesene share found in the reactor head space is indicated as 2.8% (broth titer 14.3 g/l) and 1.9% (broth titer 23.6 g/l) of the total volume generated in the fermentation (Amyris WO 2008/045555).

It should be remarked that in specific cases other lipophilic media additives than dodecane are reported, like decane (Yao et al. 2020; You et al. 2017; Wang et al. 2011) or isopropyl myristate (Ye et al. 2022).

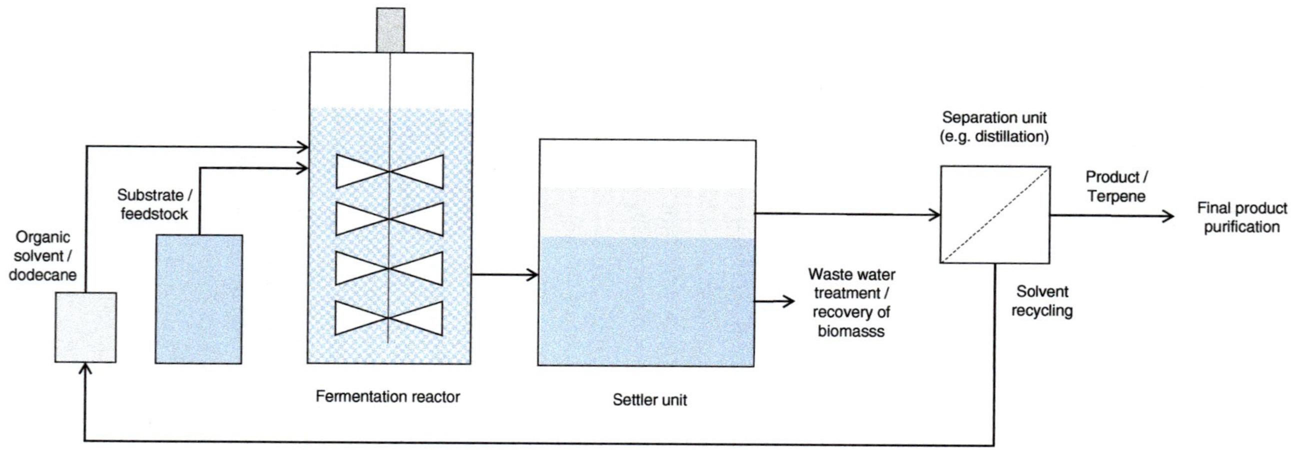

Figure 23.8 Process flow diagram for the fermentative production of β-farnesene. Source: Adapted from Scalcinati et al. (2012).

23.6 Industrial Application

It is known that the commercial production of β-farnesene was initially started by the company Amyris with the yeast strain *S. cerevisiae* and sugar cane syrup as feedstock. The experiments were performed in bioreactors with a design volume of 200 m^3, and the cells were not negatively affected by the resulting pressure at the bottom of the reactor. The obtained β-farnesene was then converted into farnesane via hydrogenation in order to qualify as a fuel additive (Milker and Holtmann 2021). The realized performance parameters may be according to their central publication in Nature and achieve a titer of 130 g/l and a productivity of 2.24 g/l× h, corresponding to a batch time of 58 h. The published carbon yield amounts to 17.3% (Meadows et al. 2016).

However, the commercialization of β-farnesane as fuel additive – after the hydrogenation of β-farnesene – was not successful. This was caused by mainly two obstacles: Amyris could not fully achieve the target cost structure for β-farnesene, and the changed market conditions caused a significant cost and price reduction, respectively, for petrochemically derived jet fuel. These circumstances enforced Amyris to adjust its company strategy, and the business focus switched to specialties with a higher value perspective. John Melo, the CEO of Amyris at that point in time, summarized the strategic shift from low value and high-volume products (i.e. fuel additives) to high value and low-volume products (i.e. specialty chemicals): "The breakthrough has really been the maturity of the technology [...] We are now able to move a molecule from target to scale in a year or less. For a partner, that is really critical - you make your investment and then you can confidently plan a development cycle around the molecule. That definitively was not possible ten years ago [...] I think our most challenging times were really in 2011-2012, when macro trends began turning against us [...] People had expected oil prices to be $100/barrel forever, but then shale oil become the next thing and the cost became more like $50/barrel. To this day, we still have the best jet fuel anybody has made in the world, but doing it biologically via fermentation, it was hard to get to the costs to where you had a sustainable business" (Coons 2021).

Reflecting this unpredicted market development, Amyris took advantage of its terpene platform for non-fuel products and launched in 2016 with Biossance®, a fully natural squalane skin care brand (Kennedy 2019). The synthesis may proceed via two alternative routes. The simplest approach would be to dimerize farnesane (the former fuel additive) into squalane. Second, farnesene might chemically be dimerized to squalene, which could then be hydrogenated to squalane (Leavell et al. 2016). As a third option, squalene might be derived in a one-step fermentation process in case the biochemical pathway is extended and farnesyl-pyrophosphate is efficiently converted by the squalene synthase (Li et al. 2020). The fermentative squalene would then be processed by chemical hydrogenation to derive squalane. However, an efficient secretion of squalene to achieve a titer >100 g/l might be difficult to achieve (Zhuang and Chappell 2015).

Accordingly, Amyris phased out the commercialization of β-farnesene as a jet fuel precursor and sold its production plant in Brotas (Brazil) and related IP to DSM in 2017 and its complete flavor and fragrance bio-based intermediates business also to DSM in 2021 (Ingredient networks 2021). A new business perspective for fermentative β-farnesene appeared in fine chemistry, in particular the synthesis of vitamin E or α-tocopherol. α-Tocopherol is mainly produced with chemical technology, and the synthesis may proceed via the condensation of isophytol and trimethylhydroquinone (TMH). Isophytol is usually produced

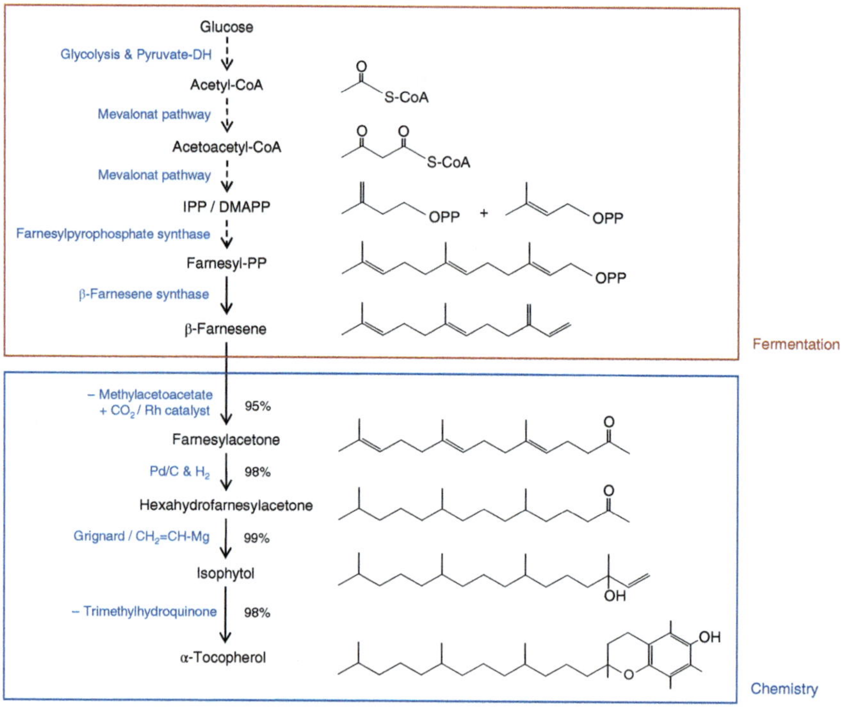

Figure 23.9 Innovative synthesis route toward α-tocopherol. Source: Adapted from Ye et al. (2022).

with organic chemistry. The Chinese Biotech company J1 Biotech realized that β-farnesene can chemically be converted into isophytol, which would leverage their existing farnesene technology. J1 Biotech and Wuhan University achieved in 2014 a β-farnesene titer of 55 g/l with an engineered *E. coli* host. The technology has originally been developed based on an in vitro reconstitution of the mevalonate pathway, which initially provided a β-farnesene titer of 1.1 g/l, which could later be further enhanced (Zhu et al. 2014). The biocatalyst was equipped with an improved β-farnesene synthase and followed a recombinant mevalonate pathway. Since Nenter & Co. had innovated the synthesis of TMH, both companies realized the synergies of their technologies and joined forces in the production of α-tocopherol. J1 Biotech provided its fermentative β-farnesene technology and the protocol for the chemical conversion of β-farnesene into isophytol. Nenter offered its proprietary TMH technology as contribution (Figure 23.9):

The yield in the three-step conversion of β-farnesene to isophytol is published by J1 Biotech and Wuhan University, and a high performance is achieved (Ye et al. 2022). The condensation of isophytol and TMH may achieve a yield of up to 98%, pending on the applied catalyst (Hasegawa et al. 2003).

Fujian Guanfu Modern Household Wares announced in 2016 that its wholly-owned subsidiary, Nenter & Co., will construct a vitamin E production plant with a capacity of 20 kt/pa (CCM Data and Business Intelligence 2016). The later publication of J1 Biotech and Wuhan University revealed that Nenter & Co. actually constructed a vitamin E plant with an annual output of 30 kt/pa in Jingzhou, Hubei Province, China (Ye et al. 2022). It is not entirely clear which β-farnesene technology, the one from Nenter or the one from Amyris, is actually used for the commercial production of vitamin E. However, J1 Biotech and Wuhan University published that Nenter has collaborated with Amyris to provide β-farnesene as the raw material for the vitamin E synthesis, which indicates that the superior Amyris technology is applied. In 2019, DSM announced that it is going to setup a 75/25 joint venture with Nenter & Co., which will include Nenter's production and related assets for vitamin E (DSM press release 2019; Kohli 2019). About 20% of the global vitamin E supply is produced with isophytol derived from fermentative β-farnesene and the β-farnesene is made in the former Amyris fermentation plant originally designed to develop the fuel additive business (Ye et al. 2022).

References

Albertsen, L., Chen, Y., Bach, L.S. et al. (2011). Diversion of flux toward sesquiterpene production in *Saccharomyces cerevisiae* by fusion of host and heterologous enzyme. *Applied and Environmental Microbiology* 77: 1033–1040.

Amyris patent (2008). Fuel compositions comparing farnesane and farnesene derivatives and method of making and using same. WO 2008/045555, PCT/US2007/021890.

Armamino, N., Charpentier, J., Flachsmann, F. et al. (2020). What's hot, what not: the trends of the past 20 years in the chemistry of odorants. *Angewandte Chemie* 59: 16310–16344. https://doi.org/10.1002/anie.202005719.

Arnaud, C. (2016). Artemisinin 'bomb' attacks many proteins. *C&EN* 9.

Bogorad, I.W., Lin, T.-S., and Liao, J.C. (2013). A, Synthetic non-oxidative glycolysis enables complete carbon conservation. *Nature* 502: 693–698.

Bomgardner, M. (2014). World cup scores bio jet fuel. *C&EN* 8.

Bruce, T.J.A., Aradottir, G.I., Smart, L.E. et al. (2015). The first crop plant genetically engineered to release an insect pheromone for defense. *Scientific Reports* https://doi.org/10.1038/srep11183.

Carsanba, E., Pintado, M., and Oliveira, C. (2021). Fermentation strategies for production of pharmaceutical terpenoids im engineered yeast. *Pharmaceuticals* 14: 295. https://doi.org/10.3390/ph14040295.

CCM Data & Business Intelligence (2016). Intense market competition: Nenter to launch 20,000 tonnes annually VE project. http://www.cnchemicals.com/Press/87555CCM:%20Intense%20market%20competition:%20Nenter%20to%20launch%2020,000%20tonnes%20annually%20VE%20project.html (accessed 6 Jan 2023).

Christianson, D.W. (2017). Structural and chemical biology of terpenoid cyclases. *Chemical Reviews* 117: 11570–11648. https://doi.org/10.1021/acs.chemrev.7b00287.

Coons, R. (2021). Amyris on the upswing. *Chemical Week* 3 (10): 42–43.

Davies, F.K., Work, V.H., Beliaev, A.S., and Posewitz, M.C. (2014). Engineering limonene and bisabolene production in wild type and a glycogen-deficient mutant of *Synechococcus* sp. PPC 7002. *Frontiers in Bioengineering and Biotechnology* 2: article 21. https://doi.org/10.3389/fbioe.2014.00021.

DSM press release (2019), DSM announces JV with Nenter to produce vitamin E. https://www.dsm.com/corporate/news/news-archive/2019/01-19-dsm-announces-jv-with-nenter-to-produce-vitamin-e.html

Dueber, J.E., Wu, G.C., Malmirchegini, G.R. et al. Synthetic protein scaffolds provide modular control over metabolic flux. *Nature Biotechnology* 27 (8): 753–759. https://doi.org/10.1038/nbt.1557.

Eiben, C.B., de Rond, T., Bloszies, C. et al. (2019). Mevalonate pathway promiscuity enables noncanonical terpene production. *ACS Synthetic Biology* 8: 2238–2247. https://doi.org/10.1021/acssynbio.9b00230.

Eisenmann, M. (2012). Holding out for reinforcement. *Nature Biotechnology* 484: S16–S18.

Gao, Y., Honzatko, R.B., and Peters, R.J. (2012). Terpenoid synthase structures: a so far incomplete view of complex catalysis. *National Product Reports, The Royal Society of Chemistry* 29 (10): 1153–1175. https://doi.org/10.1039/c2np20059g.

Gruchattka, E. and Kayser, O. (2015). *In vivo* validation of *in silico* predicted metabolic engineering strategies in yeast: disruption of α-ketoglutarate dehydrogenase and expression of ATP-Citrate lyase for terpenoid production. *PLOS One* https://doi.org/10.1371/journal.pone.0144981.

Hasegawa, A., Ishihara, K., and Yamamoto, H. (2003). Trimethylsilyl pentafluorophenylbis (trifluoromethanesulfonyl)methide as a super Lewis acid catalyst for the condensation of trimethylhydroquinone with isophytol. *Angewandte Chemie* 115: 5909–5911. https://doi.org/10.1002/anie.200352382. The yield in the synthesis of α-tocopherol is given in the German abstract: https://onlinelibrary. wiley.com/doi/10.1002/aige.200352382.

Heeres, A.S., Picone, C.S.F., van der Wielen, L.A.M. et al. (2014). Microbial advanced biofuels production: overcoming emulsification challenges for large-scale operation. *Trends in Biotechnology* 32 (4): 221–229. https://doi.org/10.1016/j.tibtech.2014.02.002.

Hommel, M. (2008). The future of artemisinins: natural, synthetic or recombinant? *Journal of Biology* 7: article 38. 10.1186.jbiol1101.

IHS Chemical Week (2012). Amyris, Total sign biofuel deal worth up to $82 million, p. 4.

IHS Chemical Week (2017). Growing renewables: harvesting performance benefits, p. 19–21.

Ingredient networks (2021). DSM acquires Amyris's bio-based flavors and fragrances business. https://www.ingredientsnetwork.com/dsm-acquires-amyris-biobased-flavors-and-news111594.html#:~:text=DSM%20has%20already%20acquired%20Amyris%E2%80%99%20Farnesene%20%E2%80%93%20a,where%20DSM%20produces%20several%20of%20Amyris%E2%80%99%20flavor%20products accessed 6.1.2023.

Karim, A.S., Dudley, Q.M., and Jewett, M.C. (2017). Cell-free synthetic systems for metabolic engineering and biosynthetic pathway prototyping. In: *Industrial Biotechnology. Microorganisms* (ed. C. Wittmann and J.C. Liao), 125–148. Wiley-VCH.

Kemper, K., Hirte, M., Reinbold, M. et al. (2017). Opportunities and challenges for the sustainable production of structurally complex diterpenoids in recombinant microbial systems. *Beilstein Journal of Organic Chemistry* 13: 845–854. https://doi.org/10.3762/bjoc.13.85.

Kennedy, H. T., New arrival in biobased baby care, but Amyris isn't only one turning to skin care, www.biofuelsdigest.com, 2019,

Kohli, K. (2019). DSM ups stake in food ingredients firm, forms vitamin E JV, in China. *Chemical Week* 15.

Kozak, B.U., van Rossum, H.M., Luttik, M.A.H. et al. (2014). Engineering acetyl coenzyme A supply: functional expression of a bacterial pyruvate dehydrogenase complex in the cytosol of *Saccharomyces cerevisiae*. *mBio* 5 (5): e01696-14.

Kung, S.H., Murarka, A., McPhee, D., and Paddon, C.J. (2018). Approaches and recent developments for the commercial production of semi-synthetic artemisinin. *Frontiers in Plant Science* 9: article 87 https://doi.org/10.3389/fpls.2018.00087.

Kuzuyama, T. and Seto, H. (2012). Two distinct pathways for essential metabolic precursors for isoprenoid biosynthesis. *Proceedings of the Japan Academy, Proceedings Series B* 88: 41–52. https://doi.org/10.2183/pjab.88.41.

Lan, E.I. and Liao, J.C. (2012). ATP drives direct photosynthetic production of 1-butanol in cyanobacteria. *PNAS* 109 (16): 6018–6023. https://doi.org/10.1073/pnas.1200074109.

Leavell, M.D., McPhee, D.J., and Paddon, C.J. (2016). Developing fermentative terpenoid production for commercial usage. *Current Opinions in Biotechnology* 37: 114–119. https://doi.org/10.1016/j.copbio.2015.10.007.

Lecourt, M. and Antoniotti, S. (2020). Chemistry, sustainability and naturality of perfumery biotech ingredients. *ChemSusChem* 13: 1–12. https://doi.org/10.1002/cssc.202001661.

Li, T., Liu, G.-S., Zhou, W. et al. (2020). Metabolic engineering of *Saccharomyces cerevisiae* to overproduce squalene. *Journal of Agricultural and Food Chemistry* 68: 2132–2138. https://doi.org/10.1021/acs.jafc.9b07419.

Liao, P., Hermmerlin, A., Bach, T.J., and Chye, M.-L. (2016). The potential of the mevalonate pathway for enhanced isoprenoid production. *Biotechnology Advances* 234: 697–713. https://doi.org/10.1016/j.biotechadv.2016.03.005.

Liebsch, A. (2014). Renewable hydrocarbons from sugarcane. *BMC Proceedings* 8 (supplement 4) O35, www.biomedcentral.com/1753-6561/8/S4/O35: .

Lin, P.P., Jaeger, A.J., Wu, T.-Y. et al. (2018). Construction and evolution of an Escherichia coli strain relying on nonoxidative glycolysis for sugar catabolism. *Proceedings of the National Academy of the Unitaed States of America* 115 (14): 3538–3546.

Liu, Z., Moradi, H., Shi, S., and Darvishi, F. (2021). Yeats as microbial cell factories for sustainable production of biofuels. *Renewable and Sustainable Energy Reviews* 143: article 110907. https://doi.org/10.1016/j.rser.2021.110907.

Liu, Y., Zhang, J., Li, Q. et al. (2022a). Engineering *Yarrowia lipolytica* for the sustainable production of β-farnesene from waste oil feedstock. *Biotechnology for Biofuels* 15: 101. https://doi.org/10.1186/s13068-022-02201-2.

Liu, Y., Wang, Z., Cui, Z. et al. (2022b). Progress and perspectives for microbial production of farnesene. *Bioresource Technology* 347: https://doi.org/10.1016/j.biortech.2022.126682.

Mackenzie, A. (2013). Synthetic biology and the technicity of biofuels. *Studies in History and Philosophy of Biological and Biomedical Sciences* 44: 190–198. https://doi.org/10.1016/j.shpsc.2013.03.014.

Martin, V.J.J., Pitera, D.J., Withers, S.T. et al. (2003). Engineering a mevalonate pathway in *Escherichia coli* for production of terpenoids. *Nature Biotechnology* 21 (7): 796–802.

Maury, J., Asadollahi, M.A., Møller, K. et al. (2008). Reconstruction of a bacterial isoprenoid biosynthetic pathway in *Saccharomyces cerevisiae*. *FEBS Letters* 582: 4032–4038. https://doi.org/10.1016/j.febslet.2008.10.045.

McGarvey, D.J. and Croteau, R. (1995). Terpenoid metabolism. *The Plant Cell* 7: 1015–1026.

Meadows, A.L., Hawkins, K.M., Tsegaye, Y. et al. (2016). Rewriting yeast central carbon metabolism for industrial isoprenoid production. *Nature* 537: 694–697. https://doi.org/10.1038/nature19769.

Milker, S. and Holtmann, D. (2021). First time β-farnesene production by the versatile bacterium *Cupriavidus necator*. *Microbial Cell Factories* 20: 89. https://doi.org/10.1186/s12934-021-01562-x.

Nguyen, N.-P.-T., Raynaud, C., Meynial-Salles, I., and Soucaille, P. (2019). Reviving the Weizmann process for commercial *n*-butanol production. *Nature Communications* 9: 3682. https://doi.org/10.1038/s41467-018-05661-z.

Okamura, E., Tomita, T., Sawa, R. et al. (2010). Unprecedented acetoacetyl-coenzyme A synthesizing enzyme of the thiolase superfamily involved in the mevalonate pathway.

Proceedings of the National Academy of the Unitaed States of America 107 (25): 11265–11270. https://doi.org/10.1073/pnas.1000532107.

Osterath, B. (2014). Im Tonnenmaβstab gegen Malaria. *Nachrichten aus der Chemie* 62: 125–127.

Pang, Y.R., Hu, Z.H., and Xiao, D.G. (2018). Advances in metabolic engineering for the microbial production of naturally occurring terpenes limonene and bisabolene: a mini review. *Chinese Journal of Biotechnology* 34 (1): 24–33. https://doi.org/10.13345/j.cjb.170150.

Peralta-Yahya, P.P. and Keasling, J.D. (2010). Advanced biofuel production in microbes. *Biotechnology Journal* 5: 147–162. https://doi.org/10.1002/biot.200900220.

Peralta-Yahya, P.P., Ouellet, M., Chan, R. et al. (2011). Identification and microbial production of a terpene-based advanced biofuel. *Nature Communication* 2: 483. https://doi.org/10.1038/ncomms1494.

Phelan, R.M., Sekurova, O.N., Keasling, J.D., and Zotchev, S.B. (2015). Engineering terpene biosynthesis in *Streptomyces* for production of the advanced biofuel precursor bisabolene. *Synthetic Biology* 4: 393–399. https://doi.org/10.1021/sb5002517.

Rodriguez, S., Kirby, J., Denby, C.M., and Keasling, J.D. (2014). Production and quantification of sesquiterpenes in *Saccharomyces cerevisiae,* including extraction, detection and quantification of terpene products and key related metabolites. *Nature Protocols* 9 (8): 1980–1996. https://doi.org/10.1038/nprot.2014.132.

van Rossum, H.M., Kozak, B.U., Pronk, J.T., and van Maris, A.J.A. (2016). Engineering cytosolic acetyl-CoA supply in *Saccharomyces cerevisiae*: Pathway stoichiometry, free-energy conservation and redox-factor balancing. *Metabolic Engineering* 36: 99–115. https://doi.org/10.1016/j.ymben.2016.03.006.

Scalcinati, G., Partow, S., Siewers, V. et al. (2012). Combined metabolic engineering of precursor and co-factor supply to increase α-santalene production by *Saccharomyces cerevisiae*. *Microbial Cell Factories* 11: 117. www.microbialcellfactories.com/content/11/1/117.

Schadeweg, V. and Boles, E. (2016a). n-Butanol production in *Saccharomyces cerevisiae* is limited by the availability of coenzyme A and cytosolic acetyl-CoA. *Biotechnology for Biofuels* 9: 44. https://doi.org/10.1186/s13068-016-0456-7.

Schadeweg, V. and Boles, E. (2016b). Increasing n-butanol production with *Saccharomyces cerevisiae* by optimizing acetyl-CoA synthesis, NADH levels and trans-2-enoyl-CoA reductase expression. *Biotechnology for Biofuels* 9: 257. https://doi.org/10.1186/s13068-016-0673-0.

Schäfer, B. (2014). Artemisinin. *Chemie in unserer Zeit* 48: 134–145. https://doi.org/10.1002/ciuz.201400645, and: Ein neuer Wirkstoff gegen eine alte Krankheit: Artemisinin, Chemie in unserer Zeit, Teil 2, 2014, Vol. 48, p. 216-225.

Shiba, Y., Paradise, E.M., Kirby, J. et al. (2007). Engineering of the pyruvate dehydrogenase bypass in Saccharomyces cerevisiae for high-level production of isoprenoids. *Metabolic Engineering* 9: 160–168.

Tippmann, S., Ferreira, R., Siewers, V. et al. (2017). Effects of acetoacetyl-CoA synthase expression on production of farnesene in Saccharomyces cerevisiae. *Journal of Indian Microbiology and Biotechnology* 44: 911–922. https://doi.org/10.1007/s10295-017-1911-6.

Turconi, J., Griolet, F., Guevel, R. et al. (2013). Semisynthetic artemisinin, the chemical path to industrial production. *Organic Process Research & Development* https://doi.org/10.1021/op4003196.

Vickers, C.E., Bongers, M., Bydder, S.F. et al. (2016). Protocols for the production and analysis of isoprenoids in bacteria and yeast. In: *Hydrocarbon and Lipid Microbiology* (ed. T.J. McGenity), 23–52. Springer Procols Handbooks https://doi.org/10.1007/8623_2015_107.

Vickers, C.E., Williams, T.C., Peng, B., and Cherry, J. (2017). Recent advances in synthetic biology for engineering isoprenoid production in yeast. *Current Opinion in Chemical Biology* 40: 47–56. https://doi.org/10.1016/j.cbpa.2017.05.017.

Vögeli, B., Engilberge, S., Girard, E. et al. (2018). Archaeal acetoacetyl-CoA thiolase/HMG-CoA synthase complex channels the intermediate via a fused CoA-binding site. *Proceedings of the National Academy of the Unitaed States of America* 115 (13): 3380–3385. https://doi.org/10.1073/pnas.1718649115.

Wang, C., Yoon, S.-H., Jang, H.J. et al. (2011). Metabolic engineering of *Escherichia coli* for α-farnesene production. *Metabolic Engineering* 13: 648–655. https://doi.org/10.1016/j.ymben.2011.08.001.

Wegner, S.A., Chen, J.-M., Ip, S.S. et al. (2021). Engineering acetyl-CoA supply and *ERG9* repression to enhance mevalonate production in *Saccharomyces cerevisiae*. *Journal of Industrial Microbiology and Biotechnology* 48: https://doi.org/10.1093/jimb/kuab050.

Yao, P., You, S., Qi, W. et al. (2020). Investigation of fermentation conditions of biodiesel by-products for high production of β-farnesene by an engineered *Escherichia coli*. *Environmental Science and Pollution Research* 27: 22758–22769. https://doi.org/10.1007/s11356-020-08893-z.

Ye, Z., Shi, B., Huang, Y. et al. (2022). Revolution of vitamin E production by starting from microbial fermented farnesene to isophytol. *The Innovation* 3 (3): 100228. https://doi.org/10.1016/j.xinn.2022.100228.

Yoshikuni, Y., Ferrin, T.E., and Keasling, J.D. (2006). Designed divergent evolution of enzyme function. *Nature Letters* 440: 1078–1082. https://doi.org/10.1038/nature04607.

You, S., Yin, Q., Zhang, J. et al. (2017). Utilization of biodiesel by-product as substrate for high-production of β-farnesene via relatively balanced mevalonate pathway in *Escherichia coli*. *Bioresource Technology* 243: 228–236. https://doi.org/10.1016/j.biortech.2017.06.058.

You, S., Chang, H., Zhang, C. et al. (2019). Recycling strategy and repression elimination for lignocellulosic farnesene production with an engineered Escherichia coli. *Journal of Agricultural and Food Chemistry* 67: 9858–9867. https://doi.org/10.1021/acs.jafc.9b03907.

Zhou, Y.J., Kerkhoven, E.J., and Nielsen, J. (2018). Barriers and opportunities in bio-based production of hydrocarbons. *Nature Energy* 3: 925–935. https://doi.org/10.1038/s41560-018-0197-x.

Zhu, F., Zhong, X., Hu, M. et al. (2014). In vitro reconstitution of mevalonate pathway and targeted engineering of farnesene overproduction in *Escherichia coli*. *Biotechnology and Bioengineering* 111 (7): 1396–1405.

Zhuang, X. and Chappell, J. (2015). Building terpene production platforms in yeast. *Biotechnology and Bioengineering* 112 (9): 1854–1864. https://doi.org/10.1002/bit.25588.

Index

Pathway Design for Industrial Fermentation, First Edition. Walter Koch.

b

m

r